Proceedings of the Social Sciences International Research Conference, 2025

Ephrem Habtemichael Redda
Editor

Proceedings of the Social Sciences International Research Conference, 2025

Social Sciences Research in the AI Era: Opportunities, Challenges and Threats

Springer

Editor
Ephrem Habtemichael Redda
Vanderbijlpark Campus
North-West University
Vanderbijlpark, South Africa

ISBN 978-981-95-9488-7 ISBN 978-981-95-9489-4 (eBook)
https://doi.org/10.1007/978-981-95-9489-4

This Springer imprint is published by the registered company Springer Nature Singapore Pte Ltd.
The registered company address is: 152 Beach Road, #21-01/04 Gateway East, Singapore 189721, Singapore

Foreword

The Social Sciences International Research Conference (SSIRC) 2025, hosted by the Faculty of Economic and Management Sciences at North-West University, exemplifies the power of institutional collaboration in advancing research and knowledge exchange. The conference took place from 6–7 November 2025 at the Radisson Blu Hotel in Umhlanga, Durban, South Africa, and was held in person.

A highlight of SSIRC 2025 was the celebration of the conference's 10th anniversary, during which the conference co-chairs, Prof Babs Surujlal and Prof Herman van der Merwe, took delegates on a nostalgic journey through the event's history. A commemorative video featuring well-wishes from past keynote speakers added to the sense of reflection and celebration.

The conference featured a keynote address by Prof Leon de Beer from the Norwegian University of Science and Technology, who shared his perspective on "When work thinks back: The impact of artificial intelligence on work-life and the human condition".

The SSIRC 2025 Organising Committee extends its sincere appreciation to the scientific committee, subject chairs, external reviewers, session chairs, keynote speaker, panellists, panel facilitator, and administrative staff for their invaluable contributions to the success of the conference.

With the theme "Social Sciences Research in the AI era: Opportunities, Challenges and Threats", SSIRC 2025 inspired meaningful dialogue and fostered impactful research collaborations, ensuring a rich and productive experience for all participants. We trust that this conference sparked new ideas and made lasting contributions to the future of social sciences research.

On behalf of the organising committee

Prof. J. Surujlal
SSIRC Co-Chair, Executive Dean:
Faculty of Economic and Management Sciences
North-West University
Vanderbijlpark, South Africa

Acknowledgements

We hope this conference creates a wonderful network platform for you and that you will enjoy all the presentations with us.

The Organising Committee of the SSIRC expresses its sincere gratitude to all individuals and institutions whose support made this event possible.

We wish to acknowledge:

- The North-West University for their generous support and partnership.
- INSETA for their generous sponsorship and partnership.
- The Radisson Blu for their excellent service and beautifully kept venue.
- The keynote and plenary speakers, whose insights have enriched this conference.
- Subject chairs, session chairs, reviewers, and authors for their commitment to academic excellence.
- The organising committee for their dedication and hard work throughout the year.
- Delegates from around the world for their participation and engagement.

Please feel free to visit our SSIRC website for more information: https://commerce.nwu.ac.za/SSIRC

For more information about the North-West University: https://www.nwu.ac.za/welcome

For more information on the Faculty of Economic Management Sciences: https://studies.nwu.ac.za/undergraduate-studies/fields-study-economic

Conference Organiser and Chair of the Scientific Committee

Prof Ephrem (E.H.) Redda, WorkWell Research Unit, Faculty of Economic and Management Sciences, North-West University

Track Chairs/Subject Chairs

Dr. Aletta Neethling, Acting Faculty Manager, Faculty of Business and Management Sciences, CPUT, South Africa, Accounting and Finance
Dr. Kekebogile Paadi, Acting Deputy Director: Human Resources Management, North-West University, South Africa, Human Resource Management
Dr. Wilbert Manyanga, North-West University, Business Resilience and Sustainability
Prof. Ankit Katrodia, Associate Professor, School of Management Sciences, North-West University, South Africa, Marketing/ Sport and Leisure
Prof. Liandi van den Berg, Professor and Director, TELIT-SA, North-West University, South Africa, Service Excellence in Higher Education
Prof. Luzaan Hamilton, Associate Professor, School of Management Sciences, North-West University, South Africa, Business Management
Prof. Marinda Pretorius, Senior Lecturer and Programme Coordinator, University of Johannesburg, South Africa, Economics
Prof. Martin Chanza, Associate Professor, North-West University, Emerging Technology and Management
Prof. Precious Mncayi, Associate Professor, School of Economic Sciences, North-West University, South Africa, Economics

Advisory Committee

Dr. R. Tomar, King Fahd University of Petroleum and Minerals, Saudi Arabia
Prof. M. Allahyari, Islamic Azad University, Iran,
Prof. P. H. Chaskar, University Putra Malaysia, Malaysia
Dr. D. Kumar, University of Sassaram, India
Prof. R. Mishra, Institute of Public Enterprise, India
Prof. C. Mafini, Vaal University of Technology, South Africa
Prof. M. Fekete-Farkas, Szent István University, Hungary
Prof. N. Meye, University of Johannesburg, South Africa
Prof. H. van der Merwe, North-West University, South Africa
Prof. J. Surujlal, North-West University, South Africa
Prof. W. Suherman, University Negeri Yogyarta, Indonesia
Prof. A. Larion, Romania Ovidius University of Constanta
Prof. Dr. Magda Róbert, Hungary
Dr. Jana Majerova, AMBIS University, Czech Republic
Prof. D. Dabija, Babes-Balyai University Clj-Nopoca, Romania
Prof. M. Asif Khan, University of Kotli, Pakistan
Prof. M. Subban, University of KwaZulu-Natal, South Africa
Dr. C. Schachtebeck, University of Johannesburg, South Africa
Prof. V. Leendertz, North-West University, South Africa
Prof. L. van den Berg, North-West University, South Africa
Prof. S. Kot, Czestochowa University of Technology, Poland
Prof. R. Nunkoo, University of Mauritius, Mauritius
Dr. V. Tandrayen-Ragoobur University of Mauritius, Mauritius
Dr. S. Fauzel, University of Mauritius, Mauritius
Dr. D. Ramdhony, University of Mauritius, Mauritius
Dr. U. Subadar-Agathee, University of Mauritius, Mauritius
Prof. R.P. Gunputh, University of Mauritius, Mauritius
Dr. R. Mahadew, University of Mauritius, Mauritius
Prof. K. Khedo, University of Mauritius, Mauritius
Dr. R. Babajee, University of Mauritius, Mauritius
Dr. J. Matadeen, University of Mauritius, Mauritius
Dr. V. Ramlugun, University of Mauritius, Mauritius
Prof. A. Henrico, North-West University, South Africa
Prof. L. Van Staden, North-West University, South Africa
Prof. Roland Goldberg, North-West University, South Africa
Dr. V. Ramiah, University of Wollongong, Australia

Review Process

The first call for papers went out early on 11 December 2023, with a due date of abstract submissions by 28 March 2025. A total of 269 abstracts were submitted, and abstract acceptance was communicated by 18 April 2025. The abstracts were blind peer-reviewed by appointed subject chairs who are experts within their respective fields. Authors with accepted abstracts were invited to submit a full paper with a deadline of 20 June 2025. A total of 125 full papers were received. These papers were subjected to a double-blind peer review process managed by the chair of the reviewers' committee according to South Africa's Department of Higher Education and Training (DHET) refereeing standards. Subject chairs within the different tracks of Business Management, Economics, Finance, Public Relations, Human Resource Management, Marketing, Sport and Leisure and Technology Integration identified reviewers unknown to the authors and from a different institution for the review process. Likewise, the identity and the institution of the reviewer remained anonymous.

A database of possible reviewers was compiled by the SSIRC organising committee, indicating each reviewer's area of expertise. Each subject chair drew from this database within their subject field or included new additions to the list of reviewers. Some of the reviewers were authors of papers; however, several reviewers were from international institutions. The list of the scientific committee and reviewers is provided within the proceedings. At least two anonymous reviewers reviewed each paper submission, and where discrepancies occurred, a third reviewer was used, or the head of the reviewers performed the third review.

Papers were reviewed according to the following criteria:

- Significance of the paper
- Originality
- Content includes the abstract, quality literature review, and/or research question
- Appropriateness of research methodology
- Adequacy of results and discussion
- Technical, including language, spelling and references

Authors received a reviewer report on the above-mentioned aspects and attended to comments and recommendations where needed by submitting a change log when required by reviewers. Eight papers were rejected after review. The second manuscript submissions, if needed, were assessed by the head of reviewers when papers were returned, or alternatively reviewed for a second round by the reviewer according to their comments. In total, 98 full papers were included in the conference proceedings.

Table 1 Institutional and weighted author contributions to proceedings

No	Name of Institution	SSIRC Proceeding (NWU)	SSIRC Proceeding (Springer)	Total SSIRC Proceeding	Institutional %
1	Botho University		1	1	1,69
2	Central University of Technology		1	1	1,69
3	Independent contractor	1		1	1,69
4	Nelson Mandela University	3		3	5,08
5	North-West University	3	6	9	15,25
6	Reverend Father Moses Orshio Adasu University Makurdi, Benue State-Nigeria	2		2	3,39
7	Rhodes University		2	2	3,39
8	Sol Plaatje University		1	1	1,69
9	STADIO		3	3	5,08
10	Tshwane University of Technology	1	3	4	6,78
11	University of Fort Hare		1	1	1,69
12	University of South Africa		5	5	8,47
13	University of Johannesburg	3	3	6	10,17
14	University of Mauritius	1	1	2	3,39
15	University of Mpumalanga		1	1	1,69
16	University of Pretoria	1		1	1,69
17	University of South Africa	1		1	1,69
18	University of the Free State	1	2	3	5,08
19	Vaal University of Technology	3	9	12	20,34
		20	**39**	**59**	100,00

Institutional Contributions to Proceedings

After the double-blind peer review process, a total of 59 full papers were accepted for inclusion in the proceedings. In total, 93 submissions, comprising both full papers and abstracts, were received and presented from 19 national and international institutions, as illustrated in Table 1. The host institution, NWU, contributed 15.25% of the total proceedings output, while 84.75% of the contributions originated from other institutions. This distribution aligns with the DHET requirement that more than 60% of published papers must be sourced from institutions other than the host.

It should be noted that the first volume is published through the traditional outlet hosted on the North-West University website, whereas the second volume is published in the Springer Proceedings Series, based on conference attendee preferences. Duplicate publication was not permitted; each paper was published in only one volume, in accordance with DHET guidelines.

Programme Thursday, 6 November 2025		
8:30-9:30	Registration	Conference Hall 1
9:30-9:45	**Prof Ephrem Redda and Prof Crizelle Els** Welcome and opening of the 10th SSIRC	Conference Hall 1
9:45-10:30	**Prof Babs Surujlal and Prof Herman van der Merwe** SSIRC: Celebrating a decade of impact and a future of possibilities	Conference Hall 1
10:30-11:00	Refreshment break	
11:00-12:00	**Prof Leon de Beer** (Norwegian University of Science and Technology/Nort-West University) Conference keynote address: When work thinks back: The impact of artificial intelligence on work-life and the human condition.	Conference Hall 1
12:00-13:00	Conference lunch	

Track Session 1: Thursday, 6 November, 13:00–14:30
BUSINESS MANAGEMENT—Conference Hall 1—Session chair: Prof Ephrem Redda
Edward George Lindeque (P) Risk dimensions of artificial intelligence: Navigating the unchartered waters of innovation.
Monish Lallbeeharry; **Rajesh Sannegadu (P)**; & Sharmila Pudaruth Identifying the project management success factors in the Mauritian construction industry.
Lizl Steynberg (P); & Jan Grundling Shadows of control: The hidden costs of authoritarian leadership in South African higher education.
Robert Nicky Tjano (P) A rhetoric in CEO and board chairperson's statements of South African state-owned entities: A perspective from directed thematic content analysis.
Sillas Molosiwa (P); Samuel Chingoiro; Olumide Jaiyeoba; & Mornay Roberts-Lombard Trends and drivers of private investment towards SDG 9 financing in Botswana: A case for pooled funding.
ECONOMICS—Meeting Room 1—Session chair: Dr. Ntombi Gamede
Niilonga Shetunyenga; & Lucius Botes (P) Factors contributing to poor implementation of rural development strategies in Namibia.
Nicholas Ngepah; Charles Saba; & Patience Ndidiamaka Agu (P) The impact of cross-border trade on poverty reduction in Africa.
Eric Effah Sarkodie; & **Ismael Maloma (P)** Poverty alleviation through microcredit in selected sub-Saharan African countries.

(continued)

(continued)

SERVICE EXCELLENCE IN HIGHER EDUCATION—Meeting Room 3—Session chair: Dr. Tandiwe Mdlungu
Bulelwa Nguza-Mduba (P) The impact of pedagogical approaches in the teaching of entrepreneurship education at a South African university.
Shaun Peens (P) Rethinking mentorship in teacher training: Evaluating the impact of school exposure sessions on student teachers in South Africa.
Nomzamo Mpanza Xaba (P); Green Paul; Smangele Moyane; & Jane Skinner Evaluating the viability of the CHE's 2013 four-year curriculum proposal: Bridging the articulation gap and enhancing student success at South African universities of technology.
HUMAN RESOURCE MANAGEMENT—Meeting Room 4—Session chair: Prof Ankit Katoria
Gibson Muridzi (P); & Shepherd Dhliwayo Investigating staff turnover in electricity sector in Zimbabwe.
Mosekama Osia Mokhele (P); & Simon Matome Nkgapele Can centralisation strengthen human resource management functionality in South African public sector: Recruitment of senior management services perspectives.
Pierre Joubert (P); & Adelaide Zandile Mqokozo Employee empowerment, career satisfaction and intention to stay of employees in a printing organisation in Pretoria.
Henk de Jager (P); & Julian Makgabo Nkoana Exploring issues of leadership in a selected municipality in North West Province.
Miranda Hlengiwe Mhlongo (P); & Vincent Xolani Mhlongo Artificial intelligence and ageing academics in South African higher education: Navigating technological shifts and pedagogical challenges.
Refreshment break
Track Session 2: Thursday, 6 November, 15:00-16:30
BUSINESS MANAGEMENT—Conference Hall 1- Session chair: Dr Robert Nicky Tjano
Merwe Oberholzer (P); Leonah Christine Muzangwa; & Susanna Levina Middelberg Zimbabwean Command Agriculture Input Subsidy Programme: Perceptions of producers and officials.
Lemlem Ghebreselassie Amhatsion (P); & Ndivhuwo Sirembe A systematic literature review on strategic and performance information disclosure by universities.
ECONOMICS—Meeting Room 1—Session chair: Prof Lucius Botes
Carol Moncho**; & Abraham Van der Vyver (P)** A critical analysis of the conversational strands on Twitter (X) regarding the postponement of the presentation and acceptance of the South African budget of 2025.
Michał Kot; & **Sebastian Kot (P)** Migration and employment evolution in Central European countries.
Naledi Masango (P); & Gillian Mrubula-Ngwenya BRICS in the Global South: Transformative technologies for economic resilience.

(continued)

(continued)

Ntombi Gamede (P) The determinants of youth unemployment in regions D and G of Johannesburg, Gauteng Province.
SERVICE EXCELLENCE IN HIGHER EDUCATION—Meeting Room 3—Session chair: Prof Bulelwa Nguza-Mduba
Mariette Frazer (P) Developing self-leadership through innovative pedagogies in retail education.
Sannah Molefe (P) Exploring the impact of artificial intelligence in blended learning in higher education.
Bernadette Kruger (P); & Nicolene Barkhuizen A Three Segment Pyramid framework for academic identity reconstruction in a VUCA world.
HUMAN RESOURCE MANAGEMENT—Meeting Room 4—Session chair: Prof Pierre Joubert
Bulelwa Ntunzi (P); Siphosethu Mapete (P); Avela Jojo; Sive Ben; & Themba Mjoli The role of stress, anxiety and job insecurity in organisational restructuring at a higher education institution.
Falkor Eckardt (P); & Crizelle Els Bridging academia and industry: Equipping students for the diversity-driven workplace.
Thabile Soga (P); & Doreen Morukhu Flexible work arrangements for improving work-life balance.
Kholiswa Mogoeng (P) Enhancing employee well-being through ethical leadership in the South African public sector: The mediating role of perceived organisational support.
Rosina Mavunga; & **Siphokazi Kwatubana (P)** Coping with prejudice in the context of multiple deprivations: Women in leadership positions in tertiary education in Zimbabwe.
End of conference—Day 1 (Check your 'Things to do list' for excursions)
Track Session 3: Friday, 7 November, 8:30-10:00
BUSINESS MANAGEMENT—Conference Hall 1—Session chair: Lemlem G. Amhatsion
Olusegun Adesile (P) Analysing the short-term dynamics between consumer spending, formal and informal SMEs, and employment growth in South Africa: A bound testing approach.
Angel Khululiwe Shabangu (P); & Moses Jametane Matsheke Hustling in the Hood: Challenges faced by small-scale retailers in a selected township in Gauteng, South Africa.
Edwill Mtengwa; **Lena Lainah Maregere (P)**; Ravinder Rena; & Joseph Lekunze Evaluating the extent to which an entrepreneurial orientation exists among the beef cattle small and medium enterprises in Zimbabwe.
Bernard Lama Ngota (P); **Elroy Eugene Smith (P)**; & Ayanda Pamella Deliwe Immigrant entrepreneurial ecosystems in South Africa: Unlocking the resilience of African immigrant entrepreneurs against all odds.
Marcia Lebambo (P) An analysis of South Africa's entrepreneurship policy environment in fostering youth entrepreneurship development.

(continued)

(continued)

ECONOMICS—Meeting Room 1—Session chair:
Afees Noah; & **Grobler Wynand (P)** Governance quality's role in enhancing entrepreneurship and economic growth: Evidence from African economies.
Clive Coetzee (P) Making sense of the proliferation of image classification methods and techniques using Landsat 9 satellite imagery.
Gibson Muridzi (P); & Shepherd Dhliwayo Investigating the influence of artificial intelligence on small business competitiveness – A systematic literature review.
Molepa Seabela; **Kanayo Ogujiuba (P);** & Maria Eggink South Africa in context: A structured literature review on socioeconomic determinants of income inequality.
SERVICE EXCELLENCE IN HIGHER EDUCATION—Meeting Room 3—Session chair:
Mothemba Cecilia Mokoena (P); & Gauda Johannes Maseko Relationship between entry qualification and academic performance of accountancy students at a university of technology.
Mothemba Cecilia Mokoena (P); & Gauda Johannes Maseko Determining perceptions and attitudes of non-accounting majors towards accounting: A case study of a selected universities of technology in South Africa.
Henk de Jager (P); Mark Bussin; & Calvin Mabaso Exploring student capital as a competitive advantage in teaching excellence for private higher education institutions.
Lilian Nwosu (P); Veruschka Pelser-Carstens; & Anina Marx Lecturers' perspectives on the use of ChatGPT in accounting education: Opportunities and threats.
Mfanelo Ntsobi (P); Bongani June Mwale (P); Blondel Nyamkure (P); & Vincent Zulu Evaluating the effectiveness of the secondary school improvement programme in Gauteng: A mixed-method approach analysing student outcomes and educational practices.
HUMAN RESOURCE MANAGEMENT—Meeting Room 4—Session chair:
Renier Steyn; & **Frans Lekukela Ramonyatse (P)** Mapping the landscape of leadership styles: A contemporary review.
Kuldeep Jobanputra; Sanjay Bhayani; & **Ankit Katrodia (P)** Exploring the drivers of employee retention: The roles of organisational support, work-life balance, and empowerment in job satisfaction.
Melody Mtimkulu (P); & Siphokazi Kwatubana Optimising academic success: Exploring influential factors in the school's physical environment for learners.
Melody Mtimkulu; & **Nomsa Mohosho (P)** Walls to wellness: Safe, clean, and green pathways to healthier school environments.
Refreshment break
Track Session 4: Friday, 7 November, 10:30-12:00
BUSINESS MANAGEMENT – Conference Hall 1—Session chair: Dr Olusegun Adesile

(continued)

(continued)

Lodewikus Janse van Rensburg (P) The stability-innovation paradox: How traditional companies foster intrapreneurship through organisational stability.
Vishay Heeralal; & **Bibi Zaheenah Chummun (P)** Investigating the factors that influence branding of grocery retail shopping apps in Durban.
Jeremiah Machingambi (P); & Edward M. Rankhumise Literature review on entrepreneurship education in South Africa: Towards developing a small business development framework.
Moses Jametane Matsheke (P); & Angel Khululiwe Shabangu The influence of instrumental readiness on entrepreneurial intentions of students at a selected university of technology in Gauteng.
Victoria Adekomaya (P); & Chris Schachtebeck Effect of personal traits on conflict management strategies in family business succession planning in South Africa.
BUSINESS MANAGEMENT—Meeting Room 1—Session chair:
Sithembiso Govuzela (P) Designing industrial image processing technology for warehouse management within supply chain management in South Africa.
Solly Chiloane; & **Elizabeth Chinomona (P)** Supplier risk management on supplier quality improvement and supplier commitment: A case of South African health industry.
Xolani Richard Mwandla; **Elizabeth Chinomona (P)**; & Pfanelo Nematatani Warehousing effectiveness in the automotive industry: A case of South Africa.
Molebaleso Lydia Ntshingila (P) Streamlining retail's supply chain performance using internet of things: Optimising e-warehouse and transportation management.
Ntombizodwa Matsoma (P); & Marcus Ambe The impact of supplier development and strategic partnership on performance of clothing manufacturers in South Africa.
SERVICE EXCELLENCE IN HIGHER EDUCATION—Meeting Room 3—Session chair: Prof Lilian Nwosu
Mariusz Sroka (P) Fire protection as an element of safety education at universities.
Lemlem Amhatsion (P); & Ndivhuwo Sirembe (P) Tracing the career paths of Thuthuka bursary funded students: A case study of the University of the Free State.
Tlotlo Segotso (P); Lilian Nwosu (P); Musili Oladele; & **Innocent Mashava (P)** Teaching and learning of accounting in higher learning institutions through the lens of the Fourth Industrial Revolution: A systematic literature review.
Conference lunch
Track Session 5: Friday, 7 November, 13:00-14:30
BUSINESS MANAGEMENT—Conference Hall 1—Session chair: Prof Vivence Kalitanyi
Victoria Adekomaya (P); & Chris Schachtebeck Balancing autonomy and oversight: Strategies for improving delegation in South African SMEs.

(continued)

(continued)

Annelie Steenkamp (P); & Luzaan Hamilton Fostering student entrepreneurship: A literature review on entrepreneurial ecosystems in higher education.
Grace Magama (P); & Peet Venter The role of continuous communication channels and feedback loops in private higher education institutions driving emergent strategising.
Storm Watson (P); & Miemie Struwig The development of a structured approach to scenario planning.
LAW—Meeting Room 1—Session chair:
Sannah Molefe (P) Investigation of artificial intelligence identification methods in alleviating identity fraud in Gauteng.
Godfrey Thenga (P) Shining a spotlight on human rights violations: An appraisal of the recycling business and law enforcement in South Africa.
Tinyiku David Ngoveni (P) Stakeholders crime prevention initiatives: A rural-based phenomenological study of Limpopo province (South Africa).
Marjone Van der Bank; & **Nadien Raath (P)** Workplace safety and cannabis: Lessons learned from the Enever v Barloworld Equipment South Africa, a Division of Barloworld South Africa (Pty) Ltd (JA86/22) [2024] case.
Tinyiku David Ngoveni (P) Multi-disciplinary approaches for the prevention illicit drug use and supply: A case study of University of Limpopo (South Africa).
MARKETING—Meeting Room 3—Session chair: Dr. Storm Watson
Chantal Rootman; **Sumaiya Sidat (P)**; & Asiphumeze Nkula Challenges of implementing digital marketing practices: Views of small retail businesses.
Alyssa Shawntay Williams (P) Unlocking electronic-customer experience (e-CX) excellence: The impact of electronic service quality (E-S-Qual) on post-purchase satisfaction in South African e-commerce retail organisations.
Lena Lainah Maregere (P); Nombulelo Dilotsotlhe; & Edwill Mtengwa An analysis of the impact of personal selling on SME poultry businesses in Gauteng Province of South Africa.
Thubelihle Ndlela (P); & Tidimalo Raditlhalo Examining digital marketing strategies for price-conscious young consumers on mobile grocery apps in South Africa.
BUSINESS RESILIENCE—Meeting Room 4—Session chair: Prof Gauda Maseko
Tlou Raphela (P); & Laura Mosiah Evaluating the hospital resilience to pandemics: A case of General Justice Gizenga Mpanza Regional Hospital, in South Africa.
Brad Bell; & **Mujinga Tshimanga (P)** Strategic risk prioritisation: Key threats facing large South African retailers over the 2025-2029 horizon.

(continued)

(continued)

Patient Alimonjo Mulolo; **Gauda Johannes Maseko (P)**; & Innocent Mudhombo Financial implications of Covid-19 on profitability of SMEs in the hospitality industry in Sedibeng District Municipality.
Tlou Raphela (P) Evaluating resilience strategies for pandemics in African countries: A case of healthcare facilities.
Refreshment break
Track Session 6: Friday, 7 November, 15:00-16:30
BUSINESS MANAGEMENT—Conference Hall 1—Session chair: Prof Luzaan Hamilton
Sinovuyo Noah (P); & Marcia Lebambo (P) The barriers to women's leadership progression within the retail sector in Gauteng, South Africa.
Vivence Kalitanyi (P) Influence of pro-activeness on financial and non-financial performance of retail SMEs in Tshwane Metropole.
Misheck Musaigwa; & **Vivence Kalitanyi (P)** Strategic agility and continuous innovation in the digital era.
Mccpowell Fombang (P); & Gauda Maseko Investigating accounting by small and medium size businesses in Emfuleni municipality.
MARKETING—Meeting Room 1—Session chair: Dr Thubelihle Ndlela
Storm Watson (P); & Elroy Smith The impact of demographics on customer attitudes towards zero-waste shopping.
Marina Precious Ngobeni (P); Nadine Oosthuizen; & **Alyssa Shawntay Williams (P)** Beyond the feed: How influencer self-disclosure transforms followers into customers.
FINANCE AND ACCOUNTING—Meeting Room 3—Session chair: Dr Aletta Neethling
Bouba Ismaila (P); & John Beneke Evaluating women financial inclusion level between French and English-speaking countries in sub-Saharan Africa.
Bibi Fatimah Boodoo; **Boopendra Seetanah(P);** & Sheereen Banon Fauzel. Financial inclusion as a catalyst for economic development: Evidence from Mauritius.
Kedibone Thantsha (P); Nicky Tjano; & Nthabiseng Moraka The moderating role of human capital on women directors and corporate sustainability disclosures tone in board chairs' statements of JSE-listed mining companies.
Cayla Viljoen; & **Herman Albertus Viviers (P)** Income tax treatment of assessed losses: A three-way comparative analysis based on the fundamental principles of good tax policy.
EMERGING TECHNOLOGY AND MANAGEMENT—Meeting Room 4—Session chair:
Rebecca Momanyi (P); & Hannah Dunga Evaluating the impact of e-commerce on food supply chain and quality performance for fresh food in Gauteng Province, South Africa.
Lehlohonolo Amos Masitenyane (P) Listener perceptions of university-community radio station programmes and content towards continued listening intention.

(continued)

(continued)

Bouba Ismaila (P); & John Beneke The role of AI, blockchain, and digitalisation in providing financial services to unbanked populations: A bibliometric analysis of literature.
Olukayode Lasisi; **Bisiriyu Abiodun Taleat (P)**; Kedibone Phago; & Muslim Akintola Agri-tech and livestock management: Economic and policy perspectives for sustainable agricultural development in Nigeria and South Africa.
Shepherd Dhliwayo; & **Solomon Gbaka (P)** Modelling the linear and asymmetric effect of financial innovation on climate resilience in Nigeria.
AWARD AND CLOSING CEREMONY—Conference Hall 1 **17:00 – 18:00**
Prof Crizelle Els Closing of the 10th SSIRC
Prof Bennie Linde and Prof Ephrem Redda Best Paper Award Best Student Paper Award Best Presentation Bingo prize
Final closing

Important Information

Refreshment breaks:
Arrival time: 7:30 – 8:00
Mid-morning tea: 10:00 – 10:30
Lunch: 12:00 – 13:00
Afternoon tea: 15:00 – 15:30

Presentation Guidelines

- There are laptops and data projectors in each room. Access to MS PowerPoint will be available, but other presentation software might not be available.
- The presentations are scheduled for a minimum of 15 minutes in total; 10 minutes per presentation and 5 minutes for interaction.
- Please bring a USB stick to the track session convenor (chair) to download on the laptop. We will not accept emails with the presentations.
- Multiple presenters can be used, but the time will be the same for the paper, no matter the number of presenters. Please keep within the time limitations set for the presentation.
- There are comfort breaks and lunches before each track session, where you will have time to accustom yourself to the room and download your presentation.

- It might happen that there are fewer than five presentations per track session, in which case the session convenor can allow more discussions.
- There might not be presenters (clickers) in the rooms and you will need to bring your own, should it be necessary.

Disclaimer

The editors assume no responsibility or liability for any plagiarism errors or omissions in the content of this document. The information in this document was provided by the authors and is published on an as-is basis. The only expectations are technical layout and formatting.

Please Scan for the Digital Programme:

NWU®

WorkWell Research Unit

S·S·I·R·C

Contents

Risk Dimensions of Artificial Intelligence: Navigating the Uncharted Waters of Innovation

Edward George Lindeque

Abstract As artificial intelligence (AI) evolves and integrates into society, it is set to disrupt every industry, introducing novel risks. In the financial services sector, the lack of model transparency, current ethical dilemmas, and data privacy issues present significant challenges. This study identifies potential risks encountered during AI development and deployment projects within this sector. Using a narrative literature review, risk data were collected from various secondary sources. The research developed an AI project risk breakdown structure (AIPRBS) that includes both risks common to information technology projects and those unique to AI systems. The analysis revealed 12 common project execution risk categories: project sponsorship, funding, end-user, stakeholder relations, scope, AI system requirements, planning and control, scheduling, project staffing, organisational dynamics, system development, and external dependencies. Unlike traditional software, AI-based technology risks can extend beyond a single enterprise, affecting multiple organisations and society at large. AI introduces risks not fully covered by conventional risk management frameworks, many of which are novel or heightened compared to traditional software projects. The study further groups AI development and deployment project risks into ten categories: economic, human-system collaboration, socio-economic and environmental, ethical and societal, system data, system technology, operational, AI model, legal and regulatory, and security and safety. Many of these distinctive risks are challenging to mitigate, and new, unforeseen risks are likely to emerge, necessitating ongoing vigilance and periodic reviews of risk profiles. The findings offer a completeness check for future AI project risk assessments in the financial services sector, ensuring risks are taken knowingly.

Keywords Artificial intelligence · AI project risks · AI systems risk · Financial services sector · Risk breakdown structure

E. G. Lindeque (✉)
North-West University, Vanderbijlpark, South Africa
e-mail: eddie.lindeque@outlook.com

E. H. Redda (ed.), *Proceedings of the Social Sciences International Research Conference, 2025*, https://doi.org/10.1007/978-981-95-9489-4_1

1 Introduction

Artificial intelligence (AI) is no longer considered an emerging technology, but rather an impetus to reforming industries and societies at an unparalleled pace into a 'wisdom economy' rather than a 'transaction economy' (Werner 2025). AI introduces novel risks, as it is one of the most transformative technologies since the invention of the steam engine (Mayer et al. 2025). Society is generally acknowledging that AI marks a pivotal moment in world history and is increasingly being adopted as both a business and a national strategy. Investment in AI has reached unprecedented levels, and the hype is expected to continue for the foreseeable future (Gartner 2024; Zeng et al. 2024). According to Gartner (2024), 79% of organisations are allocating new budget for AI implementation in the 2025/26 financial year, a 31% increase from the previous year. However, the financial services industry is still considered an emerging investor in AI systems.

Judging from the 2024 global AI summit in South Korea, there is still no global consensus on the risks associated with AI development and deployment. Twenty-seven countries, including the European Union, France, South Korea, the United Arab Emirates, the United Kingdom, and the United States, committed to developing proposals for evaluating AI risks at this summit. This agreement, which concluded the AI Seoul summit, marked the first time these nations had agreed to establish shared risk thresholds for advanced AI. These thresholds will help identify when AI capabilities might pose 'severe risks without proper safeguards', such as enabling AI to bypass human oversight through manipulation, deception, or autonomous replication and adaptation (Hern 2024; Wodecki 2024).

The National Institute of Standards and Technology (NIST 2023) argues that understanding and managing AI risks will enhance trustworthiness and foster public trust. An increasing number of scholarly articles are being published on various aspects of AI-related risks. However, the current literature does not cover an AI project risk breakdown structure (AIPRBS) for implementing AI systems. A risk breakdown structure (RBS) has its roots in project management. It is simply a hierarchical decomposition of project risks. It categorises risks into levels, from broad categories at the top to more specific risks at lower levels. The primary purpose is to identify and organise risks in a structured manner, making it easier to anticipate, manage, and address them (Chapman 2013; Hillson 2002; Holzmann and Spiegler 2011; PMI 2009).

An RBS provides a framework for use with most risk assessment techniques, whether qualitative or quantitative. This paper aims to develop an AIPRBS, which represents a list of potential risk sources that can serve as a completeness checklist to avoid blind spots in risk identification during the development or deployment phases of an AI project in the financial services sector. It is therefore not meant to be a taxonomy of AI risks alone. In addition, the AIPRBS only covers project risks after the feasibility stage of an AI project.

2 Literature Review

2.1 *Risk Defined*

Risk identification is the first step in mitigating and managing project risks. Project risks are typically seen as adversely affecting the project's timelines, scope, cost, or quality. International standards organisations provide various perspectives on what constitutes risk. For this study, all three definitions below are accepted:

- Risk is described as the impact of uncertainty on objectives. This definition supports the notion that risk encompasses not only negative consequences, but also positive opportunities (ISO 2018).
- Information technology (IT) risk is a type of business risk that is directly linked to the use, ownership, participation, influence, and integration of IT systems within organisational operations (COBIT 2019).
- Risk is a combined measure of the likelihood of an event happening and the extent of its consequences. The impact of an AI system can be harmful, positive, or a mix of both, leading to either risks or opportunities (NIST 2023).

2.2 *Balancing AI Project Risks*

Software projects are often challenging to manage and frequently fail to meet the project objectives. Effective project risk management is crucial for software development initiatives, as it enables planning and control throughout execution. It helps reduce the likelihood of project failure and enhances the value of other project management processes. The main goals of software project risk management are to minimise the probability and impact of potential risks while maximising opportunities. This process should be aligned with existing organisational practices and policies (Arnuphaptrairong 2011; Keil et al. 1998; PMI 2009; Tesch et al. 2007).

The World Economic Forum (WEF 2024) regards AI as central to the Fourth Industrial Revolution, characterised by merging technologies and the convergence of the physical, digital, and biological realms. AI is rapidly integrating itself into the financial services industry. Investors in AI should guarantee that all applications are responsible, meaning that they are honest, beneficial, and safe. Developing and deploying AI in alignment with responsible AI principles is crucial for enhancing 'risk-adjusted returns' and positioning companies for innovative growth (ISO 2023a).

AI shares certain risks with traditional software systems, such as security vulnerabilities; however, it also introduces unique risks. Many of these risks have been publicised, and some will be challenging to mitigate, let alone eliminate. Additionally, new and unforeseen risks are likely to emerge as AI continues to evolve. Therefore, continuous vigilance and risk assessment are essential as this technology advances (Marrie et al. 2023).

These unique risks arise from AI's essential characteristics: its capacity to learn from data, make autonomous decisions, and interact with the physical world (Mukherjee 2024). As with traditional software, the risks associated with AI-based technology can extend beyond a single enterprise, affecting multiple organisations and having societal repercussions. AI systems introduce risks not fully covered by existing risk frameworks and methodologies. Compared to traditional software project risks, these AI-specific risks are either novel or heightened (NIST 2023).

Risk management is therefore vital to the development and deployment of AI systems. The unique characteristics of AI technologies, such as autonomy, complexity, and adaptiveness, along with their rapid development, make it challenging for organisations to stay ahead of potential risks throughout a project. Consequently, organisations struggle to identify, assess, and address all potential risks and implement effective risk management strategies. The complexity of AI risks necessitates the development of new tools and methodologies to address the limitations of existing risk management frameworks (ISO 2023b; Tartaro et al. 2024).

3 Methodology

3.1 Research Design

A narrative literature review was employed to capture software project risks and risks associated with AI system development and deployment, drawing on various secondary sources. This type of review is beneficial for compiling and synthesising literature in a particular field of study, aiming to give the reader a thorough background of current knowledge (Cronin et al. 2008). The risks identified from these documents were thematically organised into risk categories or themes.

3.2 Sampling Method

A purposive sampling method was employed, focusing on data related to IT software project risks or AI risks in the financial services sector. Articles on risks relating to AI applications used in agriculture, cybersecurity, education, gaming, healthcare, lifestyle, space exploration, transportation, and robotics, in any form, were, therefore, excluded from consideration. Articles covering artificial general intelligence (AGI) and 'frontier AI models' were also excluded from the literature review.

3.3 Data Collection

Sources such as peer-reviewed articles, conference proceedings, postgraduate manuscripts, and grey literature (including official reports, industry practitioners' reports, news releases, and blogs) were considered. Industry standards (such as the International Organization for Standardization (ISO), the International Electrotechnical Commission (IEC), and National Institute of Standards and Technology (NIST)), discussion documents, and other frameworks also informed the literature review. Only English articles or translated documents were considered. Electronic databases, including EBSCOhost, Emerald Insight Journals, Google Scholar, JSTOR, arXiv, and Scopus, were searched to identify relevant articles.

3.4 Data Analysis

Thematic analysis, employing the constant comparative method, was conducted on a selection of articles, standards, conference proceedings, and other relevant documents. The results were captured in an electronic workbook. Theoretical coding saturation was reached when no new themes, categories, or insights emerged from the data (Fusch and Ness 2015). Since no additional codes were generated from the review of new information, it confirmed that the risk codes were well-defined and comprehensive. The Massachusetts Institute of Technology's AI risk repository, developed and maintained by Slattery et al. (2025), was used to triangulate the results of this study.

4 Results and Discussion

A total of 31 secondary sources were considered for the thematic analysis of the AI project risks, yielding 22 risk categories derived from 92 individual risk codes identified in the literature. Two higher-tier risk dimensions were used to classify the risk items in the AIPRBS, namely, 'AI project execution risks' and 'AI development and deployment risks'. All sources considered are cited in Tables 1 and 2. The results of this literature review are presented and discussed in aggregate by risk category.

4.1 Software Project Risks

The software project risks reflected in Table 1 provide a balanced view of all risks a financial services company may face when implementing AI. The project risks were aggregated into the following 12 shared risk categories derived from the literature:

Table 1 AIPRBS—AI project execution risks

Risk category	Risk items	Sources
Project sponsorship	Inadequate management commitment Roles and responsibilities	(Arnuphaptrairong 2011); (ISO 2023a); (Schmidt et al. 2001); (Tesch et al. 2007)
Funding	Budget omissions Cost overruns Underfunding development	(Schmidt et al. 2001); (Tesch et al. 2007)
End-user	User not committed Expectations not met Conflict Inadequate user involvement Resistance to change	(Arnuphaptrairong 2011); (ISO 2023a); (Nikolaenko and Sidorov 2023); (Schmidt et al. 2001); (Tesch et al. 2007)
Stakeholder relations	Ineffective communication Unclear objectives	(Arnuphaptrairong 2011); (ISO 2023a); (Tesch et al. 2007); (Wirtz et al. 2022)
Scope	Excessive scope New technologies Scope creep Weak business case Weak change management	(Arnuphaptrairong 2011); (ISO 2023a); (Schmidt et al. 2001); (Tesch et al. 2007); (Wirtz et al. 2022)
AI system requirements	Requirements creep Stakeholder identification failure Technical specifications lacking Omitting requirements Changing requirements Requirements not frozen	(Arnuphaptrairong 2011); (Schmidt et al. 2001); (Tesch et al. 2007)
Planning and control	Database procedures Not monitoring progress Deficient methodology Deficient technology Lacking practical skills	(Arnuphaptrairong, 2011); (NIST 2024); (Schmidt et al. 2001); (Tesch et al. 2007)
Scheduling	Key deliverable timelines Unrealistic milestones Project pre-emption Wrong development strategy	(Arnuphaptrairong 2011); (Schmidt et al. 2001); (Tesch et al. 2007)
Project staffing	Inappropriate staffing New technology adoption Technical complexity Consultant intensive	(Arnuphaptrairong 2011); (Keil et al. 1998); (Schmidt et al. 2001); (Tesch et al. 2007)

(continued)

Table 1 (continued)

Risk category	Risk items	Sources
Organisational dynamics	Corporate politics Evolving competitive landscape Organisational restructuring Impact of management change	(Arnuphaptrairong 2011); (Keil et al. 1998); (Schmidt et al. 2001)
System development	Up-front work System documentation System testing regime	(NIST 2023); (Schmidt et al. 2001)
External Dependencies	Missing deadlines Continuity of operations	(Schmidt et al. 2001)

project sponsorship, funding, end-user risks, stakeholder relations, scope, requirements, planning and control, scheduling, project staffing, organisational dynamics, system development, and external dependencies. The 12 risk categories in this dimension were derived from a thematic analysis of 46 risk items identified across the nine sources, as listed in Table 1.

Project sponsorship is important for any project, and inadequate top management buy-in, commitment, and ownership of an AI systems project remain significant risks. It is, therefore, essential to assign project roles and responsibilities accurately (Keil et al. 1998). Similarly, the budgeting and funding processes are key areas of risk. Omissions during the project budgeting process and inadequate estimations or project scoping will result in underfunding, ultimately leading to delays in project execution (Schmidt et al. 2001).

Managing end-user risk can make or break an AI system project. Risk factors, such as failing to manage end-user expectations, failing to secure their commitment to project deliverables, and inadequate user involvement throughout the project, can have numerous ramifications. It is also essential to manage conflicts between user departments promptly to avoid undue delays in deliverables and, ultimately, resistance to change (Nikolaenko and Sidorov 2023). Adversarial stakeholder relations often arise from ineffective communication of the project scope, goals, and deliverables, as well as a failure to report project progress on time (Tesch et al. 2007; Wirtz et al. 2022).

Many risks threatening software projects are related to scope and requirements matters. Misunderstanding or failing to identify project objectives and subsequently changing them will inevitably lead to project scope creep. Implementing new technologies, such as AI, presents several challenges in setting boundaries and defining project deliverables. Challenges include the need for additional training, potential integration issues with current systems and platforms, and uncertainty about the technology's performance and reliability (Schmidt et al. 2001; Wirtz et al. 2022). Schmidt et al. (2001) argue that a project can lack a proper business case if the system is being developed merely in reaction to market competitive forces. This suggests that if the

Table 2 AIPRBS—AI development and deployment risks

Risk category	Risk items	Sources
Economic	Eroding job quality job displacement competitive dynamics	(Giarmoleo et al. 2024) (Hagendorff 2024) (McLean et al. 2023) (Nah et al. 2023) (Shelby et al. 2023) (Sherman and Eisenberg 2024) (WEF 2024) (Weidinger et al. 2022) (Wirtz et al. 2022)
Human-system collaboration	Bias and homogenisation confabulation interaction risks exploitation of clients limited human involvement	(Giarmoleo et al. 2024) (Hagendorff 2024) (Kilian et al. 2023) (Martin 2024) (McLean et al. 2023) (Nah et al. 2023) (NIST 2023) (Sharma et al. 2024) (Sherman and Eisenberg 2024) (Sun et al. 2023) (Tartaro et al. 20 24) (WEF 2024) (Weidinger et al. 2022) (Wirtz et al. 2022)
Socio-economic and environmental	Energy consumption impact on environment, social and governance (ESG) factors hardware resource requirements power centralisation carbon footprint computational costs	(Chapman 2013) (Hagendorff 2024) (NIST 2023) (NIST 2024) (Saghiri et al. 2022) (Sherman and Eisenberg 2024) (Weidinger et al. 2022) (Wirtz et al. 2022) (Zeng et al. 2024)
Ethical and societal	Unintentional harm morality misalignment unexpected side effects	(Bengio et al. 2024) (Giarmoleo et al. 2024) (Hagendorff 2024) (NIST 2023) (Saghiri et al. 2022) (Sun et al. 2023) (Tartaro et al. 2024) (WEF 2024) (Wirtz et al. 2022) (Zeng et al. 2024)
System data	Data dependency dataset shifts inappropriate training data	(Giarmoleo et al. 2024) (Nah et al. 2023) (NIST 2023) (Somer and Thakmann 2023) (Zhang et al. 2022)
System technology	Technology readiness system modifications AI system lifecycle data storage constraints system hardware issues	(Giarmoleo et al. 2024) (ISO 2023a) (ISO 2023b) (NIST 2023) (Saghiri et al. 2022)
Operational	Lack of internal skills third-party exposure inadequate system maintenance	(ISO 2023a) (NIST 2023) (Sharma et al. 2024) (Somer and Thakmann 2023) (WEF 2024) (Zeng et al. 2024)

(continued)

Table 2 (continued)

Risk category	Risk items	Sources
AI model	Inadequate continual training pre-trained models transparency and explainability model failure detection model robustness reliability and accountability	(Hagendorff 2024) (Hendrycks et al. 2023) (ISO 2023a) (ISO 2023b) (NIST 2023) (Saghiri et al. 2022) (Sharma et al. 2024) (Sherman and Eisenberg 2024) (WEF 2024)
Legal and regulatory	Authorship infringements verifiability of models privacy violations inappropriate regulations	(Giarmoleo et al. 2024) (Hagendorff 2024) (ISO 2023a) (Maham and Küspert 2023) (Martin 2024) (Nah et al. 2023) (Holzmann and Spiegler 2011) (NIST 2024) (Saghiri et al. 2022) (Sharma et al. 2024) (Sherman and Eisenberg 2024) (Somer and Thakmann 2023) (Sun et al. 2023) (WEF 2024) (Weidinger et al. 2022) (Wirtz et al. 2022) (Zeng et al. 2024)
Security and safety	Cybercrime System robustness Model extraction Data poisoning Adversarial cyberattack Ingress Level of automation Information security	Giarmoleo et al. 2024) (Hagendorff 2024) (Hendrycks et al. 2023) (ISO 2023a) (ISO 2023b) (Leech et al. 2024) (Maham and Küspert 2023) (Martin 2024) (Nah et al. 2023) (NIST 2024) (Sharma et al. 2024) (Sherman and Eisenberg 2024) (Sun et al. 2023) (WEF 2024) (Wirtz et al. 2022) (Zhang et al. 2022)

primary motivation for introducing an AI system is to maintain a competitive edge in the financial markets rather than a well-defined business case, it can have significant implications for requirements, technical specifications, and estimation of required resources, such as human, hardware, and software.

Planning and control are about having the appropriate project management methodologies, technologies, and skill sets to closely monitor deviations across cost, scope, schedule, or quality. There should be sufficient procedures in place to ensure the security, integrity, and availability of the training data for AI models, as proposed by the ISO/IEC 42001:2023 Artificial Intelligence Management System (ISO 2023a). The absence of these controls represents risks.

There will always be some level of scheduling risk in any complex project. Risk factors include unrealistic milestones and failing to map key deliverables to agreed-upon timelines (Schmidt et al. 2001). It follows that choosing the wrong development strategy for a bespoke (as opposed to generic) AI model can result in unplanned time delays. It is also not unrealistic to experience a project pre-emption, where the project is postponed or hurried to make way for higher-priority or urgent projects. This can be due to redeployment of scarce resources, strategic shifts, or changes in organisational priorities.

Various risk items should be considered in the project staffing risk category, such as a lack of required knowledge, skills, and experience among project members. The introduction of new technology, such as AI, can be an unfamiliar territory for both users and IT staff. A high level of technical complexity amplifies the need for subject matter experts (Saghiri et al. 2022).

Organisational dynamics, such as restructuring or changes to the management team during the project, can create a mismatch between corporate needs and the project objectives. Changes in the competitive landscape can alter user requirements, thus creating instability in the project. More disruptive is corporate politics, which can have devastating effects on the project, such as underfunding or a project that is not intended to succeed (Arnuphaptrairong 2011; Schmidt et al. 2001).

Under the system development category, risks range from inadequate process and modelling documentation of the system to poor estimation of the time needed for up-front work. Underdeveloped and insufficient software testing standards, as well as the inability to document AI-based practices to the level expected of traditionally engineered software, are considered significant project risks (NIST 2023).

Finally, all software projects are exposed to external dependencies. Project timelines can be affected by AI vendors and project consultants, often resulting in missed deadlines or subpar quality requirements. Vendors or project consultants can also go out of business during the project (Schmidt et al. 2001).

4.2 AI Development and Deployment Risks

Scanning different sources, the common organisational risks associated with the development and deployment of AI systems are reflected in Table 2. The 10 risk categories in this dimension were derived from a thematic analysis of 46 individual risk items identified in 26 sources.

An AI system may have unintended impacts on the organisation's external environment. The economic risk category is a good example, where the deployment of AI may amplify existing labour inequalities and negatively affect job quality, as specific niche skill sets will be required. The automation of current tasks is likely to lead to workforce substitution. Competitive dynamics in the use of AI systems can lead to economic and cultural devaluation of human effort (Giarmoleo et al. 2024; Sherman and Eisenberg 2024).

Human-system collaboration risk revolves around biases and the exploitation of clients. AI may perpetuate algorithmic biases in its training data, algorithmic design, or deployment practices, leading to discrimination and unfair comparative outcomes that can compromise the trustworthiness of the AI system. Confabulation, the creation of confidently presented, yet incorrect or false, content (also referred to as 'hallucinations' or 'fabrications'), can mislead or deceive users, as generative AI (GAI) may inappropriately assign human traits to non-human entities (Hagendorff 2024; Nah et al. 2023; NIST 2024). The way humans engage with AI systems creates epistemic difficulties between AI-generated and human-generated content, resulting in automation bias, over-reliance, and emotional entanglement with GIA. Conversational agents (CAs) may amplify opportunities for client nudging, deception, and manipulation. Additionally, a lack of consideration for human-system interaction and oversight can lead to undetected errors in inputs or outputs (Weidinger et al. 2022).

Socio-economic and environmental factors revolve around issues such as the increased energy consumption required by the higher computational power needed for graphics processing units (GPUs) in deep-learning models, for both training and operational purposes. The increasing demand for precious metals and other materials required for hardware production, such as chips and devices, is also influenced by the expected economic lifespan of the hardware. Misuse of market power, monopolistic advantage, and unfair distribution of benefits can lead to economic centralisation of power by first movers in the AI system market (Zeng et al. 2024). Balancing the strain on resources and training-related carbon emissions with the potential of AI to address environmental problems should be considered by management teams when making decisions to invest in AI systems (Sherman and Eisenberg 2024). It is therefore essential to assess the impact of AI systems on the environmental, social, and governance (ESG) factors as proposed by ISO/IEC 42001 AI Management System (ISO 2023b). This also involves calculating the environmental and planetary impact of the computational costs associated with developing and deploying AI systems.

AI systems pose many ethical and societal dilemmas. Misusing AI for unintended purposes can cause unintentional harm, particularly to clients and society. Morality misalignment occurs when the project team fails to consider the shared societal attributes that define what is considered 'right' or 'wrong' across geographical areas or societies where the organisation operates (Bengio et al. 2024). The inability to foresee or identify the side effects of AI-based systems beyond statistical metrics also poses a risk to the project (NIST 2023).

System data risks can be split into three aspects. Firstly, data dependency arises from the AI system's reliance on data for training tasks and the complexity of such data. Secondly, inappropriate training data, which is the data used to develop an AI system, may not accurately or appropriately reflect the context or intended application of the system, and the actual ground truth may be non-existent or inaccessible. Thirdly, data shifts in the context of AI systems typically refer to a mismatch between the dataset with which the AI system was developed and the dataset on which it was deployed due to variations in data distribution over time (Giarmoleo et al. 2024; Nah et al. 2023; Zhang et al. 2022).

The system technology may not be ready for deployment due to unknown factors such as system limitations, boundary conditions, or performance drift. System modifications made during training, whether deliberate or accidental, can significantly impact an AI system's performance. Hardware risks include hardware errors resulting from defective components or from transferring trained machine learning models between different systems, while increased data collected by the AI system can cause memory or data storage issues. Finally, risks may arise at any stage of the AI system life cycle, such as design flaws, poor deployment, inadequate maintenance, and decommissioning (Giarmoleo et al. 2024; Saghiri et al. 2022).

The AI operational risks are threefold. Firstly, there may be a lack of internal skills to support the system, as users and client-facing staff may not yet be fully competent in using the AI systems. Secondly, another operational risk is the use of software, hardware, and data services provided by external vendors without adequate guardrails. Inadvertent intellectual property (IP) leakage into the public domain can lead to adverse legal and operational consequences. Thirdly, deployed AI systems may require more frequent maintenance and activation of corrective measures due to changes in data, models, or concepts (Sharma et al. 2024; Zeng et al. 2024).

It is reasonable to expect many risks associated with AI models. For instance, the accuracy of training models may diminish due to shifts in data patterns. Using pre-trained AI models increases uncertainty and poses challenges in managing bias, as well as issues with demonstrating validity and reproducibility. Predicting failure modes for emergent properties of large-scale pre-trained models becomes more difficult, raising concerns about whether intelligent agents can be trusted and controlled. Ensuring model robustness is crucial for the AI system to achieve its intended goals and withstand disruptions, including atypical inputs or abnormal changes to the input data. AI-generated output can be challenging to interpret, and assigning accountability for various types of harm caused by AI may be complex (Kilian et al. 2023; Sharma et al. 2024). This complexity is linked to the inability to provide transparency and explainability information to interested parties, such as regulators.

In the legal and regulatory space, it is essential to recognise that AI systems can pose ethical challenges when used in ways that infringe on copyright, patents, or other IP rights. The lack of validation, transparency, and explainability of opaque models used is a risk. AI can handle complex relationships hidden in large datasets and learn from them, suggesting that the decision-making process is based on 'black box' models, which may not be acceptable to regulators. The risk of infringing an individual's privacy rights should not be ignored, as enhanced data aggregation capability may compromise privacy through the unauthorised use or disclosure of personal or sensitive information. AI has the potential to recall private information and inadvertently disclose it (Maham and Küspert 2023; Somer and Thakmann 2023). Therefore, national policy uncertainty and an immature regulatory environment can inadvertently lead to insufficient scrutiny, potentially resulting in a lack of accountability.

If the security and safety risks are not actively managed, they can develop into systemic risks. AI systems face the following intrinsic cybersecurity risks: 'data poisoning attacks', which cause the AI to identify and miscategorise data; 'model extraction', which renders it impossible to restore model data or the entire model; and 'ingress attacks', where AI systems are misled due to specific disturbances (Martin 2024; WEF 2024). Adversarial cyberattacks, in which hackers aim to deceive the AI system by making subtle, undetectable changes to input data, should be guarded against. This type of attack seeks to cause the AI model to generate incorrect results that are mostly undetectable to human observers. The AI systems can cause data leaks due to ineffective controls between external data-sharing platforms. Security breaches can lead to significant harm, including flawed decision-making (Hendrycks et al. 2023; Sun et al. 2023). As a result, cybersecurity threats pose a significant challenge for financial regulatory authorities, as these attacks can directly affect the integrity of the financial institutions' operations.

5 Conclusion

With the emergence of AI, we are in the early stages of the most profound technological shift that humanity will ever experience. Many authors agree that the immature regulatory environment exacerbates the inherent risks of AI projects. Additionally, established risk management frameworks and methodologies are not designed to address the novel risks associated with the use of AI systems. However, the literature purports that understanding and actively managing AI risks will enhance trustworthiness and foster public trust in AI systems deployed by the financial services sector. For this reason, management teams should design innovative controls that can adequately address these dynamic risks.

This paper provides a comprehensive analysis of the risks associated with AI system projects in a financial services company, emphasising the importance of considering both AI-specific development and deployment risks and general software project risks. It is recommended that project teams and regulators consider the 92 risk items listed in the AIPRBS during AI system project risk assessments as a checklist for completeness. Due to the dynamic nature of AI risks, it is recommended that management teams continuously review their project risk registers to identify emerging and unforeseen risks. The AIPRBS is also a valuable tool for reporting and aggregating risks by using the 22 categories derived from thematic analysis.

An inherent limitation of the study is that its objective was not to propose an elaborate generic risk breakdown structure (RBS), as the AIPRBS only aims to inform project risk assessments in the financial services sector. Similar studies can be conducted to expand or modify the risk items for a tailored RBS applicable to AI projects across various industries.

References

Arnuphaptrairong, T.: Top Ten Lists of Software Project Risks: Evidence from the Literature Survey. In: Proc. International MultiConference of Engineers and Computer Scientists, Hong Kong (2011)

Bengio, Y., Hinton, G., Yao, A., Song, D., Abbeel, P., Darrell, T., Harari, Y.N., Zhang, Y.-Q., Xue, L., Shalev-Shwartz, S.: Managing extreme AI risks amid rapid progress. Science. **384**(6698), 842–845 (2024)

Chapman, R.J.: Simple tools and techniques for enterprise risk management, 2nd edn. John Wiley & Sons, Ltd (2013)

COBIT: COBIT Framework: Governance and Management Objectives. ISACA (2019)

Cronin, P., Ryan, F., Coughlan, M.: Undertaking a literature review: a step-by-step approach. Br. J. Nurs. **17**(1), 38–43 (2008)

Fusch, P., Ness, L.R.: Are we there yet? Data saturation in qualitative research. Qual. Rep. **20**, 1–9 (2015)
Gartner: Hype Cycle for Artificial Intelligence—2024. Gartner (2024)
Giarmoleo, F.V., Ferrero, I., Rocchi, M., Pellegrini, M.M.: What ethics can say on artificial intelligence: insights from a systematic literature review. Bus. Soc. Rev. **129**(2), 258–292 (2024)
Hagendorff, T.: Mapping the ethics of generative AI: a comprehensive scoping review. Mind. Mach. **34**(4), 1–20 (2024)
Hendrycks, D., Mazeika, M., Woodside, T.: An overview of catastrophic AI risks. arXiv preprint https://arxiv.org/abs/2306.12001 (2023)
Hern, A.: TechScape: what we learned from the global AI summit in South Korea. Guard. (2024)
Hillson, D.: Use a risk breakdown structure (RBS) to understand your risks. In: Proc. PMI Annual Seminars and Symposium, San Antonio, TX (2002)
Holzmann, V., Spiegler, I.: Developing risk breakdown structures for information technology organizations. Int. J. Proj. Manag. **29**(5), 537–546 (2011)
ISO: Risk management—guidelines (ISO 31000:2018). International Organization for Standardization (2018)
ISO/IEC: Information technology—artificial intelligence—management system (ISO/IEC 42001:2023a). ISO/IEC (2023)
ISO/IEC: Information technology—artificial intelligence—guidance on risk management (ISO/IEC 23894:2023b). ISO/IEC (2023)
Keil, M., Cule, P.E., Lyytinen, K., Schmidt, R.C.: A framework for identifying software project risks. Commun. ACM. **41**(11), 76–83 (1998)
Kilian, K.A., Ventura, C.J., Bailey, M.M.: Examining the differential risk from high-level artificial intelligence and the question of control. Futures. **151**, 103187 (2023)
Leech, G., Garfinkel, S., Yagudin, M., Briand, A., Zhuravlev, A.: Ten hard problems in artificial intelligence we must get right. arXiv preprint https://arxiv.org/abs/2402.04464 (2024)
Maham, P., Küspert, S.: Governing general purpose AI: A comprehensive map of unreliability, misuse and systemic risks (2023)
Marrie, N., Lakic, I., Anteghini, A., Valley, A., Rosman, B., Bekker, M., Shongwe, L.: South Africa and Artificial Intelligence. Boston Consulting Group (2023)
Martin, V.: Artificial intelligence and financial systems: opportunities and risks. In: Proc. PaKSoM 2024, Belgrade, Serbia (2024)
Mayer, H., Yee, L., Chui, M., Roberts, R.: Superagency in the workplace: empowering people to unlock AI'S full potential. McKinsey & Company. (2025)
McLean, S., Read, G.J.M., Thompson, J., Baber, C., Stanton, N.A., Salmon, P.M.: The risks associated with artificial general intelligence: a systematic review. J. Exp. & Theor. Artif. Intell. **35**(5), 649–663 (2023)
Mukherjee, A.: Beyond ISO 42001: the role of ISO/IEC 23894 in AI risk management. Medium. (2024)
Nah, F.-H.F., Zheng, R., Cai, J., Siau, K., Chen, L.: Generative AI and ChatGPT: applications, challenges, and AI–human collaboration. J. Inf. Technol. Case Appl. Res. **25**(3), 277–304 (2023)
Nikolaenko, V., Sidorov, A.: Analysis of 105 IT project risks. J. Risk Financ. Manag. **16**(1), 33 (2023)
NIST: Artificial intelligence risk management framework (AI RMF 1.0). NIST AI 100–1 (2023)
NIST: Artificial intelligence risk management framework: Generative artificial intelligence profile. NIST AI 600–1 (2024)
PMI: Practice standard for project risk management. Project Management Institute (2009)
Saghiri, A.M., Vahidipour, S.M., Jabbarpour, M.R., Sookhak, M., Forestiero, A.: A survey of artificial intelligence challenges: analyzing definitions, relationships, and evolutions. Appl. Sci. **12**(8), 3912 (2022)
Schmidt, R., Lyytinen, K., Keil, M., Cule, P.: Identifying software project risks: an international Delphi study. J. Manag. Inf. Syst. **17**(4), 5–36 (2001)

Sharma, R., Harish, V., Rana, G.: Navigating risk in the age of artificial intelligence: assessing and identifying risks with AI strategies. Educ. Adm.: Theory Pract. **30**(4), 512–528 (2024)

Shelby, R., Rismani, S., Henne, K., Moon, A., Rostamzadeh, N., Nicholas, P., Yilla-Akbari, N.M., Gallegos, J., Smart, A., Garcia, E., Virk, G.: Sociotechnical harms of algorithmic systems: Scoping a taxonomy for harm reduction. In: Proc. AAAI/ACM Conference on AI, Ethics, and Society, pp. 723–734 (2023)

Sherman, E., Eisenberg, I.: AI risk profiles: a standards proposal for pre-deployment AI risk disclosures. In: Proc. AAAI Conference on Artificial Intelligence, Vancouver, BC (2024)

Slattery, P., Saeri, A.K., Grundy, E.A.C., Graham, J., Noetel, M., Uuk, R., Dao, J., Pour, S., Casper, S., Thompson, N.: The AI risk repository: A comprehensive meta-review, database, and taxonomy of risks from artificial intelligence. arXiv preprint https://arxiv.org/abs/2408.12622 (2025)

Somer, P., Thakmann, S.: Risk management of AI in industry: a literature review. In: Proc. 15th Mediterranean Conference on Information Systems (MCIS) and 6th MENACIS, Madrid (2023)

Sun, H., Zhang, Z., Deng, J., Cheng, J., Huang, M.: Safety assessment of Chinese large language models. arXiv preprint https://arxiv.org/abs/2304.10436 (2023)

Tartaro, A., Panai, E., Cocchiaro, M.Z.: AI risk assessment using ethical dimensions. AI Ethics. **4**(1), 105–112 (2024)

Tesch, D., Kloppenborg, T.J., Frolick, M.N.: IT project risk factors: the project management professionals' perspective. J. Comput. Inf. Syst. **47**(4), 61–69 (2007)

WEF: Responsible AI playbook for investors. World Economic Forum (2024)

Weidinger, L., Uesato, J., Rauh, M., Griffin, C., Huang, P.-S., Mellor, J., Glaese, A., Cheng, M., Balle, B., Kasirzadeh, A.: Taxonomy of Risks Posed by Language Models. In, Proc. ACM Conference on Fairness, Accountability, and Transparency (2022)

Werner, J.: Imagination in action summit: pioneering the new era with AI. Imagin. Action. (2025)

Wirtz, B.W., Weyerer, J.C., Kehl, I.: Governance of artificial intelligence: a risk-and guideline-based integrative framework. Gov. Inf. Q. **39**(4), 101714 (2022)

Wodecki, B.: Countries Commit to AI Risk Threshold Sharing at Seoul AI Safety Summit. AI Business (2024)

Zeng, Y., Klyman, K., Zhou, A., Yang, Y., Pan, M., Jia, R., Song, D., Liang, P., Li, B.: AI risk categorization decoded (AIR 2024): From government regulations to corporate policies. arXiv preprint https://arxiv.org/abs/2406.17864 (2024)

Zhang, X., Chan, F.T.S., Yan, C., Bose, I.: Towards risk-aware artificial intelligence and machine learning systems: an overview. Decis. Support. Syst. **159**, 113800 (2022)

Shadows of Control: The Hidden Costs of Authoritarian Leadership in South African Higher Education

Lizl Steynberg and Jan P. Grundling

Abstract This paper explores the rise and impact of authoritarian leadership in South African higher education institutions (HEIs), where rigid hierarchies, micro-management, and secrecy have transformed universities into what the authors call 'factories of sadness'. Set against the backdrop of neoliberal reform and corporatisation, the study interrogates how dictatorial governance undermines academic freedom, faculty well-being, and institutional resilience. Using autoethnography as both method and voice, the research draws on the authors' lived experiences, institutional documents, and field notes to address three core objectives: (1) to identify the traits and behaviour of authoritarian leadership in HEIs; (2) to evaluate its effects on faculty morale, organisational culture, and academic innovation; and (3) to propose strategies for ethical, transparent, and participatory governance. The findings reveal a toxic leadership culture marked by bullying, opacity, and excessive control—conditions that erode collegiality, fuel faculty burnout and disengagement, and weaken institutional trust. The emotional and professional costs are significant, particularly in contexts where academic voices are marginalised or silenced. The study's originality lies in its deeply personal yet analytically rigorous approach, providing rare insight into the lived realities of authoritarianism in academia. It concludes with actionable recommendations to restore collegiality, promote democratic leadership, and foster inclusive, resilient academic cultures.

Keywords Academic morale · Authoritarian leadership · Faculty Well-being · Higher education institutions · Line manager dictator · Organisational culture

L. Steynberg (✉)
Tshwane University of Technology, Pretoria, South Africa
e-mail: steynbergl@tut.ac.za

J. P. Grundling
Centre for Local Economic Development (CENLED), University of Johannesburg, Johannesburg, South Africa

PASCAL International Observatory (Africa), Johannesburg, South Africa

E. H. Redda (ed.), *Proceedings of the Social Sciences International Research Conference, 2025*, https://doi.org/10.1007/978-981-95-9489-4_2

1 Introduction

Centralised authoritarian leadership in South African higher education institutions (HEIs) has transformed academic environments into rigid, controlled spaces, often referred to as 'factories of sadness'. Faculty face coercive hierarchies that prioritise conformity over collaboration, undermining academic autonomy and collegiality. These governance structures prioritise efficiency and financial metrics over innovation and intellectual exchange, leading to dissatisfaction and alienation (Harris and Vick 2023; Olabiyi et al. 2024). Opaque decision-making replaces open dialogue, eroding faculty morale and institutional culture (Brown and Patel 2021; Van Vuuren 2024). Moreover, authoritarian leadership exacerbates socio-economic inequalities and massification, thereby weakening institutional effectiveness. This paper explores its impact on academic morale, organisational culture and institutional outcomes, and offers insights for fostering healthier, more inclusive academic environments.

2 Purpose

This paper examines authoritarian governance in South African HEIs, where dictatorial leadership fosters oppressive environments known as 'factories of sad-ness'. It analyses leadership styles rooted in manipulation, control, and conformity, assessing their impact on faculty morale, institutional culture, and effectiveness. By synthesising empirical evidence, the paper explores coercive power dynamics, addressing key gaps in the literature and advocating for systemic change. Specifically, it aims to:

1. Identify characteristics and behaviours of dictatorial governance in South African HEIs and their contribution to the phenomenon of 'factories of sadness'.
2. Examine its impacts on academic morale, organisational culture, and institutional effectiveness.
3. Propose policy interventions to promote transparent, equitable leadership.

Highlight research gaps and future directions for addressing authoritarianism in academia.

Ultimately, the paper advocates for ethical, transparent, and collaborative governance to counter the pervasive sadness and institutional alienation.

3 Background to the Study

Faculty in South African HEIs increasingly face authoritarian leadership, transforming institutions into 'factories of sadness' that undermine the academic values of dialogue and respect. Corporatisation has centralised power, prioritising efficiency over innovation and intellectual exchange (Harris and Vick 2023; Venter et al. 2024). Socio-economic inequalities and massification further exacerbate these challenges.

Once hubs of critical thought, universities now operate under darkocratic governance, where opaque decision-making and rigid hierarchies restrict autonomy and collaboration. Despite being intended for efficiency, these structures erode morale, alienate faculty, and shift focus from academic missions to financial metrics, making HEIs disengaged and uninspired environments (Brown and Patel 2021; Moyo and McKenna 2021).

South African HEIs face corporatisation pressures alongside historical disparities, mass enrolments and resource limitations, fostering coercive leadership that undermines institutional effectiveness and staff well-being. This study examines the line manager dictator model, revealing how power dynamics and control mechanisms contribute to authoritarian governance. It synthesises empirical evidence to explain how coercive leadership causes the 'factories of sadness' phenomenon, and explores the systemic challenges these governance structures create for faculty morale, organisational culture and institutional effectiveness.

This research advocates for systemic change to address the toxic impacts of dictatorial leadership and to improve teaching, learning, and professional growth environments. By focusing on ethical, transparent and collaborative governance, it aims to provide actionable recommendations to restore collegiality, autonomy and innovation, reclaiming the transformative role of higher education in intellectual engagement and empowerment.

4 Literature Review—Authoritarian Leadership in Higher Education: Examining Structural Control, Toxic Behaviours, and the Erosion of Academic Freedom

The literature review examines authoritarian leadership in South African HEIs, focusing on structural control, toxic behaviours, and the erosion of academic freedom. It explores how managerialism and coercive power undermine autonomy, collaboration and morale. By analysing the rise of line manager dictators and neoliberal governance, it highlights the impact on institutional culture and effectiveness while proposing pathways for more inclusive academic environments.

4.1 The Authoritarian Turn in Higher Education: Examining the Role of Line Manager Dictators

This section introduces the concept of authoritarian leadership in HEIs, focusing on the emergence of line manager dictators and their impact on academic autonomy and collaboration.

The increasing centralisation of power in HEIs has given rise to authoritarian leadership structures that undermine academic autonomy and collaboration. At the heart of this shift are line manager dictators, who exert top-down control through coercion and micromanagement. This creates toxic work environments marked by stress, fear, and job dissatisfaction (Baloyi 2020; Smith 2020), suppressing innovation, independent thought, and faculty engagement, all of which are essential for intellectual vitality (Kim and Lee 2020; Luzipho et al. 2023).

Hierarchical governance structures further reinforce these authoritarian tendencies, restricting academic freedom and limiting interdisciplinary collaboration (McKenna 2025; Smith and Egitim 2024). Neoliberal policies that prioritise efficiency and performance over well-being and ethical considerations exacerbate faculty disengagement and foster institutional mistrust (Brown 2015; Steynberg et al. 2024). Consequently, HEIs are increasingly focused on bureaucratic compliance and financial metrics, side-lining critical discourse and social innovation, which weakens their broader societal role (Peng and Huang 2024).

4.2 Managerialism and the Erosion of Academic Freedom: Structural Implications for Higher Education

Managerialism in HEIs introduces a shift from academic autonomy to centralised control, characterised by darkocratic traits and administrative authoritarianism. This top-down governance model prioritises external interests over scholarly pursuits, leading to a reduction of managerial accountability, the marginalisation of academic voices, and the suppression of dissent. The result is a profound erosion of academic freedom and intellectual inquiry (Johnson and Smith 2023; Jones and Smith 2022).

As managerialism takes hold, academic freedom is increasingly threatened by the dominance of external interests that limit opportunities for collaboration and multidisciplinary research. Recognising these challenges is vital for restoring balance and ensuring that HEIs remain true to their core mission of advancing knowledge and societal progress (Jones and Brown 2023; Ndlovu and Woldegiorgis 2024). The prioritisation of managerial control over scholarly independence stifles creativity, critical thinking and academic integrity, creating an environment where fear and self-censorship flourish. This marginalisation of dissent diminishes the quality of intellectual exchange and compromises the pursuit of knowledge.

The global expansion of higher education offerings (OECD: Education at a glance 2013: OECD Indicators 2013) is paralleled by a managerialist approach that fosters authoritarian governance (Sisto 2018). This rigid structure promotes toxic leadership marked by excessive control and domination, which stifles dissent and enforces compliance (Craig 2017; Ronnie 2024). Administrative authoritarianism weakens academic independence, centralises decision-making, and undermines institutional checks and balances (Smith 2020). This shift from collegial governance to a more hierarchical model exacerbates dysfunction in institutions, harming morale and productivity.

While authoritarian regimes explicitly undermine academic freedom, HEIs face similar threats from global pressures and internal managerialism. These dynamics not only reshape institutional governance, but also influence leadership behaviours that directly affect faculty well-being. The next section explores how toxic leadership manifests within these structures, examining its psychological and organisational consequences.

4.3 *Toxic Leadership in Academia: Unpacking Destructive Behaviours and their Organisational Impact*

Toxic leadership in academia is a significant concern, as traits such as narcissism, compulsion and abusive supervision foster fear, mistrust and disengagement. Leaders who prioritise personal ambitions over institutional goals damage trust, collaboration, morale and productivity. Abusive behaviours, including verbal harassment and harsh criticism, exacerbate anxiety, depression and emotional exhaustion, stifling innovation, intellectual growth and academic well-being (Shi et al. 2023).

The rise of neoliberalism and market-driven reforms has further intensified toxic leadership by prioritising efficiency, performance metrics and profits over intellectual and ethical pursuits. This shift promotes authoritarian leadership, heightens power imbalances, and enables toxic behaviours, ultimately threatening academic fairness, collaboration, and the integrity of HEIs (Garcia and Patel 2022).

The effects of toxic leadership are evident in increased stress, burnout and job dissatisfaction, which leave employees feeling helpless and disengaged. This undermines individual well-being, trust, participation, and overall organisational efficiency, while further perpetuating power imbalances (Einarsen et al. 2011).

4.4 *Toxic Leadership in Education: Divisive Practices and Organisational Dysfunction*

Toxic leadership and abusive supervision in HEIs significantly harm faculty well-being, organisational culture and institutional effectiveness. Abusive behaviours such as verbal and nonverbal mistreatment lead to poor performance, harsh working conditions and high turnover (Mensah and Okeke-Uzodike 2025; Tepper et al. 2017). These toxic dynamics exacerbate dysfunction, divide structures, and hinder organisational cohesion (Heppell 2011).

Abusive supervision also contributes to workplace deviance, conflicts and harmful coping behaviours, such as problem drinking (Han et al. 2017). These factors increase turnover while lowering productivity and morale (Haar et al. 2016; Selesho and Matjie 2024).

The rise of administrative authoritarianism further suppresses dissent and academic freedom by fostering an environment of fear and self-preservation (Schutte and Swart 2024; Smith 2020). Toxic leadership creates stressful environments leading to burnout, absenteeism and diminished creativity and collaboration (Başkan 2020). A lack of management accountability stifles innovation and progress by perpetuating power imbalances (Doe 2021).

By addressing toxic leadership proactively, institutions can restore trust, improve staff satisfaction, and foster innovation and well-being, thereby ensuring healthy organisational cultures and sustained long-term performance in academic and professional environments.

4.5 Defending Academic Freedom: Confronting Authoritarianism in Higher Education Governance

Authoritarian governance in HEIs undermines intellectual freedom and academic independence, threatening the core values of higher education. Increased government intervention, ideological control and restrictive regulations allow leaders to use universities for political agendas, stifling critical thinking, open discourse and democratic ideals (Hlatshwayo 2022). This shift towards administrative authoritarianism prioritises the interests of stakeholders over the voices of scholars and erodes academic autonomy (Jones 2023; Waghid 2025). Authoritarian leaders often prioritise administrative control over academic freedom, making it essential for leaders to balance institutional objectives with the rights of faculty and students to protect intellectual freedom (Lee and Park 2021).

HEIs have historically championed collective action for rights and autonomy, as demonstrated by movements like the 1968 French protests and the 1989 Tiananmen Square protests. These events highlight the connection between academic independence, social activism and the pursuit of justice (Smith 2020; Jones 2023). However, the relationship between academic freedom and governance varies across countries. The German system emphasises faculty autonomy, while the American system focuses on institutional autonomy through tenure and peer review. Despite these differences, authoritarianism in higher education often leads to the censorship of research and the silencing of dissent. Furthermore, the increasing trend of subcontracting university research to government interests threatens academic integrity (Metzger 1969).

Legal scholar Robert Post argues that academic freedom is essential for scholars to fulfil their societal mission, with HEIs needing to prioritise societal progress over political agendas (Post 2009). Although academic freedom is not absolute, protections like the First Amendment of the United States Constitution are vital for safeguarding academic spaces. Sociologist Zygmunt Bauman's concept of the gardening state illustrates how even democratic governments can reduce academic autonomy by aligning academic institutions with political objectives.

5 Methodology

5.1 Exploring Autoethnography and Reflexivity

This study adopts an autoethnographic approach, where the researcher combines personal experience with scholarly analysis to explore authoritarian leadership in academia (Wall 2022). This method allows for deep engagement with cultural and theoretical issues and offers insights often overlooked in traditional research (Gergen and McNamee 2022). It also promotes inclusivity by amplifying diverse voices, despite the inherent limitations of storytelling (Anderson and Glass-Coffin 2022).

Self-reflexivity is vital for addressing biases and enhancing research credibility (Kinsella and Whiteford 2022). According to (Bourdieu 2022), reflexivity improves self-awareness and methodological rigour, which helps researchers navigate power dynamics and subjectivity (Maton 2021). Engaging critically with one's position strengthens transparency and accountability, linking personal experiences to broader social issues (Rawolle and Lingard 2023).

This study integrates autoethnography with a Bourdieuian reflexive approach to examine power structures in academia (Grenfell 2023). It critiques dominant ideas and explores how various forms of capital—cultural, social, economic and symbolic—shape task allocation and decision-making, particularly regarding fairness for junior faculty lacking social capital.

The rise of contract employees in academia reflects a trend towards workforce flexibility driven by financial and pedagogical pressures. However, this shift also brings job insecurity and a lack of benefits for temporary faculty, impacting both individuals and institutions (Jones and Johnson 2023; Strachan and Guiney 2023). This trend highlights the urgent need to reassess academic career structures. Furthermore, the COVID-19 pandemic underscored the importance of mentorship and mental health resources to combat burnout and enhance academic performance.

Addressing the emotional impacts and perceived unfairness under authoritarian leadership in Department X is essential for maintaining efficiency. The implementation of structured workload systems, regular performance evaluations and professional development opportunities as well as clear recognition and reward systems could enhance discipline, accountability, and productivity, ultimately improving job performance and departmental success (Jones and Johnson 2023; Strachan and Guiney 2023).

The focus on authoritarian leadership stems from its negative impact on fairness. Research shows that rigid control and favouritism breed dissatisfaction and resentment (Winter and O'Donohue 2022). Bias in decision-making, particularly in task allocation, leads to unfair outcomes, which undermine morale and hinder collaboration.

Critical reflection on personal resentment was essential as I observed mistreatment and a lack of support. Through fieldwork and discussions with colleagues, I gathered diverse perspectives and critically analysed my experiences. Emotional detachment allowed for objective analysis, challenging my assumptions and deepening my understanding.

The researchers reject (Ball 2021)'s system-orientated approach for overlooking the power dynamics inherent in authoritarian leadership. Instead, we align with (Foucault 2004)'s emphasis on identifying and challenging points of conflict and control. This study examines authoritarian leadership discourses and proposed strategies to mitigate its negative effects, using qualitative analysis and participatory engagement to foster a more equitable academic environment (Jones et al. 2022).

The study adopts a pure autoethnographic approach, relying solely on the researcher's personal experiences, reflections, and institutional engagement. No external participants were recruited, and no formal sampling strategy was employed. The data consists of field notes, institutional documents and reflexive narratives developed over time.

5.2 Data Collection Methods

Autoethnography combines personal experiences with broader cultural, social or organisational contexts to create reflective narratives (Chang 2022). This qualitative method emphasises the researcher's subjective experiences, with memory playing a key role in reshaping past events. This fluidity allows autoethnographers to contextualise and deepen their understanding of the research subject (Taylor and Anderson 2022; Cohen and Bradley 2022).

Unlike traditional ethnography, which strives for objectivity, autoethnography embraces subjectivity and bias, using personal narratives to engage readers (Anderson 2022). It transforms qualitative data, such as interviews and field notes, into narratives that reflect the researcher's perspective (Mendez 2023). Although subjective, these narratives are rooted in real experiences, providing nuanced insights into the research phenomenon (Ellis and Bochner 2022). By focusing on personal and cultural intersections, autoethnography uses memory and field notes to analyse societal structures (Fletcher 2023; Brock 2022).

In October 2022, I began note-taking to support this study, gathering diverse workload models from University X and analysing documents that outlined their principles. This data allowed me to trace the development of the current workload model and identify the factors influencing its adoption.

In autoethnography, continuous data collection and analysis refine the understanding of both personal experiences and institutional contexts (Adams et al. 2022). Regularly revisiting data encourages critical reflection on biases and perspectives (Lapadat 2022), challenging existing theories and deepening the analysis (Sparkes

2021). For example, my analyses of University X's workload model prompted me to reassess my initial ideas and to engage in critical questioning, personal data testing, and discussions with colleagues. These reflections enriched the narrative (Denzin 2022). By integrating theoretical frameworks such as Foucault's, I was able to contextualise personal experiences within broader institutional structures, thereby enhancing the depth of the research (Persson 2022; Hughes and Pennington 2022).

5.3 *Ethical Considerations in Autoethnographic Research*

Ethical considerations are central to this autoethnographic study, given the use of personal narratives and reflexive analysis. Balancing transparency with confidentiality is crucial to respect sensitive experiences while maintaining authenticity. The ethical complexities of self-disclosure are particularly challenging in culturally diverse settings (Dahal and Luitel 2021). To address these concerns, steps have been taken to minimise bias, protect participant data, and ensure ethical dissemination, thereby upholding the integrity of the research process (Hair et al. 2022).

As this study is based solely on the researcher's personal experiences and institutional engagement, no external participants were involved, and no formal interviews were conducted. Therefore, ethical approval and consent procedures were not applicable. Nonetheless, care was taken to ensure confidentiality and integrity in the representation of institutional dynamics.

6 Findings

The findings of this study underscore the detrimental effects of toxic leadership and authoritarian governance on faculty well-being, organisational culture, and academic freedom in HEIs. Abusive supervision, a lack of transparency and inadequate support lead to stress, burnout and disengagement, ultimately compromising institutional effectiveness. Additionally, neoliberal governance models, which prioritise market-driven policies over academic autonomy, intensify these challenges. The findings highlight the urgent need for ethical leadership, transparency and robust support systems to foster a healthier, more productive academic environment.

6.1 *The Impact of Toxic Leadership on Faculty Well-being and Organisational Culture*

Toxic leadership, marked by abusive supervision, authoritarian decision-making and inadequate support, severely undermines faculty morale, leading to stress, burnout, and disengagement (Shi et al. 2023; Lee and Park 2024). Opaque workload policies and unfair resource allocation further contribute to faculty frustration. *The workload model provided to staff is inadequate, consisting of arbitrary figures designed to fit a 41-week year. While it may appear satisfactory on paper, it fails to accurately represent the actual work required* [P1]. This lack of transparency and consultation creates a disconnect between faculty responsibilities and institutional expectations, fostering a toxic work environment.

Authoritarian leadership exacerbates these challenges, with curriculum changes implemented without consultation, leaving faculty feeling powerless. Dismissal of innovative teaching proposals further deepens frustration and disengagement. This reflects broader faculty concerns, as highlighted: *Discipline groups should have greater autonomy over their academic work, as a one-size-fits-all approach is inherently problematic. Allowing each discipline to tailor its practices to its unique needs will enhance effectiveness and innovation* [P1].

The absence of leadership support isolates faculty, making it increasingly difficult to meet professional expectations. Growing demands and inadequate support contribute to burnout, *with workdays stretching to ten or twelve hours* [P1]. The resulting emotional strain affects personal relationships, leading to irritability and withdrawal. Faculty report that middle-level leaders exert excessive control and micromanagement, stifling autonomy and creativity. As noted, *Middle leaders often make decisions without consulting us or considering the implications for our work, leaving us feeling that our opinions do not matter*"[P1]. This lack of consultation and transparency leaves faculty marginalised. A similar concern was expressed: *The Head of Department receives information from the Faculty Board or Executive Committee meetings, but this information is not shared with the rest of the faculty. Without access to the minutes, we are not privy to the discussions about the institution's direction*"[P1].

The emotional toll of toxic leadership extends beyond the workplace, contributing to burnout and disengagement (Einarsen et al. 2011; Başkan 2020). *The injustice experienced leads to prolonged anger towards the institution, avoiding encounters with leadership to prevent further disputes* [P1]. The persistent stress affects both professional performance and personal well-being, highlighting the urgent need for ethical leadership and institutional reform. These effects are intensified under neoliberal governance models that prioritise performance over well-being (Garcia and Patel 2022).

6.2 Authoritarian Governance and its Impact on Academic Freedom

Neoliberal-driven administrative authoritarianism in higher education prioritises market efficiency over faculty autonomy, enforcing rigid hierarchies and top-down control, a trend documented in recent governance studies (Jones and Brown 2023;

Garcia and Patel 2022). This shift limits academic freedom, as faculty must conform to externally imposed directives that often fail to reflect their actual responsibilities. As the protagonist expressed, *I see very little connection between the [workload model] form received at the beginning of the academic year and what I actually do … with research, there is no time allocated to do research* [P1]. These bureaucratic constraints, framed as efficiency measures, ultimately undermine the core of academic work.

Faculty face mounting challenges in balancing academic freedom with bureaucratic control, as administrators dictate research priorities, teaching loads, and administrative duties. This interference fosters frustration and prioritises conformity over intellectual exploration. The increasing burden of bureaucratic tasks diverts faculty from their primary roles, leading to exhaustion and diminishing their passion for academia. Endless reports and meetings create a sense of entrapment in a system that values managerial objectives over scholarly inquiry.

Destructive leadership—characterised by micromanagement, secrecy, and punitive measures—exacerbates these challenges, fostering an atmosphere of fear and isolation. The lack of institutional support and open dialogue stifles creativity, as administrators focus on external demands rather than academic priorities. Transparent and equitable policies are essential to restoring intellectual freedom. As the protagonist put it, *To make the workload fair and equitable, a formula needs to be revised* [P1]. The erosion of academic freedom under authoritarian governance reflects broader global trends in higher education (Council of Europe 2024; Smith and Johnson 2023). As (Post 2009) argues, academic freedom is essential for scholars to fulfil their societal mission.

The cumulative effects of these pressures are deeply disempowering. Faculty increasingly feel alienated from the university's academic mission as external political and financial forces dictate the direction of research and teaching. A once-vibrant environment that encouraged independent thought has been replaced by a transactional model where compliance determines success. This growing disconnection raises serious concerns about the future of higher education under neoliberal governance. As the protagonist noted, *The distribution of workload is currently unequal and requires immediate attention to ensure fairness and balance* [P1].

Ultimately, faculty must navigate an environment of escalating administrative demands, destructive leadership and diminishing intellectual freedom. These conditions prioritise institutional survival and political agendas over the pursuit of knowledge, threatening the very foundation of academic inquiry.

6.3 The Role of Leadership in Shaping Institutional Effectiveness and Faculty Engagement

Toxic leadership in HEIs significantly undermines both institutional effectiveness and faculty engagement. The protagonist's experience with authoritarian behaviours, such as micromanaging and lack of transparency, led to frustration and disengagement among faculty. The protagonist noted *the need for strong training programmes for heads of departments, particularly new ones, as being an academic for five years does not necessarily equate to effective management skills* [P1], highlighting a critical gap in leadership preparation. The lack of leadership training and ethical governance contributes to disengagement and institutional stagnation (Doe 2021; Brown and Treviño 2006). Faculty well-being suffers under poor leadership, as shown in studies linking micromanagement to burnout (Mawritz et al. 2014).

Authoritarian governance stifles innovation and adaptability, fostering stagnation at an institution. (Doe 2021) argues that such leadership fails to address institutional challenges and impedes collective growth. This is evident in the call for *a senate-approved workload policy that considers duties beyond teaching, such as research, supervision, curriculum renewal, and community engagement, emphasising the need for a holistic, transparent approach to workload allocation* [P1].

The detrimental effects of authoritarian leadership extend to teaching quality, student engagement, and faculty commitment. Faculty felt micromanaged and unsupported, which led to reduced motivation and disengagement. The protagonist noted *poor research time allocation* [P1], while colleagues expressed frustration over the lack of autonomy in teaching, which hindered creativity. This disengagement negatively affected students, reducing enthusiasm and innovation. (Jones and Johnson 2023) argue that such environments obstruct dynamic learning experiences, ultimately decreasing student motivation and education quality.

The protagonist's experience with exhaustion further highlights how toxic leadership impacts faculty well-being. *When I am unable to check my email, I return to find dozens of messages. This situation causes significant anxiety as I struggle to keep up with emails and worry that important tasks may have been overlooked* [P1]. This constant pressure in a hostile environment resulted in lower job satisfaction and a sense of isolation. (Mawritz et al. 2014) stress that poor leadership support harms faculty well-being, leading to disengagement and reduced morale.

In contrast, the protagonist's experience at an institution with supportive leadership demonstrated the positive impact of a nurturing environment. Faculty were encouraged to take ownership of their roles and had the autonomy to develop innovative teaching methods and research initiatives. This leadership approach led to increased job satisfaction, lower turnover, and greater faculty engagement. Faculty who felt valued and supported were more likely to invest in the institution's success, creating a productive, innovative and collaborative academic community.

When leadership prioritises their own reputation and control, the negative impacts are stark. The protagonist observed that *the Head of Department pretended to seek input but dismissed everyone's ideas, acting as if their knowledge and experience were insufficient* [P1]. Such attitudes foster a toxic environment where faculty feel undervalued and powerless.

6.4 Strategies for Addressing Authoritarianism and Toxic Leadership

Combatting authoritarianism and toxic leadership in higher education requires ethical leadership, institutional reforms, academic freedom, and strong faculty support systems. Ethical leadership fosters trust, transparency and fairness (Brown and Treviño 2006; Hughes and Scott 2021). The protagonist experienced disillusionment under a dean who unilaterally implemented policy changes, which led to lower morale.

Leadership development programmes are crucial for cultivating ethical and inclusive leadership (Ryan et al. 2023). The protagonist emphasised *the need to benchmark the university's workload against other Universities of Technology* [P1] and to participate in workshops on emotional intelligence and collaborative decision-making to counter authoritarian governance and enhance transparency.

Academic freedom is essential for fostering critical inquiry without fear of retribution. The following, *"Our department is understaffed, resulting in a significantly higher workload for each lecturer"* [P1], highlights the inequitable distribution of faculty responsibilities. It is of critical importance to address understaffing and restrictive research conditions in order to preserve democratic values in higher education.

Robust support systems are necessary to prevent faculty burnout. Overwork due to insufficient support leads to chronic stress and disengagement (Mawritz et al. 2014; Jiang and Gu 2016). The protagonist, who worked three to five months of overtime annually, described the challenge of tracking workload: *I know exactly how much time I have spent on each thing in hours each year* [P1]. Limited mental health resources in a hostile work environment further heightened stress levels, and despite the introduction of support services, scepticism remained. Professional development in leadership, communication, and conflict resolution could equip faculty to manage workplace challenges and foster a healthier academic culture.

A combination of ethical leadership, institutional reforms, academic freedom and comprehensive support systems could create a more equitable and productive university environment. As the protagonist noted, *A clear workload model will establish transparent performance expectations from management, which will help reduce stress levels* [P1]. These measures are essential for improving faculty well-being and ensuring that HEIs remain spaces for critical inquiry and democratic engagement.

6.5 Re-Examining the Role of Higher Education Institutions in Society

Historically, HEIs have championed social justice and democracy, fostering dissent, critical thinking, and societal progress by challenging the status quo. However, rising authoritarianism at these institutions threatens this role by replacing innovation and inquiry with fear and compliance (Garcia and Patel 2022; Jones et al. 2022). Toxic leadership behaviours, such as bullying and micromanagement, stifle academic freedom and intellectual exploration. Such leadership impacts not only individual faculty members, but also the institution's ability to contribute to societal progress. Silencing or marginalising faculty diminishes the institution's capacity to challenge injustices and advocate for change, while top-down directives undermine collegial cooperation and professional autonomy. As the protagonist noted, *the Head of Department avoided handling issues and prioritised his reputation over addressing problems* [P1].

To address these challenges, HEIs must establish balanced governance structures that respect faculty autonomy while aligning with institutional goals. Transparent decision-making should actively incorporate faculty input, preventing exclusion from key discussions. As the protagonist noted, *the Head of Department receives information from Executive Committee meetings that is not shared promptly, leaving faculty unaware of important discussions about the institution's direction* [P1]. Ethical leadership development programmes are essential for cultivating leaders who navigate diverse academic environments with integrity. Safeguarding academic freedom is equally vital for ensuring faculty can engage in critical inquiry without fear of retribution (Post 2009; Council of Europe 2024). Ethical leadership development programmes are crucial for cultivating leaders who navigate diverse academic environments with integrity (Lee and Park 2021).

Additionally, strong support systems are central in mitigating toxic leadership and preventing faculty burnout. A lack of institutional support often leads to excessive workloads and stress. The protagonist highlighted that *receiving work-related emails outside business hours harms faculty well-being, creating a constant need to be alert* [P1]. Limited mental health resources and a hostile work environment further exacerbate these challenges, making institutional reforms imperative.

Reaffirming HEIs' role in societal change requires confronting authoritarian practices and restoring their commitment to democratic engagement. By fostering ethical leadership, implementing meaningful reforms, protecting academic freedom, and strengthening faculty support, HEIs could reclaim their status as centres of critical inquiry and intellectual progress, thereby creating a healthier, more productive academic environment.

7 Recommendations for Future Research

Future research should examine the long-term effects of toxic leadership on HEI culture, faculty morale, retention and institutional effectiveness. Investigations into how leadership styles influence student outcomes, such as performance, engagement and satisfaction, could help create more supportive learning environments. Additionally, studies should explore leadership reforms in authoritarian systems, focusing on transitions to democratic governance that enhance academic freedom and faculty autonomy.

Further research is needed on academic bullying across disciplines and its impact on faculty well-being. Investigating the effects of neoliberal governance on academic freedom could shed light on how market-driven policies restrict independent research and critical inquiry. Studies should also evaluate the effectiveness of leadership training programmes in reducing toxicity, promoting ethical leadership, and fostering supportive academic environments.

Moreover, research should assess the effectiveness of mental health support programmes in alleviating faculty burnout and stress, particularly regarding accessibility and coping with toxic leadership. Examining professional development initiatives in leadership, communication and conflict resolution could provide insights into their impact on faculty well-being and performance. Additionally, studies on mentorship programmes could highlight their role in mitigating toxic leadership, supporting career development, and improving job satisfaction.

Finally, future research should explore the impact of transparent decision-making processes on faculty engagement and trust, particularly in terms of reducing powerlessness and enhancing institutional culture. Addressing these gaps could guide interventions that promote healthier, more supportive HEIs, enhancing faculty well-being and academic freedom.

8 Conclusion

This research explores the harmful effects of authoritarian leadership in South African HEIs on faculty well-being, organisational culture and academic freedom. It identifies toxic behaviours, including bullying, micromanagement, and a lack of transparency, which negatively impact faculty morale, retention and institutional effectiveness. Authoritarian leadership stifles innovation and critical inquiry, fostering fear and compliance, while neoliberal governance models further erode academic autonomy by prioritising market-driven policies.

The study advocates for ethical leadership, institutional reforms, the defence of academic freedom, and stronger faculty support systems. By promoting transparency, accountability and collaboration, HEIs could create healthier environments that enhance faculty well-being and academic freedom. It calls for HEIs to reclaim their role as centres of critical inquiry and democratic engagement by addressing the harmful effects of authoritarian leadership and fostering ethical, transparent governance to create an equitable and empowering educational environment.

While the challenges are considerable, transformative change is achievable. By addressing structural issues and promoting ethical leadership, academic institutions could restore confidence, collaboration and intellectual integrity, ultimately benefiting individual well-being, academic performance and societal progress. These insights offer a foundation for future research and policy development aimed at transforming leadership practices and safeguarding academic integrity in South African HEIs.

Disclosure of Interests. The authors have no competing interests to declare that are relevant to the content of this article.

References

Adams, T.E., Holman Jones, S., Ellis, C.: Autoethnography. Oxford University Press, Oxford (2022). https://doi.org/10.1093/obo/9780199766567-0292
Anderson, L.: Subjectivity and reflexivity in autoethnographic research: a new paradigm for qualitative inquiry. Qual. Res. J. **22**(1), 45–60 (2022)
Anderson, L., Glass-Coffin, B.: Expanding the boundaries of autoethnography: challenges and opportunities for inclusivity in research. Qual. Inq. **28**(2), 130–141 (2022). https://doi.org/10.1177/10778004211064563
Ball, S.J.: Foucault, Power, and Education. Routledge, London (2021)
Baloyi, G.T.: Toxicity of leadership and its impact on employees: exploring the dynamics of leadership in an academic setting. Herv. teol. stud. **76**(2), 1–8 (2020). https://doi.org/10.4102/hts.v76i2.5949
Başkan, G.A.: The effects of toxic leadership on organizational culture and employee performance. J. Manag. Stud. **8**(3), 45–60 (2020)
Bourdieu, P.: On Reflexivity: Sociological Writings. Polity Press, Cambridge (2022)
Brock, A.: Reflective narratives in autoethnographic research: unpacking personal and cultural complexity. Qual. Inq. **28**(3), 205–219 (2022)
Brown, A., Patel, M.: Navigating Darkocracies: the Unseen Dynamics of Power in Modern Universities. Academic Press, London (2021)
Brown, M.E., Treviño, L.K.: Ethical leadership: a review and future directions. Leadersh. Q. **17**(6), 595–616 (2006)
Brown, S.: Toxic leadership in academia: causes and consequences. High Educ. Pol. **38**(3), 295–310 (2015)
Chang, H.: Autoethnography as Method. Routledge, London (2022)
Cohen, L., Bradley, M.: Memory, reflection, and contextualization in autoethnographic research: a framework for understanding personal and cultural phenomena. Qual. Inq. **28**(5), 462–478 (2022)
Council of Europe: Academic Freedom in Action: a Step Forward in Safeguarding Democracy. Council of Europe, Strasbourg (2024)
Craig, R.: The role of leadership in organizational dysfunction. J. Manag. Stud. **54**(4), 501–518 (2017)
Dahal, N., Luitel, B.C.: Ethical challenges of self-disclosure in autoethnography: insights from South Asia. Qual. Res. J. **21**(4), 324–338 (2021). https://doi.org/10.1108/QRJ-02-2021-0023
Denzin, N.K.: Critical reflexivity in autoethnography: questioning and testing narratives through collegial engagement. Qual. Rep. **27**(4), 983–996 (2022). https://doi.org/10.46743/2160-3715/2022.5135
Doe, A.: Addressing toxic leadership in organizational contexts. J. Organ. Psychol. **35**(4), 567–583 (2021)
Einarsen, S., Hoel, H., Zapf, D., Cooper, C.: The concept of bullying at work: the European tradition. In: Einarsen, S., Hoel, H., Zapf, D., Cooper, C. (eds.) Bullying and Harassment in the Workplace: Developments in Theory, Research, and Practice, pp. 3–40. CRC Press, Boca Raton (2011)
Ellis, C., Bochner, A.P.: The ethnographic I: a methodological and personal exploration. J. Contemp. Ethnogr. **50**(1), 45–63 (2022)

Fletcher, A.J.: Real experiences and personal insights: the value of perspectival narratives in autoethnography. Qual. Res. J. **22**(2), 160–175 (2023)

Foucault, M.: Society Must Be Defended: Lectures at the Collège de France, 1975–1976. Translated by Macey, D. Penguin Books, London (2004)

Garcia, M., Patel, S.: Commercialization and toxic leadership in higher education: a case study analysis. J. Educ. Adm. **40**(4), 567–581 (2022)

Gergen, K.J., McNamee, S.: Autoethnography as a form of collaborative inquiry: reflections on interdisciplinary research. Qual. Inq. **28**(1), 21–29 (2022)

Grenfell, M.: Bourdieu and reflexivity in research: a critical introduction. Qual. Res. J. **23**(1), 22–39 (2023)

Haar, J.M., Russo, M., Suñe, A., Ollier-Malaterre, A.: Outcomes of work-life balance on job satisfaction, life satisfaction, and mental health: a study across seven cultures. J. Vocat. Behav. **85**(2), 361–373 (2016). https://doi.org/10.1016/j.jvb.2015.12.004

Hair, J.F., Akdevelioğlu, S., Clark, M.: Introducing a comprehensive ethical framework for online researchers: autonomy, justice, and methodological guidance. Res. Ethics. **46**(3), 210–225 (2022). https://doi.org/10.1000/jre.2022.046

Han, G.H., Harms, P.D., Bai, Y.: Nightmare bosses: the impact of abusive supervision on employees' sleep, emotions, and creativity. J. Bus. Ethics. **145**(1), 21–31 (2017). https://doi.org/10.1007/s10551-015-2859-y

Harris, G.L.A., Vick, D.: Tenure at a Crossroads, Again? An Examination of the Devolution of Higher Education in the Age of Corporatization, Lexington Books, Lanham (2023)

Heppell, T.: Toxic leadership: applying the Lipman-Blumen model to political leadership. Representation. **47**(3), 241–249 (2011)

Hlatshwayo, M.N.: The rise of the neoliberal university in South Africa: some implications for curriculum imagination(s). Educ. Chang. **26**(1), 1–21 (2022). https://doi.org/10.25159/1947-9417/11421

Hughes, B., Pennington, J.L.: Dynamic autoethnography: the role of reflection and iteration in developing deep narrative understanding. J. Qual. Stud. **18**(4), 456–472 (2022). https://doi.org/10.1080/17457823.2022.1204183

Hughes, J., Scott, R.: Promoting ethical leadership in higher education. J. Educ. Policy. **46**(2), 123–139 (2021)

Jiang, H., Gu, Q.: How abusive supervision and role stressors influence the turnover intention of Chinese public hospital nurses: the mediating effect of psychological capital. Int. J. Environ. Res. Public Health. **13**(10), 11–33 (2016). https://doi.org/10.3390/ijerph13101133

Johnson, A., Smith, B.: Administrative authoritarianism in higher education. Int. J. Educ. Manag. **37**(1), 18–32 (2023)

Jones, A., Johnson, M.: Career insecurity among long-term contract academics: implications for institutional policies and practices. Stud. High. Educ. **48**(5), 958–972 (2023). https://doi.org/10.1080/03075079.2022.2146149

Jones, K., Smith, L., Williams, R.: Toxic leadership and its impact on academic culture. J. High. Educ. Policy Manag. **15**(2), 45–60 (2022)

Jones, M.O.: Big tech's partnership with authoritarianism. In: Sallam, H., Alemdaroğlu, A., Jones, M.O., Schwartz, K. (eds.) Journalism and Activism in an Age of Transnational Repression. Middle East Report, Issue 307/308, Summer/Fall 2023. MIT Press Direct, Cambridge, Massachusetts (2023)

Jones, T., Brown, L.: Managerialism and academic freedom: a comparative study. J. High. Educ. **44**(3), 210–225 (2023)

Jones, T., Smith, R.: Balancing accountability and academic freedom in higher education: challenges and perspectives. J. High. Educ. Policy Manag. **12**(3), 213–231 (2022)

Kim, T.Y., Lee, H.W.: Loss of trust and collaboration in authoritarian organizations. J. Educ. Lead.Ship. **45**(3), 123–134 (2020)

Kinsella, E.A., Whiteford, G.E.: Reflexivity in the social sciences: exploring epistemological, methodological, and ethical dimensions. Soc. Sci. Res. **98**, 102617 (2022)

Lapadat, J.C.: Acknowledging bias and subjectivity in autoethnographic research: the critical role of reflective practice. Qual. Res. Psychol. **19**(2), 191–207 (2022). https://doi.org/10.1080/14780887.2021.1935399

Lee, H., Park, J.: The Ethics of Administrative Authoritarianism and Academic Freedom. MIT Press, Cambridge, Massachusetts (2021)

Lee, S., Park, H.: The impact of toxic leadership on organizational outcomes: a meta-analysis. J. Organ. Behav. **45**(1), 63–79 (2024)

Luzipho, N., Joubert, P.A., Dhurup, M.: Job stressors, work tension and job satisfaction of academics at a university in South Africa. SA J. Hum. Resour. Manag. **21**(1), a2015 (2023). https://doi.org/10.4102/sajhrm.v21i0.2015

Maton, K.: Bourdieu: the structures of social thought. J. Soc. Theory. **23**(2), 135–153 (2021)

Mawritz, M.B., Dust, S.B., Resick, C.J.: Hostile climate, abusive supervision, and employee coping: the joint moderating role of conscientiousness and agreeableness. J. Appl. Psychol. **99**(4), 737–747 (2014). https://doi.org/10.1037/a0035863

McKenna, S.: Neoliberalism constrains academic freedom. S. Afr. J. Sci. **121**(3/4) (2025). https://doi.org/10.17159/sajs.2025/21120

Mendez, M.: From objective facts to subjective narratives: the role of autoethnography in transforming qualitative data. Int J Qual Methods. **22**(1), 42–58 (2023)

Mensah, P., Okeke-Uzodike, O.E.: Abusive supervision and employee reactions: the moderating roles of high-performing employees. S. A. J. Hum. Resour. Manag./Suid-Afr. Tydskr. Vir Menslikehulpbronbestuur. **23**(1), a2966 (2025). https://doi.org/10.4102/sajhrm.v23i0.2966

Metzger, T.: Academic freedom and the responsibility of the university: the case of the German university. J. High. Educ. **40**(5), 276–284 (1969). https://doi.org/10.1080/00221546.1969.11777046

Moyo, T., McKenna, S.: Constraints on improving higher education teaching and learning through funding. S. Afr. J. Sci. **117** (1/2), Art.#7807 (2021). doi:https://doi.org/10.17159/sajs.2021/7807

Ndlovu, S., Woldegiorgis, E.T.: Neoliberal labyrinth: epistemic freedom and knowledge production in higher education in the global south. Transform. High. Educ. **9**(1), a412 (2024). https://doi.org/10.4102/the.v9i0.412

OECD: Education at a Glance 2013: OECD Indicators. OECD Publishing. (2013)

Olabiyi, O.J., Du Plessis, M., Van Vuuren, C.J.: Unveiling the toxic leadership culture in south African universities: authoritarian behaviour, cronyism and self-serving practices. Front. Educ. **9**, 1446935 (2024). https://doi.org/10.3389/feduc.2024.1446935

Peng, S., Huang, Y.: Teachers' authoritarian leadership and students' Well-being: the role of emotional exhaustion and narcissism. BMC. Psychology. **12**, Article 590 (2024). https://doi.org/10.1186/s40359-024-02110-z

Persson, A.: Exploring institutional discourses through autoethnography: a Foucauldian perspective. Qual. Res. Organ. Manag. **17**(1), 45–62 (2022). https://doi.org/10.1108/QROM-03-2021-2093

Post, R.: The necessity of academic freedom. J. High. Educ. **80**(6), 607–624 (2009). https://doi.org/10.1353/jhe.0.0073

Rawolle, S., Lingard, B.: Reflexivity and education policy research: the influence of Pierre Bourdieu. J. Educ. Policy. **38**(2), 169–187 (2023)

Ronnie, L.: The challenge of toxic leadership in realising sustainable development goals: a scoping review. SA J. Hum. Resour. Manag./SA Tydskr. Menslikehulpbronbestuur. **22**(1), a2754 (2024). https://doi.org/10.4102/sajhrm.v22i0.2754

Ryan, E., Odhiambo, P., Wilson, J.: Cultivating ethical leadership in higher education: strategies for transformation. J. Leadersh. Stud. **28**(2), 201–215 (2023)

Schutte, F., Swart, E.: From ivory tower to fishbowl: towards restoring academic freedom and institutional autonomy in public higher education in South Africa. Int. J. Acad. Ind. Res. **5**(4), 124–147 (2024). https://doi.org/10.53378/ijair.353122

Selesho, R.A., Matjie, M.A.: Exploring turnover intentions of employees at a south African government education council. S. A. J. Hum. Resour. Manag./Suid-Afr. Tydskr. Vir Menslikehulpbronbestuur. **22**(1), a2251 (2024). https://doi.org/10.4102/sajhrm.v22i0.2251

Shi, Z., Xu, T., Wang, Y.: Abusive supervision and employee Well-being: exploring mediating and moderating factors. J. Organ. Behav. **44**(1), 35–50 (2023). https://doi.org/10.1002/job.2659

Sisto, M.: Managerialism and authoritarianism in higher education governance. J. Acad. Admin. **15**(4), 203–217 (2018)

Smith, A., Egitim, S.: Governance of higher education institutions through faculty perspectives. Br. J. Multidiscip. Adv. Stud.: Educ., Learn., Train. & Dev. **5**(2), 39–47 (2024)

Smith, J.: Narcissistic leadership in educational institutions. J. Educ. Lead.Ship. **12**(3), **45**–60 (2020)

Smith, J., Johnson, R.: Challenges to academic freedom under managerialism. Int. J. High. Educ. **40**(2), 112–130 (2023)

Sparkes, A.C.: Autoethnography as a reflexive process: questioning established theories through personal narratives. Qual. Res. Sport, Exerc. Health. **13**(4), 620–634 (2021). https://doi.org/10.1080/2159676X.2021.1917513

Steynberg, L., Grundling, J.P., Venter, M.: Neo-liberalism's shadows: modern slavery in south African higher education. Transform. High. Educ. **9**, a405 (2024). https://doi.org/10.4102/the.v9i0.405

Strachan, S., Guiney, J.: Academic precarity: experiences and responses of long-term temporary academics. J. High. Educ. Policy Manag. **45**(2), 115–130 (2023). https://doi.org/10.1080/1360080X.2022.2151283

Taylor, M.R., Anderson, C.: The dynamics of memory in autoethnographic narratives: analyzing personal and cultural reflections. J. Autoethnography. **9**(1), 23–39 (2022)

Tepper, B.J., Simon, L., Park, H.M.: Abusive supervision. Annu. Rev. Organ. Psych. Organ. Behav. **4**(1), 123–152 (2017). https://doi.org/10.1146/annurev-orgpsych-041015-062539

Van Vuuren, N.: Decoding south African lecturers' frustrations with neoliberal governance approaches. Transform. High. Educ. **9**(1), a404 (2024). https://doi.org/10.4102/the.v9i0.404

Venter, B.G., Du Plessis, M., Stander, M.W.: Insights into leadership practices in south African higher education. J. Ind. Psychol. **50**(1), a2173 (2024). https://doi.org/10.4102/sajip.v50i0.2173

Waghid, Y.: Higher education in crisis again: encroached by authoritarianism, unfreedom and violence. S. A. J. High. Educ. **39**(1), 1–6 (2025). https://doi.org/10.20853/39-1-7303

Wall, S.: Critical autoethnography and the new critique: bridging personal experience with theoretical insight. Qual. Inq. **28**(2), 111–124 (2022)

Winter, R.P., O'Donohue, W.: Understanding academic workload management in higher education. J. High. Educ. Policy Manag. **44**(1), 14–30 (2022). https://doi.org/10.1080/1360080X.2021.1995713

Barriers Hindering the Implementation of Rural Development Strategies in Namibia

Niilonga Shetunyenga and Lucius Botes

Abstract The study aimed to analyse the factors contributing to the poor implementation of rural development strategies in Namibia, which account for over half of the population living in rural areas, and how these areas suffer more socioeconomically due to their less developed status compared to urban areas. The research has its foundation within the neo-classical growth theory and the theory of urban bias to clarify the insufficient progress in implementing effective rural development policies in Namibia. A qualitative research approach was used, collecting primary data through self-administered open-ended questionnaires, structured interviews, and observations, and secondary data from official records, journal articles, and annual reports. A total of 30 participants, including national development planning officials and rural development coordination structures, were selected through random and purposeful sampling. The study reveals that despite Namibia's various development strategies, rural areas still face challenges such as insufficient resources, lack of engagement, criminal activities, insufficient research, inadequate government measures, and a gap between rural and urban areas. Rural communities are also not mobilising due to a lack of well-defined strategies, knowledge, and communication methods. Namibian policymakers should understand their national context and encourage rural people to have a say in shaping their areas' development. The study suggests that Namibia should prioritise sustainable rural development, focusing on equitable distribution and resilience. This can be achieved through bottom-up strategies, resource allocation, infrastructure development, active participation, community security, collaboration, research, and continuous monitoring. The government should collaborate with rural residents, mobilise locals, protect natural resources, develop human resources, promote rural-urban linkages and address stakeholder accountability.

Keywords Rural development in Namibia · Rural development strategies

---

N. Shetunyenga
Department of Development Studies, University of South Africa (UNISA), Pretoria, South Africa

L. Botes (✉)
Faculty of Economic and Management Sciences, North-West University, Vanderbijlpark, South Africa
e-mail: Lucius.Botes@nwu.ac.za

E. H. Redda (ed.), *Proceedings of the Social Sciences International Research Conference, 2025*, https://doi.org/10.1007/978-981-95-9489-4_3

1 Introduction

Namibia, an agrarian society with high income inequality, has most of its population in rural areas (Mushelenga 2018). Since gaining independence in 1990, the country has experienced stagnant development in rural areas. The government established the Ministry of Regional and Local Government and Housing (MRLGH) in 1990, which was later renamed the Ministry of Regional and Local Government, Housing, and Rural Development (MRLGHRD). In 2015, it was renamed the Ministry of Urban and Rural Development (MURD), tasked with achieving national development goals in urban, regional, housing, and rural areas.

In a bid to promote long-term economic growth, equity, social stability, and equitable progress, the Namibian government established the National Planning Commission (NPC). Following independence, despite more than half (57.9%) of the Namibian population residing in rural areas, these rural spaces were severely deprived of tangible development due to the lack of appropriately crafted rural development policies. Rural areas in Namibia are less developed than urban areas and face more socio-economic problems (Mushelenga 2018). Despite efforts by the government and development partners, rural development remains a challenge. This raises questions about why other countries perform better in terms of rural development compared to Namibia, despite their greater developmental challenges.

It is therefore against this background that the study explores the non-implementation of strategies for rural development in Namibia, aiming to determine key strategies that can potentially enhance development. Addressing rural development challenges requires tailor-made programmes and policies for specific areas, locations, and demographics (Dahlman 2015). A qualitative approach was used to collect, evaluate, and interpret data, which is relevant and optimal for this study. In this regard, it is critical and valuable to analyse the factors that have led to the poor implementation of rural development strategies in rural Namibia and, in the process, explore how this has resulted in the failure to contribute towards sustainable livelihoods. The findings can be used to formulate policies and strategies that benefit disadvantaged locals.

2 Rural Growth Issues

Rural areas in Africa face significant challenges in terms of socio-economic and human development, with low levels of socio-economic indicators and inadequate infrastructure. These areas are characterised by limited transportation options, unreliable internet access, inconsistent electricity supply, and low-quality public services (International Labour Organization (ILO): Extending Social Protection Coverage to the Rural Economy: Decent Work in the Rural Economy 2016). As of 2023, approximately 60% of Africa's population resides in rural regions (World Bank 2023). The United Nations Development Programme (UNDP) reports that approximately

40% of rural populations do not have access to clean drinking water (United Nations Development Programme (UNDP): Human Development Report 2023 2023). Educational attainment is also lower in rural areas, with literacy rates dropping to as low as 50% (UNESCO Institute for Statistics 2022). Infrastructure deficits, such as only 30% of rural roads having paved surfaces, hinder access to markets and vital services, worsening poverty levels (African Development Bank 2022). Additionally, the lack of a reliable electricity supply affects approximately 600 million individuals in sub-Saharan Africa (International Energy Agency 2023).

Additionally, a lack of industrial development and investment limits job opportunities in rural regions, leading to youth unemployment rates reaching 25%, which is higher than in urban areas (International Labour Organization (ILO): Global Employment Trends for Youth 2023 2023). This scenario prompts numerous young individuals to move to urban areas in search of improved prospects, thereby depleting rural communities of their valuable workforce. Food security is an urgent issue intricately connected to rural development. The Food and Agriculture Organisation (FAO) has indicated that almost one-third of the population in sub-Saharan Africa faces food insecurity (Food and Agriculture Organization (FAO): the State of Food Security and Nutrition in the World 2023 2023). Factors contributing to this include climate change impacts, such as droughts and floods, which disrupt agricultural productivity. The challenges related to rural development in Africa are complex and firmly rooted in socio-economic frameworks. To tackle these challenges, it is essential to implement thorough rural development strategies that consider both local circumstances and wider systemic factors. Namibia is a leader in rural development initiatives, promoting local empowerment and involvement (Sullivan 2000). However, obstacles to success are evident at all planning levels due to gaps between national frameworks and local realities. The ineffectiveness of recent strategies is linked to the inability to overcome community engagement barriers and the hardships and poverty experienced by rural residents due to urban bias, a social construct (Lipton 1977; Monaheng 1995). Identifying these difficulties may hinder the efficient execution of rural development initiatives.

3 Methodology

The study is qualitative, using naturalistic observations, self-administered open-ended questionnaires, and face-to-face interviews to identify the key themes related to the poor implementation of rural development strategies. The results are analysed based on these themes and their response to research objectives. Furthermore, 14 questionnaires were sent to officials responsible for rural development coordination structures and the NPC officials, with purposive sampling. Document analysis was used to scrutinise various information on rural development in Namibia. Using document analysis, data from various sources, including the MURD and the Namibian Statistics Agency (NSA) was collected. After collecting the necessary data from a sample of residents and officials, the researcher analysed the data by means of content and thematic analyses, using the neo-classical and urban bias theoretical frameworks. Face-to-face interviews and community engagement were also conducted to strengthen the study's outcome with 21 community members. A combination of primary and secondary data was used to measure and analyse implementation strategies and their effectiveness. Reviews of Namibian development plans, including continental (Africa) plans, were also considered. The study aimed to understand the factors contributing to the poor implementation of rural development strategies in Namibia.

The study used qualitative data, collected directly from residents and officials, and analysed it using the content analysis method and Creswell's six-step conceptualisation. The analysis was conducted using the neo-classical and urban bias theoretical frameworks, identifying patterns, themes, assumptions, and meanings. The data was organised into manageable chunks, categories, and units, and conclusions and recommendations were drawn based on the collected data. Ethics refer to the researcher's conduct regarding the rights of participants and those impacted by the study. The researcher made sure that the participants would not be subjected to any emotional, bodily, or psychological harm. Furthermore, the researcher obtained ethics clearance from UNISA, followed the University of South Africa's Code of Research Ethics, and obtained permission from MURD, NPC, and UNISA.

4 Findings

The following section presents the findings from the data collected for the study, beginning with an examination of field observations and subsequently engaging in a discussion centred on the factors that lead to the poor implementation of rural development strategies.

4.1 Observations

During data collection, the following observations were made in the field:

There appears to be a lack of community engagement in matters pertaining to rural development. The lack of community engagement in rural development matters is evident, with poor turnout in local gatherings and no youth meetings or forums, except for church-related matters. Most meetings are attended by senior citizens, who are retired and receiving pensions. Younger people only participate in the end of meetings with the intention of selling their products or services. Rural communities are often busy with activities at home or at cuca shops, and poor network coverage worsens the challenge due to erratic information dissemination. The observation established that if participants were interested in responding to interviews, they may have time and interest in participating in rural development initiatives. To encourage participation and engagement, it is crucial to motivate people to engage with their leaders and stakeholders, as community involvement is crucial in rural initiatives.

The quality of rural development varies from one constituency to another. The researcher observed the existence of a disparity in the quality of rural development and the level of development across various constituencies in the regions of Namibia. Constituencies located farther from urban centres tend to encounter greater difficulties and lack access to essential services such as electricity and network connectivity, in comparison to those situated closer to urban areas.

Regarding the appointment of councillors, there is a discrepancy in credentials and skills. The appointment of councillors based solely on their political party involvement may result in a discrepancy between their qualifications and experiences, ultimately leading to the inadequate execution of rural development initiatives.

Tensions between community leaders and local inhabitants. It has been observed that there are instances of factionalism among the headmen, councillors, rural communities, and businesspersons in certain regions. The lack of communication between village development committees (VDCs), business individuals, or village leaders poses a significant challenge to effective decision-making processes. Consequently, this creates separation and exerts an adverse effect on the implementation of rural development strategies.

Members of rural communities express nervousness towards their leaders. During meetings, it was observed that participation is mostly limited to the constituency councillors and headmen, while other attendees remain passive. However, after the meetings, these individuals engage in mini meetings where they express their grievances.

Rural areas are far away from urban areas. Some of the villages are located so far away from urban settings that they must use bad gravel roads to travel to town for services. It is a sad situation, as some rural residents travel more than 160 km a day, going and coming back to their respective habitats. Some rural community members use only one car in the village and travel once to town. Some of these are contributing factors to migration to urban areas, ultimately resulting in an absence of involvement in the decision-making processes and restricted communication with local leaders regarding these challenges.

4.2 Barriers Hindering the Implementation of Namibian Rural Development Strategies

The need for policy implementation is becoming more widely acknowledged as a means of closing the gap between policy objectives, anticipated outcomes, and intended consequences. However, the process itself is dynamic, complicated, and challenging, and it is still not fully understood (Matsieli and Sooryamoorthy 2022). The sustainability of rural development is often hindered by various constraints that have been identified through careful analysis. The formulation of pertinent strategies serves as a valuable framework to foster innovation. However, it is important to note that the implementation of successful innovative practices within regional contexts, particularly in rural areas, has been stifled by several challenges stemming largely from intricate rural governance dynamics, limited human capacity, and various socio-economic issues. Additionally, the researcher sought to determine the relevance of the neo-classical and urban bias theories in explaining these factors within the Namibian context. The study established the following:

Top-down vs Bottom-up Approaches. The findings established that the bottom-up approach to rural development in Namibia is a significant factor contributing to underdevelopment, as rural dwellers continue to distance themselves from the process of formulating and implementing development plans, in the process hindering their effectiveness. From the study, it also emerged that rural communities are not actively engaged in the formulation of rural development strategies. Rural areas are also observed to suffer from a lack of attention from the central government (Participant 04). Therefore, the exclusion of rural residents is a critical factor that militates against the attainment of development goals. In line with the insights provided by (Setokoe et al. 2019), the lack of community engagement in development initiatives can be attributed to the inability of hierarchical institutions to give up their authority, the absence of genuine ownership, and the dearth of pertinent information regarding the local communities. Therefore, rural communities lack the authority and mandate to engage in the implementation of rural development strategies and hold their villages accountable, as the regional government is seemingly not supporting them.

The study also determined that rural communities' lack of engagement is also a result of the lack of attention from the regional government. The delays in the implementation of programmes and projects might be attributed to insufficient community involvement (Shetunyenga 2023). Rural development strategies are traditionally designed by government entities, but insufficient attention from the central government and limited community discourse hinder their effective and successful implementation. Insufficient community engagement and inadequate stakeholder engagement hinder the success of these strategies. A clear understanding of appropriate involvement methods is crucial for the effective and successful implementation of development strategies and agendas.

The implementation of rural development strategies in Namibia requires involvement from rural communities. This involves identifying their specific needs, formulating and implementing appropriate strategies, and involving them in the evaluation process. This approach will lead to comprehensive and sustainable strategies. However, rural communities in Namibia often receive policies without considering their perspectives, resulting in a top-down approach and poor implementation of rural development strategies.

Lack of Resources. The study indicated that rural development strategies are being used for self-serving purposes, with insufficient capital projects and financial constraints hindering their successful implementation. Participant 02 asserts that resources allocated for the purpose of implementing rural development strategies are being used for self-serving purposes. Furthermore, a constituency councillor was accused of allocating land to himself—land that had been earmarked to assist communities with development projects (The Namibian 2019). Additionally, a constituency councillor was accused of selling government-provided seeds and fertilisers, which were meant to provide support to rural farmers, on the black market (New Era 2020). Consequently, regional government officials face challenges in addressing challenges culminating from insufficient resources, leading to a disjointed and unpredictable implementation process. The misuse of resources for rural development in Namibia has a detrimental impact on efforts to alleviate poverty and reduce inequality. When there are inadequate resources, this will limit a rural community's capacity to successfully impact and shape policies (Shetunyenga 2023). To effectively address this issue, it is crucial for the government and stakeholders to take responsibility and ensure that rural areas are provided with the necessary resources to enhance the quality of life for their rural populations.

Furthermore, the challenge of limited funding resulting from budget cuts by the central government mitigates against engaging in extensive research in partnership with esteemed institutions (Participant 30). In certain instances, significant documents, such as reports on the national population, could not be produced, thereby impeding the dissemination of valuable information through printed media. The persistent reduction of the development budget is gradually diminishing the potential for future expansion and the provision of public services. Moreover, the adoption of contemporary technology incurs a substantial expense. Resultantly, incomplete capital projects or financial constraints are contributing factors to the challenges of effectively achieving rural development strategies, as evidenced by poor network coverage in certain areas of the region (Participant 14).

According to Participant 29, regional government officials are in closer proximity to communities. However, they face the challenge of lacking the necessary resources from the central government to address their insufficient capacity to effectively implement development strategies. Insufficient resources pose a significant challenge to the successful implementation of rural development strategies, resulting in a disjointed and unpredictable process, as the absence of resources impedes progress (Merrell 2022). supports these findings and submits that rural areas often experience a lack of local resources, which leads to a greater dependence on intergovernmental fiscal transfers or external donations. The national government helped regions to reduce their workload and perform tasks more efficiently by creating logical opinions and providing them with well-crafted policies to implement in their respective areas (Participant 30). However, there are not enough resources available to facilitate the effective implementation of these policies and strategies.

The results reveal a concerning lack of essential amenities in rural areas, such as healthcare, limited network connectivity, poor road infrastructure, and unreliable electricity access. The study uses technocratic, radical, and reformist approaches to analyse Namibia's poor rural development strategies. It links Monaheng's methods, focusing on infrastructure, resource distribution, and technological shortcomings. However, these approaches may overlook social aspects, face practical difficulties, and fail. This results in challenges for rural communities in maximising benefits from development initiatives, hindering their ability to engage in community gatherings.

The slow provision of infrastructure development and maintenance contributes to the poor implementation of rural development strategies in Namibia.

Lack of Well-Defined Rural Development Strategies. The findings indicate that the lack of clearly defined rural development strategies hinders implementation. According to Participant 25, there exists a limited understanding regarding certain legislative measures, notably the Public Procurement Act of 2015 (Act 15 of 2015), the Access to Information Act of 2022 (Act No. 8 of 2022), and the Regional Councils Act of 1992 (Act 22 of 1992). Insufficient explanations of the legal structure frequently serve as a significant impediment to the successful execution of rural development strategies, necessitating enhanced unity and advancements in the leadership proficiencies of local communities. Participant 28 reported a decrease in the effectiveness of policies aimed at supporting rural communities. The participant also observed that, while the central government is making efforts, the capacity of regional councils to lead rural development planning is inadequate. The alignment between rural development strategies and the organisational goals and objectives of certain regional councils is not comprehensive. Therefore, rural development strategies need to be delivered at the right geographical scale.

Criminal Activities. The study exposed that the criminal activity in question involves the theft of copper wires, which are supposed to facilitate access to electricity, and are taken from connection towers. To exacerbate the situation, certain rural constituencies are observed to be without police sub-stations, as indicated by the study conducted by Shetunyenga (Shetunyenga 2023). The theft of resources that are meant for rural development and corruption are some of the factors contributing to the poor implementation of rural development strategies. The lack of passion for public interest by constituency councillors and Community Development Committees (CDCs) is concerning, as they appear to prioritise their own enrichment over addressing the needs and concerns of rural communities (Participant 23). To improve the effectiveness of resource use, it is important to establish transparency within national, regional, and local government.

Lack of Knowledge. In Namibia's rural areas, there is insufficient knowledge of rural development processes, strategies, and policies among community members. The study established that insufficient knowledge within Namibia's rural communities hinders government officials and rural communities from effectively implementing rural development strategies. The main challenge is that community engagement in the implementation process is hindered by a lack of comprehensive understanding among community members regarding the appropriate means of involvement (Participant 11). Inevitably, community engagement is hindered by poor understanding and capacity, and the lack of channels for individuals to articulate their viewpoints and engage with governmental entities.

Regarding these findings, and as argued by the OECD (OECD 2016), rural communities are ignorant of the positive aspects of development efforts made by governments. On the other hand, officials have indicated that resources are being wasted because village residents are unaware of numerous scientific advancements. According to Shetunyenga (Shetunyenga 2023), there is inadequate awareness among rural communities in Namibia regarding their capacity to actively engage in the legislative and policy-making processes.

These findings concur with Mukwena (Mukwena 2004), who opines that it is evident that the current capabilities of regional councils to effectively lead regional development planning and establish and manage rural areas are inadequate. Therefore, it is of utmost importance to empower rural communities to effectively articulate their socio-economic, political, cultural, and spiritual aspirations and necessities. Additionally, it entails the establishment of institutions and platforms that facilitate the expression and interaction of rural communities with the government. These mechanisms play a crucial role in enabling active participation, as well as in guiding and shaping rural development (Ministry of Regional and Local Government 2012).

The study further established that cultural practices inhibit innovation and, as such, contribute to the poor implementation of the processes and strategies for developing rural areas. To accelerate rural development, it is pertinent that cultural and social norms, values, and beliefs align properly with rural development strategies. Some traditional beliefs and practices can hinder community participation in rural development. For instance, some cultural beliefs may discourage inhabitants from participating in development initiatives, while others may discourage people from collaborating with those outside their own community. It is important for elected officials to possess a comprehensive understanding of the cultural backgrounds of the populations within their various constituencies (Helao 2005).

Furthermore, the lack of capacity in rural communities in Namibia hinders the effective implementation of rural development strategies. This lack of capacity stifles efforts to promote rural development. Additionally, there is a lack of communication channels for individuals to express their viewpoints and engage with government officials. Open communication channels are crucial for fulfilling the rural development mandate. It is of utmost importance to consider that individuals in rural areas understand their local circumstances and the ability to tailor development strategies to suit their unique context.

Inadequate Communication Methods. Inhabitants lamented the absence of effective and open communication channels between the government and the rural communities. The lack of effective and open communication often breeds distrust between community leaders and rural communities, and, as such, tends to hinder effective community participation within rural communities in Namibia. The issue of mistrust is observed to often emanate from past experiences or cases where constituency councillors made some promises but failed to deliver on them. This tends to promote high levels of apathy among community members, especially when it comes to a call for participation in government initiatives, as community members evince a lack of interest in such development activities. Effective communication between elected representatives and the electorate is crucial for participatory democracy, governance, accountability, and effective public service delivery (Helao 2005). In line with the same argument, Helao (Helao 2005) indicated that elected officials ought to establish channels of engagement with local communities at the grassroots level.

One additional concern that arises is the restricted availability and inadequate accessibility of telecommunication services in rural areas. According to Participant 15, the inadequacy of information and communication technology (ICT) and electricity infrastructure continues to impede the progress of rural development in rural communities. The effective transfer of information and the provision of conducive conditions for learners and students are crucial aspects of human and skills development, which are often facilitated by ICT connectivity and access to electricity (Participant 15). The successful implementation of rural development strategies is often impeded by the lack of consistent access to electricity and expenses associated with deploying ICT infrastructure across vast rural landscapes (National Planning Commission (NPC) 2017). Therefore, conducting community visits or consultative meetings and government intervention geared towards solving challenges faced by rural communities will assist in the effective implementation of rural development strategies.

Insufficient Research, Data Reliability, and Monitoring and Evaluation Processes. The study revealed that insufficient research, unreliable data, and outdated regional councils' website information hinder the implementation of rural development strategies. Participant 20 believes that the existing monitoring and evaluation processes are ineffective. The lack of proper and effective monitoring and evaluation has therefore impeded efforts towards the efficient implementation of rural development strategies. Insufficient research and unreliable data are factors that contribute to the poor implementation of rural development strategies, as stated by Participant 26. Moreover, there is a lack of current information available on the websites of the regional councils. In agreement with Amadhila (Amadhila 2019), research is not prioritised; additionally, actions and activities concerning these concerns are based on incomplete information. Consequently, there is also no reputable source of data within Namibia that can be used for better development, monitoring, and assessment (Republic of Namibia 2015).

The findings revealed that there are no statistics after the year 2011. Data on many other key development areas is also lacking due to financial constraints. Therefore, no evidence-based statistics have been provided in Namibia to influence future policy within the past 13 years. Furthermore, it has come to the study's attention that various ministries are currently withholding data and not cooperating with the NPC in terms of data sharing and compilation. To facilitate the sustainable development of rural areas in Namibia, it is important to have access to accurate and comprehensive demographic information (Ministry of Regional and Local Government 2012).

Resultantly, the link between regional and central government initiatives and establishing a robust monitoring framework for constituencies through regional councils is crucial for sustainable rural development. Evaluators should conduct comprehensive skills audits for officials responsible for implementing rural development strategies to ensure they possess the necessary skills to effectively address rural communities' developmental needs. Poor monitoring and evaluation mechanisms contribute to the poor implementation of rural development strategies.

Insufficient Government Proactive Measures for Climate Change Response and Other Crises. Based on the findings of the study, the central government's lack of a proactive response to natural disasters such as disease outbreaks and adverse climatic conditions such as floods, COVID-19, drought, veld fires, and rural-urban migration, has occasionally derailed the implementation processes for rural development strategies (Participant 27). Furthermore, it has been observed that, in most instances, local businesses and international donors have been at the forefront in responding to the effects of natural disasters and crises. This is quite positive and can be the solid ground on which the government, through its officials, can build on in implementing rural development strategies. This is so, as local development initiatives supported or spearheaded by local businesses and communities have a higher likelihood of success than initiatives undertaken by the government alone.

It is, ultimately, important to adopt a comprehensive approach to disaster management that encompasses a well-structured and organised strategy to enhance the efficiency of preparedness, response, and recovery efforts. In the absence of proper measures, the occurrence of natural disasters can significantly delay the successful implementation of rural development strategies. This can result in preventable delays and hinder the overall progress of implementation. An additional observation made during this study was that climate change has led to increased exposure to severe droughts and shortages of water. Therefore, communities and all levels of government need to be prepared for environmental shocks by practicing risk mapping and planning.

Lack of Coordination and Coherence. The study has identified that challenges in rural development strategy implementation arise from a lack of coordination between the central government, community members, and other stakeholders. Participant 13 blamed the lack of inadequate engagement between stakeholders and community members of rural constituencies as a critical factor impeding development. The absence of effective coordination among stakeholders has led to the incompleteness of compiled reports. The inadequate involvement of civic organisations, residents, and pertinent stakeholders in the development, planning, execution, oversight, and assessment of rural development strategies has hampered progress (Participant 17). Additionally, Participant 21 identified a significant challenge in the planning and implementation of rural development strategies in the regions, which they observed are being influenced by the lack of coordination and coherence among the central government, community members, and other stakeholders.

The study submits that there is an absence of coordination and integration among sectors because of the absence of a framework of laws and regulations that guarantees proper planning, management, and governing (National Planning Commission (NPC) 2017). This is consistent with the findings of Beland et al. (2016), who opine that a contributing factor to the failure of policy implementation is the absence of shared ideas among key stakeholders responsible for carrying out implementation duties. Limited organisational capabilities and a widespread reluctance to collaborate might be blamed for the lack of coordinating mechanisms and dialogue platforms (MRLGHD 2013).

Development requires continuous, timely input (Westoby and Botes 2020). Poor collaboration and networking between stakeholders in rural development is a contributing factor to the poor implementation of rural development strategies.

In terms of rural growth, it is important to recognise that no realistic goal can be achieved when stakeholders act in isolation. As Adam Smith wisely states, no island can thrive independently. Coordination facilitates strategic intervention by using existing resources efficiently, minimising programme duplication, and effectively targeting individuals in need of specific forms of assistance (Jackson and Cobb 2013; Nawab and Sharar 2022). The successful implementation of rural development strategies cannot be attained without a coordinated approach. Osmani et al. (Osmani et al. 2022) support this belief and submit that the successful implementation of development initiatives can face significant challenges within contexts where institutions lack the necessary capacity to effectively carry out their core functions. These functions, namely cooperation, commitment, and coordination, perform a vital function in ensuring the smooth execution of development initiatives.

Gaps Between Rural and Urban Areas. Rural communities suffer from a phenomenon named population neglect (Kawamura 2021). The study revealed that rural areas providing sustenance for the poor are often replaced by towns, despite their significance and relevance to people's needs. The implementation of rural development strategies is being abused by a favoured minority in urban areas (Participant 12). Because of this, the difference between urban and rural locations has increased in an alarmingly rapid manner. Participants 16, 19, 22, and 24 submitted that the local resources are not being fully used due to an absence of public ownership and involvement. In many cases, rural areas that are meant to provide sustenance for the rural poor are being replaced by towns that continue to mushroom. This is even though the land holds significant meaning and relevance to the needs and aspirations of rural people, and the development of towns impoverishes them. The persistent disparities between rural and urban regions persist, with many rural areas still grappling with the potential for social, economic, and political marginalisation (Kawamura 2021).

Observations from this study reveal that there is a huge difference between rural and urban areas in Namibia. One of the key challenges faced in rural areas is the absence of sports facilities, which can serve as a significant attraction for young individuals to remain in these regions. Moreover, it is indeed disheartening to observe that throughout the vast rural landscapes of Namibia, there is a notable absence of libraries that possibly offer useful information resources to the rural population.

Based on these observations, it can be deduced that there exists a significant concern pertaining to the reliance of rural communities on urban areas. This reliance stems from the lack of essential services available in rural regions, which compels the rural inhabitants to migrate towards urban centres in search of these services. Ultimately, this will result in a significant depletion of talent and untapped potential within rural communities.

5 Recommendations

The factors contributing to the poor implementation of rural development strategies cannot be ignored. As such, this requires contemporary scholarship to provide suggestions and recommendations that are prerequisites for the effective implementation of sustainable rural development in Namibia. Based on discussions, analyses and interpretations of the factors that have led to poor implementation of rural development strategies in Namibia conducted in the study, the following recommendations are made, with the hope that they are able to foster effective implementation strategies for the achievement of functional rural development processes:

5.1 Recommendations for Community Members

The implementation of rural *development strategies should be a collaborative effort.* Rural residents should be part of the planning, implementation, and evaluation of rural initiatives. The government should not be solely responsible for rural development. Therefore, the implementation of rural development strategies should involve both the government and rural residents. In this regard, constituency councillors should be crucial in promoting implementation and efficiently mobilising resources, as designated by the central government.

Self-mobilisation of local inhabitants. If locals are not involved in the activities intended to improve their development, their voices will not be heard. To influence policymakers in the implementation of government programmes on rural development, locals should have a strong commitment and a strong will to provide the necessary information that can help in the formulation of effective rural development strategies. Rural residents will be able to establish their own initiatives and planning processes because these would have been formulated at grassroots levels. Furthermore, it is crucial for the local community to safeguard their resources. Ultimately, it is important that locals are granted the necessary rights to use, oversee, and possess their natural resources. By so doing, they can effectively ensure the sustainable management of these resources for the long-term benefit of their communities.

5.2 Recommendations for Implementers

Rural development strategy should incorporate a multiplicity of policies and approaches. The development of human resources, research and monitoring programmes, infrastructure development, country-specific policies, encouraging stakeholder accountability, capital stock investments, and technical investments are all essential for a strong and thriving rural community. Any rural development strategy should incorporate and activate the community, and decision-making must take community development plans into consideration.

Establishing stronger connections between the agricultural, industrial, and service sectors in rural areas. Namibia needs stronger policies to manage and coordinate agri-food supply chains, as many rural projects are already agriculture-based. Entrepreneurial training for residents interested in agriculture is crucial for creating business opportunities and cross-sector skills. Using contemporary technologies will increase productivity, modernise agriculture, and foster skills related to the agriculture value chain. The government should also consider incorporating strategies that make the market work for the poor into rural development plans. This initiative aims to create a favourable business environment for local businesses, mitigating market failures, and effectively addressing challenges faced by these businesses and enhancing their productivity.

Improving the coordination of rural and urban policies. Implementation will go well if there are fairness, better connections, and better living conditions in rural areas. Connecting rural roads to inner-urban major roads, as well as ports and logistics facilities, will improve access to rural areas. For the provision of rural services, development policy rhetoric needs to be harmonised and turned into actual implementation. Responsible authorities should plan and implement services that are typically provided in urban areas in rural areas, assisting in rural revitalisation and rural-urban integration. This will contribute to bridging the disparity between urban and rural areas and reviving rural development.

Encourage empowerment in rural communities. The central government should actively involve rural communities in policy initiation and implementation to develop comprehensive strategies that address regional needs. By involving indigenous communities in the formulation and implementation of development strategies, the government can better support these communities and improve their standards of living. Implementers should also involve beneficiaries in the process of development, conception, and implementation to create effective strategies. By consciously striving to recognise and comprehend their programme, programme implementers can create effective solutions. It is crucial to include members of the target population at every stage of the process to promote participation and reduce social stigmas. By doing so, rural communities can be at the forefront of addressing regional needs and promoting overall development.

Monitoring and evaluation. Policy monitoring and evaluation are crucial for understanding the short-term and long-term effects of various projects and programmes. This process involves assessing the impact of the implementation process, key stakeholders, and populations for which the policies are formulated. Policymakers and planners should constantly check and evaluate the impact of alternative intervention programmes on communities and society. Monitoring and evaluation research can be conducted for various purposes, including improving programmes, defending ongoing programmes, auditing them to identify failures, abolishing them when necessary, or changing implementers.

Policy monitoring and evaluation are crucial for understanding the short-term and long-term effects of various projects and programmes. This process involves assessing the impact of the implementation process, key stakeholders, and populations for which the policies are formulated. Policymakers and planners should constantly check and evaluate the impact of alternative intervention programmes on communities and society. Monitoring and evaluation research can be conducted for various purposes, including improving programmes, defending ongoing programmes, auditing them to identify failures, abolishing them when necessary, or changing implementers.

Centrality of rural populations/people. It is important for local inhabitants to be active in activities taking place in their own communities. In pursuit of this, the regional government and various stakeholders need to support communities in shaping and determining their own future. There is a need to adopt a bottom-up approach.

Develop policies to assist rural youth. From the observations conducted, it is evident that there is a lack of participation by the youth in rural development initiatives. It is advised for the young people to have youth forums at the constituency level. They should meet annually, specifically in August or December, as these are the periods when most youths are within the rural areas for holidays. Such policies in social inclusion will greatly benefit rural youth through active participation in decision-making processes that affect their lives. The youth play a crucial role in shaping a nation's future, especially in sustainable rural development. To support their growth and community development, the government should provide support and assistance, such as community library services and sports facilities. Allocating resources towards remote working infrastructure is essential to retain skilled individuals and attract mobile talent to rural areas. The absence of accessible libraries for youths is a negative situation. Establishing libraries in every constituency can facilitate research and discourage rural youth from seeking services in urban areas, as libraries offer more services than reading books. Overall, supporting the youth is crucial for achieving sustainable rural development.

Urban-rural linkages. Rural communities are highly dependent on urban communities for services. It is important to consider improving services within rural areas, mostly infrastructure. This is to promote rural-urban linkage, and curtail rural-urban migration, which has historically robbed rural communities of many people who were supposed to assist with the implementation of rural development strategies, when they migrate from their rural jurisdictions to urban areas in search of better services. By fostering connectivity and inclusion, it becomes easier to develop more effective strategies and initiatives that promote equitable development and bridge the urban-rural divide.

Research, consultancy, and technological innovation. The researcher strongly supports the promotion of new research initiatives focused on rural development priorities. It is crucial for regional councils to collaborate with institutions of higher learning, research funders, and research organisations to effectively address the needs and challenges of rural communities. In addition, it is crucial to conduct thorough research prior to implementing strategies, and to conduct evaluations at least one year after the policy has been implemented to assess its impact. This approach will assist in enhancing the standard of living in rural areas by promoting inclusive social, environmental, technological, and economic development. Furthermore, the Namibian government should view rural development through the lens of the fourth industrial revolution, recognising technological advancements that alter landscapes and present new opportunities for progress. Innovation is crucial for sustainable growth in emerging nations such as Namibia, which can help achieve national development goals and pave the way for Namibia's advancement towards becoming a knowledge-based society. Therefore, research and technological advancement are vital for the present and future success of any developing nation.

Sufficient capability to implement. As elucidated, a gap in skills is one of the contributing factors to the poor implementation of rural development strategies. Therefore, capacity development must be primary before the formulation and implementation of any development strategies. The government and community members can be drivers of rural strategies, but only if they are well capacitated. If the residents lack the necessary skills, they will not be able to fully and effectively participate in rural development initiatives. Therefore, rural communities need to identify priority gaps on their own.

Analyse the stakeholders. Perform a comprehensive stakeholder analysis, delineating the respective contributions and expectations of the stakeholders towards the MURD in terms of rural development. It is important to disseminate the lessons learnt by relevant stakeholders to regional councils and communities.

5.3 Recommendations for Stakeholders

Facilitation of local inhabitants and constituency councillors. Stakeholders should be facilitating local inhabitants and constituency councillors to assist the central government in capacity development, research of rural communities' needs, and evaluations of rural development strategies. This will promote the actionable and effective implementation of rural development strategies.

6 Conclusions

The study concludes that the factors contributing to the poor implementation of rural development strategies emanate from internal and external strategy and policy challenges that mitigate against them achieving the objective that policymakers set out to accomplish. The poor implementation results primarily from the ineffective top-down approach, lack of resources, the lack of well-defined rural development strategies, criminal activities, the lack of knowledge, inadequate communication methods, insufficient research, data reliability, monitoring and evaluation processes, insufficient government proactive measures for climate change response and other crises, the lack of coordination and coherence, and the gap between rural and urban areas. Such factors give rise to discrepancies between the planned targets and the results achieved within the stipulated periods.

References

African Development Bank: African Economic Outlook 2022. African Development Bank Group, Abidjan (2022)

Amadhila, F.K.: Evaluating the Effect of the Micro Finance Scheme Programme for Rural Entrepreneurs in Oshikoto Region, Namibia. Master's thesis, Stellenbosch University, Stellenbosch (2019)

Dahlman, C.: A New Rural Development Paradigm for Developing Countries for the 21st Century. OECD Development Centre, Brussels (2015)

Food and Agriculture Organization (FAO): the State of Food Security and Nutrition in the World 2023. FAO, Rome (2023)

Helao, T.: The Role of Regional Councillors in Consultation and Communication Regarding Rural Service Delivery in the Oshana Region of Namibia. Master's dissertation, University of the Western Cape, Cape Town (2005)

International Energy Agency: World Energy Outlook 2023. IEA Publications, Paris (2023)

International Labour Organization (ILO): Global Employment Trends for Youth 2023. ILO, Geneva (2023)

International Labour Organization (ILO): Extending Social Protection Coverage to the Rural Economy: Decent Work in the Rural Economy, ILO, Geneva (2016)

Jackson, K., Cobb, P.: Coordinating professional development across contexts and role groups. In: Evans, M. (ed.) Teacher Education and Pedagogy: Theory, Policy and Practice, pp. 80–99. Cambridge University Press, Cambridge (2013)

Kawamura, H.: In situ urbanization key to leaving no one behind. UN DESA Policy Brief No. 104, United Nations Department of Economic and Social Affairs, New York (2021)

Lipton, M.: Why Poor People Stay Poor: a Study of Urban Bias in World Development. Temple Smith, Australian National University Press (1977)

Matsieli, M., Sooryamoorthy, R.: Why is public policy implementation difficult in developing countries? Analysis of e-government implementation in Lesotho. J. Stud. Soc. Sci. Hum.Ities. **8**(3), 303–317 (2022)

Merrell, I.: Blockchain for decentralised rural development and governance. Blockchain: Res. Appl. **3**(3), 100086 (2022)

Ministry of Regional and Local Government: Housing and Rural Development (MRLGHRD): National Rural Development Policy. MRLGHRD, Windhoek (2012)

Monaheng, T.: Rural Development and Community Participation in Lesotho. Doctoral dissertation, University of South Africa, Pretoria (1995)

Mukwena, R.M.: The role of decentralisation in reducing regional inequalities in Namibia. Reg.Nal Dev. Dialogue. **25**, 108–128 (2004)

Mushelenga, P.: Rural Development Planning and Review [Speech]. Official Opening of the 2018 Annual Rural Development Planning and Review Workshop, Eembaxu RDC, Namibia, 25 June 2018

National Planning Commission (NPC): Namibia's fifth national development plan. Government of the Republic of Namibia, Windhoek (2017)

Nawab, A., Sharar, T.: Coordination in planning and implementing professional development programs for teachers: a much-needed component in rural areas. SAGE Open. **12**(1), 1–9 (2022)

New Era: Namibia: Councillor accused of selling government seeds, fertilizers. New Era, 20 May (2020)

OECD: FDI in figures. OECD Publishing, Paris (2016)

Osmani, M., Kolaj, R., Borisov, P., Arabska, E.: Why agricultural policies fail and two cases of policy failures in Albania. Agric. Resour. Econ.: Int. Sci. E-J. **8**(2), 86–104 (2022)

Republic of Namibia: National Policy on Micro, Small and Medium Enterprises in Namibia. Government Gazette, Windhoek (2015)

Setokoe, T.J., Ramukumba, T., Ferreira, I.W.: Community participation in the development of rural areas: a leaders' perspective of tourism. Afr. J. Hosp., Tour. Leis. **8**(2), 1–15 (2019)

Shetunyenga, N.N.: An exploration of the barriers to community participation in development within Ompundja constituency. Open J. Soc. Sci. **11**(9), 399–414 (2023)

Sullivan, S.: Rural Planning in Namibia: State-Led Initiatives and some Rural Realities. IIED, London (2000)

The Namibian: Namibia: Land allocation row rocks Ohangwena. The Namibian, 10 September (2019)

UNESCO Institute for Statistics: Global Education Monitoring Report 2022. UNESCO Publishing, Paris (2022)

United Nations Development Programme (UNDP): Human Development Report 2023. UNDP, New York (2023)

Westoby, P., Botes, L.: Does Community Development Work? Stories and Practice for Reconstructed Community Development in South Africa. Practical Action Publishing, Rugby (2020)

World Bank: World Development Indicators. World Bank, Washington, DC (2023)

Warehousing Effectiveness in the Automotive Industry: A Case of South Africa

Xolani Mwandla, Elizabeth Chinomona, and Pfanelo Nematatani

Abstract Warehousing is an important component of the supply chain and one of the key factors in the automotive industry for efficiently operating a warehouse to ensure sustainable competitive advantage and improve performance. With the South African automotive industry facing many challenges, such as the unreliability of rail transport and rail capacity problems, the high cost of South African ports, the cost of replacing outdated technology, and compliance with government preferential procurement policies, warehousing effectiveness and efficiency are one of the comprehensive solutions which could bring much-needed relief. In addition, the industry faces the transformational dynamics of the growth of digitisation, increasing demands for automation, and new business models. The purpose of this study was to investigate the effectiveness of warehousing in the automotive industry in South Africa. To achieve the research objectives and answer the research questions, a qualitative approach that incorporated exploratory and descriptive qualitative research designs was used. A purposive sampling technique was applied to identify the unit of analysis and select the population for this study. As a unit of analysis, supply chain professionals from selected automobile manufacturers based in Gauteng Province were used as the population for the study. For data collection, the study used a qualitative data collection method of a structured interview, and 15 participants were interviewed. The findings were recapitulated, themes and subthemes were generated, and the key insights derived from the data were highlighted. The data generated from the interview feedback were imported into the ATLAS.ti Qualitative Data Analysis Software (QDAS) programme for analysis purposes. The study recommends that strategies such as optimisation through automation, integration of digitisation and digital technologies, industry 4.0 and circular economy framework, and investment in quality management best practices should be implemented in order to enhance warehousing effectiveness in the automotive industry.

X. Mwandla · E. Chinomona (✉)
Vaal University of Technology, Vanderbijlpark, South Africa
e-mail: elizabethc@vut.ac.za

P. Nematatani
University of South Africa, (UNISA), Pretoria, South Africa
e-mail: nematp@unisa.ac.za

E. H. Redda (ed.), *Proceedings of the Social Sciences International Research Conference, 2025*, https://doi.org/10.1007/978-981-95-9489-4_4

Keywords Automotive Industry · Efficiency · Effectiveness · Warehousing · South Africa

1 Introduction

The automotive industry is a significant contributor to the South African economy, with a gross domestic product (GDP) of 6.8 per cent and employs around 110,000 people across the vehicle and component manufacturers (DTI 2018). Modern cars, perceived as a fast, comfortable, safe, flexible, and affordable mode of transport, have become a status symbol or means to reflect identity (ILO 2021). Consumer preferences determine the current styles, reliability, safety, and performance standards of vehicles (Lee et al. 2019). This has propelled the industry to continue improving manufacturing productivity to provide high-quality service to customers at the right time (Harvey 2020). As a result, the production of reliable cars and the modernisation of production systems have been growing immensely (Tellini et al. 2019). Over the past 20-plus years, the changing global motor vehicle industry has enabled the development of a vibrant automotive industry from an era of providing traditional transportation services to an intelligent era (Ji et al. 2020). Due to its development, the automotive industry is one of the most successful industries of the past century.

2 Problem Statement

Despite the popularity of automotive as a concept and on a global scale, the automotive industry has experienced strong competition in a highly competitive market affected by globalisation, stiff competition, fluctuating market demands and customer requirements (Altinisik and Hugul 2020; Ambe and Badenhorst-Weiss 2011). This situation also holds for the South African situation (Havenga et al. 2012). These challenges, including the supply chain complexity for automakers, mean the automotive industry is dominated by limited large multinational enterprises (MNEs) (Lampón et al. 2021). Small and medium-sized enterprises (SMEs) are much more prevalent among the second and third tiers of the industry's supply chains (Tolmay and Badenhorst-Weiss 2018; ILO 2021). There are areas within warehousing that have not been critically researched, at least in the South African context. When it comes to the latest advanced technology, South Africa is still lagging behind other first-world countries. In addition, debatably, there have not been many studies conducted on warehousing effectiveness in the South African automotive industry. Africa presents unique challenges and there has not been sufficient research covering the topic in Africa compared to Asia and South America.

2.1 Research Questions

There are two types of research questions, namely, the main research question and secondary research questions.

2.1.1 Main Research Questions

- What are the key factors and strategies that influence warehousing effectiveness in the automotive industry in South Africa?

2.1.2 Secondary Research Questions

- What are the challenges and risks faced by the selected automobile manufacturers concerning warehousing?
- What must be done to make relationships between the selected automobile manufacturers and their customers more effective?
- How can warehousing effectiveness be improved in the selected automobile manufacturers?
- What are the warehousing processes in the automotive industry?

1. **Literature Review.**

A review of the literature is done on the theory and variables used in this study.

2.2 Effectiveness-Efficiency Theory

Effectiveness-efficiency theory grounds the study through constituting an operational and strategic concern for organisations as they seek to create value for customers and reduce costs, while striving to maximise competitive advantage (Bambrick et al. 2023). In general, effectiveness and efficiency are the most extensively used as measures of performance. Researchers such as Hidayat et al. (2023) describe effectiveness as making the proper judgements and ensuring the successful implementation of those judgements or decisions to accomplish the proposed objectives. Meanwhile, a general description of efficiency by Amirrudin et al. (2023) refers to a measurement that is typically stated as a percentage of the actual output to the standard production predicted. In addition, Amirrudin et al. (2023) stated that the effectiveness of warehouse efficiency can be measured by a collection of metrics that primarily quantify warehouse space capacity and efficient usage of available warehouse space.

2.3 Automotive Industry

The official statement from the United Nations (UN) International Standard Industrial Classification of all Economic Activities (ISIC), in its Revision 4, defines the automotive industry as the manufacture of motor vehicles, trailers and semi-trailers, including the manufacturing of parts and accessories for motor vehicles. According to the Ports Regulator of South Africa (2021), the automotive sector in South Africa plays a critical role in the country's economy, and numerous cities sustain themselves through this industry. Altinisik and Hugul (2020) believe that automotive manufacturers are facing stronger competition due to increasing customer expectations and aggressive delivery deadlines. However, the automotive industry has evolved over the years due to increasing demand for electric vehicles worldwide and governments of different countries announcing that the new electric vehicles will be included in the programmes of government subsidies for purchasing eco-friendly cars (Dmytriiev et al. 2019).

2.4 Supply Chain Management

Supply chain management (SCM) is such a vast topic that people often define it differently based on their personal experiences or perspectives (Harland 2024). Harland (2024) further identified three main groupings of SCM definitions and conceptualisations: those focusing on material planning and control, those focusing on buyer-supplier relationships and those taking a holistic approach, such as value chain, distribution channels, and networks. However, SCM is defined by Francis et al. (2021) and Patrucco et al. (2023) as a process of planning, identifying, selecting, and coordinating all activities involved in sourcing and developing suppliers to realise supply chain performance than competitors. In the automotive industry, substantive pieces of literature by Sharma et al. (2017) and Alsmairat and Aldakhil (2022) confirm that the automotive supply chain role-players are encouraged to constantly seek solutions to optimise their supply chains and to ensure sustainability. Errico et al. (2022) argue that participating in global supply chains has been a driver of several diversified benefits, productivity and innovativeness. Richards (2017) concludes that warehouses are regarded as a crucial link in modern supply chains. To achieve this, SCM was adopted as a best practice model. Drawing from the above, the adoption of SCM served as an impetus to investigate the effectiveness of warehousing in the automotive industry in South Africa.

2.5 Inventory Management

Scholars such as Macas et al. (2021) refer to inventory as a physical stock of finished goods that a firm is offering for sale, a stock of work in progress, and the maintenance of sufficient stock to meet the expected demands as and when they arise. Saliji (2021) describes inventory as an important strategic key factor in the automobile industry and a vital asset that ensures the high competitiveness of automobile manufacturers. This is corroborated by Decampos et al. (2022), who point out that effective inventory management has consistently been a focus point of the automotive industry for decades.

2.6 Effectiveness and Efficiency

The concepts of efficiency and effectiveness cut across various fields of study; the term effectiveness means doing the right thing, while efficiency means doing the right thing right. In other words, effectiveness focuses on the activity's results and efficiency focuses on the activity's performances (Saliji 2021). According to Florina (2017), the interpretation of efficiency and effectiveness refers to the relationship between inputs and outputs, the achieved results, and the expected ones, and the increase in productivity. Meanwhile, Mago and Hofisi (2014) and Zidane and Olsson (2017) indicate that efficiency and effectiveness are often used interchangeably. The significance of this research was premised on determining the effectiveness of warehousing in the automotive industry.

2.7 Warehouse and Warehousing Management

A warehouse is defined by Custodio and Machado (2020) as a storage place for products, storage, shipment and order-picking. Warehousing is described by Polten and Emde (2021) as the central element of basically all supply chains, accounting for about 20 per cent of all logistical costs. Wilson (2020) adds that warehousing and logistics are complex business processes that require flexibility and innovative solutions. Meanwhile, Tutam (2022) emphasises that warehousing is a significant component of the supply chain and plays a crucial role in contributing to a company's success. According to Sivakumar and Ruthramathi (2019), warehousing is the segment of enterprise logistic functions that are responsible for the storage space and handling of inventory from the receiving point to the consumption point. Lyu et al. (2020) argue that although excellent models and methodologies have been proposed to improve warehousing performance, large proportions of warehousing operations are still not value-adding.

3 Methodology

For this study on the effectiveness of warehousing in the automotive industry in South Africa, a research topic that remains relatively unexplored in the South African context, a qualitative research method was deemed appropriate to close the empirical gap. Salmons (2021) defines qualitative research as an umbrella term used to describe the ways of studying behaviour, perceptions, or experiences, or through their visual or verbal actions, writings, or expressions. This section introduced and explained the rationale for the research methods chosen for the study. To address the void in the empirical literature, a qualitative research methodology was deemed relevant to explore and describe the warehousing effectiveness in the automotive industry in South Africa.

3.1 Data Collection

To assess the effectiveness of warehousing in the automotive industry in South Africa, a purposive sampling technique was deemed relevant to identify the unit of analysis and select the population for this study (Niemann et al. 2018). The reason for choosing purposive sampling is that it is a better match of the sample to the aims and objectives of the research, thus improving the rigour of the study and trustworthiness of the data and results (Campbell et al. 2020). For this study, a purposive sampling technique, also called judgement sampling, was deemed to be more suitable, as it allowed the researcher to deliberately choose participants due to the qualities those participants possess and their willingness to provide the information through the knowledge and experience they possess. Equally, the age, longevity and experience in their current position, as well as qualifications, were used as set criteria to select participants in major automobile manufacturing working environments.

4 Details of Transcripts

For this study, a total of 15 participants were interviewed from automobile manufacturers based in Gauteng Province. A Qualitative Data Analysis Software (QDAS) programme, ATLAS.ti, was used to support analyses of data.

4.1 ATLAS.Ti Data Exports

Like NVivo, ATLAS.ti is a QDAS programme used to support analyses of data. ATLAS.ti and NVivo software can be used to conduct different types of studies

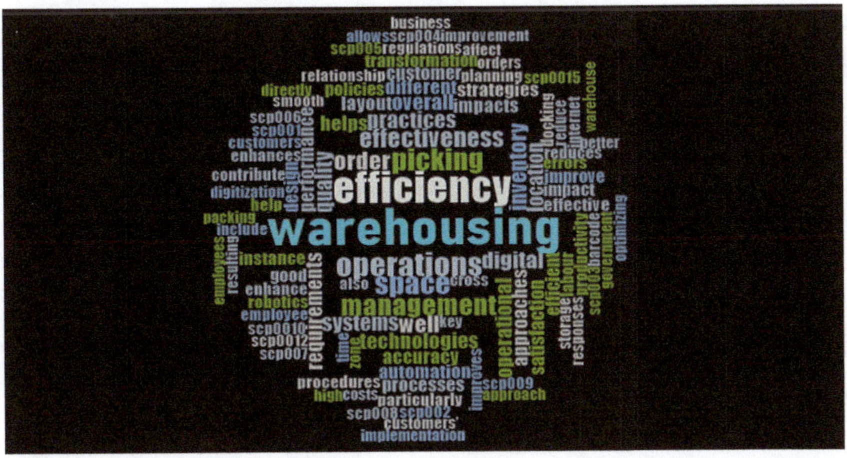

Fig. 1 ATLAS.ti-generated word query. *Source* ATLAS.ti 8

and can support a diverse array of research methods, methodologies, and analytical approaches (Woods et al. 2016). Figure 1 shows the generated word query exported from the ATLAS.ti.

Software packages such as ATLAS.ti are considered useful for qualitative data analysis, and they are furthermore powerful and sufficiently flexible to serve as the central component of a robust digital research workflow (Paulus 2023). In this study, an ATLAS.ti version of the software was applied to process the data collected through the administration of questionnaires with 15 research participants drawn from an automobile manufacturer. Table 1 displays the themes and their references (frequency of appearance or use in questionnaire transcripts) exported from ATLAS.ti.

4.2 *Themes and Subthemes*

Themes are described by Braun et al. (2023) as the key constructs that are used as elements, attributes, concepts, and descriptors. It is an implied topic that arranges a group of repeating ideas and enables researchers to answer the study question. Themes contain codes that have a common point of reference and have a magnitude of overview that combines ideas regarding the subject of enquiry (Pilcher and Cortazzi 2024). Braun et al. (2023) further pointed out that each theme may have some subthemes as subdivisions to get a comprehensive understanding of the data and reveal a pattern in the participants' responses.

The identification of themes and patterns that appear in otherwise unstructured qualitative data is one of the basic forms and an integral component of qualitative research because it provides an entry point into analysing qualitative data (Odiadi 2024). A study by Mishra and Dey (2022) implies that themes are abstract constructs

Table 1 Presentation of main themes and subthemes

No	Themes	Subthemes
1.	Warehousing effectiveness	Inventory availability, flexibility and reliability; supply chain efficiency; cost reduction
2.	Warehousing efficiency	Process improvement and standardisation; implementation of advanced technologies; operational efficiency and improved productivity
3.	Location planning	Proximity to manufacturing plants; access to major transportation networks; Markert proximity
4.	Space optimisation and requirements	Space utilisation efficiency; safety measures and compliance; reduction of excess inventory
5.	Order-picking systems	Order fulfilment speed; enhanced order accuracy; consistency and reliability
6.	Warehouse layout and design	Simulation-based optimisation approach; process flow optimisation; enhanced safety and productivity
7.	Quality management	Continuous improvement and reliability; increased customer satisfaction; value for money
8.	Digital transformation technologies	Robotics and warehousing automation; fourth industrial revolution (4IR); warehouse management systems (WMS)
9.	Lead time and customer satisfaction	Timeliness of delivery; customer communication and transparency; just-in-time system approach
10.	Government regulations and support	Subsidies and incentives; labour policies and environmental regulations; local content and productions

Source Author's compilation.

which researchers may identify before data collection through a literature review, during data collection, and after data collection. In this study, themes and subthemes were identified during the literature review and after data collection by sorting out and scrutinising the participants' responses.

5 Thematic Analyses and Discussion

Thematic analysis is commonly used in qualitative research and emphasises examining themes or patterns of meaning within data (Braun et al. 2023). This focuses on both the organisation and description of the data set, and the theoretically informed interpretation of meaning. It emphasises on analysing, identifying, and interpreting patterns of meaning or themes within qualitative data. Braun et al. (2023) further stated that thematic analysis can be used to analyse most types of qualitative data, including data collected from surveys, focus groups, or interviews.

5.1 Effectiveness and Efficiency

Although the ATLAS.ti identified ten main themes, and out of those ten, only two are the most relevant for this research, namely effectiveness and efficiency. Effectiveness and efficiency incorporate and interlink with the other eight themes, which are order-picking, quality management, space optimisation, government regulations and support, quality management, digital transformation, warehouse layout and design, and lead time.

These themes are elements and factors that enhance and improve the efficiency and effectiveness of warehousing in the automotive industry. Therefore, the thematic analyses and discussions that are anchored in this research are efficiency and effectiveness. This becomes pertinent given that the focus of the study is warehousing effectiveness and efficiency in the automotive industry. While effectiveness focuses on the results derived from certain activities, efficiency focuses on the activities' performances (Saliji 2021). In other words, the interpretation of efficiency and effectiveness refers to the relationship between inputs and outputs, between the achieved results and the expected ones, and the increase in productivity (Florina 2017). In addition, efficiency speaks to the employees' productivity, which in turn contributes to the effectiveness of warehouse operations and performance. Also, effectiveness speaks to the operations of the warehouse.

Efficiency is also defined as the use of minimal resources, which is measured in terms of cost and inventory turnover (Abu et al. 2018). Verma and Venkatesan (2023) added that effectiveness involves providing a competitive advantage, agility, profitability, innovation, improved product safety and quality, operational efficiency, improved customer experience and other social and environmental benefits. These views capture the key issues examined in this research and the outcomes derived regarding different elements of effectiveness in an automobile warehouse. The next section explores the themes of efficiency and effectiveness through the lens of research participants concerning key areas of the research questions administered to them.

5.2 *The Impacts of Quality Management on Warehousing Effectiveness and Efficiency*

The research question also examined the aspect of better quality in automobile warehousing, looking at the key best practices in quality assurance. Most of the participants indicated their awareness of quality management practices, including the establishment of clear guidelines and proper procedures for all warehouse operations, including tasks such as receiving, picking, and packing. The quality practice also involves radical alternatives for preventing poor quality in the manufacturing of vehicles and the effective application of a knowledge-driven, intelligent, quality problem-solving system. The participants indicated that investing in the basics and providing good quality offer value for money and cost-effectiveness across the spectrum of warehouse operations.

A major finding that cuts across is that there is a close relationship between quality management practices and warehouse efficiency because the latter feeds on the former. All participants highlighted the importance of quality management practices to warehouse efficiency, indicating their positive impacts on key areas such as inventory management, process standardisation, employee engagement, and customer satisfaction. For instance, participants indicated that *implementing quality management practices helps to improve inventory accuracy and visibility as well as* to *simplify operations procedures, which allows the warehouse to operate more smoothly and efficiently.*

One of the major contributions of quality management practices to warehouse inventory management is the reduction of stockouts and overstocks. Also, through the standardisation of warehouse operations processes, *errors, damaged products, and ineffective processes are easily identified and dealt with.* Overall, quality management practices have positive impacts on every aspect of warehouse operations. Generally, they help to maintain high standards and develop and sustain the culture of maintenance and continuous improvement across warehouse operations.

5.3 *The Impacts of Government Regulations and Support on Warehousing Effectiveness and Efficiency*

The research question addressed relevant areas of government regulations and support, such as policies on environmental issues and local content in South African-assembled vehicles and transformation, as well as support initiatives such as the South African Automotive Master Plan (SAAM). Participants indicated that government policy that enforces local content in South African-assembled vehicles would not cause the industry to become competitive. In other words, enforcement of local content policy does not positively affect automobile warehouse competitiveness. For instance, participants suggested that *policies that prioritise the promotion of local content over international ones tend to weaken the industry against the international market*. However, all the participants agreed that a variant of the local content policy contained in the government support through the SAAM would improve industry competitiveness levels to that of leading international competitors. This opinion indicates one of the statements that *governmental policy interventions such as the SAAM can enhance competitiveness in the industry and contribute to its growth and sustainability*. Although still in its early stages of implementation, authors such as Black, Makundi and McLennan (2017) have observed the impacts of the government-initiated automotive industry development strategy— the SAAM 2035. According to Black et al. (2017), the plan has increasingly exposed South Africa's automotive industry to international competition as the government has sought to make it more competitive and encourage exports and a more rational industry structure.

Some of the participants responded that government policies do not necessarily target warehouse efficiency. This is indicated by statements such as *Government policies are not as effective as they could be in terms of warehouse operations efficiency*. However, participants maintained that government regulations around safety, environmental protection, and labour practices within warehouses can improve efficiency and effectiveness. They cited examples of labour practices that can help with fair labour practices, employee engagement, and efficiency if there is strict compliance. A participant noted that *labour-related regulations such as minimum wage, working-hour restrictions, and employee benefits can improve warehouse operations and performance, resulting in high productivity and employee satisfaction.*

The work of Bhavani and Sharavan (2015) and Knox and Marin-Cadavid (2023) highlight the importance of employee engagement as a strategic approach for driving improvement and organisational change. Knox and Marin-Cadavid (2023) also collaborated with this research finding in the sense that employee engagement can play a key role in reducing wastage in productivity. Also, warehouse adherence to government policies on environmental issues can have safety implications for the workplace and reduce environmental impact. However, participants argued that certain environmental policies negatively harm job creation and growth in the automotive industry. However, some participants identified green manufacturing as one of the key challenges faced by automobile manufacturers in achieving profit optimisation and sustainable development.

6 Conclusions of the Research Finding

In the analysis of data collected, the study anchored the thematic analyses and discussion on effectiveness and efficiency because the main focus is the effectiveness and efficiency of automobile warehousing. Given this, different thematic areas of the data were analysed to determine their impacts on warehousing effectiveness and efficiency. The key areas included location, space optimisation, warehouse layout and design, order-picking, lead time, quality management, digital transformation, and government regulations and support. The contributions of this study are both theoretical and practical. Theoretical contributions concern the benefits and applicability of the cited literature, while practical contributions are aligned with the benefits of the study results for supply chain professionals and other warehousing and automotive supply chain stakeholders.

Theoretically, this study contributes to the existing body of information and knowledge since it is an addition to the available literature on warehousing effectiveness and efficiency in the automotive industry. The study further contributes to the existing literature on the conceptualisation, drivers, and key factors to warehousing effectiveness, such as location, space optimisation, order-picking, warehouse layout design, digital transformation, quality management, lead time, and government regulation and support. Practically, the results provided in this study are beneficial to supply chain professionals in the automotive industry regarding the enhancement of warehousing operations to ensure effectiveness and efficiency. The study revealed that key factors such as space optimisation, quality management, and shorter lead time contribute to warehousing efficiency and customer satisfaction. Furthermore, SCM professionals in the South African automotive industry can take note and ensure that warehousing automation processes are efficiently implemented in their automotive manufacturing value chain to improve productivity and reduce costs. The study recommended the following: digitisation and digital technologies should be integrated into warehousing knowledge management and organisational culture; warehousing should be optimised through an automation strategy; the inclusion of a circular economy framework, cost-effective location planning, as well as investment in the industry's best quality management practices and a review of existing government policies around warehousing. Other recommendations border on further study exploring warehousing in the era of electric vehicles and a multi-provincial study of warehousing effectiveness in two or more provinces, comprising the Eastern Cape and Kwa-Zulu-Natal.

References

Alsmairat, M., Aldakhil, A.: Modelling the interrelationships among environmental forces, organizational capabilities and supply chain sustainability. Uncertain Supply Chain. Manag. **10**(1), 117–124 (2022)

Altinisik, A., Hugul, O.: The seven-step failure diagnosis in the automotive industry. Eng. Fail. Anal. **116**, 104702 (2020)

Ambe, I.M., Badenhorst-Weiss, J.A.: South African automotive industry: trends and challenges in the supply chain. J. Contemp. Manag. **8**, 337–362 (2011)

Amirrudin, A.H., Kamaruddin, N.S., Salehuddin, N., Ibrahim, S.: Improving warehouse efficiency through effective inventory management practices. Soc. Manag. Res. J. **20**(1), 173–187 (2023)

Bambrick, S., Vafadarnikjoo, A., Reid, I., Bamford, D.: Unpacking critical success factors to improve supply chain effectiveness, efficiency and performance: a 7Vs framework for consideration. Prod. Plan. Control. **34**, 1–15 (2023)

Braun, V., Clarke, V.: Toward good practice in thematic analysis: avoiding common problems and becoming a knowing researcher. Int. J. Transgender Health. **24**(1), 1–6 (2023)

Campbell, S., Greenwood, M., Prior, S., Shearer, T., Walkem, K., Young, S., Walker, K.: Purposive sampling: complex or simple? Research case examples. J. Res. Nurs. **25**(8), 652–661 (2020)
Custodio, L., Machado, R.: Flexible automated warehouse: a literature review and an innovative framework. Int. J. Adv. Manuf. Technol. **106**, 533–558 (2020)
Decampos, H.A., Rosales, C.R., Narayanan, S.: Supply chain horizontal complexity and the moderating impact of inventory turns: a study of the automotive component industry. Int. J. Prod. Econ. **245**, 108377 (2022)
Department of Trade: Industry and Competition: South Africa's Automotive Industry Master Plan to 2035. DTIC, Pretoria (2018)
Dmytriiev, I.A., Shevchenko, I.Y., Kudryavtsev, V.M., Lushnikova, O.M., Zhytnik, T.S.: World experience and a unified model for government regulation of automotive industry development. Public Policy Adm. **18**(3), 46–58 (2019)
Errico, M., De Noni, I., Teodori, C.: SMEs' financial risks in supply chain trade with large companies: the Italian automotive component industry. J. Gen. Manag. **47**(2), 126–137 (2022)
Florina, P.: Elements on the efficiency and effectiveness of the public sector. Ovidius Univ. Ann., Econ. Sci. Ser. **17**(2), 313–319 (2017)
Francis, J.R., Mairose, B.M., Tichy, E.M.: The year the world was awakened to the importance of supply chain management. Mayo Clin. Proc.: Innov., Qual. & Outcomes. **5**(1), 187–194 (2021)
Harland, C.: Supply Chain Management: Concepts, Challenges and Future Research Directions. Springer, Cham (2024)
Harvey, M.: Warehouse Process Performance Evaluation Aligned with Supply Chain Performance Aspirations. Master's thesis,. Stellenbosch University (2020)
Havenga, J.H., King, D., Simpson, Z.P., Goedhals-Gerber, L.L., De Bod, A.: Container terminal spatial planning: a 2041 paradigm for the Western Cape Province. J. Transp. Supply Chain. Manag. **6**(1), 180–191 (2012)
Hidayat, N.S., Saidah, D., Ghifari, H.A., Deritantoro, S.: Impact of layout, material handling and WMS on warehouse management effectiveness. Adv. Transp. Logist. Res. **6**, 626–638 (2023)
International Labour Organization: The Future of Work in the Automotive Industry. ILO, Geneva (2021)
Ji, B., Zhang, X., Mumtaz, S., Han, C., Li, C., Wen, H., Wang, D.: Survey on the internet of vehicles: network architectures and applications. IEEE Commun. Stand. Mag. **4**(1), 34–41 (2020)
Lampón, J.F., Perez-Elizundia, G., Delgado-Guzmán, J.A.: Cooperation in financing the automobile industry's supply chain: reverse factoring. J. Manuf. Technol. Manag. **32**(5), 1094–1112 (2021)
Lee, U., Kang, N., Lee, I.: Selection of optimal target reliability in RBDO for electric vehicle design. Struct. Multidiscip. Optim. **60**(3), 949–963 (2019)
Lyu, Z., Lin, P., Guo, D., Huang, G.Q.: Towards zero-warehousing smart manufacturing from zero-inventory JIT production. Robot. Comput. Integr. Manuf. **63**, 101932 (2020)
Macas, C.V.M., Aguirre, J.A.E., Arcentales-Carrión, R., Peña, M.: Inventory management for retail companies: a literature review. In: In: Proc. 2nd Int. Conf. On Information Systems and Software Technologies, pp. 71–78 (2021)
Mago, S., Hofisi, C.: Conceptualising microfinance for effective smallholder farming in Africa. Int. Bus. & Econ. Res. J. **13**(6), 1437–1446 (2014)
Niemann, W., Kotzé, T., Jacobs, D.: Collaborative planning, forecasting and replenishment implementation: a south African retailer case. S. Afr. **29**(4), 1–16 (2018)
Patrucco, A.S., Luzzini, D., Krause, D., Moretto, A.M.: Purchasing strategy fit under environmental uncertainty. Int. J. Phys. Distrib. & Logist. Manag. **53**(9), 1043–1072 (2023)
Paulus, T.M.: Using qualitative data analysis software to support digital research workflows. Hum. Resour. Dev. Rev. **22**(1), 139–148 (2023)
Polten, L., Emde, S.: Scheduling automated guided vehicles in very narrow aisle warehouses. Omega. **99**, 102204 (2021)
Richards, G.: Warehouse management: a complete guide, 2nd edn. Kogan Page, London (2017)
Saliji, M., Effective Inventory Management in the Automotive Industry. Master's thesis, Mälardalen University (2021)

Salmons, J.: Doing Qualitative Research Online, 2nd edn. Sage, London (2021)
Sharma, S.K., Bhat, A., Kumar, V., Agarwal, A.: Path analysis model for supply chain risk management. Int. J. Inf. Syst. Supply Chain Manag. **10**(2), 21–41 (2017)
Sivakumar, V., Ruthramathi, R.: Challenges and features of warehousing operations in Chennai. Int. J. Innov. Technol. Explor. Eng. **9**(1), 3304–3308 (2019)
Tellini, T., Silva, F.J.G., Pereira, T., Morgado, L., Campilho, R.D.S.G., Ferreira, L.P.: Improving in-plant logistics flow by physical and digital pathways. Procedia Manuf. **38**, 965–974 (2019)
Tolmay, A.S., Badenhorst-Weiss, J.A.: Customer–supplier relationship dynamics in south African automotive supply chains. Int. J. Value Chain. Manag. **9**(2), 149–165 (2018)
Tutam, M.: Warehousing 4.0 in logistics 4.0. In: İyigün, İ., Görçün, Ö.F. (eds.) Logistics 4.0 and the Future of Supply Chains. Springer, Singapore (2022)
Wilson, G.: The future of logistics warehouses. Supply Chain. Digit. (2020)
Woods, M., Paulus, T., Atkins, D.P., Macklin, R.: Advancing qualitative research using QDAS. Soc. Sci. Comput. Rev. **34**(5), 597–617 (2016)
Zidane, Y.J.T., Olsson, N.O.: Defining project efficiency, effectiveness and efficacy. Int. J. Manag. Proj. Bus. **10**(3), 621–641 (2017)

Supplier Risk Management on Supplier Quality Improvement and Supplier Commitment: A Case of South African Health Industry

Solly Donald Chiloane and Elizabeth Chinomona

Abstract The South African health industry encounters considerable challenges in managing supplier relationships and supplier risk management, which have a direct impact on supplier quality improvement and supplier commitment. Effective supplier risk management, quality improvement initiatives, and fostering supplier commitment are crucial in addressing these challenges. This study delved into the key principles of supplier relationship management (SRM), focusing on how these practices influence supplier quality improvement and supplier commitment within the South African healthcare sector. By emphasising risk mitigation, continuous quality improvements, and strong supplier partnerships, the research sought to highlight strategies that can enhance procurement outcomes in this complex and critical industry. A total of 250 questionnaires were distributed to the identified sample of public health hospitals in Gauteng Province. The collected quantitative data were analysed using SPSS and SMART-partial least squares (SMARTPLS 3.0) structural equation modelling procedure. The findings indicate that strong supplier relationships are pivotal in ensuring timely delivery, cost efficiency, and the overall quality of healthcare supplies. Given that a study of this nature has not been performed before among South African public healthcare sectors, the results are an essential addition to the existing body of literature within the area of supplier commitment in public health industries in developing countries such as South Africa. The study concludes by suggesting recommendations for limiting the impact of the identified challenges on supplier commitment. The study highlights the need for health organisations to adopt effective SRM strategies to enhance procurement outcomes, improve supplier commitment, and address challenges within the industry. Recommendations for improving SRM practices in the South African health sector are provided.

Keywords Supplier risk management · Supplier quality improvement and supplier commitment · Supplier relationship theory

S. D. Chiloane · E. Chinomona (✉)
Vaal University of Technology, Vanderbijlpark, South Africa
e-mail: elizabethc@vut.ac.za

E. H. Redda (ed.), *Proceedings of the Social Sciences International Research Conference, 2025*, https://doi.org/10.1007/978-981-95-9489-4_5

1 Introduction

The health sector in South Africa is one of the most vital components of the country's economy and well-being; yet it faces considerable procurement challenges, especially in supplier management. The effectiveness of procurement strategies in the health industry directly affects the quality and availability of medical supplies, pharmaceuticals, and equipment. In this context, supplier relationship management (SRM) plays a critical role in optimising procurement performance, ensuring reliability in supply chains, and fostering long-term partnerships with suppliers. Supplier commitment is a crucial element of SRM, influencing the continuity, cost-effectiveness, and efficiency of procurement processes. However, limited research has focused on how specific SRM principles, such as supplier risk management and supplier quality improvement, influence supplier commitment in South Africa's healthcare industry. This research aims to address this gap by investigating how SRM practices are implemented in healthcare organisations and their effect on supplier commitment. The South African healthcare sector is marked by a diverse range of suppliers, varying service levels, and complex logistical challenges that create significant obstacles for procurement professionals. With mounting pressure on both the government and private sectors to enhance healthcare services while managing costs, it becomes imperative that procurement strategies are tightly aligned with effective supplier management principles. This study, therefore, aimed to investigate the role of SRM in improving supplier quality and supplier commitment within the South African healthcare industry (Smith & Jones 2024; Van der Merwe & Thompson 2025).

2 Problem Statement

The issue of improving organisational performance remains a significant challenge for many firms, with SRM emerging as a potential solution to enhance operational and financial outcomes. However, procurement processes often suffer from inefficiencies due to non-adherence to established procedures, a long-standing concern for both academics and practitioners (Gelderman et al. 2024). In underdeveloped nations, especially within healthcare, there is a reputation for poor resource management, inadequate personnel training, and low compensation (Ghaz & Al-Weshah, 2013; Raja et al. 2024). In South Africa, healthcare procurement departments are particularly challenged by a lack of expertise in key procurement processes and their related outcomes. The increasing reliance on suppliers for items and services previously managed in-house has allowed firms to focus on their core competencies. Despite extensive studies on procurement performance in the private sector (Hamza et al. 2017; Nawi et al. 2016), public procurement, especially in developing nations, remains a subject of heightened scrutiny (Oyugi & Kamaara, 2023). E-procurement, although recognised for its advantages, has been slow to be adopted. This highlights the growing recognition of procurement's role in organisations, especially in Southern Africa, which remains underexplored. Moenga's (2011) study of procurement practices in Kenya's tea sector highlighted challenges such as supply chain visibility, customer demands, and cost containment. However, a few studies have focused on South Africa's public health sector, particularly the link between supplier risk management (SRM), supplier quality improvement (SQI), and supplier commitment. This study aimed to fill this gap by investigating the influence of SRM and supplier quality improvement on supplier commitment within the Gauteng Province's healthcare sector.

3 Literature Review

A review of the literature is done on the theory and variables used in this study.

3.1 Supplier Relationship Management Theory

This is a critical theory that emphasises the importance of managing partnerships and collaborations between an organisation and its suppliers. Unlike traditional procurement methods that focus on quantitative contractual obligations, SRM prioritises the creation and nurturing of long-term, value-driven relationships (Wagner & Bode 2014). This approach underscores the significance of communication, mutual benefits, and aligning strategic objectives between buyers and suppliers (Schönsleben 2016). By fostering these partnerships, organisations can drive better procurement performance and enhance supplier commitment, ensuring reliable product delivery that meets both quality and cost standards. The theory of SRM asserts that supplier relationships are fundamental to the procurement process, especially considering that suppliers have direct influence over the price, quality, delivery reliability, and availability of goods and services (Sonmez 2016). This relationship involves several components, including supplier segmentation, which allows businesses to categorise suppliers based on their strategic importance and performance capabilities (Abdul 2013). Such segmentation helps organisations identify which suppliers should be engaged in deeper partnerships and which should be managed in a more transactional manner (Acevedo 2016). This is particularly relevant in sectors such as healthcare, where the delivery of timely and quality goods is critical. Furthermore, the importance of performance management in supplier relationships cannot be overstated. According to Carter et al. (2015), performance management involves setting mutual expectations with suppliers regarding operational measures and then continuously monitoring and improving those measures to ensure consistent performance. In the context of the South African healthcare industry, ensuring that suppliers meet specific performance metrics such as timely delivery, product quality, and cost control is essential for maintaining high standards in healthcare provision. Applying SRM theory to the South African healthcare industry can significantly enhance procurement performance and supplier commitment. Healthcare organisations in South Africa face unique challenges, such as limited resources, regulatory requirements, and the need for consistent and high-quality product delivery. By focusing on strategic supplier relationships, procurement departments can improve supplier performance and mitigate supply chain risks, ultimately improving patient care and outcomes. SRM offers the potential for better collaboration, shared goals, and mutual problem-solving, which is critical in the complex and demanding healthcare sector. In the specific case of South Africa, the role of supplier relationships in procurement performance is particularly vital. Suppliers in the healthcare sector are responsible for providing everything from medical equipment to pharmaceuticals, each of which requires careful coordination to ensure the right quantity, quality, and timeliness. As highlighted by Lysons and Farrington (2013), managing supplier relationships effectively in this context directly impacts procurement outcomes and organisational success. With SRM, procurement departments can ensure that their suppliers are not only reliable but also committed to long-term collaboration and innovation, which is crucial in a rapidly evolving healthcare environment.

Conceptual Framework and Hypothesis Development. The conceptual framework presented in Fig. 1 conceptualises the relationship between SRM, supplier quality improvement, and supplier commitment. SRM and SQI are the predictor variables, with supplier commitment as the outcome variable.

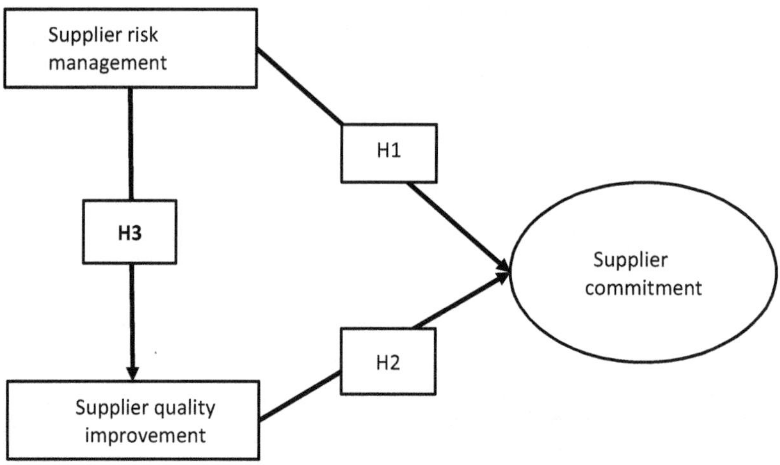

Fig. 1 Conceptual framework

3.2 Supplier Risk Management and Supplier Commitment

SRM plays a crucial role in mitigating supply chain risks and enhancing supplier commitment. Risk perception, both at the personal and organisational levels, significantly influences decision-making processes, especially within the context of supply chain management (Annamalah & Pertheban 2016; Al-Shboul 2024). From an organisational perspective, the concept of risk has been studied in various fields by many scholars. For this reason, risk in the context of supply chain management is more closely associated with supplier failure (Breuer, et al., 2013). Supplier failures and disruptions, commonly viewed as primary sources of risk, can destabilise organisations, leading to poor service delivery (Kauppi et al. 2016). SRM, through risk identification, assessment, and management, serves as a framework to address these challenges, with key components such as knowledge development and risk reduction proving essential for minimising negative impacts (Manuj, et al. 2014). Therefore, it can be concluded that SRM positively impacts supplier commitment, procurement performance, and risk mitigation, supporting organisations in achieving a competitive advantage while ensuring operational efficiency and reliability.

H1: Supplier risk management has a significant and positive relationship with supplier commitment.

3.3 Supplier Quality Improvement and Supplier Commitment

Supplier quality improvement (SQI) is a critical concept that enhances the operational efficiency of organisations, but its implementation often proves challenging. Supplier quality improvement can be viewed as an integration of strategic practices, and such practices need to stretch across inter-organisational boundaries to satisfy both existing and new customers, as well as enhance current operational processes (Quigley et al. 2018). While it may appear straightforward, many firms struggle to adopt effective supplier quality management (SQM) practices, leading to operational failures (Quigley et al., 2018). SQI integrates strategic practices that extend across inter-organisational boundaries, aiming to satisfy both current and future customers while improving operational processes (Quigley et al. 2018). From a supplier development perspective, it is viewed as a series of activities undertaken by a buying firm to identify, measure, and improve supplier performance, with a focus on continuous enhancement of value (Handley & Gray 2013). Ariesty (2016) explains that commitment is a very important aspect of the success of the supply relationship. Commitment is defined as the confidence of one of the parties that fostering good relations with other parties influences the optimal benefit gained by both sides. Consistent quality improvement is closely linked to supplier commitment, with longer relationships typically leading to reduced uncertainties about supplier performance (Kitheka et al. 2013). As such, this study hypothesises that SQI positively influences supplier commitment, with long-term relationships and continued investments in supplier development reducing uncertainties and fostering enhanced supplier performance. Effective SQM can, therefore, be seen as essential for building commitment and achieving sustained improvements in operational processes.

H2: Supplier quality improvement has a significant and positive relationship with supplier commitment.

3.4 Supplier Risk Management and Supplier Quality Improvement

SRM and SQI are integral to building strong, competitive, and efficient supply chains. SRM focuses on fostering long-term, mutually beneficial relationships with suppliers, while SQI aims to improve the quality of goods and services supplied. Both concepts are crucial for achieving operational excellence and mitigating risks within organisations (Manuj et al. 2014; Quigley et al. 2018). In the context of SRM, effective management of supplier relationships plays a pivotal role in reducing supply chain disruptions and enhancing procurement performance (Kakwezi & Nyeko 2013; Kauppi et al. 2016). The SRM framework emphasises the importance of communication, collaboration, and risk management in strengthening the connection between a buying firm and its suppliers (Annamalah & Pertheban 2016). This approach helps ensure that suppliers meet performance expectations, which is particularly important when considering supply chain risks such as supplier failure (Breuer et al. 2013). A key link between SRM and SQI is the role of long-term relationships in enhancing supplier commitment to quality improvement. Consistent collaboration between organisations and their suppliers often leads to more significant investments in quality management practices and better communication, ultimately driving quality improvements (Kitheku, et al., 2013:73). In addition, the degree of uncertainty associated with a supplier's quality performance decreases with the length of the relationship and the trust built over time (Makau 2017:17).

H3: Supplier risk management has a significant and positive relationship with supplier quality improvement.

4 Methodology

A research design is defined as a plan or framework for formulating and addressing research objectives and hypotheses (Creswell 2014; Gray 2017). It enables a researcher to develop a specific structure to solve a particularly growing research problem, question or opportunity (McDaniel & Gates, 2013). The study used a quantitative (Causal) research design, specifically a single, cross-sectional study, because data were collected only once from the sample elements (Burns et al. 2018). The single cross-sectional study is quantitative, meaning that it seeks to understand the sample structure and thereafter to recommend a final course of action (Polit & Beck 2018).

4.1 Research Design

A research design is a master plan that specifies the methods or procedures for collecting and analysing information needed for a research study (Gray 2017). Furthermore, a research design outlines guidelines for the researcher in the process of collecting, analysing and interpreting data. In research, a researcher can choose among three methods, which are qualitative, quantitative, and mixed methods. For the purpose of the study, a quantitative research approach was adopted since the study establishes the influence of the antecedents' factors on procurement performance. The following section discusses the research approach employed in this study.

4.2 Path Analysis

Path analysis was the next phase in the SEM-based data analysis. Path modelling describes how observed or measured variables relate to a theoretical concept and assesses the structural routes of a research model (Guenther, et al., 2023). An SEM approach was used to illustrate and assess the theoretical underpinnings of the investigation, as well as the interaction between model constructs, regression coefficients and p-values were examined to evaluate the study's structural model (Gray 2017). Both the prediction abilities and the standardised regression coefficients must be thoroughly explained in order to use path modelling (Polit & Beck, 2028). The results of the structural model for the suggested hypotheses are shown in Fig. 2.

For the study variables in Fig. 2, factor loadings are greater than 0,5, which suggests that the items reliably measure the latent factor, contributing to the construct validity of the measurement model.

Cronbach's alpha was used in this study to measure the reliability of all constructs; the reliability of the constructs is determined by a higher level of the Cronbach's coefficient alpha. Fornell and Larcker (1981) stated that reliability of a construct must be above 0.7 in order to enhance the internal consistency of the construct. Based on Table 1, all items are above the threshold of 0.7 and therefore reliable. According to Tseng (2014), a composite reliability index value that is higher than 0.7 proves the adequacy of the internal consistency of the research construct. An analysis of the data in Table 1 illustrates that all research constructs are above 0.7, which affirms the internal validity of all constructs measured as reliable. Average value extracted (AVE) should be greater than 0.5, as Fornell and Larcker (1981) postulated. Based on data from Table 1, all average values extracted range above 0.6, adhering to the required acceptability value; thus, all constructs met the required degree of acceptability (Polit & Beck, 2018).

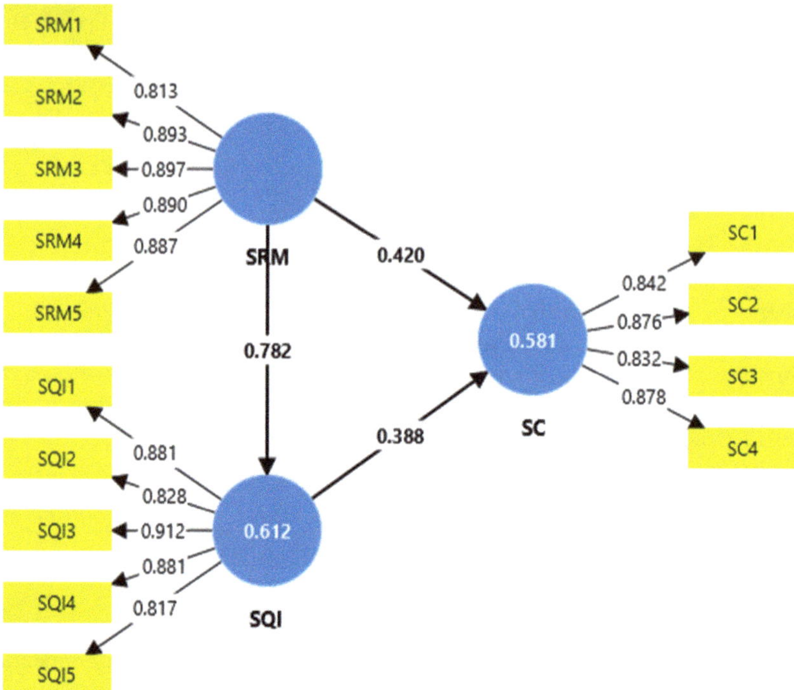

Fig. 2 Path coefficients generated by the SMARTPLS software. Source: AMOS (27.0)

5 Results and Discussion

The results acquired from testing hypothesis one confirm the correlation between SRM and supplier commitment. The hypothesis proves that there is a significant relationship between SRM and supplier commitment, and the relationship is supported. The $p < 0.001$ and the path coefficient is $\beta = 0.420$. Furthermore, the t-statistic value is observed at 8.006. The primary objective of SRM is to enhance organisational competitiveness by fostering stronger supplier relationships, reducing costs, and ensuring efficient material flow, which leads to improved operational performance (Tseng 2014; Wahu et al. 2015; Kauppi et al. 2016). In high-risk sectors like healthcare, where service continuity is critical, the implementation of SRM is especially beneficial in preventing and mitigating disruptions (Shou et al. 2018).

The results obtained from testing Hypothesis Two show that there is a positive relationship between supplier quality management and supplier commitment. The $p < 0.001$ and the path coefficient is $\beta = 0.388$. Besides, the t-statistic value is observed at 4.255. SQI involves several managerial efforts that integrate supplier capabilities into operational processes, including supplier selection, development, integration, and audits (Kitheku, et al., 2013). These activities help firms evaluate the current capabilities of their suppliers, such as through plant visits and audits, and implement

Table 1 Accuracy analysis statistics

Research construct	Scale item						
		Mean	Std. deviation	Cronbach's alpha	CR	AVE	Factor loadings
Supplier risk management	SRM1	0.461	0.636	0.843	0.844	0.673	0.813
	SRM2						0.893
	SRM3						0.897
	SRM4						0.890
	SRM5						0.887
Supplier quality improvement	SQI1	4.566	0.592	0.871	0.871	0.755	0.881
	SQI2						0.828
	SQI3						0.912
	SQI4						0.881
	SQI5						0.817
Supplier commitment	SC1	4.492	0.655	0.800	0.800	0.621	0.842
	SC2						0.876
	SC3						0.832
	SC4						0.878

Note SRM= Supplier Risk Management; SQI = Supplier Quality Improvement; SC = Supplier Commitment

Table 2 Results of hypotheses testing

Path coefficients	Hypothesis	Path coefficient	P value	T-Statistics	Decision
Supplier risk management - > supplier commitment	H1	0.420	***	8.006	Significant and supported
Supplier quality improvement - > supplier commitment	H2	0.388	***	4.255	Significant and supported
Supplier risk management- > supplier quality improvement	H3	0.788	***	12.910	Significant and supported

Significant Level < 0.05 *; significant level < 0.01 ***; significant level < 0.001 **, *Note* SRM = Supplier Risk Management; SQI = Supplier Quality Improvement; SC = Supplier Commitment

interventions to improve supplier quality, like training and allocating resources to strengthen suppliers' technical capabilities (Handley & Gray 2013; Otieno 2017). Also, a two-stage decision-making process is needed to determine the optimal investment in SQI activities, which may depend on the degree of uncertainty surrounding the supplier's capabilities (Makau 2017).

The results of this study prove that the relationship between SRM and supplier quality improvement is significant and supported (H3). The $p < 0.001$ and the path coefficient is $\beta = 0.788$. Additionally, the t-statistic value is observed at 12.910. This illustrates that SRM in the healthcare sector influences SQI in a more significant way than the other two relationships. Given the close interrelationship between these variables, SRM practices should contribute to improving supplier quality by fostering trust, collaboration, and commitment (Tseng 2014; Wahu et al. 2015).

On the other hand, SQI involves the continuous improvement of supplier performance and capabilities, typically through strategic practices such as supplier development, training, and audits (Kitheku, et al., 2013). According to Handley and Gray (2013), SQI activities focus on identifying and improving supplier performance while ensuring that quality standards meet or exceed expectations. These efforts are critical in aligning suppliers' capabilities with the purchasing organisation's operational needs and customer requirements. For example, investments in training and technical support help suppliers enhance their processes and reduce variability in product quality (Otieno 2017).

6 Conclusions of the Research Finding

The findings of this research offer statistical proof that a connection exists among enhancing supplier quality, managing supplier risks, ensuring supplier commitment, and improving procurement performance within the public healthcare sector of South Africa. The present study demonstrates that procurement performance is at its lowest in the South African public healthcare sector, with supply chain risks still prevalent. The risks involve a lack of skilled personnel who are either unenthusiastic about their roles or lack the required skills, knowledge, and qualifications, leading the public sector to depend on consultants to fill positions. This process may be ineffective when information is lacking, delayed, or slow to arrive. Additionally, supplier commitment is hindered by risks such as corruption and supply chain complications. Moreover, the research indicates that enhancing supplier commitment can be achieved by effectively managing specific risks outlined in this paper, resulting in a better overall procurement performance.

References

Abdul, S.: Supplier selection in the Malaysian telecommunications industry. International Journal of Procurement Management. 6(2), 139–158 (2013)

Acevedo, M.: Supplier segmentation and management strategies in procurement. International Journal of Procurement and Supply Chain Management. 8(3), 895–910 (2016)

Al-Shboul, M.A.: Investigating the quality of the relationship, supply risk mitigation on medium and large-sized manufacturing firms' supply chain performance in developing countries: the

moderating effect of supplier involvement. The Total Quality Management Journal. 36(4), 1162–1192 (2024). https://doi.org/10.1108/TQM-01-2023-0014
Annamalah, S., Pertheban, T.R.: The effectiveness of supply risk management in purchasing and operational performances of SMEs. J. Internet Bank. Commer. 17(30), 1–25 (2016). https://doi.org/10.2139/ssrn.2859871
Breuer, C., Siestrup, G., Haasis, H.D., Wilderbrand, H.: Collaborative risk management in sensitive logistics nodes. Team Performance Management. 19(7), 331–351 (2013). https://doi.org/10.1108/TPM-11-2012-0036
Burns, A., Groove, B.: The Practice of Nursing Research: Conduct, Critique and Utilization, 4th edn. W. B, Saunders Company (2018)
Burns, A.C., Veeck, A., Bush, R.T.: Marketing Research, 8th edn. Pearson Education, England (2018)
Carter, J.R., Monczka, R.M., Mosconi, L.: Performance management and supplier relationships. Journal Purch. Supply Manag. 21(2), 751–764 (2015)
Creswell, J.W.: Research Design: Qualitative, Quantitative and Mixed Methods Approaches, 4th edn. Sage, Lincoln (2014)
Fornell, C., Larcker, D.F.: Evaluating structural equation models with unobservable variables and measurement error. Journal Mark. Res. 18(1), 39–50 (1981). https://doi.org/10.1177/002224378101800104
Gelderman, C.J., Doves, D., Semeijn, J.: Caniëls, M.C.J.: sustainable supplier development within healthcare: an embedded case study in a hospital setting. Sustainability Letters. (2024) https://doi.org/10.62513/sl219060
Gray, D.E.: Doing Research in the Business World. Sage, London (2017)
Guarnieri, P., Cerqueira-Streit, J.A., Batista, L.C.: Reverse logistics and the sectoral agreement of packaging industry in Brazil towards a transition to circular economy. Resour. Conserv. Recycl. 153, 104541 (2020). https://doi.org/10.1016/j.resconrec.2019.104541
Hamza, S.B., Gerbi, A., Ali, S.H.: Factors affecting procurement performance in the case of Awassa textile share company. Global Journal of Management and Business Research. 16(2), 21–34 (2017)
Handley, S.M., Gray, J.V.: The role of supplier development in improving supply chain performance. Int. J. Oper. Prod. Manag. 33(3), 276–303 (2013). https://doi.org/10.1108/01409171311314061
Kakwezi, P., Nyeko, S.: Procurement Processes and Performance: Efficiency and Effectiveness of Procurement Function. Unpublished MBA thesis,. University of Nairobi, Kenya (2013)
Kauppi, K., Longoni, A., Caniato, F., Kuula, M.: Managing country disruption risks and improving operational performance: risk management along integrated supply chains. International Journal of Production Economics. 182, 484–495 (2016)
Kerzner, H.: Project Management: a Systems Approach to Planning, Scheduling, and Controlling. John Wiley & Sons (2017)
Kitheka, S., Jonathan, M., Muli, W., Ruth, N., Tabby, G.: The effect of supplier quality management on organizational performance: a survey of supermarkets in Kakamega town. International Journal of Business and Commerce. 3(1), 71–82 (2013)
Lysons, K., Farrington, B.: Procurement and supply chain management, 9th ed. Pearson Education (2013)
Makau, C.: Factors Affecting Procurement in Independent Electoral and Boundaries Commission of Kenya. Management University of Africa, Kenya, Research Project (2017)
Manuj, I., Esper, T.L., Stank, T.P.: Supply chain risk management approaches under different conditions of risk. Journal of Business Logistics. 35(3), 241–258 (2014). https://doi.org/10.1111/jbl.12051
McDaniel, J.R.C., Gates, R.: Marketing Research, 7th edn. Wiley, Danvers (2013)
Nawi, M.N.M., Roslan, S., Salleh, N.A., Zulhumadi, F., Harun, A.N.: The benefits and challenges of e-procurement implementation: a case study of Malaysian company. International Journal of Economics and Financial Issues. 6(7), 329–332 (2016)

Otieno, A.R.: An Assessment of the Effects of Supplier Quality Management on Procurement Performance: a Case Study of Sony Sugar Ltd in Migori County. Kisii University, Kenya, Diploma research project (2017)

Polit, D.F., Beck, C.T.: Essentials of Nursing Research: Appraising Evidence for Nursing Practice, 9th edn. Wolters Kluwer, Philadelphia (2018)

Quigley, J., Walls, L., Demirel, G., Maccarthy, B.L., Parsa, M.: Supplier quality improvement: the value of information under uncertainty. European journal of operational research. 264, 932–947 (2018). https://doi.org/10.1016/j.ejor.2017.05.044

Raja, M.W., Mahesar, A.J., Chanar, P.H., Shaikh, S.: Examining the impact of supplier selection on procurement performance of pharmaceutical firms in Pakistan. Remittances Review. 9(1), 354–367 (2024)

Schönsleben, P.: Integrated production logistics and supply chain management. Spring. (2016). https://doi.org/10.1007/978-3-642-35950-7_16796-2

Shou, Y., Hu, W., Kang, M., Li, Y., Park, Y.W.: Risk management and firm performance: the moderating role of supplier integration. Industrial Management & Data Systems. 118(7), 1327–1344 (2018). https://doi.org/10.1108/IMDS-09-2017-0427

Smith, R., Jones, L.: The role of SRM in healthcare procurement: a south African perspective. Journal of Health Procurement. 12(4), 100–115 (2024)

Sonmez, A.: The impact of supplier relationship management on procurement performance. J. Bus. Res. 69(3), 1245–1254 (2016)

Tseng, M.L.: The impact of knowledge management capabilities and supplier relationship management on corporate performance. International Journal of Production Economics. 154, 39–47 (2014). https://doi.org/10.1016/j.ijpe.2014.04.009

Van der Merwe, A., Thompson, C.: Aligning procurement strategies with supplier management in south African healthcare. South African Journal of Procurement Studies. 16(1), 89–102 (2025)

Wagner, S.M., Bode, C.: Supplier relationship management: theory and practice. Supply Chain Management Review. 18(5), 125–140 (2014)

Wahu, K.E., Namusonge, G.S., Mungai, C., Chilion, O.: Factors affecting performance of the procurement function in Kenyan public secondary schools: a case study of Gatundu District. International Journal of Academic Research in Business and Social Sciences. 5(4), 273–285 (2015). https://doi.org/10.6007/IJARBSS/v5-i4/1572

Zimbabwean Command Agriculture Input Subsidy Programme: Perceptions of Producers and Officials

L. C. Muzangwa, S. L. Middelberg, and M. Oberholzer

Abstract The Zimbabwean producers struggle to secure funding for farming. The government reintroduced the Command Agriculture Programme to mitigate the decline in agricultural productivity. This study aims to distinguish and corroborate the perceptions of these producers and officials regarding the effectiveness of the state-financed input subsidy programme (ISP). Quantitative data were obtained from 99 farmers, using an online questionnaire, and were then analysed by using principal component analysis and descriptive statistics. The results were compared with qualitative data gathered by interviewing six officials from the Commercial Bank of Zimbabwe and the Department of Agricultural Research Extension of the government. The results revealed that producers perceived several challenges with the ISP, including inadequate inputs, delayed payments from the Grain Marketing Board, and delayed input distribution. However, they also recognised the benefits of the programme, particularly in helping them secure input contracts and access cheaper inputs. The interviewed officials echoed the perceptions of the producers. Despite these challenges, farmers and officials remained optimistic about the programme's future. Additionally, producers identified commercial banks and agricultural companies as potential alternative sources of funding. It is recommended that the government strengthen internal controls to ensure the timely distribution of inputs and payments to producers, as well as to address political interference and corruption.

Keywords Agricultural finance · Command agriculture · Input subsidy programme · Agricultural producers · Zimbabwe

L. C. Muzangwa · S. L. Middelberg · M. Oberholzer (✉)
North-West University, Potchefstroom, South Africa
e-mail: merwe.oberholzer@nwu.ac.za

S. L. Middelberg
University of Lincoln, Lincoln, UK

E. H. Redda (ed.), *Proceedings of the Social Sciences International Research Conference, 2025*, https://doi.org/10.1007/978-981-95-9489-4_6

1 Introduction

ISPs are government schemes that promote food security and reduce production costs for producers (Mwale et al. 2022; Walls et al. 2022). Agricultural inputs such as seeds, fertilisers and fuel are provided at subsidised prices, with the option of extending these benefits to all producers, or exclusively to smallholder farmers (World Bank 2019). Although it is considered a contributor to agricultural development in Africa, ISPs remain contentious (Jayne et al. 2018; Seleka & Mmopelwa 2020). Empirical investigations of such programmes made available to farmers are critical for establishing how ISPs affect their financing decisions (Jonga et al. 2022).

In Zimbabwe, the government's controversial fast-track land redistribution programme (FTLRP), implemented in 2000, led to a severe economic downturn and the withdrawal of agrarian capital. Consequently, this has created a situation where producers lack the necessary financing and have limited access to the resources required for farming (Mazwi et al. 2019; Mafirakurewa, 2023). Most producers were allocated land through the government's FTLRP, classifying farms as A1 or A2. The A1 model is a communal arrangement that allocates small plots of at least six hectares for growing crops and grazing land to landless or poor farmers, where A2 farms range from 50 to 1500 hectares (Dube 2021; NZSA, 2019).

The prevailing national economic decline has left producers with a thin base upon which to conduct their financing decisions and practices, exacerbated by their inability to provide the necessary security to access agricultural loans from commercial banks (Mazwi et al. 2019; Chigunhah et al., 2020; Nguyen et al. 2023; Kpabitey et al. 2024). Another consideration reported by the World Bank (2019) is the adverse consequences of climate change on the African agricultural sector that affect producers' harvests, leading to financial losses and failure to settle agricultural loans. During the agricultural season of 2016–2017, Zimbabwe faced a very high food import bill, owing to economic hyperinflation. The government then decided to play a more prominent role in agricultural financing by reintroducing and funding the Command Agriculture ISP in 2016, a policy that continues to this day (Mazwi et al. 2019; Mafirakurewa et al. 2023; Musonza & Hlungwani 2024).

The Command Agriculture programme, also known as Operation Inala/Maguta, was initially introduced in 2005 to boost cereal production in Zimbabwe, which was at an all-time low (Mazwi et al. 2019). In 2008, the programme faced challenges related to the effects of 'side marketing' and was consequently abandoned (Moyo & Chambati 2013). In agriculture, side marketing or side selling refers to situations in which parties breach contractual agreements, and this can occur when producers sell farm products to other merchants, or when companies buy from sources with which they have not entered into with contracts.

With the reintroduction of the Command Agriculture programme, changes were made to address the challenges of side marketing. Several stakeholders from both the private and public sectors were included (Dube 2021), namely the state-controlled Ministry of Agriculture, the Reserve Bank of Zimbabwe (RBZ), the Lands and Rural Development Board, the parastatal Grain Marketing Board (GMB), privately owned institutions such as the Commercial Bank of Zimbabwe (CBZ) and a petroleum company, Sakunda Holdings (Mazwi et al. 2019). The Department of Extension Services (AGRITEX) of the Ministry of Agriculture was appointed to play the monitoring and supervisory role. The suitability of producers was determined by AGRITEX officials, based on farmers' proof of land and asset ownership, as well as production history (Mazwi et al. 2019). The primary expenses associated with the revived programme pertain to a special maize programme that mandates private businesses supply production inputs to the GMB for subsequent distribution to farmers. The programme aims to mobilise financial resources from the private sector, by gathering private companies to finance the importation and distribution of inputs on credit to programme-registered farmers (World Bank 2020). The GMB contracts the latter to deliver harvested maize. Producers then pay the difference between the value of their products and input loans. In turn, the GMB reimburses private business suppliers for the input. Although the programme did not officially include a subsidy on inputs –since participating farmers were required to repay the input loan –the GMB effectively subsidised output prices. The GMB purchased agricultural produce from farmers at prices higher than those charged to domestic processors, thus incurring significant costs (World Bank 2020). The redesign of the programme brought a new dimension to the ISP, even though the state remains the key facilitator of the programme (World Bank 2019; Mazwi et al. 2019).

Prior studies on ISPs, which is the focus of this study, produced mixed results. Musonza and Hlungwani (2024) assessed the effectiveness of the programme to increase food security by increasing sustainable production. Their findings showed that, although food production has improved, the programme does not meet the country's demand –due to weak programme monitoring, patronage, and corruption. Although the study identified key issues, it focused only on communal farmers in one district (Murehwa) of Zimbabwe. Muzeza et al. (2023) analysed the factors affecting A1 smallholder farmers' technical efficiency under the ISP. They found that socio-economic factors, such as age, education, and marital status, were critical determinants of technical efficiency. Data from 240 households were collected; however, the focus was only on the Chegutu and Zwimba Districts of Zimbabwe. Chisango (2018) investigated the challenges and prospects of the programme by focusing on participants from Bindura District in Mashonaland, Central Province. The conclusions emphasised that the programme is perceived as benefiting only a select group of elite government officials and faces numerous challenges. The strengths of the study are the inclusion of a variety of stakeholders in the data collection, namely producers, agricultural extension officers, and government officials. However, focusing on a single region (Bindura District) could limit generalizability. Dube's (2021) literature review of the programme, built on Chisango's (2018) findings, asserted that it exacerbated farmers' tenure insecurity and vulnerability to drought while distorting their access to credit. Mazwi et al. (2019) investigated the programme by collecting data from 75 purposively selected farmers in the Zvimba District in Mashonaland, West Province. The authors concluded that the ISP was an innovative governmental financing approach that provided much-needed input to farmers; however, there were biases in the treatment of farmers. Moreover, several production challenges faced by farmers have been identified, including labour shortages, insufficient inputs such as seeds and fertilisers, and water scarcity. The strengths of the study are as follows: a mixed-methods approach was followed with data collected during two agricultural seasons (2016–2017 and 2017–2018). However, the focus was on only one district, limiting the generalizability of the findings.

While several studies (Chisango 2018; Dube 2021; Mazwi et al. 2019; Musonza et al. 2024; Muzeza et al. 2023) have been conducted on Zimbabwe's Command Agriculture ISP, these investigations have largely been limited to specific districts in the country or narrow dimensions of the programme, such as the effectiveness of the programme in increasing food security, or determinants of technical efficiency. However, an empirical study that distinguishes and validates the perceptions of both producers and programme officials across different regions of Zimbabwe is yet to be conducted. Additionally, limited attention has been paid to operational challenges and their effects on the programme's overall effectiveness. This study addresses these gaps by adopting a mixed-methods approach.

This research is particularly significant, given Zimbabwe's reliance on agriculture for food security and economic stability. The reintroduced Command Agriculture ISP plays a critical role in addressing financing gaps and mitigating the impact of the country's economic downturns. This study offers recommendations to enhance the effectiveness of the programme by providing insights into the perceptions of producers and government officials.

The research objectives of this exploratory study were to distinguish and validate the perceptions of (1) producers and (2) institutional officials involved in the Command Agriculture ISP in Zimbabwe, and to draw conclusions that could potentially shape government policies for programme enhancement. Furthermore, the credibility and trustworthiness of this study were strengthened by applying mixed research methods for both data collection and analysis. Three objectives have been established for this purpose. First, to obtain the perceptions of farmers regarding their assessment of the programme's effectiveness (including its benefits and drawbacks), to explore the possibilities of alternative financing sources, and to obtain their views on the future of the programme by collecting quantitative data. Second, qualitative data were collected from AGRITEX and CBZ officials regarding their perceptions of the strengths, weaknesses, opportunities, and threats associated with the programme, as well as its future outcomes. Third, we validated the perceptions of farmers (quantitative data) and compared them with the observations made (qualitative data) by AGRITEX and CBZ officials by using triangulation. These findings contribute to the body of empirical knowledge on the reintroduced the Command Agriculture programme in Zimbabwe and, most importantly, make recommendations to address the shortcomings and weaknesses of the ISP.

2 Methodology

2.1 Data Sources

The key stakeholders of the Zimbabwean Command Agriculture ISP include producers (farmers), the AGRITEX, and CBZ officials. Empirical data were collected using a convergent parallel mixed-methods research design, after which the quantitative and qualitative results were triangulated. Quantitative data were gathered from the farmers by using a newly developed online questionnaire, and administered using the online survey tool, QuestionPro. These questions were developed on the basis of an extensive literature review. Qualitative data were gathered from semi-structured interviews with AGRITEX and CBZ officials. The main aim was to obtain information from all stakeholders engaged in the daily operations of the ISP. This included gathering perspectives from farmers, programme management, and financing officials. For the quantitative segment, a snowball non-probability sampling method was used to select participants. The initial participants were identified, and they, in turn, recruited others. This chain reaction continued until 99 online questionnaires were successfully completed, providing a sufficient dataset for statistical analysis. The sample size was in line with similar studies by Musonza and Hlungwani (2024) and Chisango (2018) on the Zimbabwean Command Agricultural Programme, both of which included 100 participants. The similarity in themes covered by the questionnaire and interview guide facilitated a comparison between quantitative and qualitative data.

The sample for the qualitative data collection phase included three officials from AGRITEX and three individuals employed by CBZ, the programme financier. A non-probability purposive sampling technique was used to select the interviewees. Yilmaz (2013) advised that a relatively small number of participants using this sampling technique in qualitative research is appropriate for gathering in-depth knowledge. Although six may appear limited, the participants were selected based on their unique perspectives and expertise, which was considered adequate to investigate the core challenges and benefits of ISP. Moreover, in qualitative research the emphasis is on data saturation, the point at which additional interviews yield no new insights or information, rather than the number of participants (Fusch & Ness, 2015). The study adhered to the ethical guidelines of North-West University (ethics number: NWU-00883-21-A4). Participation was voluntary, and informed consent was obtained before data collection commenced. Participants were assured of confidentiality, anonymity, and their right to withdraw at any stage without consequence. All data were stored securely and are accessible only to the research team.

Quantitative Data Collection and Analysis. The questionnaire included five sections that consisted of closed-ended questions, checklists, and 5-point Likert scale questions, where 1 indicated 'disagreement to a very large extent' and 5 indicated 'agreement to a very large extent.' In a few instances, the participants failed to indicate their preferences. These missing data points were excluded to ensure that only the remaining data were subjected to statistical analysis.

Quantitative data were collected from April to October 2021 by distributing the questionnaires online. Principal component analysis (PCA) and descriptive statistics were used to describe quantitative data. PCA is a type of factor analysis that is used to reduce the number of items (variables) to a smaller set (factors). It was also employed to provide evidence of the construct validity of self-reporting scales in the questionnaire. Maree and Pietersen (2023) maintained that factor analysis is also used to identify items that should be grouped together, as questions yielding similar responses are indicative of measuring the same underlying dimension or factor. These authors also asserted that a sample size of 99 participants is sufficient, which should be five to ten times the number of items in a construct (Maree & Pietersen 2023). In this study, the largest construct consisted of seven items, namely section C in the questionnaire. PCA was used to condense the original set of items into a more concise set of factors and to elucidate their commonalities. The questions and statements in sections B, C, D, and E of the questionnaire were factorised. Section B contained questions regarding the benefits of Command Agriculture as a source of finance, whereas section C comprised questions about its drawbacks. Section D encompassed questions concerning programmes as possible sources of agricultural funding, and section E included questions regarding the future and possible outcomes of the programme. The five items (benefits) in section B were included in one factor, and the drawbacks (items) in section C were incorporated into one factor. However, other constructs that addressed the four alternative sources of agricultural funding (section D) and the two statements concerning the future of the ISP (section E) were not factorised, because the statistics could not be kept as single factors. The constructs were reported using descriptive statistics. The group of farmers (99 participants) was sufficient to employ descriptive statistics, requiring only 15 participants per group (Maree & Pietersen 2023). Internal reliability was measured using Cronbach's alpha, where a value of at least 0.6 is appropriate, indicating that the research instrument is reliable regarding the extent to which it is repeatable and consistent (Maree & Pietersen 2023). To determine whether the data were suitable for factor analysis, the Kaiser-Meyer-Olkin (KMO) test and Bartlett's test of sphericity were conducted. The significance of the latter must be less than 0.05 ($p < 0.05$), while the KMO test should be more than 60% acceptable (Maseko et al. 2020).

Qualitative Data collection and Analysis. The interview guide consisted of three sections. The first section collected demographic data and asked general questions. The second section requested participants to identify the programme's perceived internal strengths and weaknesses, as well as its external opportunities and threats. The last section questioned officials about their possible outcomes and the future.

In October 2021, three CBZ officials were interviewed in a semi-structured format at the CBZ Agroyield Offices in Harare. One of the planned three interviews with AGRITEX officers took place at their offices in Harare, whereas the other two were conducted by telephone from South Africa, as these AGRITEX officers work primarily out of the office visiting farmers. The initial participant count was set at six, but was contingent on the concept of data saturation –the point at which no new information was gathered. Data saturation was achieved with the fifth participant, as their responses aligned with those of the previous participants. However, the responses of the sixth interviewee were consistent with those of other interviewees.

Qualitative data were analysed thematically. Manual coding was employed to accurately reflect participants' responses and to avoid misinterpreting phrases taken out of context. Each theme was derived from the interview transcripts, guided by the research objectives and literature review.

Triangulation was applied to ensure the convergence, corroboration, and consistency of the findings obtained from producers via the questionnaire, and from CBZ and AGRITEX officials through interviews. An assessment was conducted to determine how the results complemented each other.

3 Results and Discussion

3.1 Research Objective 1: Quantitative Data Analysis and Discussion

Please The first research objective was to obtain the perceptions of farmers regarding their assessment of the effectiveness of the Command Agriculture programme (including benefits and drawbacks), to explore the possibilities of alternative financing sources, and to obtain their views on the future of the programme by collecting quantitative data. The demographic information indicated that 26% of the study population had used the programme as a funding source for four years, 22% for two years, 19% for three years, and 16% and 17% for one year and five years, respectively. Since its launch, the ISP has been in operation for approximately five years. Two-thirds of the beneficiaries were landowners, whereas the remainder (34%) were managers, suggesting that most producers became landowners through the government's FTLRP, classifying farms as either A1 or A2 models. Confidence can be placed in the responses of the majority of landowners, as they had direct experience with the programme. Additionally, the inclusion of managers allowed for the diverse perceptions of individuals who interacted with the programme. More than half of the participants (52%) owned approximately six hectares of land (small farms), followed by those (30%) whose land exceeded six hectares (medium-sized to large farms). The remaining participants (18%) worked on less than six hectares of land. The study population consisted primarily of smallholder farmers, comprising 70% of the participants. The incorporation of large-scale producers provided an essential perspective for this study. Notably, 60% of the participants were male, indicating a well-balanced gender representation. Slightly more than half (51%) obtained a higher education institution qualification, followed by those with an ordinary level (25%) or an advanced level education at school (18%). Only 6% of the participants had received only a primary school education. These findings suggest that most farmers pursued additional education, indicating that individuals with higher levels of education increasingly chose farming as their profession. This also instils confidence in the quality of their responses, given that a significant proportion of the participants were relatively highly literate. The largest group consisted of individuals aged between 30 and 50 years (45%), followed by those younger than 30 years (39%). Only 16% of the study population were aged >50 years.

The remaining quantitative data (Sections B-E) were analysed using PCA and descriptive statistics. The results, as presented in Figs. 1 and 2, were based on the rotation of the items in Sections B and C of the questionnaire. This item rotation helped establish intercorrelations among the five benefits. Correlation analysis was translated into eigenvalues. The five items (benefits) yielded a single factor, as only one (out of a possible five) eigenvalue was higher than 1 (Maree & Pietersen, 2023). A factor-loading matrix was created. Only items with loadings higher than the cut-off value of 0.5 were allocated to the factor. Fig. 1 presents the five benefits from section B of the questionnaire, ranked by their importance and determined by factor loading.

The factor loadings for the five items in Fig. 1 exceeded 0.5. Among these, 'securing input contracts' exhibited the highest loading at 0.799, while 'availability of buyers' had the lowest loading, measuring 0.672. These factor loadings resulted in a factor mean of 3.232, which slightly exceeded the midpoint of the 1–5 scale. This indicates that the participants perceived the benefits of using the ISP to a 'moderate extent.' The standard deviation was 0.752, indicating that the observations were moderately variable around the mean. 56% of the variance can be attributed to these five items, surpassing the 50% threshold recommended by Maree and Pietersen (2023). As previously mentioned, the correlation analysis yielded only one factor with an eigenvalue of 2.8, which exceeded the common threshold of 1 (Maree &

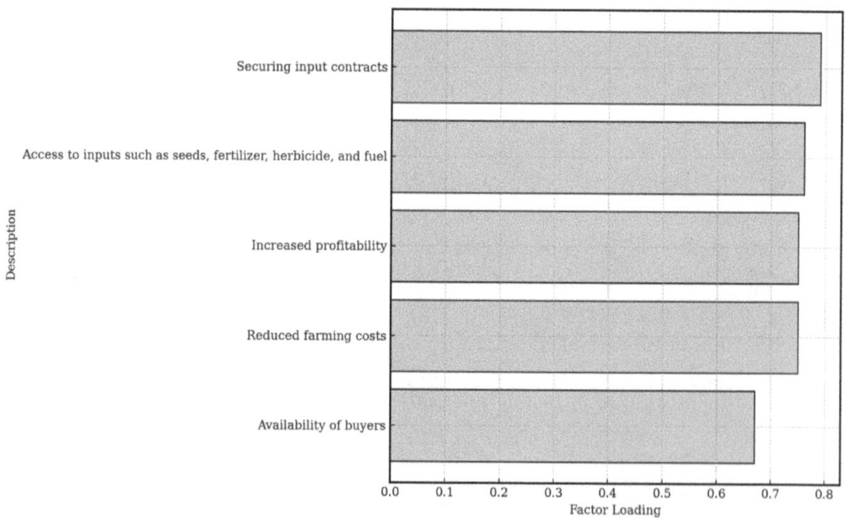

Fig. 1 Factor loadings assessing perceived benefits of the command agriculture ISP

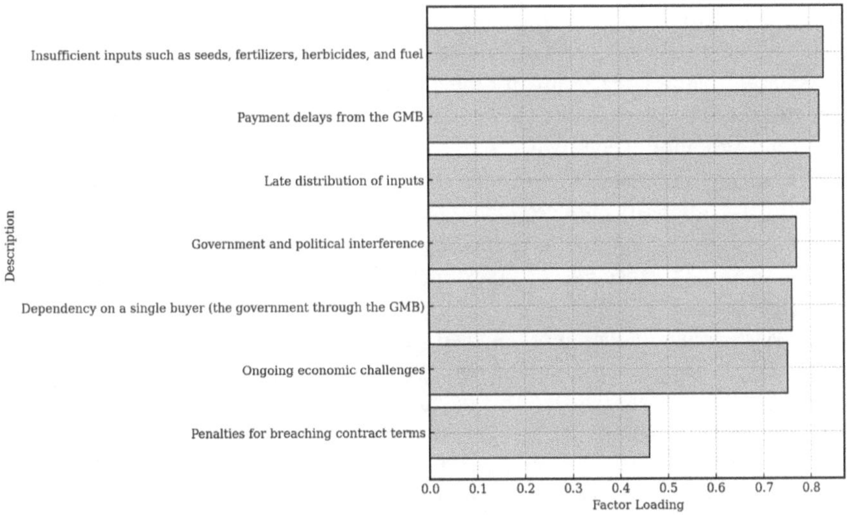

Fig. 2 Factor loadings assessing perceived drawbacks of the command agriculture ISP

Pietersen 2023). Additionally, this factorisation led to an acceptable Cronbach's alpha of 0.801, which implies that when this measuring instrument is employed at different times with participants from the same population, the same findings should be obtained. A KMO test yielded an acceptable value of 0.771 (exceeding the 0.6 threshold), and a Bartlett's significance of 0.000, which is below the threshold of 0.05).

Figure 2 presents the reported drawbacks of the ISP in terms of their factor loadings. Of the seven items, six exhibited factor loadings that exceeded 0.5; 'insufficient inputs such as seeds, fertilisers, herbicides, and fuels' stood out with the highest factor loading at 0.827, while 'ongoing economic challenges' had the lowest at 0.753.

However, 'penalties for breaching contract terms' fell below the 0.5 threshold, registering a loading of 0.460. This suggests that it has played a relatively unimportant role in describing the drawbacks of the ISP. A standard deviation of 0.897 and mean factor of 3.331 were both within acceptable ranges for consolidating the mentioned drawbacks into a single factor. Furthermore, the eigenvalue exceeded 1 at 3.952. Similar to the perceived benefits of the programme, the results highlighted that 56% of the total variance surpassed the threshold of 50%. Notably, Cronbach's alpha (0.865), the KMO test result (0.875), and the significance level of Bartlett's test of sphericity (0.000) conformed to the recommended thresholds outlined in sect. 2.1. The factor was deemed statistically acceptable for retention and significantly shaped producers' perceptions of the drawbacks associated with the Command Agriculture ISP. This factor and its related items hinder the attainment of the programme's objectives and desired outcomes.

Table 1 displays the descriptive statistics for the questions concerning alternative financing sources and opinions on the programme's outcome and future (section

D). These statistics were generated using central tendency and dispersion measures, including the means and standard deviations of the constructs.

As Table 1 shows, the results for the alternative sources of agricultural financing were diverse. The participants agreed that commercial banks with a mean score of 3.169, and agricultural companies with a score of 3.624, could be viable alternatives for agricultural financing. Conversely, there was disagreement among participants regarding the suitability of microfinance firms (2.341) and development finance firms (2.833) as sources of agricultural funding. However, all the standard deviations in Table 1, section D, exceeded 1, suggesting a substantial variation in the responses of the participants.

The participants moderately believed (mean = 3.081) that the ISP would ultimately reduce hunger regarding the outcome and future of the programme (Table 1, section E). In addition, the standard deviation, which measured the consistency of the participants' perceptions, was less than one, indicating a generally consistent viewpoint among them. By contrast, the participants moderately disagreed (mean = 2.657) with the notion that the programme would be phased out. Additionally, the standard deviation exceeded 1, signifying significant variability in their responses to alternative financing sources and opinions on the programme's outcome and future.

The policy implications drawn from the quantitative results emphasise the strengthening of the programme's monitoring and depoliticisation. The observed operational inefficiencies can be addressed by establishing an independent oversight body to pilot and strengthen ISP monitoring, particularly regarding the timely distribution of inputs and payments from the GMB to producers. This oversight body could comprise diverse memberships, including specialists from civil society organisations, academia, the private sector, and farmers' associations, ensuring impartiality and minimising political interference. Political interference can lead to preferential treatment by farmers based on political affiliations. Depoliticizing the Command Agricultural Programme would help ensure fair treatment of all farmers, regardless of political considerations.

Table 1 Descriptive statistics on alternative sources of agricultural financing and opinions on programme outcome and future

No	Section of questionnaire	N	Mean	Std. deviation
1	Section D			
	Commercial banks	83	3.169	1.238
	Microfinance	82	2.341	1.239
	Agricultural companies	93	3.624	1.160
	Development finance	84	2.833	1.260
2	Section E			
	The ISP will reduce hunger	99	3.081	0.976
	The ISP might be phased out	99	2.657	1.061

3.2 *Research Objective 2: Qualitative Data Analysis and Discussion*

The second research objective was to gather qualitative data from AGRITEX and CBZ officials regarding their perceptions of the strengths, weaknesses, opportunities, and threats associated with the programme and its future outcomes. The initial segment of the interview focused on assessing the familiarity of officials with the programme and the dynamics within its key stakeholder network. All participants demonstrated a clear and extensive understanding of the ISP objectives. They noted that the Zimbabwean programme was a joint effort between the government and banks, essentially constituting a public-private partnership. The impetus behind launching the ISP stemmed from the necessity to revive agricultural activities. All participants highlighted that each stakeholder had a distinct role, and that these roles complemented one another to ensure the programme's success. Furthermore, the relationships among key stakeholders were deemed crucial for the programme's effectiveness. Most participants described the relationship between the programme facilitators and farmers as ranging from satisfactory to good. However, a few participants noted that while the relationship was positive, it remained work in progress, requiring further efforts to enhance the programme's outcomes and future sustainability. One participant suggested that including farmers from larger farms could improve these relationships.

Two themes were developed from the qualitative data in Sections B and C of the interview guide, grounded in the research objectives, interview guide, and interviewees' responses. Theme 1 related to internal strengths and weaknesses as well as ISP external opportunities and threats. All the participants concurred that the ISP possessed numerous strengths and offered significant opportunities. Among the prominent strengths identified were its secured financial and political support from the government, as noted by Participant A1: '*The strength of the ISP lies in political and financial power because the government is the main stakeholder.*' The establishment of partnerships between the state and agricultural input suppliers was highlighted by Participant B1: '*The strengths of the ISP include its partnerships with new suppliers and the fact that the government being the main stakeholder means there is a backup.*' In addition, the programme attracted producers owing to its assurance of crop buyers, accessible input procurement, and lowered production costs.

The ISP has several opportunities at its disposal, including targeting producers of different sizes, as alluded to by Participant A1: '*They should also target medium- to large-scale farmers.*' Participant A2 added that farming infrastructure could be offered: '*The government should install irrigation facilities for farmers and offer more implements for the farmers to rent.*' The formation of supply chain partnerships with agricultural companies and expanding support for a range of agricultural products were identified as opportunities.

Participants unanimously noted the presence of numerous internal weaknesses and external threats that could potentially hinder the programme's success. Among these weaknesses are the delayed distribution of inputs, selective distribution influenced by politics and corruption, farmer defaults, the impact of drought due to climate change, and incidents of theft. Participant A2 remarked, *Not every farmer is getting access to the inputs because of political factors. Also, some farmers are failing to pay back the government because of factors like drought and theft.* Participant A3 added: *The programme has several weaknesses, which are, delayed input distribution, political involvement (corruption). Some of the farmers will have to struggle to get the inputs or Command Agriculture ISP contracts because of corruption.*

Theme 2 encompassed three subthemes pertaining to the outcomes and future prospects of the ISP. These subthemes were: 1) the achievements of the ISP, 2) the rationale behind targeting smallholder farmers, and 3) the anticipated future thoughts of what will happen to the programme.

All participants agreed that the ISP successfully achieved its intended outcomes, such as enhancing food security and improving smallholder producers' quality of life. Participant A3 emphasised that: '*The smallholder producers are getting inputs at subsidised prices on credit and they will have to pay back after harvesting and selling their produces. This seems to be profitable and the production/farming costs are quite low.*' These combined outcomes contributed to an overall enhancement of farm profitability.

All participants agreed that the programme's emphasis on smallholder farmers provided significant opportunities and benefits, including enhancing their quality of life and strengthening the country's food security. This was accomplished by empowering smallholder farmers to increase their production capacities. However, despite its successes, the Zimbabwean programme faced drawbacks such as delays in input distribution, shortages of inputs, banking errors, and late payments. While some issues were resolved, others were referred to different facilitators, with a few remaining unresolved. Notably, one participant remarked that grievance resolution was still an ongoing process.

The third sub-theme, within the broader context of the outcome and future of the ISP, focused on the views of the participants regarding the programme's future. All participants acknowledged the continual improvement of the programme, which encompassed advancements in modern farming techniques, streamlined administration, and a well-defined delineation of stakeholder roles. Participant A2, for example, observed that *Improvements have been made to the Command Agriculture ISP. There have been efforts to make smallholder farmers use more modern farming techniques and implements so as improve efficiency and effectiveness.* Considering the above, most participants were confident that the ISP would and should continue. However, the interview responses highlighted that the programme's sustainability depends on the overall performance of the country's currently unfavourable economy. Consequently, most interviewees expressed doubts regarding its long-term viability.

3.3 *Research Objective 3: Triangulation of Results*

The third research objective was to validate the perceptions of farmers (quantitative data) and compare them with observations made (qualitative data) by AGRITEX and CBZ officials, using triangulation. CBZ and AGRITEX officials highlighted strengths such as convenient access to affordable inputs and reduced production costs, which were corroborated by farmers who acknowledged these benefits. This aligns with Mazwi et al.'s (2019) findings that the Command Agricultural Programme improved farmers' access to production inputs and extension services. However, Jayne and Rashid (2013) contended that there was a lack of empirical evidence to substantiate the positive effects of ISPs on agricultural growth and poverty reduction.

Producers in the programme faced several challenges, including insufficient inputs such as seeds, fertilisers, herbicides, and fuel. They also encountered delays in payments from the GMB and observed a selective distribution of inputs to farmers influenced by politics and corruption. These issues partially align with the weaknesses identified by officials of CBZ and AGRITEX. Mazwi et al. (2019) corroborated the tendency of the GMB to make delayed payments to the maize producers. Farmers also highlighted political interference in the programme, which was confirmed by government officials during the interviews. For instance, they pointed to the preferential treatment of farmers with ties to particular political groups through the allocation of input contracts, and priority access to services. This was seen as a clear example of how political affiliations influenced programme implementation –a sentiment echoed in the findings of Musonza and Hlungwani (2024), and Chisango (2018). It points towards a lack of political will to eradicate state involvement in the Command Agricultural Programme, as this finding aligns with Dube's (2021:1628–1629) assertion that '… government officials have reserved for themselves decision-making powers on agricultural inputs, outputs, and timing which undercut the peasant incentive.' Despite its drawbacks, both producers and officials from CBZ and AGRITEX believed that the ISP has the potential to alleviate hunger, improve food security, enhance livelihoods, and support sustainable operations in the foreseeable future. This perspective is partly corroborated by Mazwi et al. (2019), who found that 'many farmers' supported the programme despite its challenges.

4 Conclusion and Recommendations

In 2016, Zimbabwe reintroduced the Command Agriculture ISP, a state-driven agricultural financing model. This study aimed to distinguish and validate the perceptions of both producers and institutional officials participating in the programme, to explore alternative funding sources, and to draw conclusions that could inform government policies. The study revealed a prevailing sense of optimism among farmers and officials regarding the programme. However, challenges persist, including insufficient inputs such as seeds, fertilisers, and fuel; delayed payments from the GMB;

and ongoing political interference by officials affiliated with the ruling party. These findings were corroborated by both quantitative and qualitative data, emphasising the overlap in perceptions between producers and institutional officials. The policy implications for the Zimbabwean government, as the primary participant, include strengthening robust internal controls. This can be achieved by the government conducting a resource audit to identify gaps in the supply chain and to improve timely access to sufficient inputs and payments for farmers through the GMB. Furthermore, efforts should be made to depoliticise the programme by establishing an independent oversight body, comprising members of the academia, civil society organisations, the private sector, and farmers' organisations to oversee resource allocation and programme implementation. In addition, transparent and equitable criteria for participation in the programme should be developed to minimise political favouritism. The state should also establish and oversee a formal grievance process to address complaints from producers and institutional officials, ensuring accountability and improving relationships between stakeholders. Finally, to broaden programme participation, targeted campaigns are necessary to attract more producers, particularly large-scale producers, as one interviewee noted that they hold the majority of the land in Zimbabwe.

This study has certain limitations that require further clarification. First, a small sample of producers and programme officials were included, primarily because of the sensitivity of the topic, which made potential interviewees hesitant to participate. Second, the benefits and drawbacks of the programme, as presented to producers in the questionnaire, were derived from the existing literature and may not comprehensively cover all aspects. To address these limitations, future research should expand its sample size to include a broader range of stakeholders. Additionally, there is a need to explore customised internal control systems designed specifically for agricultural ISPs. This study contributes to the limited empirical research on the reintroduction of the Command Agriculture Program by focusing on the perceptions of the directly involved stakeholders.

References

Chigunhah, B.R., Svotwa, E., Munyoro, G., Mabvure, T.J., Govere, I.: Private capital formation activities and bank credit access among farmers in Zimbabwe. International Journal of Economics and Financial Issues. **10**(3), 225–235 (2020)

Chisango, F.F.T.: Challenges and prospects of Zimbabwe's command farming in unlocking the country's smallholder agricultural economy. International Journal of Agricultural Economics. **3**(4), 76–82 (2018)

Dube, L.M.: Command agriculture and food security: an interrogation of state intervention in the post-fast track land redistribution era in Zimbabwe. Journal of Asian and African Studies. **56**(7), 1626–1643 (2021)

Fusch, P.I., Ness, L.R.: Are we there yet? Data saturation in qualitative research. Qual. Rep. **20**(9), 1408–1416 (2015)

Jayne, T.S., Mason, N.M., Burke, W.J., Ariga, J.: Taking stock of Africa's second-generation agricultural input subsidy programs. Food Policy. **75**, 1–14 (2018)

Jayne, T.S., Rashid, S.: Input subsidy programs in sub-Saharan Africa: a synthesis of recent evidence. Agricultural economics. **44**(6), 547–562 (2013)

Jonga, R.H., Delin, H., Masona, C., Belford, C.: The effect of agricultural technology transfer on Zimbabwe's economic development: a dynamic global trade analysis project approach. Agrekon. **61**(4), 433–448 (2022)

Kpabitey, S., Chitose, A., Kusadokoro, M.: Provision of loans and credit by cocoa buyers under non-price competition: cocoa beans market in Ghana. Open Agriculture. **9**(1), 20220256 (2024)

Mafirakurewa, T., Mushunje, A., Zantsi, S.: Factors influencing smallholder farmers' decisions to participate in loan-based farming in Mutare district, Zimbabwe: a double-hurdle model approach. Agriculture. **139**(12), 2225 (2023)

Maree, K., Pietersen, J.: Sampling. In: Maree, K. (ed.) First Steps in Research, 3rd edn. Van Schaik, Pretoria (2023)

Maseko, G.J., Oberholzer, M., Middelberg, S.L.: Perspectives of university students on their student debt accountability. Southern African Journal of Accountability and Auditing Research. **22**(1), 27–38 (2020)

Mazwi, F., Chemura, A., Mudimu, G.T., Chambati, W.: Political economy of command agriculture in Zimbabwe: a state-led contract farming model. Agrarian South: Journal of Political Economy. **8**(1–2), 232–257 (2019)

Moyo, S., Chambati, W.: Land and Agrarian Reform in Zimbabwe: beyond White-Settler Capitalism. CODESRIA, Dakar (2013)

Musonza, F., Hlungwani, P.M.: An evaluation of command agriculture and food security among communal farmers in rural Zimbabwe. Journal of Land and Rural Studies. **12**(2), 155–183 (2024)

Muzeza, N.T., Taruvinga, A., Mukarumbwa, P.: Analysis of factors affecting technical efficiency of A1 smallholder maize farmers under command agriculture scheme in Zimbabwe: the case of Chegutu and Zvimba districts. Cogent Economics & Finance. **11**(1), 2163543 (2023)

Mwale, M.L., Kamninga, T.M., Cassim, L.: The effects of the Malawi farm input subsidy program on household per-capita consumption convergence. Dev. Pract. **32**(3), 336–348 (2022)

Nguyen, L., Russ, J., Triyana, M.: The effect of agricultural input subsidies on productivity: A meta-analysis. Policy Research Working Paper 10399. World Bank, Washington, DC (2023)

Seleka, T.B., Mmopelwa, D.: Effects of input subsidies on cropland allocation and diversification in Botswana's subsistence economy. Agrekon. **59**(3), 337–353 (2020)

Walls, H., Johnston, D., Matita, M., Chirwa, E., Mazalale, J., Quaife, M., Kamwanja, T., Smith, R.: How effectively might agricultural input subsidies improve nutrition? A case study of Malawi's farm input subsidy Programme (FISP). Food Secur. **15**, 21–39 (2022)

World Bank: Zimbabwe Public Expenditure Review: with a Focus on Agriculture. World Bank, Washington, DC (2019)

World Bank: Agriculture Subsidies for Better Outcomes: Options for Zimbabwe. World Bank, Washington, DC (2020)

Yilmaz, K.: Comparison of quantitative and qualitative research traditions: epistemological, theoretical, and methodological differences. European journal of education. **48**(2), 311–325 (2013)

Streamlining Retail's Supply Chain Performance Using Internet of Things: Optimising E-Warehouse and Transportation Management

Molebaleso Lydia Ntshingila

Abstract The fast-paced, integrated channel retail world requires supply chain models that are adaptable to e-commerce growth and evolving consumer expectations. The Internet of Things (IoT) offers a solution by connecting physical assets, enabling enhanced visibility and control. While South African retail has adopted some technology, critical implementation gaps remain in supply chain optimisation, particularly in e-warehouse and transportation management. This study investigated the potential of IoT to enhance retail supply chain performance, aiming to streamline operations through these two critical functions in Gauteng Province. A quantitative methodology was employed, using a structured questionnaire for the collection of primary data from retail stakeholders. Hypotheses were tested using structural equation modelling (SEM), with confirmatory factor analysis (CFA) performed using SPSS version 29.0. The study's results statistically confirm that the best-practice implementation of IoT solutions in e-warehouse and transportation management provides a significant advantage in predicting the competitiveness of South African retailers. The findings demonstrate that integrating smart technologies into core logistical functions delivers tangible performance benefits. To fully unlock the transformative potential of IoT and foster a more efficient, competitive retail landscape based on total quality management (TQM), retailers must prioritise allocating budget and resources for strategic pilot projects in e-warehouse and transportation management.

Keywords E-Warehousing · Internet of things (IoT) · Supply chain performance

1 Introduction

The evolution of the Internet of Things (IoT) is significantly driven by large-scale industries that leverage their predictive and tracking capabilities to optimise supply chain operations (Sharma et al. 2020). To meet the needs of a wider customer base

M. L. Ntshingila (✉)
Vaal University of Technology, Vanderbijlpark, South Africa
e-mail: lydian@vut.ac.za

E. H. Redda (ed.), *Proceedings of the Social Sciences International Research Conference, 2025*, https://doi.org/10.1007/978-981-95-9489-4_7

with increasingly specific requirements, the transportation industry needs to adapt and diversify its services. Logistics has become an important part of all companies due to the great demands of an increasingly competitive market. For better customer service and lowest cost, all functions that contribute must be managed efficiently (Enes & Silva 2020). IoT-based applications offer significant benefits to industries by tracking coded objects, reducing errors and theft, speeding up processes, and enabling more efficient and adaptable organisational systems (Sharma et al. 2020). Production and distribution are crucial to deliver on time and minimise costs in a low finished product inventory. However, to reduce disruptions, strong supplier relationships, careful planning, and efficient processes are required (Orjuela-Castro et al. 2017). Transportation costs often have a ripple effect that affects other areas of logistics, such as warehousing and inventory management. Hence, a reduction in transportation is significant to improve a company's overall logistics efficiency and profitability (Parkhi et al. 2014).

To meet customer demand and a competitive advantage in today's fast-paced business environment, warehouse operations are important as they enable rapid fulfilment across various retail, web-based, and catalogue channels (Sainathuni et al. 2014). To optimise the operation and improve the level of warehouse service, the application of IoT in warehouses enhances the effectiveness and accuracy of the warehouse processes, such as stock-taking, product sorting, and inbound and outbound (Sharma et al. 2020). Warehouses used to operate as a supplement to a firm's main operations. To increase efficiency and cost savings, other manufacturers and retailers find it advantageous to partner with logistics specialists, while others maintain their own facilities (Bowen 2008). For retailers need constant, exceptional logistical performance, they are forced by limited trade margins in the retail landscape. To establish a competitive advantage, retailers must focus on optimising in-store replenishment from distribution centres (DCs). Retailers also need to develop efficient vehicle route strategies and optimal delivery policies (Holzapfel et al.2016). There is, therefore, a need for warehouse management system (WMS) implementation, real-time delivery tracking, and driver training programmes as additional considerations for enhanced efficiency.

Integrating warehousing and transportation systems is a strategic investment that creates a more streamlined and synchronised supply chain (Wanjari, 2020). The suitability of a particular warehousing system for a specific industry sector is primarily within the logistics landscape and is driven by the characteristics of the customer orders being processed (Boysen et al. 2021). Modern warehouses serve as the nerve centre of efficient goods distribution, providing the functions of accumulation, processing, and distribution of goods between manufacturers and end users. They have exceeded their role as static storage facilities by developing into experienced logistics hubs, integrating a multitude of interconnected and complementary functions (Marchuk et al. 2020). One key consideration that businesses are constantly striving to optimise in their operations is the geographic distribution of warehouses, which offer the potential for a wider inventory range (*Mishra* et al. 2024). Furthermore, a strategic integration of innovative technologies and solutions is a critical driver of a continuous adaptation that is recognised by global corporations in today's competitive business landscape (Marchuk et al. 2020). Businesses generate faster revenue when clients receive their product on time, hence it is important to optimise the supply chain on both transportation and warehousing.

2 Literature Review

A variety of literary works were reviewed, including books, journals, completed theses, and other online resources. The author aims to gain a clear understanding of the interrelated concepts involved in this study.

E-Warehouse Management. Modern warehouses are the backbone of warehousing logistics; they are technical structures that integrate various elements of the logistics system to perform important functions such as accumulating, processing, and distributing goods to the end users (Marchuk et al. 2020). Warehouse capacity is crucial information that is typically measured by the number of storage locations categorised by unit load and rack type (Manzini et al. 2015). Warehousing management practices involve the planning, implementation, and control of strategies, processes, and techniques for optimal warehouse operations (Rotich & Ndeto 2024). Warehousing significantly impacts logistics efficiency and cost (Wanjiru & Ngugi 2020). Within the retail outlet, warehouses play a crucial role in consolidating and storing a diverse range of products. To enhance efficiency, this strategic approach helps reduce transportation costs and lead times (Kembro & Norrman 2022).

To increase warehouse efficiency, a WMS is used because it is a database information technology that coordinates integrated storage activity and maintains accurate inventory by recording every transaction to support storage operations (Istiqomah et al. 2020). To help businesses track and trace the inventory effectively, WMS is one of the key logistics and supply chain applications that can assist them to have a competitive advantage over those that do not use the application (Alajkovic et al. 2022). For a competitive market that facilitates product visibility, organisations have adopted advanced technology to form an adaptive and flexible warehouse design. Businesses can access continuous location and data updates on their goods by providing real-time data storage and sharing (Sahara & Aamer 2022).

Transportation Management. Transportation is a frequently overlooked source of competitive advantage. Effective transportation performance relies on adherence to delivery schedules, safeguarding product quality, and delivering a consistently satisfactory service. A sustainable improvement can be achieved through transport coordination (Kadłubek et al. 2022; Kmiecik et al. 2022). Effective transportation planning is crucial for designing efficient internal material flow systems within and across facilities (Bekrar et al. 2021). Today, the Internet and transportation have revolutionised global connectivity, enabling unprecedented freedom and economic efficiency in the movement of people and goods (Kwon 2020). Transportation is a critical logistics operation with a substantial environmental impact. Effective coordination can significantly improve sustainability (Kmiecik 2022). The maximum cost reduction is possible if the advanced transportation mode is independent of retailers' demand. Hence, less environmental transport is important in the retail sector.

To improve the final warehouse operation, transport optimisation and tracking are a crucial part of WMS implementation (Žunić et al. 2018). Mohsen (2023) states that although transportation management systems (TMS) may eliminate the need for manual operations like juggling spreadsheets, scattered files and databases, and phone calls for every shipment, they are only used by approximately one-third of businesses to move goods.

Competitive Advantage. Competitive advantage is a distinctive characteristic that provides an organisation with a significant advantage over competitors, leading to increased profits and sustained long-term success (Al-Khawaldah et al. 2022). The resource-based view posits that competitive advantage stems from a firm's unique bundle of resources and capabilities, which must be valuable, rare, inimitable, and non-substitutable to create a sustainable competitive edge. The capability-based view highlights the role of firm-specific capabilities but cannot be easily replicated or acquired in achieving competitive advantage (Herden 2020; Saragih et al. 2020). The firm's ability to create value for its customers is through effective competitive advantage (Al-Khawaldah et al. 2022). When the execution and the benefits of the same strategy are not duplicated by competitors, competitive advantage is considered sustainable (Sousa et al. 2021). In response to intense global market competition, organisations have shifted their focus to core competencies. Therefore, organisations strive to gain a competitive advantage by effectively managing supplier relationships and purchasing activities (Saragih et al., 2020).

Total Quality Management. TQM is defined as a management system focusing on continuous improvement through the application of tools, techniques, and a culture of shared values. The enhancement of customer satisfaction is the goal of TQM by improving the quality of products and services with the least consumption of resources. To ensure customer satisfaction in terms of quality, quantity, price, and delivery, TQM focuses on continuous improvement across all organisational aspects (Abbas 2020). To achieve customer satisfaction through zero-defect products, organisations can improve product quality and quality control techniques by focusing on the performance of the product's specification (Riyadi 2020).

Internet of Things. IoT is a framework connecting physical objects to the Internet to be digitally represented and accessible. Furthermore, in IoT, a 'thing' is any device equipped with sensors that collects and transmits data automatically over a network (Mouha 2021). The involvement of information and communication technology has become a necessity since supply chains are getting riskier, more complex, and more costly. The implementation of IoT in supply chain management provides real-time visibility across all stages, enabling the seamless flow of goods and information (Khan et al. 2022). IoT as a pathway to operational excellence has assumed a critical role in several aspects of SCM. Moreover, it has gained increased popularity and attracted the attention of the SCM community since the advent of wireless technology (Rejeb et al. 2020).

Supply Chain Performance. Performance is a valuable tool for measuring overall organisational efficiency and enabling proactive corrective action (Islam et al. 2021).

Supply chain performance is the ability to consistently meet customer demands through efficient operations and timely fulfilment (Zaman et al., 2023). Process-based approaches, perspective-based approaches, and hierarchical-based approaches are the three key approaches included to assess supply chain performance (Aityassine et al. 2022). Kurdi et al. (2022) and Mondol (2021) indicate that inventory turnover rate, shipping schedules, and transactional duration are factors that can be used to assess supply chain performance. According to Hanaysha and Alzoubi (2022), the resource collection process is required in the supply chain to create and deliver the products to the end customers. Therefore, the provision of information is crucial in measuring the supply chain's performance.

2.1 Conceptual Framework

Conceptual Framework and the Formulation of Hypotheses. The study tested the conceptual framework presented in Fig. 1, which identifies the four retail supply chain management variables, which are e-warehousing, transportation management, total quality management, and competitive advantage as the predictor variables that exert an influence on the IoT, which is the mediating variable. The framework further shows that IoT also exerts an influence on the business performance (outcome variable) of retail businesses.

E-warehouse and IoT. Storage and retrieval operations, organisational structure, materials handling equipment, storage media, and the warehouse building itself, which provides environmental protection for the stored goods, are the general

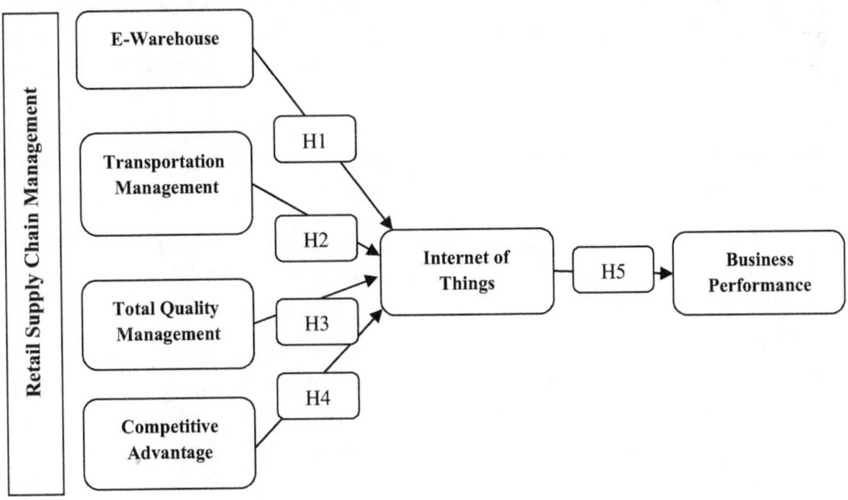

Fig. 1 Conceptual framework for internet of things, retail supply chain management, and performance

concepts of warehousing. Moreover, distribution ensures that goods reach customers from the production line timeously and economically (Wanjiru & Ngugi 2020; Maina & Wachiuri 2024). To ensure accurate and fast performance in all functions to meet customers' demands, warehouses can contain thousands of products, which is why they should be optimally used (Mostafa et al. 2019). The rise of e-commerce and demand for faster delivery have made efficient logistics crucial for competitive advantage (Maina & Wachiuri 2024). Modern warehousing facilities and logistics services support trade and manufacturing activities (Oyetoro, 2024). For a more integrated, agile, adaptive, efficient, and effective Circular Economy (CE), digitalisation, the IoT, sensors, and big data are key enablers that improve market performance in businesses (Fatimah et al. 2023). Since warehousing is considered the bridge between vendors and consumers, it directly impacts supply chain performance, affecting both lead times from vendors and delivery times to customers (Trivellas et al. 2020). Therefore, this study hypothesises that:

H1: There is a positive relationship between e-warehousing and IoT.

Transportation Management and IoT. Transportation is the most important component in a logistics system, as it is related to the transportation time in every subsection. Distribution and transportation processes are critical operational issues closely linked to customer characteristics and requirements (Trivellas et al. 2020). To adapt to new circumstances, rapid changes in the industry require strategic thinking and flexibility (De Neufville 2020). Due to the exponential growth of technology devices, immense data volumes are now generated across all sectors, including business, government, and society. The transport sector, recognising its critical role in urban development, generates and uses significant data, impacting people, processes, and profit (Iyer 2021). Advanced technologies, including blockchain, AI-powered logistics platforms, and IoT sensors, are streamlining cross-border transport management. While not replacing physical technologies, blockchain technology is increasingly used to enhance cargo traceability in cross-border trade by providing a decentralised, distributed ledger for secure data transactions and goods authentication (Koh et al. 2020). Furthermore, AI is used to provide real-time vehicle information for effective traffic management and mobility-on-demand services into a unified user interface for trip planning (Iyer 2021). This leads to the following hypothesis:

H2: There is a positive relationship between transportation management and IoT in South African retail businesses.There is a positive relationship between transportation management and IoT in South African retail businesses.

Total Quality Management and Internet of Things. TQM gains renewed importance in the context of Industry 4.0. Integrating TQM with Industry 4.0 advances the preservation of established quality principles with greater efficiency and scope through innovative technologies (Wolniak & Grebski 2023). According to Almetwally et al. (2020), industries are increasingly considering the (IoT) a necessity. Through digitalisation and smart manufacturing, Industry 4.0 transforms product creation into a personalised and customised experience (Wolniak & Grebski 2023). A smart internal supply chain was proposed when a manufacturing company in

Portugal analysed the supply chain problems. Quality-related problems were indirectly reduced through the proposed solution, which resulted in product callback or customer complaints (Sader 2020). Therefore, quality as a major driver behind the development of Industry 4.0 increases high-quality products and reduces costs. Accordingly, the following hypothesis is suggested:

H3: There is a positive relationship between TQM and IoT in South African retail businesses.

Competitive Advantage and IoT. To integrate various functional areas and processes internally and externally with suppliers and customers, organisations must adopt technologies into their operations (De Vass et al. 2021). Furthermore, new avenues have been opened through digital transformation for organisations to forge closer connections with their customers (Agustian et al. 2023). To respond to market needs quickly and cost-effectively, retailers must improve their capabilities by integrating IoT between business partners (De Vass et al. 2021). Moreover, production ecosystems strengthen a firm's prevailing competitive position by infusing digital connectivity into value chain interdependencies. To lead new competitive scenarios and dynamic, digital ecosystems, data-based competition should be incorporated into product-based competition (Subramaniam 2020). Hence, the adoption of IoT enhances the integration capabilities of retailers. These lead to the following hypothesis:

H4: There is a positive relationship between competitive advantage and IoT in South African retailers.

IoT and Business Performance. To enable real-time tracking of various performance-related parameters, big data-driven supply chain (BDDSC) provides organisations with enormous information. Management's priorities to improve the firm's performance are to reduce the order cycle time and inventory (Gawankar et al. 2020). Retail businesses, like any organisation, to sustain itself in the industry and grow in the long run, must enhance their performance. To ensure good performance, many firms have been able to successfully implement supply chain management strategies through e-commerce (Andonov et al. 2021). Moreover, by leveraging IoT data, businesses can gain insights into product usage, performance, and customer feedback (Rajan 2024). To maximise supply chain performance, retail businesses must implement supply chain integration (SCI) as it improves visibility, traceability, interoperability, and collaborative decision-making (De Vass et al. 2021). Accordingly, the following hypothesis is suggested:

H5: There is a positive relationship between IoT and business performance in South African retailers.

3 Methodology

The research sample, procedures for data collection, statistical approaches, and ethical considerations are discussed in this section.

3.1 Research Design

A quantitative research design was employed in this study to investigate the impact of IoT implementation in e-warehouse and transportation management on the supply chain performance and competitiveness of retail businesses within Gauteng Province, South Africa. The hypothesised relationships between IoT adoption and key outcomes were tested using structural equation modelling (SEM).

3.2 Sampling Method

This quantitative study examined retail businesses operating within the Gauteng Province of South Africa. The geographic scope of data collection was limited to Gauteng, given its significant retail business population and its designation as the central business region of South Africa's economy. The study targeted managers and relevant staff members working in supply chain departments of retail businesses located in Gauteng as respondents for the questionnaires. To determine the sample size, the study was established by benchmarking against historical samples of 350 to 870 respondents used in prior research (Gawankar et al. 2020, 380 respondents; Chinomona et al., 2023). Therefore, a sample size of 563 respondents and a response rate of 59, 24% was acceptable.

3.3 Research Instrument

A structured questionnaire served as the primary data collection instrument. The questionnaire was designed based on a thorough review of existing literature and consisted of closed-ended questions using a Likert scale (e.g., 1 = Strongly Disagree to 5 = Strongly Agree) to measure respondents' perceptions and experiences.

3.4 Data Collection

The questionnaire was pilot tested with a small group of retail professionals to ensure the clarity, comprehensibility, and validity of the questions before the main data collection phase. Necessary revisions were made based on the feedback received.

3.5 Data Analysis

This section presents the demographic details of participating retail businesses as well as the results and discussion of the inferential statistics. It shows that (n = 187; 33.2%) were clothing retail stores and (n = 376; 66.7%) were grocery stores.

4 Results and Discussion

EW, e-warehousing; TM, transport management; TQM, total quality management; CA, competitive advantage; IoT, Internet of Things; BP, business performance; α, alpha; CR, composite reliability; AVE, average variance explained (Tables 1 and 2).

The Item-Total Correlation column shows the correlation between each item (e.g., EW1, TM2) and the total score of its respective construct. High item-total correlations indicate that the items are measuring the same underlying construct. Values above 0.7 are generally considered acceptable, indicating good reliability. CR values above 0.7 are considered acceptable. AVE values of 0.5 or higher are generally considered acceptable, suggesting that the construct explains a substantial portion of the variance in its items. The constructs demonstrate good reliability and validity (Table 3).

EW, e-warehousing; TM, transport management; TQM, total quality management; CA, competitive advantage; IoT, Internet of Things; BP, business performance; *, Correlation is significant at the 0.01 level (2-tailed).

The correlation analysis, apart from confirming discriminant validity, served to confirm the association and direction of the association between constructs. Inter-factor correlations were positive and significant, ranging from r = 0.215 to r = 0.761. This result demonstrates that when the intensity of one construct increases, positive increases can be expected in the other constructs, and the reverse is also true.

Confirmatory Factor Analysis. Confirmatory factor analysis (CFA) is a SEM technique employed to confirm hypothesised relationships between observed indicators and latent factors (Goretzko et al. 2024). In CFA, the accuracy of measurement scales used in the study was assessed to ascertain their internal consistency, reliability, validity, and model fit. The results of the CFA are presented in Table 2. Internal consistency reliability of measurement scales was assessed by using total item correlations, Cronbach's alpha test, and composite reliability (CR) test. A measurement is

Table 1 Profile of participating retail businesses

Variable	Category	Number	Percentage
Type of business	Clothing retail stores	187	33%
	Grocery retail stores	376	67%
Total		563	100%

Table 2 Psychometric properties of the measurement scales

Research constructs	Cronbach's test Item-total correlations α value	Mean score	CR	AVE	Highest shared variance	Factor loadings
E-warehouse	EW1 0.701 0.927 EW2 0.855 EW3 0.873 EW4 0.846 EW5 0.805	0.420	0.930	0.728	0.42	0.727 0.886 0.918 0.895 0.851
Transport Management	TM1 0.675 0.936 TM2 0.802 TM3 0.810 TM4 0.875 TM5 0.860 TM6 0.861	0.461	0.934	0.722	0.18	0.680 0.815 0.827 0.921 0.927 0.931
Total quality management	TQM1 0.683 0.840 TQM2 0.663 TQM3 0.704 TQM4 0.661	0.373	0.842	0.575	0.21	0.720 0.699 0.825 0.778
Competitive Advantage	CA1 0.816 0.900 CA2 0.812 CA3 0.620 CA4 0.787 CA5 0.811	0.456	0.907	0.682	0.44	0.832 0.642 0.870 0.897 0.865
Internet of things	IoT1 0.817 0.926 IoT2 0.752 IoT3 0.804 IoT4 0.761 IoT5 0.782 IoT 6 0.811	0.468	0.931	0.680	0.34	0.860 0.806 0.786 0.843 0.800 0.853
Business performance	BP1 0.819 0.905 BP2 0.913 BP3 0.859 BP4 0.590	0.430	0.907	0.735	0.26	0.865 0.975 0.922 0.621

Table 3 Discriminant validity

Constructs	EW	TM	TQM	CA	IoT	BP
E-warehousing	1.000	0.381*				
Transport management	0.215*	1.000				
Total quality Management	0.514*	0.296*	1.000			
Competitive advantage	0.570*	0.249*	0.761*	1.000		
Internet of things	0.455*	0.413*	0.512*	0.370*	1.000	
Business performance	0.333*	0.324*	0.416*	0.472*	0.475*	1.000

deemed reliable when it produces consistent and reproducible results applied repeatedly under similar conditions (Abbas 2020). The item-to-total correlations for all measurement scale items were found to be above the established cutoff value of 0.3, which, according to Abudu-Birresborn et al. (2023), is acceptable.

Hypotheses Tests Results.

(H1) suggests that there is a positive relationship between e-warehousing and IoT. This hypothesis is accepted in this study since there is a positive and significant relationship between the two constructs ($\beta = 0.537$; $p < 0.001$). (H2) suggests that there is a positive relationship between transport management and IoT. This hypothesis is supported here since there is a positive and significant relationship between these two constructs ($\beta = 0.291$; $p < 0.001$). (H3) suggests that there is a positive relationship between TQM and IoT. This hypothesis is also supported because there is a positive and significant relationship between the two constructs ($\beta = 0.803$; $p < 0.001$). (H4) suggests that there is a positive relationship between competitive advantage and IoT. This hypothesis is supported as well because there is a positive and significant relationship between the two constructs ($\beta = 0.882$; $p < 0.001$). (H5) suggests that there is a positive relationship between IoT and retail business performance. Since this study elicited the views of owners and managers of retail businesses exclusively, future studies could be conducted that include the views of professional employees in these enterprises as well (Table 4).

Future studies could also consider other business best practices that were excluded. The first limitation is that the geographic scope of the study was restricted to retail businesses that were based in the Gauteng Province. The second limitation is that the study did not consider all IoT factors, some of which are also important in influencing retail business performance. The third limitation is that the study, like all surveys, was susceptible to response bias. However, this bias was reduced by ensuring that the questionnaire was self-administered, which provided respondents with the freedom to complete the questionnaire without being pressured by the researchers. Future studies could be conducted using larger samples composed of retail businesses drawn from different South African provinces.

Table 4 Structural equation modelling testing of hypotheses

Paths	Hypothesis	Path coefficient	Unstandardised estimates	Hypotheses result
EW → IoT	H1	0.537*	0.65	Supported
TM → IoT	H2	0.291*	0.32	Supported
TQM → IoT	H3	0.803*	0.78	Supported
CA → IoT	H4	0.882*	0.92	Supported
IoT → BP	H5	0.484*	0.61	Supported

5 Conclusion

This research aims to deliver empirical evidence regarding the effects of implementing IoT in e-warehousing and transportation management on the performance and competitiveness of retail supply chains in Gauteng Province. The results provide valuable insights for retail businesses that are either contemplating or actively engaged in IoT solutions. By contributing context-specific evidence from the South African retail sector, the study aspires to enhance the existing body of knowledge and offer practical recommendations for optimising supply chain operations through the adoption of IoT technologies.

References

Abbas, J.: Impact of total quality management on corporate green performance through the mediating role of corporate social responsibility. Journal of cleaner production. 242, 118458 (2020)

Abudu-Birresborn, D., Brennenstuhl, S., Puts, M., McCleary, L., Yakong, V., Chu, C.H., Cranley, L.: Evaluating knowledge and attitudes scales for the care of older adults among nursing students in Ghana. BMC nursing. 22(1), 47 (2023)

Agustian, K., Mubarok, E.S., Zen, A., Wiwin, W., Malik, A.J.: The impact of digital transformation on business models and competitive advantage. Technol. Soc. Perspect. 1(2), 79–93 (2023)

Aityassine, F., Soumadi, M., Aldiabat, B., Al-Shorman, H., Akour, I., Alshurideh, M., Al-Hawary, S.: The effect of supply chain resilience on the supply chain performance of chemical industrial companies. Uncertain Supply Chain Manag. 10(4), 1271–1278 (2022)

Alajkovic, A., Brcic, M., Ivandic, V., Bubalo, L., Koncic, M., Kovac, M.: Delivery pattern planning in retailing with transport and warehouse workload balancing. Croatian operational research review. 13(1), 99–111 (2022)

Al-Khawaldah, R., Al-Zoubi, W., Alshaer, S., Almarshad, M., AlShalabi, F., Altahrawi, M., Al-Hawary, S.: Green supply chain management and competitive advantage: the mediating role of organizational ambidexterity. Uncertain Supply Chain Manag. 10(3), 961–972 (2022)

Almetwally, S.A.H., Hassan, M.K., Mourad, M.H.: Real-time internet of things-based water quality management system. Procedia CIRP. 91, 478–485 (2020)

Andonov, A., Dimitrov, G.P., Totev, V.: Impact of e-commerce on business performance. Tem journal. 10(4), 1558–1564 (2021)

Bekrar, A., Ait El Cadi, A., Todosijevic, R., Sarkis, J.: Digitalizing the closing-of-the-loop for supply chains: a transportation and blockchain perspective. Sustainability. 13(5), 2895 (2021)

Bowen Jr., J.T.: Moving places: the geography of warehousing in the US. Journal of Transport Geography. 16(6), 379–387 (2008)

Boysen, N., de Koster, R., Füßler, D.: The forgotten sons: warehousing systems for brick-and-mortar retail chains. European Journal of Operational Research. 288(2), 361–381 (2021)

De Neufville, R.: Airport systems planning, design, and management. In: Air Transport Management, pp. 79–96. Routledge (2020)

De Vass, T., Shee, H., Miah, S.J.: IoT in supply chain management: a narrative on retail sector sustainability. International Journal of Logistics Research and Applications. 24(6), 605–624 (2021)

Enes, M., Silva, Â.: Service quality improvement on warehouse-store transportation in a retail food company. International Journal for Quality Research. 14(2), 321–334 (2020)

Fatimah, Y.A., Kannan, D., Govindan, K., Hasibuan, Z.A.: Circular economy e-business model portfolio development for e-business applications. Journal of Cleaner Production. 415, 137528 (2023)

Gawankar, S.A., Gunasekaran, A., Kamble, S.: Investments in big data-driven supply chains and organizational performance. International journal of production research. 58(5), 1574–1593 (2020)

Goretzko, D., Siemund, K., Sterner, P.: Evaluating model fit of measurement models in confirmatory factor analysis. Educational and Psychological Measurement. 84(1), 123–144 (2024)

Hanaysha, J.R., Alzoubi, H.M.: IoT adoption and supply chain performance. Uncertain Supply Chain Manag. 10(2), 537–550 (2022)

Herden, T.T.: Explaining competitive advantage from analytics. Bus. Res. 13(1), 163–214 (2020)

Holzapfel, A., Hübner, A., Kuhn, H., Sternbeck, M.G.: Delivery pattern and transportation planning in grocery retailing. European Journal of Operational Research. 252(1), 54–68 (2016)

Islam, M.R., Ali, S.M., Fathollahi-Fard, A.M., Kabir, G.: Prediction of warehouse performance using a PSO-based grey model. Journal of Computational Design and Engineering. 8(2), 705–727 (2021)

Istiqomah, N.A., Sansabilla, P.F., Himawan, D., Rifni, M.: Barcode-based warehouse management system implementation. In: J. Phys. Conf. Ser. 1573, 012038 (2020)

Iyer, L.S.: AI-enabled applications for intelligent transportation. Transportation Engineering. 5, 100083 (2021)

Kembro, J., Norrman, A.: Transformation from manual to smart warehousing. The International Journal of Logistics Management. 33(5), 107–135 (2022)

Khan, Y., Su'ud, M.B.M., Alam, M.M., Ahmad, S.F., Ahmad, A.Y.B., Khan, N.: IoT applications in sustainable supply chain management. Sustainability. 15(1), 694 (2022)

Kmiecik, M.: Logistics coordination based on inventory and transportation planning. Sustainability. 14(13), 8134 (2022)

Koh, L., Dolgui, A., Sarkis, J.: Blockchain in transport and logistics. Int. J. Prod. Res. 58(7), 2054–2062 (2020)

Kurdi, B., Alzoubi, H., Akour, I., Alshurideh, M.: Blockchain and smart inventory systems in retail. Uncertain Supply Chain Manag. 10(4), 1111–1116 (2022)

Kwon, O.K.: COVID-19 impacts on global supply chains. Journal of International Logistics and Trade. 18(3), 107–111 (2020)

Maina, E., Wachiuri, E.: Distribution management practices in Nairobi logistics firms. Int. J. Soc. Sci. Manag. Entrep. 8(4), 1–15 (2024)

Manzini, R., Bozer, Y., Heragu, S.: Decision models for warehousing systems. International Journal of Production Economics. 170, 711–716 (2015)

Marchuk, V.Y., Harmash, O.M., Ovdiienko, O.V.: World trends in warehousing logistics. Intellect. Logist. Supply Chain Manag. 2, 32–50 (2020)

Mishra, A., Dutta, P., Gottipalli, N.: Optimization model for downstream supply chain networks. International Journal of Productivity and Performance Management. 73(3), 912–942 (2024)

Mohsen, B.M.: Developments in digital technologies for supply chain management. Procedia Computer Science. 220, 788–795 (2023)

Mondol, E.P.: Blockchain and smart inventory systems in retail. Int. J. Comput. Inf. Manuf. 1(1), 1–12 (2021)

Mostafa, N., Hamdy, W., Alawady, H.: IoT impacts on warehousing. Social sciences. 8(3), 84 (2019)

Mouha, R.A.R.A.: Internet of things. J. Data Anal. Inf. Process. 9(2), 77–89 (2021)

Orjuela-Castro, J.A., Herrera-Ramírez, M.M., Adarme-Jaimes, W.: Warehousing and transportation logistics of mango. Revista Facultad de Ingeniería. 26(44), 73–86 (2017)

Oyetoro, A.: The logistics sector in the Sultanate of Oman. Industry report. (2024)

Parkhi, S., Jagadeesh, D., Kumar, R.A.: Transport cost optimization in retail distribution. Journal of Supply Chain Management Systems. 3(4), 31–38 (2014)

Rajan, P.: Integrating IoT analytics into marketing decision-making. International Journal of Data Informatics and Intelligent Computing. 3(1), 12–22 (2024)

Rejeb, A., Simske, S., Rejeb, K., Treiblmaier, H., Zailani, S.: IoT research in supply chain management. Internet Things. 12, 100318 (2020)
Riyadi, S.: Technology competences and supply chain performance. International Journal of Supply Chain Management. 9(2), 452–459 (2020)
Rotich, V.C., Ndeto, C.: Warehousing management practices in Nairobi. International Journal of Social Sciences Management and Entrepreneurship. 8(4), 1–14 (2024)
Sader, S.: Total Quality Management in Industry 4.0. Master's thesis,. Szent István University (2020)
Sahara, C.R., Aamer, A.M.: IoT-based smart warehouse integration. International Journal of Pervasive Computing and Communications. 18(5), 622–644 (2022)
Sainathuni, B., Parikh, P.J., Zhang, X., Kong, N.: Warehouse-inventory-transportation problem. European Journal of Operational Research. 237(2), 690–700 (2014)
Saragih, J., Tarigan, A., Pratama, I., Wardati, J., Silalahi, E.F.: TQM and supply chain practices. Pol. J. Manag. Stud. 21(2), 384–397 (2020)
Sharma, A., Kaur, J., Singh, I.: IoT in pharmaceutical manufacturing and warehousing. SN Computer Science. 1(4), 232 (2020)
Sousa, P.H.R., Reyes Junior, E., Costa, C.C.B., Reis, A.L.N.: Innovation process and resource-based view. Iberoamerican Journal of Strategic Management. 20, 1–35 (2021)
Subramaniam, M.: Digital ecosystems and competitive strategy. J. Organ. Des. 9, 1–10 (2020)
Trivellas, P., Malindretos, G., Reklitis, P.: Green logistics and supply chain performance. Sustainability. 12(24), 10515 (2020)
Wanjari, S.S.: Warehousing operations from an integrated supply chain approach. European Journal of Business and Management Research. 5(5), 1–8 (2020)
Wanjiru, G.K., Ngugi, P.K.: Effect of e-warehousing on public health institutions. Online source (2020)
Wolniak, R., Grebski, W.: Total quality management in industry 4.0. Zesz. Nauk. Politech. Śląskiej. 185, 595–604 (2023)
Zaman, S.I., Khan, S., Zaman, S.A.A., Khan, S.A.: Grey decision-making model for digital warehouse management. Decision Analytics Journal. 100293 (2023)
Žunić, E., Delalić, S., Hodžić, K., Beširević, A., Hindija, H.: Smart warehouse management system concept. In: Proc. 14th NEUREL, pp. 1–5. IEEE (2018)

The Impact of E-Commerce on Fresh Food Supply Chains and Quality in Gauteng, South Africa

R. M. Momanyi and H. M. Dunga

Abstract The rapid expansion of e-commerce has significantly transformed traditional supply chain structures, particularly within the food sector, presenting both opportunities and challenges in the delivery of fresh, high-quality produce. This study examines the impact of e-commerce adoption on the food supply chain and fresh produce quality performance in Gauteng Province, South Africa. Specifically, it explores the influence of digital platforms on consumer purchasing behaviour, product quality, and last-mile delivery efficiency. Employing a qualitative research design, the study conducted semi-structured interviews with 12 consumers across four municipalities. A content analysis approach was used to assess participants' experiences and perceptions regarding the benefits, challenges, and quality-related concerns associated with purchasing fresh food online. The findings indicate that while e-commerce enhances accessibility, product variety, and convenience, it also introduces significant challenges in maintaining the freshness, consistency, and overall quality of perishable goods. Commonly reported issues included delayed deliveries, inadequate packaging, order inaccuracies, and the lack of pre-purchase product inspection. Participants emphasised the need for improved logistics coordination, accurate product information, and rigorous quality control mechanisms. The study concludes that the effective integration of e-commerce into the fresh food sector requires robust supply chain management practices prioritising quality assurance, customer satisfaction, and responsive service delivery. Recommendations include the development of user-friendly digital interfaces, faster and more reliable delivery systems, innovative packaging solutions, real-time tracking capabilities, and enhanced consumer education. Furthermore, the study advocates for future research incorporating advanced technologies such as blockchain, artificial intelligence (AI), and the Internet of Things (IoT) to improve traceability and demand forecasting. Broader participant representation and the inclusion of perspectives from retailers and suppliers are also recommended to enrich future analyses. This research contributes to the growing discourse on the digital transformation of food systems and offers practical insights for e-commerce platforms, policymakers, and supply chain stakeholders.

R. M. Momanyi (✉) · H. M. Dunga
Vaal University of Technology, Vanderbijlpark, South Africa
e-mail: beckymomanyi18@gmail.com

E. H. Redda (ed.), *Proceedings of the Social Sciences International Research Conference, 2025*, https://doi.org/10.1007/978-981-95-9489-4_8

Keywords Gauteng province · E-commerce · Food supply chain · Quality performance · Consumer perception · Fresh produce · Delivery logistics

1 Introduction

E-commerce has reshaped traditional purchasing behaviour, offering consumers unprecedented convenience and triggering major changes in global retail (Ntumba et al. 2023). The rise of digital marketplaces and advanced online platforms enables consumers to access a wide range of fresh food products anytime, anywhere. As digital tools are increasingly adopted to modernise food distribution, e-commerce's impact on the food supply chain intensifies (Kittipanya-Ngam and Tan 2020). China exemplifies this transformation, with platforms like Alibaba's Tmall and JD.com redefining online food retail with diverse options and seamless experiences (Hoang and Le Tan 2023). Similarly, Amazon's acquisition of Whole Foods in the US signals a significant shift, merging online and physical channels to boost customer satisfaction and operational efficiency (White 2020).

Africa also sees this digital shift. The retail food industry increasingly uses e-commerce, especially for fresh produce. Digital platforms solve issues with affordability, access, and distribution, changing food supply chains across Africa. Nigerian sites like Supermart.ng, Konga, and Jumia give urban and rural customers a variety of food choices. These firms use new delivery methods and payment options to improve access and overcome logistical barriers. Agri-tech efforts, such as Twiga Foods in Kenya and Farmcrowdy in Nigeria, show e-commerce's power to link small farmers to consumers, making supply chains efficient and fair, while offering fresher goods (Iqbal and Butt 2020). But merging e-commerce with Africa's food retail still brings big hurdles, especially for vendors handling perishables.

In South Africa, e-commerce is growing quickly. An estimated 27 million South Africans shopped online by 2022. This may grow to 37.9 million by 2027, over a 40% increase (Fortune 2024). Lower prices, more convenience, and flexible delivery have made e-commerce a popular alternative to traditional retail. Local retailers like Takea-lot, Pick n Pay, Woolworths, Checkers, and Food Lover's Market have invested in digital infrastructure to suit changing consumer needs. Yet, worries about the quality of fresh food ordered online persists. Nearly 40% of online grocery orders are reportedly returned, mainly due to quality problems. Fresh food's perishability makes online supply chain management tough, since buyers cannot inspect products before purchasing. This lack of inspection raises return rates and lowers consumer trust, while increasing costs for transport and waste management (Dube and Dube 2023).

From agricultural input to consumer delivery, the food supply chain is a complex network of interrelated parties and processes (Bhat et al. 2021). Numerous factors including socioeconomic dynamics, policy frameworks, technology, and consumer preferences influence it. Effective supply chain management is essential in this landscape, especially in ensuring the delivery of fresh, high-quality food within digital retail. There is significant potential for preserving food quality and minimising waste through technological interventions (Ciccullo et al. 2021). However, cooperation among government agencies, technology firms, and supply chain stakeholders is crucial for success. Quality performance is vital for e-commerce in food retail. It drives customer satisfaction, boosts operational effectiveness, and shapes regulatory compliance. Quality metrics throughout the supply chain set benchmarks for continuous improvement and competitive advantage (Emon et al. 2024). The South African government has invested in strengthening supply chain infrastructure, specifically to improve fresh food distribution.

Gauteng Province, South Africa's economic hub, is central to the e-commerce shift. As the population diversifies, demand for fresh, quality food increases. Yet whether supply chains meet consumer needs remains in question. It is urgent to examine how e-commerce affects the reliability and quality of fresh food delivery in this region. Retailers need this knowledge to improve service, streamline operations, and compete in a digital market (Ishfaq et al. 2022). This study explores how e-commerce adoption impacts Gauteng's food supply chain. It focuses on how e-commerce affects fresh food quality and its effects on customer satisfaction and supply chain efficiency. The paper also presents the literature review, then the method, findings, and conclusions, with recommendations based on results.

2 Literature Review

2.1 Theoretical and Empirical Perspectives

E-commerce has changed retail, especially in the food supply chain. Relying on online platforms for fresh produce now brings concerns over quality, reliability, and customer satisfaction (Liu et al. 2023). In Gauteng Province, with its strong economy and more online shoppers, understanding these issues is important for better operations and customer experiences. This study combines theory and empirical evidence to assess e-commerce's effect on fresh food supply chains and quality outcomes in Gauteng. The main goal is to see how e-commerce adoption changes food supply performance and vendor quality. As online grocery shopping grows in South Africa, especially for fruits, vegetables, dairy, and meat, the findings will inform business strategy and policy. The study first looks at how e-commerce platforms shape consumer choices, focusing on behaviours and motivations. Digital convenience, better prices, and flexible delivery drive online sales, but not being able to see produce before buying makes it harder to judge quality (Gauri et al. 2021).

Second, the study looks at how consumers judge freshness and quality in e-commerce food purchases. Because fresh food spoils quickly, keeping quality high from farm to doorstep is crucial for satisfied customers (Turan and Ozturkoglu 2022). Delivery conditions like timing, packaging, and handling are key for the consumer experience. Consumers often complain about mismatched products, delivery delays, spoiled items, and weak customer support (Ikpe and Shamsuddoha 2024). Fixing these issues helps pinpoint weak spots in online food supply chains. The study aims to give actionable recommendations to improve e-commerce operations, boost customer satisfaction, and build consumer trust. These suggestions are for platform developers and fresh produce retailers who want to raise service quality.

On the empirical side, the research listens to consumers who buy fresh produce online in Gauteng. The team used surveys and interviews with a targeted sample of online grocery shoppers. They focused on how platform features like usability, product descriptions, customer service, and delivery flexibility shape buying decisions. The main part of the analysis is how consumers rate product quality, especially perishability. Participants shared if their products met their standards for freshness, shelf life, packaging, and condition (Akram et al. 2023). These results show how well quality control works and its effect on retaining customers. The study also details top concerns for online buyers, like damaged goods, delays, wrong orders, and poor real-time support. Addressing these helps retailers and logistics firms know which areas need urgent attention.

By combining theory and evidence, the study shows how e-commerce changes Gauteng's food supply chains. Digital retail offers access and convenience, but also brings new problems in quality assurance and customer satisfaction. The findings highlight a need for better cold chain systems, technology for monitoring quality, and ways to collect customer feedback. Policymakers should foster public-private collaboration in setting and enforcing quality rules (Rao et al. 2021). Regulations must protect consumers, ensure food safety, and encourage innovation. In summary, this study offers key insights into the links between consumer behaviour, e-commerce, and supply chain management in South Africa. By joining theory and data, it builds a path for stronger, more customer-focused e-commerce strategies in fresh food. This supports retailers, logistics firms, and policymakers working to develop the digital food ecosystem in Gauteng and beyond.

3 Methodology

3.1 Research Design

This study adopted a qualitative research approach. This enables the exploration of social phenomena in natural settings, allowing for the collection of rich, descriptive data. Qualitative research is particularly suited to gaining in-depth understanding of participant experiences, attitudes, and behaviours. It offers flexibility, letting participants express their views in their own words without the constraints of predetermined response options that are typical in quantitative methods (Lim 2025). This approach was selected for a deeper and more nuanced examination of consumer perceptions and challenges in the context of online fresh food purchasing in Gauteng Province.

1.1 Sampling Method.

To recruit participants, the study employed convenience sampling, a non-probability sampling method. Individuals were selected based on their accessibility and willingness to participate. This technique was deemed appropriate due to time limitations and logistical constraints. It allowed the researcher to engage readily available respondents within the study area (Islam and Aldaihani 2022).

1.2 Research Instrument.

A total of 12 participants were selected for in-depth interviews. All were consumers of fresh food via online platforms and have bought fresh foods from eight retail shops across four municipalities in Gauteng Province. The selection was guided by ease of access instead of a random or systematic sampling procedure. Respondents answered the same open-ended questions or inquiries, following a predetermined thematic framework. This allowed vital issues to be addressed and provided order, while also giving them room to discuss their encounters, views, and experiences. The researcher was able to acquire more detailed information through follow-up questions.

3.2 Data Collection

Data were collected using semi-structured interviews. This provided the flexibility to explore key themes while allowing participants to elaborate on their experiences. Interviews were audio-recorded and transcribed. The researcher reviewed the transcripts multiple times to ensure familiarity with the data and to achieve data saturation up to the point when no new themes emerged. The initial coding process led to the identification of broad topics, which were subsequently refined into more specific sub-themes. These thematic groupings formed the foundation for the study's analysis, conclusions, and recommendations.

3.3 Data Analysis

The data was analysed using thematic analysis. This method involves systematically identifying, organising, and interpreting patterns within qualitative data (Naeem et al. 2023). To enhance rigour and transparency, the study used NVivo software for coding and theme development. This tool helped organise qualitative data and ensured consistent codes across transcripts. It contributed to the reliability and credibility of the findings.

4 Results and Discussion

This section presents the thematic outcomes from the qualitative data collected through semi-structured interviews with 12 participants across four municipalities in Gauteng Province. Thematic analysis, supported by NVivo software, was used to identify patterns in participants' responses about their experiences with delivery and quality performance of fresh food purchased via e-commerce platforms. Four key themes emerged: delivery conditions satisfaction, quality satisfaction levels, e-commerce challenges, and positive acknowledgement of quality.

Theme1: Delivery Conditions Satisfaction. This theme captures participants' diverse perspectives on satisfaction with the delivery conditions of fresh food ordered online. Many expressed satisfaction with timeliness, packaging materials, and temperature control mechanisms used by online retailers. However, some noted negative experiences, such as delays and packaging failures. Despite these issues, most participants indicated that packaging efforts generally preserved product quality during transit.

The packaging for fresh food is impressive, and temperature regulation for meat and seafood is particularly well-managed. Even a slight delay in delivery does not usually affect the quality, which significantly enhances my satisfaction with online food purchases.

I have generally been satisfied with the delivery process. The packaging materials are usually effective, and I appreciate that the food often arrives in insulated bags to maintain its temperature. However, there was a time when my order was delayed, and the melting ice packs affected the freshness of some items.

Overall, the delivery conditions have been satisfactory. Strong packaging ensures no damage during transport, and temperature control keeps my dairy products chilled. There was a rare instance of a delayed delivery affecting vegetable freshness, but I am generally very satisfied when all factors are well-handled.

Theme2: Quality Satisfaction Levels. This theme reflects participants' perceptions of the quality and freshness of fresh food purchased online. Many respondents reported high satisfaction with the freshness and overall condition of delivered products. Several also shared instances of disappointment due to damaged or overripe items. These mixed experiences suggest variability in quality control across platforms and deliveries.

I was particularly satisfied with the pork ribs I ordered for a special dinner; they were very fresh and flavourful. On the downside, the bananas I ordered were bruised and overripe upon arrival, which was disappointing as I prefer them firm.

I was extremely pleased with the freshness and sweetness of the strawberries I ordered online for an event. They were noticeably superior to those typically found in stores. However, I was very disappointed with the avocados I received on another occasion; they were overripe and unusable, which was particularly frustrating since I needed them for a recipe.

I was highly impressed with the quality of the prawns I ordered online; they arrived perfectly chilled and in pristine condition, leading to a delightful meal. Conversely, I once received tomatoes that were overly soft and damaged, which affected the quality of my dish.

Theme3: E-commerce Challenges. Despite the operational functionality of e-commerce platforms, this theme reveals challenges that affect customer satisfaction and trust. Participants reported issues like order substitutions, minimum purchase thresholds, delivery delays, and product mismatches. All of these influenced their online shopping experiences.

A major obstacle for me is the minimum purchase amount required for free delivery. There are times when I need just a few items, but I end up purchasing extra products to meet the minimum order quantity and avoid the delivery fee, which is quite inconvenient.

Substitutions for unavailable products have been problematic. On one occasion, I ordered organic apples but received regular ones instead. This didn't meet my dietary needs and forced me to buy the correct items elsewhere.

Inconsistent delivery times have been a major issue. On one occasion, my fresh meat arrived at nearly room temperature due to a delivery delay, forcing me to cook it immediately to avoid spoilage and disrupting my schedule.

Theme4: Positive Acknowledgement of Quality. This theme highlights the overall satisfaction of participants with the quality of fresh food acquired online. In many cases, participants reported that the freshness, taste, and condition of online products met or exceeded those from physical stores. Nonetheless, occasional dissatisfaction occurred, often related to baked goods or certain vegetables.

My experience has been mostly positive. I frequently receive fresh, sweet strawberries, which surpass those from local stores. However, there was a rare case where the bread wasn't as fresh as expected. The quality generally meets my expectations. Fresh seafood is a highlight, arriving cold and fresh. However, I received some overly soft bell peppers once, prompting me to switch platforms for better quality.

Overall satisfaction is high, particularly with locally sourced products. The freshness of eggs is consistently good. However, a slightly stale cake once made me con-sider baking my own instead of ordering baked goods online.

In conclusion, the thematic findings of this study reveal the complex and sometimes contradictory experiences of consumers using e-commerce for purchasing fresh food in Gauteng Province. While participants acknowledged the convenience, accessibility, and variety of online retail, their satisfaction was reduced by recurring issues. These included inconsistent delivery, product substitutions, and fluctuations in the quality of perishable goods. Such issues point to ongoing operational gaps in the digital supply chain. The study highlights a need for improvements in cold chain logistics, quality assurance protocols, and responsive customer service. Addressing these areas is crucial for enhancing the consumer experience and fostering long-term trust, loyalty, and sustainable growth in the e-commerce fresh food sector.

5 Conclusion

This study has highlighted the transformative role of e-commerce platforms in reshaping how consumers buy fresh food in Gauteng Province, South Africa. The findings show that e-commerce offers significant advantages, such as broader product selection, competitive pricing, and greater accessibility. These factors influence consumer purchasing behaviour. For many, especially students and busy professionals, online grocery shopping is a practical and time-saving alternative to traditional retail. The availability of organic and locally sourced options, detailed product descriptions, and personalised recommendations further enrich the online shopping experience. However, the study also revealed key challenges that affect satisfaction and reliability. These include inconsistent delivery, poor product substitutions, perishable items arriving in bad condition, and unresponsive customer support. Such limitations affect trust and may deter repeat purchases if not addressed.

For long-term sustainability in the fresh food e-commerce sector, this study recommends several interventions. These include investing in robust cold chain logistics, implementing strict quality control, improving user interfaces, enhancing real-time customer service, and developing flexible delivery options. Advanced technologies like the Internet of Things (IoT), blockchain, and artificial intelligence (AI) could also help improve traceability, inventory management, and satisfaction. While e-commerce is changing access to fresh food in South Africa's cities, it needs infrastructure, policy, and innovation to ensure quality, efficiency, and fairness. The research provides practical insights for retailers, developers, policymakers, and supply chain stakeholders aiming to build a more reliable, inclusive, and consumer-focused digital food ecosystem.

References

Akram, H.W., Akhtar, S., Ahmad, A., Anwar, I., Sulaiman, M.A.B.A.: Developing a conceptual framework model for effective perishable food cold supply chain management based on a structured literature review. Sustainability. **15**(6), 4907 (2023)

Bhat, S.A., Huang, N.F., Sofi, I.B., Sultan, M.: Agriculture food supply chain management based on blockchain and IoT: a narrative on enterprise blockchain interoperability. Agriculture. **12**(1), 40 (2021)

Ciccullo, F., Cagliano, R., Bartezzaghi, G., Perego, A.: Implementing the circular economy paradigm in the Agri-food supply chain: the role of food waste prevention technologies. Resour. Conserv. Recycl. **164**, 105114 (2021)

Dube, M., Dube, S.: Towards sustainable colour cosmetics packaging. Cosmetics. **10**(5), 139 (2023)

Emon, M.M.H., Khan, T., Siam, S.A.J.: Quantifying the influence of supplier relationship management and supply chain performance: an investigation of Bangladesh's manufacturing and service sectors. Braz. J. Oper. Prod. Manag. **21**(2), 2015 (2024)

Fortune, P.: Democratising Digital Advertising and e-Commerce in South Africa. Master's thesis, University of the Witwatersrand, Johannesburg (2024)

Gauri, D.K., Jindal, R.P., Ratchford, B., Fox, E., Bhatnagar, A., Pandey, A., Howerton, E.: Evolution of retail formats: past, present, and future. J. Retail. **97**(1), 42–61 (2021)

Hoang, H., Le Tan, T.: Unveiling digital transformation: investigating technology adoption in Vietnam's food delivery industry for enhanced customer experience. Heliyon. **9**(9) (2023)
Ikpe, V., Shamsuddoha, M.: Functional model of supply chain waste reduction and control strategies for retailers: the USA retail industry. Logistics. **8**(1), 22 (2024)
Iqbal, R., Butt, T.A.: Safe farming as a service of blockchain-based supply chain management for improved transparency. Clust. Comput. **23**, 2139–2150 (2020)
Ishfaq, R., Davis-Sramek, B., Gibson, B.: Digital supply chains in omnichannel retail: a conceptual framework. J. Bus. Logist. **43**(2), 169–188 (2022)
Islam, M.A., Aldaihani, F.M.F.: Justification for adopting qualitative research methods, research approaches, sampling strategy, sample size, interview method, saturation, and data analysis. J. Int. Bus. Manag. **5**(1), 1–11 (2022)
Kittipanya-Ngam, P., Tan, K.H.: A framework for food supply chain digitalisation: lessons from Thailand. Prod. Plan. Control. **31**(2–3), 158–172 (2020)
Lim, W.M.: What is qualitative research? An overview and guidelines. Australas. Mark. J. **33**(2), 199–229 (2025)
Liu, M., Jia, W., Yan, W., He, J.: Factors influencing consumers' repurchase behaviour on fresh food e-commerce platforms: an empirical study. Adv. Eng. Inform. **56**, 101936 (2023)
Naeem, M., Ozuem, W., Howell, K., Ranfagni, S.: A step-by-step process of thematic analysis to develop a conceptual model in qualitative research. Int J Qual Methods. **22**, 16094069231205789 (2023)
Ntumba, C., Aguayo, S., Maina, K.: Revolutionising retail: a mini-review of e-commerce evolution. J. Digit. Mark. Commun. **3**(2), 100–110 (2023)
Rao, M., Bast, A., de Boer, A.: European private food safety standards in global Agri-food supply chains: a systematic review. Int. Food Agribus. Manag. Rev. **24**(5), 739–754 (2021)
Turan, C., Ozturkoglu, Y.: A conceptual framework model for an effective cold food chain management in a sustainability environment. J. Model. Manag. **17**(4), 1262–1279 (2022)
White, S.: Amazon and whole foods: adventures in grocery shopping. CASE J. **16**(2), 115–153 (2020)

Trends and Drivers of Private Investments Towards SDG Financing in Botswana: A Case for Pooled Funding

Sillas Molosiwa, Samuel Chingoiro, Olumide Jaiyeoba, and Mornay Robert-Lombard

Abstract This study, carried out in Gaborone, Botswana, was done out of a need for new findings that can provide practical insights on the progress of Sustainable Development Goals (SDGs) and assessing a mechanism that can potentially aid in scaling up financial contributions from the private-sector towards SDG 9. The need arose from the challenge of the financing gap towards SDGs in general and in particular SDG 9 in Botswana. This study contributed to the area of sustainable development by assessing Botswana's progress with SDGs, then examining the relationship between features of a pooled fund and institutional investor buy-in towards SDG 9. Using a survey questionnaire and quantitative methods, the study examined 5 features of a pooled fund as predictors of institutional investor buy-in, i.e., willingness of the institutional investor to increase the rate of investment towards infrastructure. The findings indicate that out of the examined 5 features (multi-stakeholder governance structure; multiplicity of investors; concessional finance; independent fund administrator; long-term project plans), only three of these proved to have a significant influence on institutional investor buy-in. The recommendations emanating from this study include a recommendation for a special infrastructure pooled fund to be set up, characterised by some of the features indicated in this study that positively influence institutional investor buy-in. Additional recommendations include setting up a dedicated SDGs hub for tracking and consolidating SDGs data and involving institutional investors early-on during project planning for alignment and de-risking purposes as a driver or catalyst for SDG 9 financing from institutional investors.

S. Molosiwa (✉) · S. Chingoiro · O. Jaiyeoba
Botho University, Gaborone, Botswana
e-mail: sillasmolosiwa@gnosisflux.com

S. Chingoiro
e-mail: samuel.chingoiro@bothouniversity.ac.bw

O. Jaiyeoba
e-mail: Olumide.jaiyeoba@bothouniversity.ac.bw

M. Robert-Lombard
University of the Western Cape, Cape Town, South Africa
e-mail: mroberts-lombard@uwc.ac.za

E. H. Redda (ed.), *Proceedings of the Social Sciences International Research Conference, 2025*, https://doi.org/10.1007/978-981-95-9489-4_9

Keywords Institutional investment · Pooled fund · Sustainable development goal · Infrastructure financing · Botswana

1 Introduction

The United Nations' (UN's) Sustainable Development Goal 9 (SDG 9) promotes resilient infrastructure, sustainable industrialisation, and innovation. It is central to long-term economic transformation, especially in emerging economies like Botswana. Yet, substantial financing shortfalls slow progress towards this goal. Globally, developing countries face an estimated annual infrastructure financing gap of US$373 billion (World Commission on Environment and Development 1987). Africa alone requires US$130 billion to US$170 billion per year, while the annual shortfall ranges from US$50 billion to US$90 billion. Nevertheless, private-sector investment remains low. In Botswana, most infrastructure financing depends almost entirely on national budget allocations, which are already overstretched (Ozili 2023).

This challenge is acute in Botswana. The absence of innovative financing mechanisms limits progress towards SDG 9 targets. At the same time, institutional investors in Botswana held over BWP 8.8 billion (about US$663 million) in capital in 2021 (Office of the Auditor General, Republic of Botswana 2019). This raises a critical question: why does this capital remain underutilised in financing infrastructure development? The challenge lies in creating incentives for private-sector participation, particularly from institutional investors, in sustainable infrastructure projects (Lawal et al. 2024).

This study answers this gap, investigating the potential of a pooled funding mechanism as an innovative instrument to mobilise domestic private capital. Pooled funding, a blended finance tool, combines contributions from private investors, development partners, and multilateral institutions. It could unlock financing for infrastructure development in line with SDG 9. Accordingly, the study's primary aim is to assess the suitability of the pooled fund model in Botswana's context as a means of financing infrastructure development aligned with SDG 9. The study was guided by the following research questions:

- How does multi-stakeholder participation influence investment buy-in in pooled funding models?
- What role does concessional finance play in attracting private investment into the pooled fund?
- How does the presence of an independent fund administrator affect trust and participation in the pooled fund?
- How can the alignment between long-term infrastructure planning and investor commitment be strengthened through pooled funding?
- What impact do governance structures have on investor confidence in pooled funds for infrastructure development?

By addressing these questions, the study aims to provide conceptual and practical insights. It shows how innovative financing mechanisms can unlock institutional capital, strengthen infrastructure development, and accelerate progress towards SDG 9 in Botswana and similar developing economies.

2 Literature Review

This study applies a multi-theoretical framework to examine how pooled funding mechanisms can mobilise institutional investment for infrastructure aligned with SDG 9 in Botswana. Central is the three pillars model of sustainable development, rooted in the Brundtland Report and the triple bottom line, which stresses economic, social, and environmental integration (UNEP, United Nations Environment Programme Finance Initiative 2019). Sustainable finance theory further suggests institutional investors increasingly prioritise long-term environmental, social, and governance (ESG) objectives (Public Investment Corporation SOC Limited 2023). Flow of funds theory explains how private capital can be channelled into sustainable infrastructure, while the economic theory of infrastructure and commons management highlights inclusive access and societal benefit (Frischmann 2005). Collectively, these theories shape the study's focus on pooled fund features influencing investor buy-in, illustrated in Fig. 1. Recent developments, including Public Investment Corporation SOC Limited (2023)'s six sustainable finance theories and Parmaksiz et al. (2023) insights, enrich this framework by explaining how signalling, priorities, and resource endowments influence institutional capital mobilisation.

Moreover, global analysis emphasises that securitisation and structured vehicles are key to transforming illiquid development assets into investible opportunities for institutional investors, enhancing risk sharing and accessibility. Infrastructure debt funds and pooled investment vehicles offer tailored mechanisms for delivering large-scale project financing, directly aligning with the study's pooled funding focus (Schmitt 2023). United Nations Conference on Trade and Development (2025) highlighted the vital role of sovereign institutional investors, including African sovereign wealth and pension funds, in bridging infrastructure financing gaps on the continent, underlining the relevance of local institutional investors in Botswana's landscape.

As illustrated in Fig. 1, the conceptual framework of this research focuses on evaluating the impact of specific features of a pooled funding mechanism on institutional investor buy-in for infrastructure projects in Botswana. At its core, the objective is to determine if these features, derived from various pooled funds in humanitarian, developmental, and health sectors, can positively influence institutional investors' participation in infrastructure funding. Key features under investigation include flexibility, predictability, stronger risk management, cost-effectiveness, strategic decision-making, and transparency, all of which collectively provide pooled funds with their comparative advantage (United Nations 2015a).

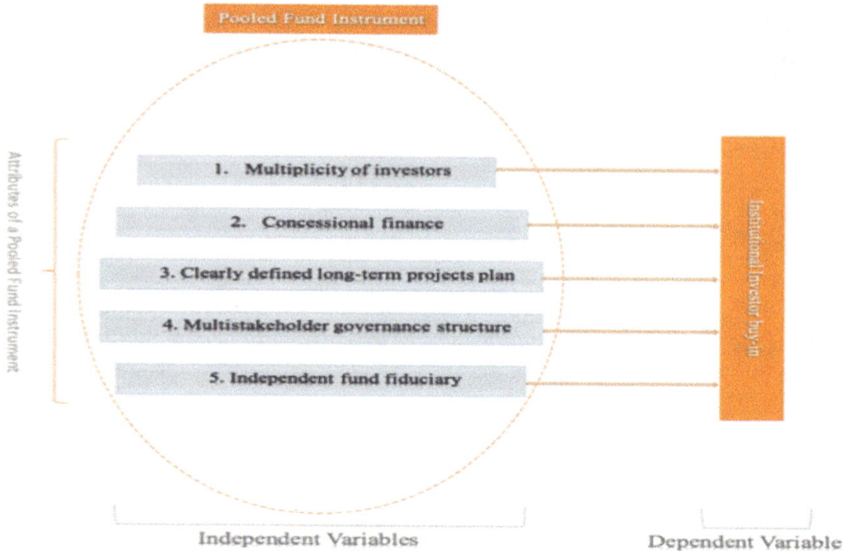

Fig. 1 A conceptual framework depicting variables

The research explores whether a pooled fund for infrastructure, with multiple stakeholders such as institutional investors, multilateral development banks (MDBs), international financial institutions, local governments, and donor governments, can attract investment. Critical elements for consideration include concessional finance from MDBs, a comprehensive long-term infrastructure investment plan, a multi-stakeholder governance structure with clearly defined roles, and an independent trustee to oversee fund administration. These features are tested through hypotheses, examining their potential to influence investor buy-in, with institutional investor participation serving as the dependent variable. As such it becomes important to note that institutional investor buy-in is significantly influenced by several key factors that enhance the attractiveness and viability of infrastructure projects. A multiplicity of investors, encompassing development banks, donors, and government actors, fosters a diversified risk profile and demonstrates a broad commitment to the project's success. This collective involvement can enhance credibility and reduce perceived investment risks. Concessional finance plays a pivotal role in de-risking infrastructure projects. By offering below-market financing terms, it lowers the cost of capital and improves the financial viability of projects, making them more appealing to institutional investors seeking stable returns (ICVA 2023). A clearly defined long-term project plan provides institutional investors with the predictability and stability they require. Such planning outlines project timelines, milestones, and expected returns, aligning with the long-term investment horizons typical of institutional investors. An inclusive multi-stakeholder governance structure ensures transparency, accountability, and effective decision-making. When various stakeholders, including public and private entities, collaborate, it mitigates governance risks and enhances project credibility (Public Investment Corporation SOC Limited 2023). Finally, an independent fund fiduciary acts as a neutral party overseeing fund management, ensuring that investments are made in the best interests of all stakeholders. This independence builds trust among institutional investors, assuring them that their investments are managed with due diligence and integrity. Collectively, these elements create a conducive environment for institutional investors, aligning their investment criteria with the project's structure and governance, thus facilitating greater buy-in.

Considering the discussion above, the study explores whether key features of pooled funding mechanisms can attract institutional investors to finance infrastructure projects in Botswana, thereby contributing to SDG 9. Institutional investors such as pension funds and insurance companies manage vast assets globally. If mobilised, even a small portion of their capital could help close the SDG financing gap. Despite this potential, institutional investor engagement in infrastructure in developing countries remains low.

The proposed hypotheses highlighted below are derived through a comprehensive analysis of Botswana's infrastructure financing challenges and the theoretical frameworks underpinning sustainable finance. Recognising the substantial infrastructure financing gap in Botswana, where national budget allocations are overstretched, the study posits that innovative financing mechanisms are essential to mobilise institutional investment for infrastructure aligned with SDG 9. Drawing from the three pillars model of sustainable development, the theory of sustainable finance, and the flow of funds theory, the study hypothesises that specific features of pooled funding mechanisms, such as a multiplicity of investors, concessional finance, clearly defined long-term project plans, multi-stakeholder governance structures, and independent fund fiduciaries, can positively influence institutional investor buy-in. These hypotheses are informed by recent developments in sustainable finance, including the Public Investment Corporation SOC Limited (2023) formulation of sustainable finance theories and the identification of key financial instruments to mobilise institutional investors in emerging markets. Furthermore, the study considers the role of sovereign institutional investors in bridging infrastructure financing gaps, as highlighted by [35]. By integrating these theoretical insights and contextual factors, the study aims to assess the suitability of the pooled fund model in Botswana's context as a means of financing infrastructure development aligned with SDG 9. Furthermore, pooled funds, commonly used in humanitarian and health sectors, offer comparative advantages including flexibility, cost efficiency, risk sharing, and strategic oversight. This study tests whether these features, when adapted for infrastructure investment, can influence investor participation (Purvis et al. 2019).

The first hypothesis addresses the presence of multiple contributors, including development banks, donors, and government actors, proposing that:

H1: A multiplicity of investors is significantly and positively associated with institutional investor buy-in.

Second, concessional finance offered by development partners to reduce investment risk may be a key incentive for private capital mobilisation. Evidence suggests that blended finance, which includes concessional elements, can help bridge the financing gap left by limited official development assistance (Public Investment Corporation SOC Limited 2023). Therefore:

H2: Concessional finance is significantly and positively associated with institutional investor buy-in.

Third, long-term project pipelines provide predictability, which is essential for large-scale investment planning. While public-private partnerships (PPPs) are commonly used, they often encounter inefficiencies and limitations, suggesting a need for innovative approaches (Infrastructure Consortium for Africa Secretariat 2022). Thus:

H3: A clear long-term infrastructure plan is significantly and positively associated with institutional investor buy-in.

Fourth, pooled funds that feature inclusive governance structures can enhance transparency, align stakeholder interests, and support coordinated decision-making. With the SDGs highlighting the importance of multi-stakeholder collaboration (Masuga 2023), the following hypothesis is proposed:

H4: A multi-stakeholder governance structure is significantly and positively associated with institutional investor buy-in.

Lastly, the presence of an independent fund administrator or fiduciary is expected to increase investor confidence through improved transparency and accountability in fund management (Bartz-Zuccala et al. 2020). Hence:

H5: An independent fund fiduciary is significantly and positively associated with institutional investor buy-in.

Together, these hypotheses aim to assess whether a pooled fund structure tailored for infrastructure investment can catalyse institutional investor participation and contribute meaningfully to sustainable development efforts in Botswana.

3 Methodology

This study adopted a quantitative research methodology to examine institutional investor engagement in infrastructure financing through pooled fund mechanisms in Botswana. The aim was to generate statistical insights into how pooled fund features influence investor participation. A structured questionnaire was chosen as the primary data collection tool, addressing the lack of prior empirical research in Botswana. The study followed a pragmatic philosophy, focused on practical solutions to the infrastructure financing gap, and applied a deductive approach guided by five hypotheses: multiplicity of investors, concessional finance, long-term project planning, multi-stakeholder governance, and independent fiduciary oversight. A census approach was employed, targeting 114 institutional investors, including pension funds, asset managers, and insurance companies regulated by the Non-Bank Financial Institutions Regulatory Authority. The small population size made a census both feasible and necessary, ensuring comprehensive coverage and enhanced reliability. Data were collected through SurveyMonkey, email, and Google Forms. The questionnaire covered demographics, investment motivations, perceived risks and opportunities, and pooled fund features, measured using a five-point Likert scale. Items were adapted from established sources such as (The Global Goals n.d.; ICVA 2024; Motshidisi 2022; United Nations 2015b; United Nations Conference on Trade and Development 2025; World Commission on Environment and Development 1987), ensuring content validity.

Data were analysed using SPSS, with weighted averages used to assess pooled fund features. Reliability was confirmed with a Cronbach's alpha of 0.801, exceeding the 0.7 threshold and demonstrating strong internal consistency. Ethical clearance was obtained from Botho University, and informed consent, confidentiality, and anonymity were upheld throughout the study. Overall, the methodology provided a systematic and reliable framework for examining institutional investor behaviour. By combining a census approach with a structured survey, the study generated actionable insights into how pooled fund mechanisms can mobilise private capital to support sustainable infrastructure and advance SDG 9.

4 Results and Discussion

This section presents the research findings in alignment with the study's objective, which is to assess the suitability of the pooled funding mechanism in mobilising domestic institutional investment towards infrastructure development with particular focus on SDG 9. The analysis includes descriptive statistics, an assessment of psychometric properties, and inferential statistics used to test the study's hypotheses.

4.1 Descriptive Statistics

Data were collected through a structured, closed-ended questionnaire distributed digitally via SurveyMonkey, Google Forms, and email. From a population of 114 institutional investors, 101 valid responses were obtained, meeting the required sample size. All three targeted subgroups were represented, with pension funds forming the majority (66.3%), followed by insurance companies (19.8%) and asset managers. Most organisations (43.6%) had been in operation for 5–10 years, while 35.6% had operated for more than 10 years. Respondents' demographics showed that 70.3% were male and 29.7% female, with the majority (72.3%) aged 30–49. Regarding education, 45.5% held undergraduate degrees, and many occupied management or director-level positions (44.5%). In terms of marital status, 60.4% were married and 32.7% single. On infrastructure investment, 35.6% currently invest,

33.7% do not, and 30.7% plan to invest in the future. These findings reflect a balanced mix of personal and organisational characteristics among institutional investors.

4.2 Indicator Analysis

This section presents the study's findings in relation to its main objective, which is to assess the suitability of the pooled funding mechanism as a driver for domestic private-sector investment in infrastructure, aligned with SDG 9.

Investment in Infrastructure and Perceived Barriers. In line with the hypothesis that private-sector investment in infrastructure is currently low in Botswana and globally, participants were asked to identify perceived barriers to such investments. This multi-response item revealed that 58.5% of respondents considered infrastructure projects too complex to invest in. Additionally, 53.5% cited the absence of viable project plans and pipeline opportunities, alongside unfavourable policy and regulatory environments. Another 50.5% highlighted misalignment between client interests and infrastructure investments, while 47.5% pointed to insufficient financial mechanisms linking capital sources to infrastructure projects.

Sustainable Development Goal 9. Respondents' views on SDG 9 were measured across six indicators. A significant proportion (39.6%, *Mean* = 3.85, *SD* = 0.95) agreed that financial return is the primary driver for infrastructure investment. Similarly, 34.7% agreed (*Mean* = 3.83, *SD* = 1.06) that several inhibiting factors limit institutional investment in infrastructure. Notably, 42.6% agreed and 28.7% strongly agreed that their organisations were willing to contribute more financial resources towards SDG 9 (*Mean* = 3.73, *SD* = 1.03). Regarding current engagement, 39.6% agreed that their organisation is significantly participating in SDG 9 financing (*Mean* = 3.90, *SD* = 0.96), while 43.6% agreed that Botswana has made notable progress since 2016 (*Mean* = 3.99, *SD* = 4.17).

Multiplicity of Investors. Findings indicated moderate support for pooled financing from multiple sources. Respondents agreed that a pooled infrastructure fund drawing from MDBs, donors, government, and the private sector would increase investment from their firms (30.7% agreement, *Mean* = 3.33, *SD* = 1.24). Furthermore, 35.6% agreed (*Mean* = 3.67, *SD* = 1.24) that firms are more likely to invest when other investors are involved. However, only 29.7% agreed that solo or partnered investments have comparable success rates to those with multiple investors (*Mean* = 3.47, *SD* = 1.24). The types of collaborating investors were also considered significant, with 32.7% agreeing that investor composition impacts project success (*Mean* = 3.47, *SD* = 1.24). Regarding concessional finance, 33.7% agreed that its availability influences their investment decisions (*Mean* = 3.88, *SD* = 0.84), and 45.5% believed their firms could contribute concessional debt/equity for socially beneficial projects (*Mean* = 4.04, *SD* = 0.73). Moreover, 31.7% strongly agreed that concessional finance would likely boost infrastructure investment (*Mean* = 4.14, *SD* = 0.80). Organisational capacity was also examined: 52.5% agreed they had adequate expertise for concessional finance projects (*Mean* = 3.68, *SD* = 0.92), while 52.5% believed concessional finance reduces financial risk.

Independent Fund Fiduciary. The role of an independent fiduciary in pooled funds was positively viewed. Respondents agreed (*Mean* = 3.99, *SD* = 0.87) that such a structure would likely encourage investment. Similarly, 33.7% strongly agreed and 29.7% agreed that co-investment is more likely when funds are managed by an independent fiduciary (*Mean* = 3.96, *SD* = 0.82). A smaller proportion (45.5%) supported the notion that an independent trustee is unnecessary (*Mean* = 3.65, *SD* = 1.00), indicating mixed views. On fund governance effectiveness, 28.7% strongly agreed that fiduciary performance influences investment decisions (*Mean* = 3.52, *SD* = 1.02). Additionally, 44.6% agreed that sound financial management by a fiduciary could attract increased investment (*Mean* = 3.77, *SD* = 1.15), whereas 42.6% believed poor management would significantly undermine project outcomes (*Mean* = 3.84, *SD* = 1.04).

Long-Term Project Plans. The availability of a long-term project pipeline was widely considered an incentive for investment. About 55.5% of respondents agreed (*Mean* = 3.95, *SD* = 0.94) that a clearly defined pipeline within a trust fund would enhance investment willingness. However, only 42.6% agreed that the clarity of national project plans does not impact investment (*Mean* = 3.68, *SD* = 1.22), indicating diverging views. The interest in co-investing in government-planned projects was lower, with only 37.6% in agreement (*Mean* = 3.25, *SD* = 1.14). Ambiguous project plans were identified as potential deterrents to investment (*Mean* = 3.24, *SD* = 1.33), while 45.5% agreed that viable long-term plans would enhance fund allocation (*Mean* = 3.39, *SD* = 1.23).

Multi-Stakeholder Governance Structure. Perceptions of governance structures were varied. Only 30.7% agreed that multi-stakeholder governance is more efficient than government—or privately led models (*Mean* = 3.57, *SD* = 1.20). Similarly, 35.7% believed that a multi-stakeholder-led trust fund would encourage investment (*Mean* = 3.34, *SD* = 1.31), while nearly 37% disagreed that such governance negatively impacts fund effectiveness (*Mean* = 2.95, *SD* = 1.19). Respondents were divided on whether stakeholder involvement enhances transparency. While 29.7% agreed that involving investors in governance promotes transparency (*Mean* = 2.98, *SD* = 1.27), 32.7% strongly agreed that it is unnecessary (*Mean* = 2.91, *SD* = 1.34). Lastly, 42.6% believed that stakeholder involvement is essential during the planning phase of projects (*Mean* = 2.88, *SD* = 1.34).

4.3 Correlation Analysis

The dependent variable in the conceptual framework is *institutional investor buy-in*, while the independent variables are *multiplicity of investors*, *concessional finance*, *independent fund fiduciary/administrator*, *clearly defined long-term project plans*, and *multi-stakeholder governance structure*. Table 1 presents the results of the

Pearson correlation analysis conducted to explore the strength and direction of relationships between the dependent and independent variables.

Table 1 shows that *institutional investor buy-in* and *multiplicity of investors* have a moderate positive and statistically significant correlation ($r = 0.306, p < 0.01$). This suggests that as the diversity of investor participation increases, institutional investors are more likely to invest or increase their level of engagement. The statistical significance indicates that this association is unlikely to have occurred by chance. Such a relationship may be attributed to the benefits associated with multiple investors, including risk diversification, greater capital availability, enhanced credibility, and alignment with broader development objectives.

Table 1 also indicates that the correlation between *institutional investor buy-in* and *concessional finance* is weak and not statistically significant ($r = 0.149$, $p = 0.137$), indicating that the relationship is neither strong nor reliable enough to infer a meaningful connection. Thus, it cannot be concluded that concessional finance significantly influences investor buy-in within the context of this study. Furthermore, a moderate and statistically significant positive relationship is observed between *institutional investor buy-in* and *independent fund fiduciary* ($r = 0.449$, $p < 0.01$). This implies that positive perceptions of the role of a fiduciary or fund administrator are likely associated with increased institutional investor participation. The finding aligns with existing literature that highlights the importance of fiduciary trust and accountability in fostering investor confidence (United Nations 2015b).

Similarly, *multi-stakeholder governance structure* demonstrates a strong and statistically significant positive correlation with *institutional investor buy-in* ($r = 0.523, p < 0.01$). This indicates that broader governance participation may contribute to enhanced investor interest. The significance of multi-stakeholder governance is echoed in studies such as Gasser et al. (2015), which highlight its capacity to synthesise diverse interests into effective collaborative frameworks. Finally, *clearly defined long-term project plans* are positively and significantly correlated with *institutional investor buy-in* ($r = 0.486$, $p < 0.01$). This finding suggests that the presence of clear and viable long-term infrastructure plans may positively influence investor engagement. The result is consistent with literature noting that a lack of well-defined investment pipelines is a key barrier to infrastructure financing (United Nations 2015a).

4.4 Regression Analysis

This section presents the results of a multiple linear regression analysis conducted to assess the effect of five independent variables namely multiplicity of investors, concessional finance, independent fund fiduciary, clearly defined long-term project plans and multi-stakeholder governance structure on the dependent variable institutional investor buy-in.

Model Summary. The model summary established that the independent variables collectively explain 56.2% of the variance in institutional investor buy-in, as indicated

Table 1 Correlation analysis

Variables		Institutional investors buy-in	Multiplicity of investors	Concessional finance	Independent fund fiduciary	Long-term project plans	Multi-stakeholder governance
Institutional investors buy-in	R	1	0.306**	0.149	0.449**	0.486**	0.523**
	p-value		0.002	0.137	0.000	0.000	0.000
Multiplicity of investors	R		1	0.120	0.188	0.243*	0.091
	p-value			0.233	0.060	0.014	0.366
Concessional finance	R			1	0.016	0.073	0.132
	p-value				0.876	0.470	0.188
Independent fund fiduciary	R				1	0.064	0.160
	p-value					0.524	0.109
Long-term project plans	R					1	0.226*
	p-value						0.023
Multi-stakeholder governance	R						1

**Correlation is significant at the 0.01 level (2-tailed)
*Correlation is significant at the 0.05 level (2-tailed)

by $R^2 = 0.562$. This suggests a moderate level of explanatory power, implying that while the model accounts for more than half of the variance, 43.8% remains unexplained and could be attributed to factors outside the current model. Furthermore, the adjusted R^2 (0.539) also supports the model's adequacy. Furthermore, the Durbin-Watson statistic (2.025) falls within the acceptable range (1.5–2.5), suggesting no serious autocorrelation in the residuals.

Analysis of the Variance. In terms of the results of the Analysis of Variance, the statistical significance of the regression model was confirmed. The F-value = 24.383 with a $p < 0.001$, indicated that the model as a whole is statistically significant and provides a meaningful prediction of institutional investor buy-in.

Coefficients and Multicollinearity Diagnostics. In terms of the coefficients for the regression analysis, it was established that the three independent variables, that is, independent fund fiduciary, long-term project plans, and multi-stakeholder governance structure, are statistically significant at the 1% level. Conversely, multiplicity of investors and concessional finance are not statistically significant. As such, multicollinearity does not appear to be an issue, as all variance inflation factor (VIF) values are below 2, well under the commonly accepted threshold of 10.

Hypothesis Testing. In terms of hypothesis testing, which evaluates the relationships between the independent variables and the dependent variable, it was established that the hypotheses concerning independent fund fiduciary (H3), long-term project plans (H4), and multi-stakeholder governance structure (H5) are supported, as their respective p-values are below 0.05 (they are all $p = 0.000$). In contrast, the hypotheses related to multiplicity of investors (H1) and concessional finance (H2) are not supported at conventional levels of significance as their p-values range between 0.105 and 0.425, which is above the cut-off level of 0.05).

5 Discussion

The identification of barriers to private-sector investment in infrastructure, such as project complexity, lack of viable plans, unfavourable policy environments, misaligned client interests, and insufficient financial mechanisms, aligns with findings from other studies. For instance, (McHugh 2021) noted that the current Public Investment Programme in Botswana is not suitable for PPPs, citing limited private-sector involvement and inadequate project information. Similarly, the Botswana Sustainable Financing Strategy (Gregory et al. 2021) acknowledges challenges in attracting private investment due to policy and institutional gaps.

Regarding SDG 9, the study's findings that financial return is a primary driver for infrastructure investment and that respondents are willing to contribute more resources towards SDG 9 financing are consistent with global trends. The UN emphasises the importance of resilient infrastructure and sustainable industrialisation in achieving SDG 9. However, the moderate engagement levels reported in this study suggest that while there is willingness, actual participation may be hindered by the aforementioned barriers (Government of Botswana 2023).

The study's exploration of pooled financing mechanisms reveals moderate support among respondents for such models. This is in line with global perspectives that advocate for pooling resources to mitigate risks and enhance investment attractiveness. However, the study also indicates that investor composition and concessional finance are not significant factors influencing investment decisions, highlighting the need for tailored approaches in Botswana's context. Furthermore, the positive perception of an independent fiduciary in pooled funds aligns with literature emphasising the role of governance structures in enhancing investor confidence. Studies by United Nations (2015b) underscore the importance of fiduciary trust and accountability in fostering investment. Similarly, the study's findings that long-term project plans and multi-stakeholder governance structures positively influence institutional investor buy-in are supported by existing literature. The United Nations (2015b) highlights that a lack of well-defined investment pipelines is a key barrier to infrastructure financing, while (Gasser et al. 2015) emphasise the effectiveness of multi-stakeholder governance in synthesising diverse interests into collaborative frameworks.

In conclusion, while the study's findings align with global perspectives on infrastructure financing, they also underscore the unique challenges and opportunities within Botswana's context. Addressing the identified barriers and leveraging the positive perceptions towards pooled funding mechanisms, independent fiduciaries, long-term planning, and multi-stakeholder governance can enhance the attractiveness of infrastructure investments and contribute to achieving SDG 9 in Botswana.

6 Conclusion, Recommendations, Limitations, and Future Research

The study's findings largely align with previous research while highlighting context-specific differences in Botswana. Respondents identified barriers to private-sector investment, such as project complexity, lack of viable plans, and regulatory challenges, consistent with (Gregory et al. 2021; McHugh 2021), which note policy gaps and limited private-sector engagement. Similarly, the positive perceptions of long-term project pipelines, independent fiduciaries, and multi-stakeholder governance reflect global evidence that well-structured governance and clear investment plans encourage institutional investor participation (Gasser et al. 2015). However, unlike some international studies, concessional finance and multiplicity of investors were not statistically significant in this context, possibly due to the small size of Botswana's investment market and limited familiarity with pooled funding mechanisms. The study also shows moderate willingness to contribute towards SDG 9 financing, suggesting that, while interest exists, structural and institutional barriers continue to constrain actual investment levels.

Based on the study's findings, several actionable recommendations can be drawn to enhance institutional investor buy-in for infrastructure projects in Botswana, directly addressing the study's research questions and aligning with SDG 9 targets.

First, the strong positive correlations for independent fund fiduciary, clearly defined long-term project plans, and multi-stakeholder governance structure suggest that these features are critical for mobilising institutional investment. It is therefore recommended that pooled funds adopt independent fiduciaries or fund administrators to enhance transparency and accountability. Such independent oversight can build trust, ensure due diligence, and signal professional fund management to potential investors.

Second, the significance of clearly defined long-term project pipelines highlights the need for strategic planning. The government and fund managers should develop comprehensive infrastructure pipelines, detailing project timelines, expected returns, risk mitigation measures, and alignment with SDG 9 objectives. This predictability allows institutional investors to match projects with their long-term investment horizons, enhancing engagement. Third, multi-stakeholder governance emerged as a strong predictor of buy-in. Pooled funds should establish governance structures that integrate public sector agencies, development banks, donors, and private investors. This facilitates transparency, mitigates governance risks, and ensures diverse perspectives in decision-making, fostering credibility and broader participation. Although multiplicity of investors and concessional finance were not statistically significant in this context, strategies can still consider these elements. Targeted awareness campaigns could highlight the benefits of co-investment and blended finance, addressing knowledge gaps and encouraging institutional investors to participate alongside other contributors. Finally, the study underscores the value of a census approach in gathering insights from all 114 institutional investors in Botswana, ensuring comprehensive representation and actionable recommendations. Collectively, these strategies, strengthening fiduciary oversight, developing long-term pipelines, implementing inclusive governance, and promoting blended finance awareness, provide a practical roadmap for leveraging pooled funds to mobilise domestic institutional capital and accelerate progress towards SDG 9.

Conclusively, the study has several limitations. First, the research was confined to the population of 114 institutional investors in Botswana, which, although comprehensively covered through a census approach, limits the generalisability of findings to other countries or larger markets. Second, the study relied on self-reported perceptions of investors, which may be influenced by social desirability bias or subjective interpretations of pooled fund features. Third, the focus was primarily on investors' willingness rather than actual investment behaviour, which may differ when real financial commitments are required. Future research could address these limitations by expanding the scope to include comparative studies across multiple emerging markets. Investigating alternative financing mechanisms beyond traditional PPPs, such as blended finance or infrastructure debt funds, would provide deeper insights into incentivising private-sector participation. Additionally, research could examine regulatory frameworks, capacity gaps, and practical challenges in structuring complex infrastructure deals in Botswana. These studies would help refine strategies to mobilise institutional investment effectively towards SDG 9 and sustainable infrastructure development.

References

Bartz-Zuccala, W., Taskin, Ö., Hos, T., Sangaré, C., Schwarz, R., Horrocks, P.: Scaling up blended Finance in developing countries. OECD (2020). https://doi.org/10.1787/2fb14da0-en

Frischmann, B.: An economic theory of infrastructure and commons management. Minn. l. reV. **89**, 917–1030 (2005)

Gasser, U., Budish, R., West, S.M.: Multistakeholder as governance groups: observations from case studies (Berkman Center Research Publication No. 2015-1). The Berkman Center for Internet & Society at Harvard University (2015). https://dx.doi.org/10.2139/ssrn.2549270

Government of Botswana: Botswana sustainable financing strategy (BSFS) 2023–2030. Integrated National Financing Frameworks (2023, May 28). https://inff.org/assets/resource/botswana-sustainable-financing-strategy-bsfs-2023.pdf

Gregory, N., Karlin, A., Lauridsen, M.L., Mutambatsere, E., Schellekens, P., Sierra-Escalante, K., Spicer, M.: Using blended concessional finance to invest in challenging markets: Economic considerations, transparency, governance, and lessons of experience. International Finance Corporation (2021). https://www.ifc.org/content/dam/ifc/doc/mgrt/ifc-blendedfinance-fin-092021.pdf

ICVA: Pooled funding at a Crossroads: a comprehensive review and analysis (2023, May). https://www.icvanetwork.org/uploads/2023/06/Pooled-Funding-at-a-Crossroads-A-Comprehensive-Review-and-Analysis.pdf

ICVA: Pooled funding models: Governance systems: Briefing Note (2024, February). https://www.icvanetwork.org/uploads/2023/12/Pooled-Funding-Models-Governance-Systems-Briefing-Note-.pdf

Infrastructure Consortium for Africa Secretariat: Infrastructure financing trends in Africa 2019–2020. Africa Development Bank Group (2022) https://www.afdb.org/sites/default/files/documents/publications/04112022ift_africa_report_2019-2020-2_english.pdf

Lawal, C.I., Friday, S.C., Ayodeji, D.C., Sobowale, A.: Strategic framework for transparent, data-driven financial decision-making in achieving sustainable national development goals. International Journal of Advanced Multidisciplinary Research and Studies. **4**(6), 1655–1678 (2024). https://doi.org/10.62225/2583049X.2024.4.6.4098

Masuga, T.: Exploring private-sector challenges in financing PPPs in Botswana (Master's Dissertation). Graduate School of Business, University of Cape Town (2023) http://hdl.handle.net/11427/40551

McHugh, C.A.: Mobilising private funding of development finance. J. Dev. Stud. **57**(12), 1979–2001 (2021). https://doi.org/10.1080/00220388.2021.1945042

Motshidisi, O.A.: NBFI Sector performance overview: 2022 Annual Report Highlights. Non-Bank Financial Institutions Regulatory Authority (2022, November 16) https://www.nbfira.org.bw/sites/default/files/CEOs%20Speech-%202022%20Annual-Report%20Highlights%20for%20Media%20Briefing%20%2816.11.2022%29.pdf

Office of the Auditor General, Republic of Botswana: Performance audit on preparedness for implementation of sustainable development goals (2019). https://www.gov.bw/sites/default/files/2020-03/REPORT%20ON%20SDG_0.pdf

Ozili, P.K.: Theories of sustainable finance. Manag. Glob. Transit. **21**(1), 5–22 (2023). https://doi.org/10.26493/1854-6935.21.5-22

Parmaksiz, K., Bal, R., van de Bovenkamp, H., Kok, M.O.: From promise to practice: a guide to developing pooled procurement mechanisms for medicines and vaccines. J. Pharm. Policy Pract. **16**(1), 73 (2023). https://doi.org/10.1186/s40545-023-00574-9

Public Investment Corporation SOC Limited: Integrated annual report 2023 (Reporting Period 1 April 2022 to 31 March 2023). Public Investment Corporation SOC Ltd, Pretoria (2023)

Purvis, B., Mao, Y., Robinson, D.: Three pillars of sustainability: in search of conceptual origins. Sustain. Sci. **14**, 681–695 (2019). https://doi.org/10.1007/s11625-018-0627-5

Schmitt, C.: Pooled Funding Models: Governance systems: a comparative study. ICVA (2023, November) https://www.icvanetwork.org/uploads/2023/12/ICVA-Pooled-Funding-Models-Governance-Systems.pdf

UNEP, United Nations Environment Programme Finance Initiative: Fiduciary duty in the 21st century: Final Report (2019). https://www.unepfi.org/wordpress/wp-content/uploads/2019/10/Fiduciary-duty-21st-century-final-report.pdf

United Nations: Addis Ababa Action Agenda of the Third International Conference on Financing for Development (2015a). https://sustainabledevelopment.un.org/content/documents/2051AAAA_Outcome.pdf

United Nations Conference on Trade and Development. Leveraging the potential of sovereign investors for infrastructure investment in Africa (2025). https://unctad.org/system/files/official-document/diae2024d2_en.pdf

United Nations: UN general assembly transforming our world: The 2030 agenda for sustainable development (a/RES/70/1) (2015b, October 21). https://docs.un.org/en/A/RES/70/1

World Commission on Environment and Development: Our common future. Oxford University Press (1987)

Rethinking Mentorship in Teacher Training: Evaluating the Impact of School Exposure Sessions on Student Teachers in South Africa

Shaun Peens

Abstract Mentorship plays a critical role in sustainable educator development, particularly in South Africa's education system, which faces socio-economic disparities in a rapidly evolving learning environment. Unlike structured mentorship models in successful international contexts, South African student teachers often lack formalised support during their teaching practice, potentially affecting their preparedness for the profession. This study examined the effectiveness of mentorship experiences during school exposure sessions, assessing student teachers' perceptions of mentorship quality and their overall readiness for professional teaching. Key objectives included evaluating the adequacy of mentorship and analysing student teachers' confidence and preparedness for teaching roles upon graduation. Using a quantitative research approach, a structured online questionnaire was administered to student teachers who had school-based teaching exposure. Findings indicated that additional mentorship in isolation may not significantly enhance current experiences, given the limited feedback and selective support from mentor teachers. Notably, few mentors actively encouraged student teachers to adopt their own teaching methods; feedback during teaching periods remained sparse, and mentorship support was inconsistent. Furthermore, sporadic responses indicated a preference for experienced mentors to guide them, highlighting a critical gap in the current mentorship framework. It is recommended that universities should collaborate with schools to create structured, well-supported mentorship programmes, like internships, to equip student teachers better for the realities of the profession.

Keywords Mentorship · Internship · Teacher development · Experienced educators · Teaching practice

---

S. Peens (✉)
Central University of Technology, Bloemfontein, South Africa
e-mail: speens@cut.ac.za

E. H. Redda (ed.), *Proceedings of the Social Sciences International Research Conference, 2025*, https://doi.org/10.1007/978-981-95-9489-4_10

1 Introduction

The study aimed to assess whether student teachers' mentorship experiences during their teaching practice periods adequately prepare them for the teaching profession upon completion of their university studies. The research explored their perceptions of mentorship quality, the practical applicability of their training, and overall readiness for professional teaching.

The objectives comprised:

1. the evaluation of the efficacy of mentorship from a student's perspective,
2. an analysis of g student teachers' confidence and preparedness to take up teaching roles after obtaining a B.Ed. degree.

Subsequent to the identification of key challenges and gaps in the school exposure experience to explore possible themes for future studies on teacher training programmes, the researcher compared the findings with models used in other countries. The research aimed to propose enhancements to teacher training programmes based on identified challenges and gaps.

The research underscores the vital role of mentorship in sustainable educator development within the context of South Africa's education system, which grapples with socio-economic disparities and an evolving learning environment. Although the researcher asserted an absence of structured mentorship programmes as a barrier to empowering educators to navigate these challenges in comparison with successful international models, more than 50% of respondents deemed the current model as sufficient. The study further explored whether a robust mentorship model modelled on international models and characterised by dedicated mentors would provide the necessary individualised support and professional development opportunities for educators as opposed to occasional guidance and occasions to lead classrooms independently during teaching practice periods.

Core principles of effective mentorship were examined, including the significance of strong mentor-mentee relationships, clear communication, and ongoing feedback. The research envisioned that well-rounded mentorship may enhance teaching skills, improve classroom management, foster leadership potential, and contribute to educators' well-being. Additionally, it emphasised the need for culturally responsive teaching practices within mentorship programmes to create inclusive learning environments.

2 Literature Review

The research emphasises the critical role of mentorship in fostering sustainable educator development in South Africa; given the challenges within the South African education system. It was therefore challenging to find a suitable theoretical framework as a combination of educational and human resource development had to be embedded. To combine both these spheres, several possible theories were considered. Social cognitive theory addressed observational learning, self-efficacy, and reciprocal determinism through mentorship, which allowed student teachers to observe experienced mentors, to learn from their practices, and to develop self-efficacy through guided experiences and feedback (Hudson 2013). The quality of observed learning and feedback mechanisms is critical; hence situated learning theory in turn posits that learning is inherently social and contextual (Clemente et al. 2023; Saeidnia et al. 2024). Teaching practice is a form of situated learning where student teachers learn by participating in the daily activities of the school, guided by mentors (Orland-Barak 2002; Orland-Barak and Wang 2020). Effective mentorship provides opportunities for legitimate peripheral participation, steadily increasing the student teacher's involvement and responsibility within a structured framework.

The views above clearly indicated that transformative learning theory which emphasises that learning involves critical reflection on one's experiences, assumptions, and beliefs, leading to a change in perspective would be the most appropriate theoretical framework where quantitative data could be absorbed into future studies, Mentorship in teaching practice offers the opportunity for student teachers to reflect on their teaching experiences critically, challenge their pre-conceived notions about teaching, and transform their understanding of what it means to be a teacher (Alexander et al. 2014).

2.1 *International Comparisons*

Since a silo approach to evaluate whether a University of Technology in South Africa is on par with its teaching practice programme would not be a feasible way to discover weaknesses and measure strengths; an assessment of international approaches towards internships for education students will be discussed below.

In the Philippines, a study assessed the readiness of education students for teaching internships, leading to the development of a student teaching internship enhancement programme. The study identified areas in which students needed more support, such as communication skills and classroom management (Guerrero 2024). Although classroom management forms part of undergraduate studies, the level and types of school exposure vary significantly, and no single theoretical model can prepare students effectively for diverse situations and to cover all possible scenarios in a single undergraduate module could not be possible either. It is therefore worthwhile for a formalised programme as developed in the Philippines, placing students in real-life situations with a dedicated mentor teacher.

Australia seems to offer prolonged opportunities to implement additional programmes on what students are learning through the intensives aiming to evaluate this knowledge with further research for additional assignments (Master of Teaching (Secondary) Internship 2023). A possible challenge with this approach could be that more theoretical content forms part of additional studies, which increases knowledge, but might not be suitable for practical implementation in classroom control and teaching effectiveness. Students could learn 'how to do on paper' but lack the required skills for effective implementation. Although the researcher agrees that knowledge is power, and indispensable in most spheres of life, education is based on a combination of proper, relevant knowledge and an even deeper understanding of how to deliver the content to a diverse group of learners.

Some universities in the United Kingdom offer Master of Teaching Internship Programmes where students spend the full duration of the course in schools, implementing what they are learning over an extended period. These internships provide extensive guidance and support from education experts (Master of Teaching (Secondary) Internship 2023). The unique challenge in the United Kingdom is that most students have limited teaching experience upon graduation (Guerrero 2024), and that the practical component requires extensive external support, which this master's degree aims to provide. Interestingly no formal period is allocated but indicated for the duration of study and upon approval of education experts.

However, the researcher could perceive similarities in his ideas and the approach currently used in the United States as many teacher education programmes incorporate clinical experiences, including internships, to allow candidates to enact practices learned in the preparation programme within an educational context (Putman and Polly 2021). Some programmes offer year-long teacher internship pilot programmes, providing tuition support and placing interns with mentor teachers in local partner schools. These internships aim to redesign the undergraduate internship experience (Year-Long Teacher Internship Pilot Programme 2022). The programme the researcher regards as the most valuable offers a two-year internship programme for individuals already working as teachers in public schools, providing a pathway to a teaching credential (Internship Program 2025). This approach is not new to South Africa either, as this approach exists within the medical field in which medical students are placed for a two-year internship and another year community service for certification to practice, as one example.

In a general perspective, internships offer a unique opportunity for students to combine theory with practice and acquire new skills. A successful internship can provide students with an opportunity for future employment by shaping their professional and personal skills (Shtëmbari and Elgün 2023).

2.2 *Effectiveness of South Africa's Current Mentorship Model*

In South Africa, the National Department of Education has implemented an Advanced Certificate in Education in School Leadership, initially targeting practising school principals, combining a content and process-rich programme involving work-based learning and employing mentoring as a development tool (Naicker et al. 2014). The researcher, however, argues that this additional support should not be limited to leadership positions, like principals, heads of departments and seasoned teachers in the system, but to all prospective teachers.

School-based learnership programmes are currently embedded in the curriculum for B. Ed courses, but function more like a tick box system instead of a professional skills development programme, without effectively contributing to classroom competence from recent graduate teachers (Van Schalkwyk 2014). This form of support could assist in-service training programmes, workshops, seminars, and mentoring to provide ongoing professional development for teachers (Uribe-Banda et al. 2023), ensuring a suitable level of skills and professionalism. These types of assistance should be provided prior to a new graduate having to take sole responsibility of any class. This approach is vital to improve teachers' professional knowledge and competencies (Assaad et al. 2019). Experiential learning in initial teacher education requires mentoring schools to provide a learning environment to equip student teachers with both subject content knowledge and pedagogical content knowledge (Thaba and Kanjere 2014). This is affirmed by research, indicating that mentoring plays a significant role in the training and development of teacher trainees, influencing their experiences and shaping them into effective teachers (Kumi-Manu et al. 2023; Ojo et al. 2017). Effective entering relationships between mentor teachers and student teachers are extremely capable of transforming both teachers and students involved, with mentoring integrated into the general approach to teaching and professionalism (Waty 2020). Mentoring has the potential to improve the quality of teaching and learning, enhance teachers' efficacy, develop their leadership skills and build capacity within schools (Sibanda and Amin 2021).

From the research it seems clear that the implementation of a mentoring programme is a definite need and should be envisaged for the improvement of our education sector. Challenges often arise when higher learning institutions do not address the immediate needs of student teachers adequately, although tertiary institutions should not be the only space where these challenges are addressed (Msila 2013). This paper indicates by means of student feedback how stakeholders may develop suitable bridging models between higher learning institutions and public education.

3 Methodology

3.1 Research Design

The research design employed a quantitative approach, using a questionnaire to gather data on the experiences of education students during their teaching practice. The questionnaire consisted of structured questions, designed to evaluate the support, guidance, and feedback received from mentors, as well as the students' self-efficacy and preparedness for the teaching profession. This method allows for the collection of standardised data from a large sample ($n = 86$) of respondents, facilitating statistical analysis and the identification of trends and patterns in the mentorship experiences of education students (Mokoele and Weyers 2021). The use of a questionnaire is appropriate for this study as it enables the researcher to collect data on a range of variables relevant to the research questions. One of the aims was to evaluate whether the support offered by mentor teachers increases when the same students return to school the following year.

3.2 Population and Sampling Method

The target population for this study comprised of education students enrolled in the Faculty of Humanities at a University of Technology in South Africa. A convenience sampling method was employed to select respondents for the study, as most of the students form part of the student group who the researcher could motivate personally on the reasons and purpose of the study. Students were encouraged to take part anonymously. In addition, this sampling technique was chosen based on its practicality and accessibility, as the researcher has direct access to the student population through their academic courses. Participation in the questionnaire was voluntary but limited to students who have completed at least 3 weeks of teaching practice to date, ensuring that all respondents have had practical experience in a school environment.

3.3 Research Instrument

The research instrument for this study, a questionnaire, is a widely used tool for collecting data, particularly in social science research (Taherdoost 2016). The questionnaire was designed to evaluate the support, guidance, and feedback received from mentor teachers, as well as the students' self-efficacy and preparedness for the teaching profession. It included structured questions, including both closed-ended and open-ended questions. Closed-ended questions will allow for the collection of quantitative data that can be statistically analysed, while open-ended questions will provide an opportunity for respondents to provide more detailed responses to their exposure during their teaching practice at schools (Labana et al. 2024).

While a pilot study is the recommended approach for assessing validity and reliability (Srinivasan and Lohith 2017), especially for questionnaires (Ranganathan et al. 2024), the following factors were considered to strengthen the responses. The study guaranteed respondents that all responses would be kept in the strictest confidence, assuring that all answers were free from bias (Apriyani et al. 2020). Only relevant aspects of the topic were included, strengthening content validity (Sullivan 2011). Questions were purposefully structured to be clear, simple, and formulated to be relevant, concise, and direct (Viruel et al. 2025). Despite this, the limitations of the questionnaire should also be acknowledged, including the potential for response bias and the possibility that respondents may interpret questions differently (Chen 2019).

3.4 *Data Analysis*

The study employed quantitative data analysis techniques to examine the relationships between mentoring-related variables and student teachers' preparedness for real-world teaching challenges. Data analysis points included descriptive statistics and correlation- and regression analysis.

The data were analysed using both correlational analysis and linear regression modelling, with the aim of understanding the strength and direction of associations between key variables, as well as identifying significant predictors of preparedness. All analyses were conducted using SPSS and an alpha level of 0.05 was used to determine statistical significance. Please note that the first paragraph of a section or subsection is not indented.

4 Correlation Analysis

A Pearson correlation matrix was computed to assess the bivariate relationships among the key variables: preparedness for teaching in diverse classrooms (Prep-Diverse), preparedness for real-world teaching challenges (Prep-Real), mentor support, feedback frequency, and mentor approach. Pearson's r values range from -1 to $+1$, indicating the strength and direction of linear relationships between variable pairs. A positive correlation indicates that as one variable increases, so does the other, while a negative correlation indicates an inverse relationship.

The correlation matrix revealed moderate to strong positive associations between preparedness variables ($r = 0.65$), as well as modest positive correlations between mentor support, mentor approach, and preparedness measures. Feedback frequency exhibited weaker correlations with preparedness variables (Table 1).

This analysis was useful for:

- Exploring potential multicollinearity among predictors.
- Informing the selection of variables for further regression modelling.

5 Linear Regression Analysis

To further investigate the predictive effects of mentoring-related factors on student teachers' preparedness, a multiple linear regression model was constructed. The dependent variable was preparedness for real-world teaching challenges (Prep-Real), and the independent variables included mentor support, feedback frequency, and mentor approach.

The linear regression analysis provided estimates of regression coefficients, standard errors, t-values, and associated p-values for each predictor. The model helped to determine the extent to which each mentoring variable significantly contributed to perceived preparedness, controlling for the influence of the others.

The results indicated that mentor support was a statistically significant positive predictor of preparedness ($\beta = 0.532$, $p = 0.017$), suggesting that higher levels of mentor support are associated with increased feelings of readiness among student teachers. In contrast, feedback frequency and mentor approach were not statistically significant predictors in this model, indicating that their influence may be limited or mediated by other factors.

6 Results and Discussion

The questionnaire included 19 questions 18 of which were closed- ended and one open-ended, allowing respondents to elaborate on their experiences, opinions, and suggestions. The open-ended question allowed respondents to provide more detail on the aspects of mentorship they found most beneficial.

6.1 Descriptive Statistics

The section uses descriptive statistics to summarise the data collected from the questionnaires. Demographic information such as gender, age, and academic levels was included to provide context for understanding the characteristics of the respondents. Interestingly, the data indicated that 73% of the respondents have dropped a subject at some time during their student career and are not first-time entry and exit students; with a 64%/36% ratio female to male and 92%/8% split between students enrolled for Economic and Management Sciences and Languages respectively.

In examining the teaching practice exposure, it was noted that 7% of the respondents have had three weeks (First Year) exposure, 15% had 9 weeks (Second Year), 48% had 15 weeks (Third Year) and the remaining 30% had 21 weeks (fourth year). Since each year group is expected to be exposed to different facets of the teaching process, this data will be applied to most of the closed-ended questions to evaluate the impact for each year group.

Overall, the support received from mentor teachers was indicated as either excellent (62%) or good (31%); however, those who indicated average and poor were mostly third- and fourth-year students who would have needed it most during their teaching practice time. It is worth noting that these students had been exposed to at least fifteen weeks of teaching practice at a school, but indicated a lack of mentorship during teaching practice, which has potentially been an issue.

Comparing the open-ended question to the closed- ended question data, there seems to be a recurring theme that constructive feedback and guidance regarding their teaching improved student confidence. The majority of respondents, namely

59%, indicated that constructive feedback was provided daily, whereas 30% indicated irregular feedback and 11% rare to none.

Responses included: *Effective communication, support and encouragement and constructive feedback, The most beneficial aspects of mentorship for me were the guidance, feedback, and support I received throughout my development and constructive feedback provided by my mentor was invaluable. It helped me reflect on my lesson planning, classroom management, and teaching strategies, allowing me to improve continuously. My mentor would often observe my lessons and then offer specific, actionable suggestions, which made a significant difference in my confidence and effectiveness.*

Regarding opportunities to lead classes independently, results indicated that 22% of respondents (First and Second Years) who were attending school as observers were indeed requested to present lessons to classes on their own. In contrast, 44% of third- and fourth-year students who were obliged to lead classes indicated they were doing so less than once a week with 1 in 5 respondents indicating no guidance on lesson planning and classroom management.

In terms of feedback provided by mentor teachers from lessons observed, 31% of students got daily feedback with 58% indicating irregular feedback and 10% indicating no feedback had been received at all. Apart from one respondent, all other students were seniors who could have benefited a great deal from feedback sessions by their mentor teachers. Interestingly, 55% of the respondents who indicated no feedback had been received also indicated that no support was given on how to deal with learner behavioural challenges (Fig. 1).

An interesting outcome, albeit noted by the researcher, was on the question enquiring how teaching methods were motivated; as only 6% of mentor teachers expected their students to follow their teaching style, with 52% encouraging them to use their own preferred style. However, 42% indicated that they were requested to follow teaching guidelines in combination with their own.

The researcher is of the opinion that each student should be assigned to a single mentor to follow the holistic processes and aspects of the teaching profession. However, results indicated that 67% of students preferred having multiple mentors to shadow their approach and methods.

Fig. 1 Frequency of lesson observance by mentor with feedback

With 92% of students indicating that mentor teachers furnished them with resource materials, only 47% agreed that they had been assisted to use such effectively, with 31% admitted to some assistance and 22% claimed to receiving the materials with little to no assistance. A good sign, nevertheless, is that 76% of respondents indicated that mentor teachers included them in professional development activities like school administrative tasks, although 62% of those not included were senior students who would have benefited from the process.

The last two questions of the questionnaire were scale and rank driven, compared to 'choose best answer-type' options in previous questions. In these questions the focus was on affective aspects, avoiding factual thinking, to assess the respondents' perceived confidence and preparedness to teach and whether possible needs might exist. The first statement indicated overall confidence that students could mimic their mentor teachers' classroom management. Only 6% of respondents were neutral, with no respondents feeling they could not manage independently. The same responses were received for statements 3 and 5, evaluating the time spent with mentor teachers sufficient to deal with real-life challenges and for preparation for their teaching career respectively. In both cases only 6% of respondents disagreed, indicating they would need more time with mentor teachers. Statement two focused on preparedness for diverse class situations, 18% of respondents agreeing to be not sufficiently prepared for diverse situations. Statement four enquired whether a single mentor proved to be more beneficial, where 37% of respondents indicated a neutral reaction or? disagreeing with the statement.

The next question had to determine the level of support the respondent would suggest improving school exposure for student teachers. The support options were listed as (a) more time needed with mentor teachers, (b) preferring a more experienced mentor teacher, (c) being assigned to a single mentor teacher and (d) no additional support needed with the focus only on obtaining their degrees. For this question 40% of students indicated they would rank more time with their mentor teachers as the first option, with 36% preferring a more experienced mentor while opting for a single mentor ranked third (12%) with 7% indicating they are only here for the degree.

6.2 *Correlation Analysis*

A correlation analysis was conducted to identify relationships between distinct aspects of the mentorship experience and the student teachers' confidence and preparedness.

Interpretation

- There is a moderate positive correlation between preparedness for diverse classrooms (Prep-Diverse) and preparedness for real-world teaching (Prep-Real) ($r = 0.65$).
- Mentor Support is positively related to both preparedness measures, more strongly with Prep-Real ($r = 0.34$) than Prep-Diverse ($r = 0.29$).

Table 1 The correlation matrix for mentorship/confidence/preparedness variables

	Prep-Diverse	Prep-Real	Mentor Support	Feedback Frequency	Mentor Approach
Prep-diverse	1.00	0.65	0.29	0.23	0.35
Prep-real	0.65	1.00	0.34	0.13	0.42
Mentor support	0.29	0.34	1.00	0.32	0.40
Feedback frequency	0.23	0.13	0.32	1.00	0.10
Mentor approach	0.35	0.42	0.40	0.10	1.00

- Mentor Approach shows relatively stronger correlations with Prep-Real ($r = 0.42$) and Prep-Diverse ($r = 0.35$), suggesting its style may influence both.
- Feedback Frequency has weaker correlations across the board, especially with Prep-Real ($r = 0.13$), indicating it might not play a strong role in perceived preparedness.

6.3 *Linear Regression Analysis*

A linear model was used to measure how well three specific predictors: mentor support, feedback frequency, and mentor approach explain or predict the level of preparedness that student teachers perceive for real-world teaching challenges. These were ranked to determine the extent of support by the mentor, how often the mentor provides feedback and the style or method the mentor uses in guiding the student and how these impact student teachers' preparedness (Table 2).

Mentor Support is a significant positive predictor of student teachers' preparedness for real-world teaching challenges ($\beta = 0.532, p = 0.017$). This coefficient indicates that for each unit increase in mentor support, the preparedness for real-world challenges increases by approximately 0.53 units. The *p*-value (0.017) indicates that this relationship is statistically significant (typically, a *p*-value <0.05 is considered significant).

Feedback Frequency ($\beta = 0.007, p = 0.926$) and Mentor Approach ($\beta = -0.007$, $p = 0.991$) were not statistically significant predictors, indicating that they may have limited influence on preparedness.

Table 2 Linear model coefficients predicting preparedness for real-world teaching challenges

Predictor	Estimate	Std. Error	t-value	p-value
(intercept)	2.170	1.515	1.43	0.161
Mentor support	0.532	0.212	2.50	0.017
Feedback frequency	0.007	0.077	0.09	0.926
Mentor approach	−0.007	0.557	−0.01	0.991

6.4 Discussion of Results

From the key objectives, the researcher aimed to gain insight into the efficacy of mentorship from a student's perspective and to analyse student teachers' confidence and preparedness to take up teaching roles after obtaining a degree.

The response rate on the open-ended questions was 73% of which a quarter were single word responses with limited insight. Although five respondents mentioned no mentorship had been received at all, most respondents argued that mentorship mainly benefited them when constructive feedback and guidance were received, which improved their confidence. Summarising the responses from the open-ended questions, the feedback does not support the researcher's initial assumption that mentorship was inadequate. Findings from closed-ended questions revealed that the researcher underestimated the exposure and experiences student teachers gain during their tenure at school during their teaching practice periods; as the findings clearly indicated that students feel comfortable and ready to teach; and therefore, objective two of this study was not supported. An aspect which was satisfactory proven, is that senior students responded with more certainty about their preparedness than First- and Second-Year students, suggesting that mentorship benefits do increase per year group.

With regard to negative aspects, only one student from the fourth-year group was in line with their academic progress, 13 have failed at least one year, and 12 have failed more than once during their time at university. From the feedback provided, 44% of senior students conducted classes independently, but only 1 in 5 got feedback, suggesting the mentor teacher shifted responsibilities without student development in mind. Looking at the open-ended question, the researcher felt demotivated, with 73% of the respondents giving single word to one liner responses, arguably only to finish the questionnaire without a real desire to contribute to the outcomes of the study.

In contrast to what the researcher believes that a single mentor over a longer period could prove more valuable, the respondents regarded more than one mentor teacher as more beneficial. The reasons for this response are not clear and could prove to be valuable in a future study. The researcher could argue that since students must be assessed during their teaching practice period on both their majors, it might influence their need for more than one mentor teacher.

7 Conclusion

On the positive side, findings indicate that no mentor teacher really encouraged student teachers to follow their teaching methods, but rather a good combination or the student's own approach. The majority of students indicated that more time at school and having an experienced mentor teacher guiding them would prove valuable.

The study ultimately appreciates the investment in well-designed mentorship programmes as a strategy for fostering sustainable educator development, drawing lessons from successful international practices to empower South African educators, as has been done in the Philippines and the United States. Further research to explore mentorship complexities in South Africa would benefit the entire education sector.

Future research could follow a qualitative or mixed-method approach to delve directly into the reasons why most respondents prefer a single mentor, why they feel more time at schools could be beneficial, and the reasons why an experienced mentor teacher may improve their school exposure experiences.

References

Alexander, G., van Wyk, M.M., Moreeng, B.: Constructing student-teacher identities via a mentorship programme initiative: a case for school-based learning. Mediterr. J. Soc. Sci. **5**(9), 405 https://www.mcser.org/journal/index.php/mjss/article/download/2652/2620(2014).

Apriyani, D., Krisnawati, M., Nurmasitah, S., Sudana, I.M., Fajri, S.N.: Standard competency gaps of vocational teachers. Int. J. Innov. Learn. **28**(1), 40 (2020). https://doi.org/10.1504/ijil.2020.108474

Assaad, R., Hendy, R., Salehi-Isfahani, D.: Inequality of opportunity in educational attainment in the Middle East and North Africa: evidence from household surveys. Int. J. Educ. Dev. **66**, 24 (2019). https://doi.org/10.1016/j.ijedudev.2019.02.001

Chen, W.: A case study on developing students' leadership skills via team work activities. Open J. Soc. Sci. **7**(10), 414 (2019). https://www.scirp.org/journal/paperinformation?paperid=96748

Clemente, F.M., Ramírez-Campillo, R., Beato, M., Moran, J., Kawczyński, A., Makar, P., Sarmento, H., Afonso, J.: Arbitrary absolute vs. individualized running speed thresholds in team sports: a scoping review with evidence gap map. Biology of sport **40**(3), 919 (2023). Termedia Publ. House. https://doi.org/10.5114/biolsport.2023.122480

Guerrero, R.: R.: readiness of education students in teaching internship basis for student teaching internship enhancement program (STIEP). Int. J. Soc. Sci. Hum. Res. **7**(5) (2024). https://doi.org/10.47191/ijsshr/v7-i05-02

Hudson, P.: What makes school leaders inspirational and how does this relate to mentoring? Open J. Lead.Ship **2**(4), 87 (2013). https://doi.org/10.4236/ojl.2013.24014

Internship Program: California state university. Los Angel. https://www.calstatela.edu/coe/edsp/multiple-single-subject-and-education-specialist-internship-program(2025).

Kumi-Manu, R.N., Acquah, S., Addo-Wuver, F.: Preservice teachers' professional experiences during internship at the University of Education, Winneba, Ghana. East Afr. J. Educ. Soc. Sci. **4**(3), 198 (2023). https://doi.org/10.46606/eajess2023v04i03.0291

Labana, R.V., Dolar, M.V., Ferrer, E.J.J., Rabago, J.R.C.: The awareness-to-agreement-to-adherence model of food safety compliance: the case of street food consumers in Manila City, Philippines. J. Public Health Emerg. **8**, 13 (2024). https://doi.org/10.21037/jphe-23-182

Master of Teaching (Secondary) Internship. University of Melbourne. https://study.unimelb.edu.au/find/courses/graduate/master-of-teaching-secondary-internship/the-experience/(2023).

Mokoele, B., Weyers, M.: Measuring the need for educational supervision amongst child protection social workers: an exploration. Soc. Work./Maatskaplike Werk **57**(4), 407 (2021). https://doi.org/10.15270/57-4-965

Msila, V.: Teacher-learners' search for relevance: lessons from a principals' leadership/management qualification in South Africa. Int. J. Educ. Sci. **5**(4), 443 (2013). https://doi.org/10.1080/09751122.2013.11890106

Naicker, I., Chikoko, V., Mthiyane, S.E.: Does mentorship add value to in-service leadership development for school principals? Evid. S. Afr. **17**(2), 421 (2014). https://doi.org/10.1080/09720073.2014.11891451

Ojo, O.A., Adu, E.O., Adu, E.O.: School experience: institutions' expectations from pre-service teachers. Anthropol. **29**(1), 19 (2017). https://doi.org/10.1080/09720073.2017.1335825

Orland-Barak, L.: What's in a case?: what mentors' cases reveal about the practice of mentoring. J. Curric. Stud. **34**(4), 451 (2002). https://doi.org/10.1080/00220270110091140

Orland-Barak, L., Wang, J.: Teacher mentoring in service of preservice teachers' learning to teach: conceptual bases, characteristics, and challenges for teacher education reform. J. Teach. Educ. **72**(1), 86 (2020). https://doi.org/10.1177/0022487119894230

Putman, S.M., Polly, D.: Examining the development and implementation of an embedded, multi-semester internship: preliminary perceptions of teacher education candidates, clinical educators, and university faculty. Peabody J. Educ. **96**(1), 99 (2021). https://doi.org/10.1080/0161956x.2020.1864250

Ranganathan, P., Caduff, C., Frampton, C.: Designing and validating a research questionnaire–part 2. Perspect. Clin. Res. **15**(1), 42 (2024). https://journals.lww.com/picp/fulltext/2024/15010/designing_and_validating_a_research_questionnaire.8.aspx

Saeidnia, H.R., Fotami, S.G.H., Lund, B., Ghiasi, N.: Ethical considerations in artificial intelligence interventions for mental health and Well-being: ensuring responsible implementation and impact. Soc. Sci. **13**(7), 381 (2024). https://doi.org/10.3390/socsci13070381

Shtëmbari, E., Elgün, R.F.: Generation Z 'life skills' acquired and enhanced through internships before and during COVID-19 pandemic. Adm. Sci. **13**(2), 38 (2023). https://doi.org/10.3390/admsci13020038

Sibanda, D., Amin, N.: The link between a mentorship programme for mathematics, science, and technology in-service teachers and professional development. SAGE Open. **11**(1) (2021). https://doi.org/10.1177/2158244020988732

Srinivasan, R., Lohith, C.P.: Pilot study—assessment of validity and reliability. In: India Studies in Business and Economics, p. 43. Springer Nature (2017). https://doi.org/10.1007/978-981-10-3590-6_6

Sullivan, G.M.: A primer on the validity of assessment instruments. J. Grad. Med. Educ. **3**(2), 119 (2011). https://doi.org/10.4300/jgme-d-11-00075.1

Taherdoost, H.: Validity and reliability of the research instrument; how to test the validation of a questionnaire/survey in a research. SSRN Electron. J. (2016). https://doi.org/10.2139/ssrn.3205040

Thaba, K., Kanjere, M.: Experiential learning in initial teacher education: do schools have mentoring capacity? Mediterr. J. Soc. Sci. (2014). https://doi.org/10.5901/mjss.2014.v5n27p537

Uribe-Banda, C., Wood, E., Gottardo, A., Biddle, J., Ghaa, C., Iminza, R., Wade, A., Korir, E.: Assessing blended and online-only delivery formats for teacher professional development in Kenya. Cogent. Education. **10**(1) (2023). https://doi.org/10.1080/2331186x.2023.2191414

Van Schalkwyk, A.: A learnership programme for student teachers: a partnership between a school and an open distance learning institution. Mediterr. J. Soc. Sci. (2014). https://doi.org/10.5901/mjss.2014.v5n20p1029

Viruel, S.R., Rivas, E.S., Ruiz-Palmero, J.: The role of artificial intelligence in project-based learning: teacher perceptions and pedagogical implications. Education Sciences. **15**(2), 150 (2025). https://doi.org/10.3390/educsci15020150

Waty, M.: The impact of individual student mentoring for teacher professionalism in English preservice teacher education. SAGA journal of English language teaching and applied. Linguistics. **1**(2) (2020). https://doi.org/10.21460/saga.2020.12.47

Year-Long Teacher Internship Pilot Program, (2022), University of Maryland. https://education.umd.edu/research/centers/maryland-pds-2025-project/components/year-long-teacher-internship-pilot-program-C2%A0

How Continuous Communication Channels and Feedback Loops Drive Emergent Strategising

Grace Magama and Peet Venter

Abstract Private higher education institutions (PHEIs) face distinctive challenges that demand adaptive, collaborative approaches to strategy. However, limited empirical research has examined how communication channels and feedback loops drive the emergence of strategies in this sector. The purpose of this study was to explore how communication channels and feedback loops in a PHEI in South Africa drive emergent strategising to navigate the rapidly changing environment effectively, and to provide a comprehensive understanding of the theory and practice of communication and feedback loops. Grounded in Bourdieu's social capital theory and the strategy-as-practice (SAP) framework, the study employed an intrinsic case study research design, and purposively sampled eighteen (18) managers at multiple levels of management. Data were collected from semi-structured interviews and written narratives, and were thematically analysed, revealing how continuous communication channels and feedback loops drive emergent strategising through six practices: continuous communication, feedback loops, social capital, technology–culture integration, facilitative leadership, and formal–informal synergy. These practices offer insights into how continuous communication channels and feedback loops drive emergent strategic decision-making that evolves from daily practices, social relationships, and organisational culture. The study recommends institutionalising cross-functional communication through regular, open forums and fostering inclusive cultures that value iterative feedback processes. Such mechanisms enable smooth information flow and timely decision-making, which are essential for maintaining agility in dynamic academic environments. For managers in higher education and beyond, the study emphasises the importance of a culture that embeds feedback and communication into daily practice to drive emergent strategising.

Keywords Continuous communication channels · Formal–informal channels · Higher education institutions · Emergent strategising · Feedback loops · Social capital

G. Magama (✉) · P. Venter
School of Business Leadership, University of South Africa (UNISA), Pretoria, South Africa
e-mail: 42073790@mylife.unisa.ac.za

E. H. Redda (ed.), *Proceedings of the Social Sciences International Research Conference, 2025*, https://doi.org/10.1007/978-981-95-9489-4_11

1 Introduction

In South Africa's evolving educational landscape, higher educational institutions (HEIS) face numerous pressures, including the need to integrate technology into teaching practices (Gobind 2024; Ohei et al. 2023) respond to the growing diversity in student demographics (Marcum 2022), and navigate shifting governmental policies (Wildschut 2020). These challenges require adaptive strategies that allow educational institutions to remain relevant and competitive in a dynamic environment (Kovalenko et al. 2021; Rusok et al. 2021). One of the critical approaches for achieving adaptability is through cross-functional collaboration across academic and administrative departments (Bidandi and Anthony 2022; de Lange 2012). In South African HEIs, where traditional hierarchies and departmental silos often hinder communication (Czerniewicz 2022), fostering collaboration has become vital for maintaining organisational agility (Žitkienė and Deksnys 2018). Hence, building on Bourdieu's theory of social capital, the study explores how educational institutions can foster continuous communication channels and feedback loops in private higher education institutions (PHEIs), driving emergent strategising (Bourdieu 1986).

Educational institutions can cultivate adaptive strategies by dismantling departmental silos (Mizuta 2022) and fostering inclusive communication across various functional areas (Chirwa and Boikanyo 2022), enabling them to quickly address internal pressures and external changes, including technological advancements and evolving student needs (Hwang and Fu 2020; Nasir and Hasan 2024; Riveira and Fadhila 2024). Feedback loops, an essential element of cross-functional collaboration, offer real-time insights that assist institutions in adjusting their strategies and ensuring alignment with long-term objectives (Brown and Duguid 2001).

Research shows that higher educational managers who prioritise collaborative decision-making and implement continuous communication channels are better equipped to handle strategic adjustments, particularly in response to technological advancements and evolving student needs (Loveday 2009). Further, cross-functional collaboration enables departments to work together, share knowledge, and drive adaptive strategies (Shams et al. 2019). In emergent strategising, collaboration and communication become even more significant. Emergent strategies are shaped by real-time adaptations and continuous feedback loops, where micro-practices—such as informal discussions, quick adjustments, and feedback from various stakeholders—play a vital role in strategy formation (Mintzberg 1994a).

In South Africa, where educational institutions face resource constraints and systemic inequalities (Marcum 2022), the need for more inclusive and collaborative approaches is even more pressing. The paper explores how continuous communication and feedback loops drive emergent strategies in South African HEIs, positioning them to respond swiftly to complex challenges (Jassawalla and Sashittal 2006). By examining these dynamics in the South African context, the paper offers insights into how continuous communication channels and feedback loops drive emergent strategising, which can enhance their capacity to adapt to rapidly evolving educational demands and societal expectations, thereby ensuring sustainable performance in the future. The following questions guided the research:

- How do continuous communication channels and feedback loops drive emergent strategies?

In the following sections, we outline the concepts and existing research, then describe the research methods. We then present the research findings, after which the article concludes.

2 Literature Review

2.1 Emergent Strategising in Higher Education

Emergent strategising refers to the organic development of strategies over time, influenced by everyday decisions and actions rather than stemming from a formal planning process (Mintzberg 1994b). In contrast to intentional strategies, which are premeditated and carried out through a top-down approach, emergent strategies evolve as organisations respond to unexpected changes in their external and internal environments (Mintzberg and Waters 1985).

The concept has gained significant attention in the strategic management literature, especially in organisational contexts like HE, where adaptability and responsiveness to external pressures are essential. In the South African context, where educational institutions face internal constraints and external demands for reform, emergent strategies allow institutions to remain flexible and responsive to these evolving challenges (Reddy and Thiven 2006; Pramjeeth et al. 2023; Ogude et al, n.d.). Traditional strategic models often emphasise intentional planning, in which organisations delineate explicit objectives and systematic methodologies (Mintzberg 1987). Emergent strategies develop from routine behaviours, impromptu decisions, and immediate problem-solving (Lynch 2015). The differentiation between deliberate and emergent strategies is crucial for comprehending how institutions may adjust to complex and unexpected settings (Whittington et al. 2006).

As indicated above, emergent strategising is especially pertinent in HEIs given the sector's dynamic nature. Technological innovations, shifting regulatory frameworks, and developing student requirements necessitate that institutions be adaptable and adaptive in their decision-making processes (Ohei et al. 2023). The micro-practices of individuals and departments influence emergent strategies as they address urgent difficulties through their actions and interactions. Research indicates that emergent strategising is more efficacious in complex and dynamic situations, enabling organisations to maintain agility and responsiveness (Farjoun 2002).

International comparisons show similarities and differences. Asian PHEIs are adopting digital platforms and e-learning systems to respond swiftly to resource and governance constraints (Ngai et al. 2019). Emergent strategies in Europe generally involve mergers and collaborations among PHEIs to adapt to competitive higher education (HE) markets (Williams 2017). Emergent strategising is relevant and context-specific for South African PHEIs, as shown by these comparisons.

Therefore, in South African HPEIs, emergent strategising is essential for sustaining competitiveness and responding to external pressures, including budget limitations, the digital transformation of education, and the necessity for socio-economic inclusiveness (Czerniewicz 2018). Further, cross-functional collaboration, wherein departments cooperate in response to changing circumstances, is essential for cultivating emergent strategies (Jassawalla and Sashittal 2006). In essence, such collaboration facilitates real-time adjustments throughout institutions, ensuring responsiveness to evolving demands.

2.2 The Role of Cross-Functional Collaboration in South African Educational Institutions

In South African HEIs, effective interdepartmental collaboration is crucial for addressing the challenges of twenty-first-century education, where technology, student diversity, and global trends necessitate innovative strategies (Czerniewicz 2018). Cross-functional collaboration is the engagement of individuals and teams from multiple departments working together to achieve shared strategic objectives (Ahmad and Boit 2023).

Hence, communication is essential for fostering cross-functional collaboration, enabling knowledge sharing and aligning departments with the institution's overarching strategic goals. Mohamed et al. (2004) assert that interdepartmental communication fosters strategic responsiveness by facilitating the dissemination of essential information throughout the organisation. In educational institutions, where decision-making necessitates input from diverse stakeholders (including academic departments, administrative offices, and external partners), the capacity to disseminate knowledge and synchronise activities is essential for cultivating emergent strategies (Mohamed et al. 2004).

Educational institutions in South Africa face challenges that make cross-functional collaboration increasingly important. PHEIs often operate with limited resources, and their competitiveness depends on the effectiveness of departmental collaboration in responding to external demands. In addition, the transition to blended learning and the increasing use of digital tools necessitate extensive collaboration among IT departments, academic personnel, and administrative teams to ensure that initiatives align with institutional objectives and address the evolving needs of students. Studies indicate that continual communication among departments enhances their ability to predict changes and respond in a coordinated and effective manner (Mohamed et al. 2004).

2.3 Adaptive Strategies Within the Context of Higher Education (HE)

Adaptive strategies enable educational institutions to respond flexibly to environmental changes by making real-time modifications to their strategic plans (Teece et al. 1997). Hence, adaptive techniques are crucial for institutions aiming to maintain relevance in a progressively competitive and uncertain environment. Further, the emergence of online education, changes in government regulation, and increasing student expectations are among the problems that necessitate that educational institutions implement flexible and adaptive strategies (Czerniewicz 2018).

International research shows that teamwork fosters adaptation. Kezar (2005) found that rethinking internal structures for cooperation improves institutional agility in European and North American institutions. Further, the Triple Helix approach in Africa emphasises partnership between universities, the business sector, and government for innovation in HE (Gachie 2020).

Given that in South Africa, where socio-economic inequalities and resource limitations pose significant challenges, adaptive strategies are essential for PHEIs. These tactics enable institutions to adapt to market needs swiftly, embrace new technologies, and manage alterations in regulatory frameworks (Cloete 2014). Implementing adaptive techniques in HE necessitates flexibility and openness to experimenting with innovative models, such as blended learning, as well as restructuring courses to accommodate a diverse student population (Czerniewicz 2018).

Adaptive strategies are most efficacious when founded on collaborative frameworks, wherein diverse departments cooperate to ensure the institution aligns with its overarching strategic objectives. Cross-functional collaboration and effective communication channels enable educational institutions to make real-time modifications, thereby maintaining agility in a rapidly evolving environment. Studies demonstrate that organisations promoting collaborative decision-making are more adept at executing adaptive strategies that address both internal and external requirements (Teece et al. 1997).

2.4 The Importance of Feedback Loops and Continuous Communication in Supporting Real-Time Strategic Adjustments

Feedback loops are a crucial element of any adaptive approach, enabling organisations to consistently assess the efficacy of their efforts and implement real-time modifications based on contemporary data (Senge 1990). In HE, feedback loops manifest as student surveys, performance evaluations, or internal audits, which furnish essential data enabling institutions to modify their strategy in response to evolving conditions (Fullan 2025).

Ongoing communication within departments is crucial for creating effective feedback mechanisms. Through departmental collaboration and information sharing, issues can be identified promptly, enabling swift responses and real-time strategic modifications. (Mohamed et al. 2004) assert that efficient communication routes facilitate the exchange of tacit information within departments, which may frequently remain unrecognised within formal reporting frameworks. Essentially, the informal exchange of knowledge facilitates expedited decision-making and the execution of adaptive methods that address both internal obstacles and external demands.

In the context of South African HE, continuous communication and feedback mechanisms are crucial for institutions seeking to remain competitive and meet the diverse needs of their student bodies (Ferreira 2005). Input from students, professors, and other stakeholders provides essential data for modifying instructional approaches, enhancing curricula, and executing technology-based initiatives. Without efficient feedback loops and communication systems, institutions may struggle to execute adaptive strategies that address evolving demands in real-time (Fullan 2025).

2.5 *Theoretical/Conceptual Framework*

Drawing on the above discussion, Fig. 1 combines and provides a conceptual framework that illustrates how feedback loops, continuous communication, technological tools, and organisational culture contribute to real-time strategic adjustments, enabling agility and responsiveness. Feedback loops allow institutions to monitor performance and implement timely adjustments, while continuous communication ensures that information flows effectively across departments. Technological tools facilitate these processes by providing real-time data, and a collaborative organisational culture promotes their use. Together, these elements empower institutions to make strategic adjustments in real-time, enhancing their capacity to adapt quickly to shifting external conditions.

2.6 *Theoretical Perspective*

Bourdieu's concept of social capital offers a valuable perspective on how relationships and networks shape the dynamics of feedback loops, ongoing communication, and strategic planning in higher education (HE). Social capital fosters trust and collaboration, enhancing both formal feedback and informal communication channels. Hence, in environments rich in social capital, technological tools are adopted more efficiently, accelerating and improving feedback loops and promoting strategic responsiveness and agility (Bourdieu 1986). By contrast, SAP concentrates on the

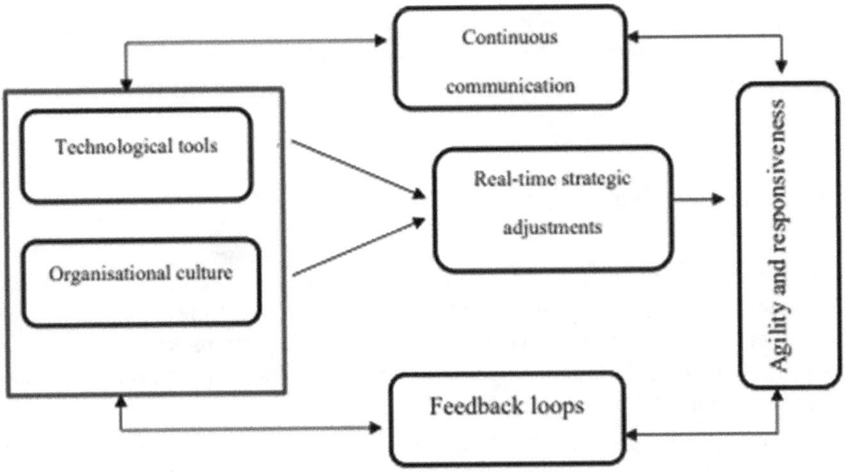

Fig. 1 Framework for real-time strategic adjustments in institutions. *Source* : Author's own compilation

micro-level actions and interactions through which strategy is enacted in daily organisational life (Whittington et al. 2006). Therefore, integrating these perspectives facilitates an understanding of how informal and formal communication networks, reinforced by trust and relational capital, enable emergent strategising through real-time responsiveness and collaborative action.

3 Methodology

3.1 Research Design

This study followed a constructivist-interpretive approach, using an intrinsic qualitative case study to provide an in-depth exploration of communication practices and feedback loops within a single South African PHEI, enabling effective navigation of rapidly changing environments. In line with the approach, Rüegg-Stürm et al. (2015) posit that constructivism is embedded in practice and that knowledge is studied in relation to the world.

3.2 Sampling Method

According to Yin (2018), there is no accepted rule for the number of participants required in qualitative research. Therefore, the study employed purposive sampling to select participants who were managers at the time of data collection and who occupied top-, middle-, and low-level management positions, ensuring a diverse range of perspectives was captured.

3.3 Research Instrument

Pathak and Intratat (2012) posit that semi-structured interviews are appropriate when the research would benefit from a more open framework. Data were collected through 18 semi-structured interviews and 18 written narratives from multi-level managers to gather interesting stories and understand their experiences and the meanings they assigned to them.

3.4 Data Gathering

The data used in this article were initially obtained as part of a PhD research study that examined microfoundational practices influencing emergent strategy-making by managers in a PHEI in South Africa. The broader study aimed to explore how strategy emerged through microfoundational practices influenced by *phronesis* and *metis*.

For the thesis, data were collected from four publicly accessible documents, 18 semi-structured interviews with managers, and 18 written narratives from the same participants. The interviews explored cognition, social interaction, the role of power, individual routines, and factors within the institution, as well as extraneous factors, while the narratives presented personal thoughts and reflections.

Ethical clearance for the project was obtained from the University of South Africa's ethics committee, and all participants provided informed consent for the use of their contributions in scholarly publications.

This study focused on a secondary examination of the original dataset, with a specific emphasis on how continuous communication and feedback loops drive emergent strategising within the PHEI. While these elements were evident in the original thesis, this essay presents a more concentrated examination of communication dynamics and feedback loops, notably the formal–informal, inclusive, and open iterative communication that enabled the institution to adopt emergent, adaptable strategies.

3.5 Data Analysis

ATLAS.ti version 25 was used for thematic analysis to identify recurring patterns related to feedback loops, communication practices, and strategic adjustments.

Coding focused on open, inclusive, and informal communication, and how these practices contributed to real-time strategy formation (Khandkar 2009).

4 Results and Discussion

The section below discusses the key findings on how continuous communication channels and feedback loops drive emergent strategising within a PHEI. Using Bourdieu's social capital theory and the SAP framework, six interconnected organisational practices emerged as enablers of emergent strategising: continuous communication, feedback loops, social capital, technology and culture, formal and informal synergy, and facilitative leadership. As such, these practices collectively demonstrate that strategy emerges not as a product of formal planning but as an evolving process embedded in communicative practices, relationships, and institutional culture.

4.1 Continuous Communication

Continuous communication became the key element of emergent strategising. Participants explained how a smooth mix of formal and informal communication ensured alignment, responsiveness, and shared awareness across departments and campuses.

A middle manager reflected: *Through the practice of open, continuous communication, feedback loops were created, and managers regularly held meetings and workshops, providing platforms for discussing challenges and brainstorming solutions.*

As one top manager reflected in a written narrative: *Weekly meetings are held with various department heads and managers to address ongoing issues, from registrations to exams. We also consult through informal channels like phone calls and Teams meetings, which allow for more flexible collaboration* (Written narrative, Middle Manager 5).

These accounts show that communication at the PHEI goes beyond structured information sharing; it functions as an adaptive mechanism that supports sense-making and collaborative decision-making. Further, digital tools like Microsoft Teams and SharePoint have become essential channels for maintaining open dialogue, especially in a post-pandemic context.

As a top manager noted: *Since COVID-19, my work setup has predominantly been remote, involving online communication with managers and colleagues* via *platforms like Microsoft Teams, Google Meet, and email* (Written narrative, Top Manager 8).

The frequency and accessibility of communication fostered a collective rhythm of engagement, in which information flowed continuously rather than episodically. This corroborates Celestin and Vanitha's (2017), findings that cross-functional dialogue enhances responsiveness and adaptability in dynamic educational environments.

Outside of formal meetings, there are informal interactions where we clarify specific issues using Teams calls (Interview, Middle Manager 3).

Therefore, continuous communication is not just an administrative instrument but a strategic approach, serving as a way for actors to build shared understanding and implement emerging strategies every day.

4.2 Feedback Loops Practice

Feedback loops served as the driving force of organisational learning, transforming communication into strategic adaptation. They established iterative cycles of reflection and action, ensuring that strategies emerged from lived realities. Participants repeatedly highlighted how formal and informal feedback systems fostered agility.

We have surveys when our students give feedback on so on, and then based on the surveys and also issues that we pick up, maybe in modules and so on. That goes on to a improvement plan that is part of those QA, quality assurance, in the school. And that is then always, you know, checked. And we try to close the loop … (Interview, Top Manager 1).

Similarly, a middle manager echoed: *It also inspired me to develop workshops based on student feedback, rather than relying solely on quantitative data. For example, I now use qualitative surveys to gather input from students before planning workshops, which allows us to address their specific needs and challenges* (Written Narrative, Middle Manager 6).

This dynamic shows that feedback is not one-way but mutual, thus flowing between leadership, staff, and students. Informal feedback also holds considerable importance. As one top manager indicated: *But we also do, I don't want to say informal but, but* ad hoc *meetings should, should the need arise if we need to do something specific..* (Interview, Top Manager 8).

Another middle manager reflected: *For instance, informal discussions often provide new perspectives that help streamline processes or develop new ideas. However, to ensure everyone is on the same page, we always formalise decisions through emails or Teams updates, reinforcing transparency and accountability* (Written narrative, Middle Manager 4).

Such examples illustrate double-loop learning as posited by Argyris (2008), where feedback not only refines actions but also reshapes assumptions underlying institutional practices. This iterative responsiveness aligns with Hendry and Seidl (2003), who found that feedback-driven cycles within organisations promote continuous adaptation and collective ownership of strategy. Similarly, Kezar and Holcombe (2017) emphasise that shared feedback mechanisms in HE facilitate participatory strategising and enhance institutional agility.

4.3 Social Capital Practice

Social capital supported the effectiveness of both communication and feedback systems. It was expressed through trust, collegiality, and shared purpose, allowing information to flow freely across hierarchical boundaries.

As one lower manager explained: *We have WhatsApp groups, and if there is anything that is come up, we share ideas. If there is an issue, that is arisen. At one campus. We have to make sure that the other campuses are aware of that. So we have a very viable information sharing and collaborative manner. So we always try to be very we always try to avoid being reactive* (Interview, Lower Manager 2).

Another elaborated, by stating that: *...we share ideas across campuses using WhatsApp groups to collaborate and ensure all campuses are updated on important matters* (Written narrative, Lower Manager 1).

These interactions illustrate Bourdieu's (1986) concept of social capital as a shared resource fostering cooperation and mutual understanding. Managers described a culture of solidarity rather than competition: *We rely on each other's expertise; there's no competition, it's about pulling together to make things work* (Interview, Middle Manager 3).

Such trust also created psychological safety: *The trust between departments means we can raise issues freely without fear—everyone knows it's about improvement, not blame* (Interview, Top Manager 6).

The relational depth of this communication system transformed everyday interaction into a strategic resource, echoing (Schaarsberg 2025), who noted that institutions rich in social capital sustain more resilient and adaptive feedback mechanisms. Within the PHEI, social capital was not incidental; it was strategically functional, shaping negotiation, learning, and alignment toward emergent goals.

4.4 Technology–Culture Practice

Digital platforms enabled continuous communication and continuity; their strategic impact depended on a culture that embraced openness, innovation, and trust, which drove emergent strategising.

A top manager explained: *Due to the dispersed nature of the institution with multiple campuses, my daily interactions with other managers primarily happen* via *Teams meetings and emails. These are the primary tools used for operational discussions and strategic updates. We also use live sheets on platforms like Microsoft SharePoint, which allow real-time updates and collaboration across teams. Formal interactions, such as workshops and planning sessions, occur periodically, particularly when developing the academic plan for the upcoming year* (Written Narrative, Top Manager 6).

However, participants recognised that technology alone was insufficient. As one middle manager stated: *Technology helps, but it only works because we've built a culture where people are willing to engage and learn new tools ... but technology also is, is a factor. And then sometimes, you know, we, we struggle with getting the [technology] skills because, you know, some, some, some, some specialisations, they, they, they are not really available* (Interview, Middle Manager 1).

This sentiment reflects the principle of sociotechnical alignment, which holds that technological infrastructure must harmonise with human and cultural systems to facilitate learning.

Another participant described how innovation had become cultural: *Technology has had a significant impact on how we strategise, particularly in terms of student engagement and assessments. The rise of AI tools like ChatGPT has changed how we approach assessments, forcing us to rethink how to develop strategies that balance the use of AI while still building essential skills in students* (Written narrative, Top Manager 1).

Further, a lower manager added: *We use technology to link across campuses and perform tasks collaboratively …* (Interview, Low Manager 1).

This intersection of digital and cultural capital reflects Monticelli et al. (2024), who found that technology-enabled communication only improves strategising when it is embedded within a participative organisational culture. As such, in this case, the PHEI's culture enhanced the communicative potential of its technology.

4.5 Leadership Facilitation Practice

Leadership emerged as an essential intermediary between communication, feedback, emergent strategising, and learning. As such, instead of taking on authoritative roles, leaders served as facilitators of dialogue, empowering others to contribute strategically.

One middle manager reflected: *I learn from them, and they learn from me. So there is that mutual respect, and whatever I do to them, I expect them to also do it back to me* (Interview, Middle Manager 1).

This indicates that leadership here is based on guidance rather than control. A lower manager echoed this view: … *sense that the CEO, the CEO is a very open door policy person.* (Lower Manager 2).

Such relational leadership reflects the SAP emphasis on leadership as an enacted, situated activity that shapes strategy through interaction. Another participant elaborated: *We do it vertically—everyone contributes upwards and downwards. It improves communication between departments and reduces silos* (Interview, Middle Manager 7).

A lower manager summarised: ... *the master brains behind this whole setup in our CEO, and the way that he thinks, in the way that he, you know that he does his business as quite easy, and it's open type of communication* (Interview, Low Manager 3).

These accounts emphasise facilitative leadership as a key practice that maintains psychological safety and inclusion, in line with Bass et al. (2025), who found that leadership practices promoting psychological safety and collaborative learning enable adaptive and emergent strategising within complex organisational systems.

4.6 Formal and Informal Synergy Practice

Finally, the synergy between formal and informal communication created a flexible environment conducive to emergent strategising. Formal structures provided accountability and coordination, while informal exchanges allowed rapid problem-solving and creativity.

As one middle manager noted: *We've got weekly campus meetings and MANCO meetings to discuss deliverables and review progress.* (Interview, Middle Manager 4).

Yet, innovation often began informally: *Sometimes the WhatsApp conversation starts an idea that ends up being formalised at MANCO*–[that is how most innovations happen] (Lower Manager 1).

Participants valued both systems equally. One top manager explained: *Formal meetings are good for accountability, but informal chats drive creativity and connection* (Interview, Top Manager 6).

Another added: *Informal discussions often provide new perspectives that help streamline processes or develop new ideas* (Middle Manager, 4).

This duality aligns with Schaarsberg (2025), who found that institutions that leverage both formal and informal communication channels are better able to maintain adaptive strategies. At the PHEI, such synergy blurred the boundary between operational coordination and strategic evolution, making strategy a living, ongoing practice.

5 Conclusion

This study examined how continuous communication and feedback loops drive emergent strategising within a PHEI in South Africa. The findings reveal that strategic adaptability is not achieved through rigid plans, but through six interconnected practices, namely, continuous communication, feedback loops, social capital, technology–culture integration, facilitative leadership, and formal–informal synergy. As such, continuous communication across departments enabled openness and collaboration, while feedback loops transformed dialogue into ongoing learning and rapid response. Together, they fostered agility by linking reflection with action. Further, social capital, through trust, shared understanding, and informal relationships, strengthened these processes, turning interpersonal networks into strategic assets that bridged hierarchies and functions. Equally, technology–culture integration showed that digital tools enhance strategising only when grounded in a culture of participation and transparency. Facilitative leadership further supported this ecosystem by promoting inclusion, dialogue, and psychological safety rather than control. Finally, a formal–informal synergy balanced structure with flexibility: formal meetings ensured accountability, while informal exchanges fostered creativity and speed.

In essence, strategy emerged not from isolated plans but from daily communicative practices and relational dynamics. For HE leaders, embedding these six practices, particularly open communication, iterative feedback, and facilitative leadership, offers a sustainable pathway to organisational agility and responsiveness in complex environments. Amidst the uncertainty, economic strain, and rapid change facing South African HE, emergent strategising anchored in continuous communication and feedback loops strengthened by social capital, facilitative leadership, technology–culture integration, and formal–informal synergy, provides a route towards institutional resilience and relevance. By embedding these practices into the fabric of daily organisational life, PHEIs can foster responsiveness, inclusivity, and innovation that endure well into the future.

References

Ahmad, T., Boit, J.: The role of cross-functional collaboration in digital transformation. Zenodo. (2023). https://doi.org/10.5281/zenodo.13325767

Argyris, C.: Double-Loop Learning and Organizational Change: Facilitating Transformational Change. Wiley, Hoboken (2008)

Bass, G.A., Duffy, C.C., Cannon, J.W., Yelon, J.A., Ferrada, P., Evans, S., et al.: Surgeon leadership in trauma resuscitation requires a competency-based multimodal training framework. Trauma Surg. Acute Care Open. **10**(4) (2025). https://doi.org/10.1136/tsaco-2025-001975

Bidandi, F., Anthony, A.: Collaboration and partnerships between south African higher education institutions and stakeholders: case study of a post-apartheid university. Discov. Educ. **1**(2) (2022). https://doi.org/10.1007/s44217-022-00001-2

Bourdieu, P.: The forms of capital. In: Richardson, J.G. (ed.) Handbook of Theory and Research for the Sociology of Education, pp. 241–258. Greenwood, New York (1986)

Brown, J.S., Duguid, P.: Knowledge and organization: a social-practice perspective. Organ. Sci. **12**(2), 198–213 (2001). https://doi.org/10.1287/orsc.12.2.198.10116

Celestin, M., Vanitha, N.: Breaking down silos: collaborative strategies that actually work. Int. J. Appl. Adv. Sci. Res. **6**(1), 391–397 (2017) https://doi.org/10.47037/ijaasr.v6i1.6732

Chirwa, M., Boikanyo, D.H.: The role of effective communication in successful strategy implementation. Acta Commer. **22**(1) (2022). https://doi.org/10.4102/ac.v22i1.1020

Cloete, N.: The south African higher education system: performance and policy. Stud. High. Educ. **39**(8), 1355–1368 (2014). https://doi.org/10.1080/03075079.2014.949533

Czerniewicz, L.: Inequality as higher education goes online. In: Networked Learning, pp. 95–106. Springer, Cham (2018). https://doi.org/10.1007/978-3-319-74857-3_6

Czerniewicz, L.: Multi-layered digital inequalities in HEIs: the paradox of the post-digital society. Educ. Inf. Technol. (2022). https://doi.org/10.1007/s40593-021-00285-9

de Lange, G.: Developing a University-Community Engagement Conceptual Framework. PhD thesis,. University of the Free State (2012)

Farjoun, M.: Towards an organic perspective on strategy. Strateg. Manag. J. **23**(7), 561–594 (2002). https://doi.org/10.1002/smj.239

Ferreira, M.: A framework for continuous improvement in the south African higher education sector. High. Educ. World Rep. (2005)

Fullan, M.: The New Meaning of Educational Change, 6th edn. Teachers College Press, New York (2025)

Gachie, W.: Higher education institutions, private sector and government collaboration for innovation within the triple helix model. Afr. J. Sci. Technol. Innov. Dev. **12**(2), 203–215 (2020). https://doi.org/10.1080/20421338.2019.1631120

Gobind, D.: The Internal Factors that Influence Incremental Technological Innovation in a South African Organisation. PhD thesis,. University of the Witwatersrand (2024)

Hendry, J., Seidl, D.: The structure and significance of strategic episodes. J. Manag. Stud. **40**(1), 175–196 (2003). https://doi.org/10.1111/1467-6486.00008

Hwang, G.J., Fu, Q.K.: Advancement and research trends of smart learning environments in the mobile era. Int. J. Mob. Learn. Organ. **14**(1), 114–129 (2020). https://doi.org/10.1504/ijmlo.2020.103911

Jassawalla, A.R., Sashittal, H.C.: Collaboration in cross-functional product innovation teams. Adv. Interdiscip. Stud. Work Teams. **12**, 1–25 (2006). https://doi.org/10.1016/S1572-0977(06)12001-4

Kezar, A.: Redesigning for collaboration within higher education institutions. Res. High. Educ. **46**(7), 831–860 (2005). https://doi.org/10.1007/s11162-004-6227-5

Kezar, A., Holcombe, E.: Shared Leadership in Higher Education. American Council on Education, Washington, DC (2017)

Khandkar, S.H.: Open coding. Qual. Res. J. (2009)

Kovalenko, M., Lomonosova, O., Rusnak, A.: Adaptive management of higher education institutions. Balt. J. Econ. Stud. **7**(2), 118–128 (2021) https://doi.org/10.30525/2256-0742/2021-7-2-118-128

Loveday, J. E.: An abstract of the dissertation of title: Working together for student success: Cross-functional collaboration at community colleges (2009)

Lynch, R.: Strategic Management, 7th edn. Pearson Education, Harlow (2015)

Marcum, J.: Education, Race, and Social Change in South Africa. University of California Press, Berkeley (2022)

Mintzberg, H.: The strategy concept II. Calif. Manag. Rev. **30**(1), 25–32 (1987). https://doi.org/10.2307/41165264

Mintzberg, H.: Rethinking strategic planning. Long Range Plan. **27**(3), 12–21 (1994a). https://doi.org/10.1016/0024-6301(94)90185-6

Mintzberg, H.: The Rise and Fall of Strategic Planning. Free Press, New York (1994b)

Mintzberg, H., Waters, J.A.: Of strategies, deliberate and emergent. Strateg. Manag. J. **6**(3), 257–272 (1985). https://doi.org/10.1002/smj.4250060306

Mizuta, M.: Silos in Higher Education Institutions. PhD thesis,. Portland State University (2022). https://doi.org/10.15760/etd.7965

Mohamed, M., Stankosky, M., Murray, A.: Applying knowledge management principles to enhance cross-functional team performance. J. Knowl. Manag. **8**(3), 127–142 (2004). https://doi.org/10.1108/13673270410541097

Monticelli, J.M., Fossatti, P., de Quadros da Silva, L., Bitencourt Soster Luz, C.: Innovative university: evidence from university management. Int. J. Educ. Manag. **38**(2), 509–524 (2024). https://doi.org/10.1108/IJEM-05-2022-0174

Nasir, M., Hasan, M.: Utilizing artificial intelligence in education to enhance teaching effectiveness. Jurnal. Serambimekkah. Ac. Id. (2024). https://jurnal.serambimekkah.ac.id/index.php/ice/article/view/1367

Ngai, C.S.B., Lee, W.M., Ng, P.P.K., Wu, D.D.: Innovating collaborative e-learning practices in higher education. Stud. High. Educ. **44**(11), 1990–2010 (2019). https://doi.org/10.1080/03075079.2018.1482266

Ohei, K.K., Mantzaris, E., Ntshangase, B.A., Olutade, E.O.: Incorporating new technologies into teaching in South Africa. Int. J. Res. Bus. Soc. Sci. **12**(6), 286–295 (2023) https://doi.org/10.20525/ijrbs.v12i6.2715

Pathak, A., Intratat, C.: Semi-structured interviews and teacher perceptions. Malays. J. ELT Res. **8**(1), 1–10 (2012)

Pramjeeth, S., Nupen, D., Jagernath, J.: Challenges impacting higher education leaders. Afr. J. Interdiscip. Stud. **5**(1), 1–13 (2023) https://doi.org/10.51415/ajims.v5i1.1178

Reddy, T., Thiven, R.: Higher education and social transformation in South Africa. Cah. Rech. Afr. **5**, 121–145 (2006). https://doi.org/10.4000/cres.1118

Riveira, C., Fadhila, A.: Technological advancements shaping future learning environments. Journal. as-Salafiyah. Id. (2024). https://journal.assalafiyah.id/index.php/sahri/article/view/121

Rüegg-Stürm, J., von Arx, W., Grand, S.: Constructivist paradigms. In: Cambridge Handbook of Strategy as Practice, pp. 78–94. Cambridge University Press, Cambridge (2015). https://doi.org/10.1017/CCO9781139681032.005

Rusok, N.H.M., Samy, N.K., Bhaumik, A.: Higher education institutions as learning organisations. Int. J. Acad. Res. Bus. Soc. Sci. **11**(7) (2021). https://doi.org/10.6007/IJARBSS/V11-I7/10566

Schaarsberg, A.: The Role of Communication Practices in Shaping Social Identity and Interdepartmental Collaboration During Organisational Growth (2025)

Senge, P.M.: The Fifth Discipline. Broadway Business, New York (1990)

Shams, Vrontis, D., Weber, Y., Tsoukatos, E., Ferraris, A.: Cross-functional knowledge management: The international landscape. Cross-functional knowledge management: The international landscape, pp. 1–210 (2019). https://doi.org/10.4324/9780429020650

Teece, D.J., Pisano, G., Shuen, A.: Dynamic capabilities and strategic management. Strateg. Manag. J. **18**, 509–533 (1997)

Whittington, R., Molloy, E., Mayer, M., Smith, A.: Practices of strategising. Long Range Plan. **39**, 615–629 (2006). https://doi.org/10.1016/j.lrp.2006.10.004

Wildschut, L.: Evaluation education in South Africa. Afr. Eval. J. **8**(1) (2020). https://doi.org/10.4102/aej.v8i1.476

Williams, J.: Collaboration, alliance and merger among Higher Education Instititions (2017). https://www.oecd.org/edu/workingpapers

Yin, R.K.: Case Study Research and Applications, 6th edn. Sage, Thousand Oaks (2018)

Žitkienė, R., Deksnys, M.: Organizational agility conceptual model. Mont. J. Eco. **14**(2), 115–129 (2018). https://doi.org/10.14254/1800-5845/2018.14-2.7

Employee Empowerment, Career Satisfaction and Intention to Stay of Employees in a Printing Organisation in Pretoria

Pierre Joubert and Adelaide Zandile Mqokoza

Abstract One of the crucial tasks of government is to build a public service that is capable of meeting the needs of South African citizens. Government will not be able to improve public service delivery without working together with the public servants. The study was intended to explore the relationship between employee empowerment and career satisfaction with a view to retain the employees of the Government Printing Works. This study followed a quantitative research approach and a structured questionnaire was used to collect data. The questionnaire consisted of four sections; A (demographic profile), B (employee empowerment), C (career satisfaction) and D (intention to stay). Three hundred and fifty questionnaires were distributed and 201 were returned and used for data analysis. The results of the correlation analysis revealed that there is a significant moderate positive relationship between employee empowerment and career satisfaction as well as a weak positive relationship between employee empowerment and intention to stay. It was recommended that conditions that may lead to employees feeling powerless should be identified and satisfactorily resolved and certain powers should be delegated to employees to enable work decisions to be taken quicker. To improve career satisfaction, it is recommended that opportunities for growth and greater responsibility be provided that will not only challenge employees but also give them a sense of reward and fulfilment. Future research can incorporate more than one organisation in the printing sector and more than one government department, as well as a mixed methods approach to gain a deeper understanding of the studied constructs and their relationships.

1 Introduction

Employee empowerment (EE) has been a topic of discussion for many years (Choi et al. 2016; Das and Moharana 2017; Prathiba 2016). In previous studies conducted on this construct, the concept of EE was examined in conjunction with participative

P. Joubert (✉) · A. Z. Mqokoza
Sol Plaatje University, Kimberley, South Africa
e-mail: pierre.joubert@spu.ac.za

E. H. Redda (ed.), *Proceedings of the Social Sciences International Research Conference, 2025*, https://doi.org/10.1007/978-981-95-9489-4_12

management, motivation, job autonomy and employee enrichment (Priyadharshany and Sujatha 2015). However, the last two decades record a growing concentration on the concept of EE as a single discourse, due to its significance in dealing with increased market competition and globalisation effects (Kohli and Sharma 2017). Conger and Kanungo (1988) define empowerment as a process whereby conditions of powerlessness in the workplace are identified and removed. It further implies the act of delegating power to enable work decisions to be taken quickly (Turkmeglou 2019). According to Choi et al. (2016), the delegation of power gives employees the sense that they are valued by the organisation and, in return, engenders commitment on their part towards organisational success. Choi et al. (2016) further add that empowerment may foster critical thinking and innovation among employees. This study will focus on the empowerment of employees within a printing organisation in Pretoria.

2 Literature Review

Srivastava and Madan (2018) state that there are two approaches to EE, namely structural/relational empowerment and psychological/motivational empowerment. While structural empowerment is concerned with management practices, psychological empowerment entails employees' psychological strengths. Additionally, structural empowerment is concerned with management's role in EE, while psychological empowerment deals with employees' perception of empowerment (Kohli and Sharma 2017). Four dimensions of psychological empowerment have been identified, namely: meaning, competence, self-determination and impact (Pandey 2016) who explained that structural empowerment encompasses elements such as opportunities to grow and develop, resources and flexibility. Mohapatra and Mishra (2019) assert that because employees spend most of their time at the workplace it is important that they are content while being there. Moreover, because EE increases a worker's authority, knowledge and motivation, employees perceive the granting of decision-making authority to be an indication that the organisation appreciates their job contributions (Mohapatra and Mishra 2019). Consequently, committed, motivated and knowledgeable employees are more likely to compete successfully and adapt more easily to organisational and market changes (Khokher and Raziq 2017). Mahmood and Sahar (2017) opine that in organisations in which employees' career needs are developed and fulfilled, staff members are more engaged with the organisation's goals. According to Takawira and Coetzee (2019), employees' career satisfaction (CS) is an important component of their level of work engagement and retention. Career satisfaction levels are likely to be increased within organisations in which the employees feel empowered. Some of the positive contributions of CS are palpable in employees' health, well-being and longevity of service. Employees with high levels of CS are more committed and contribute to organisational success (Agrawal and Srivastava 2018). According to Shahid (2018), employees leave organisations if they are not satisfied with multiple factors such as lack of recognition, unfulfilling jobs and inadequate career advancement. In a competitive world, every employer prefers to have committed, dependable, loyal and hardworking employees who remain with the organisation for a lengthy period, while employees have many opportunities to leave and join another organisation (Shanker 2014). Retention of high-quality employees is critical because turnover of such employees will negatively impact the organisation's operations and performance (Woon et al. 2017). Intention to stay (ITS) is being willing to continuously work for an organisation (Yao and Huang 2018). Mohapatra and Mishra (2019) opine that companies need to reduce the turnover rate of the employees, a situation that can be achieved by implementing EE effectively and supporting employees' career advancement. Takawira and Coetzee (2019) found that employees' CS remains important for their well-being, work engagement level and retention, while Kumar and Kumar (2017) indicate that EE is a type of motivational strategy which gives employees a sense of satisfaction towards their job and the organisation.

Studies conducted in different settings have discovered that there is a significant positive relationship between EE and employee ITS within the organisation (Pranee et al. 2020; Rozsa et al. 2019). Evidence confirms a positive relationship between CS and ITS (Ismail et al. 2018). The Government Printing Works (GPW) is a South African security printing specialist that deals with the printing of passports, visas, birth certificates, smart card identification documents and examination materials, as well as government stationery and publications, such as tender bulletins and government gazettes. The GPW operates as a self-funded organisation within the parameters of South Africa's Department of Home Affairs (DHA). Empowered employees will ensure strengthened support for the DHA and the effective functioning of the printing organisation (Janks 2014). When an organisation avails opportunities and strategies to empower employees, it is an indication that the organisation values their contribution (Mohapatra and Mishra 2019). Overall, this study aims to uncover the link between EE, CS and ITS of employees who work for the GPW (Fig. 1).

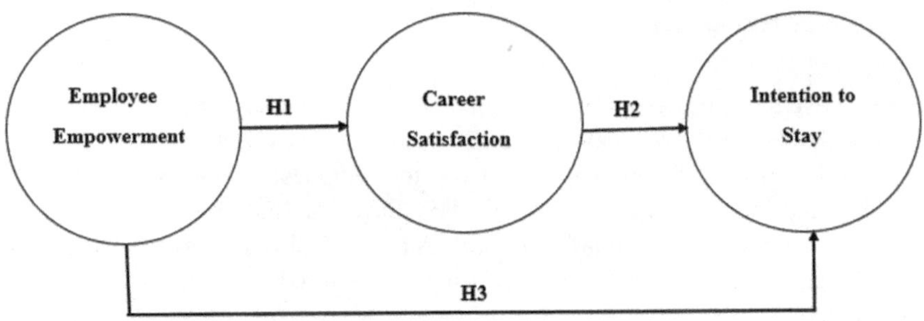

Fig. 1 Conceptual framework of the relationship between EE, CS and ITS

2.1 Theoretical and Conceptual Frameworks

The study is premised on the Reinforcement Theory for EE because when employees feel empowered by the organisation that employs them, they are likely to improve their performance and seek innovative ways of solving company problems (Srivastava and Madan 2018). The Self Concordance Theory is the underlying theory for CS in this study because both this theory and the concept of CS advocate for fulfilling individuals' intrinsic goals. Career satisfaction is the echo of the internal attitudes of individuals towards the development of their career aspirations; it forms a subjective part of career success (Ayketin et al. 2016). The concept 'intention to stay' is grounded in the Social Exchange Theory, which states that if a person receives benefits from others, the person in turn feels an obligation to compensate that party with loyalty and effort (Maliwa et al. 2023). With regard to the retention of human resources, GPW continues to lose experienced and qualified artisans and administrative personnel to the private printing sector because the latter offers more attractive remuneration packages and career growth that may contribute to CS of an individual (Mtshali et al. 2018). Although many studies in South Africa have focused on empowerment or empowerment techniques (Abdissa and Fitwi 2016), and the relationship between EE and job satisfaction (Alzalabani and Nair 2011), no comprehensive study has focused on the relationship between EE, CS and ITS of employees (Kohli and Sharma 2017). Therefore, this study aims to fill that research gap by investigating the relationship between EE, CS and ITS among the GPW's employees. Based on the literature review and problem statement, the following conceptual framework is presented for the study:

3 Methodology

3.1 Research Design

A descriptive research design that is guided by initial hypotheses and determines the relationship between variables was deemed fit for this study. A quantitative research approach based on a theoretical framework that emphasises quantification in the collection and analysis of data was used for this study. The study comprised two steps: the literature review and the empirical study. An empirical study was conducted with regard to employees' EE, CS and ITS using a structured questionnaire. The relationship between se three constructs was investigated and the results documented to contribute to the scientific body of knowledge (Leedy and Ormrod 2013). For this study, the target population ($N = 648$) consisted of employees who occupy permanent positions within the GPW and excludes seasonal workers, interns and apprentices. The study participants included both males and females who are employed by the GPW as Artisans, Printers Assistants, Deputy Directors, Assistant Directors, Clerks,

Security Officers, Administrators, Receptionists and Data Capturers. The comprehensive list of the employees who are employed at the GPW was obtained from the Human Resources Service Condition Office of the GPW. In this study the sample size was equal to $n = 324$ that is half of the population size $(648/2) = 324$ (Redelinghuys and Bothma 2016).

3.2 Sampling Method

Historical evidence was considered in determining the sample size because previous quantitative studies used sample sizes of approximately 200 participants (Daniel 2016; Park 2018; Yap et al. 2010). Subjects included in the sample were selected in the following manner: they should have worked for the organisation for not less than six (6) months. Interns, apprentices and casual workers were not included in this research.

3.3 Research Instrument and Data Collection

The survey was conducted using a fully structured questionnaire composed of close-ended questions, having various options the participants could select. The questionnaire consisted of four sections: Section A consisted of biographical information of the respondents such as gender, age, length of service in the organisation and highest academic qualification. Section B focused on EE and consisted of statements on EE utilising a five-item instrument adapted from Hanaysha (2016) which previously reported a reliability coefficient of 0.77 and was used without any modification. Section C consisted of statements on CS utilising a five-item instrument adapted from Greenhaus et al. (1990), which originally reported a coefficient of 0.84 and was used without any modification. Section D consisted of statements on ITS utilising a five-item instrument adapted from Wayne et al. (1997), using a five-point Likert scale, ranging from strongly disagree (1) to strongly agree (5). The scale previously reported a reliability coefficient of 0.89.

3.4 Data Analysis

After the completion of the data preparation process, the collected data was analysed using the Statistical Package for Social Science (SPSS) version 25 for Windows. In analysing the data, statistical analysis techniques were applied. From a total of 350 distributed questionnaires, 201 were returned and deemed valid and useful for analysis purposes. As Leedy and Ormrod (2013) states, samples of over 200 are desirable in conducting quantitative studies. The total response rate in this study was 57.43%.

4 Results and Discussion

An analysis of the demographic profile of the study sample indicates that the majority of respondents were female, 30–39 years old, with 0–5 years' work experience, and in possession of a degree. The average respondent, therefore, was in a position to understand and respond effectively to the questions posed in the questionnaire. The results of respondents' perceptions towards EE are reported in Table 1.

The mean scores ranged between 3.58 and 4.46. Item B2 recorded the highest mean score of 4.46, which suggests that the majority of the respondents strongly agree that they are confident they possess the necessary capabilities and skills to perform their jobs. The lowest mean score of 3,58 is recorded for B3, which suggests that most respondents feel that they are given authority to make their own decisions concerning their jobs. The overall mean score for EE is 3.99, confirms that the majority of respondents agree that they feel empowered by the opportunities they are presented with at the GP.

The mean scores for CS ranged between 2.83 and 3.59. Item C1 recorded the highest score of 3.59, which indicates that the majority of respondents agree that they are satisfied with the success they have achieved in their careers. Item C3 reported the lowest score of 2.83, highlighting that the majority of respondents are neutral with regard to the satisfaction about the progress they have made towards meeting their income goals. The reported overall mean score of 3.31 indicates that the majority of respondents agree that they are satisfied with their career progress.

Table 1 Mean ratings for employee empowerment

Scale description	N	Min	Max	Mean	Standard deviation
B1: I feel competent to perform the tasks required for my position.	201	1	5	4.38	0.993
B2: I am confident about my capabilities and skills to do my job.	201	1	5	4.46	0.830
B3: I have the authority to make necessary decisions in my job.	201	1	5	3.58	1.231
B4: My manager trusts me to make the appropriate decisions in my job.	201	1	5	3.75	1.123
B5: I have considerable opportunity for interdependence and freedom in how I do my job.	201	1	5	3.80	1.079
Overall mean score	3.99				

Mean scale rating: 0.0.1–1. =Strongly disagree; 1.01–2 = Disagree; 2.01–3 = Neither agree nor disagree; 3.01–4 = Agree; 4.01–5 = Strongly agree

Table 2 Mean scores for career satisfaction

Scale description	N	Min	Max	Mean	Standard deviation
C1: I am satisfied with the success I have achieved in my career.	201	1	5	3.59	1.324
C2: I am satisfied with the progress I have made towards meeting my overall career goals.	201	1	5	3.32	1.196
C3: I am satisfied with the progress I have made towards meeting my goals for income.	201	1	5	2.83	1.154
C4: I am satisfied with the progress I have made towards meeting my goals for advancement.	201	1	5	3.25	1.090
C5: I am satisfied with the progress I have made towards meeting my goals for the development of new skills.	201	1	5	3.57	1.125
Overall mean score	3.31				

Mean scale rating: 0.0.1–1.0 = Strongly disagree; 1.01–2 = Disagree; 2.01–3 = Neither agree nor disagree; 3.01–4 = Agree; 4.01–5 = Strongly agree

Table 3 Mean scores for intention to stay

Scale description	N	Min	Max	Mean	Standard deviation
D1: I will most probably stay in this company in the foreseeable future.	201	1	5	2.92	1.336
D2: I definitely intend to maintain my current relationship with this organisation.	201	1	5	3.57	1.112
D3: I have no intention of leaving this organisation.	201	1	5	2.73	1.329
D4: I want to continue working 5 years from now in my current job.	201	1	5	2.66	1.423
D5: I am not even planning on looking for a new job.	201	1	5	2.45	1.392
Overall mean score	2.86				

Mean scale rating: 0.0.1–1. =Strongly disagree; 1.01–2 = Disagree; 2.01–3 = Neither agree nor disagree; 3.01–4 = Agree; 4.01–5 = Strongly agree

The mean scores for ITS ranged between 2.45 and 3.57. Item D2 reported the highest score of 3.57, indicating that the majority of respondents agree that they intend to maintain their current relationship with the organisation. The lowest score is recorded by D5 as 2.45, highlighting that most respondents are neutral regarding planning to look for a new job. The overall mean score for ITS is 2.86, thus, confirming that the majority of respondents neither agree nor disagree that they intend to stay with the GPW. The overall mean score for EE = 3.99, the overall mean score for CS = 3.31

Table 4 Correlation coefficient scale

Value of correlation coefficient	Relationship interpretation
Less than 0.1	No correlation
0.1–0.3	Weak correlation
0.4–0.5	Moderate correlation
Above 0.5	Strong correlation

Source Apuke 2017

Table 5 Correlation analysis findings between study constructs

Correlations			
Constructs	EE	CS	ITS
EE	1		
CS	0.485**	1	
ITS	0.287**	0.535**	1

$N = 201$ **Correlation is significant at the 0.01 level (2-tailed)

and, lastly, the overall mean score for ITS = 2.86. The mean score for EE indicates that GPW employees generally feel empowered by the opportunities provided by their employee, as well as being allowed to perform their jobs independently and to make the necessary decisions when it is appropriate to do so. The mean score for CS indicates that respondents are content with the rate at which their careers are advancing as well as their progress towards the attainment of new skills. The overall mean score for ITS indicates that respondents are neutral regarding their continued employment at the GPW, employees are non-committal when expressing plans regarding their remaining with the GPW for longer periods.

The analysis of the correlations is reported in Table 5.

The findings depicted in Table 5 reveal a significant moderate positive relationship ($r = 0.485; p < 0.01$) between EE and CS, meaning that there is indeed an association between EE and CS. This result is consistent with findings reported in existing literature that EE has a positive influence on employee CS (Kumar and Kumar 2017; Takawira and Coetzee 2019). The results of the correlation analysis also reveal a weak positive relationship between EE and ITS ($r = 0.287; p < 0.01$). This finding indicates that a positive relationship exists between the two constructs, even though it is not strong. A weak relationship reflects that there is very little that the participating GPW employees associate with their willingness to continue working for the organisation. The weak relationship could also indicate that respondents do not feel sufficiently empowered to want to remain with the organisation, a fact that is consistent with the findings of Pranee et al. (2020) which state that employees will not willingly remain with the organisation if it fails to empower them in the desired manner. The results reported in Table 5 also highlight an existence of a strong relationship between CS and ITS ($r = 0.535, p < 0.01$). The results indicate that there is a strong relationship between employees' CS and their willingness to remain with the organisation. This finding is consistent with existing literature which indicates that employees stay longer with organisations when they are content with their respective career progress (Ismail et al. 2018). The more employees are satisfied with the rate at which their careers are progressing within a particular organisation, the higher the chances of their continued employment.

Linear regression analysis was performed to establish and explore predictive relationships between EE and CS, CS and ITS, and EE and ITS.

When investigating the predictive relationship between EE and CS, CS was entered into the regression equation as a dependent variable and EE was entered as an independent variable. The results highlight a statistical significance at $p < 0.001$; $p = 0.000$ (t -value = 7.830; $R^2 = 0.236$), indicating that roughly 24% of the variance in CS is accounted for by EE. The results also reveal that EE contributes positively towards predicting employee CS. Kumar and Kumar (2017) support this finding because investing in continued education and overall development of employees positively influences CS.

Model 2 above presents the regression analysis of CS and ITS in which CS was entered into the regression equation as an independent variable, and ITS was entered as a dependent variable. The results revealed that there is a statistical significance between CS and ITS at $p < 0.001$; $p = 0.000$ (t value = 8.925; $R^2 = 0.286$), indicating the existence of a significant predictive relationship. The results indicate that CS contributed positively towards employee ITS because 29% of the variance in ITS is contributed by CS. This result means that an increase in the CS of GPW employees influences predictions about their ITS with the organisation. Employees who are content with the rate at which they are advancing in their careers would decide to stay with the organisation because they feel the organisation contributes to their career progress. The results echo the findings of Ismail et al. (2018) that CS is a predicting factor in employees' continued employment with the organisation. Employees display loyalty to organisations that can contribute to their career growth.

Table 6 Regression analysis model 1: Employee empowerment and career satisfaction

Dependent variable: Career satisfaction				Collinearity statistics	
Model 1 Independent variable:	Beta	T	Sig	Tolerance	VIF
Employee empowerment	0.485	7.830	0.000	1.000	1.000

$R = 0.485$ $R^2 = 0.236$ Adjusted $R^2 = 0.232$ $p < 0.001$

Table 7 Regression analysis Model 2: Career satisfaction and intention to stay

Dependent variable: Intention to stay				Collinearity statistics	
Model 2 Independent variable:	Beta	T	Sig	Tolerance	VIF
Career satisfaction	0.535	8.925	0.000	1.000	1.000

$R = 0.535$ $R^2 = 0.286$ Adjusted $R^2 = 0.282$ $p < 0.001$

Table 8 Regression analysis Model 3: Employee empowerment and intention to stay

Dependent variable: Intention to stay				Collinearity statistics	
Model 3 Independent variable:	Beta	T	Sig	Tolerance	VIF
Employee empowerment	0.287	4.225	0.000	1.000	1.000

$R = 0.287$ $R^2 = 0.082$ Adjusted $R^2 = 0.078$ $p < 0.001$

Model 3 above highlights that statistical significance was found between EE and ITS at $p < 0.001$; $p = 0.000$ (t value = 4.225; $R^2 = 0.082$) hence a predictive relationship between EE and ITS is concluded. This finding indicates that approximately 8% of the variance in ITS is contributed by EE and is consistent with Dahou and Hacini's (2018) statement that an organisation can have a highly committed workforce if they consistently practice EE.

H1: There is a significant relationship between EE and CS

In examining the proposed hypothesis, EE was found to have a positive impact on CS hence the relationship is concluded to be statistically significant ($p = 0.000$). This fact means that empowering employees within the organisation greatly contributes to the level of contentment they feel towards their different careers. Confirmation of the hypothesis is consistent with existing literature regarding the association between EE and CS.

H2: There is a significant relationship between CS and ITS

Career satisfaction and ITS presented a statistically significant positive relationship ($p = 0.000$). This finding indicates that CS positively contributes to employees' intention to continue their employment with the organisation. An organisation that contributes towards its employees' career growth can provide sufficient reason for them to decide to continue to work for the organisation. When employees are satisfied with the rate at which their careers are progressing, they become competent, hence it is unlikely that they will terminate their employment with the organisation.

H3: There is a significant relationship between EE and ITS

A significant positive association was confirmed between EE and ITS ($p = 0.000$). The findings are consistent with existing literature which states that when employees are empowered within an organisation, they will desire to continue their employment with that organisation. This emotion results from the fact that empowering employees gives them a sense of belonging, which increases their level of commitment to the organisation.

Table 9 Hypotheses testing outcomes

Hypothesis	Relationship	Beta coefficient	*t*-value	*p*-value	Accepted/Rejected
H1	EE➔CS	0.485	7.830	0.000	Accepted
H2	CS➔ ITS	0.535	8.925	0.000	Accepted
H3	EE➔ITS	0.287	4.225	0.000	Accepted

NB: supported at $p < 0.001$

Table 10 Internal consistency findings of the measuring instrument

Sections	Scale items	Number of items	Cronbach's alpha coefficient
B	B1, B2, B3, B4, B5	5	0.856
C	C1, C2, C3, C4, C5	5	0.889
D	D1, D2, D3, D4, D5	5	0.907

Cronbach's alpha coefficient was utilised to determine internal consistency of the measuring instrument. The reliability findings are reported in Table 10.

Table 10 reflects satisfactory reliability coefficients because all the sections of the measuring instrument scored above the set minimum of 0.70 (Andrew et al. 2011). The reliability scores range from 0.86 to 0.91, highlighting a high degree of internal consistency. To establish content validity, the questionnaire was reviewed by subject experts to check and correct any technical errors. The results of the pilot study revealed that the questionnaire would be capable of evaluating the intended content when carrying out a larger-scale survey, thus, no changes were necessary. To ascertain predictive validity, regression analysis was applied to the collected data to establish and confirm predictive relationships. Relationships were established between EE and CS, CS and ITS, and EE and ITS. Construct validity was determined through the application of correlation analysis. The findings of the correlation analysis revealed the existence of positive associations between EE and CS, CS and ITS, and EE and ITS. This result indicates that there is satisfactory construct validity in confirmed associations.

5 Conclusion

The discovery of the effect of EE and CS on the employees' ITS with the GPW may assist management to use and benefit from the recommendations emanating from this research. This practice would assist GPW to retain employees and not to continually spend resources on recruitment, selection and induction processes. Through this research, the granting of a decision-making authority to lower-level employees is explored. The analysis presented conveys valuable information on the importance of supporting the career growth of employees. This study contributes to existing literature and adds value to the field of Human Resource Management. It is anticipated that its findings will be useful for future research.

Since the study results showed an association between EE and ITS, it, therefore, is necessary to improve EE. Consequently, it is recommended that GPW pays attention to the following aspects to improve EE:

- Identify conditions in GPW that may lead to employees feeling powerless and insignificant and implement practices that will resolve this problem.
- Delegate power to enable work decisions to be taken quicker and to encourage employees' innovativeness.

- Maintain a balance between full management control and the freedom of employees, that entails the capability of voice, i.e., the extent to which employees can effectively express their views, proposals and ideas, as well as having autonomous control of their working processes.
- Encourage lower-level employees and senior management to have open dialogues, during which these employees are granted opportunities to express suggestions and have their voices heard.

The study's results highlighted the existence of a strong relationship between CS and ITS, thus indicating it is necessary to increase CS in order to increase ITS. In order to improve CS among employees, it is recommended that GPW pays attention to the following aspects:

- Provide opportunities for growth and increased responsibility for employees, by providing them with jobs that will simultaneously challenge them and provide a sense of reward and fulfilment.
- Offer recognition and support for employee effort, diligence and commitment.
- Offer study opportunities and funding via bursaries as well as training and development interventions.
- Offer support to employees who are studying part-time in order to improve their careers.

The study's focus was only on the employees of GPW. Due to the sample being limited to one organisation, the results do not allow for the generalisation to other printing organisations, or other government departments. It is recommended that future research incorporate more than one organisation in the printing sector and more than one government department. Since this study employed a quantitative research design, future research on these constructs should use a qualitative or mixed methods approach to gain more understanding of such issues. This research was a once-off study, therefore, it is recommended that follow-up studies be conducted to provide a more balanced and complete view of the studied constructs.

Declaration This paper is an extract from a dissertation submitted to Vaal University of Technology: Mqokoza, A.Z. Employee empowerment, career satisfaction and intention to stay of employees in a printing organisation in Pretoria. Master's dissertation, Vaal University of Technology. (2022).

References

Abdissa, G., Fitwi, T.: Factors affecting performance of micro and small enterprises in south-West Ethiopia: the case of bench Maji, Sheka, and Kefa zones. Glob. J. Manag. Bus. Res. **16**(10), 47–64 (2016)

Agrawal, S., Srivastava, S.: Organizational commitment and career satisfaction among women employees. Indian J. Ind. Relat.S. **54**(1), 132–145 (2018)

Alzalabani, H.A., Nair, R.D.: Employee empowerment and job satisfaction: a case study of RC and SABIC in the Kingdom of Saudi Arabia. Int. Employ. Rev. **17**(1), 21–39 (2011)

Andrew, D.P.S., McEvoy, C.D., Pederson, P.M.: Research methods and design in sport management. Hum. Kinetics. (2011)

Apuke, O.D.: Quantitative research methods: a synopsis approach. J. Bus. Manag. Rev. **6**(10), 40–47 (2017)

Ayketin, I., Erdil, O., Erdogmus, N., Akgun, A.E.: Academicians' career satisfaction: the mediation effect of research productivity. Educ. Sci. Theory Pract. **16**(6), 1921–1945 (2016)

Choi, S.L., Goh, C.F., Adam, M.B.H., Tan, O.W.: Transformational leadership, empowerment and job satisfaction: the mediating role of employee empowerment. Hum. Resour. Health. **14**(73), 1–14 (2016)

Conger, J.A., Kanungo, R.N.: The empowerment process: integrating theory and practice. Acad. Manag. Rev. **13**, 471–482 (1988)

Daniel, E.: The usefulness of qualitative and quantitative approaches and methods in researching problem-solving ability in science education curriculum. J. Educ. Pract. **7**(15), 91–100 (2016)

Das, S., Moharana, T.R.: Employee empowerment and organisational effectiv,eness: an empirical study on central public sector in India. J. Indian Manag. **40-51**, 2249–7803 (2017)

Greenhaus, J.H., Parasurama, S., Wormley, W.M.: Effects of race on organisational experiences job performance evaluations and career outcomes. Acad. Manag. J. **33**(1), 64–86 (1990)

Hanaysha, J.: Examining the effects of employee empowerment, teamwork and employee training on organisational commitment. Procedia–Soc. Behav. Sci., 298–306 (2016)

Ismail, A., Nor, A.M., Mohd Nowalid, W.A.W., Abdullah, A.A.: Career administration, proactive behavior and career satisfaction: a mediating relationship. Int. J. Bus. Manag. Sci. **8**(3) (2018)

Janks, H.: Globalisation, diversity and education: a south African perspective. Educ. Forum. **78**(1), 8–25 (2014)

Khokher, N.A., Raziq, A.: The relationship between employee empowerment and subjective wellbeing. Lahore J. Bus. **5**(2), 81–99 (2017)

Kohli, A., Sharma, A.: An analysis of employee empowerment and job satisfaction: a review. Jarpuria Inst. Manag. **17**(1), 14–20 (2017)

Kumar, P.J., Kumar, A.A.: Employee empowerment: an empirical study. Glob. J. Manag. Bus. Res. **17**(4), 59–64 (2017)

Leedy, P.D., Ormrod, J.E.: Practical Research. Planning and Design, 10th edn. Pearson (2013)

Mahmood, A., Sahar, A.: Impact of psychological empowerment and perceived career support on employee work engagement with the mediating role of affective commitment. Pak. J. Commer. Soc. Sci. **11**(3), 1084–1099 (2017)

Maliwa, N., Joubert, P.A., Dhurup, M.: Job stressors, work tension and job satisfaction of academics at a university of technology. SAJ. Hum. Resour. Manag. **20** (2023)

Mohapatra, M., Mishra, S.: The employee empowerment as a key factor defining organisational performance in emerging markets. Int. J. Bus. Insights Transform. **12**(1), 48–52 (2019)

Mqokoza, A.Z.: Employee Empowerment, Career Satisfaction and Intention to Stay of Employees in a Printing Organisation in Pretoria. Masters dissertation,. Vaal University of Technology (2022)

Mtshali, E. P., Nyakala, K. S., Munyai, T.T., Ramdass, K. Identifying causes of low productivity in the printing industry. In Proceedings of the International Conference on Industrial Engineering and Operations Management, (pp. 933–944). (2018)

Pandey, J.: Structural and psychological empowerment in rural India. Indian J. Ind. Relat.S. **51**(4), 579–593 (2016)

Park, Y.: Empirical investigation on the predictors of career satisfaction. Ind. Commer. Train. **50**(4), 165–171 (2018)

Pranee, S., Pungnirund, B., Saeng-On, P., Pulphon, S.: The influence of servant leadership and CSR practices on the employee's intention of stay in the pharmacy firms: mediating role of empowerment. Syst. Rev. Pharm. **11**(3) (2020)

Prathiba, S.: A study on impact of employee empowerment and employee engagement on organisational commitment. South Indian Educ. Soc. J. Manag. **12**(2), 45–54 (2016)
Priyadharshany, A.J., Sujatha, S.: Does structural empowerment impact on job satisfaction via psychological empowerment? A mediation analysis. Sona. Glob. Manag. Rev. **10**(1), 23–43 (2015)
Redelinghuys, K., Bothma, E.: Person-environment fit, job satisfaction and intention to leave: the moderating effect of leader empowering behaviour. J. Psychol. Afr. **26**, 11–21 (2016)
Rozsa, Z., Formánek, I., Maňák, R.: Determining the factors of the employees' intention to stay or leave in the Slovak's SMEs. Int. J. Entrep. Knowl. **7**(2), 63–72 (2019)
Shahid, A.: Employee intention to stay: an environment based on trust and motivation. J. Manag. Res. **10**(4), 58–71 (2018)
Shanker, M.: Organisational commitment and intention to stay of employees in Indian companies. J. Psychosoc. Res. **8**(2), 199–208 (2014)
Srivastava, S., Madan, P.: A measure of employee empowerment in Indian work setting. Indian J Ind RelatS. **54**(2), 296–307 (2018)
Takawira, N., Coetzee, M.: Psychosocial resources, support, and career satisfaction among professional women: a south African study. J. Psychol. Afr. **29**(3), 208–216 (2019)
Turkmeglou, M.A.: Investigating benefits of benefits and drawbacks in hospitality sector. Int. Res. Bus. Stud. **12**(1), 1–13 (2019)
Wayne, S.J., Shore, L.M., Liden, R.C.: Perceived organisational support and leader-member exchange: a social exchange perspective. Acad. Manag. J. **40**, 82–111 (1997)
Woon, W., Tan, C.L., Nasurdin, A.M.: Linking organizational climate, psychological ownership and intention to stay: a proposed model. Glob. Bus. Manag. Res. **9**(1), 127–141 (2017)
Yao, C., Huang, P.: Effects of leadership style on job satisfaction and intention to stay in shipping industry. J. Coast. Res. **83**, 796–801 (2018)
Yap, M., Cukier, W., Holmes, M.R., Hannan, C.A.: Career satisfaction: a look behind the races. Ind. Relat.S J.S. **65**(4), 584–608 (2010)

Exploring Issues of Leadership in a Selected Municipality in Northwest Province

Julian Makgabo Nkoana and Henk de Jager

Abstract A municipality's primary function is to provide basic services, such as water, electricity, and infrastructure, thereby creating a conducive environment for socio-economic development. Efficient leadership from municipal officials is often cited as the main reason for the lack of these basic services. This study aimed to investigate the predominant leadership issues in a specific municipality of the Northwest province. The study employed a qualitative research approach, using a purposive sampling technique to select six senior officials who met specific inclusion criteria. The study used thematic analysis to analyse the data, adhering to all ethical protocols in the study. The study primarily identified that ethical leadership is crucial for developing trust between municipal officials and the community. For this reason, ethical leadership is key to how municipal officials execute their duties while adhering to sound governance principles, practices, and accountability. Open and honest communication with all relevant stakeholders is crucial in addressing structural shortcomings within the municipality, thereby creating a conducive environment for socio-economic development. Active community participation through all relevant platforms is strongly encouraged to develop an open and transparent environment. The importance of the equitable distribution of resources between cities and rural communities is strongly emphasised. Lastly, the application of technology to address shortcomings in various systems and procedures is critical for good governance and upholding ethical standards.

Keywords Governance · Municipalities · Service Delivery · Ethics · Leadership

1 Introduction

Leadership within a municipal environment is pivotal for driving efficiency, effectiveness, and compliance with organisational mandates. Recent scholarship highlights leadership as a multi-dimensional process involving the capacity to influence, inspire,

---

J. M. Nkoana · H. de Jager (✉)
STADIO Higher Education, Johannesburg, South Africa
e-mail: henkd@stadio.ac.za

E. H. Redda (ed.), *Proceedings of the Social Sciences International Research Conference, 2025*, https://doi.org/10.1007/978-981-95-9489-4_13

and guide teams toward achieving set objectives within ethical and strategic frameworks (Northouse 2023). For this study, leadership is conceptualised as a dynamic and adaptive process where leaders mobilise resources, foster collaboration, and drive innovative strategies to enhance performance and service delivery outcomes (DuBrin 2018).

In the context of municipalities, leadership extends beyond administrative duties to encompass the ability of executive mayors and municipal managers to mobilise communities and align them with developmental goals. This requires a synthesis of political acumen and managerial expertise to navigate complex local governance structures (Tshishonga 2013). Effective leadership is crucial in addressing challenges related to service provision, which encompasses access to critical infrastructure, public transportation, waste management, and electricity (YuSheng and Ibrahim 2019). The 2022 Auditor-General's Performance Report identifies key inefficiencies such as the over-reliance on consultants due to inadequate technical and financial expertise, particularly in infrastructure projects (Auditor-General, 2022). Vacancies and instability in critical roles, such as chief financial officer and municipal manager, exacerbate these challenges, leading to governance lapses and the institutionalisation of weak financial and performance management controls (Mamokhere 2024).

The need for a cohesive and visionary approach to service delivery underscores the role of leadership in municipal service delivery. Research indicates that leadership has a significant influence on the ability to expedite basic service delivery and improve living conditions for the disadvantaged (Dlamini et al. 2021). In this regard, Tshabalala (2020) argues that municipal leadership should prioritise inclusive governance and capacity building to address systemic inefficiencies. This is echoed by YuSheng and Ibrahim (2019), who describe service delivery as the equitable distribution of essential services to improve societal well-being. Despite three decades of democracy, the persistence of malfeasance and insufficient internal checks has prompted communities to resort to protests as a means of holding leadership accountable (Mamokhere 2024). The main objective of the research is to explore issues of leadership in a selected municipality to address some of these gaps.

2 Literature Review

2.1 *Introduction*

The significance of leadership in politics and governance underscores its role in influencing the goals and strategies of governments worldwide (Ramolobe 2023). In the political arena, leaders are often mandated by constitutional frameworks to wield authority, make decisions, and drive development. Municipal leadership, as a microcosm of governance, presents unique challenges due to the intertwining of political and administrative responsibilities.

2.2 Conceptualising Leadership

Leadership remains a multifaceted and complex concept, as noted by scholars across various disciplines (Oc 2018). Despite its complexity, Rudolph et al. (2020) identify four fundamental principles that underpin leadership. The first principle is that leadership is an activity and process, emphasising its dynamic and evolving nature as leaders guide their followers. The second principle revolves around influence, persuasion, and exemplary behaviour, where leaders inspire and motivate others by embodying the values and goals of the organisation. The third principle emphasises the leader-follower relationship, founded on trust, effective communication, and mutual respect, which fosters collaboration and shared objectives. Finally, the fourth principle evaluates leadership through the achievement of goals and objectives, with effective leaders steering collective efforts toward achieving predefined aims.

In a democratic government, effective leadership entails encouraging people to be accountable, work together, and be responsive to the diverse needs of various groups. This is especially important for addressing the challenges that South African cities and towns face in delivering services and managing their governments. Mamokhere (2024) emphasises the importance of ethical leadership in mitigating governance failures and ensuring effective service delivery, a view supported by Ramolobe (2023), who highlights the significance of strong leader-follower relationships in municipal leadership.

2.3 Conceptualising Governance

Keping (2018) characterises good governance as the foundation of effective and ethical governance practices. This perspective emphasises principles such as transparency, accountability, and responsiveness in fostering sustainable governance. Reddy et al. (2020) expand on this, highlighting governance as the interaction and participation of multiple stakeholders, including communities, civic associations, and institutions, in the formulation and implementation of public policy. Collaboration and engagement are essential to the success of policymaking and governance.

Good governance requires public participation, collective action, and collaboration. It requires meaningful involvement platforms and effective municipal service administration and delivery strategies (Keping 2018; Ramolobe 2023). This approach requires ethical leadership to prioritise community needs through equitable and inclusive policies. Governance is a dynamic process that involves stakeholders in decision-making and problem-solving (Mawuko-Yevugah and Attipoe 2021).

2.4 Conceptualising Service Delivery

YuSheng and Ibrahim (2019) define service delivery in municipalities as the distribution of government services and advantages to the community, ensuring that basic services such as water, sanitation, housing, and electricity are delivered to communities. They emphasise that when the government provides services effectively, it is fulfilling its responsibility to address societal needs and achieve its goals and objectives. Furunes and Mkono (2019) add that service delivery is measured by the satisfaction of citizens with the services they receive, highlighting the importance of municipal efforts in allocating these services to the public. What will be the enablers for good service delivery?

2.5 African Leadership and Governance

Leadership and governance challenges in Africa have remained complex and deeply rooted in the continent's socio-political fabric. Obiora (2019) asserts that these challenges are diverse and often difficult to diagnose precisely. Issues such as inadequate infrastructure, governance inefficiencies, and human rights violations have persisted, raising concerns about the commitment of African leaders to improve their nations' conditions. According to Henn (2023), unethical behaviour among public officials, often stemming from the highest levels of leadership, has created a trickle-down effect, normalising unethical behaviour and undermining governance systems. Mawuko-Yevugah and Attipoe (2021) liken this deterioration in governance to a disease that gradually erodes the foundations of societal progress.

One of the significant governance issues highlighted by Obiora (2019) is the misconception surrounding democratic practices in Africa. Many leaders have maintained prolonged tenures, often establishing political dynasties by placing family members in positions of authority. This trend has slowed the growth of democracy and led to systems of government that value loyalty over skill. Mawuko-Yevugah and Attipoe (2021) stress the need for leaders who are known for their real efforts rather than their empty words.

2.6 Implications for Leadership

Leadership is inherently tied to ethical responsibility, requiring leaders to be accountable for their actions and the behaviours they inspire in others. According to Ramolobe (2023), moral leadership involves setting an example for ethical behaviour in both personal and professional contexts, thereby promoting a culture of integrity within organisations. Ethical leaders act as moral managers, leveraging their resources and influence to encourage ethical behaviour among employees, particularly in the realm of public service delivery. In municipal governance, where service delivery is a primary mandate, leaders exert significant influence on outcomes, even if they are not directly involved, through guidance and strategic oversight. However, the ethical principles conveyed by moral leaders can sometimes become overly rigid, limiting their effectiveness. Leadership in the South African context presents additional challenges. Issues such as the continued remuneration of leaders with documented unethical behaviour highlight the need for systemic reforms to ensure accountability (Mawuko-Yevugah and Attipoe 2021).

2.7 Issues Affecting Municipal Leadership in the South African Municipalities

Mbandlwa et al. (2020) note that South Africa is home to diverse communities, each facing issues shaped by the population's needs and the government's response to them.

Inadequacy in Management, Institutional Capacity, Skills Inequality, and Incapable Staffing. South African municipalities face challenges due to a shortage of qualified staff. This makes it harder for them to meet the community's needs and fulfil their roles as service providers. This lack of sufficiency leads to widespread dissatisfaction and results in repeated service failures. For these problems to be solved and for city services to run smoothly, the skills gap must be closed, and properly trained staff must be hired (Kalonda and Govender 2021; Ramolobe 2023). Mbandlwa et al. (2020) state that South African municipalities face challenges related to administrative competency and a lack of resources, which makes it difficult for them to provide good services.

Political Interference, Budgeting Flaws, and Limited Budget. Zweni (2023) claims that local government has become increasingly vital in recent decades, providing essential services to improve the lives of its inhabitants. However, poor budgeting practices continue to limit the capacity of towns to fulfil their mandates by underfunding key services. Many of these processes are inefficient, opaque, and misallocate resources due to low funding. A lack of resources hinders staff training, infrastructure development, and maintenance, resulting in service delivery failures. These systemic concerns harm the quality of life and socio-economic growth in South African communities (Ramolobe 2023; Mbandlwa et al. 2020).

Unethical leadership harms service delivery, as evidenced by the public fund misappropriation in certain areas (Mamokhere 2024). According to Mawuko-Yevugah and Attipoe (2021), political meddling undermines local governance autonomy and efficiency, thereby complicating municipal service delivery.

Inadequate Community Consultations. Mbandlwa et al. (2020) highlight that local governments continue to struggle to recognise and assist with citizens' problems, as they do not receive adequate input. Mawuko-Yevugah and Attipoe (2021) argue that South African municipal authorities struggle to understand the needs, wants, and problems of their constituents because they do not engage effectively with community members, stakeholders, and experts. This lack of consultation often leads to choices based on insufficient information or population needs, which hinders the success of service delivery efforts.

Poverty and High Unemployment Rate. Nkhabu (2021) and Matyana and Thusi (2023) note that high unemployment and poverty rates in communities significantly impair the ability of municipal leadership to provide effective services in South Africa. Leaders must meet the demands of marginalised groups within budgetary and capability limits. Poverty and unemployment exacerbate societal tensions and instability, leading to rallies, strikes, and other service disruptions (Mamokhere 2024). Mbandlwa et al. (2020) note that community poverty and high unemployment, especially among recent graduates, sometimes lead to protests and nepotism.

2.8 Conclusions

This section covered key government and leadership topics. The aim was to enhance discipline-specific expertise by using both published and unpublished sources, which were then analysed to provide a theoretical foundation for the study. Governance, service delivery, and leadership were also examined as part of a holistic understanding of the topic. The next section discusses the research methodology and techniques employed in this study.

3 Methodology

3.1 Research Design and Approach

This study used constructivism within an ontological philosophy. Constructivism posits that knowledge is constructed through human experience and interaction rather than being based on objective truth (Hossain et al. 2024). Ontology, the study of reality, examines facts, values, behaviours, and the interplay between objectivity and subjectivity (Pham 2024). The study used a qualitative research approach. This approach was selected to consider the participants' experiences and worldviews, as it facilitates an in-depth exploration of meanings and insights that provide a comprehensive understanding of the phenomenon (Mohajan 2018).

3.2 Sampling Method

In line with a qualitative approach, the study used purposive sampling to select the participants. Purposeful sampling aims to identify participants who meet certain criteria, ensuring that the study includes a range of experts' views that are important to the research problem (Busetto et al. 2020). For this study, this method proved especially effective, as it enabled the collection of rich, detailed data aligned with the study's objectives through the purposive selection of six individuals with specific knowledge and experience (Fife and Gossner 2024). Participants selected for the study were in top management positions with at least 10 years' experience in local government and held a professional qualification.

3.3 Data Collection

Primary data were gathered through semi-structured interviews, which align with the qualitative approach adopted for this study. Semi-structured interviews were chosen as they provided the flexibility to explore participants' perspectives in depth while maintaining a focused discussion on the key aspects of the phenomenon being studied (Bell et al. 2022). This strategy collects extensive, contextual data needed to comprehend complicated social phenomena.

3.4 Data Analysis

The researcher analysed the collected data using thematic analysis, a method that is particularly effective for identifying, analysing, and describing patterns in qualitative data (Bell et al. 2022). The thematic analysis involved identifying phrases, terms, and ideas that appeared repeatedly, revealing the subjects' points of view and the study's objectives. The data were organised and entered into Excel sheets to facilitate analysis and grouping in a structured manner. It is common in qualitative research to derive codes directly from the data (Naeem et al. 2024), and this study employed coding as a primary method for analysing qualitative data.

4 Results and Discussion

Ethical leadership has a significant impact on how the municipality is managed and services are delivered, aligning the municipality with the public administration's ethical imperative. The study found that ethical leadership is crucial for developing trust, responsibility, and openness in the public sector, as well as addressing structural issues that hinder service delivery. Openness and responsibility, when properly implemented, create stakeholder trust and streamline operations. Madumo (2022) postulates that ethical leadership is necessary for long-term government, especially in places where dishonest and poor management is common. Similarly, Mamokhere (2024) suggests that ethical leadership fosters a culture of honesty and ensures that public resources are used fairly and effectively.

4.1 Participants' Profiles and Themes

The participants' ages ranged from 45 to 60 years; 57% had 10–15 years of work experience, and 43% had more than 16 years, which indicates a level of maturity and adequate experience in a municipal environment. At the same time, 86% of the participants held a master's degree, which demonstrates a sufficient level of knowledge in managing municipalities. Table 1 represents the main themes identified through thematic analysis.

Theme 1—The Significance of Leadership. Leadership is essentially about bringing constructive change, motivating others to reach their full potential, and resolving conflicts with vision and integrity. Innovation, resilience, and organisational success are all facilitated by effective leadership in a fast-paced world.

Sub-Theme 1.1. Impact of Unethical Behaviour. The study highlighted the issue of competent, ethical leadership, which is necessary to restore the city's systems. Eduzor (2024) postulates that immoral behaviour hurts both the performance of organisations and the public's trust in them. This underscores the similarity between the current findings and previous studies, highlighting the importance of ethical behaviour in government, especially when resources are limited, and the public is critical of the performance of government entities.

Table 1 Themes and sub-themes from interviews

Themes and sub-themes	Frequency
1. The significance of leadership.	45
1.1. Impact of unethical behaviour.	23
1.2. Trust deficit in leadership.	22
2. The role of leadership in the selected municipality.	108
2.1. Community participation as accountability mechanisms.	57
2.2. Collaborative structures.	51
3. The improvement of ethical leadership.	79
3.1. Stakeholder collaboration.	59
3.2. Resource allocation fairness.	20

Participants emphasised the importance of ethical leadership in limiting unethical behaviour in the selected municipality. Afzal and Tumpa (2024) agree and state that ethical leadership stops people from being dishonest and encourages a culture of fairness and responsibility.

Unethical behaviour in procurement processes leads to irregular expenditures and delays in service delivery. (Participant 3).

Some politicians use their influence to bypass established procedures, creating systemic inefficiencies. (Participant 2).

The data show that unethical behaviour permeates every aspect of municipal operations, especially procurement, where a lack of openness and accountability leads to inefficiency, delayed services, and public dissatisfaction.

Sub-Theme 1.2. Trust Deficit in Leadership. A lack of trust has emerged as an issue in the selected municipality, primarily because of unkept service delivery promises, limited public involvement, and the pervasive influence of politics on processes and decisions. In this regard, Afzal and Tumpa (2024) note that inconsistent ethical leadership breeds scepticism and undermines institutional integrity. This study revealed that favouritism, resource mismanagement, and a lack of openness in municipal leadership are eroding accountability and creating a significant trust deficit in leadership. Vogel et al. (2021) indicate that leadership by example fosters trust and promotes ethics across governance levels. Concurrently, Agarwal (2024) agrees that ethical leadership sets a moral tone and shapes organisational culture.

The public no longer trusts the municipality due to unmet service delivery promises. (Participant 5).

Ward councillors fail to communicate effectively, leading to widespread frustration. (Participant 6).

The study revealed a substantial divide between city leaders and the community, stemming from unfilled promises and a lack of communication, which leads to community scepticism and disengagement with municipal projects.

Theme 2: The Role of Leadership in the Selected Municipality. The lack of vision, openness, and decisiveness in service delivery is the result of poor managerial and political leadership. The public's opinion of the government's capacity to deliver services effectively or practise good governance is impacted by unethical behaviour and maladministration in the public sector, which are not only perceived as individual acts but as collective acts committed by the entire leadership.

Sub-Theme 2.1. Community Participation as Accountability Mechanisms. Community involvement has become a key part of holding municipal leaders accountable, but the results often tell a different story. Public forums, town hall meetings, and participatory budget consultations are meant to bring people and city officials closer together, but they have not always provided the desired results. The participants agree that, although systems exist, participation and attendance are often limited due to a lack of trust, poor communication, and logistical challenges. Communities indicate that the lack of openness and conversation between city leaders and the community undermines trust, making it harder for people to work together. Wang (2022) states that trust is a crucial aspect of effective leadership, particularly in public administration, where leaders are accountable to the people they serve. Park et al. (2022) also emphasise that open communication is essential for engaging people and fostering trust. In this regard, Mamokhere (2024) concurs that weak community participation tools aggravate municipal government issues.

Community meetings are announced but often too late for many people to attend, especially those in rural areas. (Participant 2).

The platforms for participation exist, but they're not accessible to everyone, especially those without internet or transportation. (Participant 4).

We don't see our suggestions being taken seriously or acted upon, which makes people disengage over time. (Participant 5).

Sub-Theme 2.2. Collaborative Structures. The study suggests that collaborative arrangements, particularly committees, are essential to enhance the accountability of municipalities. The reality often indicates that not all structures function as well as they should due to challenges such as unclear roles, insufficient resources, and poor coordination. Van Assen (2020) claims that oversight committees can be more effective with sufficient resources and autonomy. This study supports this view by emphasising explicit terms of reference, periodic performance reviews, and the improved integration of internal and external committees. Fernandes et al. (2022) agree that robust performance monitoring systems are necessary to match organisational goals with ethical governance.

Our committees need better funding and clear guidance on their roles to operate effectively. (Participant 3).

External committees have excellent ideas, but their input often gets lost since there's no proper system to integrate their recommendations. (Participant 6).

Collaboration between different departments and with the public is key to accountability, but we're not there yet. (Participant 6).

This study further indicates that performance monitoring systems can enhance data collection, analysis, and reporting by incorporating technology. Elmatsani et al. (2024) support the use of digital tools to promote municipal transparency and efficiency. However, technological integration alone is insufficient and must be supported by accountability and regular audits to achieve meaningful results, as the study postulates.

Theme 3: The Improvement of Ethical Leadership. Ethical leadership encourages maximum employee engagement, which improves culture, productivity, and openness. Honesty, justice, community, respect, and integrity are the tenets of ethical leadership. Madumo (2022) states that leadership training in South African municipalities typically lacks contextual relevance, leaving leaders unprepared to handle complicated ethical issues. According to Eduzor (2024), targeted leadership training can improve organisational performance and ethical decision-making. Wang (2022) indicates that although most training programmes claim to embed ethical values, they often fail to do so.

Sub-Theme 3.1. Stakeholder Collaboration. All participants agreed that strong intergovernmental relations (IGR) are needed to ensure that everyone is held accountable and to guarantee that sufficient services are delivered at the local government level. It became clear that national, provincial, and local governments need to collaborate to ensure that policies, funding, and strategies align with community needs.

However, participants identified certain challenges with the current IGR framework, including the clarity of communication among stakeholders and the definition of roles and responsibilities. According to Van Assen (2020), collaborative governance models align strategies and maximise resources. Mamokhere (2024) asserts that governance issues necessitate clear communication and coordinated actions from all parties. However, the participants indicated that political interference and resource discrepancies typically hinder this collaboration. In this regard, Fernandes et al. (2022) caution that politics can either aid or hurt collaboration.

Intergovernmental relations should not just be about compliance but about real collaboration. All tiers of government need to align their priorities to avoid duplication of efforts. (Participant 1).

There is a lack of clear communication channels between provincial and local governments. This often results in delayed projects and unmet community needs. (Participant 3).

Political agendas sometimes hinder cooperation between different levels of government, which undermines accountability and trust. (Participant 5).

If we want effective service delivery, the roles and responsibilities of each stakeholder need to be clearly defined and adhered to. (Participant 1).

Resource allocation disparities make collaboration difficult. Smaller municipalities often struggle to keep up with their provincial counterparts. (Participant 4).

The study emphasised the significance of intergovernmental interactions in fostering collective accountability. According to their contribution, successful teamwork requires efficient communication and clear role clarification. The lack of a shared accountability framework often undermines collaboration, as highlighted by the participants. Government layers' misaligned agendas are frequently the cause of duplication or resource waste.

Sub-Theme 3.2. Resource allocation fairness. The participants agreed that the issue of fair resource allocation is key to providing equitable services at the local government level. Madumo (2022) concurs, emphasising the importance of fair resource distribution for ethical government. Participants indicated that the unfair division of resources is a major challenge, frequently caused by political favouritism, geographical bias, and differences in income and social class. These actions were perceived as contradicting the principles of fairness and equitable access to public goods and services. Several participants stated that the current systems for allocating resources are unclear and do not adequately consider the specific needs of marginalised groups. As a result, areas that are not well-developed continue to struggle to access services, thereby exacerbating the gap between cities and rural areas. In line with Van Assen's (2020) assertion, the most crucial factor in ensuring review boards function effectively is to provide them with the necessary power through effective leadership and sufficient resources. Vogel et al. (2021) also discuss how ethical leadership can help make people more responsible, which in turn makes methods for monitoring and oversight more effective.

Resource allocation is often skewed toward urban areas, leaving rural communities underfunded and underdeveloped. (Participant 1).

Favouritism, whether political or economic, creates inequality in service delivery, which fuels resentment among marginalised groups. (Participant 4).

Without a clear, needs-based allocation system, smaller municipalities will always lag. (Participant 2).

Transparency in resource distribution is critical to rebuilding trust between government and citizens. (Participant 6).

The criteria for resource allocation are not always communicated to stakeholders, which leads to suspicions of favouritism. (Participant 5).

Participants noted that the lack of a needs-based approach exacerbates inequality and restricts vulnerable individuals' access to basic services. Urban communities receive greater funding due to their political and economic influence, leaving rural municipalities without sufficient funds for essential infrastructure and services.

5 Recommendations

The selected municipality must prioritise institutional improvements that strengthen ethical leadership to overcome the difficulties identified in this study. Structured, context-specific leadership training is essential. All programmes should include municipal leadership-specific modules on ethical decision-making, governance compliance, and conflict resolution. Madumo (2022) emphasises local governance training to improve leader competency. Eduzor (2024) highlights that leadership training must include practical scenarios to connect theoretical and real-world knowledge. Mamokhere (2024) cautions that training without evaluation is insufficient. The municipality must implement monitoring systems to evaluate all programmes, ensuring that training leads to improved leadership.

Stronger supervision systems are another important step toward municipal accountability and transparency. Participants emphasised the importance of providing the Municipal Public Accounts Committee (MPAC) with sufficient resources and autonomy to ensure compliance. Van Assen (2020) agrees that strong oversight bodies promote ethical government. Fernandes et al. (2022) emphasise the requirement for oversight committee performance reviews to ensure accountability. Agarwal (2024) argues that political influence often weakens all bodies. The municipality should insulate oversight committees from external influences, allowing them to operate independently and adhere to governance norms.

The municipality should use technology to enhance transparency and foster citizen participation. E-governance solutions facilitate real-time communication with stakeholders and enhance accessibility to municipal services. Digital transformation can address governance inefficiencies and regain public trust (Elmatsani et al. 2024). According to Eduzor (2024), digital tools simplify decision-making and increase responsibility. Park et al. (2022) noted that the municipality must address rural infrastructure and digital literacy deficiencies. A hybrid paradigm that blends traditional and digital involvement could assure inclusivity while maximising technology's benefits.

Comparative studies across various municipalities should be added to future research to cover a wider range of governance problems and leadership styles. Madumo (2022) says that cross-regional studies are very helpful for understanding systemic problems and best practices that can help with changing the way government works. Mamokhere (2024) also says that comparison studies can find patterns and trends that are not clear from research that is only done in one place.

6 Conclusion

This study suggests that ethical leadership is essential in addressing systemic governance challenges within the selected municipality. The findings suggest that municipal leadership requires transparency, accountability, and community engagement. Madumo (2022) recommends ethics-focused leadership reforms to address systemic inefficiencies and restore public trust. Mamokhere (2024) asserts that ethical leadership is crucial for effective resource allocation and efficient public service. This study also suggests that entrenched unethical behaviour, political power, and poor controls make ethical governance difficult. Agarwal (2024) concurs that systemic contexts with resistance to change make ethical leadership challenging.

Structured leadership training, digital tools for transparency, and oversight bodies for accountability are strongly supported as effective measures to address ethical issues in leadership. Afzal and Tumpa (2024) postulate that technology and strong institutions strengthen governance. At the same time, Eduzor (2024) supports local government-specific leadership training to improve ethical behaviour. The study ultimately postulates that ethical leadership is necessary for sustainable growth and effective public administration.

References

Afzal, F., Tumpa, R.: Exploring leadership styles to foster sustainability in construction projects: a systematic literature review. Sustainability. **16**(3), 971 (2024). https://doi.org/10.3390/su16030971

Agarwal, P.: Exploring the Foundations and Impact of Ethical Leadership: a Multidimensional Analysis, 6th edn. Sage, London (2024)

Bell, E., Bryman, A., Harley, B.: Qualitative Data Analysis, 6th edn. Oxford University Press, Oxford (2022)

Busetto, L., Wick, W., Gumbinger, C.: How to use and assess qualitative research methods. Neurol. Res. Pract. **2**, 7 (2020). https://doi.org/10.1186/s42466-020-00059-z

Dlamini, M., Adam, E., Chirima, G., Hamandawana, H.: A remote sensing-based approach to investigate changes in land use and land cover in the lower uMfolozi floodplain system, South Africa. Trans. R. Soc. S. Afr. **76**(1), 13–25 (2021). https://doi.org/10.1080/0035919X.2020.1858365

DuBrin, A.J.: Tolerating Ambiguity for Leadership and Professional Effectiveness. Page Publishing, Staten Island (2018)

Eduzor, N.C.: Leadership styles and organisational performance: examining the impact of transformational leadership on employee engagement and business success. Int. J. Dev., Sustain. Environ. Manag. **4**(1), 16–24 (2024). https://doi.org/10.5281/zenodo.10500674

Elmatsani, H.M., Widianingsih, I., Nurasa, H., Munajat, M.D.E., Suwanda, S.: Exploring the evolution of leadership in government: a bibliometric study from the e-government era into the digital age. Cogent Soc. Sci. **10**(1), 1–27 (2024). https://doi.org/10.1080/23311886.2024.2414877

Fernandes, C.I., Veiga, P.M., Ferreira, J.J., Rammal, H.G., Pereira, V.: Assessing strategic leadership in organisations: using bibliometric data to develop a holistic model. J. Bus. Res. **7**(3), 45–87 (2022)

Fife, S.T., Gossner, J.D.: Deductive qualitative analysis: evaluating, expanding, and refining theory. Int J Qual Methods. **23** (2024). https://doi.org/10.1177/16094069241244856

Furunes, T., Mkono, M.: Service-delivery success and failure under the sharing economy. Int. J. Contemp. Hosp. Manag. **31**(8), 3352–3370 (2019). https://doi.org/10.1108/IJCHM-06-2018-0532

Henn, S.J.: Complements or substitutes? How institutional arrangements bind traditional authorities and the state in Africa. Am. Polit. Sci. Rev. **117**(3), 871–890 (2023). https://doi.org/10.1017/S0003055422001137

Hossain, M.S., Alam, M.K., Ali, M.S.: Phenomenological approach in qualitative studies: data collection and saturation. ICRRD Qual. Index Res. J. **5**(2) (2024) doi.org/10.53272/icrrd.v5i2

Kalonda, J.K., Govender, K.: Factors affecting municipal service delivery: a case study of Katima Mulilo town council. Namibia. Afr. J. Public Aff. **12**(2), 1–26 (2021)

Keping, Y.: Governance and good governance: a new framework for political analysis. Fudan J. Humanit. Soc. Sci. **11**, 1–8 (2018). https://doi.org/10.1007/s40647-017-0197-4

Madumo, O.: Contextualising Leadership Challenges in Municipalities: a Developmental Perspective, 6th edn. Sage, London (2022)

Mamokhere, J.: Why do leaders fail? Issues affecting leadership in achieving sustainable service delivery in south African municipalities. J. Public Adm. Dev. Altern. **7**(3), 34–76 (2024) https://doi.org/10.55190/200BJPADA.2024.309

Matyana, M., Thusi, X.: Unemployment and poverty in South Africa: assessing the National Development Plan 2030 predictions. Int. J. Dev. Sustain. **12**(6), 212–226 (2023)

Mawuko-Yevugah, L., Attipoe, H.A.: Chieftaincy and traditional authority in modern democratic Ghana. S. Afr. J. Philos. **40**(3), 319–335 (2021). https://doi.org/10.1080/02580136.2021.1964206

Mbandlwa, Z., Dorasamy, N., Fagbadebo, O.M.: Leadership challenges in the south African local government system. J. Crit.Al Rev. **7**(13), 1642–1653 (2020)

Mohajan, H.K.: Qualitative research methodology in social sciences and related subjects. J. Econ. Dev. Environ. People. **7**(2), 23–48 (2018)

Naeem, M., Ozuem, W., Howell, K., Ranfagni, S.: Demystification and actualisation of data saturation in qualitative research through thematic analysis. Int J Qual Methods. **23** (2024). https://doi.org/10.1177/16094069241229777

Nkhabu, M.F.: The Role of Local Government in the Implementation of the Sustainable Development Goals: a Case of Tshwane Municipality. Master's thesis,. University of Pretoria (2021)

Northouse, P.G.: Leadership: Theory and Practice, 10th edn. Sage, Thousand Oaks (2023)

Obiora, S.: Impact of leadership and governance on poverty alleviation in Africa. SSRN (2019). https://doi.org/10.2139/ssrn.3576621

Oc, B.: Contextual leadership: a systematic review of how contextual factors shape leadership and its outcomes. Leadership Quarterly. **29**(1), 218–235 (2018). https://doi.org/10.1016/j.leaqua.2017.12.004

Park, J., Han, S.J., Kim, J., Kim, W.: Structural relationships among transformational leadership, affective organisational commitment and job performance: the mediating role of employee engagement. Eur. J. Train. Dev. **46**(9), 920–936 (2022)

Pham, C.H.: Qualitative data analysis. In: Advances in Educational Technologies and Instructional Design. Sage, London (2024)

Ramolobe, K.S.: Dynamics of traditional leaders' relationships with municipal councillors and service delivery. J. Local Gov.Ment Res. Innov. **4**, 95 (2023). https://doi.org/10.4102/jolgri.v4i0.95

Reddy, S., Allan, S., Coghlan, S., Cooper, P.: A governance model for the application of AI in healthcare. J. Am. Med. Inform. Assoc. **27**(3), 491–497 (2020)

Rudolph, C.W., Murphy, L.D., Zacher, H.: A systematic review and critique of research on healthy leadership. Lead.Ship Q. **31**(1), 101335 (2020). https://doi.org/10.1016/j.leaqua.2019.101335

Tshabalala, T.J.: Principals' Narratives on Balancing Leadership, Management and Instruction in South African Schools. Doctoral dissertation, University of the Free State (2020)

Tshishonga, N.: Emotional intelligence and the enhancement of humane local political leadership at local government level. J. Public Adm. **48**(2) (2013)

Van Assen, M.F.: Empowering leadership and contextual ambidexterity: the mediating role of committed leadership for continuous improvement. Eur. Manag. J. **38**(3), 435–449 (2020). https://doi.org/10.1016/j.emj.2019.12.002

Vogel, B., Reichard, R.J., Batistič, S., Černe, M.: A bibliometric review of the leadership development field. Lead.Ship Q. **32**(5), 101497 (2021)

Wang, C.J.: Mechanisms linking transformational leadership, perceived organisational support, creativity and performance in hospitality. Behav. Sci. **12**(10), 406 (2022). https://doi.org/10.3390/bs12100406

YuSheng, K., Ibrahim, M.: Service innovation, service delivery and customer satisfaction and loyalty in Ghanaian banking. Int. J. Bank Mark. **37**(5), 1215–1233 (2019). https://doi.org/10.1108/IJBM-06-2018-0142

Zweni, N.: Employee participation in training and development interventions at a municipality. SA J. Hum. Resour. Manag. 21, 1–10 (2023). doi:https://doi.org/10.4102/sajhrm.v21i0.225865

An Analysis of South Africa's Entrepreneurship Policy Environment in Fostering Youth Entrepreneurship Development

Marcia Lebambo

Abstract Entrepreneurship has been emphasised as crucial to the country's economy, particularly in light of ongoing efforts to promote development through entrepreneurship policies and supportive institutions since 1994. Despite nearly three decades of dedicated initiatives, South Africa continues to face significant challenges, including elevated unemployment rates, rising inequality, persistent poverty, and low levels of entrepreneurial activity among the youth. The unemployment rate among young individuals (aged 15–34) stands at 45.5%, and South Africa's supportive environment for entrepreneurship is rated lower than global averages. This paper examined the effectiveness of the entrepreneurship policy environment in creating a supportive environment for youth-owned businesses to develop and thrive. An interpretivist paradigm was employed to gain a deeper understanding of the social phenomenon of youth entrepreneurship. This approach allowed participants to express their experiences and beliefs in their own words. Using an exploratory qualitative design, the study involved 19 youth entrepreneurs who participated in in-depth interviews. Thematic analysis was used to interpret and analyse the transcripts from the data collected. The study concludes that while policy intentions and proclamations are positive, the lack of understanding of the unique challenges faced by youth-owned business environments renders the implementation of support institutions' programmes and initiatives ineffective. Finally, the paper provides recommendations within academic and practical contexts. Among others, the paper recommends that for policy to be impactful in the context of youth entrepreneurs in South Africa, the following are key factors: (1) assess policy effectiveness, (2) examine policy alignment, and (3) explore systemic policy factors.

Keywords Entrepreneurship · Policy environment · Youth entrepreneurship development

M. Lebambo (✉)
Central University of Technology, Free State, Bloemfontein, South Africa
e-mail: Mlebambo@cut.ac.za

E. H. Redda (ed.), *Proceedings of the Social Sciences International Research Conference, 2025*, https://doi.org/10.1007/978-981-95-9489-4_14

1 Introduction

Youth entrepreneurship has emerged as a pivotal catalyst for economic development and social inclusion on a global scale. When young individuals are empowered to innovate, create employment opportunities, and address societal challenges, they significantly contribute to both social inclusion and economic advancement (Gopala and Shekhar 2024). However, developing countries such as South Africa are facing challenging economic conditions, experiencing a double-edged sword effect where low economic growth hinders entrepreneurial activities, and the lack of entrepreneurial dynamics further militates against economic development. In addition, as a country with a larger youth population, the labour markets are unable to absorb the large influx of young graduates (Global Entrepreneurship Monitor 2023). The World Employment and Social Outlook Trends 2025 report from the International Labour Organisation highlights that employment growth has remained insufficient to address the ongoing decent work deficits worldwide significantly. Young people continue to experience disproportionately high unemployment rates, estimated at around 12.6%, with little indication of improvement on the horizon. In South Africa, the youth unemployment rate has now reached 56.3% in the first quarter of 2025. An estimated 9.2 million youths are not engaged in education, employment, or training (Statistics South Africa 2025).

In response to this challenge, the government has introduced various support mechanisms aimed at promoting the growth of youth-owned businesses, with the intention of tackling the increasing levels of unemployment and poverty. A range of strategies, including policies, support institutions, and development programmes, are all designed to inspire more young individuals to create their own income opportunities through entrepreneurship. Universities across the country are also incorporating entrepreneurial skills into their curricula, believing that entrepreneurship education can help address the various social and economic issues that arise from a declining economy, such as crime, limited entrepreneurial activity, growing inequality, and poverty. However, the GEM, 2022/2023 Report states that the entrepreneurship rate for non-graduates exceeds that of graduates in South Africa. Despite these efforts, the youth unemployment rate in South Africa remains high. Graduate unemployment has recently gained attention, with over a million university graduates facing challenges in finding work, which further intensifies the already high youth unemployment problem. Research indicates that many graduates feel unprepared to enter the job market, while entrepreneurial activities among young people are not meeting expectations (Mseleku 2024; OECD 2025; MacGinty 2024). Das (2022) argues that the misconception equating entrepreneurship with youth entrepreneurship has resulted in the latter being overshadowed by broader entrepreneurship initiatives. This has led to a one-size-fits-all approach, where policies and programmes are uniformly applied to diverse demographic groups without considering the unique needs of young individuals. Supporting this perspective, UNWTO (2013) advises that each policy entrepreneurship framework should be treated as a living document, emphasising that a single policy approach is insufficient in the realm of entrepreneurship support.

This situation presents a complex dilemma for South Africa, where there is optimism that entrepreneurship could offer a way out of unemployment. However, although considerable research effort has been devoted to understanding how entrepreneurship policies stimulate entrepreneurship, the impact of these initiatives on promoting youth entrepreneurship remains significantly underexplored. Thus, the knowledge gap that this study investigates. It is crucial to evaluate the impact of these programmes to pinpoint the factors contributing to the paradox of extensive support in the form of policies and support programmes, alongside persistently high unemployment rates and limited meaningful growth in entrepreneurial activities. The problem statement thus reads:

Although the government has introduced various entrepreneurship support structures, including policies and capacity development institutions, youth entrepreneurship activities remain low. According to StatsSA (2025), youth unemployment has consistently remained high over the past decade, increasing from 36.9% in 2025 to 46.1%, a rise of 9.2 percentage points. The 2021/2022 Global Entrepreneurship Monitor (GEM) South Africa Report indicates a higher prevalence of entrepreneurial activity among youth aged 25–34, many of whom seek self-employment due to a lack of available job opportunities. However, the GEM Report for 2023/2024 also highlights a decline in overall entrepreneurial activity in South Africa. The report states that the established business ownership of businesses more than three to five years declined by 50% from 3.5% in 2019 to 1.8% in 2022/23, although this was exacerbated by the COVID-19 pandemic. The recovery and resilience rate remains low.

Therefore, the guiding research questions for the development of this paper are:

1. Why have entrepreneurial activities among the young remained low despite the government's investment in supporting youth-owned businesses?
2. What are the views of youth entrepreneur on the role of the government in supporting their businesses to grow, become profitable and remain competitive?

2 Literature Review

2.1 *Challenges of Growing Youth Entrepreneurship*

> We all know that government does not create jobs. Business creates jobs. Around 80% of all the people employed in South Africa are employed in the private sector.—President Cyril Ramaphosa (South African Government 2022)

Ramaphosa's 2022 State of the Nation Address (SONA) received mixed reactions. Some interpreted his remarks as an indication that the president was giving up on the government's failure to create jobs, particularly for young people. Conversely, others viewed the statement as an acknowledgement of the vital role that small, medium, and microenterprises (SMMEs) play in the country's economic development. Notwithstanding the varying perspectives, the private sector is a fundamental driver of economic growth, innovation, and job creation. Consequently, it is essential for both private and public institutions to collaborate effectively in order to address and mitigate socio-economic challenges (Sibiya et al. 2023).

Notably, a vital part of the state of the nation's address by the president of South Africa was the acknowledgement that the stunted growth of SMMEs is largely due to the need to alleviate the regulatory burden placed on informal businesses. He stated, *There are many regulations in this country that are unduly complicated, costly, and difficult to comply with. This prevents companies from growing and creating jobs* (South African Government 2022). This assertion is supported by existing literature on the challenges faced by SMMEs in South Africa. For instance, Lebambo and Shambare (2020) found that a lack of appropriate business infrastructure, barriers to entering new markets, insufficient industry mentors and networking opportunities, intense competition, high compliance costs, and difficulties in accessing bank loans continue to undermine business viability in the tourism sector. Furthermore, Sibiya et al. (2023) investigated the challenges experienced by SMMEs and the effects of interventions by the South African National and Provincial Governments. Their research revealed that the regulatory and legislative frameworks impose stringent compliance requirements, which further threaten the viability of SMMEs.

One of the significant challenges faced by youth start-ups in South Africa is access to capital. According to the research conducted by Gwija et al. (2014) on the challenges and prospects of youth entrepreneurship development in a designated community in the Western Cape, a lack of start-up and expansion capital was ranked as a critical issue. The study found that access to initial capital is the primary obstacle in establishing a new business venture, especially for young individuals from less advantaged backgrounds. Youth entrepreneurs often face challenges in securing financial capital due to information asymmetries, resulting in high-interest rates and stringent collateral requirements imposed by financial institutions (Fraser et al. 2015; Manolova et al. 2019).

Awareness of government development programmes is another challenge facing the youth of South Africa (Gwija et al. 2014). Paver et al. (2019) found that although many programmes exist to assist the youth because of the lack of proper coordination, the intended beneficiaries do not take full advantage of the programmes because they are aware of their existence. This is corroborated by Lebambo (2019), who notes that poor coordination in sharing information and a lack of access to local government offices limit the effectiveness of awareness programmes; as a result, entrepreneurs in rural areas do not fully benefit from government entrepreneurship programmes. "In many cases, they are the actual object of policies. As a result, rural entrepreneurs are subjected to a one-size-fits-all policy implementation approach, which is counter-effective" (Lebambo 2019).

2.2 South Africa's State of Youth Entrepreneurship in Comparison to Other BRICS Nations

Other BRICS nations, such as China and India, have witnessed growth in the technology sector, with a surge of young entrepreneurs emerging who are eager to tackle local challenges through innovative solutions (Bhaskar and Seth 2025). In contrast, South Africa faces various environmental challenges that hinder youth access to ICTs. These challenges include limited availability of resources, insufficient infrastructure, a lack of digital knowledge and skills, affordability issues related to technology, language barriers, and elevated unemployment rates (Alao and Brink 2023). Such persistent difficulties particularly impact young individuals from townships and underserved communities. The authors highlight that the diverse digital skills required across various industries, especially in the digital era, motivate individuals and businesses to engage more frequently with ICT services and products. Therefore, investing in technological infrastructure is crucial to support youth entrepreneurship. Gopala and Shekhar (2024) found that in India, young entrepreneurs with alternative career options favoured entrepreneurship as their chosen path. Conversely, a report by GEM (2022/2023) indicated that in South Africa, despite a significant percentage of respondents believing they possess the necessary capabilities, knowledge, skills, and experience for entrepreneurship, 53.0% would refrain from starting a new business due to fear of failure. This percentage surpasses the GEM global average of 45% and the combined average for participating African countries, which stands at 44.2%. This raises an intriguing question: why do citizens of a country who recognise their knowledge, skills, and experience and see promising opportunities still harbour fears about launching a business? This question reflects an integral aspect of the broader investigation associated with this research.

2.3 Overview of South Africa's Entrepreneurship Policy Landscape

While South Africa lacks a singular, comprehensive youth entrepreneurship policy, several significant policies and initiatives focus on youth entrepreneurship. These efforts are often integrated into broader frameworks, such as the National Youth Policy and various policies aimed at supporting SMMEs. Notably, the 1995 White Paper on the National Strategy for the Development and Promotion of Small Business in South Africa was the first documented policy from the post-Apartheid government to promote entrepreneurship. The Integrated Strategy for Promoting Entrepreneurship outlines three strategic pillars aimed at fostering entrepreneurial growth. Specifically, focusing on increasing financial and non-financial support services, creating a demand for the products and services provided by the SMMEs and reducing regulatory constraints. In the wake of this groundbreaking policy, a series of complementary initiatives have been introduced to bolster the national strategy. These initiatives include, but are not limited to:

1. *National Small Business Act No. 102 of 1996*: A user-friendly legal framework with particular emphasis on the impediments faced by black entrepreneurs and others who are previously disadvantaged.
2. *National Empowerment Act No. 105 of 1998*: Promoting and supporting business ventures pioneered and run by historically disadvantaged persons.
3. *Competition Act No 89 of 1998*: To ensure that all South Africans have the same chance to participate fairly in the national economy.
4. *Preferential Procurement Act No. 5 of 2000*: Intended to create demand for small enterprise products and services through procurement opportunities in the corporate sector, state-owned enterprises (SOEs) and government.
5. *Broad-based Black Economic Empowerment Act 53 of 2003*: Amended Act 46 of 2013. promoting access to finance for black start-ups, SMMEs, cooperatives and black entrepreneurs, including those in the informal business sector; and increasing effective economic participation and black-owned and managed enterprises.
6. *The National Development Plan 2030*: South Africa aims to double its gross domestic product (GDP) by 2030. The plan includes increasing employment from 13 million in 2010 to 24 million in 2030 and reducing inequality.

In conjunction with the development of policies to support small business growth, the government, in collaboration with the private sector and various non-governmental organisations, has established several institutions. These institutions are tasked with implementing the three strategic pillars outlined in the national strategy. The details of these institutions are presented in Table 1.

The key government policies and institutional initiatives in South Africa that support youth entrepreneurship, especially through policy and institutional mechanisms, included in Table 2.

Despite the existing frameworks established to support entrepreneurship, encompassing various policies and institutional resources, South Africa's track record in promoting youth entrepreneurship remains insufficient. The South African business environment presents numerous challenges, including excessive bureaucratic obstacles, stringent compliance requirements, corruption in public procurement, and slow government response times in implementing support initiatives. The observed decline in entrepreneurial activity, despite the availability of supportive measures, raises questions about traditional theoretical frameworks that suggest a direct relationship between entrepreneurship support and increased entrepreneurial activity. This leads to a critical inquiry: in a country where entrepreneurship promotion is a governmental priority and has seen the allocation of both financial and non-financial resources, what underlying factors hinder the practical realisation of these policy goals?

Table 1 Entrepreneurship support institutions and programmes

Institution	Responsibility
Government-led	
Small Enterprise Development and Finance Agency (SEDFA). The merger of the Small Enterprise Finance Agency (SEFA) and the Small Enterprise Development Agency (SEDA)	In 2025, the SEDA merged with the SEFA and the Cooperative Banks Development Agency (CBDA) to form the SEDFA, streamlining support for micro, small, and medium enterprises (MSMEs) through a unified source for financial and non-financial assistance
The National Youth Development Agency (NYDA)	The NYDA aims to assist young south Africans between the ages of 14 and 35 to start businesses and finance existing ones
National Empowerment Fund (NEF)	The (NEF) WAS formed to offer financial and non-financial support to black empowered businesses
Cooperative Banks Development Agency (CBDA)	The agency's vision is to facilitate financial inclusion and economic transformation while ensuring a sustainable cooperative banking sector that empowers communities
Presidential Youth Employment Initiative (PYEI)	The PYEI, known as the Basic Education Employment Initiative (BEEI), has effectively reduced youth unemployment across all nine provinces. Educator assistants support teachers with class preparation and activities during and after lessons
Department of Trade, Industry and Competition (DTIC)	Facilitates access to finance, incentives, and business development services for youth entrepreneurs
Black Business Supplier Development Programme (BBSDP)	Offering grants to small black-owned enterprises to help them improve their competitiveness and sustainability
Various incubators and accelerator programmes	
Raizcorp	Various programmes aimed at creating, supporting and scaling entrepreneurship and the entrepreneurial ecosystem
Harambee youth employment accelerator	This not-for-profit social enterprise collaborates with partners to tackle youth unemployment. They concentrate on addressing the challenges faced by entry-level job seekers and assist individuals in finding their first positions in the workforce
SAB Kickstart	This programme is designed to support youth entrepreneurs by offering them coaching and loan funding to help these businesses grow, among other services

Source Developed for this research

Table 2 Government policies and strategies related to the youth

Policy/programme	Purpose and focus
National Youth Policy (NYP) 2030	Empowers youth through entrepreneurship, education, and employment. Focuses on skills development, digital inclusion, and addressing dropout rates
Youth Enterprise Development Strategy (YEDS)	They create a supportive environment for young entrepreneurs by providing financial help, mentorship, and access to markets
Youth Employment Service (yes)	Connects youth to work opportunities and supports entrepreneurship via training and funding. Encourages private sector involvement
National Development Plan (NDP) 2030	Long-term vision for inclusive economic growth, with youth entrepreneurship as a key driver of employment and innovation
Presidential Youth Employment Intervention	Targets systemic barriers to youth employment, including entrepreneurship support through hubs and digital platforms
Integrated Youth Development Strategy	Coordinates efforts across sectors to promote youth development, including enterprise support and capacity building

2.4 *Theoretical/Conceptual Framework*

The study is founded on institutional theory and resource-based view theory. Institutional theory has become an increasingly prominent lens for entrepreneurship research. It was selected for this study to clarify how both formal structures (such as laws and policies) and informal norms influence entrepreneurial behaviour. Specifically, the focus is on how these laws and policies impact the entrepreneurial actions of young people in South Africa. The term *institution* broadly encompasses formal rule sets as defined by North (1990). In the framework of this study, these rules include regulatory structures, governmental agencies, laws, courts, professions, as well as various societal and cultural practices that impose pressures on youth entrepreneurs to thrive, scale, and sustain their businesses (DiMaggio 1988). Entrepreneurs are both limited and empowered by their institutional environments. This interplay can create entrepreneurial opportunities and impact the rate and scale of new venture creation, with external factors such as favourable market incentives and access to capital playing significant roles (Bruton and Ahlstrom 2003; Aldrich 1990). The resource-based view (RBV) has developed into one of the most significant theoretical frameworks for comprehending how resources contribute to organisational growth (Alvarez and Busenitz 2007; Kellermanns et al. 2016). This has been adopted in several entrepreneurship studies to understand how resources enhance business performance. The RBV is applicable in evaluating how government programmes provide access to critical resources, including financial and non-financial resources, such as funding, grants, training, skills development, and networks, for youth entrepreneurs in South Africa.

3 Methodology

3.1 Research Design

The study employed an exploratory qualitative design to capture the views of business owners on the role of the government's entrepreneurial support initiatives in their businesses. Qualitative research enables investigators to explore participants' inner perspectives, allowing for a more profound insight into the personal significance that individuals attach to their lived experiences.

Sutton and Austin (2015) argue that qualitative research aims to elicit and interpret the lived experiences of participants, ultimately conveying their perspectives in a way that meaningfully contributes to academic discourse. As a result, this study adopted an interpretivist paradigm using an exploratory research design. The goal of exploratory research is not to provide final and conclusive answers to research questions, but rather to explore the research topic in varying depths and tends to "address new problems with little or no prior research" (Saka et al. 2023). In the context of the study, while research has examined the challenges faced by young entrepreneurs, the role of government policies and programmes in fostering the growth of youth-owned businesses remains unclear. Thus, the study adopted an interpretivism paradigm as it aims to achieve a deep understanding of the social phenomenon of youth entrepreneurship, allowing participants to use their own words to relate their experiences and beliefs. (Rashid et al. 2019).

3.2 Sampling Method, Data Collection, and Instrument

A total of 19 Gauteng-based youth entrepreneurs participated in the study through in-depth interviews. Qualitative experts recommended that qualitative studies require a minimum sample size of at least 12 to reach data saturation (Clarke and Braun 2013). *In addition, 3.4 details on how quality and trustworthiness were ensured in the study. In terms of the number of youth-owned businesses,* there is no specific statistic available on the number of youths who own businesses in South Africa. However, the 2023–2024 GEM Report reveals that 18% of individuals aged 24–34 and 10% of those aged 18–24 are actively participating in entrepreneurial endeavours. Additionally, Statistics South Africa indicates that around 21 million young people comprise 33.1% of the nation's population.

The participants' descriptions and backgrounds are presented in Table 3. Since youth-owned businesses are often advertised on social media (Sarmah et al. 2023), participants were found on LinkedIn, Instagram, and Facebook business pages. All have separated their personal accounts from those of the business. Only 5 of the 17 interviews were conducted face-to-face. The remaining number used online platforms such as Microsoft Teams and Zoom. The interviews last between 50 and 90 minutes. With consent, some of the interviews were recorded and subsequently transcribed. The interviews and quotes are presented verbatim and include participant numbers (YEP number), which identify Youth Entrepreneur Participants (YEP 1; YEP 2; YEP 3, etc.). Pseudonyms are used to ensure anonymity. An interview guide was developed as an instrument for data collection, which guided the development of questions and interviews with the participants. All businesses were based in the Gauteng province, primarily in the City of Tshwane, mainly in Pretoria, and three were based in Johannesburg, specifically at participants 1, 5, and 18. Since the study adopted the RBV, two important resources—financial and non-financial support—are used to determine the kinds of support the youth entrepreneurship sampled in this study have received from the government since the formation of their businesses.

The inclusion criteria used to select the businesses

Table 3 Description of participants

YEP	Organisation type	Years of experience	Financial support, e.g. loans/grants	Non-financial support received
1.	Tour operator	5	Yes	Yes
2.	Restaurant	4	No	Yes
3.	Internet café	2	No	No
4.	Supermarket	5	No	No
5.	Hardware	6	Yes	Yes
6.	Supermarket	4	No	No
7.	Guest house	3	Yes	Yes
8.	Business consultancy	9	No	Yes
9.	Daycare	4	No	No
10.	Social media influencer	3	No	Yes
11.	Visual media	9	Yes	Yes
12.	Food and catering	2	No	No
13.	Restaurant	7	No	No
14.	Online boutique	2	No	No
15.	Event planner	5	No	No
16.	Transport	3	No	Yes
17.	Plumbing and electrical contractor	8	No	No
18.	Daycare	6	No	Yes
19.	IT support	4	No	No

- 2 or more years of operating.
- The age of the owners falls within the youth age category, 24–35.
- Based in the Gauteng province and owner of the business.
- Business registered and falling within the Micro category in accordance with the National Small. Business Amendment Act (26 of 2003).

3.3 *Data Analysis*

Data analysis commenced with a thorough process of data preparation, which continued informally through interviews and observations, and persisted during transcription as recurring themes, patterns, and categories became apparent (Creswell 1998, 2007; Silverman 2000). The qualitative analysis was conducted using the data collected from recorded and transcribed interviews, along with observation notes. This involved sorting, organising, and condensing the data into more manageable segments based on specific subjects. As noted by Babbie (2004: 387), qualitative data analysis involves examining social research data without converting it into a numerical format. ATLAS.ti, qualitative data analysis software, was employed for the data analysis process.

3.4 *Trustworthiness*

Trustworthiness is important in qualitative research. Yin (2016) states that trustworthiness in qualitative research is concerned with demonstrating the authenticity of your work. That is, the extent to which the research can be audited by developing a case study database and by the use of quotations in the written report. The study followed Lincoln and Guba's (1985) five suggested issues to ensure trustworthiness. This includes *credibility, transferability*, *dependability*, *confirmability*, and *reflexivity*. To ensure credibility, the recordings of the interviews were used to compare the original data with what the study reported. (2) Transferability was achieved through a detailed description of the data, and thematic analysis contributed to this process. (3) Dependability ensures the robustness and consistency of the findings over time. The study incorporated the participants' insights into the evaluation of the results. This involved actively engaging them in interpreting the data and formulating recommendations based on the study's findings. (4) Confirmability was achieved by ensuring that both the data and the interpretations of the findings were clearly derived from the transcripts and notes taken during the interviews with participants, rather than being the inventions of the researcher. (5) Reflexivity, the researcher undertook thorough and critical self-reflection regarding the research process, examining personal biases, ethical considerations, and the impact of their background on the study.

4 Results and Discussion

Theme 1: Policy effectiveness. This theme discusses the stated goals of government policies and programmes targeting youth entrepreneurship.

The participants, in general, were of the view that the policies and programmes on paper look good and the objectives outlined address the needs of the young people, especially in terms of skills and taking into account the legacy of the past, where the majority of marginalised people were not given equal opportunity to participate.

> When you look at the National Small Business, the likes of the NYDA, when you look at their objectives, they were intended to address the legacy of apartheid, give young people opportunities, but have they achieved that? On paper, yes; in reality, no (Youth Entrepreneur Participant: 3)

The regulatory burdens we face are substantial... Five years later, I am still waiting for a simple consent letter from the municipality; the process is dragging on far too long... I am on the verge of giving up (Youth Entrepreneur Participant 9—a daycare owner). The participant details that the journey to have the establishment rezoned from residential to commercial property took over 5 years, primarily due to delays at the municipality. Various municipal departments—including the fire department, electricity, health, and waste management—took an excessive amount of time to process their parts of the application that were essential for rezoning and obtaining the necessary consent to operate the daycare. These delays significantly hindered the business's ability to secure funding and other support needed for growth, as municipal approval letters were crucial documents for obtaining such approvals.

These sentiments resonate with numerous studies investigating the challenges faced by SMMEs in South Africa. Research indicates that red tape and the high cost of doing business are critical factors that undermine the viability of SMMEs in the country (Sibiya 2023; Ngorora and Mago 2013; Ndabeni and Rogerson 2005; Ngcobo and Sukdeo 2015). Specifically, Lekhanya and Visser (2016) examined the risks and key factors affecting the growth of entrepreneurial ventures in South Africa and found that environmental regulatory requirements present significant challenges for SMMEs. Furthermore, stringent government policies inhibit their expansion, while public procurement regulations and licensing requirements negatively impact their growth prospects. The various taxes, levies, and compliance costs imposed on youth-owned businesses continue to threaten their growth potential.

Theme 2: Policy Alignment: Inclusivity and Equity. In line with institutional theory, this theme examines how formal rules, along with regulatory structures and governmental laws, either hinder or support the growth of businesses. It is important to assess how these formal structures account for the unique context of South Africa, particularly the historical marginalisation of communities where the majority of black individuals were not allowed to fully participate in the mainstream economy. Therefore, evaluating how entrepreneurship programmes address disparities among youth—especially in relation to socio-economic challenges faced by young entrepreneurs in townships and rural areas—is crucial. The study found that disparities persist between establishments located in rural, urban, and township communities. Rural and township areas in South Africa face unique challenges that are not encountered by their urban counterparts. According to Kim and Davidson (2024), macroeconomic and cultural factors such as negative perceptions surrounding townships, inadequate infrastructure, crime, and competition from large corporations continue to hinder township businesses today.

Youth Entrepreneur Participant 13 shared the following: *As an entrepreneur based in Soshanguve, my challenges extend beyond just load shedding. Even outside of the load shedding schedule, we still experience power supply cuts initiated by the City of Tshwane. There have been days when we have been without water and electricity. If we were in Sandton or wealthier areas, do you think the government would allow us to go days without electricity and water? It is a burden to operate a business in a township when they cut off water without notice. How can you run a restaurant without water?*

Mametja et al. (2023) investigated access to quality, affordable, and reliable ICT infrastructure for township SMMEs, finding that while access to ICT is essential for digital adoption, microenterprises in these areas face significant costs related to internet connectivity, which can be a burden for financially constrained businesses. To promote inclusivity, the government must regulate telecommunications service providers to ensure that access to internet connectivity is both affordable and accessible (Mametja et al. 2023).

The GEM 2022/2024 Report highlights that youth and non-graduates continue to face three significant challenges, including access to training and mentoring (human capital). The GEM suggests solutions such as funding programmes aimed at reducing diversity and inclusion gaps throughout the entrepreneurial journey, as well as promoting digital literacy to support opportunities for scaling up.

Theme 3: Policy and Access to Business Resources and Support. This theme addresses the provision of access to critical resources such as capital, training, and networks for youth entrepreneurs in South Africa. Specifically, the interconnectedness of physical infrastructure, human capital, institutional support, and resources.

Owoseni and Obeng (2024) examined the institutional constraints that impact small businesses in South Africa and found that the bureaucracy involved in starting a business presents significant hurdles to their development. The requirements imposed are often beyond the capacity of small businesses. Key bureaucratic constraints, including taxes, regulatory compliance, and a lack of government support, further exacerbate these challenges. The South African Minister of Small Business Development, Stella Ndabeni, spoke at a small business event in the Northwest in 2025, highlighting the significant impact of red tape and compliance issues on the growth of small enterprises. She stated, *We have identified 27 pieces of legislation that require repeal or amendment, and we are actively engaging with the relevant departments. Ultimately, much of the red tape revolves around compliance; it pertains to certain regulations imposed by those who enforce them* (Channel Africa 2025). This is despite the appointment of Mr. Sipho Nkosi to head the Red Tape Reduction (RTR) task team in the Presidency, which aims to cut red tape across government since 2023.

Other challenges mentioned by respondents include competition from established businesses. Especially in the tourism and hospitality sectors, they expressed concern that, due to infrastructure challenges in their restaurants and lodging businesses, they fear they cannot compete effectively with large, established businesses. Although competition is inevitable in a business environment, the unique systemic challenges facing SMMEs in South Africa require focused training and support on strategic planning and competitiveness. This emphasis will help SMMEs navigate unfair competition from established and larger businesses. Additionally, Islam et al. (2021) suggest that programmes supporting SMMEs in developing niche strategies and fostering innovation can empower them to compete successfully. Policies such as the Preferential Procurement Act No. 5 of 2000 aim to generate demand for products and services from small enterprises by creating procurement opportunities in the corporate sector, SOEs, and the government. This policy can be leveraged to support this initiative.

> If the roads are bad, there is no electricity, or there is not enough water, people will not want to visit businesses in areas with poor infrastructure, crime, or electricity problems. The government needs to improve its infrastructure if it wants to help and support small businesses in South Africa, especially for us in the townships ... [pause] If you are in a township and have agues house one has to bear in mind that people have options to use well know n hotel names like sun international…[pause].. so infrastructure plays a role …. [YEP 7, who owns a guest house]

The study found that although a myriad of programmes offering financial and non-financial support exist across different sectors, the majority of training and development programmes, particularly those not focused on finance, are not sector-specific. Instead, they tend to be one-size-fits-all programmes that do not consider the unique challenges each sector and business faces. Participants mentioned that they would attend courses focused on generic customer service or business management, often without considering the specific issues relevant to their industry.

For example, businesses such as tour operators, guest houses, restaurants, transportation, plumbing and electrical contracting, and daycare services (YEP 1, 7, 9, 13, 16, 17, 18) have unique and often complicated requirements regarding regulations and compliance, which need specialised support. A daycare business owner indicated that their sector has shifted from being governed by the Department of Social Development to being housed under the Department of Education. This transition has brought about new compliance and regulatory frameworks that they were not fully supported or equipped to address. Delays in getting housing plans approved, rezoning plans processed, and meeting compliance requirements have hindered their ability to operate fully compliant businesses.

Similarly, a restaurant owner mentioned that it took years to obtain the food handler certificate from the municipality certifying their establishment as suitable for food supply and handling. Plumbing and electrical contractors expressed a desire for training that focuses on advancements in plumbing and construction, rather than generic business management skills, to enable them to compete with larger construction firms. They also seek better access to markets and rely on walk-in customers. Many of them are listed in databases of various organisations registered to supply their products or services, yet due to competition, they often miss out on contracts and job opportunities.

This situation ties into the ongoing debate about access to markets for small businesses, especially for youth entrepreneurs who may lack years of operating experience and must compete with larger organisations. These findings suggest that perhaps policies have not been fully effective because the government's understanding of the needs of young entrepreneurs and the unique challenges they face is not thorough.

Basson and Omoruyi (2025) conducted an examination of the internal and external factors influencing the growth of SMMEs. Their findings indicated a notable deficiency in consistent communication between government agencies and SMMEs. As a result, the government often fails to understand the challenges faced by SMMEs, leading to ineffective methods in addressing their needs and requirements. There is an urgent need to enhance support in areas such as infrastructure provision, equipment funding, and more accessible grants. Furthermore, numerous studies indicate that policies are ineffective due to a lack of understanding of the unique challenges and circumstances faced by small businesses in South Africa. Consequently, a one-size-fits-all support approach is employed, resulting in the ineffective implementation of assistance programmes (Lebambo 2019; UNWTO 2013; Scheba and Turok 2020). Figure 1 is an ATLAS.ti visual representation of word frequencies from transcripts on the words which were repeatedly mentioned by the youth entrepreneurs on the subject of the role of policy on resources and support.

Fig. 1 ATLAS.ti word frequencies

5 Conclusion

The study highlights a range of challenges that significantly hinder the growth and development of young entrepreneurs. A significant issue is the digital divide, which continues to disadvantage young individuals in rural areas and townships due to limited access to technology. Additionally, essential entrepreneurial skills are lacking, there is an inadequate supporting business infrastructure, and a lack of adequate policies and systems. Tackling these challenges is crucial in unlocking the potential opportunities for youth entrepreneurs. With more than seven million South African youth unemployed, representing over 50% of the total unemployed, this suggests that current policy interventions are not having the desired impact. A limitation of the study was that the data collection method was qualitative, and therefore, the number of interviews may not be sufficient to draw robust conclusions from this study. However, future research can explore the same topic using a longitudinal case study to focus on the entrepreneur's journey from micro to small and medium growth path.

6 Recommendations

The paper recommends that, for policy to be impactful in the context of South Africa, the following key factors should be considered.

1. *Assess Policy Effectiveness*: Impact studies should be conducted to evaluate the outcomes of policies and the support provided by institutions. This will give a clear indication of a policy's effectiveness.
2. *Examine Policy Alignment*: It is crucial to ensure that policies are relevant and responsive to the evolving needs of youth. As these needs change due to global and regional developments, advances in technology, AI, and the socio-political landscape of South Africa, ongoing evaluation of policy relevance and responsiveness to youth is important.
3. *Explore Systemic Policy Factors*: Barriers, enablers, and institutional dynamics should be addressed. As highlighted in the literature, systematic challenges within institutional dynamics have been significant impediments to the effectiveness of policies. These systemic factors must be tackled promptly, as they represent significant barriers to promoting entrepreneurship. Challenges include bureaucratic procedures and licensing, which continue to burden small businesses. Additionally, public procurement processes need to be transformed to counter the dominance of large businesses.

7 Contributions

This study enriches the existing body of knowledge by addressing the gap in academic literature regarding the effectiveness of South African government policies on entrepreneurship and small business support since 1994. While much research on entrepreneurship focuses on the role of government policies in promoting the growth of SMMEs, this study examines explicitly the youth population, who are most affected by socio-economic challenges such as unemployment, poverty, and inequality in South Africa's underperforming economy. Furthermore, the study highlights the specific policy factors that should be prioritised when evaluating the effectiveness of entrepreneurship policies in South Africa, using the effectiveness–alignment–systemic factors assessment tool.

References

Alao, A., Brink, R.: Information and communication technology management for sustainable youth employability in underserved society: technology use for skills development of youths. Int. J. Sociotechnol. Knowl. Dev. **15**(1), 1–19 (2023)

Aldrich H.: Using an ecological approach to study organizational founding rates. Entrep. Theory Pract. **14**(3), 7–24 (1990)

Alvarez, S.A., Busenitz, L.W.: The entrepreneurship of resource-based theory. In: Cuervo, Á., Ribeiro, D., Roig, S. (eds.) Entrepreneurship, pp. 197–218. Springer, Berlin, Heidelberg (2007). https://doi.org/10.1007/978-3-540-48543-8_10

Babbie, E.: The Practice of Social Research (10th ed.). Wadsworth (2004)

Basson, N., Omoruyi, O.: Internal and external factors affecting the growth of SMMEs: a qualitative study. South. Afr. J. Entrep. Small Bus. Manag. **17**(1), a919 (2025)

Bhaskar, P., Seth, N.: Exploring differences in youth entrepreneurial intentions: a comparative analysis of India and China. AiBi J. Res., Adm. Eng. **13**(1) (2025). Available at https://www.researchgate.net/publication/392005863_Exploring_differences_in_youth_entrepreneurial_intentions_A_comparative_analysis_of_India_and_China

Bruton, G. D., Ahlstrom, D.: An institutional view of China's venture capital industry: explaining the differences between China and the West. J. Bus. Ventur. **18**(2), 233–259 (2003)

Channel Africa: Red tape holding back co-operatives, legislative changes needed: SA Minister (2025). Available at https://web.sabc.co.za/sabc/home/channelafrica/economy/details?id=3aefbfd7-1efa-4625-b2fd-0209dacea205

Clarke, V., Braun, V.: Successful Qualitative Research: A Practical Guide for Beginners. Sage, London (2013)

Creswell, J.W.: Qualitative Inquiry and Research Design: Choosing Among Five Traditions. Sage, Thousand Oaks (1998)

Creswell, J.W.: Qualitative Inquiry and Research Design: Choosing Among Five Approaches. Sage, London (2007)

Das, N.: The role of youth entrepreneurship in economic growth and social inclusion in India. J. Policy Options. **5**(1), 22–29 (2022)

DiMaggio, P.J.: Interest and agency in institutional theory. In: Zucker, L.G. (ed.) Institutional Patterns and Organizations, pp. 3–21. Ballinger, Cambridge, MA (1988)

Fraser, S., Bhaumik, S.K., Wright, M.: What do we know about entrepreneurial finance and its relationship with growth? Int. Small Bus. J. **33**(1), 70–88 (2015)

Global Entrepreneurship Monitor: GEM South Africa 2023 report. GEM Consortium (2023). Available at https://www.gemconsortium.org/report/20222023-global-entrepreneurship-monitor-global-report-adapting-to-a-new-normal-2

Gopala, K.N., Shekhar, B.: Youth entrepreneurship in India: driving economic growth and social inclusion. Educ. Adm. Theory Pract. **30**(11), 1926–1930 (2024)

Gwija, S.A, Eresia-Eke, C., Iwu, C.G.: Challenges and prospects of youth entrepreneurship development in a designated community in the Western Cape, South Africa. J. Econ. Behav. Stud. **6**(1), 10–20 (2014)

Islam, A.K.M.H., Sarker, M.R., Hossain, M.I., Ali, K., Noor, K.M.A.: Challenges of small- and medium-sized enterprises in business growth: a case of the footwear industry. J. Oper. Strateg. Plan. **4**(1), 119–143 (2021). https://doi.org/10.1177/2516600X20974121

Kellermanns, F., Walter, J., Crook, T.R., Kemmerer, B., Narayanan, V.: The resource-based view in entrepreneurship: a content-analytical comparison of researchers' and entrepreneurs' views. J. Small Bus. Manag. **54**(1), 26–48 (2016)

Kim, S., Davidson, A.: Entrepreneurship in South Africa's townships: Ecosystem challenges and recommendations for the path forward. Aspen Netw. Dev. Entrep. Available at https://andeglobal.org/publication/south-africa-townships-challenges-and-recommendations/ (2024)

Lebambo, M.: The role of entrepreneurial policies in developing rural tourism entrepreneurship in South Africa. Afr. J. Hosp. Tour. Leis. **8**(3), 1–21 (2019)

Lebambo, M., Shambare, R.: Entrepreneurship development framework for small-scale rural tourism establishments in South Africa. Afr. J. Hosp. Tour. Leis. **9**(1), 1–19 (2020)

Lekhanya, L.M., Visser, K.: Risks and factors contributing towards rural entrepreneurial orientation growth of business in an emerging economy. J. Risk Gov.Ance & Control.: Financ. Mark. & Inst.S. **6**(4), 72–83 (2016)

Lincoln, Y.S., Guba, E.G.: Naturalistic Inquiry. Sage Publications, Inc., Beverly Hills, CA (1985)

MacGinty, H.: Graduate unemployment in South Africa [Master's thesis, Stellenbosch University]. Labour market intelligence (2024). Available at https://lmi-research.org.za/publication/dissertation-2024-graduate-unemployment-in-south-africa

Mametja, T.R., Lebambo, M.M., Tichaawa, T.M.: The adoption of digital technologies by women-owned tourism micro-enterprises. Afr. J. Hosp. Tour. Leis. **12**(2), 717–734 (2023)

Manolova, T.S., Edelman, L.F., Shirokova, G., Tsukanova, T.: Youth entrepreneurship in emerging economies: can family support help navigate institutional voids? J. East-West Bus. **25**(4), 363–395 (2019)

Mseleku, Z.: Transitioning from higher education to the labour market: the role of graduate internships on youth graduate employability. Cogent Educ. **11**(1) (2024)

Ndabeni, L., Rogerson, C.M.: Entrepreneurship in rural tourism: the challenges of South Africa's Wild Coast. Afr. Insight. **35**(4), 130–141 (2005)

Ngcobo, S., Sukdeo, R.: Challenges facing SMMEs during their first two years of operation in South Africa. Corp. Own.Ship & Control. **12**(3–5), 505–512 (2015)

Ngorora, G.P.K., Mago, S.: Challenges of rural entrepreneurship in South Africa: insights from Nkonkobe municipal area in the Eastern Cape Province. Int. J. Inf. Technol. Bus. Manag. **1**(16), 1–11 (2013)

North, D.C.: Institutional Change and Economic Performance. Cambridge University Press, Cambridge (1990)

OECD: Education and skills today (2025). Available at https://oecdedutoday.com/youngpeoples-career-expectations-arent-meeting-labour-market-realities-but-we-can-change-this/

Owoseni, T.A., Obeng, F.: Institutional constraints affecting small businesses in South Africa. Br. J. Multidiscip. Adv. Stud. **5**, 26–44 (2024)

Paver, R., Rothmann, S., van den Broeck, A., de Witte, H.: Labour market interventions to assist the unemployed in two townships in South Africa. SA J. Ind. Psychol. **45**(1), 1–14 (2019)

Rashid, Y., Rashid, A., Warraich, M.A., Sabir, S.S., Waseem, A.: Methods: a step-by-step guide for business researchers. Int. J. Qual. Methods. **18** (2019)

Saka, R., Osademe, G.C., Ononokpono, N.: Exploratory research design in management science: a review of literature on conduct and application. Int. J. Res. Innov. Soc. Sci. **7**(51), 1384–1395 (2023)

Sarmah, R., Khatun, A., Singh, A.: Perception and awareness of youth toward social advertising campaigns run by private brands. Int. J. Asian Bus. Inf. Manag. **14**(1), 1–20 (2023)

Scheba, A., Turok, I.: Strengthening township economies in South Africa: the case for better regulation and policy innovation. Springer Neth. **21**(1), 77–94 (2020)

Sibiya, A., van der Westhuizen, J., Sibiya, B.: Challenges experienced by SMMEs and interventions by the South African national and provincial government: A literature review. Afr J Inter/ Multidiscip Stud. **5**(1), 1–11 (2023)

Silverman, D.: Doing Qualitative Research: A Practical Handbook. Sage, London (2000)

South African Government: President Cyril Ramaphosa: 2022 state of the nation address (2022). Available at https://www.gov.za/news/speeches/president-cyril-ramaphosa-2022-state-nation-address-10-feb-2022

Statistics South Africa: Statistical release P0211 quarterly labour force survey (2025). Available at https://www.statssa.gov.za/publications/P0211/P02114thQuarter2025.pdf

Sutton, J., Austin, Z.: Qualitative research: data collection, analysis, and management. Can. J. Hosp. Pharm. **68**(3), 226–231 (2015)

United Nations World Tourism Organization: Annual Report (2013). Available from http://www.slideshare.net/chafikyahou/world-tourism-organization-annual-report-2013

Yin, R.K.: Qualitative Research from Start to Finish, 2nd edn. Guilford Press, New York (2016)

Employees' Lived Experiences in Major Organisational Restructuring in a Higher Education Institution

Bulelwa Ntunzi, Siphosethu Mapete, Sive Ben, Avela Jojo, and Themba Mjoli

Abstract Higher education institutions routinely use organisational restructuring to improve operational performance, adjust to shifting national policies, and manage budgetary constraints. These reorganisation initiatives, however, typically have unanticipated negative consequences, especially for academic and administrative employees, who often experience increased stress, role uncertainty, job insecurity, and emotional strain. This study explored the experiences of academic and administrative employees during organisational restructuring in higher education institutions. The research found increased stress, uncertainty, job insecurity, and emotional strain resulting from unclear role expectations, inconsistent leadership communication, and a lack of support systems. Suggested solutions include increased communication transparency, increased employee involvement in decision-making, revised job descriptions, and mental health support services. The study emphasises the importance of inclusive, employee-centred tactics, putting psychological health first, specifying post-restructuring duties precisely, and establishing organised support networks.

Keywords Experiences · Restructuring · Higher Education

1 Introduction

According to Nicholson et al. (2023), higher education institutions are essential for training people for the workforce and promoting sustainable development. Unpredictable funding, heightened competition, and the quickening speed of globalisation are just a few of the growing difficulties these institutions have faced recently. Due

B. Ntunzi · S. Mapete (✉) · S. Ben · T. Mjoli
University of Fort Hare, Dikeni, South Africa
e-mail: smapete@ufh.ac.za

A. Jojo
iYunivesithi Walter Sisulu, Butterworth, South Africa

E. H. Redda (ed.), *Proceedings of the Social Sciences International Research Conference, 2025*, https://doi.org/10.1007/978-981-95-9489-4_15

to these considerations, institutions must critically re-examine their strategic planning and governance frameworks to stay flexible in a rapidly evolving environment (Vlachopoulos 2021). In particular, the COVID-19 pandemic magnified pre-existing structural vulnerabilities, forcing academic institutions to rectify governance and resource allocation flaws (Doidge and Doyle 2020; Jayasuriya 2020). In this context, organisational restructuring is a purposeful, strategic process that involves changing roles, processes, and structures to adjust to internal and external challenges (Jerab and Mabrouk 2023). In higher education, this restructuring is essential to a more extensive organisational transformation to improve overall institutional effectiveness, optimise processes, and define responsibilities.

Organisational restructuring and workforce reductions are frequently positioned as essential strategies for reducing costs and ensuring long-term viability, as Harney et al. (2018) argue. However, these large-scale changes are hardly without consequence. De Jong et al. (2016) highlight the significant psychosocial risks that emerge during such transitions; employees often endure considerable emotional and psychological stress. The unpredictability inherent in organisational change tends to amplify anxiety and uncertainty, contributing to elevated workplace stress levels. Harney et al. (2018) show how organisational restructuring has adverse effects on employee well-being during times of transition by explicitly linking it to higher stress levels. According to Trchalíková (2023), high stress levels at work are not recent occurrences; they frequently undercut motivation, impair performance, increase turnover, and encourage misconduct, poor communication, and increased conflict. Notwithstanding these risks, employee opinions are frequently disregarded during the transition process, which heightens resistance to new initiatives and feelings of job insecurity. Employers must thus actively seek out and consider employee input before restructuring. Not only is employee involvement advantageous but it is also necessary to promote easier transitions and lower resistance, which will help the business achieve its goals.

Since few, if any, of the studies cited above were conducted in South Africa, the purpose of this study was to investigate the extent to which findings in other parts of the world apply to South Africa. Due to the cultural diversity of the South African population, this country is often referred to as 'the world within a country.' Hence, it is necessary to investigate the extent to which findings in other cultural contexts also apply to the South African context. More specifically, the aim is to explore and understand how organisational restructuring influences the experiences of employees' stress, anxiety, and job insecurity in South African institutions of higher education in the same way as elsewhere in the world.

2 Literature Review

Organisational restructuring is a multi-faceted issue that arises when a company aims to adjust its structure, procedures, and values to more effectively meet the needs and expectations of both internal and external stakeholders (Elsaman 2024). According to Yang (2024), organisational restructuring is a challenging issue that entails making specific changes to the business structure. These changes may include reevaluating relationships among various elements, adding or removing components, and adjusting corporate control. Furthermore, this procedure can cause internal disarray and heighten stress levels for employees (Niedhammer and Bertrais 2024).

Organisational restructuring has a significant psychological and emotional influence on workers, particularly with relation to stress and anxiety. Organisational change may be a significant source of stress, upsetting employees' feelings of security and well-being, according to studies by Backhaus et al. (2024) and Contreras and Gonzalez (2021). Employees experience higher anxiety during restructuring because of unclear job roles, increased workloads, and uncertainty about their future within the organisation (Jerab and Mabrouk 2023). Employees find it difficult to adjust to environmental changes with inadequate support and communication mechanisms, which worsens this stress. Empirical research by Kivimäki et al. (2001) and others has shown how important employees' opinions on job security are throughout the restructuring stages.

Recent empirical studies have confirmed that employee anxiety, decreased work satisfaction, and decreased organisational commitment are all strongly impacted by uncertainty regarding employment permanence. For instance, An et al. (2023) discovered that job instability is intricately linked to elevated levels of stress and anxiety in white-collar workers, which are further intensified by exposure to social media. Similarly, Wang et al. (2024) demonstrate that working adults' mental health outcomes are negatively correlated with perceived job insecurity, which increases concerns about their future employability and financial stability. In addition to lowering morale, research by Park et al. (2020) highlights that job insecurity predicts withdrawal behaviours, including decreased work effort and an increased desire to leave. These recent findings add to the understanding of the widespread effects of job insecurity during restructuring and support previous findings by Tumelo and Donald (2025) and Mathisen et al. (2023). They also underscore the urgent need for strategic organisational interventions that align employees' emotional and psychological well-being with operational goals.

Effective communication techniques are essential to reduce the adverse effects. According to Smollan's (2017) literature analysis, subpar or inconsistent leadership communication increases employee stress and breeds mistrust. When communication is ambiguous, slow, or one-way, employees often express irritation because it makes it challenging to comprehend the reasoning behind changes and prepare for new tasks or responsibilities. During times of transition, Rosenberg (2018) highlights that open and honest communication reduces conflict and fosters trust. Recent research further supports these findings. Klewes et al. (2017) emphasise that digital transformation needs effective communication to combat uncertainty and disinformation. Additionally, Akinwalere (2023) shows that open and constant communication lowers resistance and favours employee engagement. Hence, effective communication that is open, timely, and inclusive is crucial for building trust among employees, minimising opposition, and facilitating seamless transitions during periods of organisational change. The study's main objectives are to (i) examine employees' first-hand experiences of organisational restructuring, focusing on their stress and anxiety levels and (ii) explore employees' perceptions of job security during restructuring.

- Evaluate management's communication tactics during the restructuring process.
- Identify employees' coping mechanisms to deal with stress and anxiety during organisational change.

3 Methodology

3.1 Research Philosophy and Design

This research is fundamentally interpretivist in its philosophical orientation, prioritising an understanding of subjective meanings and lived experiences within specific social contexts (Pulla and Carter 2018). This epistemological position is especially pertinent to organisational research, since meaningful analysis relies heavily on the intricacies of interpersonal interactions and the complexity of human conduct (Lyu 2024). The study investigated how employees perceive and react to organisational restructuring through an interpretive lens to gather rich qualitative data representative of individual experiences. A comprehensive examination of topics like communication, psychological impact, and work insecurity was made possible by the intimate alignment of this philosophical approach with the qualitative methodology used.

The research used a qualitative approach. It relied on semi-structured interviews to gain a detailed understanding of participants' opinions. The researcher explored emerging topics and concerns of specific importance using this approach's significant flexibility (Regmi 2024). The qualitative approach was particularly well-suited when examining sophisticated organisational phenomena, where quantitative measures might not have adequately reflected the complex reality of employee experience (Daněk and Urgošíková 2024). The investigation aimed to create a diverse set of employees from multiple departments and organisational levels to fully understand the multi-faceted effects of restructuring.

3.2 Sampling Method

The researchers opted for purposive sampling, intentionally selecting participants who had experienced organisational restructuring (Stratton 2024). This approach ensured that the data collected would be both relevant and insightful. Participants represented a variety of departmental backgrounds, enabling the study to capture a broad spectrum of perspectives and individual experiences. Data collection concluded with 17 participants, as thematic saturation was reached and additional interviews were unlikely to yield new information or perspectives.

3.3 Research Instrument

The interview guide was developed by thoroughly reviewing existing literature on organisational change, job insecurity, and employee well-being. Key themes identified in previous studies informed the creation of open-ended questions, which aimed to elicit rich qualitative data.

3.4 Data Collection

Semi-structured interviews were used to gather data from employees from different organisational departments. The methodology used a methodical approach to guarantee data integrity and complied with strict ethical guidelines. Internal email correspondence was used to invite participants, stressing the value of informed consent obtained before the interviews and the voluntary nature of participation. All interviews took place in private settings to protect participants' privacy and create a comfortable atmosphere. Each session lasted 30–45 min and was audio recorded with the participants' express consent to enable accurate transcription. The recordings were transcribed verbatim to capture the nuances and complexities present in the participants' narratives and maintain the reliability of their responses. All participants consented to the study, and the university's Inter-Faculty Human Research Committee granted ethical clearance.

3.5 Data Analysis

This study examined the subjective experiences of higher education staff members with organisational restructuring, focusing on the emotional, psychological, and practical ramifications. The investigation was structured around three principal research questions: (1) How do employees experience organisational restructuring? (2) What are the psychological and emotional effects of restructuring on staff members? and (3) How effective have internal communication and institutional support systems been during the restructuring process? Qualitative data were collected through semi-structured interviews with seventeen participants, as shown in Table 1. The participants in this study ranged in age from 28 to 57 years, with professional experience spanning 2–15 years. Among the 17 respondents, 10 were members of the academic staff (58.8%), while the remaining 7 were administrative personnel (41.2%). The group included academic and administrative staff from five faculties, including Social Sciences and Humanities, Science and Agriculture, Health Sciences, Management, and Commerce, representing various departments.

The dataset was subjected to thematic analysis using ATLAS.ti with the six-phase model proposed by Braun and Clarke (2006). This process encompassed stages of

Table 1 Demographic information of participants

Participant ID	Gender	Age	Employment Type
P1	Female	33	Support staff
P2	Female	32	Support staff
P3	Female	42	Support staff
P4	Male	35	Academic staff
P5	Female	28	Academic staff
P6	Male	49	Support staff
P7	Male	36	Academic staff
P8	Female	31	Academic staff
P9	Female	57	Academic staff
P10	Male	34	Academic staff
P11	Female	36	Support staff
P12	Female	34	Support staff
P13	Female	35	Academic staff
P14	Male	36	Academic staff
P15	Female	33	Support staff
P16	Male	30	Academic staff
P17	Female	40	Academic staff

familiarisation with the data, initial coding, development of themes, review and refinement of these themes, their definitive delineation, and subsequent reporting. The analytical procedure resulted in the identification of seven principal themes: increased stress and anxiety, role ambiguity, fear of retrenchment, workload concerns, poor communication, lack of formal institutional support and exclusion from decision-making.

4 Results

The interviews provided detailed, personal stories that describe what individuals experienced as they went through changes within the university setting. This diversity enabled a more thorough analysis of how organisational restructuring influenced various departments and functional units across the institution.

4.1 Experiences of Stress and Anxiety During Restructuring

During the restructuring process, employees expressed that they experienced significant stress levels. They noted that the changes led to disruptions in their work environment and uncertainties, adversely affecting their psychological welfare.

Heightened Stress and Anxiety. Numerous participants expressed heightened anxiety during the restructuring phase because of the ambiguity and the absence of clear information regarding the expected changes. For example, one employee explained: *Over time, it has become more stressful, and I've become more anxious about it than when it started* (P10). Another noted, *I think that hearing that with this restructuring, there will also be retrenchment. That is why I'm anxious or have anxiety* (P1). These responses suggest that stress was not a short-term reaction but persisted and worsened over time.

Negative Health and Well-being Effects. Employees also associated stress with adverse physical and emotional outcomes. Numerous people indicated that their work schedules were interrupted and that they felt overwhelmed. As one respondent reflected, *The stress that this restructuring has brought to our department, it's too much* (P1). Another shared: *One is anxious about the job security* (P11). These narratives emphasise the impact on the mental well-being of employees, which was worsened by the lack of supportive structures within the organisation.

4.2 Perceptions of Job Security

Throughout the interviews, worries about job security were prevalent. Participants indicated ongoing concerns about retrenchment and uncertainty about their future.

Uncertainty and Job Insecurity. Employees described uncertainty regarding their employment as a constant source of anxiety. For example, one participant shared, *I was so anxious, I did not know the security of my employment, of my job, and what was gonna happen* (P7). Another echoed this sentiment: *There is that fear of uncertainty to say what is going to happen to me, what is going to happen to other colleagues in the future* (P8). Even staff with permanent contracts expressed scepticism about their long-term security: *Although I was having a permanent position, there's often been a lingering sense of uncertainty about how long I can sustain this position* (P17).

Fear of Retrenchment. The threat of retrenchment was a recurring concern. Participants stated that rumours and incomplete information deepened their anxiety. For instance, one employee noted: *What really made me more anxious was I was not sure about my future in the organisation* (P9). Another stated: *Yeah, you cannot be comfortable when they are introducing new change, and then you learn of certain things like retrenchment* (P1). These comments demonstrate how restructuring fostered a culture of fear, leaving workers feeling insecure and unappreciated.

Communication and Employee Engagement. During restructuring, management's communication was perceived as insufficient, lacking consistency, and frequently exclusionary. This pattern highlights how ineffective communication tactics eroded trust and heightened employee anxiety.

Lack of Transparent Communication. Participants repeatedly emphasised the absence of clear and timely communication. Staff often learned about restructuring decisions only after implementation, which left them confused and frustrated. For example, *There was no communication, we were just told now there are two departments, and this is your position, that's it* (P2). Another respondent remarked: *There's not much communication, if I may say so, and that also on its own brings the frustration to us* (P3). Poor communication contributed to widespread rumours, further heightening employees' insecurity.

Exclusion from Decision-Making. A strong sense of exclusion emerged from participants' narratives. Many expressed frustrations with being sidelined from decisions that directly affected their work. *We were never consulted. We were only informed once everything was done, and then you are told, 'This is your new position now* (P3). Others felt their knowledge and experience were disregarded: *We are the ones doing the work every day, but when it comes to decisions that affect us, they don't involve us'* (P7). Employees described this exclusion as disempowering and morale-depleting, with one participant stating: *If you're going to restructure, at least hear from the people who are doing the job. It felt like we were just numbers, not people* (P5).

4.3 Coping Mechanisms During Organisational Change

Despite the challenges, employees devised several ways of coping with anxiety and stress. These ranged from informal peer networks to individual therapy activities, although the respondents also remarked on the absence of formal institutional support.

Peer Support and Informal Sharing. During the restructuring, coworkers offered essential emotional support. Peer support was a key way to handle the situation when formal services were unavailable: *It is peer support. I think that peer support has been the way forward* (P9). This informal support fostered solidarity, although participants acknowledged it could not replace formal institutional interventions.

Personal Coping Strategies. Several employees employed customised methods to manage stress, such as using external wellness resources. One participant explained, *I see a therapist, and I did go to wellness about my mental health, so I do those things that are at my disposal to deal with mental health issues* (P4). Specific individuals have been observed to prioritise their work tasks to shift their focus away from feelings of anxiety.

Lack of Institutional Support. One prevalent concern that emerged was the absence of structured psychological support. Participants expressed discontent over the absence of counselling services and corporate wellness programmes to support restructuring efforts. For instance, *How I wish that HR had introduced this to us, how I wish that we had been called for counselling to prepare us as employees* (P1). Another added: *We haven't been offered any counselling services, or even been suggested counselling services, so it's just voicing our concerns and fears to each other* (P10). The lack of attention made employees feel deserted during a crucial time of transition.

5 Discussion

The findings of this study reinforce a substantial body of existing research linking organisational restructuring to increased psychological distress among employees. Participants' elevated stress and anxiety levels are in line with earlier studies showing that organisational restructuring can significantly affect workers' mental health, increasing their risk of anxiety, depression, and burnout (An et al. 2023; Contreras and Gonzalez 2021). The results of this study support previous research emphasising the value of open communication in reducing adverse psychological effects by indicating that ambiguous processes and poor communication can worsen stress (Clarito 2024; Harney et al. 2018).

The study highlighted the significance of job insecurity, aligning with existing literature that identifies insecurity as a key factor in the detrimental impact of restructuring on well-being. Further supporting this association and supporting other studies on the link between work insecurity, low morale, and decreased productivity are the participants' worries about possible layoffs and uncertainty about their job stability. This conclusion aligns with the research conducted by Jerab and Mabrouk (2023), Ohu et al. (2019), Harney et al. (2018), and Wang et al. (2024).

The literature also strongly supports the issue of role ambiguity found in the results. Other studies have demonstrated that this uncertainty lowers morale and performance, and participants expressed bewilderment because of unclear duties and outdated job descriptions (Verlinden et al. 2022; Mañas et al. 2018). To lessen uncertainty and its negative impacts, it is underlined how important it is to have clear job descriptions, maintain constant communication, and provide specialised organisational assistance throughout transitions (Cengiz et al. 2021).

Furthermore, the participants reported increasing workload, which is in line with research that links higher workloads to job dissatisfaction and burnout, especially when there is a lack of proper staffing and recognition (Ullah and Noël 2023; Farrell et al. 2024; Umar 2023). To preserve organisational stability and employee well-being, participants suggested implementing practical advice from the literature, such as openly discussing task expectations and hiring more staff (Gupta and Dhingra 2024).

Communication challenges emerged as direct stressors and drivers of mistrust and disengagement in this study. This supports studies showing that employees' fear and resistance increase when leadership communicates poorly, ambiguously, or unidirectionally (Klewes et al. 2017; Ophelia and Hidayat 2021). According to Akinwalere (2023), the participants' experiences of depending on unofficial channels and rumours indicate the acknowledged hazards of communication failures during organisational transformation. On the other hand, effective communication techniques that are inclusive and transparent are well known for their capacity to increase trust, lower uncertainty, and promote flexibility (Packard 2021; Rosenberg 2018).

The lack of institutional psychological support identified in the results corresponds with broader findings that support networks offering emotional and psychological resources are crucial in buffering the negative impacts of restructuring (Buick et al. 2018; Smollan 2017; Day et al. 2017). According to the research, the participants' dependence on outside treatment and peer support underlines the lack of official counselling and wellness programmes, which are essential elements to combat restructuring-related suffering (Heyns et al. 2021; Georgalis et al. 2014).

Finally, studies showing that employee participation in change processes promotes resilience, ownership, and effective adaptation align with the concept of exclusion from decision-making. Employee resistance and disengagement tend to rise when excluded, exacerbating emotional suffering and lowering morale (Niesen and Randall 2012; Smollan 2017; Dibra 2021; Sun and Uzapi 2023). The importance of organisation-level participatory interventions to mitigate job insecurity and its repercussions is underscored by longitudinal and systematic reviews that show increased job demands and decreased autonomy during restructuring worsen employee mental health (Potter et al. 2024; Harney et al. 2018; Abildgaard et al. 2018).

6 Conclusion

This study contributes theoretically and practically to our knowledge of organisational restructuring in higher education, especially in South Africa. Integrating employees' lived experiences of stress, anxiety, job insecurity, and position ambiguity improves the conceptualisation of psychological and emotional effects during restructuring and conceptually extends existing models of organisational transformation. The findings highlight the need for clear position descriptions, open communication, and employee participation to mitigate negative consequences. They also support and broaden ideas about organisational behaviour, job insecurity, and change management.

Additionally, using an interpretivist qualitative methodology, the study highlights the subjective reality of academic and administrative personnel, bringing to light contextual and cultural subtleties that are sometimes missed in research from other countries that is primarily quantitative. The study offers insightful, practical advice for institutional leaders and legislators hoping to conduct restructuring projects in a more efficient and compassionate manner. Important practical contributions include the identification of specific stresses that institutions need to address to protect employees' well-being and morale, such as ambiguous communication routes, a lack of psychological support, and exclusion from decision-making.

The research promotes open, mutually beneficial communication techniques, updating job descriptions to accurately reflect responsibilities, formally acknowledging and compensating for extra work, and setting up organised support networks, such as wellness and counselling services. By implementing these strategies, employers may lower resistance, improve employee engagement, and create an institutional culture that is adaptable enough to adjust to changing circumstances and outside influences. The research contains limitations despite its significant contributions. Initially, the study was conducted in a single university, which would restrict the findings' applicability to other academic institutions, fields, and cultural contexts. Although the qualitative methodology and small sample size are suitable for investigating depth of experience, they also limit the capacity to draw general statistical conclusions. Furthermore, because the data are cross-sectional, it only records employee experiences at a particular stage of the restructuring process, ignoring long-term psychological and occupational effects.

Future research should thus pursue longitudinal designs to better understand how employees' stress, job perceptions, and coping strategies evolve post-restructuring. To enhance generalisability and for comparison assessments of contextual factors impacting restructuring results, the study should be expanded to various institutions in South Africa and abroad. The rich qualitative insights presented here might be combined with quantitative measurements in mixed-methods techniques to assess specific hypotheses and offer more substantial proof of the efficacy of interventions. Additionally, filling the gaps noted by Chen et al. (2023) and others would involve investigating how new workplace dynamics like digital transformation and artificial intelligence integration affect restructuring experiences.

In conclusion, future research might examine the efficacy of tailored psychological support therapies and participative decision-making in organisational change scenarios to provide best practice guidelines for universities. The study's conclusion highlights the need to address organisational restructuring as a holistic process that considers not just operational and financial aspects but also employees' psychological health and active involvement. Promoting employee well-being in the face of continuous change and cultivating sustained institutional transformation requires integrating adaptable support systems, transparent communication, and inclusive change management techniques.

References

Abildgaard, J.S., Nielsen, K., Sverke, M.: Can job insecurity be managed? Evaluating an organizational-level intervention addressing the negative effects of restructuring. Work Stress. **32**(2), 105–123 (2018). https://doi.org/10.1080/02678373.2017.1367735

Akinwalere, I.: Transformation in organisational communication. J. Public Adm. Finance Law **29**, 238–249 (2023). https://doi.org/10.47743/jopafl-2023-29-21

An, H., Gu, X., Obrenovic, B., Godinic, D.: The role of job insecurity, social media exposure, and job stress in predicting anxiety among white-collar employees. Psychol. Res. Behav. Manag. **16**, 3303–3318 (2023). https://doi.org/10.2147/PRBM.S416100

Backhaus, I., Lohmann-Haislah, A., Burr, H., Nielsen, K., di Tecco, C., Dragano, N.: Organizational change: challenges for workplace psychosocial risks and employee mental health. BMC Public Health **24**(1), 2477 (2024). https://doi.org/10.1186/s12889-024-19815-w

Braun, V., Clarke, V.: Using thematic analysis in psychology. Qual. Res. Psychol. **3**(2), 77–101 (2006). https://doi.org/10.1191/1478088706qp063oa

Buick, F., Blackman, D., Johnson, S.: Enabling middle managers as change agents: why organisational support needs to change. Aust. J. Public Adm. **77**(2), 222–235 (2018)

Cengiz, A., Yoder, L.H., Danesh, V., Danesh, V.: A concept analysis of role ambiguity experienced by hospital nurses providing bedside nursing care. Nurs. Health Sci. **23**(4), 807–817 (2021). https://doi.org/10.1111/nhs.12888

Chen, J.S., Khiêm, M.Đ, Tsou, H.T.: Digital organisational restructuring on business value creation in SMEs during the COVID-19 pandemic. Technol. Anal. Strateg. Manag. **36**(12), 4036–4051 (2023). https://doi.org/10.1080/09537325.2023.2239378

Clarito, M.J.R.: Organizational change and employees' work-life balance in ABC semiconductor company. World J. Adv. Res. Rev. **21**(3), 1332–1341 (2024). https://doi.org/10.30574/wjarr.2024.21.3.0837

Contreras, S., Gonzalez, J.A.: Organizational change and work stress, attitudes, and cognitive load utilization: a natural experiment in a university restructuring. Personnel Rev. **50**(1), 264–284 (2021). https://doi.org/10.1108/PR-06-2018-0231

Daněk, A., Urgošíková, M.: Qualitative research design in education: tools, challenges, and practical implications. Ad Alta **14**(2), 385–389 (2024). https://doi.org/10.33543/j.1402.385389

Day, A., Crown, S.N., Ivany, M.: Organisational change and employee burnout: the moderating effects of support and job control. Saf. Sci. **100**, 4–12 (2017). https://doi.org/10.1016/j.ssci.2017.03.004

De Jong, T., Wiezer, N., de Weerd, M., Nielsen, K., Mattila-Holappa, P., Mockałło, Z.: The impact of restructuring on employee well-being: a systematic review of longitudinal studies. Work Stress. **30**, 91–114 (2016). https://doi.org/10.1080/02678373.2015.1136710

Dibra, M.: Employee participation rights during company restructuring: lessons learned from the 'capability for voice' in British law. Revue Générale de Droit **51**(1), 245–292 (2021). https://doi.org/10.7202/1081842ar

Doidge, S., Doyle, J.: Australian universities in the age of COVID-19. Educ. Philos. Theory **54**(6), 668–674 (2020). https://doi.org/10.1080/00131857.2020.1804343

Elsaman, H.: Conceptualising the model of creating shared value, organisational performance and the mediating role of change management. SA J. Hum. Resour. Manag. **22**, 1–8 (2024). https://doi.org/10.4102/sajhrm.v22i0.2766

Farrell, C., Hassard, J., Morris, J.: Organizational restructuring, precarious employment and work intensification: women managers' experience of work under neoliberalism. Econ. Ind. Democracy (2024). https://doi.org/10.1177/0143831X241306232

Georgalis, J., Samaratunge, R., Kimberley, N., Lu, Y.: Change process characteristics and resistance to organisational change: the role of employee perceptions of justice. Aust. J. Manag. **40**(1), 89–113 (2014). https://doi.org/10.1177/0312896214526212

Gupta, K., Dhingra, K.: Navigating change: A comprehensive analysis of corporate restructuring strategies. Int. J. Multidiscip. Res. 7(3) (2024). https://doi.org/10.36948/ijfmr.2024.v06i02.15802

Harney, B., Fu, N., Freeney, Y.: Balancing tensions: buffering the impact of organisational restructuring and downsizing on employee well-being. Hum. Resour. Manag. J. **28**, 235–254 (2018). https://doi.org/10.1111/1748-8583.12175

Heyns, M.M., McCallaghan, S., Senne, O.W.: Supervisor support and work engagement: the mediating role of psychological safety in a post-restructuring business organisation. J. Psychol. Afr. **31**(2), 140–144 (2021). https://doi.org/10.1080/14330237.2021.1903180

Jayasuriya, K.: COVID-19, markets and the crisis of the higher education regulatory state: the case of Australia. Globalizations **18**(4), 584–599 (2020). https://doi.org/10.1080/14747731.2020.1815461

Jerab, D.A., Mabrouk, T.: Strategies for effective organizational restructuring: a comprehensive guide. SSRN Electron. J. (2023). https://doi.org/10.2139/ssrn.4584665

Kivimäki, M., Vahtera, J., Pentti, J., Ferrie, J.E.: Downsizing, changes in work, and self-rated health of employees. J. Occup. Organ. Psychol. **74**(4), 394–408 (2001)

Klewes, J., Popp, D., Rost-Hein, M.: Digital transformation and communications: how key trends will transform the way companies communicate. In: Digital Transformation, pp. 7–31. Springer, Cham (2017). https://doi.org/10.1007/978-3-319-41845-2_2

Lyu, Y.: Shifting research paradigms in educational research: from positivism to interpretivism. Sci. Technol. Soc. Dev. Proc. Ser. **2**, 11–14 (2024). https://doi.org/10.70088/bmq18v35

Mañas, M.A., Díaz-Fúnez, P.A., Pecino, V., López-Liria, R., Padilla, D., Aguilar-Parra, J.M.: Consequences of team job demands: role ambiguity climate, affective engagement, and extra-role performance. Front. Psychol. **8**, 2292 (2018). https://doi.org/10.3389/fpsyg.2017.02292

Mathisen, G.E., Tjora, T., Bergh, L.I.V., Jain, A., Leka, S.: The differential impact of organizational restructuring and downsizing on the psychosocial work environment and safety climate in the petroleum industry. Saf. Sci. **166**, 106255 (2023). https://doi.org/10.1016/j.ssci.2023.106255

Nicholson, D.T., Vargas, V.R., Skelly, G.: Enquiry-based skills education for sustainable development as a catalyst for organisational change. Int. J. Sustain. High. Educ. **24**(8), 1897–1914 (2023). https://doi.org/10.1108/IJSHE-07-2022-0225

Niedhammer, I., Bertrais, S.: Organisational changes and psychosocial work factors: prospective findings from the national French working conditions survey. Saf. Health Work **15**(4), 496–499 (2024). https://doi.org/10.1016/j.shaw.2024.10.002

Nielsen,K.,&Randall,R.(2012).Theimportanceofemployeeparticipationandperceptionsofchangesin proceduresinateamworkingintervention.Workandstress, **26**(2),91–111. https://doi.org/10.1080/02678373.2012.682721

Niesen, W., Van Hootegem, A., Vander Elst, T., Battistelli, A., De Witte, H.: Job insecurity and innovative work behaviour: a psychological contract perspective. Psychol. Belg. **57**(4), 174–189 (2018). https://doi.org/10.5334/pb.381

Ohu, E.A., Spitzmueller, C., Zhang, J., Thomas, C.L., Osezua, A., Yu, J.: When work–family conflict hits home: parental work–family conflict and child health. J. Occup. Health Psychol. **24**(5), 590–601 (2019). https://doi.org/10.1037/ocp0000145

Ophelia, A., Hidayat, Z.: Leadership communication during organizational change: internal communication strategy in a multinational company. Acad. J. Interdiscip. Stud. **10**(2), 24 (2021). https://doi.org/10.36941/ajis-2021-0035

Packard, T.: Organization redesign. In: Organizational Change for the Human Services. Oxford University Press (2021). https://doi.org/10.1093/oso/9780197549995.003.0016

Park, K.H., Youn, S.J., Moon, J.: The effect of workforce restructuring on withdrawal behavior. J. Asian Finance Econ. Bus. **7**(7), 413–424 (2020). https://doi.org/10.13106/JAFEB.2020.VOL7.NO7.413

Potter, R., Afsharian, A., Richter, S., Neser, D., Zadow, A., Dollard, M., Lushington, K.: Longitudinal investigation of restructuring, psychosocial safety climate and burnout in Australian universities. J. Ind. Relat. **67**(2), 224–248 (2024). https://doi.org/10.1177/00221856241247577

Pulla, V., Carter, E.: Employing interpretivism in social work research. Int. J. Soc. Work Hum. Serv. Pract. 6(1), 9–14 (2018). https://doi.org/10.13189/ijrh.2018.060102

Regmi, B.P.: Qualitative research design: a discussion on its types. Res. J. **9**(1), 37–45 (2024). https://doi.org/10.3126/rj.v9i1.74415

Rosenberg, A.: Taking apart structural change: the constitutive role of communication in relieving tensions. Int. J. Organ. Anal. **26**(2), 368–381 (2018). https://doi.org/10.1108/IJOA-04-2017-1156

Smollan, R.K.: Learning to cope with stressful organisational change. Int. J. Work Organ. Emot. **8**(2), 148 (2017). https://doi.org/10.1504/IJWOE.2017.10007614

Stratton, S.J.: Purposeful sampling: advantages and pitfalls. Prehosp. Disaster Med. **39**(2), 121–122 (2024). https://doi.org/10.1017/S1049023X24000281

Sun, Z., Uzapi, H.: Employee involvement and the effectiveness of organizational change in Côte d'Ivoire. Adv. Econ. Manag. Res. **5**(1), 265 (2023). https://doi.org/10.56028/aemr.5.1.265.2023

Trchalíková (Vydrželová), P.: The impact of stress on work performance and employee turnover. Acta Acad. Karvin. 23(2), 73–88 (2023). https://doi.org/10.25142/aak.2023.016

Ullah, R., Noël, F.: Downsizing, workload and interpersonal conflict. Rev. Gest. Ressour. Hum. **127**(1), 51–71 (2023). https://doi.org/10.3917/grhu.127.0051

Tumelo, T., Donald, F.: Perceived job insecurity, facades of conformity, emotional exhaustion and disengagement. SA J. Ind. Psychol. **51**, 1–9 (2025). https://doi.org/10.4102/sajip.v51i0.2221

Umar, M.A.: Corporate restructuring: a strategy for improving organizational performance. Int. J. Strateg. Decis. Sci. **14**(1), 1–11 (2023). https://doi.org/10.4018/IJSDS.319974

Vlachopoulos, D.: Organizational change management in higher education through the lens of executive coaches. Educ. Sci. **11**(6), 269 (2021). https://doi.org/10.3390/educsci11060269

Verlinden, S., Wynen, J., Kleizen, B., Verhoest, K.: Blurred lines: exploring the impact of change complexity on role clarity in the public sector. Rev. Public Pers. Adm. (2022). https://doi.org/10.1177/0734371X221093573

Yang,Y.(2024).Strategicplanningofstructuraltransformationswithinthecontextofcorporateorganizational development.ActualProblemsofEconomics, **1**(274),14–21. https://doi.org/10.32752/1993-6788-2024-1-274-14-21

Wang, M.L., Narcisse, M., Togher, K., McElfish, P.A.: Job flexibility, job security, and mental health among US working adults. JAMA Netw. Open **7**(3), e243439 (2024). https://doi.org/10.1001/jamanetworkopen.2024.3439(2016)

The Stability-Innovation Paradox: How Traditional Companies Foster Intrapreneurship Through Organisational Stability—Latest

Lodewikus Johannes Janse van Rensburg

Abstract Innovation-oriented companies champion entrepreneurial thinking yet may inadvertently constrain intrapreneurial behaviours through their structures. This study investigated whether companies recognised for innovation excellence foster stronger intrapreneurial cultures than traditional firms. I analysed 14,009 employee reviews from Glassdoor using GPT-4 to extract five intrapreneurial dimensions: innovation orientation, autonomy, risk-taking, proactiveness, and empowerment. Our matched-pairs design compared 11 innovation-oriented companies with 11 traditional controls across the USA, Europe, and South Africa. Results revealed a paradox: innovation-oriented companies showed only marginally higher innovation orientation ($d = 0.089$, $p < .001$) and risk-taking ($d = 0.051$, $p < .01$), while traditional companies demonstrated significantly greater proactivity ($d = -0.074$, $p < .001$) and trending higher autonomy. Effect sizes remained small (<0.1), suggesting minimal practical differences between company types. Geographic variations proved substantial, with US patterns matching theoretical expectations while European and South African contexts showed reversed relationships. Linguistic analysis uncovered distinct conceptualisations: innovation-oriented firms emphasised 'disruption' and 'cutting-edge', while traditional companies stressed 'continuous improvement' and 'work-life balance'. These findings challenge fundamental assumptions about organisational entrepreneurship. I propose that organisational stability, rather than constraining innovation, may enable certain intrapreneurial behaviours through psychological safety and resource predictability. I introduce 'intrapreneurial complementarity' to explain how different organisational contexts cultivate distinct but equally valuable entrepreneurial behaviours. This research demonstrates the potential of AI-powered cultural analysis while revealing the culturally contingent nature of intrapreneurship. Organisations need not pursue disruptive transformation to foster entrepreneurship; stability and innovation can paradoxically enable each other.

Keywords Intrapreneurship · Organisational culture · Computational analysis · GPT-4 · Innovation paradox

L. J. J. van Rensburg (✉)
University of Johannesburg, Johannesburg, South Africa
e-mail: leonjvr@uj.ac.za

E. H. Redda (ed.), *Proceedings of the Social Sciences International Research Conference, 2025*, https://doi.org/10.1007/978-981-95-9489-4_16

1 Introduction

The modern business landscape presents organisations with an existential challenge: maintaining competitive advantage through continuous innovation while operating within established organisational structures. This tension has sparked intense interest in intrapreneurship, defined as the practice of entrepreneurial behaviour within existing organisations, as a potential solution to the innovation imperative (Antoncic and Hisrich 2003; Neessen et al. 2019). However, a fundamental question remains inadequately addressed: do companies celebrated for innovation excellence cultivate stronger intrapreneurial cultures than their traditional counterparts?

This question carries profound implications for organisational strategy and resource allocation. Globally, companies invest over $1.7 trillion annually in innovation initiatives (UNESCO 2021), with a significant portion directed toward fostering intrapreneurial capabilities. Management consultancies promote innovation-oriented transformation as essential for survival, while business schools teach aspiring managers that innovative companies represent the gold standard for entrepreneurial culture. However, empirical evidence supporting these assumptions remains surprisingly limited, and its findings are methodologically constrained by small samples and survey biases.

This research challenges this conventional wisdom by investigating whether innovation-oriented companies, specifically those recognised in rankings like Fast Company's Most Innovative Companies, genuinely foster stronger intrapreneurial cultures than traditional firms. I conceptualised intrapreneurial culture as encompassing five key dimensions from established frameworks: innovation orientation, autonomy, risk-taking, proactiveness and empowerment (Covin and Slevin 1989; Lumpkin and Dess 1996). Rather than relying on limited survey data or case studies, I employed a novel computational approach, analysing 14,009 employee reviews using GPT-4 to extract nuanced cultural insights at an unprecedented scale.

The timing of this investigation proved particularly salient given three converging trends reshaping organisational life. First, democratising employee voice through platforms like Glassdoor provided unprecedented windows into organisational cultures, offering authentic, unsolicited perspectives that complemented traditional research methods. Second, advances in large language models enabled sophisticated analysis of unstructured text data, moving beyond simple sentiment analysis to extract complex cultural dimensions. Third, the post-pandemic reconfiguration of work intensified debates about organisational culture, autonomy and innovation, making cultural assessment more challenging and critical.

The findings revealed a striking paradox that challenges foundational assumptions in innovation management. While innovation-oriented companies demonstrated marginally higher innovation orientation and risk-taking behaviours, traditional companies unexpectedly showed stronger proactiveness and a trend toward higher autonomy. These statistically significant differences were minor in magnitude ($d < 0.1$), suggesting that the gap between innovation-oriented and traditional companies may be far narrower than commonly believed. More intriguingly, these patterns varied dramatically by geographic context, with US companies conforming to expectations while European and South African contexts showed reversed relationships.

This paradoxical finding, namely that traditional companies may foster certain intrapreneurial behaviours more effectively than their innovation-oriented counterparts, demands theoretical explanation. I propose that organisational stability, often dismissed as antithetical to entrepreneurship, may paradoxically enable certain forms of intrapreneurial behaviour. In stable environments with clear structures and predictable resources, employees may feel more empowered to take calculated risks and pursue incremental innovations. Conversely, the relentless pace and constant disruption characteristic of innovation-oriented companies may, counterintuitively, constrain certain entrepreneurial behaviours by creating environments too chaotic for sustained intrapreneurial initiatives.

This research makes several significant contributions to organisational scholarship and practice. Theoretically, I introduce the *intrapreneurial complementarity* concept, suggesting that different organisational contexts cultivate distinct but equally valuable forms of entrepreneurial behaviour. Methodologically, I demonstrate the potential of AI-powered analysis for large-scale cultural assessment while maintaining interpretive depth. Practically, I provide evidence suggesting that organisations need not mimic Silicon Valley archetypes to foster meaningful intrapreneurial behaviour, potentially liberating traditional companies from costly and disruptive transformation initiatives.

The remainder of this paper unfolds as follows. I first review relevant literature on intrapreneurship, organisational culture and computational methods, developing our theoretical framework. I then detail our methodology, including our innovative use of GPT-4 for cultural dimension extraction. Our results reveal complex patterns in intrapreneurial culture across organisational types and regions. I discuss theoretical and practical implications before concluding with future research directions.

2 Literature Review and Theoretical Framework

2.1 The Evolution and Dimensions of Intrapreneurship

Intrapreneurship, referring to entrepreneurial behaviour within established organisations, emerged as a distinct concept when scholars recognised that innovation need not be confined to startups (Pinchot 1985). Sharma and Chrisman (1999) defined it as 'the process whereby an individual or group of individuals, in association with an existing organisation, create a new organisation or instigate renewal or innovation within that organisation'. This definition captures the dual nature of intrapreneurship: new venture creation and organisational transformation through innovation.

The field draws from multiple theoretical traditions. Strategic management focuses on competitive advantage through internal entrepreneurial activities (Covin and Miles 1999). Entrepreneurship theory offers concepts of opportunity recognition and value creation in the face of uncertainty (Shane and Venkataraman 2000). Organisational behaviour research highlights the human dimensions, identifying antecedents such as management support, work discretion and resource availability that enable intrapreneurial behaviour (Hornsby et al. 2002).

Three major frameworks dominate intrapreneurship research. Antoncic and Hisrich (2003) emphasised outcome-oriented dimensions: new business venturing, innovativeness, self-renewal and proactiveness. Lumpkin and Dess's (1996) entrepreneurial orientation framework takes a behavioural approach with five dimensions: autonomy, innovativeness, risk-taking, proactiveness and competitive aggressiveness. Most recently, Kuratko et al. (2019) shifted their focus to organisational enablers, emphasising the ecosystem that supports intrapreneurial activities.

For this study, I synthesised these frameworks into five dimensions particularly relevant to employee experiences: innovation orientation (encouraging novel ideas), autonomy (decision-making freedom), risk-taking (tolerance for experimentation), proactiveness (initiative-taking), and empowerment (providing resources and support). These dimensions consistently appear across frameworks and are directly observable by employees in their daily work.

2.2 *Measuring Intrapreneurial Culture: From Surveys to Digital Traces*

Traditional measurement relies heavily on survey instruments, such as the Corporate Entrepreneurship Assessment Instrument (CEAI), which faces limitations including response bias, low response rates and access constraints (Hornsby et al. 2002). Alternative approaches, which use innovation outputs (such as patents and R&D spending), provide objectivity but overlook cultural and processual aspects (Zahra 1991). These measures also represent lagging indicators, limiting their diagnostic value.

Recent developments have significantly advanced measurement capabilities. Lajçi (2025) conducted a comprehensive scoping review, revealing a fragmented landscape of assessment tools with limited cross-validation, which reinforces the need for new approaches. The intersection of digital transformation and intrapreneurship has become crucial, as noted by D'Angelo et al. (2023) showing how digital capabilities reshape corporate entrepreneurship, requiring new measurement frameworks.

Employee review platforms represent 'digital voice', meaning voluntary public expressions of workplace experiences (Briscoe and Gupta 2016). These platforms offer unique advantages: anonymity reduces fear of retaliation, public visibility creates accountability, and temporal currency provides real-time insights. Green et al. (2019) validated that Glassdoor ratings predict firm performance, with employee satisfaction changes preceding stock returns. Luo et al. (2016) found that reviews correlate with turnover, confirming they capture genuine sentiments.

Recent work has expanded the validation of employee reviews for cultural analysis. Gupta et al. (2022) used topic modelling on employee reviews to identify organisational culture constituents, finding consistent discussion of innovation, autonomy and management support, all themes central to intrapreneurship. Lee and Choi (2025) revealed how cultural contexts shape the expression of workplace experiences, which presaged our regional findings.

2.3 Natural Language Processing for Cultural Analysis

The development of transformer-based language models represents a paradigm shift for organisational research. GPT-4 and similar models can comprehend context, sentiment and complex linguistic patterns that traditional methods often overlook (Brown et al. 2020). For cultural research, this enables the detection not just of keyword mentions but also of contextual meanings, which are crucial for complex constructs like intrapreneurship.

Recent applications demonstrate this potential. Rita et al. (2025) combined sentiment analysis and topic modelling to extract actionable insights from employee feedback. Lee and Choi (2024) demonstrated how organisational culture moderates R&D investment returns by using Neuro-Linguistic Programming in employee reviews. Their finding that culture mediates strategic outcomes concurs with our discovery that innovation recognition does not directly translate to intrapreneurial culture.

2.4 Regional and Cultural Context

Cross-cultural research reveals the contextual nature of intrapreneurship. While US literature emphasises radical innovation and disruption, European contexts focus more on incremental improvement and sustainability (Bosma et al. 2010). Urban and Wood (2017) demonstrated that South African firms use intrapreneurship to address social challenges, not just for competitive advantage. Recent work by Lima de Santana and Pinheiro (2024) on public sector intrapreneurship confirms that cultural values fundamentally shape entrepreneurial expression within organisations.

Digital transformation creates new intrapreneurial contexts. Wu and Li (2024) found that the digital-innovation relationship between Chinese corporations is nonlinear and context-dependent. Fan (2024) demonstrated that digital capabilities facilitate sustainable development through corporate entrepreneurship, challenging the assumption that intrapreneurship must pursue disruption. These findings support our argument against universal intrapreneurship models.

2.5 Theoretical Framework and Hypotheses

Building on this literature, I position innovation-oriented companies (those recognised for innovation outcomes) as distinct from intrapreneurial culture (our construct of interest). External innovation recognition may reflect concentrated R&D efforts or acquisitions rather than pervasive cultural characteristics. By analysing employee reviews, I capture the lived experience of organisational culture rather than external perceptions or managerial aspirations.

Our framework addresses critical gaps: the lack of scalable culture measurement tools (Lajçi 2025), the need to understand digital-era intrapreneurship (D'Angelo et al. 2023) and calls for contextual theorising (Zahra et al. 2014). Figure 1 presents our theoretical framework for measuring intrapreneurial culture using AI-powered employee reviews.

This leads to our primary hypothesis:

H1: Innovation-oriented companies will demonstrate significantly higher scores across all five intrapreneurial dimensions than traditional companies.

However, recognising cultural contingencies, I also propose:

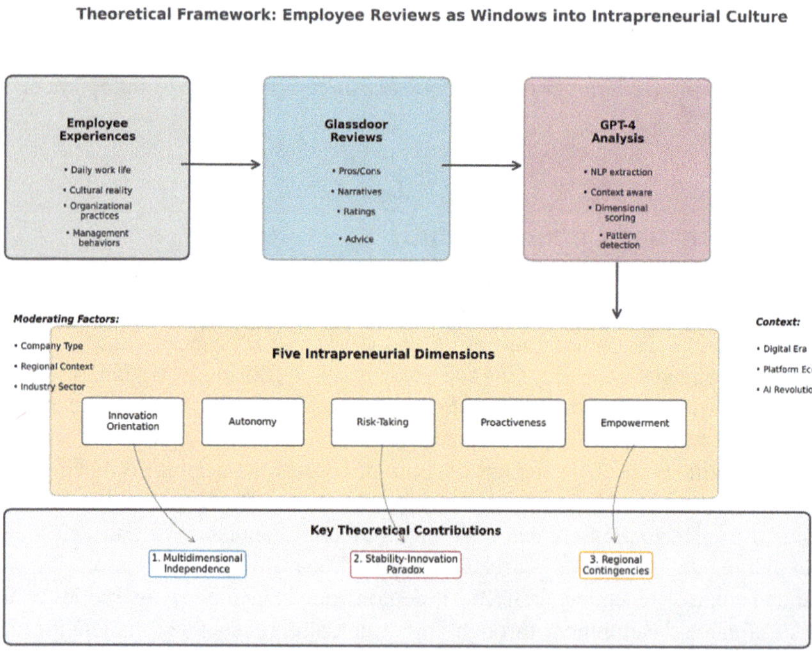

Fig. 1 Theoretical framework for AI-powered measurement of intrapreneurial culture through employee reviews. The author's own conceptual synthesis is based on Sharma and Chrisman (1999), Lumpkin and Dess (1996), Antoncic and Hisrich (2003), Kuratko et al. (2019), D'Angelo et al. (2023), Gupta et al. (2022), Lee and Choi (2025), and Lajçi (2025)

H2: The relationship between company type and intrapreneurial dimensions will vary significantly across geographic regions.

3 Methodology

3.1 Research Design

The researcher employed an explanatory sequential mixed-methods design grounded in pragmatist philosophy (Creswell and Plano Clark 2017; Morgan 2007). This approach combines computational pattern detection across thousands of reviews with qualitative analysis of exemplar texts, aligning with the principles of computational grounded theory (Nelson 2020). The pragmatist orientation enabled us to move seamlessly between theoretical frameworks and empirical patterns, proving particularly valuable in confronting unexpected findings.

3.2 Sample Selection

Our matched-pairs design compared innovation-oriented companies with traditional controls to isolate the effect of innovation orientation on intrapreneurial culture. Innovation-oriented companies were identified through the triangulation of multiple sources, including Fast Company's Most Innovative Companies (2020–2024), Forbes Innovation Rankings and regional innovation indices. Companies needed to appear in at least two rankings during our study period to qualify, resulting in 11 firms: Amazon, Apple, Google, Microsoft, Nvidia, Tesla (US); Novo Nordisk, SAP (EU); and Old Mutual, Standard Bank (SA).

Control company selection followed a rigorous matching protocol. For each innovation-oriented company, I identified matches using four sequential criteria: (1) alignment on 2-digit Standard Industrial Classification codes for industry comparability, (2) annual revenue within 50% range to control for size effects, (3) substantial operations in the same geographic region for cultural context, and (4) absence from any innovation rankings during the study period. While broad, the 50% revenue range proved necessary, given market concentration in sectors such as technology. This process yielded: General Motors, Johnson & Johnson, Verizon (US); Bayer, Beiersdorf, Capgemini, Thyssenkrupp (EU); ABSA, Sanlam, Sasol, Telkom SA (SA).

Table 1 presents the distribution of companies and reviews across geographic regions and company types. Review counts excluded 2218 reviews from companies with ambiguous innovation status. The imbalanced distribution reflects natural variation in company sizes and employee review participation rates across regions.

Table 1 Distribution of companies and reviews by geographic region and company type

Region	Innovation-oriented		Control		Total
	Companies	Reviews	Companies	Reviews	Reviews
United States	6	3586	3	3270	6856
Europe	2	1543	4	2778	4321
South Africa	2	819	4	2013	2832
Total	11	5948	11	5843	11,791

Data Collection

Data collection took place in June 2025, in accordance with institutional ethical guidelines for the use of public data. Three research assistants, all graduate students in organisational behaviour, underwent 2 weeks of training covering platform navigation protocols, data quality criteria, ethical considerations, and inter-rater reliability testing. The team achieved 98% agreement on 100 sample reviews before beginning collection.

Our systematic protocol involved navigating to each company's Glassdoor page using standardised search procedures, filtering for English-language reviews posted between January 2020 and June 2025 and extracting complete review text (including pros, cons and management advice) along with metadata (job title, location, employment status and posting date). Quality assurance included independent verification of 10% of reviews through random sampling. This process yielded 14,130 raw reviews, from which language detection algorithms excluded 121 non-English texts, resulting in 14,009 reviews for analysis.

The temporal scope is deliberately inclusive of the COVID-19 pandemic period, recognising that excluding this disruption would compromise ecological validity. Sensitivity analyses revealed stable patterns across years, suggesting our results reflect enduring cultural characteristics rather than temporary responses.

3.3 *Feature Extraction Using GPT-4*

Development and Validation. I developed the extraction approach through iterative prompt refinement, selecting a temperature of 0.3 to strike a balance between consistency and nuanced interpretation. Prompts were refined to capture contextual meaning beyond surface keywords.

To establish face validity, I conducted a preliminary assessment where two organisational behaviour researchers independently reviewed GPT-4's outputs for a subset of 50 reviews, confirming that the extracted dimensions aligned with theoretical expectations. While a comprehensive validation study comparing GPT-4 outputs with human coding across all dimensions would strengthen our findings (with a target inter-rater reliability of $\kappa > 0.75$), this represents a key limitation and a direction for future research.

Implementation. Our validated prompt instructed GPT-4 to analyse each review across five intrapreneurial dimensions, scoring each on a 0–1 continuous scale while considering context and tone beyond keywords. Processing parameters included the GPT-4-0613 model (most stable at the time of the study), a temperature of 0.3, a batch size of 10 reviews and rate limiting at five requests/minute.

I acknowledge that systematic reliability testing comparing GPT-4 outputs with human coding represents an important direction for future research.

Limitations. I acknowledge two key validation limitations. First, GPT-4's extracted dimensions lack validation against established psychometric instruments, such as ENTRESCALE or CEAI. Second, I established face validity through researcher review, but did not conduct systematic inter-rater reliability testing between GPT-4 and human coders. This positions our research as exploratory, demonstrating the potential of AI while requiring future validation studies. I operationally defined concepts like *stability* as the absence of explicit innovation-promoting behaviours, recognising this needed theoretical refinement.

3.4 Statistical Analysis

Primary analyses employed independent samples' *t*-tests to compare company types on each dimension, with a Bonferroni correction setting significance at $\alpha = 0.01$ to control the family-wise error rate. With 5948 innovation-oriented and 5843 control reviews, I achieved 80% power to detect effect sizes $d \geq 0.063$. I calculated Cohen's d with 95% confidence intervals, interpreting within Chen et al.'s (2010) framework, recognising the' potential accumulation of minor effects in organisational contexts.

Statistical assumption testing revealed expected violations. Shapiro–Wilk tests rejected normality for all dimensions ($p < 0.001$), unsurprising given bounded 0–1 scales. Levene's tests indicated heterogeneous variances for innovation orientation, autonomy and risk-taking ($p < 0.01$). Most critically, the nested structure of reviews within companies violated independence, with intraclass correlations ranging from 0.08 to 0.12.

To ensure robustness, I conducted multiple parallel analyses. Mann–Whitney U tests provided non-parametric alternatives unaffected by normality violations. Welch's t-tests accommodated unequal variances. Multilevel models explicitly accounted for company-level clustering, treating the company as a random effect. Remarkably, all approaches yielded consistent direction and approximate magnitude results, strengthening confidence in our findings.

Regional variations were examined through two-way Analysis of Variance (ANOVA) testing of interactions between Company Type and Region. Significant interactions prompted post-hoc analyses using Tukey's Honestly Significant Difference (HSD) to identify specific patterns while controlling multiple comparisons.

3.5 *Qualitative Analysis*

Reviews scoring ≥ 0.8 on respective dimensions underwent thematic analysis following Braun and Clarke (2006). The lead researcher systematically analysed themes emerging from these high-scoring reviews. Linguistic pattern analysis using term frequency-inverse document frequency (TF-IDF) identified vocabulary differences between company types, revealing distinct conceptualisations rather than merely different levels of intrapreneurial behaviour.

4 Results

4.1 *Sample Characteristics and Data Quality*

Our analytical sample comprised 14,009 employee reviews, distributed across innovation-oriented companies ($n = 5948$, 42.5%) and control companies ($n = 5843$, 41.7%). I excluded 121 non-English reviews and omitted 2,218 reviews from companies with ambiguous innovation classifications from the primary analyses. Current employees authored 58% of reviews, while former employees contributed 42%. This distribution remained consistent across both company types, mitigating potential bias from differential reviewer populations.

The temporal distribution demonstrated remarkable stability, with the following percentages: 2020 (18%), 2021 (22%), 2022 (21%), 2023 (20%), 2024 (16%), and 2025 (3%). This consistency across years, including the pandemic, suggests our findings reflect enduring cultural patterns rather than temporary disruptions. Sensitivity analyses confirmed no significant temporal trends in dimensional scores.

4.2 *Descriptive Patterns and Dimensional Independence*

Initial examination revealed compelling patterns in how employees perceive intrapreneurial opportunities. Across all companies, empowerment emerged as the highest-scored dimension (M = 0.412, SD = 0.227), suggesting that modern organisations recognise the importance of employee agency, regardless of their innovation orientation. Conversely, risk-taking registered lowest (M = 0.220, SD = 0.172), indicating persistent organisational caution toward failure tolerance, despite rhetoric about embracing experimentation.

Correlation analysis revealed nuanced dimensional relationships that challenge simplistic conceptualisations of intrapreneurial culture. Innovation orientation showed the strongest associations with risk-taking ($r = 0.43$, $p < 0.001$) and empowerment ($r = 0.38$, $p < 0.001$), confirming theoretical expectations that innovative behaviours require both failure tolerance and organisational support. However, autonomy demonstrated surprisingly weak correlations with other dimensions (r = 0.12–0.21), suggesting it operates independently rather than as an integrated component. This finding implies that organisations may grant employee autonomy without necessarily fostering other entrepreneurial behaviours, or conversely, may promote innovation within structured environments.

4.3 Hypothesis Testing: Unveiling the Innovation Paradox

Table 2 presents our primary findings, comparing intrapreneurial dimensions across company types and reveals patterns that both support and challenge theoretical expectations.

Innovation-oriented companies demonstrated statistically significant innovation orientation and risk-taking advantages, aligning with H1 predictions. However, these advantages proved modest (d = 0.089 and 0.051, respectively), falling well below conventional thresholds for practical significance. More surprisingly, control companies exhibited significant advantages in proactiveness (d = −0.074), directly contradicting our hypothesis. Though not significant after correction, the trending advantage in autonomy for control companies reinforces this unexpected pattern.

I compared these effect sizes with meta-analytic benchmarks from organisational culture research to contextualise them. Hartnell et al. (2019) reported typical culture-outcome effect sizes ranging from 0.15 to 0.30. Our observed effects fall substantially below this range, suggesting that despite statistical significance, practical differences between innovation-oriented and control companies remain minimal.

Table 2 Means, standard deviations, and effect sizes for intrapreneurial dimensions by company type

Dimension	Innovation-oriented	Control	t	p	Cohen's d	95% CI
	M (SD)	M (SD)				
Innovation	0.297 (0.213)	0.278 (0.209)	4.91	<0.001***	0.089	[0.053, 0.125]
Autonomy	0.379 (0.208)	0.388 (0.203)	−2.22	0.027	−0.044	[−0.080, −0.008]
Risk-taking	0.225 (0.176)	0.216 (0.168)	2.97	0.003**	0.051	[0.015, 0.087]
Proactiveness	0.308 (0.199)	0.324 (0.204)	−3.84	<0.001***	−0.074	[−0.110, −0.038]
Empowerment	0.416 (0.229)	0.408 (0.225)	1.98	0.047	0.037	[0.001, 0.073]

Note Innovation-oriented n = 5,948; Control n = 5,843. CI = confidence interval for Cohen's d. Significance tests use a Bonferroni-adjusted α of.01 to control the family-wise error rate
*p < 0.01. **p < 0.001

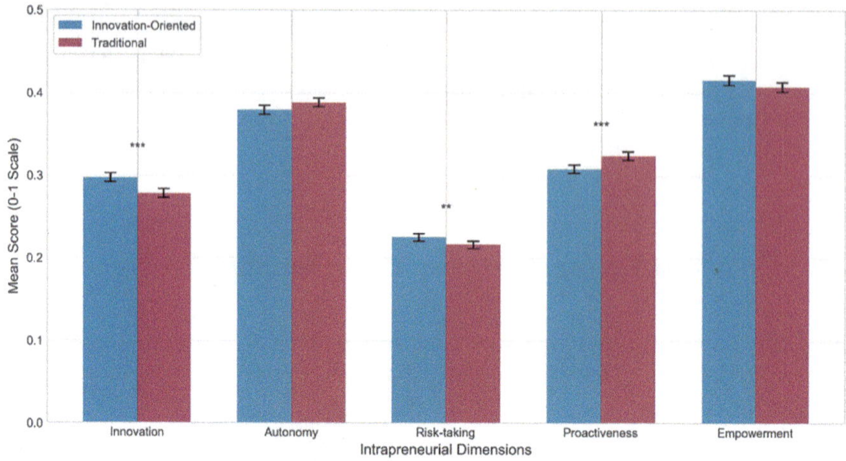

Note. Error bars represent 95% confidence intervals. ***p < .001, **p < .01, ns = not significant (α = .01)

Fig. 2 Intrapreneurial dimension scores by company type with 95% confidence intervals

Power analysis revealed important interpretive constraints. While I had excellent power for detecting innovation orientation differences (98.7%) and good power for proactiveness (84.6%), other dimensions showed concerning limitations: autonomy (26.9%), risk-taking (42.8%) and empowerment (17.0%). These power limitations suggest some null findings may represent Type II errors rather than genuine similarity. Figure 2 visualises these main results with confidence intervals.

4.4 Robustness Checks and Multilevel Considerations

Recognising the nested structure of our data, I conducted multilevel modelling to ensure robustness. Intraclass correlation coefficients ranging from 0.08 to 0.12 indicated that 8–12% of variance occurs at company level, confirming meaningful but modest clustering. Table 3 presents the results of the multilevel model.

Table 3 Multilevel model results for company type effects on intrapreneurial dimensions

Dimension	Fixed effect (Company type)	*SE*	*p*	ICC
Innovation	0.021	0.008	0.009**	0.11
Autonomy	−0.009	0.007	0.197	0.08
Risk-taking	0.011	0.006	0.067	0.09
Proactiveness	−0.018	0.007	0.010**	0.10
Empowerment	0.008	0.009	0.374	0.12

Models include the company as a random effect. Fixed effect represents the difference between innovation-oriented and control companies (innovation-oriented coded as 1). ICC = intraclass correlation coefficient, indicating the proportion of variance at company level. SE = standard error. *p < 0.01.

The multilevel models' fixed effects closely mirrored our t-test results: innovation orientation ($\beta = 0.021$, $p = 0.009$) and proactiveness ($\beta = -0.018$, $p = 0.010$). Other dimensions showed similar patterns but reduced significance, likely due to decreased power. Non-parametric Mann–Whitney U tests and Welch's t-tests for unequal variances yielded virtually identical conclusions, with all three analytical approaches converging on the same substantive findings. This convergence across methods with different assumptions strongly supports the validity of our results.

4.5 *Geographic Contingencies: Culture Trumps Strategy*

Regional analysis revealed that geographic context fundamentally moderates the relationship between company type and intrapreneurial culture, providing strong support for H2. Table 4 presents the ANOVA results for Region × Type interactions.

Significant interactions emerged for innovation ($F(2{,}14{,}003) = 12.43, p < 0.001$), autonomy ($F = 8.92$, $p < 0.001$), risk-taking ($F = 6.21$, $p = 0.002$) and proactiveness ($F = 15.87, p < 0.001$). Post-hoc analyses revealed striking regional patterns. In

Table 4 Analysis of variance for region × company type interactions on intrapreneurial dimensions

Dimension	Region × type		Post-hoc comparisons
	$F(2, 14{,}003)$	p	Significant regional differences
Innovation	12.43	<0.001***	US: Innov > Control***; EU: Control > Innov*; SA: Control > Innov*
Autonomy	8.92	<0.001***	US: ns; EU: Control > Innov**; SA: Control > Innov**
Risk-taking	6.21	0.002**	US: Innov > Control**; EU: ns; SA: ns
Proactiveness	15.87	<0.001***	US: Control > Innov*; EU: Control > Innov***; SA: Control > Innov***
Empowerment	3.76	0.023*	No significant regional differences after Bonferroni correction

Innov = Innovation-oriented companies. Post-hoc comparisons use Tukey's HSD with Bonferroni adjustment. Only significant differences are shown. Ns = not significant
$p < 0.05$. *$p < 0.01$ **$p < 0.001$

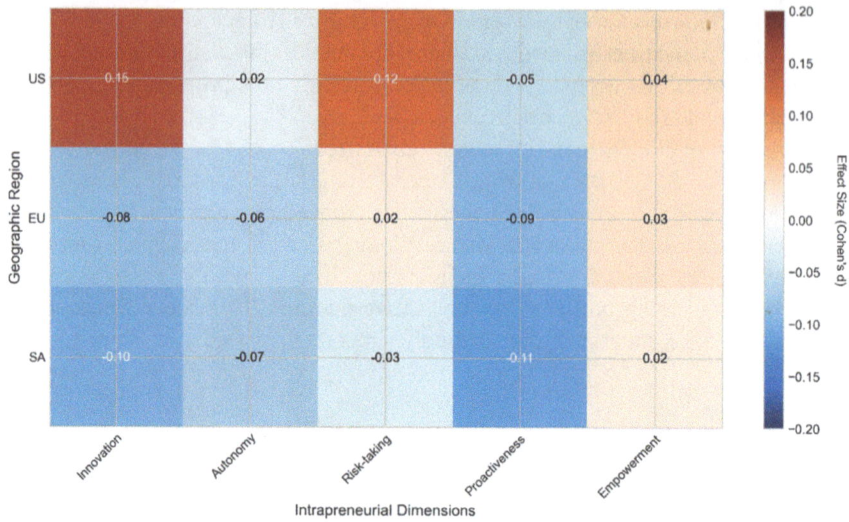

Note. Positive values indicate innovation-oriented > traditional; negative values indicate traditional > innovation-oriented.

Fig. 3 Heatmap of intrapreneurial dimensions by geographic region and company type

the United States, results largely conformed to theoretical expectations: innovation-oriented companies scored significantly higher on innovation orientation and risk-taking while showing no differences in autonomy or empowerment. This alignment suggests American conceptualisations of innovation culture, which emphasise disruption and rapid experimentation, accurately characterise US innovation-oriented firms.

However, the European and South African contexts revealed dramatically different patterns. In Europe, control companies often equalled or exceeded innovation-oriented firms on traditionally 'innovative' dimensions. South African traditional firms demonstrated advantages across multiple dimensions. Given their consistency and magnitude, these reversals cannot be dismissed as statistical artefacts, suggesting fundamental cultural differences in how intrapreneurship manifests globally. Figure 3 presents these regional variations through a heatmap visualisation.

4.6 *Qualitative Insights: The Language of Innovation*

Analysis of high-scoring reviews ($\geq$0.8) revealed distinct linguistic patterns explaining our quantitative findings. The vocabulary of innovation-oriented companies emphasised speed and disruption, with terms like 'cutting-edge', 'fast-paced', and 'disruptive' appearing more frequently. A Google employee exemplified this orientation: *The pace of innovation here is breathtaking. I am constantly experimenting with new technologies and approaches, even if they might fail.*

The language of control companies emphasised sustainability and gradual improvement, with terms such as 'continuous improvement', 'work-life balance', and 'stable environment' appearing more frequently. A Johnson & Johnson employee highlighted how stability enables autonomy: *I have complete flexibility in structuring my work. As long as I meet objectives, nobody micromanages how I achieve them. This trust is empowering because I can pursue process improvements that might take months to show results.*

The proactiveness advantage for control companies emerged through descriptions of incremental innovation. A Capgemini employee explained: *We are encouraged to identify and implement process improvements. It is not flashy innovation, but we are constantly improving things.* This suggests stability provides the temporal continuity necessary for sustained entrepreneurial initiatives rather than constraining them.

4.7 Synthesis: Reconceptualising Intrapreneurial Excellence

Our comprehensive analysis reveals that H1 received only partial support. While innovation-oriented companies show advantages in innovation orientation and risk-taking, these effects are smaller than anticipated and reversed for other dimensions. H2 received strong support, with regional variations fundamentally altering the relationship between company type and intrapreneurial culture.

The dimensional independence I observed, particularly autonomy's weak correlations with other dimensions, suggests intrapreneurship represents a multi-faceted phenomenon rather than a unitary construct. Organisations may excel in specific dimensions while lagging in others, with different configurations potentially equally effective. The distinct vocabularies between company types reveal not just different levels but fundamentally different conceptualisations of intrapreneurial behaviour, suggesting multiple legitimate pathways to fostering organisational entrepreneurship.

5 Discussion

Our investigation into intrapreneurial cultures across innovation-oriented and traditional companies has revealed a paradox that challenges fundamental assumptions in organisational theory. The finding that traditional companies demonstrate equal or superior performance on specific intrapreneurial dimensions demands theoretical explanation and practical reconsideration of how organisations cultivate entrepreneurial behaviour.

5.1 Theoretical Implications: The Stability-Innovation Paradox

Our results necessitate reconceptualising the relationship between organisational innovation orientation and intrapreneurial behaviour. Traditional frameworks rooted in contingency theory posit that companies must choose between stability and innovation, with the latter requiring cultural transformation toward risk-taking and disruption (March 1991; O'Reilly and Tushman 2013). Our findings suggest this dichotomy is false, or at least incomplete.

I propose *intrapreneurial complementarity* to explain how different organisational contexts cultivate distinct but equally valuable entrepreneurial behaviours. In innovation-oriented companies, intrapreneurship manifests as radical experimentation and technological disruption, aligning with popular conceptions of entrepreneurship. In traditional companies, it emerges through incremental innovation, process optimisation and sustainable improvements. Both forms contribute to organisational competitiveness through different mechanisms and timescales.

The stability-innovation paradox extends beyond simple complementarity. Our data suggest that organisational stability serves as an enabling condition for certain intrapreneurial behaviours, particularly those requiring sustained effort and long-term thinking. In environments characterised by constant disruption, which are common in innovation-oriented firms, employees may lack the temporal stability or resource predictability necessary for meaningful entrepreneurial initiatives. The higher proactiveness scores in traditional companies support this interpretation, suggesting that stable structures provide psychological safety and resource security, which empowers the taking of calculated risks.

Figure 4 illustrates this stability-innovation paradox, showing how organisational characteristics traditionally viewed as opposing may enable different forms of entrepreneurial behaviour.

This paradox aligns with emerging research on the 'dark side' of innovation (Anderson et al. 2014) and the benefits of organisational slack (Cyert and March 1963). While innovation-oriented companies excel at generating novel ideas, execution may suffer from initiative overload and change fatigue. Traditional companies, emphasising operational excellence, may create conditions more conducive to implementing entrepreneurial ideas, a possibility supported by their employees' emphasis on 'continuous improvement' over 'disruption'.

5.2 Methodological Contributions and Limitations

Our use of GPT-4 for extracting cultural dimensions from employee reviews represents both a methodological advancement and a limitation that requires careful consideration. Traditional organisational culture research relies heavily on surveys, which suffer from social desirability bias, low response rates and an inability to

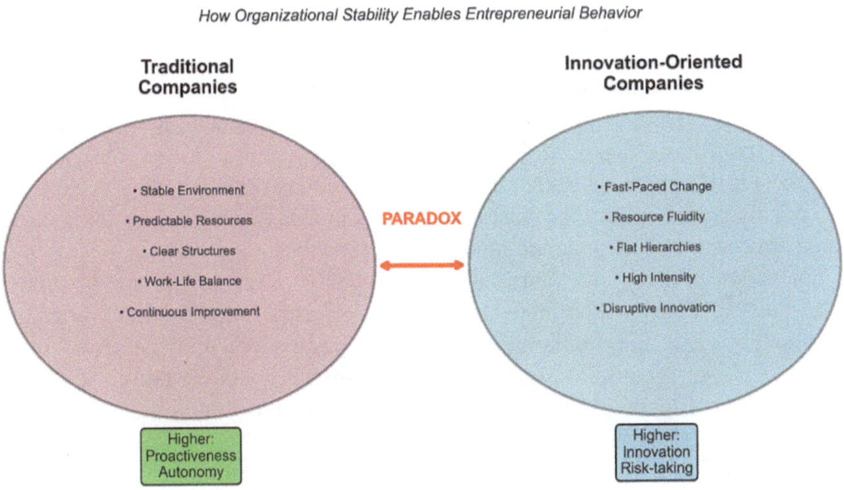

Fig. 4 The stability-innovation paradox: how organisational stability enables entrepreneurial behaviour

capture nuance (Podsakoff et al. 2003). Our approach demonstrates how large language models can analyse naturally occurring organisational data at scale while maintaining interpretive sophistication.

While our preliminary face validity assessment suggests that GPT-4 can identify complex cultural constructs, the lack of systematic inter-rater reliability testing with human coders represents a significant limitation. Future research should conduct comprehensive validation comparing GPT-4 outputs with human coding (targeting $\kappa > 0.75$) and established psychometric instruments. Future research should bridge the gap between traditional, validated measures and AI-extracted dimensions, potentially through hybrid approaches that combine survey data with computational text analysis.

Our multilevel analytical approach contributes rigour to big data organisational research. By acknowledging and modelling the nested structure of reviews within companies, we avoid the atomistic fallacy common in large-scale text analysis. The modest ICC values (0.08–0.12) reveal meaningful company-level clustering, while confirming substantial within-organisation variance, an important finding that suggests organisational culture is more heterogeneous than often assumed.

5.3 *Practical Implications: Rethinking Innovation Strategy*

The minimal practical differences between innovation-oriented and traditional companies (effect sizes < 0.1) carry profound implications for organisational leaders and consultants. Expensive cultural transformation initiatives that aim to mimic Silicon Valley archetypes may be misguided. Instead, organisations should consider how existing cultural strengths might support different but equally valuable forms of intrapreneurial behaviour.

For traditional companies, our results offer liberation from narratives demanding self-disruption. Higher proactiveness scores and trending autonomy advantages suggest that these organisations already possess cultural assets that support entrepreneurial behaviour. Rather than a wholesale transformation, they might focus on recognising and amplifying existing intrapreneurial activities that are aligned with stable structures and incremental innovation approaches.

Innovation-oriented companies may reconsider whether an extreme disruption orientation fully serves their innovation goals. Linguistic analysis reveals terms like 'burnout', 'chaos' and 'initiative overload' suggesting potential negative consequences. These organisations might benefit from introducing stability elements enabling more effective idea development and implementation without sacrificing creative energy.

The dramatic regional variations demand localised approaches to intrapreneurship cultivation. Multinational corporations cannot assume that innovation practices successful in Silicon Valley will translate globally. Our findings suggest that traditional organisational structures may provide superior platforms for intrapreneurial behaviour in collectivist cultures or those emphasising work-life balance. Organisations must develop culturally sensitive approaches aligned with local values while pursuing global innovation objectives.

5.4 *Boundary Conditions and Future Research*

Several boundary conditions constrain the generalisability of our findings. The exclusive focus on English-language reviews may mask important cultural variations in how intrapreneurship is expressed. Glassdoor's user base likely overrepresents technology-savvy, white-collar workers while underrepresenting manufacturing or service employees. The study period, encompassing the COVID-19 pandemic, introduces temporal complexity; however, our sensitivity analyses suggest stable patterns.

Most critically, small effect sizes raise questions about practical significance. While statistically significant, given our large sample, differences of $d < 0.1$ may not translate to meaningful behavioural variations. However, I argue that even minor cultural differences can compound over time to produce substantial outcomes in complex organisational systems, a proposition requiring longitudinal investigation.

Future research should pursue several directions. Validating AI-extracted dimensions against established instruments, such as ENTRESCALE, represents a pressing need. Longitudinal studies could illuminate whether different intrapreneurial forms represent evolutionary stages or stable alternatives. Ethnographic research within matched pairs could reveal mechanisms underlying the stability-innovation paradox.

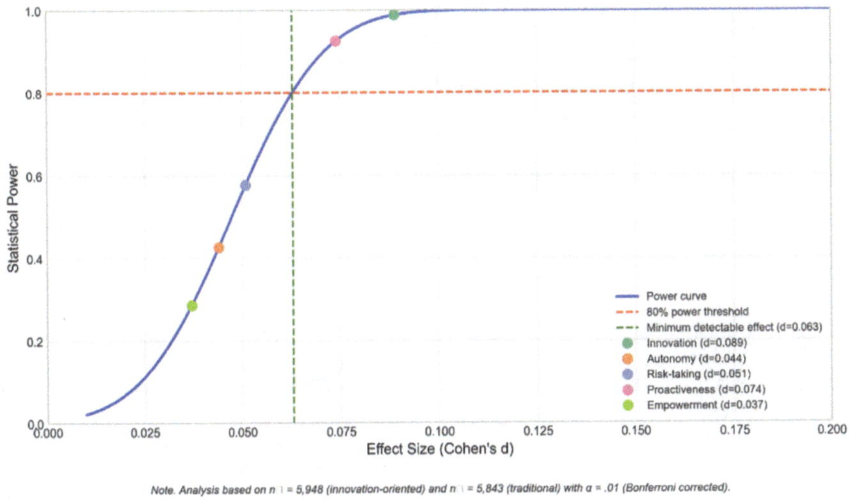

Fig. 5 Statistical power analysis for intrapreneurial dimensions

Cross-cultural research extending beyond English contexts is essential for determining whether our findings represent universal patterns or Western-specific phenomena.

The power analysis visualisation in Fig. 5 illustrates the statistical constraints on our findings. It highlights the need for even larger samples or more sensitive measures to detect subtle cultural differences.

6 Conclusion

Our discovery of the stability-innovation paradox challenges simplistic narratives about organisational intrapreneurship while opening new avenues for theory and practice. By demonstrating that traditional companies can equal or exceed innovation-oriented firms on key intrapreneurial dimensions, I advocate for a more nuanced, culturally sensitive understanding of organisational entrepreneurship. The future of innovation may lie not in universal disruption but in recognising and cultivating diverse forms of intrapreneurial excellence suited to specific organisational contexts. As organisations navigate tensions between innovation imperatives and employee well-being, our findings suggest these need not be opposing forces, as stability and innovation can paradoxically enable each other in the complex ecology of modern organisations.

References

Anderson, N., Potočnik, K., Zhou, J.: Innovation and creativity in organisations: a state-of-the-science review, prospective commentary, and guiding framework. J. Manag. **40**(5), 1297–1333 (2014)

Antoncic, B., Hisrich, R.D.: Clarifying the intrapreneurship concept. J. Small Bus. Enterp. Dev. **10**(1), 7–24 (2003)

Bosma, N., Stam, E., Wennekers, S.: Intrapreneurship: An International Study. Utrecht University School of Economics, Utrecht (2010)

Braun, V., Clarke, V.: Using thematic analysis in psychology. Qual. Res. Psychol. **3**(2), 77–101 (2006)

Briscoe, F., Gupta, A.: Social activism in and around organisations. Acad. Manag. Ann. **10**(1), 671–727 (2016)

Brown, T.B., Mann, B., Ryder, N., Subbiah, M., Kaplan, J., Dhariwal, P., et al.: Language models are few-shot learners. Adv. Neural Inf. Process. Syst. **33**, 1877–1901 (2020)

Chen, G., Ployhart, R.E., Thomas, H.C., Anderson, N., Bliese, P.D.: The power of momentum: a new model of dynamic relationships between job satisfaction change and turnover intentions. Acad. Manag. J. **54**(1), 159–181 (2010)

Covin, J.G., Miles, M.P.: Corporate entrepreneurship and the pursuit of competitive advantage. Entrep. Theory Pract. **23**(3), 47–63 (1999)

Covin, J.G., Slevin, D.P.: Strategic management of small firms in hostile and benign environments. Strateg. Manag. J. **10**(1), 75–87 (1989)

Creswell, J.W., Plano Clark, V.L.: Designing and Conducting Mixed Methods Research, 3rd edn. Sage, Thousand Oaks, CA (2017)

Cyert, R.M., March, J.G.: A Behavioral Theory of the Firm. Prentice-Hall, Englewood Cliffs, NJ (1963)

D'Angelo, S., Cavallo, A., Ghezzi, A., Di Lorenzo, F.: Understanding corporate entrepreneurship in the digital age: a review and research agenda. RMS **17**(8), 2883–2915 (2023)

Fan, Y.: Digital transformation and corporate sustainable development: the role of corporate entrepreneurship. Sustainability **16**(2), 456 (2024)

Green, T.C., Huang, R., Wen, Q., Zhou, D.: Crowdsourced employer reviews and stock returns. J. Financ. Econ. **134**(1), 236–251 (2019)

Gupta, A., Maranas, C.D., Albert, R.: Elucidation of organizational culture constituents through topic modeling of employee reviews. Hum. Resour. Manag. J. **32**(4), 987–1008 (2022)

Hartnell, C.A., Ou, A.Y., Kinicki, A.J., Choi, D., Karam, E.P.: A meta-analytic test of organizational culture's association with elements of an organization's system and its relative predictive validity on organizational outcomes. J. Appl. Psychol. **104**(6), 832–850 (2019)

Hornsby, J.S., Kuratko, D.F., Zahra, S.A.: Middle managers' perception of the internal environment for corporate entrepreneurship: assessing a measurement scale. J. Bus. Ventur. **17**(3), 253–273 (2002)

Kuratko, D.F., Fisher, G., Audretsch, D.B.: Unraveling the entrepreneurial mindset. Small Bus. Econ. **57**(4), 1681–1691 (2019)

Lajçi, R.: Intrapreneurship assessment in organizations: a scoping review of tools and frameworks. Int. J. Entrep. Behav. Res. **31**(1), 45–72 (2025)

Lee, J., Choi, Y.: Organizational culture moderates the relationship between R, D investment and firm performance. Res. Policy **53**(2), 104923 (2024)

Lee, S., Choi, H.: Cultural contexts in workplace experience expression: a cross-national analysis of employee reviews. Cross Cult. Strat. Manag. **32**(1), 78–95 (2025)

Lima de Santana, J., Pinheiro, R.: Public sector intrapreneurship: a systematic literature review. Int. Public Manag. J. **27**(3), 412–438 (2024)

Lumpkin, G.T., Dess, G.G.: Clarifying the entrepreneurial orientation construct and linking it to performance. Acad. Manag. Rev. **21**(1), 135–172 (1996)

Luo, N., Zhou, Y., Shon, J.J.: Employee satisfaction and corporate performance: mining employee reviews on glassdoor.com. Proceedings of the 37th International Conference on Information Systems, 1–16 (2016)

March, J.G.: Exploration and exploitation in organizational learning. Organ. Sci. **2**(1), 71–87 (1991)

Morgan, D.L.: Paradigms lost and pragmatism regained: methodological implications of combining qualitative and quantitative methods. J. Mixed Methods Res. **1**(1), 48–76 (2007)

Neessen, P.C., Caniëls, M.C., Vos, B., De Jong, J.P.: The intrapreneurial employee: toward an integrated model of intrapreneurship and research agenda. Int. Entrep. Manag. J. **15**(2), 545–571 (2019)

Nelson, L.K.: Computational grounded theory: a methodological framework. Sociol. Methods Res. **49**(1), 3–42 (2020)

O'Reilly, C.A., Tushman, M.L.: Organizational ambidexterity: past, present, and future. Acad. Manag. Perspect. **27**(4), 324–338 (2013)

Pinchot, G.: Intrapreneuring: Why You Don't Have to Leave the Corporation to Become an Entrepreneur. Harper, Row, New York (1985)

Podsakoff, P.M., MacKenzie, S.B., Lee, J.Y., Podsakoff, N.P.: Common method biases in behavioral research: a critical review of the literature and recommended remedies. J. Appl. Psychol. **88**(5), 879–903 (2003)

Rita, P., António, N., Afonso, A.P.: Combining sentiment analysis and topic modeling to extract actionable insights from employee feedback. Int. J. Hum. Resour. Manag. **36**(2), 234–259 (2025)

Shane, S., Venkataraman, S.: The promise of entrepreneurship as a field of research. Acad. Manag. Rev. **25**(1), 217–226 (2000)

Sharma, P., Chrisman, J.J.: Toward a reconciliation of the definitional issues in the field of corporate entrepreneurship. Entrep. Theory Pract. **23**(3), 11–28 (1999)

UNESCO: UNESCO Science Report 2021: The Race Against Time for Smarter Development. UNESCO Publishing, Paris (2021)

Urban, B., Wood, E.: The innovating firm as corporate entrepreneurship. Eur. J. Innov. Manag. **20**(4), 534–556 (2017)

Wu, J., Li, S.: Digital transformation and corporate innovation: evidence from Chinese listed companies. Technol. Soc. **76**, 102423 (2024)

Zahra, S.A.: Predictors and financial outcomes of corporate entrepreneurship: an exploratory study. J. Bus. Ventur. **6**(4), 259–285 (1991)

Zahra, S.A., Wright, M., Abdelgawad, S.G.: Contextualization and the advancement of entrepreneurship research. Int. Small Bus. J. **32**(5), 479–500 (2014)

Evaluating Women's Financial Inclusion in French and English-Speaking Countries in Sub-Saharan Africa

Bouba Ismaila and John Beneke

Abstract In the past two decades, many governments, especially in the Global South, have acknowledged the importance of financial inclusion. Arguably, the African continent has been left behind in this regard. However, with research showing that about three-quarters of the population now own phones with internet connections, which are used for e-commerce, mobile payment, and more, the continent has begun to catch up. The purpose of this paper was to determine whether financial inclusion among women differs between French and English-speaking countries in Sub-Saharan Africa. The study used the 2021 Global Financial Index data, collected by the World Bank for its report on financial inclusion worldwide. Binary logistic regression using Chi-Square tests and cross-tabulations was employed to analyse data. The results indicate that women in English-speaking countries enjoy a better overall financial inclusion rate. The study findings imply that French-speaking countries should do more to raise their population's financial inclusion by creating more favourable conditions to make this a reality for most, if not all. The originality of this study lies in its examination of the differences in women's financial inclusion based on countries' colonial heritage, an approach which has not been widely explored in African studies.

Keywords Colonial legacy · Financial inclusion · English-speaking countries · French-speaking countries · Mobile money · Sub-saharan africa · Women

---

B. Ismaila (✉) · J. Beneke
Vaal University of Technology, Vanderbijlpark, South Africa
e-mail: boubai@vut.ac.za

J. Beneke
e-mail: johnb@vut.ac.za

E. H. Redda (ed.), *Proceedings of the Social Sciences International Research Conference, 2025*, https://doi.org/10.1007/978-981-95-9489-4_17

1 Introduction

1.1 Background

Advancements in innovation and technology have hit at the heart of the African population, much like other populations worldwide, as the proportion of individuals owning internet-enabled mobile phones for purposes such as e-commerce, mobile payments, and related activities continues to grow exponentially (Ahmad et al. 2020). Over the past two decades, many governments, particularly in the Global South, have recognised the importance of financial inclusion. However, the African continent has lagged in this regard, with levels of inclusion varying significantly across regions and depending, in part, on differing colonial legacies. Several studies have probed the question of financial inclusion in Africa over the recent past years (Avom et al. 2021; Evans 2018; Ezzahid and Elouaourti 2022; Girón et al. 2021; Ifediora et al. 2022; Kouladoum et al. 2022; Magale and Yengeni 2024; Njanike and Mpofu 2024). The number of people who have an account with a credit union, microfinance institution, post office, mobile money service provider, or a bank—defined by the World Bank Global Findex 2021 as holding a regulated institution account, either individually or jointly—is the primary indicator of financial inclusion (World Bank 2023) used in this study.

These studies examined several key factors that influence Africans' financial inclusion and their contribution to economic progress, including microeconomic issues, the role of the state, and the impact of technology. For example, research by Tinta et al. (2022) shows that an individual's personality, financial literacy, formal accounting restrictions, and inventiveness all affect their selection of traditional versus mobile accounts. Similarly, financial services can be used to control risk, lessen the effects of financial shocks, encourage the establishment of new companies, enable individuals to make investments in their education and health, and ultimately boost human development (Matekenya et al. 2021). Most of these writers have praised the startling increase in the use of digital and mobile technology in Africa for helping to increase financial inclusion in various regions of the continent. Furthermore, based on ICT indicators such as the subscription rates for fixed and mobile phone users, fixed broadband, internet users, and a composite indicator of digitalisation, Kouladoum et al. (2022) found that digital technology has a positive and significant impact on financial inclusion at all levels.

Looking more specifically at women's financial inclusion, Ahamadou and Agada, (2023) found several major obstacles to financial inclusion in Sub-Saharan Africa, with the most financially excluded being women, young people under 25, the poor, and those with low levels of education. Regardless, financial products and services enable women to build enterprises, survive financial shocks, amass wealth, and exercise greater financial autonomy, just like they do for men (Tiwari et al. 2019). Yet, women often encounter major obstacles when trying to acquire credit, own property, or participate in official financial systems, contributing to Africa's long-standing gender inequities in access to financial services (Ojo 2024). While evidence shows that women in Sub-Saharan Africa are less financially included, the benefits of their inclusion are noteworthy. For instance, a study by Ajide et al. (2024) concludes that financial inclusion strengthens political empowerment of women by granting financial accessibility for civil liberties, civil society participation, and participation in politics.

A study on finance for sustainable development in Africa by Zawaira et al. (2023) demonstrates that women are more likely to turn to the informal sources compared to men, a concern that needs to be resolved at the policy level. They add that improving access to finance is at the centre of improving gender equality. Increasing the economic freedoms and opportunities for women, will enable them to contribute to their families and societies. This may be validated in the face of a finding such as 'being male and/or married are positive determinants of financial inclusion for Central Africa and Africa' (Soumaré et al. 2016).

Likewise, results of a study by Ojo (2023) showed that tackling socio-spatial issues, including gendered social and physical mobility, digital and financial access, improves women's involvement in entrepreneurship. Sadly, despite the rapid growth of digital finance, many African women—particularly those living in underprivileged and marginalised communities—lack access to digital financial services, as well as the digital assets, such as smartphones and internet access, and the digital literacy necessary to conduct online transactions (Nghargbu and Jumare 2024)

1.2 Problem Statement

Over the past 20 years, many nations, especially those in the Global South, have realised how important financial inclusion is. Despite notable advancements over the past 10 years, the African continent still appears to be lagging, with women disproportionately affected. Sub-Saharan Africa's overall financial inclusion rate was slightly over 23% in 2011, though it increased to over 43% in 2017 and beyond, with digital financial services accounting for a significant portion of this growth (Ajide 2020). The level of financial inclusion, however, seems to differ from region to region and even from one colonial legacy to the next. It is also reasonable to assume that account holding (financial or mobile money) is influenced by income level and employment status. Thus, this study aims to examine the role of colonial heritage in shaping the financial inclusion of women in French-speaking and English-speaking Sub-Saharan African countries, and to further explore the potential impact of income and employment status, if any.

1.3 Research Aim

The aim of this paper is to establish whether there is any difference in women's financial inclusion between French and English-speaking countries in Sub-Saharan Africa, and whether income level and employment status have any influence.

2 Literature Review

Over the past 20 years, financial inclusion—particularly in emerging economies—has emerged as a key area of research. Since 2011, the World Bank has conducted extensive research and published studies on this specific issue every four years, encompassing more than 70% of all nations worldwide (World Bank 2023). Numerous writers have reported on various facets of financial inclusion in various parts of the world using their database (Girón et al. 2021; Ifediora et al. 2022; Matsebula and Yu 2020; Nsiah and Tweneboah 2023; Osabutey and Jackson 2024; Tinta et al. 2022). As financial inclusion has moved up the global reform agenda, an increasing number of policymakers, regulators, researchers, market practitioners, and other stakeholders have become interested in it, according to the World Bank (2014). This suggests that promoting financial inclusion may provide an opportunity to reduce poverty and foster shared financial growth (Mushonga et al. 2018).

Although some African countries advanced in financial inclusion earlier, overall, Africa has fallen behind. One notable example is South Africa, where the inclusion rate rose from 63% in 2005 to 81% in 2012, driven largely by increased use of banking services, with women benefiting particularly from the new award system introduced by the South African Social Security Agency (SASSA). Both macro and micro factors can influence financial inclusion. For instance, Tinta et al. (2022) found that in Africa, the likelihood of holding a traditional account increases with age, income, education, urban residence, marital status, and employment. This argument is corroborated by Girón et al. (2021), whose findings reveal that many factors are important for financial inclusion in Asia and Africa, including levels of education, income, trust, accessibility to financial institutions, lack of documentation, and gender.

At macro level, different institutions—private and public—play an important and maybe bigger role in influencing financial inclusion. As evidence of this, a study by Nsiah and Tweneboah (2023a) shows that the money supply, institutional quality, and GNI per capita all have a major impact on financial inclusion in Africa. It also suggests that these characteristics may help lower barriers to financial inclusion on the continent. Similarly, Kwenda and Chinoda (2019) found that financial inclusion throughout the continent was positively impacted by institutional quality and governance.

When it comes to gender balance, it may be contended that economically empowering women could not only increase their personal agency but also contribute substantially to the growth of society, promoting economic resilience and lowering poverty. Technology has unquestionably played a pivotal role in financial inclusion for millions of Africans, particularly with the arrival of mobile money in much of the continent in the past decade. It may also help address gender inequality, at least in part.

Given the difficulty of achieving financial inclusion in Sub-Saharan Africa and commercial banks' profit-driven limitations (Philip 2003), mobile money has proven to be an effective, often superior alternative due to its affordability and ease of use. Improving the selection, worth, and accessibility of financial services and products for underserved, unserved, and financially ineligible populations should be the goal here.

One would think that the increasing number of smartphone users and internet users on the continent, along with additional factors such as stringent bank account opening criteria or the younger generation's acceptance of technology, would encourage mobile money accounts' opining. According to one study, technologically savvy generations, and easily accessible services like phone credit recharge, income, and education are some of the primary variables driving Ghana's adoption of mobile money (Amoah et al. 2020). Reducing both income inequality and enhancing mobile money innovations simultaneously could lead to more inclusive outcomes in terms of gender equality in financial inclusion (Asongu et al. 2024). Furthermore, Asongu and Le Roux (2023) revealed that mobile money innovation has an overall positive effect in transforming unemployed women into self-employed women. Likewise, enhancing financial accessibility is essential to advancing gender equality and expanding women's economic liberties and opportunities to support their families and communities (Zawaira et al. 2023).

Findings by Yeyouomo et al. (2023) demonstrate that by addressing the gender disparity in financial service access and use, fintechs help close the gender gap in financial inclusion. However, these authors note that their findings raise concerns about the potential of fintech developments alone to close this gap and suggest the necessity of specific legislative measures to accomplish that. On the other hand, although necessary, the recent tightening of mobile phone regulations in the Global South has upset the multilevel and formal/informal-hybrid infrastructures of money movement in these communities. Tougher regulations may create a new digital divide that makes it more difficult for the poor to access financial services (Stark 2021).

Although the authors found no prior research explicitly comparing financial inclusion—especially among women—across nations with different colonial backgrounds, several studies' findings may offer some insight. For example, at a regional level, data from the Central Bank of West African States (2020), Bank of Central African States (2020) and African Development Bank (2020) show that the levels of banking penetration are, respectively, 19% in West Africa, 15% in Central Africa, 35% in East Africa, 70% in Southern Africa, and 40% in North Africa (Tinta et al. 2022). At national level, according to these authors, and the Global Financial Index (The World Bank 2023), the financial inclusion rates for a few of these nations are as follows: South Africa (90%) and Nigeria (64%), Uganda (66%), Benin (49%), Côte d'Ivoire (51%), and Cameroon (47%). These latter figures indicate a notable improvement within just a couple of years.

3 Research Method

The results of this study were derived from secondary data and used a quantitative design. Market research, economics, and the social sciences are just a few of the fields that apply quantitative design. According to Watson (2015), the goal of quantitative research is to measure data and validate the measures used, analyse it for trends and connections, as in the case of this study. The World Bank's 2021 Global Financial Index data, collected for its global report on financial inclusion, was used in the study. The sample was made up of the populations of 33 Sub-Saharan nations with data available, divided into those that spoke English (17) and French (16). The World Bank provided the sample and data gathering process, which is briefly covered in the following two subsections.

3.1 Sampling Procedure

In economies where in-person surveys were used, selecting primary sampling units was the first step in the sampling process. One or more sampling stages were used to produce clustering, and these units were stratified by population size, geography, or both. Simple random sampling was employed in other cases, and sample selection was based on probabilities proportional to population size where population information was available. Additional information can be found here:https://microdata.worldbank.org/index.php/catalog/4607#study_desc1684332928200

3.2 Data Collection Method

In economies where fewer than 80% of the population had telephone access or where in-person surveying was the standard approach, surveys were carried out in person. However, in several of these economies, in-person interviews were not feasible in 2021 due to persistent mobility limitations caused by COVID-19. Therefore, 67 economies that had been polled in person in 2017 were subjected to phone-based questionnaires. These 67 economies were chosen for inclusion based on factors such as population size, COVID-19 infection rates, phone penetration rates, and the viability of using phone-based techniques in place of in-person data collection, while adhering to all government-issued guidelines during the interviewing process (World Bank 2023).

3.3 Data Analysis

Data were analysed using SPSS's binary logistic regression with cross-tabulations and Chi-Square tests. Because there are two options for financial inclusion—yes (if respondents have one or both account types) or no (if they have neither account type)—binary logistic regression was judged adequate. The association between respondents' language and various accounts was examined using cross-tabulation, and the significance (Sig.) of the discrepancies was assessed using the Chi-Square test. Both the adjusted and unadjusted odds ratios (OR) for income and employment status were included in the analysis.

The binary logistic regression equation is as follows.

$$\text{Logit(P)} = \beta 0 + \beta 1 * \text{Language} + \beta 2 * \text{Income} + \beta 3 * \text{Employment}$$

where:

P is the probability of having an account (having an account =1, not having = 0), and β0, β1, β2, and β3 are coefficients to be determined.

3.4 Hypotheses

Independent variables: language (main variable); employment status and income level (additional variables).

Dependent variables: having neither account, having a bank account, having a mobile money account.

The following hypotheses were formulated for the study:

Ho (Null Hypothesis): There is no significant difference in the likelihood of financial inclusion (having a bank or mobile money account, or not) between women in French-speaking and English-speaking countries in Sub-Saharan Africa.

H1 (Alternative Hypothesis 1): Women in French-speaking countries in Sub-Saharan Africa are less likely to be financially included compared to those in English-speaking countries.

H2 (Alternative Hypothesis 2): Women in English-speaking countries in Sub-Saharan Africa are less likely to be financially included compared to those in French-speaking countries.

4 Results and Discussions

Descriptive statistics analysis of the variables is presented below, and the results are discussed. First, the different accounts' ownership per group (language) of countries is examined.

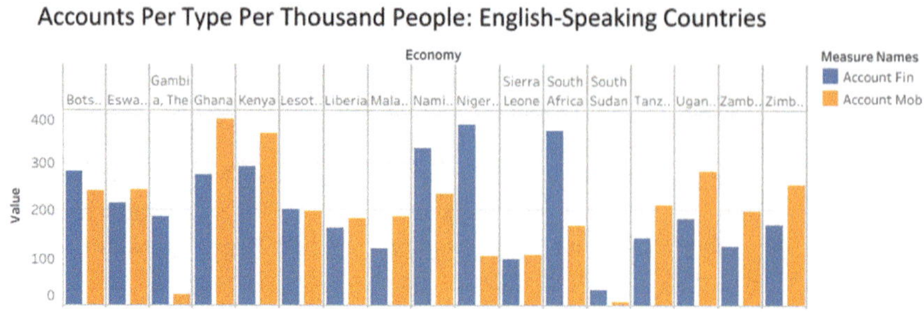

Fig. 1 Account ownership for English-speaking countries

4.1 Account Ownership in English-Speaking Countries

Figure 1 depicts the results for women in English-speaking countries.

This result demonstrates that a slim majority of countries in this group have a higher number of mobile money accounts than financial institution accounts, except for South Africa, Botswana, Nigeria, The Gambia, South Sudan, and Namibia, showing the opposite, with Lesotho having more or less the same numbers.

4.2 Account Ownership in French-Speaking Countries

The different accounts' ownership for French-speaking countries is illustrated in Figure 2.

The figure reveals that, except for Mauritius, Niger, Comoros, and Chad, the results for this group of nations show that the vast majority of them have a significantly higher number of mobile money accounts than banking institution accounts. Given that mobile phones impact the lives of billions of people worldwide, especially in Sub-Saharan Africa, the results in Figures 1 and 2 support the findings of Kouladoum et al. (2022), which found that the rate of financial inclusion in Sub-Saharan Africa

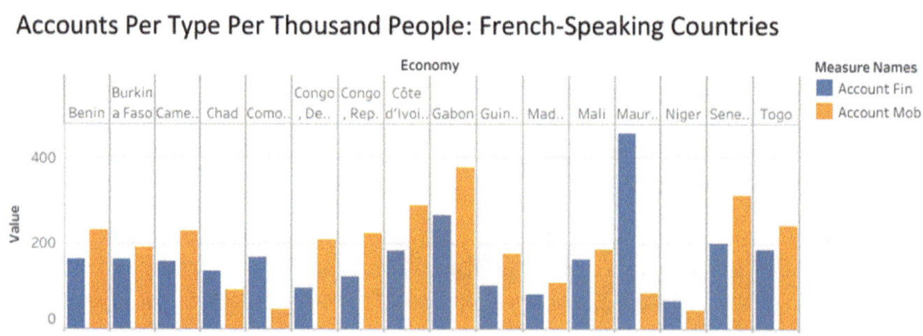

Fig. 2 Account ownership for French-speaking countries

rises as digital technologies develop. Regarding the effects of mobile money, Ahmad et al. (2020) report that mobile penetration increased more than five times outside of Sub-Saharan Africa, however, the 'miracle' occurred there, where it rose by 3200% from a baseline that was clearly fairly low. Corroborating this, Ezzahid and Elouaourti (2022) posit that mobile money accounts have proliferated across Sub-Saharan Africa since 2014, which is one reason why financial inclusion is increasing in several countries in the region.

4.3 Language Spoken and the Different Account Ownership Relationships

The relationship between the language spoken and the different accounts that respondents had is illustrated in Figure 3.

English-speaking nations have a slight edge in total account ownership, according to the data in Figure 3. The ownership of financial institution accounts is 9% greater in English-speaking nations, whereas the ownership of mobile accounts is 4% higher. Nevertheless, these account ownership rates remain lower than those in other emerging regions, regardless of the group (Kwenda and Chinoda 2019; Louis Ntu et al. 2024; Warsame and Abdalla 2024). The explanation for this might be that branch penetration in Africa is still low outside of major cities, especially in low-income or sparsely inhabited areas. This is especially true for financial institution accounts (Beck and Cull 2013). However, the incredible proliferation of mobile money accounts throughout the continent between 2011 and 2021 has resulted in significant advancements (Ahmad et al. 2020; Evans 2018; Magale and Chiwenga 2024; Shipalana 2019). Consequently, Donou-Adonsou and Sylwester (2016) found

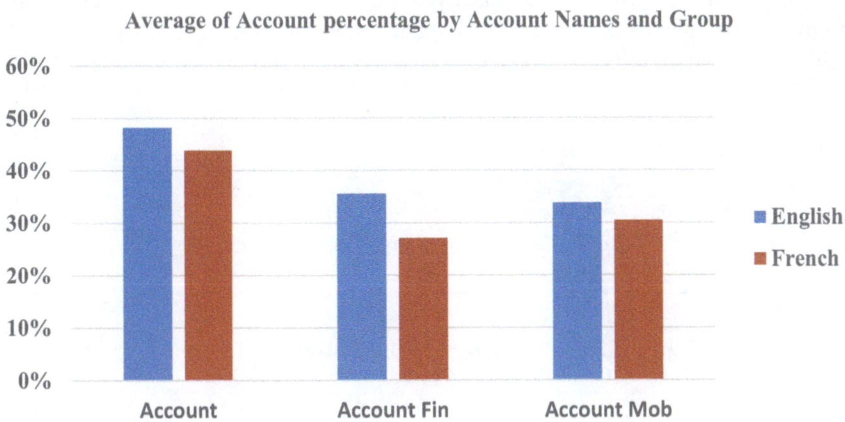

Fig. 3 Financial inclusion comparison

that wider access to bank/mobile accounts often reduces poverty, making this an advantage for the continent.

4.4 Logistic Regression Results

Further statistical analysis of the results with logistic regression is shown in summary in Table 1.

Table 1 shows that a sizeable proportion of women in English-speaking nations are financially included, while many women in French-speaking nations are not. Additionally, when language alone is considered, women in English-speaking nations are 1.66 times more likely to experience financial inclusion than those in French-speaking nations (p <.005). After adjusting the odds ratio for income and employment, the likelihood increases slightly, with women in English-speaking countries being 1.74 times more likely to be financially included (p <.005). Therefore, the null hypothesis is rejected in favour of the alternative hypothesis 1 (H_1), which states that women in French-speaking countries are less likely to be financially included than those in English-speaking countries.

Given these findings, it is important to note that several research studies on financial inclusion (Mhlanga 2021; Njanike and Mpofu 2024; Nsiah and Tweneboah 2023; Philip 2003) have primarily examined the concept in terms of having a bank account or mobile money account, or more broadly, having a financial account as defined earlier in this paper. In South Africa, for instance, the introduction of the Mzansi account in October 2004 to provide financial services to small wage workers and others living on financial earnings resulted in the creation of 1.5 million new bank accounts by May 2006 (Kirsten 2006). This raised the country's financial inclusion rate—as per the definition of financial inclusion by account ownership. Yet, it seems that many of these accounts were inactive, with little to no transactivity. Of the six million Mzansi accounts that had been opened, only an estimated 3.5 million were active by the end of 2008; the remaining accounts were inactive (World Bank 2014).

Table 1 Summary of logistic regression test analysis

Variable	Financial exclusion	Financial inclusion	Unadjusted		Adjusted*	
	(*N* = 13764)	(*N* = 17297)	OR (95% CI)	*p* -value	OR (95% CI)	*p* -value
Language						
English	5705 (37.9%)	9336 (62.1%)	1.66 (1.583-1.733)	<0.001	1.74 (1.664-1.828)	<0.001
French	8059 (50.3%)	7961 (49.7%)	ref		ref	

* Adjusted for income and employment status.

Therefore, the question is: should financial inclusion be limited to owning a bank account or another type of financial account?

4.5 Implications

The study's main implication is that French-speaking nations in Sub-Saharan Africa need to recognise that they are falling behind in women's financial inclusion in a region already lagging behind others and should begin taking more prompt and decisive action. They should begin by learning from the superior performance of their English-speaking neighbours first, and then look to other developing nations that are performing better. For people to properly integrate into the nation's financial system, focus should be placed on ways to increase women's participation in both financial institutions and mobile money accounts.

4.6 Conclusions, Limitations, and Future Research

The purpose of this study was to determine whether the level of financial inclusion among women in French- and English-speaking nations in Sub-Saharan Africa differs, and to assess the likelihood of each group being more or less financially included. The World Bank team's secondary data were analysed using binary logistic regression in SPSS. Although the region still lags behind other growing regions of the world, the results reveal that both categories have made impressive improvements in terms of account ownership, particularly in mobile money accounts. More significantly, it has been shown that English countries had a 1.66-fold higher likelihood of financial inclusion than French countries, and a 1.74-fold higher likelihood when income and employment status adjustments (OR) were made. However, the discussion of these nations' financial inclusion should go beyond simply having bank or mobile money accounts.

Although opening a bank account is a crucial initial step, it should only be seen as the beginning of true financial inclusion. Consideration should also be given to how these accounts are used and how they could be better used to achieve successful financial inclusion. It became evident that financial inclusion alone would not eradicate poverty; other developmental initiatives, such as education and the establishment of adequate physical and digital infrastructure, are equally, if not more important (Shipalana 2019). Kouladoum et al. (2022) add that more money should be invested in the human capital sector, as well as in the financial and technological infrastructures, since financial literacy may be crucial in promoting financial stability and financial inclusivity in Africa.

Furthermore, although this was not considered in this specific study, gender parity and youth are additional factors that need to be addressed for effective financial inclusion in Sub-Saharan Africa. The topic of gender balance and youth integration in financial systems has been explored in numerous studies, including those mentioned above, to achieve meaningful financial inclusion in the region, as well as in other emerging economies. For example, the findings by Girón et al. (2021) indicate that two demographic groups are excluded from the financial system: women and young people. Considering this, Ncube (2023) recommends expanding women's access to formal financial institutions, promoting their financial literacy, and enacting legislation that specifically addresses the challenges faced by female entrepreneurs.

Lastly, it has been demonstrated that within the majority of the two groups under study, the number of people who own mobile money accounts is higher than that of financial accounts, especially in French-speaking nations (Tinta et al. 2022). However, Ifediora et al. (2022) exposed that mobile money accounts and transactions do not significantly foster economic development. This suggests that financial education policies that can help Africans better understand the potential benefits of using traditional banking services should be implemented.

This study's accomplishments include a focus on the distinctions among nations based on their colonial past, an area that is rarely examined by researchers studying financial inclusion in Africa. Practically, more needs to be done by French-speaking nations to increase the number of women who have access to financial institutions and mobile money accounts to ensure that their populations are properly integrated into the financial system.

This study's limitations include its focus solely on financial inclusion defined by account ownership, while ignoring other factors that can influence both account ownership and effective financial inclusion, such as infrastructure, gender, location, and education. To gain a more comprehensive understanding of the region, future research should consider these factors, either individually or in combination.

References

Ahamadou, M., Agada, D.B.: Adopting FinTech to promote financial inclusion: evidence from West African Economic and Monetary Union. Int. J. Appl. Econ. Finance Account. **17**(1), 135–145 (2023). https://doi.org/10.33094/ijaefa.v17i1.1090

Ahmad, A.H., Green, C., Jiang, F.: Mobile money, financial inclusion and development: a review with reference to African experience. J. Econ. Surv. **34**(4), 753–792 (2020)

Ajide, F., Osinubi, T., Ojeyinka, T., Kudaisi, B.V.: The interactive effects of financial inclusion and women political empowerment on health outcomes in Africa. In: In: The Role of Female Leaders in Achieving the Sustainable Development Goals, pp. 136–159 (2024). https://doi.org/10.4018/9798369318348.ch008

Ajide, F.M.: Financial inclusion in Africa: does it promote entrepreneurship? J. Financ. Econ. Policy. **12**(4), 687–706 (2020)

Amoah, A., Korle, K., Asiama, R.: Mobile money as a financial inclusion instrument: what are the determinants? Int. J. Soc. Econ. (2020). https://doi.org/10.1108/IJSE-05-2020-0271

Asongu, S.A., Agyemang-Mintah, P., Nnanna, J., Ngoungou, Y.E.: Mobile money innovations, income inequality and gender inclusion in sub-Saharan Africa. Financial Innovation. **10**, 11 (2024). https://doi.org/10.1186/s40854-023-00553-8

Asongu, S.A., le Roux, S.: The role of mobile money innovations in transforming unemployed women to self-employed women in sub-Saharan Africa. Technol. Forecast. Soc. Change. **191**, 122548 (2023). https://doi.org/10.1016/j.techfore.2023.122548

Avom, D., Bangaké, C., Ndoya, H.: Measuring financial inclusion in African countries. Econ. Bull. **41**(3), 867–881 (2021)

Donou-Adonsou, F., Sylwester, K.: Financial development and poverty reduction in developing countries: new evidence from banks and microfinance institutions. Rev. Dev. Finance. **6**(1), 82–90 (2016)

Evans, O.: Connecting the poor: the internet, mobile phones and financial inclusion in Africa. Digit. Policy Regul. Gov. **20**(6), 568–581 (2018)

Ezzahid, E., Elouaourti, Z.: Financial inclusion, financial frictions, and economic growth: evidence from Africa. J. Afr. Bus. **23**(3), 731–756 (2022)

Girón, A., Kazemikhasragh, A., Cicchiello, A.F., Panetti, E.: Financial inclusion measurement in the least developed countries in Asia and Africa. J. Knowl. Econ. **13**(2), 1198–1211 (2021). https://doi.org/10.1007/s13132-021-00773-2

Ifediora, C., Offor, K.O., Eze, E.F., Takon, S.M., Ageme, A.E., Ibe, G.I., Onwumere, J.U.: Financial inclusion and its impact on economic growth: empirical evidence from sub-Saharan Africa. Cogent Econ. Finance. **10**, 2060551 (2022)

Kirsten, M.: Policy initiatives to expand financial outreach in South Africa (2006)

Kouladoum, J.-C., Wirajing, M.A.K., Nchofoung, T.N.: Digital technologies and financial inclusion in Sub-Saharan Africa. Telecommun. Policy. **46**(9), 102387 (2022)

Kwenda, F., Chinoda, T.: The impact of institutional quality and governance on financial inclusion in Africa. J. Econ. Financ. Sci. **12**(1), 1–9 (2019)

Louis Ntu, S., Egwu, B., Tambi, M.: The effect of financial services on youth involvement in agribusiness entrepreneurship in Fako Division. Cameroon. Asian J. Agric. Food Sci. **11**, 38–47 (2024). https://doi.org/10.24203/ajafs.v11i2.7183

Magale, E.G., Chiwenga, E.: Informal financial services in the financial inclusion matrix: gendered perspectives from Kenya and Zimbabwe. In: In: Sustainable Development Goals Series, pp. 11–19 (2024). https://doi.org/10.1007/978-3-031-53337-2_2

Magale, E.G., Yengeni, S.E.: Leaving no woman behind? Interrogating financial inclusion strategies for women in South Africa and Kenya. In: Sustainable Development Goals Series, pp. 109–118 (2024). doi:https://doi.org/10.1007/978-3-031-53333-4_11

Matekenya, W., Moyo, C., Jeke, L.: Financial inclusion and human development: evidence from sub-Saharan Africa. Dev. South. Afr. **38**(5), 683–700 (2021)

Matsebula, V., Yu, D.: An analysis of financial inclusion in South Africa. Afr. Rev. Econ. Finance. **12**(1), 171–202 (2020)

Mhlanga, D.: Factors that matter for financial inclusion: evidence from sub-Saharan Africa, the Zimbabwe case. Acad. J. Interdiscip. Stud. **10**(6), 48–59 (2021). https://doi.org/10.36941/ajis-2021-0152

Mushonga, M., Arun, T.G., Marwa, N.W.: Drivers, inhibitors and the future of co-operative financial institutions: a Delphi study on a South African perspective. Technol. Forecast. Soc. Change. **133**, 254–268 (2018)

Ncube, P.: Gender inequality and financial inclusion in South Africa: lessons from India. J. Law Soc. Dev. **10**, 1–20 (2023)

Nghargbu, R., Jumare, F.: Digital financial inclusion for women in Africa: prospects and challenges. In: Sustainable Development Goals Series, 115–122 (2024). https://doi.org/10.1007/978-3-031-53337-2_10

Njanike, K., Mpofu, R.T.: Factors influencing financial inclusion for social inclusion in selected African countries. Insight Africa. **16**(1), 93–112 (2024). https://journals.sagepub.com/doi/10.1177/09750878231194558

Nsiah, A.Y., Tweneboah, G.: Determinants of financial inclusion in Africa: is institutional quality relevant? Cogent Soc. Sci. **9**, 2184305 (2023)

Ojo, T.A.: Exploring digital and financial divide and its effects on South African women's engagement in entrepreneurship. Adv. Sci. Technol, Innov (2023)

Ojo, T.A.: Gendered finance: inclusion and transformation. In: Sustainable Development Goals Series, pp. 1–10 (2024). doi:https://doi.org/10.1007/978-3-031-53337-2_1

Osabutey, E.L., Jackson, T.: Mobile money and financial inclusion in Africa: emerging themes, challenges and policy implications. Technol. Forecast. Soc. Change. **202**, 123339 (2024)

Philip, K.: Co-operatives in South Africa: Their Role in Job Creation and Poverty Reduction. South African Foundation, Johannesburg (2003)

Shipalana, P.: Digitising Financial Services: A Tool for Financial Inclusion in South Africa. South African Institute of International Affairs (2019)

Soumaré, I., Tchana Tchana, F., Kengne, T.M.: Analysis of the determinants of financial inclusion in Central and West Africa. Transnatl. Corp. Rev. **8**(4), 231–249 (2016). https://doi.org/10.1080/19186444.2016.1265763

Stark, L.: Mobile money and the impact of mobile phone regulatory enforcement among the urban poor in Tanzania. Hum. Technol. **17**(1), 22–44 (2021). https://doi.org/10.17011/ht/urn.202106223977

Tinta, A.A., Ouédraogo, I.M., Al-Hassan, R.M.: The micro determinants of financial inclusion and financial resilience in Africa. Afr. Dev. Rev. **34**(2), 293–306 (2022)

Tiwari, J., Schaub, E., Sultana, N.: Barriers to 'last mile' financial inclusion: cases from northern Kenya. Dev. Pract. **29**(8), 988–1000 (2019). https://www.tandfonline.com/doi/full/10.1080/09614524.2019.1654432

Warsame, M., Abdalla, Y.: Impact of mobile financial services on financial inclusion: empirical insights from Kenya. Int. Rev. Econ. (2024). https://link.springer.com/article/10.1007/s12232-024-00456-9

Watson, R.: Quantitative research. Nurs. Stand. 29 (31, 2015).

World Bank: Financial Inclusion. International Bank for Reconstruction and Development/World Bank (2014).

World Bank: Global Financial Inclusion (Global Findex) Database (2023).

Yeyouomo, A.K., Asongu, S.A., Agyemang-Mintah, P.: Fintechs and the financial inclusion gender gap in sub-Saharan African countries. Women's Stud. Int. Forum. **97**, 102695 (2023). https://doi.org/10.1016/j.wsif.2023.102695

Zawaira, T., Chisadza, C., Clance, M., Gupta, R.: Financial inclusion and gender inequality in African countries. In: Finance for Sustainable Development in Africa: Evolution, Impact and Policy Implications, pp. 233–251 (2023). https://doi.org/10.4324/9781003215042-17

Unlocking Electronic Customer Experience (e-CX) Excellence: The Impact of Electronic Service Quality (E-S-QUAL) on Post-purchase Satisfaction in South African E-Commerce Retail Organisations

Alyssa Shawntay Williams

Abstract E-commerce has become an integral part of modern retail, with South African customers increasingly relying on online platforms for their shopping needs. The growth of the e-commerce industry depends not only on the variety of products offered but also on the quality of the online experience, which plays a pivotal role in shaping customer satisfaction. This paper explores the influence of electronic service quality (E-S-Qual) on post-purchase satisfaction among South African e-commerce customers, aiming to provide valuable insights for e-commerce retailers in the region. A quantitative research approach was used, with data collected from 405 respondents via an online survey. Confirmatory Factor Analysis (CFA) and Structural Equation Modelling (SEM) were applied to assess the relationships between E-S-Qual dimensions (efficiency, system availability, fulfilment, and privacy) and customer satisfaction. The findings confirm that the model demonstrated high reliability, and the E-S-Qual dimensions impacted post-purchase satisfaction. However, the results revealed that the efficiency dimension of E-S-Qual was a significant driver of customer satisfaction. Based on the findings, it is recommended that e-commerce retail organisations focus on improving the efficiency of their online platforms by implementing user-friendly interfaces, fast-loading pages, and streamlined transaction processes. While shopping online, these enhancements will ensure a positive electronic customer experience (e-CX), directly contributing to customer satisfaction and fostering long-term customer retention.

Keywords Post-purchase outcomes · Consumer satisfaction · Digital commerce · Customer Relationship Marketing (CRM) · Online shopping · Consumer behaviour

---

A. S. Williams (✉)
Rhodes University, Makhanda (Grahamstown), South Africa
e-mail: a.williams@ru.ac.za

E. H. Redda (ed.), *Proceedings of the Social Sciences International Research Conference, 2025*, https://doi.org/10.1007/978-981-95-9489-4_18

1 Introduction

The rapid growth of e-commerce has revolutionised how customers engage with retail organisations (Haleem 2026; Sagar 2024). For an e-commerce business to succeed, ensuring a seamless electronic customer experience (e-CX) through an organisation's website is essential. This experience plays a critical role in determining customer satisfaction with both the organisation and the products or services it offers (Vakulenko et al. 2019; Pappas et al. 2014). Despite the potential for an enriching e-CX, website usability and service delivery challenges often arise, which may detract from a customer's satisfaction (Mamakou et al. 2024; Williams and Oosthuizen 2021). The concept of electronic service quality (E-S-Qual) has emerged as a comprehensive framework for understanding and measuring the quality of digital service delivery in e-commerce environments (Williams and Oosthuizen 2021). E-S-Qual encompasses key dimensions such as efficiency, system availability, fulfilment, and privacy. These elements serve as a guiding compass throughout the customer's journey, ensuring that their experience is not only visually appealing but also streamlined and effective (Mamakou et al. 2024; Çelik 2021; Parasuraman et al. 2005). For South African e-commerce retail organisations, optimising E-S-Qual can be crucial in enhancing post-purchase customer satisfaction, ultimately influencing loyalty and repeat business.

The significance of this research extends beyond academic inquiry to address pressing practical challenges faced by South African e-commerce retailers. As competition intensifies and customer expectations continue to evolve, organisations must strategically allocate their limited resources to service quality improvements that yield the most significant impact on customer satisfaction (Vakulenko et al. 2019). Without empirical evidence of which E-S-Qual dimensions most significantly influence post-purchase satisfaction in South Africa, retailers risk misdirecting their investments and failing to achieve desired customer outcomes.

2 Literature Review

The literature review explores the critical factors that shape E-S-Qual and its direct impact on customer satisfaction. E-S-Qual is a framework used to assess how users perceive and interact with online services, enabling organisations to meet customer expectations better and improve satisfaction (Rita et al. 2019; Parasuraman et al. 2005). It measures digital services' quality based on efficiency, reliability, responsiveness, security, ease of use, information accuracy, and customer support. E-S-Qual emerged in response to the rapid evolution of the e-commerce sector, requiring an adaptive research approach to meet customer expectations in a digital environment (Zehir and Narcıkara 2016; Parasuraman et al. 2005; Santos 2003). Service quality is inherently perceptual, based on comparing customers' expectations and their actual service experiences (Blythe and Martin 2019).

Enhancing service quality is essential for profitability and long-term success (Jameel et al. 2023; Jinliang et al. 2023). Failures to meet expectations, or 'service gaps', lead to dissatisfaction (Lewis and Mitchell 1990; Lamb et al. 2019), while aligning service with expectations improves satisfaction and competitive advantage (Paul et al. 2016). Service quality influences competition, purchase decisions, customer retention, and organisational profitability (Biscaia et al. 2023; Lin et al. 2023; Kaya et al. 2019).

In e-commerce, service quality is increasingly critical (Zehir and Narcıkara 2016; Santos 2003). Measuring E-S-Qual ensures that organisations meet evolving customer expectations (Zehir and Narcıkara 2016; Akinci et al. 2010). Unlike traditional models such as SERVQUAL, which focus on physical facilities and face-to-face interactions, E-S-Qual is suited for virtual environments (Tabaeeian et al. 2023; Palese and Usai 2018; Rossiter 2007).

Early tools like SITEQUAL often relied on convenience sampling, limiting generalisability (Boshoff 2007). E-S-Qual, in contrast, considers technical website quality, customer satisfaction dimensions, and digital tool effectiveness, although research remains limited (Williams 2024; Raza et al. 2020; Akinci et al. 2010). By bridging the gap between customer expectations and digital offerings, E-S-Qual helps organisations enhance service quality, build customer relationships, and maintain a competitive edge in the evolving e-commerce landscape.

2.1 *Definition of Electronic Service Quality (E-S-Qual)*

E-S-Qual measures the quality of services delivered via electronic channels, such as websites and mobile applications, considering factors like technical performance, ease of use, and responsiveness (Lamb et al. 2019; Boshoff 2007). Like traditional service quality models, it aims to meet customer needs and ensure satisfaction (Ahmed et al. 2020; Zehir and Narcıkara 2016). High E-S-Qual enhances customer relations, retention, competitive advantage, and long-term profitability (Çelik 2021; Mujinga 2020; San et al. 2020; Botha 2019; Zehir and Narcıkara 2016), making it critical for organisational success in the digital landscape (Roberts-Lombard 2024; San et al. 2020; Zehir and Narcıkara 2016; Reichheld and Schefter 2000).

E-S-Qual helps organisations align services with customer expectations through technology, improving digital service delivery and overall satisfaction (Nandankar et al. 2023; Mukherjee 2019; Parasuraman et al. 2005; Santos 2003). Today, customers expect seamless user experiences, personalised interactions, and efficient website functionality, which shape service delivery expectations (Benazic 2023; Zehir and Narcıkara 2016; Madu and Madu 2002; Zeithaml et al. 2002). Website service quality is influenced by functionality, information usefulness, usability, aesthetics, privacy, security, and personalisation (Jeon and Jeong 2016; Jeon 2009), with content, design, responsiveness, and security highlighted as key determinants (Kim and Lee 2005; Yoo and Donthu 2001).

Parasuraman et al. (2005) E-S-Qual model is widely regarded as the most effective tool for assessing online service quality. Studies demonstrate that E-S-Qual significantly impacts customer satisfaction, loyalty, and repurchase intentions (Zia et al. 2022), though further evaluation in diverse contexts and countries is recommended to test its reliability and robustness (Theodosiou et al. 2019; Rafiq et al. 2012).

2.2 *Importance of Electronic Service Quality (E-S-Qual)*

E-S-Qual is a critical tool for assessing and improving the quality of digital services. Its ability to measure alignment with customer expectations in a fast-paced digital world makes it invaluable for organisations looking to enhance customer satisfaction, build long-lasting relationships, and remain competitive (Ulkhaq et al. 2019). The rapid growth of online services means that providing a high-quality online experience has become essential (Spurer and Legentil 2023). The E-S-Qual framework offers a better understanding of how organisations can improve their service quality compared to other models (Ghosh 2018; Parasuraman 2005). Although grounded in the SERVQUAL model, which was initially designed for traditional brick-and-mortar service settings, E-S-Qual adapts to the digital environment, ensuring its relevance to online service quality (Zia et al. 2022).

E-S-Qual has been applied across various industries, including banking, higher education, fashion, and retail, providing valuable insights into the factors that impact digital service quality (Zia et al. 2022; Çelik 2021; Bhatt 2020; Kapalu 2020). By quantifying the impact of each dimension, E-S-Qual enables e-commerce organisations to prioritise their resources and optimise service delivery (Le 2015).

The role of technology in shaping e-CX has been highlighted in many studies, demonstrating that an efficient website can significantly enhance customer satisfaction and loyalty (Ulkhaq et al. 2019). E-commerce organisations that understand and leverage these insights can build stronger customer relationships and gain a competitive advantage in the market.

2.3 *Dimensions of Electronic Service Quality (E-S-Qual)*

As highlighted, E-S-Qual comprises four key dimensions encompassing all customer interactions with an e-commerce retail organisation, both before, during, and after a transaction (Le and Park 2023; Rafiq et al. 2012). Parasuraman et al. (2005) developed a 22-item scale to measure these four dimensions, which are summarised below:

Efficiency. Efficiency refers to how easily customers navigate a website, find information, and complete transactions quickly (Dwivedi et al. 2021; Parasuraman et al. 2005). An efficient website allows users to locate what they need and complete transactions smoothly, directly influencing satisfaction (Zanzalari 2024; Mujinga 2020; Botha 2019; Leonnard 2019; Duarte et al. 2018).

System Availability. System availability refers to a website's technical performance, including uptime and page load speeds (Subramanian 2024; Mujinga 2020; Ulkhaq et al. 2019; Boshoff 2007; Parasuraman et al. 2005). Studies show a significant positive relationship between system availability and customer satisfaction (Santouridis and Krikoni 2016; Botha 2019).

Fulfilment. Fulfilment refers to how well a website delivers on product availability and delivery promises (Parasuraman et al. 2005). It is measured by timely delivery, product availability, and accurate order execution (Rita 2019; Blut 2016). Meeting these expectations directly impacts customer satisfaction (Mwiya et al. 2022; Zehir and Narcıkara 2016).

Privacy. Privacy involves safeguarding customers' personal and financial information during online transactions (Securities and Exchange Commission 2024). Ensuring privacy is essential for building trust and secure transactions in the digital environment (Mwiya et al. 2022; Mofokeng 2021; Ghosh 2018). Organisations that protect customer data are more likely to foster trust and loyalty (Shaw et al. 2022; Makhitha and Ngobeni 2021).

2.4 *Post-Purchase Outcome of Customer Satisfaction*

Customer satisfaction is the result of comparing a customer's pre-purchase expectations and the post-purchase performance of the product or service (Lamb et al. 2019; Duarte et al. 2018). A positive customer experience is crucial for achieving high satisfaction levels, as it directly influences a customer's intention to reuse the product or service (Ahmad et al. 2022). This suggests that an e-commerce retail organisation that creates a positive e-CX is more likely to ensure customer satisfaction and foster long-term retention.

As previously discussed, the factors of E-S-Qual have the potential to impact customer satisfaction positively in e-commerce organisations, with various studies supporting its critical role in shaping customer perceptions and behaviours (Mamakou et al. 2024; Jain et al. 2022; Olofsson et al. 2022; Çelik 2021; Mujinga 2020; Raza et al. 2020; Botha 2019; Lamb et al. 2019; Ulkhaq et al. 2019; Amin 2016; Santouridis and Krikoni 2016; Akinci et al. 2010).

The following section will outline the problem statement.

3 Problem Statement

The relationship between E-S-Qual and customer satisfaction underscores the importance of ensuring a seamless, high-quality online service to foster positive post-purchase outcomes. In summary, the literature reveals that E-S-Qual is pivotal in shaping customer satisfaction in the e-commerce environment. E-S-Qual dimensions are essential factors influencing customers' perceptions of service quality.

Meeting or exceeding customer expectations through effective service delivery leads to positive post-purchase outcomes, fostering customer loyalty and repeat business. As e-commerce continues to grow in South Africa and globally, e-commerce retail organisations must prioritise enhancing their digital service quality to maintain a competitive edge.

As the e-commerce sector in South Africa continues to grow, understanding the role of E-S-Qual in shaping post-purchase customer satisfaction is essential. However, there is a lack of comprehensive studies focusing on how E-S-Qual dimensions influence customer satisfaction in South African e-commerce retail organisations. Therefore, this study addresses the question: *What key E-S-Qual dimensions significantly influence the post-purchase outcome of customer satisfaction in South African e-commerce retail organisations?*

4 Research Objectives

The primary objective of this study is to determine the E-S-Qual dimensions which significantly influence the post-purchase outcome of customer satisfaction in South African e-commerce retail organisations. The following secondary research objectives have been formulated:

- Assess the impact of various E-S-Qual factors on customer satisfaction in e-commerce retail organisations;
- Identify which E-S-Qual factors are statistically significant in shaping customer satisfaction in e-commerce retail organisations; and
- To provide actionable recommendations for e-commerce retail organisations to enhance their E-S-Qual strategies and improve customer satisfaction.

This paper is structured as follows: Sect. 2 reviews E-S-Qual and its impact on e-commerce. Sections 3–5 cover the problem statement, research objectives, and hypotheses. Section 6 outlines the research methodology, while Sect. 7 presents the results and findings. Section 8 offers practical recommendations, Sect. 9 discusses insignificant relationships, and Sect. 10 concludes by summarising key findings, contributions, and directions for future research.

5 Hypotheses

The variables that can possibly influence customer satisfaction are divided into a set of hypotheses (see Table 1).

The following section will outline the research methodology used.

Table 1 Proposed hypotheses

Hypothesis number		Hypotheses
E-S-Qual	H_1	There is a positive and statistically significant relationship between efficiency and *customer satisfaction*
	H_2	There is a positive and statistically significant relationship between system availability and *customer satisfaction*
	H_3	There is a positive and statistically significant relationship between fulfilment and *customer satisfaction*
	H_4	There is a positive and statistically significant relationship between privacy and *customer satisfaction*

6 Research Methodology

The following outlines the approach and procedures employed to investigate the impact of E-S-Qual on post-purchase satisfaction among South African e-commerce customers:

Research Approach and Design. This study adopts a positivistic research paradigm with a mono-method quantitative design (Camilleri 2021). Grounded in secondary literature to define the research problem and objectives (Bell et al. 2022; Wiid and Diggines 2020), the approach emphasises empirical observation and measurement (Bell et al. 2022) and the application of quantitative methods in business research (Wiid and Diggines 2020). Primary data were collected from South African e-commerce consumers, focusing on Generation Y (Millennials), a technologically engaged group with a strong workforce presence (Yildirim and Korkmaz 2020).

Data Collection Techniques. Data was gathered through an online survey. The survey contained specific instructions to ensure respondents met the necessary criteria: they had to be active e-commerce customers and part of Generation Y.

Sample. The sample consisted of South African e-commerce users from Generation Y, chosen for their digital familiarity and market influence. Non-probability sampling, including volunteer and snowball methods, was employed (Saunders 2019). A minimum of 384 respondents was considered adequate (Krejcie and Morgan 1970). Of 617 responses, 405 (65.6%) were valid after cleaning, sufficient for robust statistical analyses (Ayuni 2019; Wu and Hsu 2014).

Data Analysis. The data was cleaned using Microsoft Excel and analysed with IBM's Analysis of Moment Structures (AMOS) version 27.0. A Confirmatory Factor Analysis (CFA) was conducted to assess construct validity and reliability. Cronbach's Alpha coefficients assessed the reliability of the measures (Hair et al. 2017), and face and content validity were ensured through expert review during the item development phase. All of these aligned with the positivistic methodology employed in this study (Afzal et al. 2024; Mani and Tripathi 2020; Duffett 2017). SEM and regression analyses examined relationships between independent variables and customer satisfaction (Afzal et al. 2024; Mani and Tripathi 2020; Duffett 2017; Guerra-Tamez et al. 2024).

Ethics. Ethical approval for this study was granted by the Rhodes University Human Ethics Committee (RU-HEC), reference number 2021–5128-6437

7 Results and Findings

The results and findings section presents the data analysis outcomes, highlighting the key relationships between the dimensions of E-S-Qual and post-purchase satisfaction and the statistical significance of these findings in the context of South African e-commerce customers.

7.1 Descriptive Statistics and Participant Profile

The final sample included 405 validated South African e-commerce customers, all from Generation Y, mostly born in 1992, reflecting strong engagement with digital technologies and e-commerce. Their regular online shopping behaviours made them suitable for examining the relationship between E-S-Qual dimensions and post-purchase satisfaction.

7.2 Reliability and Validity Assessment

CFA results confirmed the four-factor structure of E-S-Qual, supporting the theoretical framework proposed by Parasuraman et al. (2005). Table 2 summarises the goodness-of-fit results before considering modification indices (MIs). Dimensions highlighted in green indicate a good model fit and did not require modifications. In contrast, the dimensions highlighted in purple required targeted modifications, which will be discussed in detail below.

From Table 2, MIs greater than four were considered for inclusion in the factor structure. The higher MI values were prioritised for linking items, ensuring meaningful and theoretically sound adjustments. The targeted modifications were as follows:

- MIs for covaried items—Efficiency: The analysis revealed high agreement with efficiency-related statements, such as: *The website made it easy to find what I needed* (Q76) and *The website made it easy to get anywhere on the website* (Q77). Three pairs of items showed strong correlations: Q76/Q77, Q77/Q78, and Q80/Q82 indicating that respondents who found the website easy to navigate were more likely to report quick transaction completion and fast page loading times.

Table 2 Goodness-of-fit results before modification indices (MIs)

Factor	Parameters	MI	χ^2 (df, p)	CMIN/df	SRMR	CFI	RMSEA
Customer Satisfaction	All significant	144.788	(27, 0.000)	5.363	0.044	0.936	0.104
Efficiency	All significant	110.012	(20, 0.000)	5.501	0.0454	0.933	0.106
System Availability	All significant	37.669	(2, 0.000)	18.834	0.0611	0.91	0.21
Fulfilment (E-S-Qual)	All significant	7.865	(2, 0.020)	3.932	0.0295	0.986	0.085
Privacy	All significant	0.000	(0, 0.000)	-	0.000	1.000	-

These relationships implied that ease of finding information, navigation, and fast transaction processing were interconnected.

- MIs for covaried items—System Availability: The CFA model for system availability revealed a strong correlation between the statements *The website is always available for business* (Q84) and *The website launches and runs right away* (Q85). This correlation suggested that respondents who perceived the website as consistently available were likely to report it launching and running immediately, emphasising the importance of continuous system availability for customer satisfaction.

The CFA and the application of MIs ensured a thorough evaluation of each primary dimension's measurement properties and underlying structure. The adjustments made to the model based on the MI results helped to reinforce the validity and reliability of the dimensions of E-S-Qual and their impact on customer satisfaction, further supporting the study's findings in the context of South African e-commerce retail organisations.

The statistical results presented in Table 3 indicate that the factor structures used in the CFA exhibit reasonable validity within the study context.

Below is a breakdown of the key indicators:

- CMIN/df: Values below 3.00 are considered good, and values below 5.00 are acceptable. Most dimensions, except for 'System Availability', fall within this range, indicating a reasonable model fit.
- SRMR: Values below 0.08 are considered good fits. All dimensions meet this criterion, indicating a good fit between the observed and predicted data.
- CFI: Values above 0.90 indicate a good fit. Most dimensions exceed this threshold, except for the overall E-S-Qual model, which is slightly below the ideal value (0.947).

Table 3 Goodness-of-fit results after modification indices (MIs) taken into account

Factor	Parameters	$\chi^2(df, p)$	CMIN/df	SRMR	CFI	RMSEA
Customer satisfaction	All significant	71.192 (24, 0.000)	2.966	0.0316	0.974	0.07
Efficiency	All significant	47.119 (17, 0.000)	2.772	0.0291	0.978	0.066
System availability	All significant	4.989 (1, 0.026)	4.989	0.0163	0.99	0.099
Fulfilment	All significant	7.865 (2, 0.020)	3.932	0.0295	0.986	0.085
Privacy	All significant	0.000 (0, 0.000)	-	0.000	1.000	-
E-S-Qual	All significant	305.684 (144, 0.000)	2.123	0.0509	0.947	0.053

- RMSEA: Values below 0.10 are considered a good fit. All dimensions, except for 'System Availability', meet this criterion. 'System Availability' is slightly above 0.099, which is still considered acceptable.

These findings demonstrate that the model fits well and is suitable for further analysis and interpretation. The results provide a solid foundation for understanding the constructs of customer satisfaction, efficiency, system availability, fulfilment, and privacy in South African e-commerce, highlighting the key factors contributing to customer satisfaction.

Cronbach's alpha coefficients were calculated to assess the reliability of each factor. All factors showed strong internal consistency, with customer satisfaction (0.906), efficiency (0.877), system availability (0.752), fulfilment (0.758), privacy (0.744), and overall E-S-Qual (0.909) exceeding the 0.7 threshold. These results confirm that the factors are reliable and suitable for further analysis, supporting their use in addressing the study's objectives.

7.3 Structural Equation Modelling (SEM) Results

Table 4 presents the results from the regression analysis of the SEM model. The dimensions that were found to be statistically insignificant are highlighted in red and subsequently removed from the model.

Further, the analysis shows that efficiency significantly impacts customer satisfaction, as indicated by the high CR value (4.113) and p-value less than 0.001. In contrast, system availability, fulfilment, and privacy did not significantly affect customer satisfaction, with p-values greater than 0.05. These findings suggest that efficiency is the key factor influencing customer satisfaction among the e-CX factors in this study.

Table 4 Results of regression analysis of SEM model

			Estimate	SE	CR	P
Customer satisfaction	←	Efficiency	0.157	0.038	4.113	***
Customer satisfaction	←	System availability	0.003	0.046	0.055	0.956
Customer satisfaction	←	Fulfilment	0.057	0.039	1.454	0.146
Customer satisfaction	←	Privacy	0.017	0.029	0.594	0.552

***$p < 0.001$

7.4 Key Findings

The most significant finding of this study is that 'efficiency' emerged as the primary driver of customer satisfaction among South African e-commerce customers. This suggests that customers prioritise ease of navigation, quick access to information, and streamlined transaction processes above other service quality dimensions. While all four E-S-Qual dimensions significantly influenced post-purchase satisfaction, efficiency's prominence indicates that South African customers particularly value time-saving and user-friendly e-CX.

These findings contribute to the growing body of literature on E-S-Qual in emerging markets and provide empirical evidence for the applicability of the E-S-Qual framework in the South African context, supporting the recommendations of Theodosiou et al. (2019) and Rafiq et al. (2012) for cross-cultural validation of the model.

8 Practical Managerial Implications and Recommendations

The findings of this study provide a clear strategic direction for South African e-commerce retail organisations seeking to enhance customer satisfaction (Roberts-Lombard 2024; Çelik 2021). The most significant finding—that 'efficiency' emerged as the only statistically significant driver of customer satisfaction (CR = 4.113, $p < 0.001$)—provides a focused roadmap for resource allocation and strategic investment, which means that e-commerce retailers should prioritise investments in website optimisation (Mujinga 2020; Botha 2019; Leonnard 2019). This focused approach includes the following recommendations:

- User interface design: Implement intuitive navigation systems that allow customers to easily locate products and information, as ease of navigation directly impacts customer satisfaction (Zanzalari 2024; Duarte et al. 2018).
- Search functionality: Develop robust search engines with filtering options to help customers quickly find desired items, supporting the streamlined experience that customers expect (Mamakou et al. 2024).

- Check-out process: Streamline the purchasing process by minimising the steps required to complete transactions, which is crucial for ensuring convenience and time-saving experiences (Parasuraman et al. 2005).
- Mobile optimisation: Ensure websites are fully responsive and optimised for mobile devices, given the increasing prevalence of mobile commerce among Generation Y customers (Yildirim and Korkmaz 2020).

While the study confirmed the reliability of all E-S-Qual dimensions (Cronbach's $\alpha > 0.7$), the SEM results revealed that system availability ($p = 0.956$), fulfilment ($p = 0.146$), and privacy ($p = 0.552$) did not significantly influence customer satisfaction in this South African context. Therefore, a brief discussion of the insignificant variables follows.

9 Discussion on Insignificant Relationships

The non-significant system availability, fulfilment, and privacy relationships diverge from much of the international E-S-Qual literature. Studies by Mwiya et al. (2022), Shaw et al. (2022), and Santouridis and Krikoni (2016) found these dimensions to be significant satisfaction drivers in their respective contexts. This divergence may reflect unique South African e-commerce market characteristics, Generation Y customer preferences, or cultural factors influencing service quality expectations.

The finding that privacy was not a significant predictor contrasts with increasing global concerns about data security (Makhitha and Ngobeni 2021; Ghosh 2018), suggesting that South African customers may view privacy protection as a baseline expectation rather than a satisfaction differentiator.

However, the finding suggests several important managerial implications:

- System Availability: Baseline Expectation: The non-significant link between system availability and satisfaction ($\alpha = 0.752$) suggests technical functionality is a baseline expectation rather than a differentiator (Parasuraman et al. 2005). Retailers should maintain acceptable uptime and performance while investing in cost-effective infrastructure that meets basic customer expectations.
- Fulfilment: Context-Specific Priority: The non-significant effect of fulfilment ($\alpha = 0.758$) may reflect local market or Generation Y behaviours. Retailers should monitor evolving fulfilment expectations and prioritise improvements only if competitive pressures or customer feedback indicate a need.
- Privacy: Expected Standard: The non-significant relationship for privacy ($\alpha = 0.744$) suggests customers assume basic data protection. Organisations should ensure compliance with the Protection of Personal Information Act (POPIA) and focus investments on regulatory standards rather than attempting to enhance satisfaction.

10 Conclusions, Limitations, and Future Research

This study investigated E-S-Qual dimensions and their influence on post-purchase satisfaction among South African e-commerce customers using data from 405 respondents. The most significant finding was that only 'efficiency' demonstrated a statistically significant positive relationship with customer satisfaction ($CR = 4.113$, $p < 0.001$), while system availability, fulfilment, and privacy showed no significant relationships. This singular focus on efficiency challenges established the E-S-Qual theory and provides unique insights into South African customer behaviour, suggesting that local customers prioritise simplicity, speed, and ease of use above other service quality considerations. The finding contrasts with international literature where all four dimensions typically drive satisfaction, indicating that E-S-Qual dimensions may have varying significance across different markets and cultural contexts.

The study makes significant theoretical and practical contributions to service quality literature. Theoretically, it provides context-specific E-S-Qual validation for emerging markets. It suggests that dimensions may operate hierarchically, with 'efficiency' serving as a primary satisfaction driver while others function as baseline expectations. Practically, the research offers clear resource allocation guidance for South African e-commerce organisations, demonstrating the need for locally adapted strategies focused on efficiency improvements rather than universally applied international best practices. The findings enable competitive differentiation through targeted efficiency enhancements while maintaining baseline performance in other dimensions, providing a strategic advantage in the South African market.

Several limitations affect the study's generalisability and scope. The use of non-probability sampling methods and focus on Generation Y customers limits applicability to the broader population and other demographic groups. The cross-sectional design prevents the establishment of causal relationships and understanding evolving preferences over time, while reliance on self-reported measures may introduce response bias. The study did not explicitly differentiate between e-commerce sectors, examine cultural variables, or account for post-COVID-19 digital adoption patterns, representing significant gaps in understanding the underlying mechanisms driving the observed relationships.

Future research should prioritise replication studies across other African countries, longitudinal analysis of changing E-S-Qual importance, and multi-generational studies to understand age-related differences. Extended research directions include cultural mediation studies, sector-specific analysis, infrastructure impact studies, and qualitative exploration of why efficiency dominates satisfaction. The findings suggest that the E-S-Qual theory needs refinement for emerging markets through contextual models that recognise varying dimensional importance and hierarchical frameworks that distinguish primary drivers from baseline expectations. As South African e-commerce evolves, continued research into cultural, economic, and technological factors shaping service quality expectations will be crucial for both theoretical understanding and practical application.

References

Afzal, B., Xiao, W., Nazir, A., Junaid, D., Silva, L.J.O.: Analysing the impact of social media influencers on consumer shopping behaviour: Empirical evidence from Pakistan. Sustainability (2024). https://doi.org/10.3390/su16146079

Ahmad, F., Mustafa, K., Hamid, S.A.R., Khawaja, K.F., Zada, S., Jamil, S., Qaisar, M.N., Vega-Muñoz, A., Contreras-Barraza, N., Anwer, N.: Online customer experience leads to loyalty via customer engagement: Moderating role of value co-creation. Front. Psychol. **13**, 897851 (2022)

Ahmed, R.R., Romeika, G., Kauliene, R., Streimikis, J., Dapkus, R.: ES-QUAL model and customer satisfaction in online banking: Evidence from multivariate analysis techniques. Oeconomia Copernicana **11**(1), 59–93 (2020)

Akinci, S., Atilgan-Inan, E., Aksoy, S.: Re-assessment of ES-Qual and E-RecS-Qual in a pure service setting. J. Bus. Res. **63**(3), 232–240 (2010)

Amin, M.: Internet banking service quality and its implication on e-customer satisfaction and e-customer loyalty. Int. J. Bank Mark. **34**(3), 280–306 (2016)

Ayuni, R.F.: The online shopping habits and e-loyalty of Gen Z as natives in the digital era. J. Indonesian Econ. Bus. (2019). https://doi.org/10.22146/jieb.39848

Bell, E., Bryman, A., Harley, B.: Research methodology: Business and management contexts, 6th edn. Oxford Univ. Press, New York (2022)

Benazic, D.: Perceived mobile banking service quality by users in Croatia. In: Tipuric, D., Maric, M. (eds.) 94th Int. Sci. Conf. on Econ. and Social Dev., Dubrovnik, Croatia, pp. 93–103 (2023)

Bhatt, H.R.D.V.: A study on impact of e-service quality dimensions of online shopping platforms on overall service experience. Alochana Chakra J. **9**(6), 1066–1088 (2020)

Biscaia, R., Yoshida, M., Kim, Y.: Service quality and its effects on consumer outcomes: A meta-analytic review in spectator sport. Eur. Sport Manag. q. **23**(3), 897–921 (2023)

Blut, M.: E-service quality: development of a hierarchical model. J. Retail. **92**(4), 500–517 (2016)

Blythe, J., Martin, J.: Essentials of Marketing, 7th edn. Pearson Education, Harlow (2019)

Boshoff, C.: A psychometric assessment of ES-QUAL: A scale to measure electronic service quality. J. Electron. Commer. Res. **8**(1), 101–114 (2007)

Botha, B.: Online service quality and its influence on customer satisfaction: A descriptive study. Bachelor's thesis, The Independent Institution of Education (2019)

Camilleri, M.A.: E-commerce websites, consumer order fulfilment and after-sales service satisfaction: The customer is always right, even after the shopping cart check-out. J. Strategy Manag. (2021). https://doi.org/10.1108/jsma-02-2021-0045

Çelik, K.: The effect of e-service quality and after-sales e-service quality on e-satisfaction. Bus. Manag. Stud.: Int. J. **9**(3), 1137–1155 (2021)

Duarte, P., e Silva, S.C., Ferreira, M.B.: How convenient is it? Delivering online shopping convenience to enhance customer satisfaction and encourage e-WOM. J. Retail. Consum. Serv. 44, 161–169 (2018)

Duffett, R.G.: Influence of social media marketing communications on young consumers' attitudes. Young Consum. (2017). https://doi.org/10.1108/yc-07-2016-00622

Dwivedi, Y.K., Ismagilova, E., Hughes, D.L., Carlson, J., Filieri, R., Jacobson, J., Jain, V., Karjaluoto, H., Kefi, H., Krishen, A.S., Kumar, V.: Setting the future of digital and social media marketing research: Perspectives and research propositions. Int. J. Inf. Manag. **59**, 102168 (2021)

Ghosh, M.: Measuring electronic service quality in India using E-S-QUAL. Int. J. Qual. Reliab. Manag. **35**(2), 430–445 (2018)

Guerra-Tamez, C.R., Flores, K.K., Serna-Mendiburu, G.M., Robles, D.C., Cortés, J.I.: Decoding Gen Z: AI's influence on brand trust and purchasing behaviour. Front. Artif. Intell. (2024). https://doi.org/10.3389/frai.2024.1323512

Hair Jr, J.F., Black, W.C., Babin, B.J., Anderson, R.E.: Multivariate Data Analysis (MVDA), 7th edn. Pearson Education, Harlow (2017)

Haleem, A.: U.S. ecommerce sales mark fourth straight year of single-digit growth (2026). https://www.digitalcommerce360.com/article/us-ecommerce-sales/. Last accessed 2026/05/19

Jain, M., Dixit, S., Shukla, A.: Role of e-service quality, brand commitment and e-WOM trust on e-WOM intentions of millennials. Int. Rev. Retail Distrib. Consum. Res. **33**(1), 23–43 (2022)

Jameel, A.S., Hamdi, S.S., Karem, M.A., Ahmad, A.R.: E-service quality, e-satisfaction and e-loyalty among online shoppers: Evidence from Iraq. In: Al-kadhum 2nd Int. Conf. on Modern Appl. ICT, Baghdad, Iraq. AIP Publ., pp. 1–10 (2023)

Jeon, M.M.: Impact of perceived website service quality on customer e-loyalty on a lodging website. PhD thesis, Iowa State Univ., Ames (2009)

Jeon, M.M., Jeong, M.: Influence of website quality on customer perceived service quality of a lodging website. J. Qual. Assur. Hosp. Tour. **17**(4), 453–470 (2016)

Jinliang, W., Chau, K.Y., Baei, F., Moslehpour, M., Nguyen, K.L., Nguyen, T.T.H.: Integrated perspective of eco-innovation, green branding, and sustainable product: A case of an emerging economy. Econ. Res. **36**(3), 1–23 (2023)

Kapalu, N.M.: Electronic service quality and customer satisfaction in higher education nexus: a case of the Copperbelt University. Bachelor's thesis, Cavendish Univ. (2020)

Kaya, B., Behravesh, E., Abubakar, A.M., Kaya, O.S., Orús, C.: The moderating role of website familiarity in the relationships between e-service quality, e-satisfaction and e-loyalty. J. Internet Commer. **18**(4), 369–394 (2019)

Kim, W.G., Lee, H.Y.: Comparison of web service quality between online travel agencies and online travel suppliers. J. Travel Tour. Mark. **17**(2–3), 105–116 (2005)

Krejcie, R.V., Morgan, D.W.: Determining sample size for research activities. Educ. Psychol. Meas. **30**(1), 607–610 (1970)

Lamb, C.W., Hair, J.F., McDaniel, C., Boshoff, C., Terblanche, N., Elliott, R., Klopper, H.B.: Marketing, 6th edn. Oxford Univ. Press South Africa, Cape Town (2019)

Le, H.T.P.M., Park, J.: What leads customers to stay with an online transportation service in emerging markets? Int. J. Bus. Innov. Res. **30**(3), 323–351 (2023)

Le, T.H.: Service quality in luxury ecommerce in China market. Master's dissertation, Aalto Univ. Sch. Bus. (2015)

Leonnard: Exploring the relationship among e-service quality, e-trust, e-satisfaction and loyalty at higher education institutions. J. Eff. Responsib. Educ. Sci. **12**(4), 103–110 (2019)

Lewis, B.R., Mitchell, V.W.: Defining and measuring the quality of customer service. Mark. Intell. Plan. **8**(6), 11–17 (1990)

Lin, X.Q., Chen, Y.C., Liu, C.H., Li, Y.: Measuring creativity: Role of service quality management, knowledge sharing and social interaction. Total Qual. Manag. Bus. Excell. **34**(1–2), 1–18 (2023)

Madu, C.N., Madu, A.A.: Dimensions of e-quality. Int. J. Qual. Reliab. Manag. **19**(3), 246–258 (2002)

Makhitha, K.M., Ngobeni, K.M.: The impact of risk factors on South African consumers' attitude towards online shopping. Acta Commercii **21**(1), 1–10 (2021)

Mamakou, X.J., Zaharias, P., Milesi, M.: Measuring customer satisfaction in electronic commerce: The impact of e-service quality and user experience. Int. J. Qual. Reliab. Manag. **41**(3), 915–943 (2024)

Mani, A.P., Tripathi, V.M.: Does education affect the online impulse buying in millennials? Int. J. Recent Technol. Eng. (IJRTE) https://doi.org/10.35940/ijrte.b1019.0982s1219(2020)

Mofokeng, T.E.: The impact of online shopping attributes on customer satisfaction and loyalty: Moderating effects of e-commerce experience. Cogent Bus. Manag. **8**(1), 1968206 (2021)

Mujinga, M.: Online banking service quality: A South African E-S-QUAL analysis. In: Hattingh, M., Matthee, M., Smuts, H., Pappas, I., Dwivedi, Y., Mäntymäki, M. (eds.) Responsible Design, Implementation and Use of ICT. 19th IFIP WG 6.11 Conf. on e-Business, e-Services, and e-Society (I3E 2020), Skukuza, South Africa. Springer, Cham, pp. 228–238 (2020)

Mukherjee, S.P.: Quality: Domains and dimensions. Springer Nature, Singapore (2019)

Mwiya, B., Katai, M., Bwalya, J., Kayekesi, M., Kaonga, S., Kasanda, E., Munyonzwe, C., Kaulungombe, B., Sakala, E., Muyenga, A., Mwenya, D.: Examining the effects of electronic service quality on online banking customer satisfaction: Evidence from Zambia. Cogent Bus. Manag. **9**(1), 2143017 (2022)

Nandankar, S., Sachan, A., Adhikari, A., Mukherjee, A.: Developing and validating e-marketplace service quality model in B2G e-commerce settings: A mixed-methods approach. Int. J. Oper. Prod. Manag. **43**(12), 1809–1840 (2023)
Olofsson, L., Andersson Karlström, L.: The effect of e-service quality on customers' e-satisfaction in online apparel stores: Within the Swedish context. Bachelor's thesis, Jönköping Univ., Jönköping (2022)
Palese, B., Usai, A.: The relative importance of service quality dimensions in e-commerce experiences. Int. J. Inf. Manag. **40**, 132–140 (2018)
Pappas, I.O., Pateli, A.G., Giannakos, M.N., Chrissikopoulos, V.: Moderating effects of online shopping experience on customer satisfaction and repurchase intentions. Int. J. Retail Distrib. Manag. **42**(3), 187–204 (2014)
Parasuraman, A., Zeithaml, V.A., Malhotra, A.: ES-QUAL: A multiple-item scale for assessing electronic service quality. J. Serv. Res. **7**(3), 213–233 (2005)
Paul, J., Mittal, A., Srivastav, G.: Impact of service quality on customer satisfaction in private and public sector banks. Int. J. Bank Mark. **34**(5), 606–622 (2016)
Rafiq, M., Lu, X., Fulford, H.: Measuring internet retail service quality using ES-QUAL. J. Mark. Manag. **28**(9–10), 1159–1173 (2012)
Raza, S.A., Umer, A., Qureshi, M.A., Dahri, A.S.: Internet banking service quality, e-customer satisfaction and loyalty: The modified e-SERVQUAL model. TQM J. **32**(6), 1443–1466 (2020)
Reichheld, F.F., Schefter, P.: The economics of e-loyalty (2000) https://hbswk.hbs.edu/archive/the-economics-of-e-loyalty
Rita, P., Oliveira, T., Farisa, A.: The impact of e-service quality and customer satisfaction on customer behaviour in online shopping. Heliyon **5**(10), e02690 (2019)
Roberts-Lombard, M.: Marketing management. In: Nieuwenhuizen, C. (ed.) Bus. Manag.: An African Perspect. 2nd edn. Juta Publ., Cape Town, pp. 527–573 (2024)
Rossiter, J.R.: Toward a valid measure of e-retailing service quality. J. Theor. Appl. Electron. Commer. Res. **2**(3), 36–48 (2007)
Sagar, S.: The impact of digital transformation on retail management and consumer behaviour. J. Bus. Manag. **26**(1), 6–14 (2024)
San, W.H., Von, W.Y., Qureshi, M.I.: The impact of e-service quality on customer satisfaction in Malaysia. J. Mark. Inf. Syst. **3**(1), 46–62 (2020)
Santos, J.: E-service quality: A model of virtual service quality dimensions. Manag. Serv. Qual. Int. J. **13**(3), 233–246 (2003)
Santouridis, I., Krikoni, E.: Extending E-S-QUAL with hedonic dimensions to measure the quality of internet travel services and its effects on perceived value, loyalty and switching barriers. Int. J. Technol. Mark. **11**(3), 301–318 (2016)
Saunders, M., Lewis, P., Thornhill, A.: Research Methods for Business Students, 8th edn, Pearson Education, Harlow (2019)
Securities and Exchange Commission: Regulation S-P: Privacy of consumer financial information and safeguarding customer information (2024) https://www.federalregister.gov/documents/2024/06/03/2024-11116/regulation-s-p-privacy-of-consumer-financial-information-and-safeguarding-customer-information
Shaw, N., Eschenbrenner, B., Baier, D.: Online shopping continuance after COVID-19: A comparison of Canada, Germany and the United States. J. Retail. Consum. Serv. **69**, 103100 (2022)
Spurer, E., Legentil, L.: How does customer satisfaction impact the performance of an e-commerce company? Bachelor's thesis. Tampere Univ. Appl. Sci, Tampere (2023)
Subramanian, K.: Uptime vs. availability: How to measure and improve reliability (2024) https://www.pluralsight.com/resources/blog/tech-operations/uptime-availability-metrics-app-reliability
Tabaeeian, R.A., Mohammad Shafiee, M., Ansari, A.: No title. Int. J. Retail Distrib. Manag. **51**(4), 444–464 (2023)

Theodosiou, M., Katsikea, E., Samiee, S., Makri, K.: A comparison of formative versus reflective approaches for the measurement of electronic service quality. J. Interact. Mark. **47**(1), 53–67 (2019)

Ulkhaq, M.M., Rabbani, M., Rachmania, B.A., Wibowo, A.T., Ardi, F.: Integrating importance–performance analysis into E-S-QUAL and E-RecS-QUAL scales for assessing electronic service quality. In: Int. Conf. Energy, Nat. Resour. Electr. Power, Jeju Isl., South Korea. IOP Publ., pp. 012002 (2019)

Vakulenko, Y., Shams, P., Hellström, D., Hjort, K.: Online retail experience and customer satisfaction: The mediating role of last mile delivery. Int. Rev. Retail Distrib. Consum. Res. **29**(3), 306–320 (2019)

Wiid, J., Diggines, C.: Marketing Research, 4th edn, Juta, Cape Town (2020)

Williams, A.S.: The influence of electronic-consumer experience (e-CX) on consumer post-purchase outcomes of South African e-commerce retail organisations. PhD thesis, Rhodes University, Makhanda/Grahamstown (2024)

Williams, A.S., Oosthuizen, N.: A framework of electronic customer experience (e-CX) factors necessary for ensuring customer satisfaction of e-commerce retail organisations. In: Social Sciences International Research Conference, North-West University, North West, pp. 239–251 (2021)

Wu, C., Hsu, C.: How to improve e-satisfaction and e-loyalty and strengthen the links between them: Value from regulatory fit. Hum. Factors Ergon. Manuf. Serv. Ind. (2014). https://doi.org/10.1002/hfm.20549

Yildirim, N., Korkmaz, Y.: Challenge of millennials in project management: Insights on attitudes and perceptions of Generation Y in software development projects. In: Information Resources Management Association (ed.) Five generations and only one workforce: How successful businesses are managing a multi-generational Workforce. IGI Global, Hershey, pp. 255–282 (2020)

Yoo, B., Donthu, N.: Developing a scale to measure the perceived quality of an Internet shopping site (SITEQUAL). Q. J. Electron. Commer. **2**(1), 31–46 (2001)

Zanzalari, D.: Advantages of e-commerce: E-commerce can work for businesses and customers. (2024) https://www.thebalancemoney.com/advantages-of-ecommerce-1141610

Zehir, C., Narcıkara, E.: E-service quality and e-recovery service quality: Effects on value perceptions and loyalty intentions. Procedia Soc. Behav. Sci. **229**, 427–443 (2016)

Zeithaml, V.A., Parasuraman, A., Malhotra, A.: Service quality delivery through web sites: A critical review of extant knowledge. J. Acad. Mark. Sci. **30**(4), 362–375 (2002)

Zia, S., Rafique, R., Rehman, H.U., Chudhery, M.A.Z.: A comparison between E-TailQ and ES-Qual for measuring e-service quality in the retail industry: An emerging economy case. TQM J. **35**(8), 2228–2254 (2022)

The Role of AI, Blockchain, and Digitalisation in Providing Financial Services to Unbanked Populations: A Bibliometric Analysis of Literature

Bouba Ismaila and John Beneke

Abstract New prospects for financial inclusion have been made possible by the growing use of blockchain, artificial intelligence (AI), and digital currencies, particularly among unbanked people. The role of these technologies in delivering financial services to underbanked communities is rapidly developing. These technologies provide creative ways to close the gap between official financial services and the unbanked, fostering economic growth and financial inclusion. The review aimed to evaluate the current state of research on the role of AI, blockchain, and digitalisation of unbanked populations. This assessment of the literature looks at the field's present situation, major trends, prospects, difficulties, and potential future paths. The review revealed that digital currencies, blockchain, and AI have enormous potential to revolutionise financial services for underbanked people. The quick rise in research on this topic indicates that people are becoming more interested in these technologies. The review also cautioned that, although there are significant prospects, the effective deployment and uptake of these technologies depend on resolving technological, privacy, and regulatory issues. The goal of future studies and legislative initiatives should be to establish a supportive atmosphere that encourages creativity while maintaining safety and diversity.

Keywords Artificial Intelligence · Blockchain · Digital currencies · Financial inclusion · Underbanked communities

---

B. Ismaila (✉) · J. Beneke
Vaal University of Technology, Vanderbijlpark, South Africa
e-mail: boubai@vut.ac.za

J. Beneke
e-mail: johnb@vut.ac.za

E. H. Redda (ed.), *Proceedings of the Social Sciences International Research Conference, 2025*, https://doi.org/10.1007/978-981-95-9489-4_19

1 Introduction

The use of blockchain, artificial intelligence (AI), and digital currencies to offer financial services to underbanked populations is a quickly developing area. It is perceived that these technologies can promote financial inclusion and economic development by providing creative ways to close the gap between the unbanked and formal financial services. This review aims to explore the status of current research on the potential of finance digitalisation, blockchain, and AI to provide financial services to underbanked or unbanked individuals. This assessment examines the current state of research in the field now, as well as important trends, opportunities, challenges, and potential paths forward.

There is a case to be made that it becomes necessary to close the gap between innovation and conventional financial institutions as the world adopts digital currencies, blockchain, and AI technologies. For instance, Mhlanga (2023) found that blockchain technology can support digital financial inclusion in a variety of ways, such as through its use in financial transactions as a tool to increase savings, in the granting of credit, and in the provision of insurance. Credit scoring is a tool that traditional financial institutions use primarily to qualify clients for credit. This often unilaterally exclude consumers who do not have a credit record.

AI has the potential to help resolve this issue by extending credit access to include previously unbanked populations, by providing alternative credit assessment such as borrowers' social interactions as data source (Garcia et al. 2024; Hoang et al. 2020; Muñoz-Cancino et al. 2023). In the case of small businesses, these authors posit that these businesses' interactions with their owners, suppliers, consumers, and other businesses create a business ecosystem that offers new information that helps financial institutions improve their evaluation of creditworthiness.

The effects of digitalisation are evident in the increased accessibility of financial services in remote locations, facilitated by the expansion of digital banking solutions. This trend is particularly pronounced in developing countries, especially in Africa and Asia, where mobile money services have grown substantially and are now used for nearly all types of transactions. In this regard, beyond enabling unbanked individuals to open accounts, mobile money has also helped many rural communities in Zambia escape poverty (Mwila and Kunda 2022). Furthermore, Chitimira and Ncube (2023) reported that to date, mobile money services have significantly increased the majority of Zimbabweans' financial inclusion.

2 Research Method

A narrative review process was used to locate and examine research and literature on the evolution, patterns, and potential effects of blockchain, AI, and digital currencies on unbanked and underbanked populations. The search approach entailed locating a pertinent database, in this case Scopus. Because most of the articles found on this search engine are also found on other search engines like Google Scholar or ResearchGate, and because it offers a comparable flexible connection with a wide range of searching capabilities across a vast number of online scholarly published papers, it was judged sufficient for the purpose of this study. Books, journal articles, conference papers, dissertations, and theses were among the many forms of literature that were considered in the review. No time frame was considered because of the authors' perception of the topic being relatively new in research, yet the results turned out to be 2009–2025 period. Because of its exploratory nature, the review did not adopt a more inclusive approach to article selection such as what is used for a systematic review.

The search technique used free words or expressions linked with Booleans, such as OR and AND, and included artificial intelligence, AI, financial inclusion, blockchain, block chain, block-chain, digital currency, mobile money, unbanked populations, and underbanked populations. The result of the search produced 113 articles and papers. All studies deemed irrelevant to this review or lacking all three main keywords or their variations (i.e. artificial intelligence or AI; financial inclusion, blockchain or block chain, or block-chain; digital currency/currencies or digitalisation) were excluded from the review. A total of 87 papers were finally considered for the review. Bibliometrix was then used to analyse the papers for insights.

3 Results and Discussion

The findings of the review are presented and discussed in this section.

3.1 Key Trends

Most Relevant Keywords. The most relevant words that appeared in the papers reviewed are depicted in Fig. 1.

The frequency of the most pertinent keywords is displayed in the chart, offering useful quantitative support for the research on blockchain and AI in developing nations. The top ten most pertinent keywords taken from the review article are displayed in this graphic; each keyword stands for a thematic pillar in the discussion of financial inclusion for underbanked or unbanked groups. Keywords that indicate the main themes are:

- The most frequently used keyword is finance (15 occurrences), highlighting the review article's primary focus on financial systems, services, and access, a crucial component when talking about the underprivileged.
- Electronic money (12 references) and AI (13 mentions) both stand out as important facilitators. This implies that AI plays a key role in credit assessment, risk modelling, and customised financial solutions. Likewise, for people living outside

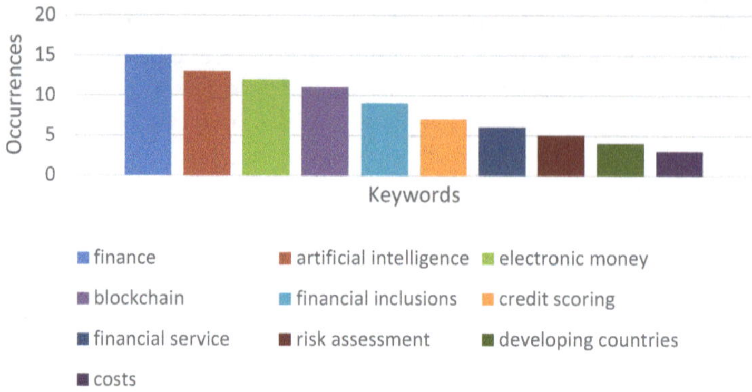

Fig. 1 Keyword frequency

of established banking networks, electronic money is essential for facilitating cashless transactions and remote banking access.

- Blockchain (11 references) is prominent, demonstrating its significance in developing decentralised, transparent, and safe financial ecosystems. Its coexistence with AI and electronic money demonstrates how the two technologies work together in: improving cross-border remittances for unbanked groups; lowering transaction costs; and reducing fraud.
- The theme term *financial inclusion* (9 times) supports the article's goal of examining technological advancements meant to increase access to financial systems. This is closely related to using blockchain and AI to reach groups that are economically disenfranchised, geographically remote, or lack access to digital technology.
- Credit scoring (7 mentions) and risk assessment (5 mentions) both emphasise how AI might be used to create alternative credit models for people without official credit histories; and improve lending decision-making to reach underrepresented populations with microloans, insurance, and savings plans.
- The financial service (6 mentions) attests that these technologies are not merely theoretical but are being implemented in practical fintech solutions.
- The terms poverty (3 mentions) and developing countries (4 mentions) ground the conversation in actual socio-economic contexts where financial exclusion is most acute.

These keywords imply that the technological solutions under review have a particularly significant impact in low-income and rural areas, where traditional banking is either non-existent or insufficient.

Word Cloud. The most relevant keywords that came out of the literature from the authors whose papers were considered for the review are shown in the Word Cloud in Fig. 1. Some words or phrases that are spelled differently (e.g. blockchain, block-chain, block chain) or that are related to the same topic (e.g. large language

Fig. 2 Word cloud

model, AI, artificial intelligence) have been combined for the purpose of clarification and showing a clearer picture (Fig. 2).

In this figure, the bigger the keyword, the more frequent its appearance in the reviewed papers. It may be noticed that the initial keywords of our search are among the top five in the cloud words. Those most prevalent topics are finance, AI, and electronic money, playing a major role. This implies a major emphasis on digital transactions and financial services powered by AI. Additionally, blockchain is widely used, highlighting its significance for safe transactions, decentralised finance (DeFi), and financial confidence. Another important word that denotes research on AI-based credit evaluation and alternative lending models is credit scoring. A focus on AI and fintech risk management is seen in the many mentions of risk assessment.

Furthermore, financial services and economic development demonstrate the wider influence of digital finance on economic expansion. Increased use of blockchain-based financial inclusion solutions and mobile banking is suggested by mobile money and decentralised finance. Also, words like poverty alleviation, rural areas, and emerging nations suggest a focus on financial inclusion for marginalised groups. Likewise, the existence of learning systems and human–computer interaction points to an expanding role for AI in accessible digital financial services. Online social networking may reveal the ways in which digital platforms impact financial practices. A thorough analysis of the literature points to further research initiatives that combine financial inclusion with AI trends.

Keyword Frequency Over Time. The review's keyword frequency over time is illustrated in Fig. 3, showing how these words/phrases evolved in research.

Insights from the figure:

Dominant and emerging themes over time: the keywords with the largest growth, particularly from 2018 onward, are finance and electronic money. This implies that studies on financial inclusion should place a high priority on electronic payments, digital transactions, and financial services. After 2017, blockchain began to emerge and continues to grow steadily. This is consistent with decentralised finance (DeFi) solutions for financial inclusion and the growth of cryptocurrencies. After 2018, artificial intelligence and credit scoring gained popularity, suggesting a greater emphasis on alternative credit models and AI-driven risk assessment.

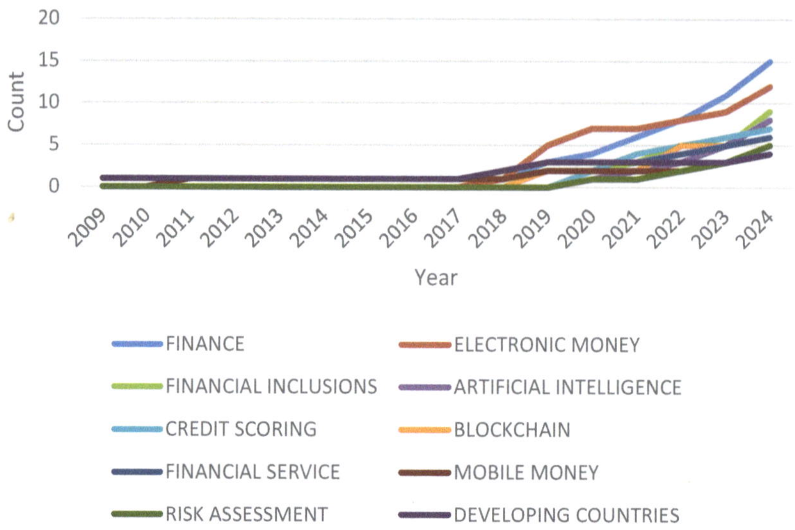

Fig. 3 Keyword frequency over time

Financial inclusion and developing countries: after 2019, there has been an upward trend in financial inclusion, which is consistent with international initiatives to connect unbanked people to the financial system. The persistent presence of developing countries indicates that research on financial inclusion in emerging economies is current and ongoing.

Changes driven by technology (mobile money, risk assessment): one of the major trends of the changes is mobile money. Research on mobile-based financial solutions is likely to increase, as evidenced by the growth of mobile money starting from 2018. This is in line with its actual acceptance in South Asia, Latin America, and Africa. Additionally, risk assessment has grown in importance. The growing use of 'risk assessment' draws attention to issues with trust in digital banking, financial crime, and AI bias.

Publications Over the Years. Notably, research on AI, blockchain, and financial inclusion has increased significantly in recent years, particularly in 2023 (13 papers) and 2024 (34 papers), which together accounted for more than half of the papers published between 2009 and 2025. However, early research in this area was inconsistent, with only one to two studies published between 2009 and 2017, suggesting that interest in the field is expanding. The following chart (Fig. 4) shows these trends.

This figure shows that there was little to no interest in the topic until 2018, followed by a marked increase in 2023 and 2024. There is no data yet for 2025, as this review was conducted at the beginning of the year.

Co-occurrences of Keywords. The co-occurrence of keywords in the reviewed papers and their clusters is depicted in Fig. 5.

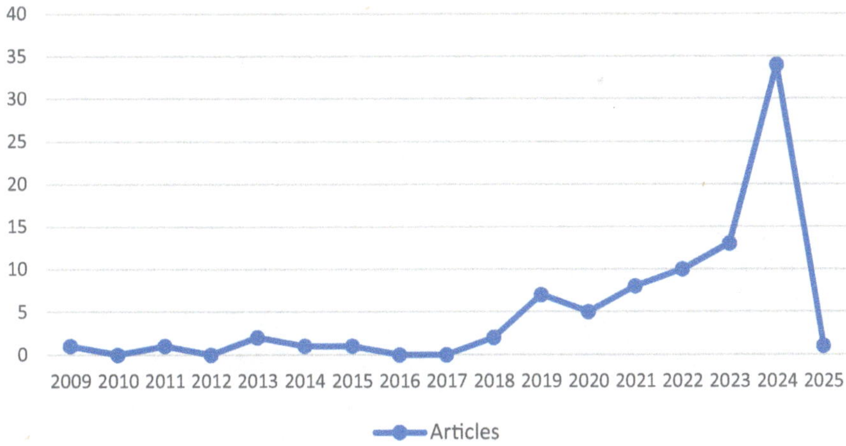

Fig. 4 Annual production over the years

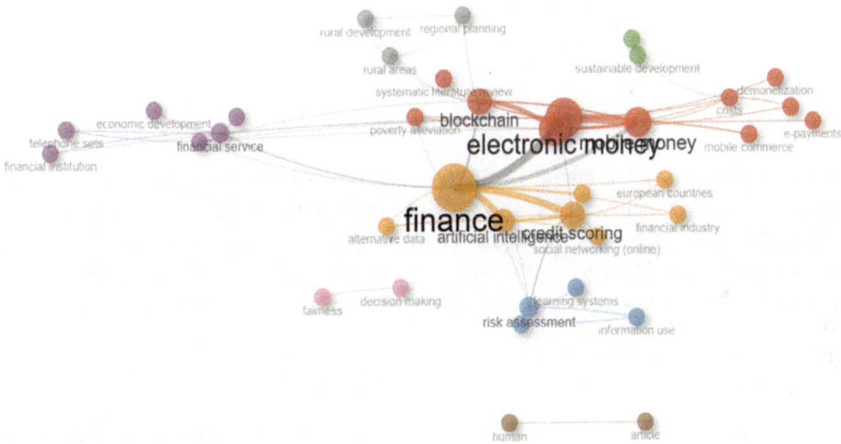

Fig. 5 Keywords co-occurrence

The main theme here is finance, which is closely related to credit scoring, blockchain, mobile money, and electronic money. This implies that when talking about AI and blockchain for financial inclusion, the main issues are financial transactions, payments, and credit accessibility. The terms *finance* and *electronic money* have huge font sizes, which suggests that they are highly important and often used in the research. AI's role in financial decision-making is demonstrated by its tight ties to risk assessment and credit rating. AI facilitates alternative credit scoring models for people without conventional credit histories and aids in the assessment of financial risks. Blockchain's usefulness in safe, decentralised financial transactions is demonstrated by its association with mobile payments and electronic money. This suggests

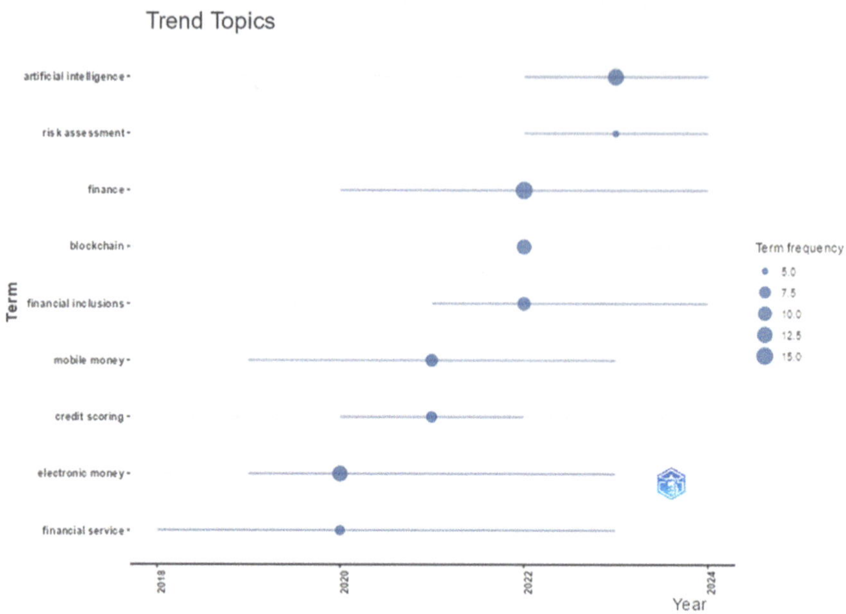

Fig. 6 Trend topics

that blockchain-based financial inclusion solutions—such as digital wallets, smart contracts, and cryptocurrencies—are being investigated.

Trend Topics. Figure 6 shows the growth of keyword trends in financial inclusion research from 2018 to 2023, with a focus on the usage of digital currencies, blockchain, and AI to reach underserved and unbanked populations.

Insights from the trend topics plot.

Development of risk assessment and AI (2022–2023):

i. AI (2022–2023): it has recently become a hot issue with a high phrase frequency (big bubble), implying that AI is being used increasingly in fraud detection, credit scoring, and customised financial solutions. Moreover, by automating decision-making for underrepresented groups, AI is viewed as a scalable solution for improving financial inclusion.
ii. Risk evaluation (2022–2023): this trend, which occurred at the same time, illustrates how AI is being used to assess non-traditional creditworthiness, loan default risk, and underwriting for insurance in underprivileged areas.

Blockchain and finance as key themes (2021–2022):

i. Finance (2021–2022): two years of consistent focus demonstrates ongoing initiatives to use digital innovation to make traditional banking systems more inclusive.

ii. Blockchain (2022): a promising performance suggests that attention is moving towards decentralised finance (DeFi) platforms using smart contracts to facilitate clear transactions and identifying management for those without bank accounts.

Early facilitators of mobile and electronic money (2019–2022):

i. Mobile money (2019–2022): a long-term trend that is becoming steadily more important, particularly in developing nations with inadequate financial systems, enabling bill payments, savings, and peer-to-peer transfers.
ii. Electronic money (2020–2022) suggests the foundation for digital inclusion was laid with the introduction of e-wallets, digital payment platforms, and agent banking.

Newly applied subjects: financial services and credit scoring:

i. Credit scoring (2020–2021) shows early interest in using AI to develop alternative credit models for consumers without credit histories.
ii. Financial service (2020–2023): a persistent but infrequent word, this one suggests that although digital financial services are the ultimate objective, the focus has been on the technologies that make this possible rather than the services themselves.

Lastly, from 2020 to 2022, financial inclusion became a cross-cutting issue. Financial inclusion (2020–2022) has been a persistent issue over several years, demonstrating that it is the primary obstacle hindering the integration of all other technologies. It further highlights how e-money, blockchain, and AI are being used specifically to combat inequality and exclusion.

4 Opportunities and Challenges

4.1 Opportunities

Blockchain and AI can offer scalable and affordable ways to serve unbanked people by providing services like insurance, savings accounts, and credit (Mhlanga 2023). Echoing that, Carè et al. (2024) findings show that mobile money and blockchain can improve financial access in developing nations, and digital banking can lower transaction costs and increase access to credit. Corroborating the same argument, Lu (2024) agrees that the fintech revolution has greatly benefited financial institutions and their customers by lowering service costs, increasing operational effectiveness, and making premium product options possible with AI and blockchain. Furthermore, for financial institutions, AI presents a great opportunity for new credit scoring models, using alternative data sources to improve access to credit for unbanked populations (Tigges et al. 2024). This presents a great opportunity for individuals without traditional banking records. Likewise, Saravanakrishnan et al. (2024) assert that by delivering alternative financial services, including lending, borrowing, and trading directly on blockchain networks, Decentralised Finance (DeFi) is upending established banking structures and achieving impressive results in reaching previously underserved and unbanked people. Moreover, blockchain technology can be used to reduce fraud and boost trust in financial institutions by providing transparent and safe transaction records (Rosati and Čuk 2019). Additionally, a study by Mhlanga (2023) established that if blockchain technology can be effectively applied to increase financial inclusion, sustainable development can be guaranteed on numerous levels.

4.2 Challenges

Despite the opportunities that AI, digital finance, and blockchain offer, there are also many challenges in their use and implementation. For example, AI-based financial solutions need user data, which raises worries regarding data misuse in terms of data privacy and security. Supporting this argument, Lu (2024) states that in terms of data security, consumer protection, and financial stability, fintech presents several legal, moral, and regulatory issues for regulators and financial authorities. With respect to regulatory barriers, regulations in different nations are impeding the use of blockchain and AI in financial services (Bowers et al. 2019). Liu et al. (2024) add that financial regulators frequently concentrate on reducing risks in the financial services industry and might not give enough consideration to the negative consequences of their policies.

Another challenge derives from technological barriers in many regions of the developing world. To that effect, many unbanked people find it difficult to use digital financial services due to limited access to technology and low levels of digital literacy in rural and underserved locations (Msweli and Mawela 2021). Scalability and cost of fintech are factors not to be ignored either, as the high implementation costs of blockchain and AI solutions present a problem, particularly in areas with low incomes. Furthermore, Modiba et al.'s findings emphasised a wide range of obstacles that hinder rural residents' ability to engage in the digital economy, including the availability of a steady supply of electricity, inadequate or non-existent infrastructure needed to access digital goods, and gender-related problems where women's gendered roles prevent them from using digital goods.

5 Way Forward

5.1 Policy and Regulatory Frameworks

Regarding policy and regulations, many reviewed papers emphasised the need for creating thorough regulatory frameworks to handle the operational, security, and privacy risks related to blockchain and AI to balance innovation and security; and governments and regulatory bodies worldwide have been responding to that. It has been reported that global think-tanks, policymakers, and IT giants are becoming more eager to promote agendas that support AI for the Planet or AI for Good (Galaz 2024). However, Lu (2024) contends that, when it comes to data security, consumer protection, and financial stability, fintech has presented legislators and financial authorities with several legal, moral, and regulatory issues. To address these challenges, Monye (2022) argued that achieving comprehensive, accessible, and sustainable financial inclusion requires legislative reforms, along with identifying gaps in the institutional and legal frameworks for digital financial services and AI, as well as the barriers that contribute to financial exclusion.

5.2 *Infrastructure Development and Digital Literacy Programmes*

Okello et al. (2024) advocate that governments in developing countries use research findings to advocate for the design of appropriate digital infrastructures to reach remote areas and ensure the 'last mile' connectivity for users of digital financial services. They add that countries like Uganda should support women, youth, and other vulnerable groups—especially in rural communities—to understand and use fintechs. This can be achieved through digital literacy initiatives that encourage the adoption of digital financial services and the use of platforms such as mobile money. Additionally, a suggestion under consideration is to explore the potential of using blockchain technology to update—or even replace—the current infrastructure for international payments and remittances, such as correspondent banking, ensuring that transactions can be validated and recorded on a distributed ledger (Mhlanga 2023).

5.3 *Partnerships*

Some of the reviewed papers proposed different types of partnerships that can be formed to enable better financial inclusion for underserved populations using AI and blockchain in secure environments. These partnerships could be private–private, private–public, private–NGOs, or public–NGOs partnerships. Ravikumar et al. (2024) report that digital lending start-ups use AI-powered alternative scoring methods and personalised digital loans to help the underbanked and unbanked population. Considering that, one would imagine that a partnership among fintech start-ups, traditional financial institutions, mobile money providers, and NGOs, for example, would benefit even more the underserved populations. In Nigeria, a study by Ayo et al. (2011) proposed fostering collaborations between governments, financial institutions, and technology companies to develop and deploy inclusive financial solutions that integrates both account-based and electronic money systems. This concept, which uses a two-factor authentication system using a phone and a national ID card, would assist to accelerate the adoption of mobile-to-ATM, bank-to-ATM, and ATM-to-bank money transfers in Nigeria and improve access to cash (ibid).

6 Future Research

Digitalisation, blockchain, and AI have enormous potential to promote financial inclusion and economic growth by delivering financial services to unbanked people, especially in developing nations. Even though there has been significant progress, there are still several gaps that require further study, particularly in areas like Africa where a high percentage of the population do not have bank accounts and where financial inclusion is vital.

6.1 Solutions for African Context

Localised AI models: future studies ought to concentrate on creating AI models that are specific to the socio-economic and cultural circumstances of African nations. This involves using alternative data sources (such as utility payments and mobile phone usage) for financial assessments and credit scoring, which are especially important in areas with limited access to traditional banking data. Additionally, there is blockchain for remittances and cross-border payments. Because many African nations rely heavily on remittances, research should examine how blockchain may be optimised to lower transaction costs and boost cross-border payment efficiency, both of which are essential for financial inclusion.

6.2 Addressing Infrastructure and Digital Literacy Gaps

Infrastructure Development: to enhance internet connectivity and digital infrastructure in underserved and rural areas, research should investigate scalable and affordable solutions. Examining how public–private partnerships may increase connectivity and digital services should be part of this. Future research should further concentrate on creating and assessing digital literacy initiatives that address the unique requirements of unbanked groups, especially women and the elderly, who frequently encounter more obstacles when trying to use digital financial services.

6.3 Regulatory and Policy Frameworks

Harmonised regulatory approaches: especially regarding AI and blockchain technologies, research should examine the creation of harmonised regulatory frameworks that strike a compromise between innovation and consumer protection. Building confidence and guaranteeing the secure uptake of digital financial services depend on this.

Furthermore, for policy impact assessments, to facilitate the use of AI and blockchain technology, future research should deeply evaluate how current regulations affect financial inclusion and pinpoint areas in which regulatory changes are required.

6.4 Sustainability and Scalability

Sustainability and long-term impact: future research should assess the long-term viability of digital financial services as well as their influence on economic growth, specifically regarding the creation of jobs and the alleviation of poverty. Likewise, with regard to scalable AI and blockchain solutions, research ought to concentrate on exploring the creation of scalable blockchain and AI solutions that can be implemented in various geographical locations with differing degrees of technological preparedness. This involves investigating open-source, inexpensive technologies that are readily adaptable to regional settings.

6.5 Partnerships and Collaborations

Public–private partnerships: studies should investigate how these collaborations promote financial inclusion, especially when it comes to using the advantages of both industries to create and implement inclusive financial products. Similarly, cross-sector collaboration should be emphasised. Future research should examine how non-governmental organisations (NGOs), fintech companies, and traditional financial institutions might work together to speed up the adoption of digital financial services among unbanked populations.

6.6 Success and Impact Measurement

Success stories and best practices: to give policymakers and practitioners useful information, future research should compile and examine success stories and best practices from various locations, especially from Africa, South Asia, and South

America. Moreover, impact metrics should be considered. Research should concentrate on creating reliable metrics to assess how digitalisation, blockchain, and AI affect financial inclusion, especially regarding insurance, savings, and credit availability. This comprises studies that follow the development of financial inclusion programmes over an extended period.

7 Conclusion

The review results indicate that Africa has been at the forefront of research on financial inclusion from 2011 to 2024, with the greatest focus in 2022 and 2024. This could be since this continent suffers from the lowest financial inclusion among all continents (Chitimira and Ncube 2023; Gwala and Mashau 2024; Mhlanga 2023; Msweli and Mawela 2021). In Africa, mobile money services have a particularly significant impact; for instance, countries such as Somalia and Uganda have adopted and heavily relied on them for everyday transactions and remittances (Elmi and Ngwenyama 2019). Asia and Latin America have shown minimal research activity on the topic compared to Africa, which is surprising given that they rank next after Africa in having the lowest financial inclusion rates. It has been reported that following demonetisation, digital payments and mobile banking have grown significantly in India similar to trends in Africa, with fintechs playing a crucial role in advancing financial inclusion (Pal et al. 2020). Europe has only reported research focus on financial inclusion between 2021 and 2024, which is understandable since this has not been an issue of concern there, compared to the other regions.

In conclusion, future research should focus on addressing the specific challenges faced by unbanked populations in the developing world, particularly in Africa, through the adoption of AI, blockchain, and digitalisation, which hold transformative potential for financial inclusion. By prioritising context-specific solutions, infrastructure development, regulatory frameworks, and ethical considerations, researchers can help maximise the impact of these technologies on financial inclusion and economic development in the region.

Acknowledgements The authors acknowledge the use of OpenAI's ChatGPT, Google Gemini, and Deepseek for assistance in the analysis of more complicated plots from Bibliometrix analysis, such as 'co-occurrence of keywords' and generating insights from them for some of the discussion in this paper. AI tools, such as large language models (LLMs), may have biases due to the vast datasets on which they were trained. These biases include data bias, which may over-represent certain geographical contexts or cultural norms; representational bias, which may perpetuate societal stereotypes about gender roles; and algorithmic/confirmation bias, which can limit the exploration of novel findings. To mitigate these biases, AI tools are used for data organisation, thematic identification, and initial drafting, rather than independent writing or critical analysis.

References

Ayo, C.K., Adewoye, J.O., Oni, A.A.: Framework for mobile money implementation in Nigeria. J. Afr. Res. Bus. Technol. **2011**(1), 1–8 (2011)

Bowers, J., Reaves, B., Sherman, I., Traynor, P., Butler, K.: Regulators, mount up! Analysis of privacy policies for mobile money services. In: Proceedings of the Thirteenth Symposium on Usable Privacy and Security (SOUPS), 97–114 (2019)

Carè, R., Boitan, I.A., Stoian, A.M., Fatima, R.: Exploring the landscape of financial inclusion through the lens of financial technologies: A review. Financ. Res. Lett. **58**, 106500 (2024)

Chitimira, H., Ncube, M.: The prospects and challenges for mobile money regulation and the promotion of financial inclusion in Zimbabwe. In: Ius Gentium, 106, 125–140. https://doi.org/10.1007/978-3-031-23863-5_6 (2023)

Elmi, M., Ngwenyama, O.: The role of mobile money in Somalia's remittance system. In: Lecture Notes in Computer Science, Springer, Cham (2019)

Galaz, V.: Dark machines: How Artificial Intelligence, digitalization and automation is changing our living planet. Routledge, London (2024). https://doi.org/10.4324/9781003317814

Garcia, A.C.B., Garcia, M.G.P., Rigobon, R.: Algorithmic discrimination in the credit domain: What do we know about it? AI & Soc. **39**(4), 2059–2098 (2024). https://doi.org/10.1007/s00146-023-01676-3

Gwala, R.S., Mashau, P.: Digitalisation of banking as a catalyst for inter-country money transfers in Africa. In: Examining the Rapid Advance of Digital Technology in Africa, 162–184 (2024). https://doi.org/10.4018/978-1-6684-9962-7.ch009

Hoang, M.D., Le, L., Nguyen, A.T., Le, T., Nguyen, H.D.: Federated Artificial Intelligence for unified credit assessment. In: International Conference on Human-Computer Interaction, LNCS, 44–56. Springer, Cham (2020)

Liu, Y., Li, X., Zheng, Z.E.: Consequences of China's 2018 online lending regulation and the promise of PolicyTech. Inf. Syst. Res. **35**(3), 1235–1256 (2024). https://doi.org/10.1287/isre.2021.0580

Lu, L.: Global Fintech revolution: Practice, policy, and regulation. Oxford University Press, Oxford (2024)

Mhlanga, D.: Blockchain technology for digital financial inclusion in Industry 4.0 towards sustainable development. Frontiers in Blockchain 6, 1035405 (2023). https://doi.org/10.3389/fbloc.2023.1035405

Monye, O.: Digital financial inclusion and regulation (2022). https://www.scopus.com/inward/record.uri?eid=2-s2.0-85193369338

Msweli, N.T., Mawela, T.: Financial inclusion of the elderly: Exploring the role of mobile banking adoption. Acta Informatica Pragensia **10**(1), 1–21 (2021). https://doi.org/10.18267/J.AIP.143

Muñoz-Cancino, R., Bravo, C., Ríos, S.A., Graña, M.: On the combination of graph data for assessing thin-file borrowers' creditworthiness. Expert Syst. Appl. **213**, 118809 (2023). https://doi.org/10.1016/j.eswa.2022.118809

Mwila, R., Kunda, D.: Predictive model to determine the growth of mobile money transactions in Zambia using data mining techniques. Int. J. Data Min. Bioinform. **27**(1–3), 139–170 (2022). https://www.inderscienceonline.com/doi/abs/10.1504/IJDMB.2022.130330

Okello, C.B.G., Akol, M.C., Manzi, A.A., Balinda, R.: Recalibrating the scope of financial inclusion through financial technologies in the digital age: The role of digital literacy as a moderator in rural Uganda. Information Technology and People (2024)

Pal, A., De', R., Herath, T.: The role of mobile payment technology in sustainable and human-centric development: Evidence from the post-demonetization period in India. Information Systems Frontiers **22**(3), 607–631 (2020). https://doi.org/10.1007/s10796-020-09982-7

Ravikumar, T., Murugan, N., Suhashini, J.: Role of digital lending in serving financially excluded individuals and small medium businesses in India. In: Perspectives in Finance and Digital Transformations in Business, 211–217 (2024). https://www.researchgate.net/publication/382011158_Role_of_Digital_Lending_in_Serving_Financially_Excluded_Individuals_and_Small_Medium_Businesses_in_India

Rosati, P., Čuk, T.: Blockchain beyond cryptocurrencies. In: Palgrave Studies in Digital Business and Enabling Technologies, pp. 149–170. Springer, Cham (2019). https://doi.org/10.1007/978-3-030-02330-0_10

Saravanakrishnan, V., Nandhini, M., Palanivelu, P.: DeFi's transformative influence on the global financial landscape. In: Digital Technologies, Ethics, and Decentralization in the Digital Era, 99–120 (2024). https://www.igi-global.com/gateway/chapter/338868

Tigges, M., Mestwerdt, S., Tschirner, S., Mauer, R.: Who gets the money? A qualitative analysis of fintech lending and credit scoring through the adoption of AI and alternative data. Technol. Forecast. Soc. Change **205** (2024). https://doi.org/10.1016/j.techfore.2024.123491

Beyond the Feed: How Influencer Self-Disclosure Transforms Followers into Customers

Marina P. S. Ngobeni, Alyssa Shawntay Williams, and Nadine Oosthuizen

Abstract In the digital marketing era, social media influencers have emerged as new storytellers, incorporating personal stories into their content to connect with their followers. This paper investigates the role of influencer self-disclosure in fostering parasocial relationships (PSRs) with influencers and its subsequent impact on purchase intentions (PIs) among young adults in South Africa. The study demonstrates that influencers who engage in self-disclosure, such as sharing personal opinions, emotions, and daily experiences, can create a sense of intimacy, thereby strengthening the PSR. Using a quantitative approach, online and paper-based questionnaires were used to collect data from 311 participants, and structural equation modelling (SEM) was employed for hypothesis testing. The results revealed that influencer self-disclosure significantly enhanced PSR ($\beta = 0.42$, $p < 0.01$), which in turn had a positive impact on PI ($\beta = 0.67$, $p < 0.01$). These results suggest that followers are more likely to develop emotional connections with influencers who they believe disclose personal information, increasing their PSR and purchase intention. The study contributes to the influencer marketing literature by highlighting the effectiveness of self-disclosure in creating and strengthening bonds between influencers and followers. Moreover, this study has practical implications for both organisations and influencers to leverage self-disclosure as a strategic tool to enhance the effectiveness of influencer marketing campaigns.

Keywords Influencers · Parasocial relationships · Self-Disclosure · Purchase intentions · Influencer marketing

M. P. S. Ngobeni
Vaal University of Technology, Vanderbijlpark, South Africa
e-mail: marina.ngobeni@eduvos.com

A. S. Williams (✉) · N. Oosthuizen
Rhodes University, Makhanda (Grahamstown), South Africa
e-mail: a.williams@ru.ac.za

N. Oosthuizen
e-mail: n.oosthuizen@ru.ac.za

E. H. Redda (ed.), *Proceedings of the Social Sciences International Research Conference, 2025*, https://doi.org/10.1007/978-981-95-9489-4_20

1 Introduction

The proliferation of social media usage has increased consumer engagement, thus opening doors for endorsement partnerships between organisations and influencers (Koay et al. 2023). Influencers are individuals who have built a substantial following on social media platforms and possess the ability to affect the purchasing decisions of others because of their authority, knowledge, position, or relationship with their audience (De Veirman et al. 2017). Social media platforms have allowed individuals to share personal content more widely than ever (Leite and Baptista 2022; Zhang and Lu 2023). As a result, influencers engage in various self-disclosure activities, including sharing biographical information, emotions, thoughts, desires, values, and beliefs with others on social media platforms (Leite et al. 2024). Influencers gain and retain followers by showcasing their lifestyles, sharing their experiences, and expressing opinions about various products and services. While Instagram has emerged as the dominant platform for influencer marketing (Chen, 2024), other social media platforms such as TikTok, YouTube, and Facebook are considered conduits for influencers, thus making influencer marketing a multibillion-dollar industry with over 23 000 organisations partnering with influencers worldwide (Garg and Bakshi 2024). Influencers share rapid and widespread (viral) content, such as status updates, images, and videos, on social media, which allows them to influence followers' opinions of brands, products, and services (Listiawati et al. 2024). They can be seen as opinion leaders because of their personality, knowledge in a particular field, relatability, and relationships with their followers (Abell and Biswas 2023).

Earlier research has revealed that influencers' personal stories can emotionally impact followers, which can affect their intentions to purchase products endorsed by influencers (Koay 2023; Leite et al. 2024). Through such self-disclosure, followers can develop parasocial relationships (PSRs) with influencers, giving them the illusion of a personal relationship with influencers. Meanwhile, Liu and Zheng (2024) attest that followers are more likely to purchase products endorsed by influencers with whom they have developed a PSR. However, there is a dearth of research into the mediating role of PSRs in the relationship between influencer self-disclosure and purchase intentions (PIs). The importance of research into consumers' PI has been confirmed by several scholars who agree that understanding PI may lead to increased return on investment for marketing activities such as influencer marketing campaigns (Lee and Lee 2022; Chen 2024; Gao et al. 2024). Therefore, this study contributes to the existing literature by examining the impact of influencer self-disclosure on PSR with influencers and PI. This study also contributes to organisations' attainment of influencer marketing goals by recommending strategies for leveraging followers' PSR with influencers to increase their PI.

2 Literature Review

2.1 Risk Defined

This literature review examines the key constructs that underpin the relationship between influencer marketing and consumer behaviour, specifically focusing on influencer self-disclosure, PSRs, and purchase intention among young adult consumers.

2.2 Theoretical Foundation

The study of influencer marketing and consumer behaviour is anchored in several foundational theories. Social Cognitive Theory (Bandura 1986) provides insight into how individuals learn and adopt behaviours through observing role models, making it particularly relevant for understanding influencer-follower dynamics. The theory's emphasis on observational learning, self-efficacy, and reciprocal determinism helps explain how influencers shape consumer attitudes and behaviours (Gleason et al. 2017; Milaković 2021).

Media Dependency Theory (Ball-Rokeach and DeFleur 1976) further explains the intensity of relationships between audiences and media figures, suggesting that dependency increases when media fulfil important goals and when alternative sources are limited (Patwardhan and Yang 2003). This theoretical framework is particularly relevant for understanding the development of strong PSRs with influencers (Kim 2020).

The Theory of Planned Behaviour suggests that PIs are determined by attitudes towards the behaviour, subjective norms, and perceived behavioural control (Ajzen 1991). In the influencer marketing context, these components translate to consumer attitudes towards influencer-endorsed products, social pressures from peer groups and influencer communities, and perceived ability to make purchases.

2.3 Influencer Self-Disclosure

Influencer self-disclosure refers to the voluntary sharing of personal information with followers (Zhang and Lu 2023; Leite et al. 2024). Influencers typically use self-disclosure to build trust, commitment and intimacy with their followers (Chung and Cho 2017; Leite and Baptista 2022). In addition to sharing product reviews and recommendations, influencers provide their followers with a wide range of topics relating to their personal lives, such as details about their family, friends, professional life, moods, feelings, and opinions (Leite and Baptista 2022). Such self-disclosure builds trust, intimacy, and commitment and proves their authenticity (Chung and

Cho 2017; Koay et al. 2023). Moreover, self-disclosure can be embedded in influencers' content as a strategy to develop organisations' brands and attract followers (Leite et al. 2024). Strategic self-disclosure allows influencers to differentiate themselves in a crowded marketplace while maintaining authenticity (Leite et al., 2024). However, research also highlights the risks of excessive self-disclosure, including privacy concerns, potential backlash, and audience fatigue (Duffy and Hund 2015; Koay et al. 2023). By fostering a friendly and realistic environment, influencers can influence followers' feelings about themselves and the brands they endorse (Zhang and Lu 2023). This discussion highlights the essential role of self-disclosure in establishing and developing relationships with followers. However, there is a paucity of literature directly linking self-disclosure to PSR. This gap justifies the introduction of self-disclosure as an influencer characteristic that can influence PSR with influencers in this research. Therefore, this study formulated and tested the following hypothesis:

H1: There is a significant relationship between influencer self-disclosure and PSR among young adults.

2.4 *Parasocial Relationships with Influencers*

PSR refers to a one-sided relationship where a follower develops an illusion of intimacy between them and an influencer they follow (Liu et al. 2023). Unlike traditional media, social media platforms enable more interactive and seemingly personal communication, intensifying the potential for PSR development (Chung and Cho 2017). PSR with influencers can provide a sense of social proof and trust, making followers more likely to consider the recommendations of influencers. Moreover, the likelihood of followers purchasing an influencer-endorsed product is typically higher among influencers to whom followers are attached (Ju and Lou 2022). Therefore, the present study proposes that PSR is a pertinent indicator of PI.

Marketing scholars have shown interest in the role of PSR in shaping PI. For example, Taher et al. (2022) revealed that strong PSR positively influences the PI of beauty products. In addition, various surveys have been conducted to support this further, involving participants from different industries and contexts. These studies include the work of Hwang and Zhang (2018) with Chinese individuals aged 18 to 40 years who follow celebrities on social media, Sokolova and Kefi's (2020) research on beauty bloggers and influencers on YouTube, and Lee and Lee's (2022) survey of Korean women in their 20 s who watch beauty vlogs on YouTube, confirming the significant impact of PSR on PI. Considering these results, this study offers a South African context by examining the development of PSR between influencers and young adults of South Africa. Therefore, the following hypothesis was formulated and tested:

H2: There is a significant relationship between PSR and PI among young adults.

2.5 Purchase Intention

PI refers to the planned action in which consumers desire to purchase a product or service (Chung and Cho 2017; Gao et al. 2024). Several scholars agree that PI is complex and is frequently explained in terms of consumer behaviour, perceptions, and attitudes (Kotler et al. 2017; Yones and Muthaiyah 2023; Chen 2024). Scholars have highlighted that PI are developed due to normative, external, and psychological influences that a consumer may or may not be aware of. Although numerous factors can influence a consumer's decision to purchase, PI is a valuable tool for predicting purchasing behaviour (Chen 2024).

In this study, PI is defined as consumers' willingness to purchase a product after receiving the desired product/service exposure from an influencer on social media. The stronger an individual's intention to engage in a specific behaviour, the higher the possibility they may act upon it (Yones and Muthaiyah 2023). This implies that a consumer's PI can be transformed into actual purchase activity based on their experiences at a particular time. Therefore, organisations must understand how consumers evaluate product/service decisions to present the desired value proposition and increase brand awareness and stimulate PI (Chen 2024).

3 Methodology

This section outlines the procedure used to collect and analyse the data for this study.

3.1 Research Design

A well-planned research design is essential for ensuring that a research project is executed efficiently, producing valid and reliable results (Hair et al. 2021). This study used a descriptive research design, which facilitated the exploration of the five key questions concerning the study's who, what, when, where, and how aspects, as outlined by Zikmund et al. (2019). Accordingly, the sample consisted of university students (who), focusing on identifying specific characteristics of influencers (what). Furthermore, the research was conducted in May 2023 (when), within the geographic context of the Eastern Cape (where), employing a survey strategy (how) as specified within the research design.

3.2 Sampling Method

The target population for this study were young adults between 18 and 25 years old. This study employed a non-probability sampling technique, in particular, convenience sampling was used to draw respondents from a university within the researcher's residence. The decision to employ the convenience sampling procedure was based on the researcher's location, time, and budget constraints, making it impractical for the researcher to collect data across the country. While convenience sampling is prone to sampling bias (Malhotra et al. 2017), the researcher attempted to eliminate sampling biases by establishing inclusion criteria for the study: the participants had to be between the ages of 18 and 25 years, have active social media accounts and had engaged with influencer content in the last six months. In addition, the researchers had no direct contact with the participants and employed a third party to distribute the questionnaire to students across all faculties at the selected institution.

3.3 Research Instrument

This study collected primary data using a survey research strategy. The survey strategy was selected for several reasons, including its cost-effectiveness, efficiency, standardisation, time efficiency, and flexibility. Consistent with the survey research strategy (Malhotra et al. 2017), this study employed a structured questionnaire as the primary data collection instrument. Specifically, an in-person and online, self-administered questionnaire was deemed appropriate for empirical investigation and the testing of relationships among the variables of this study. The questionnaire for this study comprised structured questions with closed-ended responses. Structured questions were well-suited as respondents' responses could be quantified and adapted into numerical data that could be summarised and analysed using statistical techniques (Malhotra et al. 2017).

For this study, questions in Sections A, B, and C were developed by the researcher in line with the current study's objectives, while scale items in Sections D, E, and F were adapted from previous scholars who reported high internal consistency among scale items. This was observed by Cronbach's alpha (CA) coefficient values ranging between 0.7 and 0.95 (Malhotra et al. 2017). Section A comprised questions relating to the study's inclusion criterion, while Section B elicited information about the demographic profile of the respondents. Section C comprised questions about respondents' engagement with influencers. Furthermore, Section D comprised indicator statements that sought to elicit information about the influencer's effectiveness characteristics: influencer credibility, influencer attractiveness, influencer self-disclosure, and influencer-product congruity. Section E canvassed information on respondents' PSR with the influencers they follow, while Section F pertained to respondents' PI.

3.4 Data Collection

Initially, an invitation to participate was sent to the students at the selected university via their student email addresses. The email included an invitation, a consent form, and a link to the Google Forms questionnaire. To promote the study, posters were placed around the selected campus, which enabled students to read about the study and then scan a QR code for the questionnaire. In addition, paper-based questionnaires were distributed to students in dining halls with the assistance of a field worker.

3.5 Data Analysis

Once data had been collected, it was captured on a Microsoft Excel spreadsheet to facilitate initial data preparation, which involved data cleaning, editing, and coding. Subsequently, data were analysed using the Statistical Package for the Social Sciences (SPSS) and Analysis of Moment Structures (AMOS). Quantitative data analysis methods, such as descriptive and inferential statistics, were adopted. Descriptive statistics help show the general pattern of participants' responses for each scaled variable (Brown et al. 2018). The descriptive statistics used in this study included percentages, frequency distribution, measures of central tendency (mean, mode, median), and measures of variability (standard deviation). These descriptive statistics helped summarise the sample's demographic characteristics and influencer engagement information. Moreover, descriptive statistics were used to summarise and describe the responses from the self-administered questionnaire.

Furthermore, inferential statistics entails drawing conclusions and generalising results about a population based on sample analysis (Malhotra et al., 2017). In this study, Pearson product-moment correlation and multiple regression analysis were used to analyse data and determine whether there is a correlation between the study's variables. Furthermore, structural equation modelling (SEM) was used to test the proposed hypotheses and transform the hypothesised model into a validated model (Hair et al. 2019).

4 Results

This section outlines the results obtained from empirical research. In particular, the section discusses the sample descriptive statistics, validity and reliability analysis, correlation analysis, and SEM results.

4.1 Descriptive Statistics

Three hundred and eleven (311) young adults participated in this study. A majority of the participants were undergraduate students (n = 266) between the ages of 18 and 21 years (n = 206). This cohort uses social media to draw lifestyle inspiration and as a connection for social interaction (Lim et al. 2022). In addition, most of the participants were female (n = 234), which corresponds with Lokithasan's et al. (2019) assertion that women are more prone to following and engaging with social media influencers. Moreover, influencers are more prevalent in creating beauty, fashion, and lifestyle content, which women tend to follow. Furthermore, the language and ethnic distribution indicated a significant black (n = 275) and Xhosa speaking (n = 110) majority.

To gain deeper insight into the impact of influencer self-disclosure on PSR and PI, it was essential to explore where, how, and why young adults engage with influencers. While TikTok emerged as the most frequented platform for general use, Instagram stood out as the primary space for influencer engagement. This suggests that TikTok is favoured for entertainment, whereas Instagram is the leading platform for product endorsements and influencer–follower interactions. Consequently, Instagram's formats, such as posts, reels, and stories, are particularly well-suited for endorsement content, enabling self-disclosure and fostering the development of PSR. Furthermore, the inclination towards lifestyle content highlights the study's focus on self-disclosure. Lifestyle content includes self-disclosure through influencers' daily routines, emotional narratives, and behind-the-scenes content. These results show that respondents engage with influencers who share personal anecdotes (Leite et al. 2024), which explains the relationship between influencer self-disclosure and PSR.

In addition, respondents were asked to indicate the information they sought from influencers. The results suggest that respondents sought product options, quality, and usage information. This aligns with the notion that followers consider influencers as valuable sources of product information (Abell and Biswas 2023). The influencers' PI are influenced by their authentic opinion leadership, which is regarded as more credible than traditional advertisements (De Veirman et al. 2017; Lou and Yuan 2019). The demographic profile of this study offers valuable insight into the consumer segment targeted by influencer marketing campaigns. The young adult cohort is considered early adopters of mobile technologies and is the first generation to grow up with smartphones and social media (Rosenberg et al., 2025). Moreover, this generation spends the most time on social media networks, thus giving them more access to influencer marketing content (Eastman et al. 2021). Organisations need to be aware of their consumption patterns as they are considered the largest group of prospective income earners and are approaching the age of making their own purchase decisions.

4.2 Validity and Reliability of the Measuring Instrument

The study's independent variable, influencer self-disclosure, demonstrated strong reliability with a CA coefficient of 0.879 and factor loadings between 0.668 and 0.836. The measurement model also indicated satisfactory fit indices with acceptable absolute fit indices (CMIN/DF = 3.449, SRMR = 0.038, RMSEA = 0.090) and incremental fit indices (IFI = 0.980, NFI = 0.972, CFI = 0.979). Similarly, the PSR variable demonstrated reliability, with a CA coefficient of 0.88. Although one item (D27 – 'Influencer's social media profile reveals his/her lifestyle preferences, i.e. activities, restaurants, etc.') had a factor loading slightly below the threshold, it was retained based on overall fit indices. (CMIN/DF = 3.70, SRMR = 0.05, RMSEA = 0.09) and incremental fit indices (IFI = 0.90). For PI, despite two items (F4: *I consider purchasing a product/ service that influencer endorses because he/ she is credible*) and (F7: *I consider purchasing a product/ service that influencer endorses because he/ she is a good match for the product/ service*) with relatively low factor loadings, they were retained for theoretical relevance and did not disrupt the factor structure.

PI exhibited strong reliability with a CA coefficient of 0.89. The measurement model demonstrated adequate fit indices (CMIN/DF = 3.75, SRMR = 0.06, RMSEA = 0.09) and incremental fit indices (IFI = 0.95, NFI = 0.93, CFI = 0.95). Ultimately, the validity and reliability of the study's measurement model were considered acceptable due to the careful methodological procedure and strong CA coefficient values.

4.3 Correlation Analysis

This study employed Pearson's Product-Moment Correlation to assess the strength and direction of linear relationships between variables (Hair et al. 2019; Malhotra et al. 2017). The significant positive correlations are shown in Table 1.

The correlation analysis provided valuable insight into the relationships among the variables of this study (Malhotra et al. 2017). Table 1 demonstrates that all

Table 1 Pearson's correlation results

Correlation analysis			
	Self-disclosure	Parasocial relationships	Purchase intentions
	1	0.416**	0.365**
		1	0.666**
			1

**Correlation is significant at the 0.01 level (2-tailed)
*Correlation is significant at the 0.05 level (2-tailed)

correlation estimates were significant at $p < 0.01$, and a linear relationship exists between the study's variables. The strongest correlation was between the intervening and dependent variables, PSR and PI ($r = 0.67$; $p < 0.01$). These results highlight the importance and effectiveness of influencers in stimulating PSR and PI.

4.4 Structural Equation Modelling

In this study, testing the indirect structural model was necessary as the model includes an intervening variable between the independent and the dependent variables (Hair et al. 2019). Therefore, the independent (self-disclosure), intervening (PI), and dependent variables (PI) were considered for the hypothesised model to be tested during the SEM procedure.

After running the hypothesised model, it became evident that it was suitable as it yielded p-values less than 0.01. When evaluating the outcomes of the structural model and the subsequent testing of hypotheses, three statistics were observed for this study: the direction and strength of the Beta coefficients, the statistical significances represented by the critical ratios (CR) and p-value, and the square multiple correlations (SMC) values (Malhotra et al. 2017; Hair et al. 2019). While the standardised path coefficients (β) indicate the strength and direction of the relationships between the variables in this study, the CRs are used to determine the significance of the relationship at the significance level of 0.01. The results of the structural model and the hypotheses testing results are presented in Table 2.

The results of the structural model presented in Table 2 indicate a certain degree of influence of the independent variable on the intervening variable ($\beta = 0.084$, $CR = 2.603$). Furthermore, the SMC values validate the relationships between the variables investigated in this study. Specifically, the SMC value of 0.83 suggests that the independent variable contributes towards explaining 83 per cent of the variance in overall PSR with influencers. Meanwhile, the SMC value 0.70 (70% of the variance) suggests that PSR yields good explanatory power in stimulating PI among young adults. These results imply that outstanding variances can be explained by other factors that were not considered in the current study.

Table 2 Results from the structural model

Path	Beta Coeffi-cient (β)	Critical Rario (CR)	Square Multiple Correlations (SMC)	Signifi-cance ($p < 0.01$)	Decision
PSR < Self-disclosure	0.084	2.603	0.827	***	Accept H^1
PI < PSR	0.836	7.248	0.699	***	Accept H^2

5 Practical Managerial Implications and Recommendations

The study's results show that influencer self-disclosure has a statistically significant, positive relationship with the intervening variable, PSR. This implies that the more influencers self-disclose, the more likely followers are to develop PSR with influencers. This is supported by existing literature (Farivar et al., 2021; Koay et al. 2023), which proposes that influencer self-disclosure plays an integral role in developing PSR. Influencer self-disclosure tends to elicit engagement from followers; as followers become more invested in the influencer due to their self-disclosure, they may develop emotional attachments. Moreover, Farivar et al. (2021) attest that influencers' sharing of their thoughts, emotions, and experiences can provide a sense of intimacy that enhances followers' PSR with influencers.

Based on these results, the study recommends that organisations encourage influencers to integrate self-disclosure in their content through storytelling using words, images, videos, and other media to convey an engaging narrative that invokes emotional or intellectual responses from followers. Storytelling has been shown to play an essential role in building deeper connections with audiences. Therefore, influencers can share their feelings, opinions, or anecdotes related to the products they endorse. For example, a gaming influencer can share a live stream video and talk about their experience and game aspects while playing. Incorporating authentic self-disclosure into content can strengthen followers' PSR relationships with influencers as they may relate to them personally.

Additionally, while self-disclosure is valuable in enhancing followers' PSR with influencers, organisations should ensure that influencer self-disclosure is relevant and consistent. Organisations should encourage their influencers to engage in self-disclosure activities pertinent to the products they endorse, as irrelevant self-disclosure may not contribute to PSR. For example, a skincare influencer shares before and after pictures, using the endorsed products to describe their skincare routine and sharing challenges and benefits observed. Moreover, since followers are likely to form PSR with influencers who self-disclose, influencers must maintain consistency in their self-disclosure practices and engage with their followers about their self-disclosure by responding to comments and participating in discussions.

6 Conclusion

Through the literature and empirical results, this study contributes to the body of knowledge on influencer marketing by providing a holistic view of PSR for PI in the South African context, thus bridging the gap in South African literature. Furthermore, this study offers a viewpoint on the role of influencers in stimulating PI among their followers. Practically, this study contributes to organisations' attainment of influencer marketing goals by recommending strategies for leveraging followers' PSR with influencers to increase PI. Organisations can refine the influencer selection criteria by identifying influencer characteristics that contribute to the development of PSR and subsequently enhance PI. In addition, organisations can encourage the selected influencers to optimise their content by emphasising self-disclosure to reinforce PSR.

References

Abell, A., Biswas, D.: Digital engagement on social media: how food image content influences social media and influencer marketing outcomes. J. Interact. Mark. **58**(1), 1–15 (2023)

Ajzen, I.: The theory of planned behavior. Organ. Behav. Hum. Decis. Process. **50**(2), 179–211 (1991). https://doi.org/10.1016/0749-5978(91)90020-T

Ball-Rokeach, S.J., DeFleur, M.L.: A dependency model of mass-media effects. Commun. Res. **3**(1), 3–21 (1976)

Bandura, A.: Social foundations of thought and action: A social cognitive theory. Prentice-Hall, Englewood Cliffs (1986)

Brown, T.J., Suter, T.A., Churchill, G.A.: Basic marketing research, customer insights, and managerial action, 9th edn. Cengage Learning, Boston (2018)

Chen, S.: The impact of brand awareness on purchase intention. Highl. Bus. Econ. Manag. **30**(1), 239–246 (2024)

Chung, S., Cho, H.: Fostering parasocial relationships with celebrities on social media: implications for celebrity endorsement. Psychol. Mark. **34**(4), 481–495 (2017)

De Veirman, M., Cauberghe, V., Hudders, L.: Marketing through Instagram influencers: the impact of number of followers and product divergence on brand attitude. Int. J. Advert. **36**(5), 798–828 (2017)

Duffy, B.E., Hund, E.: 'Having it all' on social media: entrepreneurial femininity and self-branding among fashion bloggers. Soc. Media Soc. **1**(2), 1–11 (2015)

Eastman, J.K., Iyer, R., Eastman, K.L., Gordon-Wilson, S., Modi, P.: Reaching the price-conscious consumer: the impact of personality, generational cohort and social media use. J. Consum. Behav. **20**(4), 898–912 (2021)

Farivar, S., Wang, F., Yuan, Y.: Opinion leadership vs. parasocial relationship: key factors in IM. J. Retail. Consum. Serv. **59**, 102371 (2021)

Gao, Y., Yuan, S., Lei, X.: How livestreamers trigger female purchase intention? The role of parasocial romantic relationships. Curr. Psychol. **43**(12), 11080–11091 (2024)

Garg, M., Bakshi, A.: Exploring the impact of beauty vloggers' credible attributes, parasocial interaction, and trust on consumer purchase intention in influencer marketing. Humanit. Soc. Sci. Commun. **11**(1), 1–14 (2024)

Gleason, T.R., Theran, S.A., Newberg, E.M.: Parasocial interactions and relationships in early adolescence. Front. Psychol. **8**, 255 (2017)

Hair, J.F., Black, W.C., Babin, B.J., Anderson, R.E.: Multivariate data analysis, 8th edn. Cengage Learning (2019)

Hair, J.F., Oritinau, D.J., Harrison, D., Celsi, M.W., Bush, R.P.: Essentials of Marketing, 5th edn. McGraw-Hill, New York (2021)

Hwang, K., Zhang, Q.: Influence of parasocial relationship between digital celebrities and their followers on followers' purchase, electronic word-of-mouth intentions, and persuasion knowledge. Comput. Hum. Behav. **87**, 155–173 (2018)

Ju, I., Lou, C.: Does influencer–follower relationship matter? exploring how relationship norms and influencer–product congruence affect advertising effectiveness across product categories. J. Interact. Advert. **22**(2), 157–177 (2022)

Kim, Y.C.: Media system dependency theory. Int. Encycl. Media Psychol. **3**, 1–17 (2020)

Koay, K.Y., Lim, W.M., Kaur, S., Soh, K., Poon, W.C.: How and when social media influencers' intimate self-disclosure fosters purchase intentions: the roles of congruency and parasocial relationships. Mark. Intell. Plan. **41**(6), 790–809 (2023)

Kotler, P., Kartajaya, H., Setiawan, I.: Marketing 4.0: Moving from traditional to digital. Wiley, New Jersey (2017)

Kursan Milaković, I.: Purchase experience during the COVID-19 pandemic and social cognitive theory: the relevance of consumer vulnerability, resilience, and adaptability for purchase satisfaction and repurchase. Int. J. Consum. Stud. **45**(6), 1425–1442 (2021)

Lee, M., Lee, H.H.: Do parasocial interactions and vicarious experiences in the beauty YouTube channels promote consumer purchase intention? Int. J. Consum. Stud. **46**(1), 235–248 (2022)

Leite, F.P., Pontes, N., Septianto, F.: To share or not to share: when is influencer self-disclosure perceived as appropriate? J. Consum. Behav. **23**(5), 2585–2598 (2024)

Leite, F.P., Baptista, P. de P.: The effects of social media influencer self-disclosure on behavioural intentions: the role of source credibility, parasocial relationships, and brand trust. J. Mark. Theory Pract. **30**(3), 295–311 (2022)

Lim, M.S., Molenaar, A., Brennan, L., Reid, M., McCaffrey, T.: Young adults' use of different social media platforms for health information: insights from web-based conversations. J. Med. Internet Res. **24**(1), 23656 (2022)

Listiawati, L., Ahmad, Y.Z., Amelia, K., Zaelani, A.B.: Analysing the influence of endorsers on brand image formation in social media: a comparative study of celebrities, influencers, and emerging trends. Kampret J. **4**(1), 34–44 (2024)

Liu, X., Zheng, X.: The power of social media influencers on brand credibility and purchase intention. Humanit. Soc. Sci. Commun. **11**(1), 1–12 (2024)

Lokithasan, K., Simon, S., Jasmin, N.Z.B., Othman, N.A.B.: Male and female social media influencers: the impact of gender on emerging adults. Int. J. Mod. Trends Soc. Sci. **2**(9), 21–30 (2019)

Lou, C., Yuan, S.: Influencer marketing: how message value affects consumer trust of branded content on social media. J. Interact. Advert. **19**(1), 58–73 (2019)

Malhotra, N., Nunan, D., Birks, D.: Marketing research: An applied approach, 5th edn. Pearson Education, London (2017)

Patwardhan, P., Yang, J.: Internet dependency relations and online consumer behavior. J. Interact. Advert. **3**(2), 57–69 (2003)

Rosenberg, H., Blondheim, M., Sabag-Ben Porat, C.: Who in the world Is generation Z? The rise of mobile natives and their socio-technological identity. Societies **15**(11), 314 (2025). https://doi.org/10.3390/soc15110314

Ruggiero, T.E.: Uses and gratifications theory in the 21st century. Mass Commun. Soc. **3**(1), 3–37 (2000)

Sokolova, K., Kefi, H.: Instagram and YouTube bloggers promote it, so why should I buy? How credibility and parasocial interaction influence purchase intentions. J. Retail. Consum. Serv. **53**, 101742 (2020)

Taher, S.S., Chan, T.J., Zolkepli, I.A., Sharipudin, M.N.S.: Mediating role of parasocial relationships on social media influencers' reputation signals and purchase intention of beauty products. Rom. J. Commun. Public Relat. **24**(3), 45–66 (2022)

Yones, P.C.P., Muthaiyah, S.: EWOM via the TikTok application and its influence on the purchase intention of some products. Asia Pac. Manag. Rev. **28**(2), 174–184 (2023)

Zhang, A.L., Lu, H.: Scientists as influencers: the role of source identity, self-disclosure, and anti-intellectualism in science communication on social media. Soc. Media Soc. **9**(2), 1–16 (2023)

Zikmund, W.G., Quinlan, C., Griffin, M.: Business research methods, 2nd edn. Cengage Learning, Boston (2019)

Investigation of Artificial Intelligence Identification Methods in Alleviating Identity Fraud in Gauteng

S. N. Molefe

Abstract In Gauteng, South Africa, identity fraud is becoming a bigger problem as scammers use technology to get over security defences. The statistics of fraud in Gauteng Province are staggering. As digital financial transactions and remote identity verification have grown in popularity, artificial intelligence (AI) has become a vital tool in the fight against fraud. Financial institutions and government organisations have embraced AI-driven identity verification techniques, such as biometric authentication and behavioural analytics, to improve security and identify questionable activities. Identity fraud has resulted in significant financial losses on a global scale. Countries such as the United States and the United Kingdom have successfully implemented AI-based fraud detection systems. In South Africa, digital banking fraud has increased, particularly in Gauteng, where identity theft poses a major risk to both people and businesses. The deployment of AI-driven solutions in Gauteng is still hampered by infrastructure problems, data privacy concerns, and poor technology adoption, despite the potential of these solutions. By analysing its effectiveness and implementation challenges through case studies and expect interviews, this study employs a qualitative methodology to examine the role of AI in identifying fraud detection. The findings indicate that while A enhances fraud detection abilities, increased stakeholder involvement and regulatory support are necessary for its efficacy. The report recommends strengthening AI governance, increasing awareness of AI-driven fraud prevention strategies, and fostering partnerships between the public and private sectors in order to optimise Gauteng's fraud detection efforts.

Keywords Identity fraud · Artificial Intelligence · Impersonation

S. N. Molefe (✉)
Vaal University of Technology, Vanderbijlpark, South Africa
e-mail: nthabisengm2@vut.ac.za

E. H. Redda (ed.), *Proceedings of the Social Sciences International Research Conference, 2025*, https://doi.org/10.1007/978-981-95-9489-4_21

1 Introduction

This study focuses on how artificial intelligence (AI) is used to underpin efforts to alleviate identity fraud. In the past, identity theft mainly involved tangible crimes. The perpetrators focused on acquiring physical objects, like pilfered purses or postal deliveries to impersonate others. The Social Security Number (SSN) gradually become a central part of a person's identity (Soban 2023). Soban (2023) believes that if a deceptive individual were to obtain sensitive personal information such as one's legal designation, domicile, or identification number, replicating their persona becomes exceedingly simple, potentially resulting in unauthorised transactions and fraudulent loans. Identity fraud protection emerged to track this core data. But in the digital age. 'fraud' now includes online details like passwords, username, PIN, and device identifications (Soban 2023). Protecting financial stability, consumer confidence, and the smooth operation of economic function relies heavily on robust strategies designed to deter fraud the monetary consequences of fraud activities can be detrimental to the organisations and individuals in Gauteng resulting in substantial financial meltdown, causing damage to their reputation (Karpoff 2021; Mandal 2023). It has become a lot easier for fraudsters to commit their crimes because of computers and online tools. They have found some danced ways to take advantage of the digital world (Caldow 2024). Impersonation is now a prominent form of insurance fraud. This is largely due to AI-powered deepfake technology, which is used to manipulate and circumvent verification systems during the application process. This technology enables the creation of fraudulent identities that appears authentic. The subsequent section discusses how AI has transformed over time to help prevent identity fraud (Caldow 2024). These crimes typically involve the illegal use or abuse of another person's identity or related information. Kalvet et al. (2019) note that identity fraud not only violates individuals' privacy but also causes significant stress and potential financial and legal consequences.

Breakdown of Financial Crime Application Fraud (South Africa Banking Sector) (See Fig. 1).

2 Literature Review

2.1 Theoretical Framework

The study adopted the Technology Acceptance Model (TAM) which is introduced by Davis in his Ph.D. dissertation in 1989. Davis alludes that TAM is rooted in two theories reasoned action and planned behaviour. The TAM was introduced 25 years ago; however, it still plays a vital role in clarifying the users' behaviour in terms of technology usage (Marangunic and Granic 2015). The Financial institutions and government institutions have braced and accepted AI-driven identity verification techniques, such as biometric authentication and behavioural analytics,

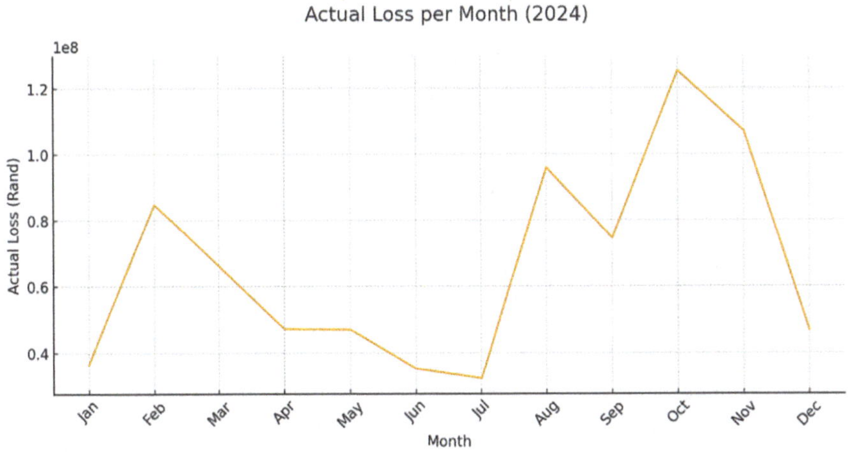

Fig. 1 Sabric 2024

to improve security questionable activities (Onesi-Ozigagun et al. 2024; Marakalala 2025; Mandru 2022).

2.2 *The Principles that Guide the Study Are the Following:*

- Regulation: Regulating the ethical use of AI (Knochel et al. 2024),
- Data security: Inhabitation of contravening private or sensitive data (Knochel et al. 2024)
- Quality control: Verification processes ought to be documented and transparent (Knochel et al. 2024),
- Originality: Deviate from using an AI-generated version for publication (Knochel et al. 2024),
- Bias mitigation: The government institutions and financial institutions should deviate from utilisation of undesirable biases when using AI-driven identity verification. Birhane et al. (2024) indicate the following as examples of existing biases: existing gender and racial biases when undertaking investigations,
- Accountability: AI systems should be used responsibly by experts; AI cannot make conscious decisions. AI model output should be evaluated (Resnik and Hosseini 2024)
- Transparency: Acknowledging and reporting the usage of AI tools promotes accountability, fosters trust, and facilitates verification and replication. AI tools are constantly being updated, but changes may not necessarily be reflected in the model, so providing dates of use is good practice (Hosseini et al. 2023)

- Broader Impact: The government institutions and financial institutions should be aware of the social and environmental impacts of using AI (Bolte and Van Wynsberghe 2024).

2.3 *Evolution of Artificial Intelligence*

The increasing application of AI to advance economic growth and social empowerment draws attention to present problems and lays out opportunities (Duan et al. 2019; Kopka and Grashof 2022). AI's impact on the economy has already been covered in detail. Digital technologies like AI and immense data.

Prior discourse has extensively examined the economic impact of AI. Such digital technologies, including AI and big data, have disrupted conventional commercial prototype, increased throughput, decreased inefficiencies, and made it possible for organisations to become more flexible. Consequently, these advancements improve the experiences of all parties involved (Chauhan et al. 2022). The emergence of AI has ushered in a new era of identity fraud. Identity fraud, which was labour-intensive, manual endeavour, has developed into a complex, AI-enabled enterprise with far-reaching effects (Ansaldo 2024). Deepfake now happen every five minutes, and digital document forgeries have increased by 244%, according to the 2025 Identity Fraud Report and phishing scam to the most intricate schemes like deepfake and identity fraud, the strategies employed by these individuals have become increasingly difficult to detect. The fraud landscape in Africa is changing significantly. The continent's overall identity fraud rate dropped marginally to 3.42% in the first quarter of 2025 from 3.50%, in the same period the previous year, according to new data from the global identity verification platform Sumsub. This slight improvement, however, conceals a more traditional document forgery is drastically decreasing.

Many African nations are making progress in lowering traditional fraud techniques, especially document forgery, according to Sumsub's Q1 2025 Identity Fraud Report (Kanali 2025).

However, at the same time, cybercriminals are increasingly using more sophisticated scams that are more difficult to spot, as well as AI-generated fake ID's (Kanali 2025). For example, overall fraud rates in South Africa decreased by more than 26%, and document forgery decreased by more than 73%. Kenya also made a significant stride, reducing document forgery incidents by 45% and overall fraud by over 15% (Kanali 2025). Nigeria offered a more nuanced picture; although the overall fraud rete increased marginally, certain types of fraud, like document forgery, saw notable declines with an estimate 80% decrease. However, Tanzania saw a 9.6% rise in total fraud, indicating regional difficulties in fending off the changing threats (Kanali 2025).

2.4 Definition of Identity Fraud

It has been acknowledged that identity theft and identity fraud are often used in please of each other, seeing the interconnection of their characteristics, however they are not similar in terms of action and their legal definition (Newman and McNally 2005).

Identity fraud is termed as the deliberate use of someone's private identification data, without their permission it is considered illegal (Federal Trade Commission 1998). Theft of someone's identity data (name, SSN, account number, etc.) it is considered as identity fraud. In contrast, identity fraud consists of the misuse of that information to obtain illicit financial or other benefits. While most frauds involve direct communication between the victim and the perpetrator as a necessary condition, identity fraud is unique in that it usually does not include any contact or relationship between the victim and the perpetrator (Merdovic and Jovanović 2024). The definition of identity fraud implies theft of individual private data and associated frauds. The lengthy process of finding the identity document. The victim lose courage to report the matter to the law enforcers (Randa and Reyns 2020).

2.5 Types of Artificial Intelligence

According to Kumar (2025), 'AI can be broadly classified into several types based on capabilities, functionalities, and technologies. Here is an overview of the different types of AI.':

Based on Technologies. Kumar (2025) discusses AI based on Technologies.

Machine Learning. This type of 'AI' has capacity generate software that is capable individually to study through retrieving and using information. Machine learning (ML) is able function by itself exclusive of direct programming.

Deep Learning. A subset of ML involving many layers of neural networks. It is used for learning from large amounts of data and is the technology behind voice control in consumer devices, image recognition, and many other applications.

Natural Language Processing Machines. 'This AI technology enables machines to understand and interpret human language. It is used in chatbots, translation services, and sentiment analysis applications.'

Robotics. 'This field involves designing, constructing, operating, and using robots and computer systems for controlling them, sensory feedback, and information processing.'

Computer Vision. 'This technology allows machines to interpret the world visually, and it's used in various applications such as medical image analysis, surveillance, and manufacturing.'

Expert Systems. 'These AI systems answer questions and solve problems in a specific domain of expertise using rule-based systems' (Kumar 2025).

AI's Double Role: Weapon and Shield. Ansaldo (2024) points out that AI is not just a tool for attackers; it is also a critical line of defence for those fighting fraud. While criminals use AI to enhance their tactics, businesses are deploying it to detect and prevent fraud with greater precision (Ansaldo 2024). AI-driven defences, such as biometric verification and motion-based detection, have made significant strides. Biometric verification systems now analyse unique characteristics, such as typing patterns or physical gestures, to verify identities. Motion-based detection can assess subtle user behaviours, such as the angle at which someone holds a device, to flag potentially fraudulent activity. These tools make it more difficult for criminals to circumvent traditional security measures (Ansaldo 2024). To combat emerging challenges from AI-driven attacks, organisations must leverage AI-powered tools that can provide granular, real-time environment visibility and alerting. Incorporating autonomous or human-in-the-loop responses can also accelerate the detection of never-before-seen threats (Ansaldo 2024). By using these technologies effectively, businesses can detect and prevent sophisticated fraud attempts that might otherwise evade traditional security measures (Data Dome 2025). Facial recognition is one of the AI technologies that has been around for a while. This biometric authentication method analyses unique facial features to verify an individual's identity, securing it from identity fraud and unauthorised account access. PricewaterhouseCoopers (2023) states that ML techniques are generally already embedded in our fraud detection systems, reducing false positive rates and driving efficiency in our investigation teams while also improving our ability to spot suspicious activity.

Regional Insights and AI Global Trends. Ansaldo (2024) notes that the global nature of AI-driven fraud adds another layer of complexity. National ID cards, tax IDs, and passports are the most frequently targeted documents, thanks to their high value on the black market. Fraudsters often use these stolen identities in cross-border schemes, exploiting differences in regulatory environments (Ansaldo 2024); the globalisation of fraud calls for a unified response. Industry leaders are advocating for stricter standards in document verification and cross-border data sharing. These efforts aim to close the gaps that fraudsters currently exploit, making it harder for them to operate on an international scale (Ansaldo 2024).

3 Challenges of AI

High False Positives: Legitimate transactions are often flagged as fraudulent, leading to customer dissatisfaction and requiring manual reviews by the company's team.

Lack of Adaptability: Fraudsters continually refine their tactics, rendering regular rules ineffective.

Slow Detection: If the legacy system triggers manual review, it becomes a slow process that delays fraud detection. This way, fraudsters get more time to exploit vulnerabilities in accuracy (Ansaldo 2024).

3.1 Prevention and Mitigation of Identity Fraud

According to Jain et al. (2011:2), Technological Solutions provides the follows:

Biometric Systems. Biometric recognition offers a robust and intuitive approach to personal identification. Using identities intrinsic to the individual, biometrics traits are inherently resistant to manipulation, sharing, or forgetfulness. Consequently, they provide a strong and relatively stable connection between an individual and their identity. However, intra-user variability exists, encompassing; (a) variations in fingerprint patterns streaming from differences in finger position and orientation during capture; (b) variations in facial imagery resulting from alterations in pose; and (c) variations in iris images attributable to pupillary dilation and gaze direction (Jain et al. 2011:2).

Encryption and Secure Authentication. Discuss encryption protocols, multi-factor authentication (MFA), and blockchain technologies in securing personal data.

Encryption safeguards data by keeping it confidential. We use it to protect data stored on devices and data sent online. Encryption prevents unauthorised access, changes, exposure, and fraud. It transforms readable data (plaintext) into an unreadable format (ciphertext). Decryption reverses this process, turning ciphertext back into plaintext. We employ encryption for secure emails, website browsing, and secure messages (Jackson 2021). This post brings together the concept of biometrics, a versatile type of data, and its application in authentication. Biometric authentication has grown increasingly feasible and dependable due to the distinctiveness of physical attributes—each individual embodies a password. Nonetheless, since these 'passwords' are immutable, there has arisen a collective apprehension about the potential compromise or fraud of biometric data, which could lead to irreversible privacy concerns. Enter blockchain, offering a remedy to address this challenge (Qazafi 2023). Blockchain has special encryption and peer-to-peer communication methods. A blockchain is formed in sequential order; changing any one of the blocks requires the permission of other blocks. The blocks communicate by a peer-to-peer method, where each node in the network maintains a copy of the entire blockchain. This decentralised approach means that there is no central authority or single point of control over the blockchain (Qazafi 2023).

3.2 Preventative Measures that Should Be Followed to Limit Security Breaches

According to Charandura (2022), preventative measures are the following:

Information Security Policies and Security Standards. Cyber Security framework and strategy that provides a framework for an assured cyber security environment using a risk-based approach should be formally developed. Furthermore, security standards should be documented, specifying expected security configurations and system parameters. These serve as security templates and ensure secure, consistent, and standardised configurations.

IT Capacity Management and Planning. This will assist the university management in establishing current and future processing needs to optimise the IT infrastructure that makes up the framework of the business.

Software Upgrades. There should be an exercise to evaluate all unsupported software (i.e. Windows operating systems, end-user applications, databases, etc.) within the IT environment. The outcome of this exercise should be a software upgrade roadmap that will assist the University IT management with rolling out the necessary upgrades.

Periodic Monitoring of Patch Management. Several external and internal network hosts suffer from patch management issues, where Microsoft and non-Microsoft patches had not been timeously applied. There should be ensuring that required vendor patches are installed when necessary.

Continuous Vulnerability Management. Vulnerability management is how vulnerabilities in the IT environment are identified, and the risks of these vulnerabilities are evaluated. This evaluation leads to correcting the vulnerabilities and removing the threat. Universities should implement a process to regularly identify, classify, remediate, and mitigate internal and external network vulnerabilities.

Monitoring and Early Detection. Implementation of a Security Incident and Event Management (SIEM) solution, which includes activity monitoring to facilitate the detection of patterns which may indicate successful compromise or misuse. This should be configured to perform log aggregation, correlation, alerting, dashboard, and reporting for all IT assets, including servers, databases, applications, firewalls, routers, and switches (Charandura 2022).

Private–Public Partnerships. Examine collaboration between governments, financial institutions, and technology companies to prevent and mitigate identity fraud.

The term *public–private partnership* describes a range of possible relationships among public and private entities in the context of infrastructure and other services. Other terms used for this type of activity include private sector participation (PSP) and privatisation. While the three terms have often been used interchangeably (Asian Development Bank. 2024).

3.3 The Protection of Personal Information Act (no. 4 of 2013) (POPIA)

The Protection of Personal Information Act (No. 4 of 2013) (POPIA) supports or complements Sect. 14 of the Constitution of the Republic of South Africa (Act 108 of 1996) and its protection of every citizen's right to privacy. The Protection of Personal Information Act (POPIA or POPIA Act) is South Africa's data protection law. The purpose of the POPIA is to protect people from harm by protecting their personal information. To prevent their money from being stolen, to protect their identity, and to safeguard their privacy, which is a fundamental human right.

3.4 The General Data Protection Regulation and Businesses in South Africa

The POPIA (No. 4 of 2013) (POPIA) works hand in hand with the General Data Protection Regulation (GDPR). The GDPR is a legal framework that detects how personal information from individuals is collected and used. It is important to understand that GDPR applies not just to companies located within the European Union (EU), but also to those outside the EU that offer goods or service to EU residents or track their online activity. The GDPR came into effect on May 25, 2018, established by EU to consolidate existing data protection laws. This means it protects the data privacy rights of everyone residing in the EU. EU laws generally apply across all EU members states. A key point is that any business processing data of EU residents must adhere to GDPR requirements. While GDPR is not South African law—the Protection of Personal Information Act (POPIA) governs data protection in South Africa—certain aspects of GDPR can impact South African businesses if they fall under the GDPR' scope.

This means that South African businesses which participate in business within any UN state or have a partnership with an EU business will fall within the scope of the GDPR. Businesses in South Africa that have a presence in the EU will therefore need to be alert of the new requirements under the GDPR to endure their businesses in a data protection compliant manner. A matter of concern is the enforcement date, which was the 25th of May 2018.

3.5 Methodology

This qualitative study employed an exploratory research design to investigate the use of AI in combating identity fraud in the Gauteng. Qualitative research is particularly well-suited for exploring complex issues in depth and holistically, providing rich, detailed data that capture the nuances of participants' experiences and perspectives

(Dane 2007). An exploratory design was chosen to gain fresh insights (Creswell 2013). AI offer significant potential in combating crime within Gauteng Province. To effectively explore this potential, employing a robust research methodology is essential. According to Bonginkosi Dhlamini, MEC in the Gauteng Department of e-Government, real-time monitoring and advanced AI capabilities have boosted crime-fighting efforts in the province since the launch of the partnership with Vumacam (Mzekandaba 2024). Several studies have used qualitative methods, particularly semi-structured interviews, to gather in-depth insights from police officials. Qualitative Research with Semi-Structured Interviews is defined as one of the research instruments used in qualitative research (Ruslin et al. 2022) The study examined the integration of AI within financial and government institutions. The study employed qualitative methods, conducting semi-structured interviews with four government and six financial institutions in Gauteng.

Research Design. Qualitative research emphasises exploring and understanding the meaning that a person or group of people ascribe to a social or human problem (Creswell 2014). Qualitative research has roots in disciplines like anthropology sociology, the humanities, and programme evolution. This research approach uses open-ended questions to generate data that is not easily measured numerically. Qualitative research allows researchers to understand issues by examining them within their natural setting and exploring the meanings individuals ascribe to them (Denzin and Lincoln 2005). Its main focus is to generate meaning, purpose, or reality from opinions and experiences of participants (Merriam 2009). This study adopts a qualitative exploratory research design to investigate how AI identification methods are used to alleviate identity fraud in selected government and financial institutions. The exploratory design is suitable because the research seeks to gain an in-depth understanding of a relatively underexplored area—the practical and ethical dimensions of using AI-based identification systems to prevent and detect identity-related crimes. Ethical considerations include ensuring informed consent, confidentiality, data protection, and fairness in the use of AI to prevent bias and misuse of information. The present research undertaken is anchored in the interpretivist paradigm, which assumes that knowledge is constructed through participants' lived experiences and perceptions. To achieve this goal, Berryman (2019:273) believes that for interpretive researchers to find answers to qualitative questions, they need to structure their research questions in a way that focuses on understanding how and why. This paradigm enables the researcher to interpret how professionals within institutions understand, implement, and evaluate AI-driven identification technologies in combating identity fraud.

3.6 Research Paradigm

Research philosophy and research paradigm are terms that tend to be used loosely, even interchangeably (Jansen 2023). Jansen (2023) continues to state that broadly speaking, they both refer to the set of beliefs, assumptions, and principles that underlie the way you approach your study (whether that is a dissertation, thesis or any other sort of academic research project). Honebein (1996) describes the constructivist philosophical paradigm as an approach that asserts that people construct their own understanding and knowledge of the world through experiencing things and reflecting on those experiences. When applied through a constructivist lens, phenomenology is a powerful tool for exploring how individuals make sense of their lived experiences. Both phenomenology and constructivism prioritise the subjective nature of knowledge and the role of individual perception in shaping reality (Johnson 2024). This study adopted a participant-focused framework, driven by a data collection method that captured primary, real-world insights on identity fraud from police investigators, in their familiar settings, expressed in their own terms and shaped by their lived experiences and views on AI in alleviating identity fraud.

Research Approach. Exploratory research was employed in this study, providing valuable insights that aid researchers in forming hypotheses and questions to gain a deeper understanding of their customers. It is essential for researchers conducting exploratory studies to stay open-minded and be ready to adjust their research methods and focus based on new findings (Saha 2024).

According to Saha (2024), some common applications of exploratory research for consumers are: Identifying emerging trends or issues in a particular field, 2. Gaining a deeper understanding of a complex phenomenon:

- Formulating new research questions and hypotheses
- Evaluating the feasibility of a larger-scale research project

This approach enabled the collection of detailed perspectives on the challenges and opportunities associated with implementing AI. Participants were purposively sampled, and data were analysed thematically to identify key themes and insights. Qualitative Approach with Semi-Structured Interviews: In exploring the role of AI in alleviating identity fraud, a study used a qualitative research method, including semi-structured interviews with police officials in Gauteng. Purposive sampling was employed to select participants, and thematic analysis was used to interpret the data. AI, which analyses unique physiological and behavioural characteristics, has become a pivotal tool in alleviating identity fraud in Gauteng. AI enhances the capabilities of law enforcement agencies by providing accurate and efficient methods for identifying individuals involved in identity fraud.

3.7 Study Findings and Discussions

In responding to the objective of this study, about four government institutions and six financial institutions detective members were purposively selected to offer their experiences on this subject (the use of AI in alleviating fraud).

3.8 *The Use of AI in the Alleviation of Identity Fraud in the Gauteng Province.*

To achieve the main objective, the following question was posed to the identified participants: What is the role of AI in combating identity fraud? All selected participants were asked this question, and their respective responses were transcribed. The targeted sample are as follows: Government institutions: A1, A2, A3, A4 and Financial institutions: B1, B2, B3, B4, B5, B6.

Participant A1 and Participant B3 replied that the fingerprints identification method is used to identify a person. However, one model is not enough other modals is surveillance cameras that are installed in the street of Gauteng for facial reignition. In other areas of Gauteng, the surveillance cameras installation is still not instituted.

Participant B6 replied that they use identification methods, such as facial recognition, fingerprint scanning, iris, and voice recognition, are widely used to verify identities and prevent impersonation. Lourie (2025) states that President Cyril Ramaposa has reiterated South Africa's dedication to using technology to improve law enforcement. He announced that the smarter policing methods to modernise its approach to fight crime. In his State of the Nation Address, President Ramaphosa emphasised the importance of AI and digital forensics in tackling financial crimes, corruption, and violent offences. He stated, *We have seen the value of technology in fighting crime*, citing AI's use in fraud risk detection and verification.

Participant A2 replied as deepfakes become an increasing threat, the need for robust countermeasures is more critical than ever. One promising defence is facial biometrics with liveness detection, which helps verify that biometric input is from a real person rather than a recorded or synthetic video. However, the rapid growth of AI-generated media poses a challenge to the reliability of these biometric systems. Gauteng Premier Panyaza Lesufi unveiled at least 28 CCTV surveillance cameras in Soshanguve township on Thursday as part of the province's crime-fighting initiatives to ensure safety and security for resident (Moatshe 2025).

Participant B4, A3, and A4 replied that AI monitors user behaviour, such as login times, device use, and keystroke patterns, to detect anomalies: The analysis of interviews with 10 participants from four government institutions and six financial institutions revealed several key themes related to the use of AI in alleviating identity fraud. Participants shared insights from their professional experiences, highlighting the practical applications, benefits, and challenges of AI identification methods.

Participant B1, B2, and B5 gave a similar reply stating that AI has become a transformative tool in investigating identity fraud by enhancing the accuracy, speed, and efficiency of identification processes. According to Rousan and Benedetto (2020), MFA can improve system security by using two or more kinds of security types, system as passwords and cards, but this is not an ideal security scheme. Passwords may be forgotten or inputted incorrectly, or the identification cards may be stolen be stolen, which speaks to traditional manual verification methods are often vulnerable to human error and manipulation, whereas AI systems employ advanced algorithms that analyse large datasets to authenticate individuals with greater precision.

3.9 Discussion

Identity fraud is a growing global concern, fuelled by the increasing digitisation of services and the sophistication of cybercriminals. As traditional security measures become insufficient, AI has emerged as a powerful tool in the fight against identity fraud. AI offers the ability to detect, prevent, and respond to identity-related crimes in ways that are faster, more accurate, and more scalable than human-led processes. The collaboration between the Department of e-Government, Community Safety, and the government institutions. It is intended to complement and enhance all other available crime-fighting initiatives and the trajectory of investigations in South Africa. The financial institutions are expected to benefit from these initiatives involving surveillance cameras on the streets, as criminals will be identified and linked to cases of identity fraud. In the financial sector, AI-driven identification methods are instrumental in verifying customer identities during account creation, online transactions, and credit applications. Automated systems continuously monitor user activity, flagging irregularities such as unusual login locations or transaction patterns. By integrating AI into fraud detection systems, financial institutions and financial institutions can respond to threats in real time, thereby reducing losses and maintaining consumer trust. Similarly, government institutions employ AI identification systems to secure national databases, prevent duplication of identity records, and detect false document submissions (Save 2025; Gondaliya 2025). AI identification systems to secure national databases, prevent duplication of identity records, and detect false document submissions. AI can cross-reference data across multiple agencies, improving the accuracy of identity validation and reducing administrative bottlenecks. Such applications contribute to public confidence in digital governance systems (Obiokafor 2024; Pheiffer and Naude 2024).

AI serves as a critical enabler in strengthening identity fraud defences. When combined with human oversight, regulatory frameworks, and user awareness, AI can significantly reduce the success rate of identity fraud schemes. Its ability to adapt, learn, and scale gives institutions an edge in an ever-evolving threat landscape. However, its effectiveness hinges on regular updates, transparency, ethical use, and fairness. A balanced approach that blends technology and human judgement will be essential in fully realising AI's potential as a tool against identity fraud. Clear legal and ethical guidelines are necessary to ensure compliance with regulations such as the GDPR or the POPIA. Identity fraud presents a multifaceted challenge that permeates various aspects of contemporary society, significantly affecting individuals, businesses, and governments (Pheiffer and Naude 2024). This pervasive issue underscores the necessity for a comprehensive understanding of its nature and extent. The absence of a universally accepted definition of identity fraud hampers effective research, policy-making, and international cooperation, as noted by Newman and McNally (2005). Addressing this definitional gap is a crucial first step towards devising effective intervention. Empowering individuals with knowledge about safeguarding their personal information and recognising potential threats can significantly reduce the incidence of identity fraud. In conclusion, a multifaceted approach that combines comprehensive research, innovative technological solutions, robust legal frameworks, and public education is imperative to combat identity fraud effectively.

By addressing definitional ambiguities and leveraging international cooperation, stakeholders can develop strategies that protect individuals' personal information, reduce the prevalence of identity fraud, and mitigate its far-reaching consequences Obiokafor 2024. The use of AI in identity verification often involves collecting sensitive personal data, which raises concerns about data protection, consent, and surveillance. All government institutions and financial institutions must adhere to the GDPR (Ross 2020) and the POPIA (Raaff et al. 2022).

3.10 Conclusion and Recommendations

AI serves as a critical enabler in strengthening identity fraud defences. When combined with human oversight, regulatory frameworks, and user awareness, AI can significantly reduce the success rate of identity fraud schemes. Its ability to adapt, learn, and scale gives institutions an edge in an ever-evolving threat landscape. However, its effectiveness hinges on regular updates, transparency, ethical use, and fairness. A balanced approach that blends technology and human judgement will be essential in fully realising AI's potential as a tool against identity fraud. Clear legal and ethical guidelines are necessary to ensure compliance with regulations such as the GDPR or the POPIA. Identity fraud presents a multifaceted challenge that permeates various aspects of contemporary society, significantly affecting individuals, businesses, and governments. This pervasive issue underscores the necessity for a comprehensive understanding of its nature and extent. The absence of a universally accepted definition of identity fraud hampers effective research, policymaking, and international cooperation, as noted by Newman and McNally (2005). Addressing this definitional gap is a crucial first step towards devising effective intervention. Empowering individuals with knowledge about safeguarding their personal information and recognising potential threats can significantly reduce the incidence of identity fraud. In conclusion, a multifaceted approach that combines comprehensive research, innovative technological solutions, robust legal frameworks, and public education is imperative to combat identity fraud effectively. By addressing the definitional ambiguities and leveraging international cooperation, stakeholders can develop strategies that protect individuals' personal information, reduce the prevalence of identity fraud, and mitigate its far-reaching consequences. The use of AI in identity verification often involves collecting sensitive personal data, which raises concerns about data protection, consent, and surveillance. All government institutions and financial institutions must adhere to the GDPR or the POPIA to avoid civil claims.

References

Ansaldo, M: AI's role in the evolution of identity fraud. https://securitybuzz.com/cybersecurity-news/ais-role-in-the-evolution-of-identity-fraud (2024). Accessed 24 June 2025

Asian Development Bank: Public–private partnership handbook. https://www.adb.org/sites/default/files/institutional-document/31484/public-private-partnership.pdf (2024). Accessed 1 November 2025

Birhane, A., Steed, R., Ojewale, V., Vecchione, B., Raji, I.D.: AI auditing: the broken bus on the road to AI accountability (2024). Accessed 25 January 2024. https://doi.org/10.48550/arxiv.2401.14462

Bolte, L. van Wynberghe. A.: Sustainable AI and the third wave of AI ethics: a structural turn. AI Ethics (2024). https://doi.org/10.1007/s4381-024-00522-6

Caldow, G: Using artificial intelligence to prevent fraud in South Africa. https://cms.law/en/zaf/publication/using-artificial-intelligence-to-prevent-fraud-in-south-africa (2024). Acessed 20 June 2025

Charandura, K: Cybersecurity in the education industry. https://www.grantthornton.co.za/Newsroom/cybersecurity-in-the-education-industry (2022). Accessed 10 October 2025

Chauhan, C., Parida, V., Dhir, A.: Linking circular economy and digitalisation technologies: A systematic literature review of past achievements and future promises. Technol. Forecast. Soc. Chang. **177**, 121508 (2022). https://www.sciencedirect.com/science/article/pii/S0040162522000403?via%3Dihub

Creswell, J.W.: Research design: Qualitative, quantitative, and mixed methods approaches. SAGE Publications, Thousand Oaks (2014)

Creswell, J.W.: Research design: Qualitative, quantitative, and mixed methods approaches, 4th edn. SAGE Publications, London (2013)

Denzin, N.K., Lincoln, Y.S.: Introduction: The discipline and practice of qualitative research. In: Denzin, N.K., Lincoln, Y.S. (eds.) The SAGE handbook of qualitative research, 3rd edn, pp. 1–32. SAGE Publications, Thousand Oaks (2005)

Duan, Y., Edwards, J.S., Dwivedi, Y.K.: Artificial intelligence for decision making in the era of big data: Evolution, challenges and research agenda. Int. J. Inf. Manage. **48**, 63–71 (2019). https://doi.org/10.1016/j.ijinfomgt.2019.01.021

Federal Trade Commission: Annual report of the federal trade commission for the fiscal year ended September 30, 1988. U.S. Federal Trade Commission (1988)

Gondaliya, D: How AI flagging systems detect and prevent fraud in 2025. https://www.biztechcs.com/blog/how-ai-flagging-system-is-used-for-fraud-detection/ (2025). Accessed 15 October 2025

Honebein, P.C.: Seven goals for the design of constructivist learning environments. In: Wilson, B.G. (ed.) Constructivist learning environments: Case studies in instructional design. Educational Technology Publications, Englewood Cliffs (1996)

Hosseini, M., Resnik, D.B., Holmes, K.: The ethics od disclosing the use of artificial intelligence tools in writing scholarly manuscripts Res. ETH **19**, 449–465 (2023)

Jackson, A: Authenticating vs. encryption: What is the difference? https://www.venafi.com/blog/authentication-vs-encryption-whats-the-difference (2021). Accessed 20 June 2025

Jansen, D: What is research philosophy? https://gradcoach.com/research-philosophy (2023). Accessed 9 June 2025

Johnson, K: Phenomenology through the lens of constructivism. https://www.phoenix.edu/research/news/2024/phenomenology-through-the-lens-of-constructivism.html (2024). Accessed 10 June 2025

Kanali, N: Africa sees drop in traditional identity fraud but rise in synthetic scams. https://africabusinesscommunities.com/tech-24/africa-sees-drop-in-traditional-identity-fraud-but-rise-in-synthetic-scams-report (2025). Accessed 25 June 2025

Karpoff, J.M.: The future of financial fraud. J. Corp. Finan. **66**, 101694 (2021). https://doi.org/10.1016/j.jcorpfin.2020.101694

Kalvet, T., Tiits, M., Ubakivi-Hadachi, P.: Risks and societal implications of identity fraud. In: Identity in the Information Society. Springer. https://doi.org/10.1007/978-3-030-13283-5_6 (2019)

Knochel, T., Schweizer, K., Aar, O.A., Akil, A.M.: Principles for Responsibility AI Usage in Research. https://doi.org/10.312334/osf.io/g3m5f (2024)

Kopka, A., Grashof, N.: Artificial intelligence: Catalyst or barrier on the path to sustainability? Technol. Forecast. Soc. Chang. **175**, 121318 (2022). https://doi.org/10.1016/j.techfore.2021.121318

Kumar, A.: Types of artificial intelligence. https://www.simplilearn.com/tutorials/artificial-intelligence-tutorial/types-of-artificial-intelligence (2025). Accessed 20 June 2025

Lourie, G.: Ramaphosa says smart policing and AI to drive South Africa's fight against crime. https://techfinancials.co.za/2025/02/06/ramaphosa-says-smart-policing-and-ai-to-drive-sas-fight-against-crime(2025). Accessed 1 October 2025

Mandal, A. Fathoming fraud: Unveiling theories, investigating pathways and combating fraud. J. Financ. Crime. (2023)

Mandru, S.K.: How AI can improve identity verification and access control processes. J. Artif. Intell. & Cloud Comput. (2022). https://doi.org/10.47363/JAICC/2022(1)E101

Marakalala, M.C.: An exploration of biometrics payments technology to combat online banking fraud and digital identity in the South African Banking Industry. OIDA Int. J. Sustain. Dev. **18**(12), 1261–1261–1276, (2025). Accessed 29 January 2026

Marangunic, N., Granic, A.: Technology Acceptance Model: A Literature Review from 1986 to 2013'. Univ. Access Inf. Soc. **14**(1), 81–95 (2015). https://doi.org/10.1007/s10209-014-0348-1

Merdovic, B., Jovanović, B.: Understanding identity fraud and fraud. Kultura Polisa. https://doi.org/10.51738/Kpolisa2024.21.2r.17mj (2024). Accessed 27 June 2025

Merriam, S.B.: Qualitative research: A guide to design and implementation. John Wiley & Sons, San Francisco (2009)

Moatshe, R.: Premier Lesufi unveils 28 CCTV cameras to boost safety in Soshanguve (2025). Accessed 1 October 2025

Mzekandaba, S.: Tech-enabled crime-fighting yields gains for Gauteng Province. ITWeb. https://www.itweb.co.za/article/tech-enabled-crime-fighting-yields-gains-for-gp/LPp6VMrBGQ8MDKQz (2024). Accessed 29 October 2025

Newman, G.R., McNally, M.M.: Identity fraud literature review. https://www.ojp.gov/ncjrs/virtual-library/abstracts/identity-fraud-literature-review (2005). Accessed 27 June 2025

Obiokafor, I.N.: Strategies to mitigate cyber identity theft in Africa's digital transformation. J Afr Stud Sustain Dev Vol. **7**(4), (2024)

Onesi-Ozigagun, O., Ololade, Y., Eyo-Udo, N.L., Ogundipe, D.O.: AI-Driven biometrics for secure fintech: Pioneering safety and trust. 24 April 2024. International Journal of Engineering Research Updates 6(2), 001–012. DOI:10.5343O/ijeru.2024.6.2.0023

Pheiffer, D.C., Naude, C.J.: An exploration of the use of software technology to combat identify theft in South Africa. Int. J. Soc. Sci. Res. Rev. **7**(5), 191–202 (2024). https://doi.org/10.47814/ijssrr.v7i5.2043

PricewaterCoopers: The impact of AI on fraud and scam. http://www.pwc.co.uk/forensic-services/assets/impact-of-ai-on-fraud-and-scammers.pdf (2023). Accessed on 28 January 202. Accessed 20 June 2025

Qazafi, M.: The future of personal digital records: Unlocking security and efficiency through blockchain and smart contracts. https://aws.amazon.com/blogs/database/the-future-of-personal-digital-records-unlocking-security-and-efficiency-through-blockchain-and-smart-contracts/ (2023)

Raaff, E., Rothwell, N., Wynne, A.: Aligning South African data and cloud policy with the POPI Act. In: Proceedings of the International Conference on Cyber Warfare and Security. 17(1):279–287. (2022). https://doi.org/10.34190/iccws.17.1

Randa, R., Reyns, B.W.: The physical and emotional toll of identity fraud victimization. Deviant Behav. **41**(10), 1290–1304 (2020). https://www.researchgate.net/journal/Deviant-Behavior-1521-0456

Resnik, D.B. & Hosseini, M.: Ethics of using artificial intelligence in scientific research: new guidance needed for a new tool, vol 5, pp. 1499–1521, 2025. (2024). Accessed 27 January 2026

Riley, D.R.: DataDome adds LLM detection and intent-based AI models to enhance fraud protection (2025). Accessed 15 April 2025

Ross, A.: The European Union's General Data Protection Regulation (GDPR) and its implications for South African data privacy law: An evaluation of selected 'content principles' Comparative and International Law Journal of South Africa. Vol. **53 No. 3** (2020). https://doi.org/10.25159/2522-3062/7985

Ross, D.: POPIA manual. Prepared in accordance with section 51 of the protection of personal information act of 2013 (2020)

Rousan, M.A.I., Benedetto, I.: A comparative analysis of biometrics types: Literature review. J. Comput. Sci. **16**(12), 1778–1788 (2020). https://doi.org/10.3844/jcssp.2020.1778.1788

Ruslin, R., Mashuri, S., Rasak, M. S. A., Alhabsyi, F., Syam, H.: Semi-structured interview: A methodological reflection on the development of a qualitative research instrument in educational studies. IOSR J. Res. Method Educ. **12**(1), 22–29 (2022). https://doi.org/10.9790/7388-120105 2229

Saha, S.: What is exploratory research, and how to conduct it? https://www.entropik.io/blogs/what-is-exploratory-research (2024). Accessed 26 June 2025

Save: Value summit. "Value methodology for strengthening the pillars of the economy." New Delhi, India (2025)

Soban, K.: History of identity fraud and identity fraud measures. https://facia.ai/blog/history-of-identity-fraud-and-identity-fraud-measure (2023). Accessed 22 June 2025

Republic of South Africa.: Protection of Personal Information Act (No. 4 of 2013). Government Printer, Pretoria. (2013)

The Barriers to Women's Leadership Progression Within the Retail Sector in Gauteng, South Africa

Sinovuyo Noah and Marcia Lebambo

Abstract The retail sector in Gauteng, South Africa, employs a large number of women, yet their representation in senior leadership roles remains disproportionately low. Despite the presence of gender equity legislation and organisational diversity initiatives, many women continue to face barriers that hinder their advancement into top management. This systematic review explores the personal, organisational, and societal factors that contribute to this leadership gap. Drawing on glass ceiling theory and social role theory, the study examines 70 scholarly and policy-based sources published between 2010 and 2024. Using the Preferred Reporting Items for Systematic Reviews and Meta-Analyses (PRISMA) framework, literature was sourced from academic databases and institutional repositories and analysed through thematic synthesis. The findings uncover a complicated network of challenges: personal challenges (such as a lack of balance between work and life and imposter syndrome); organisational hurdles like restricted access to mentorship, biased advancement practices, and leadership cultures dominated by men; as well as societal influences including gender stereotypes and overlapping discrimination. Emerging trends also indicate ineffective execution of equity policies and shifting leadership expectations among younger demographics. These barriers are intricately linked, frequently amplifying each other across various levels of both the workplace and society. The review emphasises that achieving genuine progress in gender equity necessitates more than just policies; it calls for fundamental changes, cultural shifts, and focused leadership development. Key recommendations include integrating mentorship initiatives, establishing clear promotion pathways, and ensuring that organisations are held responsible for inclusivity results. The insights from this review provide practical guidance for stakeholders and institutions, such as the Commission for Gender Equality, to foster a more inclusive leadership landscape in South Africa's retail industry.

S. Noah (✉) · M. Lebambo
Tshwane University of Technology, Pretoria, South Africa
e-mail: 213479741@tut4life.ac.za

M. Lebambo
e-mail: lebambomm@tut.ac.za

E. H. Redda (ed.), *Proceedings of the Social Sciences International Research Conference, 2025*, https://doi.org/10.1007/978-981-95-9489-4_22

Keywords Gauteng · Gender barriers · Glass ceiling · Retail sector · Social role theory women in leadership · South Africa

1 Introduction

The retail sector is the second-largest employer in South Africa, playing a pivotal role in shaping the country's economic and workforce dynamics (Teutenberg 2021). Despite women constituting a significant portion of this workforce, their presence in senior leadership remains disproportionately low. In Gauteng, the economic heart of South Africa, retail giants such as Shoprite, Pick n Pay, and Woolworths dominate the commercial landscape. Yet, women occupy fewer than 25% of senior executive roles, a figure that has shown minimal progress in recent years (W&RSETA 2021; USDA Foreign Agricultural Service 2023). This ongoing gap persists despite the introduction of progressive gender policies and global efforts to foster inclusive leadership, indicating deeper systemic and structural issues.

This systematic review examines the barriers that hinder women's progression into leadership roles within the retail sector of Gauteng province of South Africa. The review explores how organisational practices, socio-cultural norms, and specific industry dynamics collectively impede women's career progression. While international literature frequently highlights recurring issues such as gender stereotypes, work–life imbalance, and discrimination (Edwards 2016), these issues remain underexplored in South Africa's unique socio-economic context. The Commission for Gender Equality (CGE 2021) reports that gender disparities in retail leadership have increased since 2017, raising urgent questions about the underlying causes, whether institutional, cultural, or both.

The objectives of this review are fourfold:

- To investigate the barriers hindering women's leadership progression in the retail sector in Gauteng, South Africa.
- To categorise and analyse these barriers across personal, organisational, and societal levels.
- To explore how cultural expectations, workplace practices, and policy implementation affect women's advancement.
- To provide a foundation for future research and practical interventions to promote gender equity in retail leadership.

The scope of this review is geographically limited to Gauteng and sectorally focused on the retail industry. It draws from studies published between 2010 and 2024, with a specific emphasis on South African research while incorporating relevant international comparisons to enrich the analysis. Key theoretical concepts underpinning the review include the glass ceiling theory, an invisible yet pervasive institutional barrier that hinders women's advancement (Loden 1978), and social role theory, which explores how gender norms shape perceptions of leadership suitability (Eagly 1987).

This review is structured thematically. It begins with a discussion of relevant theoretical frameworks, followed by a synthesis of empirical findings organised into personal, organisational, and societal barriers. The paper concludes with implications for future research and practice.

The findings contribute to national gender equity efforts, including those led by the Commission for Gender Equality (CGE) (2021), and aims to support the development of inclusive leadership pipelines within South Africa's retail industry.

2 Literature Review

2.1 Overview

Research on women's leadership has expanded considerably in the last two decades, with increasing attention paid to gender disparities in management, career progression, and organisational culture (Hill et al. 2016; Eagly and Chin 2010). Despite these advances, the retail sector, particularly in South Africa, continues to exhibit persistent leadership gaps. This literature review explores the historical and contemporary landscape of women's leadership in retail, the barriers that continue to hinder progression, and the theoretical lenses through which these barriers are commonly understood.

The review draws from both South African and international literature, with a particular focus on empirical and theoretical work relevant to Gauteng's retail sector. The findings are organised thematically into personal, organisational, and societal barriers to leadership, followed by a synthesis of debates and gaps in the literature.

2.2 Theoretical/Conceptual Framework

Two core theories underpin this study: the glass ceiling theory and social role theory, both of which provide critical insights into the structural and cultural dynamics that inhibit women's career progression in leadership.

Glass Ceiling Theory. The term *glass ceiling*, first introduced by Marilyn Loden (1978), refers to the subtle yet persistent institutional barriers that hinder women and other marginalised groups from rising to top leadership positions, regardless of their qualifications (Babic and Hansez 2021). These barriers often remain unacknowledged in male-dominated industries but are deeply ingrained in promotion policies, hiring practices, workplace cultures, and evaluation measures for leadership. In the South African retail sector, women dominate frontline and junior management roles yet remain underrepresented in executive leadership (W&RSETA 2021). Despite compliance with equity legislation, women in Gauteng often remain 'stuck' at mid-management levels without clear progression pathways (Commission for Gender

Equality [CGE] 2021; Rdebe 2023). Research confirms that glass ceiling effects manifest through unequal pay, biased succession planning, limited mentoring, and exclusion from strategic opportunities (Broadbridge 2007; Bombuwela and De Alwis 2013; Shongwe 2019).

Social Role Theory. *Social role theory*, developed by Eagly (1987), posits that gender roles are learned through cultural and socialisation processes. These roles inform normative expectations about how men and women 'should' behave. Leadership, often associated with assertiveness, autonomy, and strategic authority, aligns with stereotypically masculine traits, while women are socially expected to be nurturing, communal, and deferential (Eagly and Karau 2002; Alqahtani 2019).

These gendered expectations shape both employer decisions and women's self-perceptions in the workplace. In retail settings, particularly, women who demonstrate agentic traits may be perceived as unfeminine or threatening, while communal leadership styles are undervalued (Edwards 2016; Lloyd 2023). This theory helps explain why qualified women may be overlooked for leadership roles—not due to lack of merit, but due to their incongruence with dominant leadership norms.

Conceptual Framework. The model adapted from Sharma and Kaur (2019) will be used to investigate and analyse factors contributing to and perpetuating the challenge of women's limited progression within the retail sector. The model was modified according to the reviewed literature and resulted in the four propositions (P1–P4) outlined below (Fig. 1):

P1: Work–life balance is a personal barrier that impedes the empowerment of women in the retail sector.

P2: Discrimination is an organisational barrier that impedes the empowerment of women in the retail sector.

P3: Mentorship is an organisational barrier that impedes the empowerment of women in the retail sector.

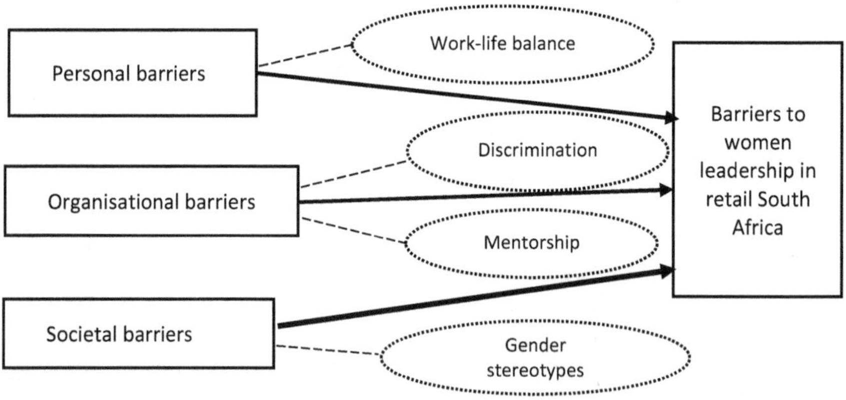

Fig. 1 Sharma and Kaur (2019) model adapted by the author to analyse factors contributing to women's limited progression within the retail sector

P4: Gender stereotypes are a societal barrier that impedes the empowerment of women in the retail sector.

2.3 Barriers to Women's Leadership in Retail

Personal Barriers. Personal barriers often arise from the tension between professional demands and domestic responsibilities. In South Africa's retail sector, women disproportionately shoulder caregiving roles while working irregular hours, including evenings and weekends (Mgaguli 2019). Lloyd (2023) notes that emotional fatigue, burnout, and a lack of time for family are cited by middle managers in fashion retail as key deterrents to career progression.

Additionally, psychological barriers such as imposter syndrome (the internalised doubt of one's own competence despite success) are prevalent among women in leadership trajectories (Clance and Imes 1978). Rdebe (2023) found that female engineers in fast-moving consumer goods (FMCG) settings often downplay their achievements, reinforcing self-doubt. While these issues originate internally, they are often exacerbated by external bias, poor managerial support, and exclusion from strategic roles (Burton and Robertson 2025).

Organisational Barriers. Organisational culture and structure remain the most persistent barriers to women's leadership progression. Broadbridge (2007) found that retail companies often reward traditionally masculine behaviours such as assertiveness, competitiveness, and presence, while undervaluing relational and democratic leadership traits more commonly associated with women.

Exclusion from informal mentorship networks, lack of transparent promotion policies, and superficial diversity initiatives are widely documented (Edwards 2016; Foster et al. 2007; Kele et al. 2022). Although mentorship is a frequently cited solution, studies show it is often poorly implemented without long-term strategy or senior management involvement (Atwood 2022; Hill et al. 2016).

Tokenism also remains an issue. In many cases, women are appointed to visible roles for representational purposes but remain excluded from key decision-making (Broadbridge 2010; Kele et al. 2022).

Societal and Cultural Barriers. Societal norms and cultural expectations heavily influence how leadership is perceived and rewarded. Social role theory highlights how women in leadership often face a double bind: if they lead assertively, they are penalised for not conforming to feminine norms; if they lead collaboratively, they are seen as weak (Eagly and Karau 2002; Alqahtani 2019).

In South African retail, traditional gender roles still expect women to prioritise home and caregiving, placing them at odds with the demands of senior leadership (Leidner 1993). This dynamic is intensified in communities where conservative values dominate workplace culture.

Black South African women, in particular, face intersectional barriers, dealing simultaneously with racial, gender, and class-based discrimination (Kele et al. 2022).

Although employment equity policies exist, cultural biases in both recruitment and promotion practices continue to hinder real transformation (Commission for Gender Equality (CGE) 2021; Shongwe 2019).

2.4 Emerging Trends and Ongoing Debates

Recent literature highlights several shifts in how women's leadership progression is being understood:

Intersectionality is gaining prominence as a framework to understand overlapping systems of oppression. Kele et al. (2022) emphasise how ethnicity, religion, and language add further complexity to leadership trajectories in retail. Okafor et al. (2011) similarly found in Nigeria that women are stereotyped based on emotionality and domestic roles, reducing their chances of being viewed as credible leaders.

Effectiveness of diversity policies remains contested. Although many companies adopt formal employment equity frameworks, researchers note that these often function as window dressing, failing to address deeper structural exclusion (Edwards 2016; Commission for Gender Equality (CGE) 2021).

Generational leadership expectations are changing. Millennial leaders in retail value authenticity, work–life integration, and emotional intelligence over traditional hierarchical leadership (Atwood 2022). Burton and Robertson (2025) found that career fulfilment in retail is increasingly linked to personal values and psychological safety, suggesting that younger cohorts may challenge existing norms.

2.5 Gaps in the Literature

Despite growing scholarship, several important gaps persist:

Context-specific research in the South African retail sector is limited. Most studies emphasise broader corporate contexts, public institutions, or focus on Western geographies (Khosa 2022; Rota 2021). Few empirical studies explore the lived experiences of retail workers and managers in Gauteng.

Longitudinal research is notably absent. Existing literature tends to be cross-sectional, missing the opportunity to track the progression and attrition of women over time (Broadbridge and Simpson 2011; Kossek et al. 2017).

Intersectional approaches remain underutilised. While the term is acknowledged, few studies in the retail context fully engage with how race, age, class, and sexuality influence leadership outcomes (Crenshaw 2013; Kele et al. 2022).

Addressing these gaps is essential for developing robust, inclusive, and effective interventions that genuinely support women's leadership development in South Africa's retail industry.

3 Methodology

3.1 Research Design

This study adopts a systematic review design to identify, evaluate, and synthesise existing literature on the barriers to women's leadership progression in the retail sector in Gauteng, South Africa. A systematic review enables the consolidation of fragmented findings across diverse contexts, offering a structured and replicable means to answer a focused research question (Tranfield et al. 2003).

Given the interdisciplinary nature of gender studies, leadership theory, and retail management, this design provides methodological rigour while allowing both thematic categorisation and critical appraisal. The process was guided by the Preferred Reporting Items for Systematic Reviews and Meta-Analyses (PRISMA) framework, which promotes transparency and replicability in literature reviews (Moher et al. 2009).

3.2 Inclusion and Exclusion Criteria

To ensure relevance and consistency, the review focused on peer-reviewed journal articles, doctoral theses, policy reports, and grey literature. A clear set of inclusion and exclusion criteria was applied. Studies were selected if they met the following inclusion criteria:

- Focused on barriers to women's leadership, particularly within the retail sector.
- Based in South Africa, with priority given to studies referencing Gauteng.
- Included international literature where theoretical or structural parallels with retail were evident (e.g., FMCG, hospitality).
- Employed qualitative, quantitative, or mixed-method approaches.
- Published between 2010 and 2024 and written in English.

Studies were excluded if they:

- Focused on unrelated sectors (e.g., agriculture or education) with no structural relevance to retail).
- Addressed general gender issues without a specific focus on leadership.
- Lacked empirical or theoretical grounding (e.g., opinion pieces or blog posts).

3.3 Search Strategy and Data Sources

The search was conducted using a multi-pronged strategy across academic databases and grey literature repositories. Databases included Google Scholar, EBSCOhost, and JSTOR; theses and dissertations from South African universities; policy reports from organisations such as the CGE and Wholesale and Retail Sector Education and Training Authority (W&RSETA); and peer-reviewed journal articles from global and regional sources.

Key search terms included:

- 'Women in retail leadership South Africa'
- 'Glass ceiling in retail'
- 'Barriers to women's leadership Gauteng'
- 'Gender stereotypes and career progression'
- 'Mentorship and women executives retail'.

Manual reference tracking and citation chaining were used to locate under-cited or seminal studies. After screening over 148 documents, approximately 70 articles met the inclusion criteria and were used in the final synthesis. These comprised peer-reviewed journal articles, doctoral theses, government reports, and policy briefs.

This PRISMA diagram shows the flow of information through the systematic review process. A total of 148 records were identified and screened, with 70 articles meeting the inclusion criteria (Fig. 2).

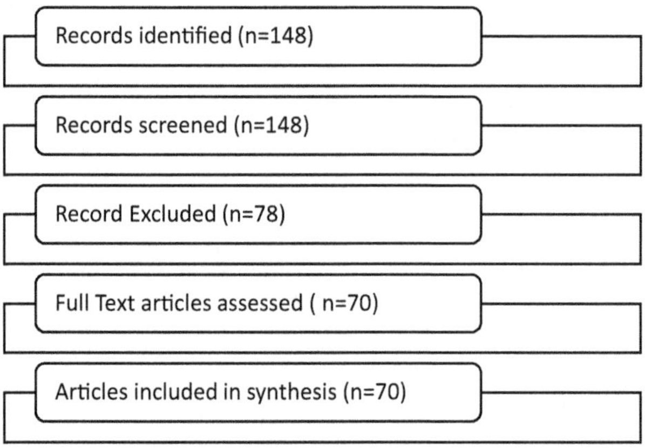

Fig. 2 PRISMA flow diagram outlining the identification, screening, and inclusion process

3.4 Data Collection and Analysis

A standardised data extraction matrix was used to record the following information from each source:

- Author(s) and publication year.
- Country and sectoral focus.
- Methodology and sample size.
- Theoretical framework(s) applied.
- Type of barriers discussed.
- Key findings and relevance to South African retail.

The analysis employed a thematic synthesis approach suitable for integrating findings from both qualitative and quantitative studies (Thomas and Harden 2008). Themes were coded and organised around the adapted conceptual framework developed from Sharma and Kaur (2019), with three main barrier categories:

1. Personal barriers (e.g., work–life conflict, burnout, imposter syndrome).
2. Organisational barriers (e.g., mentorship gaps, pay inequity, exclusion from leadership pipelines).
3. Societal barriers (e.g., gender norms, stereotypes, and intersectional discrimination).

These thematic insights were then interpreted through the lens of glass ceiling theory and social role theory to highlight structural, cultural, and individual-level influences.

3.5 Trustworthiness and Limitations

To ensure trustworthiness, triangulation was used by comparing patterns across different sectors, regions, and study types. The application of well-established theories provided interpretive consistency and strengthened analytical depth.

However, certain limitations are acknowledged:

- The review focused on English-language publications, potentially excluding relevant local-language studies.
- As a desk-based study, the review relies on published materials, which may not reflect recent post-pandemic workplace shifts.
- Although international literature was included for comparative depth, the primary focus on South African studies may limit broader generalisability.

4 Results and Discussion

This section presents the thematic synthesis of the literature on barriers to women's leadership in Gauteng's retail sector. The findings are categorised into three primary categories: personal, organisational, and societal barriers. These are followed by an analysis of emerging debates and implications, with an emphasis on how these challenges intersect and reinforce one another. The themes draw from both South African and international studies to offer a well-rounded and comparative perspective of the problem.

4.1 Personal Barriers

Personal barriers refer to internal or individual-level challenges that women encounter throughout their leadership journeys. The most frequently reported include work–life conflict, burnout, and imposter syndrome. In the retail sector, long and irregular working hours, especially during peak trading periods, are incompatible with caregiving duties that disproportionately fall on women (Mgaguli 2019). This 'always-on' culture often discourages women from pursuing or sustaining leadership roles (Lloyd 2023). Retail's inflexible schedules also exacerbate stress and limit women's ability to balance professional and domestic expectations (Ahmed et al. 2022).

Psychological barriers further constrain progression. Imposter syndrome, defined as the internal belief of being undeserving of success, is common among women in male-dominated leadership spaces (Clance and Imes 1978; Rdebe 2023). Women often hesitate to apply for promotions or seek development opportunities due to feelings of self-doubt and performance pressure (Huecker et al. 2022). Burnout and emotional fatigue triggered by high workloads, inadequate recognition, and isolation were prevalent among middle managers (Lloyd 2023). These accumulated pressures create a cycle of disengagement and underrepresentation at senior levels.

4.2 Organisational Barriers

Organisational barriers remain the most visible and systemic constraints to women's leadership progression in the retail sector.

A major challenge is the dominance of masculine leadership norms, which associate authority with assertiveness and control traits traditionally attributed to male leaders (Broadbridge 2007). Women who adopt collaborative or empathetic leadership styles are often perceived as weak or unsuitable for advancement (Edwards 2016; Lloyd 2023). Many women encounter a 'promotion cliff', where progression into senior roles slows or stops altogether despite strong performance at lower levels

(Commission for Gender Equality (CGE) 2021). This barrier is reinforced by biased evaluation criteria, informal succession planning, and a lack of gender-sensitive leadership models (Alqahtani 2019).

Mentorship and sponsorship gaps significantly limit women's visibility and development. Women often lack access to influential mentors and are excluded from key informal networks that shape career trajectories (Edwards 2016; Atwood 2022). This absence of female role models in senior positions also weakens leadership pipelines (Burton and Robertson 2025).

Tokenism was reported in several studies, where women are appointed to leadership positions without decision-making power or adequate support structures (Broadbridge 2010; Kele et al. 2022). Such appointments may satisfy equity metrics but fail to address the underlying institutional biases that hinder real progress.

Lastly, pay inequity persists in many retail settings. Women often receive lower compensation than their male counterparts in comparable positions, reinforcing perceptions of their undervaluation and leading to attrition from the leadership pipeline (Commission for Gender Equality (CGE) 2021).

4.3 *Societal and Cultural Barriers*

Societal norms and cultural expectations continue to exert a strong influence on women's leadership experiences, particularly in the South African context.

Social role theory explains how leadership traits like decisiveness and authority are typically viewed as masculine, while women are expected to display nurturing and communal behaviours (Eagly 1987). These gendered expectations are deeply embedded in workplace cultures and customer attitudes, affecting hiring and promotion decisions (Leidner 1993; Alqahtani 2019). Women who do seek leadership positions frequently encounter a double bind: they are seen as too weak if they conform to feminine standards, or too aggressive if they adopt assertive tactics (Eagly and Karau 2002; Rdebe 2023).

In South Africa, the concept of intersectionality introduces further complexity. Gender intersects with factors such as race, class, and language, creating additional barriers for Black women (Kele et al. 2022; Khosa 2022). In multinational or high-end retail environments, women from historically disadvantaged backgrounds report greater scrutiny and stereotyping, often being perceived as less competent or promotable.

4.4 *Ongoing Debates and Emerging Trends*

The literature reflects important shifts in how women's leadership challenges are conceptualised.

There is a growing consensus that merely having policies is not enough to combat deep-rooted inequalities. Although many retailers have adopted employment equity policies, these are frequently implemented without meaningful structural reform (Commission for Gender Equality (CGE) 2021). As a result, compliance becomes symbolic rather than transformative (Edwards 2016).

Intersectionality, originally framed by Crenshaw (2013), is increasingly acknowledged but rarely operationalised in South African retail research. While many studies recognise the importance of race, age, and class, few fully integrate these dimensions into their analyses (Kele et al. 2022). Generational values also represent a turning point. Millennial and Gen Z leaders are reshaping leadership expectations, prioritising authenticity, psychological safety, and work–life balance (Atwood 2022; Burton and Robertson 2025). These values challenge the dominance of hierarchical and performance-driven models traditionally associated with leadership.

Digital transformation may offer new opportunities through flexible work and remote management. However, technology could also heighten exclusion for frontline retail workers, as many are women who remain tied to physical roles without similar flexibility (Ahmed et al. 2022).

4.5 Synthesis

The analysed literature highlights that barriers do not operate in isolation. Instead, personal, organisational, and societal factors intersect in complex ways that create a cumulative disadvantage for women. For example, feelings of imposter syndrome may originate from a lack of access to mentorship networks, which in turn is reinforced by societal stereotypes regarding women's leadership capabilities. Theoretical models like glass ceiling theory and social role theory remain highly relevant in explaining how these barriers are constructed and sustained. At the same time, the increasing emphasis on intersectionality, generational change, and the limitations of policy-driven solutions points to a need for deeper structural reform and cultural transformation.

5 Conclusion

5.1 Summary of Key Findings

This systematic review has integrated over 70 sources of literature on the barriers to women's leadership in Gauteng's retail sector. Despite women comprising a large share of the retail workforce, they remain critically underrepresented in senior leadership positions. This disparity is not due to a lack of capability or ambition, but rather stems from a complex mix of personal, organisational, and societal obstacles that systematically hinder women's advancement. On a personal level, challenges such as work–life imbalance, burnout, and imposter syndrome continually undermine women's leadership aspirations and confidence. These issues are exacerbated by the unpredictable and high-pressure nature of retail work, particularly for women who balance professional and caregiving responsibilities.

At the organisational level, women continue to encounter discriminatory promotion practices, limited access to mentorship and sponsorship, and superficial inclusion. These conditions are underpinned by masculine leadership norms, informal gatekeeping, and workplace cultures that undervalue relational or democratic leadership styles, which are commonly associated with women.

At the societal level, gender stereotypes, cultural expectations, and intersectional discrimination remain deeply embedded in perceptions of leadership suitability, particularly for Black South African women—race, class, and language compound gender-based exclusion and marginalisation in leadership pipelines.

The application of glass ceiling theory (Loden 1978) and social role theory (Eagly 1987) helped frame these findings, offering insight into how structural barriers and gender role expectations interact to maintain leadership inequality. While South Africa has enacted progressive policies and legislation, such as the Employment Equity Act, the implementation often lacks accountability, cultural integration, and enforcement, and thereby limiting transformative potential.

5.2 *Implications for Practice*

The findings of this review highlight several practical and urgent recommendations for retail organisations, HR practitioners, and policymakers:

Mentorship and Sponsorship Pipelines. Institutions should implement formal, resource-backed mentorship and sponsorship programmes. The programmes should include mechanisms to identify and support high-potential women into leadership roles actively—not merely guide them.

Gender-Sensitive Leadership Development. Training programmes must challenge traditional leadership norms, redefine what effective leadership looks like, and equip women with strategic visibility and board-level readiness.

Cultural Audits and Accountability Structures. Retail organisations must go beyond diversity statements and engage in organisational self-assessment. Regular cultural audits and measurable performance indicators should be mandated to evaluate real progress on inclusion.

Structured Flexibility. Flexible work policies are essential in retail. These must be formalised and equitably accessible across all levels, including shift-based and frontline roles, to alleviate the work–life conflicts that disproportionately impact women.

Intersectional Inclusion. All interventions must reflect the experiences of Black, working-class, and young women, recognising that gender equity cannot be addressed in isolation from race, class, or age dynamics.

5.3 Implications for Future Research

Several gaps emerged across the reviewed literature, pointing to a clear agenda for future inquiry:

Longitudinal research. More career-tracking studies are needed to assess the real-world impact of leadership interventions and policy changes on women's long-term progression in retail.

Post-COVID work structures. The pandemic has reshaped retail work. Future studies should assess how remote work, digital transformation, and hybrid models either enable or constrain leadership access for women.

Deep intersectional analysis. While intersectionality is widely referenced, few studies engage with it empirically. Research should more critically explore how race, class, language, and geography intersect with gender in the South African retail context.

Cross-sectoral comparisons. Comparative studies across retail, FMCG, and hospitality could help determine whether specific industry dynamics uniquely hinder women's leadership, supporting more targeted reforms.

5.4 Concluding Statement

Achieving gender equity in leadership within the retail industry extends beyond mere representation; it necessitates the dismantling of institutional, cultural, and ideological barriers that continue to exclude women from decision-making spaces. This review has shown that while policies and programmes have created formal opportunities, deep-seated social norms and organisational practices still obstruct true transformation. Leadership in retail needs to shift from inflexible, masculine models to inclusive, adaptable, and values-oriented approaches that represent the diversity within South Africa's workforce. For this transformation to occur, organisations must implement a comprehensive strategy that focuses on structural reform, cultural evolution, and accountability in leadership. By identifying these obstacles and highlighting the systemic trends, this review contributes to the growing demand for leadership justice, not just the right to lead, but the right to lead authentically, safely, and with legitimacy.

References

Ahmed, A.M.M., Khaled, A.S., Thabit, S.A.A.: Retail sector: a progressive path for women. Int. J. Res. Rev. **9**(11), 294–301 (2022). https://doi.org/10.52403/ijrr.20221139

Alqahtani, T.: Barriers to women's leadership. Granite J.: Postgraduate Interdisciplinary J. **3**(2), 34–41 (2019)

Atwood, P.J.: *Millennial Leadership in the Retail Industry* (Doctoral Dissertation). Walden University (2022)

Babic, A., Hansez, I.: The glass ceiling for women managers: antecedents and consequences for work–family interface and Well-being at work. Front. Psychol. **12**, 618250 (2021). https://doi.org/10.3389/fpsyg.2021.618250

Bombuwela, P.M., De Alwis, A.C.: Effects of glass ceiling on women's career development in private sector organizations: case of Sri Lanka. J. Competitiveness. **5**(2), 3–18 (2013). https://doi.org/10.7441/joc.2013.02.01

Broadbridge, A.: Dominated by women: managed by men? The career development process of retail managers. Int. J. Retail Distrib. Manag. **35**(12), 956–974 (2007)

Broadbridge, A.: Social capital, gender and careers: evidence from retail senior managers. *Equality divers. Incl. Int. J.* **29**(8), 815–834 (2010)

Broadbridge, A., Simpson, R.: 25 years on: reflecting on the past and looking to the future in gender and management research. British. J. Manag. **22**(3), 470–483 (2011)

Burton, M., Robertson, C.M.: Authentic, challenged, and balanced: an analysis of perceived career and life fulfilment for women working in retail. Proc. Int. Textile and Apparel Assoc. Annu. Conf. **81**(1) (2025)

Clance, P.R., Imes, S.A.: The imposter phenomenon in high-achieving women: dynamics and therapeutic intervention. *Psychother.: theory res. Pract.* **15**(3), 241–247 (1978)

Commission for Gender Equality (CGE): Transformation in the public and private sector. CGE, Johannesburg (2021)

Crenshaw, K.: Demarginalizing the intersection of race and sex: a black feminist critique of antidiscrimination doctrine, feminist theory and antiracist politics. In: Feminist Legal Theories, pp. 23–51. Routledge, London (2013)

Eagly, A.H.: Sex differences in social behaviour: a social-role interpretation. Erlbaum, Hillsdale (1987)

Eagly, A.H., Chin, J.L.: Diversity and leadership in a changing world. Am. Psychol. **65**(3), 216–224 (2010)

Eagly, A.H., Karau, S.J.: Role congruity theory of prejudice toward female leaders. Psychol. Rev. **109**(3), 573–598 (2002). https://doi.org/10.1037/0033-295X.109.3.573

Edwards, J.: *Barriers and inequalities to women progressing to management in the retail sector*. Doctoral dissertation. National College of Ireland (2016)

Foster, C., Whysall, P., Harris, L.: Female career progression in retailing. Int. J. Retail Distrib. Manag. **35**(12), 975–981 (2007)

Hill, C., Miller, K., Benson, K., Handley, G.: Barriers and Bias: the status of women in leadership. American Association of University Women, Washington, DC (2016)

Huecker, M.R., Shreffler, J., McKeny, P.T., Davis, D.: Imposter phenomenon. In: StatPearls. StatPearls Publishing, Treasure Island (2022)

Kele, J.E., Cassell, C., Ford, J., Watson, K.: Intersectional identities and career progression in retail: the experiences of minority-ethnic women. Gender Work Organ. **29**(4), 1178–1198 (2022)

Khosa, D.: Leadership development framework for women in policing in South Africa: overcoming leadership barriers. Int. J. Res. Bus. Soc. Sci. **11**(7), 110–120 (2022)

Kossek, E.E., Su, R., Wu, L.: 'Opting out' or 'pushed out'? Integrating perspectives on women's career equality for gender inclusion and interventions. J. Manag. **43**(1), 228–254 (2017). https://doi.org/10.1177/0149206316671582

Leidner, R.: Fast food, fast talk: Service work and the routinization of everyday life. University of California Press, Berkeley (1993)

Lloyd, L.A.: *Career growth challenges faced by female middle managers: Case of the fashion retail sector* (Doctoral Dissertation). California State Polytechnic University, Pomona (2023)

Loden, M.: Implementing diversity. McGraw-Hill, New York (1978)

Mgaguli, B.: *Barriers encountered by women in career development towards senior management positions in the management industry* (Doctoral dissertation). Cape Peninsula University of Technology (2019)

Moher, D., Liberati, A., Tetzlaff, J., Altman, D.G.: Preferred reporting items for systematic reviews and meta-analyses: the PRISMA statement. BMJ. **339**, b2535 (2009). https://doi.org/10.1371/journal.pmed.1000097

Okafor, E.E., Fagbemi, A.O., Hassan, A.R.: Barriers to women leadership and managerial aspirations in Lagos, Nigeria: an empirical analysis. Afr. J. Bus. Manag. **5**(16), 6717–6726 (2011)

Rdebe, N.Z.: *Barriers to promotion faced by female engineers at an FMCG firm in Johannesburg, South Africa* (Master's Thesis). University of the Witwatersrand (2023)

Rota, K.G.: *Exploring the barriers and inequalities faced by women from the millennial generation in the workplace* (Doctoral Dissertation). National College of Ireland (2021)

Sharma, S., Kaur, R.: Glass ceiling for women and work engagement: the moderating effect of marital status. FIIB Bus. Rev. **8**(2), 132–146 (2019). https://doi.org/10.1177/2319714519845770

Shongwe, M.N.: The unequal status afforded to women in South African universities: a journey towards inclusion. University of Johannesburg (2019)

Teutenberg, S.: Retail sector report 2021. Labour Research Service (2021) Accessed 23 Nov 2021

Thomas, J., Harden, A.: Methods for the thematic synthesis of qualitative research in systematic reviews. BMC Med. Res. Methodol. **8**, 45 (2008). https://doi.org/10.1186/1471-2288-8-45

Tranfield, D., Denyer, D., Smart, P.: Towards a methodology for developing evidence-informed management knowledge by means of systematic review. Br. J. Manag. **14**(3), 207–222 (2003). https://doi.org/10.1111/1467-8551.00375

USDA Foreign Agricultural Service: Retail foods: South Africa. United States Department of Agriculture (2023) Accessed 2023

Wholesale and Retail Sector Education and Training Authority (W&RSETA): Wholesale and retail sector skills plan 2021/2022. W&RSETA (2021)

A Three-Segment Pyramid as a Framework for Academic Identity Reconstruction in the VUCA World

Bernadette Kruger and Nicolene Barkhuizen

Abstract Academics in higher education must present innovative approaches and operate optimally in response to the AI phenomenon within an increasingly short timeframe, signifying the volatile, uncertain, complex, and ambiguous (VUCA) environment. VUCA is intensified by the paradox of not only keeping pace with rapid transformation but also innovating teaching and research. Underlying processes facilitate academic identity reconstruction. Thus far, there has been little evidence of these processes that academic staff engage in to repeatedly redesign their identity and little is known about how academics navigate while they are in a perpetual state of constant change. This study applied a qualitative approach, with semi-structured interviews of qualified staff, with academic titles such as Doctor or Professor ($N = 15$), within selected private higher education institutions. The data was analysed using Atlast.ti (version 2024). The findings showed a fluid academic identity illustrated in a three segment pyramid. The first segment is *self-centred vertical leadership development*, the second is *reflective practice*, and third is *horizontal leadership development*. Self-centred vertical leadership development encompasses self-development and self-regulated learning and the premises of vertical development. Reflective practice can enable academics to leverage their experiences to learn about themselves, and their work. Horizontal leadership is a style of leadership that depends on reciprocity and trust rather than the structured corporate hierarchy. Reciprocal knowledge sharing is identified as exceptionally valuable. It is recommended that academics intentionally engage in vertical leadership development and reflective practice, as these processes facilitate the reciprocal sharing of knowledge and experience. This, in turn, supports the continuous shifting role identities of academics within a VUCA environment.

Keywords VUCA · Academic identity · Three-segment pyramid · Vertical leadership · AI phenomena

B. Kruger (✉) · N. Barkhuizen
UFS Business School, University of the Free State, Bloemfontein, South Africa
e-mail: KrugerB1@ufs.ac.za

E. H. Redda (ed.), *Proceedings of the Social Sciences International Research Conference, 2025*, https://doi.org/10.1007/978-981-95-9489-4_23

1 Introduction

AI and other technological advancements result in constant changes to academia, requiring academics to adapt or be left behind. The era of the AI phenomenon in higher education not only generates opportunities for innovation but also continues to drive constant and disruptive change globally. As a result of the rapidly changing, unpredictable conditions and the academic who is compelled to transition between roles at a pace that often outstrips traditional modes, the volatile, uncertain, complex, and ambiguous (VUCA) world is created. While the existence of multiple roles is not new, the ongoing need to integrate and perform them effectively underscores the complexity and adaptability required within the profession of academia. Such a VUCA environment unleashes frequent disruption, leading to a cascade of changes and complicating the navigation process (Jengwa and Pellissier 2022). For example, these disruptors may give rise to resistance, thereby undermining rather than fostering resilience. While the identity is reconstructed, it is common for academics to experience emotions such as frustration or overwhelm, leading up to phenomena such as quiet quitting, imposter syndrome, isolation, or withdrawal. Therefore, this paper focuses on academics in higher education, examining how they navigate and reconstruct their professional identities within this shifting landscape. The paper introduces a three-segment pyramid as a framework for understanding academic identity reconstruction and to pay closer attention to its implications for higher education in South Africa (Hedding et al. 2020). Despite the unprecedented pace of change, academics have been keeping ahead—continually adapting, leading innovation, and integrating new technologies into research and pedagogical practices (Mah et al. 2025). The emergence of AI has flagged the increasingly complex and evolving role of academia, particularly within the private higher education sector (Sthapit 2020). Regrettably, certain facets are often neglected during such a fast-paced transitioning process, for example, the transformation of digital platforms and educational delivery methods, which affects not only academics but also students. While academics grapple with transitioning to change, it is essential to acknowledge the greater struggle of students in keeping pace with often abrupt and unanticipated modifications (Shakeeb 2020). All individuals display a distinct identity in both their personal and professional capacities. The central challenge for academics and institutional leaders in today's volatile world lies in the limited time available to adapt swiftly while integrating and accommodating AI into their core responsibilities of teaching, research, and service. They continually have to ask themselves critical questions about the strategies they adopt, the actions they take, and the paths they choose to remain effective. Change creates a paradox for academics when, for instance, they must adapt to new technology or a different teaching platform, representing ambiguity. Ambiguity can be defined as having or *expressing* more than one *possible meaning,* sometimes *intent ionally* On the one hand, it is good to implement new technology, and the software increases efficiency (Oncioiu and Bularca 2025). On the other hand, when staff need to learn and become accustomed to the new software and how to operate it, it takes up a significant portion of their time. Extra hours are spent on the process, and just as the new skills have been acquired, another change is introduced or implemented due to a new technological advancement, and the cycle starts again. This amounts to an exercise that can be seen as futile in the academic role. Of course, this could also cause a digital divide among academic staff members (Mhlanga and Moloi 2020). It would be advantageous to academics if a structured way could be outlined to manage this equivocality.

2 Research Objective

To establish a framework for academic staff to navigate academic identity reconstruction in a VUCA world.

2.1 Proposition

Amid the time pressures of a volatile world, academics and leaders face the urgent challenge of integrating technological change into their core roles. There is an evident need for research-based pathways, such as the proposed three-segment pyramid, to guide academic identity reconstruction in the VUCA era.

3 Literature Review

3.1 VUCA Framework

Includedletter, VUCA environment, the paradox of the academic task is highlighted. Academics must strike a balance between autonomy and teamwork to avoid catastrophic outcomes (Sinha and Sinha 2020). The unpredictable nature of information and shifting pedagogical demands driven by AI and changing student needs require constant evolution in teaching methods. Therefore, the VUCA framework, widely recognised for its value in helping organisations navigate dynamic and rapidly evolving environments (Petrie 2015), serves as the foundation for this study's exploration of academic identity reconstruction. The issues to be addressed that cause the start of reconstruction of the identity include contending with AI education and policy development, regulating AI use, and addressing ethical concerns (Eryenyu et al. 2024). It drives the academic to continuously nurture critical thinking skills in teaching and learning.

Volatility—Rapid and Unpredictable Changes. The VUCA environment is mirrored in the volatility of situations and the resultant lack of confidence, predictiveness, and clarity that academics have to contend with. Furthermore, they have to maintain a delicate balance between keeping the traditional academic standards and operations and embracing AI and global academic trends and practices.

Several unforeseen and evolutionary forces impact higher education: changes in the perception of what constitutes education, the role of academic staff, and the environment in which educational institutions operate. Such haphazard changes are often associated with catastrophe theory as a framework for describing instability.

Uncertainty—Limited Predictability and Lack of Clarity. Characteristics of uncertainty are that it is difficult to make a forecast of the future, and that there is a lack of knowledge. Uncertainty is linked to factors such as the unknown, long-term impact of technology, evolving industry demands, and unexpected disruptions. The rapid push to adopt AI-driven teaching tools, for example, has plunged academic staff into a state of uncertainty and professional disorientation. Despite recognising the benefits of AI—such as enhanced student engagement and personalised learning—62% of lecturers in a recent study (Mbanjwa 2025) reported feeling unprepared for integration, mainly due to a lack of training, inadequate infrastructure, and limited institutional support. Only 8% felt fully equipped. This abrupt technological shift, coupled with ethical concerns and fears of job displacement, has left many lecturers struggling to redefine their roles in a digital teaching environment. The compressed timeframe for adaptation, without sufficient guidance or upskilling opportunities, underscores the psychological and professional strain staff endure in navigating AI adoption within higher education (Mbanjwa 2025).

Complexity—Interconnectedness and Multiple Variables. Complexity theory can be viewed as a theory of change and adaptation that details how change occurs in systems and academic staff's principles and role identities to flourish in turbulent environments (Barnes et al. 2021). It involves situations such as restructuring, lack of adequate resources, and the development of specialists. As an example, the new World of Work is demanding high-quality online teaching, and reorganising teaching processes in compliance with this demand depends on the academic's ability to apply various online technologies and tools. To achieve effective teaching management, it is necessary to use tools such as learning management system platforms, live streaming platforms, social networking software, and online questionnaires or meetings. In the online teaching process, academics are not only knowledge spreaders, but they also assume the roles of builders, leaders, promoters, and facilitators within the learning community. The list of new roles involuntarily imposed on the academic grows ever longer (Martin et al. 2020). In the VUCA environment of private education, platform thinking can be a valuable approach to navigating challenges and uncertainties, as putting platform-based mechanisms at the core of transformation can foster collaboration, innovation, and agility.

Ambiguity—Conflicting Interpretations and Multiple Meanings. Conflicting interpretations and multiple meanings concerning research and teaching exist in the PHE domain. There can be, for example, a lack of consensus on mission and values, different interpretations of data, and changing student expectations.

3.2 Reconstruction of Academic Identity

Due to the VUCA environment, the PHE academic role is constantly evolving to align with constant change, causing the academic identity to be continually reconstructed and therefore remain in a state of flux (Garraway 2020). Huang and Guo (2019) described the construction, deconstruction, and reconstruction of the academic identity as a dynamic process of formation, negotiation, and/or renegotiation. Therefore, academic identity is fluid rather than fixed. Academic identity will shape how a new idea is understood, the priority it is given, and the emotional and professional responses it stimulates. New ideas will be adopted to the extent they are perceived to align with and further professional identity. The current study undertook to examine how academics in South Africa navigate and reconstruct their professional identities within this shifting landscape, and to present a structured framework for understanding academic identity reconstruction.

4 Methodology

4.1 Research Design

A qualitative design was the most suitable in this instance. Qualitative research aligns well with the constructivist interpretivist philosophy and focuses on the viewpoints of participants as well as their feelings and thoughts (Schurink et al. 2021).

4.2 Sampling

The sampling method employed was convenience sampling, a category of non-probability sampling. This sampling was used to take advantage of the researcher's access to academics in private higher education institutions focused on distance learning delivery mode.

4.3 Research Instrument

The semi-structured interview schedule comprised questions posed to the sample ($N = 15$) of seasoned and mid-level academic staff members (see demographics in Table 1) about their perspectives and experiences on navigating constant change. The main question related to their view of the new role identity of the academic in private higher education, while related questions were asked about their experience of the VUCA context in higher education and how they view the new role identity.

Table 1 Demographic details of participants

Title and no		Age	Qualification	Years of experience
Dr1		50–60	PhD public admin	21
Prof2		50–60	PhD business leadership	18
Dr3		50–60	DBA	20
SA4		60–70	MCom	15
Dr5		60–70	PhD	12
Dr6		50–60	PhD	6
Dr7		40–50	DBL	4
Prof8		40–50	DCom marketing	12
Dr9		60–70	PhD history	15
Dr10		50–60	PhD entrepreneurship, PhD leadership	15
Prof11		50–60	PhD education	5
Dr12		50–60	PhD	13
SA13		40–50	BCom marketing	19
SA14		40–50	M.Admin	18
SA15		40–50	M.Admin	19

SA, Senior Academics. *Source* Author's own

4.4 Data Collection

Interviews were scheduled at convenient times for the academic participants. The interviews took the form of an informal conversation but were guided by the interview schedule. In semi-structured interviews, asking one question, getting a response, and then moving on to the next topic is rare. Structure was provided for the participants, so that they were not expected (or encouraged) to recite an entire story. However, on another level, the researcher used prompts (keywords) under specific conditions, which allowed for a flexible approach. Each participant revealed something interesting, new, and relevant to the research study that had to be explored or elaborated upon (DeJonckheere and Vaughn 2019). Although all participants were in the role of academic, the evolution of their roles resulted in some of them occupying additional positions such as CEO, Head of Campus, supervisor, moderator, part-time academic, and independent consultant in private higher education.

4.5 Data Analysis

Thematic analysis was employed, which is a flexible approach not tied to a particular philosophical position. This allowed for searching for themes or patterns that occur across the data sets of the series of 15 interviews conducted in this case. The thematic analysis involved the initial coding of the data sets related to the research question of this study (Saunders et al. 2016). Constant comparison ensured that all data were systematically compared to all other data in the data set, and that all data produced were analysed rather than potentially disregarded on thematic grounds. This method of analysis was primarily used because the experiences of seasoned academics in a VUCA environment of PHEIs were not well understood.

Saturation, as described by Saunders et al. (2016:714), is 'the stage when any additional data collected provides few, if any, new insights.' The initial codes were refined and reduced after it became apparent that no more new codes were derived from the interviews.

This refinement process also made use of the Atlas.ti feature, which considered the groundedness of each code. The groundedness in Atlas.ti of a category code is the number of quotations coded by all of its sub-codes. Therefore, 'grounded' shows how many times a code was applied throughout the texts. During this stage, codes with low groundedness were either consolidated or placed into a separate folder, while codes with high groundedness were subjected to potential splitting or re-coding to ensure that the highest grounded codes formed the basis for the next level of categories.

5 Findings and Discussion.

The findings are presented in Table 2, delineating alignment with the research objective, the interview question (IQ), and the sub-codes derived from the excerpts of the responses provided by the academic interviewees. This systematic arrangement effectively captures the essence of the theme that emerged from the participants' dialogues and served as a comprehensive reference for the development of the three-segment pyramid.

5.1 Theme.1: Academic Identity Reconstructed

5.2 Discussion of the Sub-codes and Illustrative Extracts

Self-regulated learning—defined as self-directed thoughts, emotions, and actions aimed at personal growth (Schunk and Ertmer 2000)—emerges as a key code from responses by academics to the pressures of the VUCA environment, often in contexts lacking institutional support due to the speed and intensity of change. Dr. 10 states; *It would be to have a definitive which is one of the strategies in your staff development*. When SA14 above mentions; *Maybe it's what I put on myself, the added burden* the burden of taking on the responsibility to care deeply for the student's success, the construct of intrinsic motivation is derived. She draws from a state of introspection that conscientiously recognises the impact of their actions on the student (Le et al.

Table 2 Theme 1: Academic identity reconstructed—sub-codes. Sub-codes: Self-development, individual, horizontal development, reflexive practice in academia

Interview question: How do you view the new role identity of the academic?	
Sub-codes identified	Illustrative extracts from interviews
Self-development Self-regulation	*So we should be deliberate, we need to implement actions where we can assist them to develop themselves, current KPIs is a tick box exercise. Prof 8* *Essential support is needed from management in any transition process, or it can have a detrimental effect. Academic freedom is important, as having time to explore and experiment with what is most important and enjoyable in an academic's role.* Dr. 9 *So if you want to educate yourself (autonomy) and be on top of your game with the latest technology and you have to attend webinars and those things, sometimes you have to get up at 2:00 o'clock in the morning to join for an hour. So you just have to because we are living in a global village.* Prof 2 *Maybe it's what I put on myself the added burden of knowing when that individual walks out the door and they go in a blind, play around or test the ideas that I've shared. I have a direct impact on their business, that business of this now fail or succeed. When it succeeds, I can't take credit direct, but when it fails, I blame myself for that as well. SA 14*
Individual in the institution	*But that comes back to the whole teaching philosophy and what we do and remember, research does not bring in money. Not for us but they want us to do it, because they want the esteem. But not willing to pay for it. Dr 7* *Downshifting among academics is an issue, and the fact that VUCA is a wicked problem that can either hinder or benefit communication and the sharing of initiatives.* Prof 11 *But as an academic, I think you have enough time to research to investigate to familiarise yourself to get to grips with the situation and to be able to absorb what is happening and, therefore, to adapt to the changes. Dr 12* *I think it depends on the age of the academics. Or sometimes, if you're used to doing things, the way you're doing it, you're not open to change. SA 13*
Horizontal leadership development Reflective practice	*So you must have the bottom-up, horizontal level, and peer-to-peer change management, where people are actually sharing the best practice, service and understand you know, that it can be a positive experience and can be well managed. Prof 8* *It would be to have a definitive which is one of the strategies in your staff development strategy. Provision should be made, resource wise, time wise. Maybe have workshops, other developmental interventions to expose the academics to these things, and if it's not the expertise internally to do that, then get the expertise externally. And we'll share it around all these things of how education is changing. Dr 10* *I think it's very rare that people talk about the project based on project management knowledge. And we never speak to each other on that; on the foundation of our project management knowledge, we never interact. Dr 3*

2024). Prof 8 states; *So we should be deliberate, we need to implement actions.* With each change academics face, their academic identity (professional personality) is altered, and therefore, it is continuously reconstructed. Emotions are involved, and constant change can lead to unstable emotions (Apostolidis et al. 2022). Dr. 9 highlights; *Essential support is needed from management* institutional support from management or leadership remains unclear in terms of whose responsibility it is to support the academic in a constantly changing environment and up to what point. Managers can, in this instance, attest that they do offer training and provide leave, which is the correct approach in a VUCA environment. Prof 2 states, *So if you want to educate yourself [autonomy]…* At what level, then, do you allow the academic to intervene and take responsibility for themselves (Maddock 2023)? Horizontal leadership development propagates the sharing of knowledge learned from initial experiences. Dr. 3 reflects on his experience; *We never interact.* Prof 8 also heeds *where people are actually sharing the best practice, service and understanding.* If this is not happening, feelings of isolation and fatigue are amplified, which in turn influences the reconstruction process of the academic identity (Apostolidis et al. 2022). The sub-codes and illustrative extracts informed the development of the three-segment framework discussed below.

5.3 Academic Identity—Three-Segment Pyramid

The academic identity engages in continuous reconstruction essentially due to additional roles added with a very short timeframe to adapt. The three-segment pyramid (Fig. 1) illustrates essential activities that academics naturally engage in during the involuntary reconstruction of their academic identity due to a VUCA environment, as reflected in the illustrative quotes in Table 2.

This three-segment pyramid signifies the reconstructed academic identity and the underlying processes that facilitated the construction process in three segments. The academic identity remains in a state of flux because it continually redesigns itself due to persistent exposure to constant change. The three segments of the pyramid represent distinct processes that contribute to this evolution, as evident from the illustrative extracts presented in Table 2. At the pinnacle of the pyramid is Self-Centred Vertical Leadership Development, denoting that personal development occurs within the individual and progresses vertically. This process entails focusing on individual growth and leadership skills, enabling academics to navigate the changing landscape effectively.

The second segment is dedicated to reflective practice, emphasising the significance of introspection regarding the learning experiences encountered during the exposure to change. Amidst rapid transitioning due to continuous change, this reflective practice aids in refining and enhancing current academic practices.

The third and final segment represents Horizontal Leadership Development, signifying growth in a lateral direction, where knowledge gained is shared collaboratively

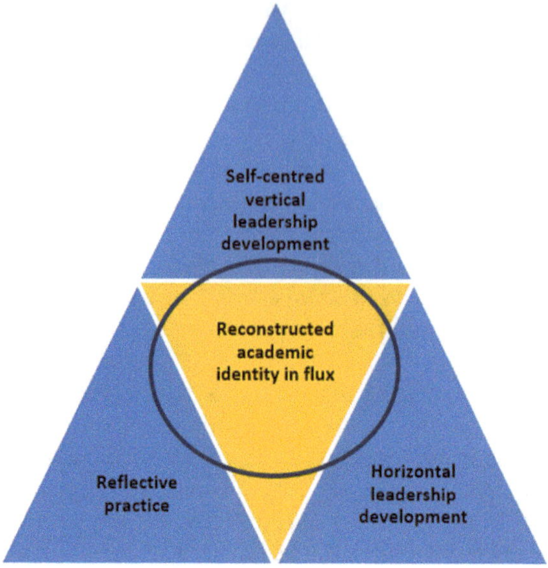

Fig. 1 Three-segment pyramid

with peers. This involves engaging with various stakeholders, including institutional leaders, staff members, and students, to foster a culture of horizontal collaboration.

This three-segment pyramid has been created to display the three essential segments that form part of the academic identity reconstruction process. It is based on the concept of vertical development of role identity and is related to vertical leadership development. The concepts behind vertical leadership development concentrate on expanding an individual's thinking capacity and ability to manage change (Petrie 2015), positioning it as a unique approach to helping leaders and organisations navigate the VUCA environment. The topic of vertical leadership development is engendering a developing body of research. The concept and related research findings, however, have not yet been introduced or applied in most traditional personal development contexts (Petrie 2015).

The three segments that are part of the academic identity reconstruction process have been identified as self-centred vertical leadership development, horizontal leadership development, and reflective practice. If these three aspects are developed concomitantly, the individual will remain resilient and be continuously engaged in improving the professional identity of academics.

Segment 1—Self-Centred Vertical Leadership Development. Self-centred vertical leadership development encompasses self-development, self-regulated learning, and the premises of vertical development. The broad meaning of self-development is the process in which a person grows or changes and becomes more advanced through their own efforts. This is also in line with personal development, as shown by Coopersmith (Coopersmith 2021). Self-development leads to self-organising in the workplace. For an academic to access the next stage of growth and improvement and to grow into an academic leader, the key is vertical development. These stages of development, building on experiences, and learned lessons, form the foundation for elevating from one stage to the next, which requires complete paradigm shifts (Jones et al. 2020).

Vertical development is about growing through successive stages to broaden and expand your capacity to see the world and shift your mindset to take in more complexity. This is not a new concept, but it is still fairly new in the context of the VUCA environment, specifically considering three important elements that are directly aligned with a complex and volatile environment. *Heat Experiences* refer to incidents where professionals must face complex situations that disrupt and disorient their habitual way of thinking. *Colliding Perspectives* are found where professionals have an opportunity to challenge their existing mental models when they are exposed to others representing different worldviews, opinions, backgrounds, and training. The third element is *Elevated Sensemaking*. As professionals process and make sense of new perspectives and experiences, they enter an elevated stage of vertical development (Ellison 2020). These elements resonate with the VUCA framework, and serves as a driving force and as shown from the responses in this study, that academics engage in self-development and self-regulated learning in a complex and uncertain environment that enables them to be better equipped for the future. One of the most difficult transitions that professionals make is to shift from one role to another, and often revert again to a previous role while at the same time learning new skills, as when transitioning from being an independent contributor to being a manager (Jones et al. 2020). It takes a certain mentality to succeed in the role of a teacher, for example, but it requires a whole different paradigm to succeed as a manager. A certain mentality is required for a professional to succeed in a constantly changing environment, where roles are added and expectations are imposed, not only by oneself but also by the leadership of the institution. In the typical sense of vertical leadership, the rationale concerns the improvement of communicating vertically and growing and learning, but in the case of an academic VUCA environment, it is required of professional academics to vertically grow in themselves, by accepting change and reforming the academic identity continuously at a rapid pace (Coopersmith 2021). This paper demonstrates that seasoned academics continually grow and learn within a VUCA environment, adopting a leadership identity.

If academic leaders are to develop vertically, it is essential that their focus shifts to the internal mechanisms that govern how they see, process, and operate, that is, their mindsets with each changing demand. Vertical development interventions prepare leaders to continuously learn and develop in response to the changing demands of the environment, thereby increasing their self-awareness. Vertical leadership development requires a philosophy that shifts from focusing on what leaders know to understanding how leaders make sense of the knowledge they acquire (Jones et al. 2020).

Segment 2—Reflective Practice. Reflection in PHE is particularly useful when events require an examination of personal experience and an in-depth consideration of situations beyond one's control (Faller et al. 2020). Critical reflection takes reflective practice to a deeper level, and reflective practice can enable practitioners to exploit their experience to learn about themselves, their work, and how they relate to the wider society and culture. It assists in developing strategies to bring matters into the open and frame appropriate and searching questions never asked before. Reflection and self-learning occur in parallel. It is essential that academic staff members take the time to reflect on their experiences, examine their assumptions, and evaluate their outcomes. For example, in this study, the rapid transitioning due to the VUCA environment is recognised, and also the consequence, namely that scholars may occasionally *get it wrong* or misalign the precise standards set forth by governing bodies such as the Council for Higher Education (CHE) or their prescriptions. However, it is essential to acknowledge that these experiences of *error* and subsequent learning are vital components of academic growth. By encountering and learning from such challenges, academics can refine their understanding and improve their future performance in fulfilling academic tasks. Although reflective development is crucial, it is not happening in the fast-paced VUCA environment, or not enough time is devoted to this crucial segment. Moreover, current expectations of constant change and 'busyness' make reflection a luxury. Without proper attention to the why and how of intentional reflection, adult learning processes will suffer in a VUCA environment, and little knowledge will be gained from different perspectives (Faller et al. 2020).

Reflection can lead to greater adaptability, resilience, and flexibility, which are critical for success in the rapidly changing and unpredictable world of private higher education. There are various tools that academics can use to facilitate reflection on their experiences and responses to unprecedented changes. Arguably, the most effective of these tools is peer discussion or group reflection sessions. By sharing experiences and perspectives with others in similar situations, academics can gain new insights into their responses to change. They can learn from the experiences of others. This can be particularly valuable in the private higher education sector, where academics may feel isolated in their roles and lack opportunities to engage with colleagues.

This segment also highlights the challenges of rapid change in which individuals constantly receive instructions to add to current processes without taking the time to learn from past experiences (Suphasri and Chinokul 2021). Instead of building something innovative based on reflective learning, the focus has shifted to continuous improvement using new processes, structures, and technology, but without critical self-assessment. Consequently, academics are not building upon their reflections, they are merely responding to the institution's leadership perceptions of what is not working. Thus, the key to growth is to not only maintain sustainability and momentum but also to consider their past efforts and their current identity in this dynamic process (Faller et al. 2020).

Segment 3—Horizontal Leadership Development. Horizontal leadership is a style of leadership that depends on reciprocity and trust rather than the structured corporate hierarchy. This type of reciprocal knowledge sharing is very valuable. When knowledge and experience are shared and reciprocated after the academic has engaged with vertical leadership development and reflection, it contributes to the ever-evolving academic identity in a VUCA environment (Yin et al. 2024).

The power of knowledge is enhanced when knowledge is shared, and knowledge sharing has arguably been the foundation of many knowledge management programmes (Al-Kurdi et al. 2020). It is critical to make the sharing of experiences intentional, as when participating in a workshop and international or national conference to discuss and share experiences to build innovative collaboration (Mills et al. 2023). This emphasises the growth and learning that can stem from collective collaboration among colleagues and, more widely, across institutions, disciplines, and social media. Collaboration is a very important mechanism for support amidst the current constant change and transition and leads to long-term connections and mutual growth that could, in turn, bring about deeper insights and relationships, increased creativity, and more innovative results. The continuous change calls for a new understanding of what it means to be educators, particularly online educators (Fox et al. 2021). More creative solutions could be brainstormed in the VUCA environment when two or more academics collaborate and share ideas. Collaboration among academics could lead to new interpretations of dealing with unprecedented change.

Innovative ideas and the concept of initiatives are possible if collaboration is prioritised among academics. The traditional approach to professional development, however, continues to hold academics back from adopting this approach.

6 Conclusion

In conclusion, the VUCA context underscores the disruption of traditional and routine academic practices, while the three-segment pyramid illustrates that, despite limited institutional support and the pressures of responding appropriately to rapid change, academics engage—often unconsciously—in self-development aligned with their personalities and the transitions required. The usability of the three-segment pyramid lies in its ability to serve as a guiding framework for academics and higher education leaders to better understand how adaptation occurs in practice and how it can be supported more deliberately. The implications are significant: by recognising and fostering these adaptive processes, institutions can more effectively prepare academics for shifting role identities and enhance resilience in the face of ongoing disruptions such as the AI revolution. Furthermore, the three-segment pyramid provides academics with a conceptual lens through which to cultivate creative responses and demonstrate agility in navigating disruptive change.

References

Al-Kurdi, O.F., El-Haddadeh, R., Eldabi, T.: The role of organisational climate in managing knowledge sharing among academics in higher education. Int. J. Inf. Manag. **50**, 1–10 (2020). https://doi.org/10.1016/j.ijinfomgt.2019.05.018

Apostolidis, C., Devine, A., Jabbar, A.: From chalk to clicks: the impact of (rapid) technology adoption on employee emotions in the higher education sector. Technol. Forecast. Soc. Change. **182**, 121860 (2022). https://doi.org/10.1016/j.techfore.2022.121860

Barnes, N., du Plessis, M., Frantz, J.: Perceived career management challenges of academics at a south African university. SA J. Hum. Resour. Manag. **19**, 1–13 (2021). https://doi.org/10.4102/sajhrm.v19i0.1515

Coopersmith, K.: Personal development planning and vertical leadership development in a VUCA world. J. Values-Based Leadersh. **15**(1), 1–15 (2021). https://doi.org/10.22543/0733.151.1380

DeJonckheere, M., Vaughn, L.M.: Semistructured interviewing in primary care research: a balance of relationship and rigour. Fam. Med. Community Health. **7**(2), e000057 (2019). https://doi.org/10.1136/fmch-2018-000057

Ellison, S.: Understanding vertical development. Academia. **1–21** (2020)

Eryenyu, C., Owomugisha, G., Biira, S., Waako, P.: The Challenges and Opportunities of Artificial Intelligence Use in Higher Education: the Case of Busitema University. J. Res. Innov. Implic, Educ (2024). https://doi.org/10.59765/upr492jr

Faller, P., Lundgren, H., Marsick, V.: Why and how does reflection matter in workplace learning? Adv. Dev. Hum. Resour. **22**(3), 248–263 (2020). https://journals.sagepub.com/doi/10.1177/1523422320927295

Fox, B., Bearman, M., Bellingham, R., North-Samardzic, A., Scarparo, S., Taylor, D., Thomas, M.K.E., Volkov, M.: Longing for connection: university educators creating meaning through sharing experiences of teaching online. Br. J. Educ. Technol. **52**(5), 2077–2092 (2021). https://doi.org/10.1111/bjet.13113

Garraway, J.: Academics' learning in times of change: a change laboratory approach (2020). https://doi.org/10.1080/0158037X.2020.1792436

Hedding, D.W., Greve, M., Breetzke, G.D., Nel, W., van Vuuren, B.J.: COVID-19 and the academe in South Africa: not business as usual. S. Afr. J. Sci. **116**(7–8), 1–10 (2020). https://doi.org/10.17159/sajs.2020/8298

Huang, Y.T., Guo, M.: Facing disadvantages: the changing professional identities of college English teachers in a managerial context. System **82** (2019). https://doi.org/10.1016/j.system.2019.02.014

Jengwa, E., Pellissier, R.: An operational excellence strategy implementation model for growth in a volatile, uncertain, complex, and ambiguous environment. Acta Commer. **22**(1), a960 (2022). https://doi.org/10.4102/ac.v22i1.960

Jones, H., Chesley, J., Egan, T.: Helping leaders grow up: vertical leadership development in practice. J. Values-Based Leadersh. **13**(1), 1–16 (2020). https://doi.org/10.22543/0733.131.1275

Le, A.T.T., Tran, T.V., Tran, T.M., Phan, T.H.: Intrinsic and extrinsic factors as motivation roles in scientific research activities of professors at Vietnamese universities. SAGE Open. **14**(1), 1–14 (2024). https://doi.org/10.1177/21582440241230838

Maddock, L.C.: Academic middle leaders, middle leading and middle leadership of university learning and teaching: a systematic review of the higher education literature. J. High. Educ. Policy Manag. (2023). https://doi.org/10.1080/1360080X.2022.2160888

Mah, D.K., Knoth, N., Egloffstein, M.: Perspectives of academic staff on artificial intelligence in higher education. Front. Educ. **10**, 1484904 (2025). https://doi.org/10.3389/feduc.2025.1484904

Martin, L., Lord, G., Warren-Smith, I.: Juggling hats: academic roles, identity work and new degree apprenticeships. Stud. High. Educ. **45**(3), 524–537 (2020). https://doi.org/10.1080/03075079.2018.1550478

Mbanjwa, S.T.: Lecturers' readiness and challenges in integrating AI-driven teaching tools at Mangosuthu University of Technology. Stud. J. Health Res. Afr. **6**(6), 1–12 (2025). https://doi.org/10.51168/sjhrafrica.v6i6.1676

Mhlanga, D., Moloi, T.: COVID-19 and the digital transformation of education in South Africa. Educ. Sci. **10**(7), 180 (2020). https://doi.org/10.3390/educsci10070180

Mills, A., Bali, M., Eaton, L.: Responding to generative AI in education. J. Appl. Learn. Teach. **6**(1), 1–12 (2023). https://doi.org/10.37074/jalt.2023.6.1.34

Oncioiu, I., Bularca, A.R.: Artificial intelligence governance in higher education. Societies 15(6), 144 (2025). https://doi.org/https://doi.org/10.3390/soc15060144

Petrie, N.: The how-to of Vertical Leadership Development. Cent. Creat. Lead.Ship, Greensboro (2015)

Saunders, M., Lewis, P., Thornhill, A.: Research Methods for Business Students, 7th edn. Pearson Education, Harlow (2016)

Schunk, D.H., Ertmer, P.A.: Self-regulation and academic learning. In: Handbook of Self-Regulation, pp. 631–649. Elsevier, Amsterdam (2000). https://doi.org/10.1016/B978-012109890-2/50048-2

Schurink, W.J., Schurink, E.M., Fouché, C.B.: Research at Grass Roots, 5th edn. Van Schaik, Pretoria (2021)

Shakeeb, S.: Remote online teaching in response to COVID-19. Maldives Natl. J. Res. **8**(1), 1–15 (2020)

Sinha, D., Sinha, S.: Managing in a VUCA world. J. Technol. Manag. Grow. Econ. **11**(1), 17–21 (2020). doi:https://doi.org/10.15415/jtmge.2020.111003

Sthapit, A.: The VUCA world: what management is for. J. Bus. Soc. Sci. Res. **5**(1), 1–10 (2020). doi:https://doi.org/10.3126/jbssr.v5i1.30194

Suphasri, P., Chinokul, S.: Reflective practice in teacher education. PASAA **62**, 1–22 (2021)

Yin, K., Niu, J., Dong, N., Zhang, K., Ashok, M.: A meta-analysis of horizontal leadership and knowledge sharing. Curr. Psychol. **43**(17), 15507–15521 (2024)https://doi.org/10.1007/s12144-023-05503-2

Multi-Disciplinary Approaches for the Prevention of Illicit Drug Use and Supply: A Case Study of University of Limpopo (South Africa)

Tinyiku David Ngoveni

Abstract The use of illicit drugs is regarded as one of the global challenges, with the Higher Education Institutions (HEIs) newly becoming a hub of drug use, supply, and distributions, currently touted as the 'drug Den', this is largely believed to sustain levels of anxiety and depression of university students. South Africa, just like other countries is not immune from this pressing problem, the provincial focus indicates that by the early 2025, the use and supply of illicit drugs abuse and supply at University of Limpopo, South Africa, was on public domain, where members of the community were raising various concerns. The qualitative research approach was applied and supported by the case study research design to accomplish the objective of this study. Three South African Police Service (SAPS), three Community Policing Forums (CPFs) executive committee members, and three Student Representative Council (SRC) were selected, using the non-probability: Purposive sampling to collect rich data from the knowledgeable and participants who witnessed this scourge. The semi-structured key informant interviews (KIIs) were used. The inductive Thematic Content Analysis (TCA) was used to analyse the gathered data.The findings of this study showcase the shortage and presence of law enforcement agencies (LEAs) around University of Limpopo (UL) premises and surrounding areas, these are selectively indicated as some of the motivational factors relating to drug supply and used by university students. As a result, the multidisciplinary approaches, involving the newly launched (2024) UL Campus Community Safety Forum should initiate proactive and reactive drug use and supply techniques to offer effective prevention measures.

Keywords Crime · Illicit Drug Prevention · Illicit Drug Use and Supply · South Africa · University of Limpopo

T. D. Ngoveni (✉)
University of South Africa (UNISA), Pretoria, South Africa
e-mail: ngovetd@unisa.ac.za

E. H. Redda (ed.), *Proceedings of the Social Sciences International Research Conference, 2025*, https://doi.org/10.1007/978-981-95-9489-4_24

1 Introduction

As the SAPS use its own Intelligence Unit, with the support of other legitimate structures, such as CPFs, Private Security officers, continue applying the hands-on approach in addressing the challenge of various crimes, including illicit drug use and supply, as per the Constitutional mandate (1996) and SAPS Act (No. 68 of 1995). Furthermore, South Africa continues experiencing the brunt of high crime rates, which does not discriminate and affect everyone across the country and negatively dent the economy largely, resulting in almost South African HEIs students living in fear of their lives (Governance and Society, [Sa]; Lancaster and Gould, 2025), and this resulted in certain members losing their trust to the SAPS officers, further escalating to taking the law into their own words (vigilantism); however, these actions can be prevented through the application of the National Crime Prevention Strategy (NCPS) of 1996 (Muchave, 2023).

By 2024, crime in South Africa is still occupying first position (1) in Africa with 74.8%, followed by Nigeria with 66.7%, and in number five world-wide, with 75.4% in 2025, below Venezuela, Papua New Guinea, Afghanistan, and Haiti, which is something worrying; most (67%) of South Africans showed concerns on persisting crime rates and violence (World Review Population, 2025). As already stated, the use of illicit drugs, like psychoactive substances, production, sale, or use is prohibited (Republic South Africa, 2023), and South Africa is regarded by Arria et al. (2017) as one of the global challenges, with the (HEIs included); as the university students found themselves in an independent private spaces, which are regarded as the hub of drug Den for the use and supply of drugs, which is largely believed to sustain levels of their anxiety. It is also reported that many university students who are already addicted to drugs, resort to criminal activities in order to finance their habitual addictions, leading to an increase in theft, burglary at residential premises, and even violent crimes in communities (universities), where drug abuse is prevalent (RehabHelper 2025).

This situation fosters challenging criminal activities, as experienced by the various institution within the country. Various operations by various criminal justice institutions led to the discovery of drugs related criminal activities. In 2024, the Directorate for Priority Crime Investigation (DPCI), commonly known as the HAWKS discovered drugs manufacturing laboratory that worth 2 billion of rands in Limpopo Province, South African Broadcasting Corporation [SABC] (2025). The South African Revenue Service (SARS) customs officers discovered cocaine worth R25 million (SARS, 2025). During March 2025, two suspected drug mules arrested at OR Airport (SABC, 2025). In April 2025, police uncovered drug laboratories in Mpumalanga Province (Newsroom Africa, 2025). Likewise, Fig. 1 indicates the perceptions of citizens on activities that causes crime across the country, from Fig. 1 Ipsos (2024) reveals that there are various activities such as poverty and unemployment, drugs and alcohol abuse, ineffective law enforcement, breakdown of traditional values, lack of education, corrupt political environment, which if left unattended, will contribute to crime to members of the community at large including within the tertiary

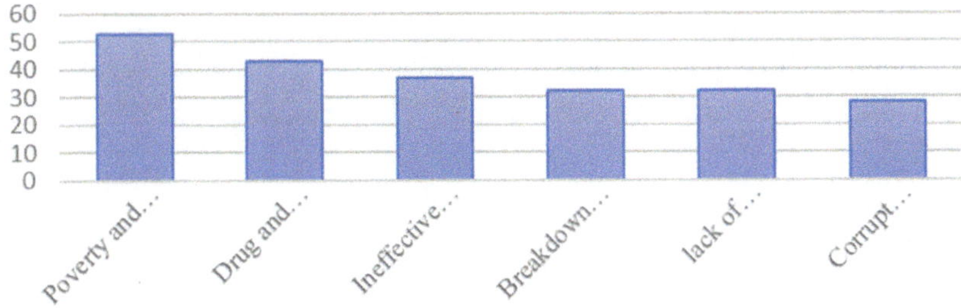

Fig. 1 Citizens perceptions on activities that causes crime in South Africa. Source: Ipsos (2024)

environment. This is supported by the survey conducted placing drugs and alcohol abuse into number two with 43%, behind poverty and unemployment 53% as the activities that lead people in committing crime.

This study discovered that students from colleges and universities use drugs with marijuana (dagga) the most commonly used drug in every year of the study, with the highest annual prevalence estimates in year three-3 (Arria, Caldeira, Allen, Bugbee, Vincent & O'Grady, 2018). The activity that is associated with diverse consequences for student health, mental health, and future job opportunities and if not taken care of, the excessive might lead to substance use disorders. The South African Community Epidemiology Network on Drugs Use (SACENDU) (2024) study, specifically for Northern Region (NR) from 11 centres that comprises eight in Mpumalanga and three in Limpopo, recorded total number of 772 candidates drug use admission, where cannabis, heroin/optiates, over the counter/prescription-medicine (OTC/PRE) identified as the main dominant drugs (SACENDU, 2024). Cannabis popularly known as weed, codeine accessed through over the counter Benylin and alcohol in the form of spirits identified as the commonly abused drugs on campuses (Mwanza & Mwale 2023). The condition, if not attended to, can spread in high volume to the space of higher learning. The following historical statistics in Table 1 depicts the use of Cannabis at UL, South Africa:

Source Peltzer, Ramlagan, and Mafuya (2010).

2 Preliminary Literature Review

2.1 Geographical Focus

The environment in which students find themselves it contributes to the choices they made in relation to drugs or alcohol. The environment created opportunities for peers to influence others, provoke students' curiosity and desire to experiment (Mwanza & Mwale 2023). In order to deal with matters such as crime of illicit drugs use and supply that impact specific country or HEIs area of concern, various issues such as the, socio-economic, even the culture are important to considered first. Rogers et al., (2017)

Table 1 The use of cannabis at UL, South Africa

Year/Location of study	Grade/Age in years	Number	Male	Female	Total
1998/urban secondary schools, Polokwane, Limpopo Province	M = 19.1	191	15.1	9.5	12.0
1998/rural secondary schools, Limpopo Province	M = 19.4	209	11.2	2.6	6.2
2004/ community survey (Limpopo Province, Johannesburg, Cape Town, East London, Pretoria)	16–17	800	12.4	1.3	6.9
2000/UL students	M = 20.1	799	10.7	2.9	6.6
2005/national, 25 higher education 3rd/ fourth year students, South Africa	18–24 (47.9%)25–29 (22.9%)30+ (10.9%)	1056	3.6	2.2	2.6

highlights that the predisposing factors (desire) such as environmental dimensions that might be available to individual to commit crime that needs consideration and attention thereafter. As already referred, crime result from various activities, like the socio-economic factors, such as poverty and unemployment, drugs and alcohol abuse, ineffective law enforcement, breakdown of traditional values, lack of education, and corrupt political environment, which if left unattended, will continues affecting community members and tertiary environment (Ipsos 2024). It is, therefore, necessary for all multidisciplinary stakeholders involved to response to illegal drug activities by university students to be familiar with the area of jurisdiction for the informed approach and productive outcome.

The geographical location, size, and population of the municipality, where universities are located, and the basic socio-economic indicators referring to the Mankweng policing area, while UL considered. The geographical setting (UL) is regarded as the study location, positioned midway between Polokwane and Tzaneen and UL benefits from close proximity to some of South Africa's best agricultural sites and mining operations. This location is not only beneficial for academic pursuits, but also offers university students, as a unique blend of urban and rural experiences. In terms of crime, the Mankweng police station, where UL is situated is a worrisome and unsafe environment and certain members of the community are concern with the service from the local SAPS (Phaladi 2023). From July 2023 to September 2024, Mankweng policing area shows concerning levels of crime activities, with contact-related crimes remaining a significant issue. The property-related crime, which saw an increase to 1772 cases from 1748 the previous year, indicating a 1.4% rise (Sekhwela 2025). Negatively, the student SRC welcomed the arrest of two suspects (23- and 33-year-old) selling drugs (like *Nyaope,* crystal meth, cocaine) to the UL premises thorough the disbanded Gate one and active Gate two (Phaladi 2025).

2.2 *Drugs Roadmap and the Associated Impacts*

The abuse of illicit substance abuse among youth (university students included) is becoming a global scourge, affecting virtually every country, not excluding South Africa in general and Limpopo Province specifically, though the extent and characteristics vary (Republic of South Africa, 2013), and this have detrimental effects; harmful to the safety and security measures, further contributing negatively to individual's health and livelihoods (Kabbash et al. 2022). The high abuse and use of illicit drugs, including alcohol in South African communities is historical challenging, as once highlighted by the first democratic President of the country, the late Nelson Mandela during the Parliament opening address, he reiterated that the individual behaviour needs proactive intervention, in reference to drug abuse and crime occurrence, while hitting on various challenges. According to the South African Drug Advisory Board report (1999), drug abuse and supply kept on gaining momentum (Peltzer et al. 2010). According to the Global Organised Crime Index (2023), South Africa, just like other countries in the Southern and West Africa, have a direct link to South Asian networks, which are seen as evolving place for the drugs organised criminal groups, and South Africa is viewed as the critical trade-in international country, with Afghanistan producing heroin and initiate smuggling rings to Europe, Africa, and Australia, including the Pakistan or Iranian heroin Global Organised Crime Index (2023). Moreover, South Africa is regarded the largest market for illicit drugs within sub-Saharan Africa (Nel 2003). Annually, the trade drugs such as heroin generates billions of rand, followed by cannabis (Africa Organised Crime Index 2023).

The peer pressure, genetic factors, curiosity, and depression are suggested as the other additional factors contributing to drug abuse among university students (Mwanza & Mwale 2023). Just like other crimes in general, the negative consequences of drug abuse and use affect not only individuals who abuse drugs, but also their immediate families and friends, various businesses, communities, and government resources (National Drugs Intelligence Centre, 2006). The abuse of drugs such as marijuana, methamphetamines, and opioids are reportedly increasing in South Africa, leading to crime increases, poverty, and health issues (Rehabhelper 2025).

The study by Montesh et al. (2015) identify the trending drug known as *'Nyaope* or *Whoonga'*, depending on the geographical location in South Africa, explained by Mthembi et al. (2018) as a mixture of low-grade heroin, cannabis products, antiretroviral drugs and other materials added as cutting agents as the latest version attracting school children and extremely young first-time drug users (Montesh et al. 2015).

The survey study by Peltzer, Ramlagan, and Mafuya (2010) South African perspective, indicates that the use of cannabis ranged among adolescents from 2% to 9% and among adults 2%, cocaine/crack (0.3%), Mandrax/sedatives (0.3%), club drugs/amphetamine-type stimulants (0.2%), opiates (0.1%) and hallucinogens (0.1%). While the primary illicit substance at admission to South African drug treatment centres was cannabis 16.9%, Methamphetamine (Tik) 12.8%, crack/cocaine 9.6%, cannabis and Mandrax 3.4%, heroin/opiates 9.2%, and prescription and OTC 2.6%. According to Mosel (2024), the 2024 Monitoring the Future Survey, in 2023, established that young American adult aged 19–30 had historically high prevalence levels of cannabis, vaping, cannabis, nicotine, and hallucinogen use. The various drugs that are dominating the country that is highly available to the members of the community including in tertiary institution where students have enough freedom: cannabis/dagga, cocaine, Mandrax/methaqualone, crack/cocaine, heroin/opiates, methamphetamine/tik, Ecstasy, Methcathinone (CAT), inhalant/solvent, poly-substance use (more than one substance use), (OTC/PRE) (SACENDU, 2024). South Africa disguising street names as: Tik, tuk, crystal, straws, globes, *ganja*, coke, snackies, white lady, acid, and crack (Montesh et al. 2015).

Mutai et al. (2024), the illicit drug use results in considerable global morbidity, but there is little data on its trends and factors associated with it in sub-Saharan Africa, some of the contributory factors to this scourge refers to the socio-demographic groups of young rural and urban men and women, including individuals' multiple problems, such as unemployment, hazardous alcohol use, high-risk sex, Intimate Partner Violence (IPV), psychological distress, and poor educational attainment. Further, investment in evidence-based interventions aimed at delaying drug use initiation could also alleviate the impact of drug use in South Africa.

The South African students in Higher and Tertiary Education Institutions (HTEIs) are not immune to alcohol, drugs, and other psychoactive substances abuse (Bullock 2004). The most obvious effects of drug abuse, which are manifested in the individuals who abuse drugs include ill health, sickness and, ultimately, death (National Drug Intelligence Centre 2006). The abuse of cocaine is a burden to many communities in many countries world-wide, frequently overloading welfare, treatment, and law enforcement agencies (LEAs) (United Nations Office on Drugs and Crime 1995). University student perspective, substance abuse is associated with disrupted study and sleep patterns, difficulty in concentration and in effectively handling stress which negatively affects academic performance, academic failure, difficulty in maintaining a social life, and inadequate coping mechanisms to manage these compounding consequences (Ramdhani 2025). Mwanza and Mwale (2023) concur indicating that students consider poor academic performance in some cases with drugs and alcohol abuse. Mphahlele (2020), focusing the use and impacts of illicit drugs UL perspective, highlights the wide range of negative consequences such as aggression,

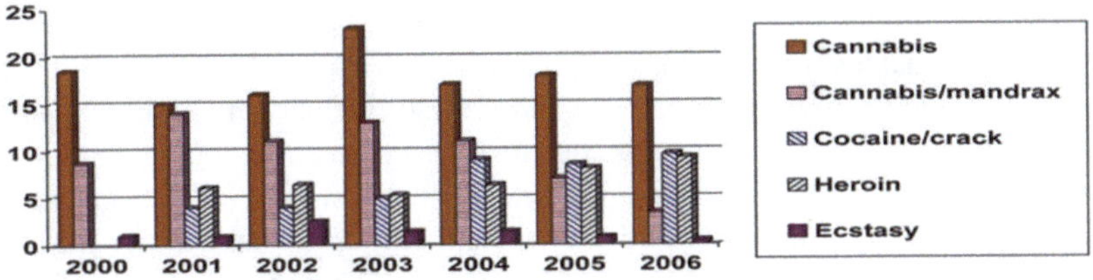

Fig. 2 The treatment demand for illicit primary drugs in South Africa for seven-year period. *Source* Peltzer, Ramlagan, and Mafuya (2010)

vandalism, acquaintance rape, risky sexual behaviour(s), social dysfunctionality, and decreased academic performance that might lead to criminal activities, arrest, and study dropouts. As a result of drug use, the graph indicates the treatment demand for illicit primary drugs in South Africa in a seven-year period with cocaine leading:

2.3 *The Prevention Illicit Drug Use and Supply, Multidisciplinary Approaches*

The South African national government, counteract legislation against the abuse of illicit drugs led to existing illicit drugs being reintroduced with new names such as *'Nyaope/Whoonga'* a disguising name (Mthembi, Mwenesongole & Cole, 2018). The name that sometimes makes it difficult for police official to present and make a case in Court of law. The situation that led to the amendment of the Drugs and Drug Trafficking Act (No. 140 of 1992) to criminalise the dominate *Nyaope* (Republic of South Africa, 2013). The prevention of illicit drugs and supply, defined as a psychoactive substance, the production, sale, or use of which is prohibited needs a proactive multidisciplinary approach which ought to include government, private sector, communities, and civil society organisations to minimise it among the youth in Limpopo Province (Republic of South Africa 2013). The study by Limpopo Department of Social Development (DSD) in collaboration with UL discovered academic advancement as a one of the tools assists in reducing the likelihood of the youth to indulge in substance use; therefore, the university space must be proactively protected by all means (Republic of South Africa 2013).

Taking into consideration the number of SAPS versus community ratio, it is evident that police cannot effectively address illicit drug and other crime prevention on their own (Ngoveni, 2024). The safety of students at the University of Limpopo who reside off-campus, once again came into the spotlight after three students were targeted after social event (bash) (Sekhwela, 2023). Therefore, to be successful in crime prevention and promoting safety and security within the country at large legally so, the government must try to move beyond the mode of crisis management and reaction and be organised and apply a proactive approach (Ngoveni 2020). It is Constitutional of the country of 1996, complimented by the NCPS, 1996, SAPS

Act, 1995, and other established legislative frameworks, policies, and documents introduced to enables multidisciplinary stakeholders' approaches that will directly involve community in promoting safety and security within their areas of jurisdiction. The communities with non-existing stakeholders, involving multidisciplinary approaches, find themselves in a position to combat, investigate, and prevent various forms of crime within their respective areas and end-up taking the law into their own hands, while violating the human rights of alleged suspect(s), as a result, they become crime perpetrators, instead of acting as crime busters (Sithuga & Rakubu, 2023), while in other South African areas, like the Westbury in Gauteng Province avoid the importance of stakeholders collaborative approaches and find itself fighting with SAPS, while dealing with crime within their area using their own approaches, eNews Central Africa [eNCA] (2025).

In addressing the crime challenging issues within their cities respectively, United States of America and Australia adopted the principles of new-urbanism and engaging in integrated Public-Private Partnerships [PPP] (Tsoriyo, Ingwani, Chakwizira & Bikam, 2021). To realise the impact of stakeholders' collaborative approaches of crime prevention, the togetherness must be promoted through other activities, such public awareness campaigns and educational programmes (Navarrete, 2023). This is emphasised by Kole (2015), by indicating that legal relationship with positive buy-ins by various crime prevention stakeholders, such as the SAPS, CPFs, private security, business forum, traditional leaders, and the media houses can play different roles of reducing crime in the country. The case study of using 'patrolling' by members of (CPFs) and CSFs in Dan Village around Tzaneen during identified peak crime periods leading to the dismantling of the Boko Haram gangs, serving as the multidisciplinary approach proved to be effective (Ngoveni 2022).

The network expansion by the SAPS by partnering with metropolitan municipalities and communities to fight crime in the priority provinces where crime is particularly high was embraced by the current (2025) President of South Africa, honourable Cyril Ramaphosa. Concurring with the importance of multidisciplinary approach the Gauteng Provincial Government, in April 2025, in a significant move to bolster municipal law enforcement and enhance public safety, to continue promoting collaborative approach in combating crime, the Gauteng Provincial Government hosted a Gauteng Safety Agreement Signing Ceremony on 'cross border law enforcement operations' at the Soweto Campus Stadium of University of Johannesburg (UJ), the event that included and supported by their Vice Chancellor, Professor Lethlokwa Mpedi (Motsila 2025). As part of proactive multidisciplinary approach, UL invited safety specialists from the provincial police service to interact with students about ways to ensure that they are at all times aware of their surroundings, to avoid being targeted (Sekhwela, 2023). In 2024, a male student was stabbed to death by the three men for attempted a phone robbery who were later arrested (Luvhengo 2024). In terms of using of illicit drugs, by 2025, UL experiencing challenges resulting in multidisciplinary approach involving police, members of the community, SRC, local drug fighting organisations playing their part leading to arrest of two suspects in possession of various types of drugs, while the NPO dealing with drug abuse assisting other students with rehabilitation (Phaladi 2025).

2.4 Theoretical Underpinning

Four this study, three theories, namely, the Broken Window Theory (BWT), Risk Management Theory (RMT), and Participative Theory (PT) were chosen among the existing ones and briefly explained: The BWT is based on the assumption that disorder and crime in the area are linked in a sequence of development. The theory herein, in line with the study focus, is researcher's assumption that, in a situation where in a use window gets broken and the one in charge decide to ignore the situation, there is a likelihood that all remaining windows that are intact are at a high risk of being broken thereafter. This theory has been taken into consideration as it assists in highlighting and cautioning all stakeholders under the CPF umbrella and the SAPS in relation to the continuous illicit drugs use and supply including other crimes taking place within UL and Mankweng policing area in general and the risk of ignoring the challenge, reactional instead of applying the collective proactive approach. Theory realises the importance of proactive action in fighting the use and supply of the illicit drugs and other crimes and the fact that all structures must be involved and take part in fighting the scourge (Lombardo & Lough 2007:122).

From the illustrations of BWT on this subject, the RMT is applied to offer prevention of drug use and supply at UL. Balzacq (2005) defines RMT as the process that can be developed by all stakeholders developing and maintaining crime information security within UL. According to the RMT, corporate risk research and assessment must be conducted first with the aim of quantifying vulnerabilities and threats the students might experience in relation to drugs. The findings of the study could be used to establish information security needs and risk management strategies within the university. The goal is to reduce the level of crime risk in general the students might be exposed to an acceptable level, which is 'zero' drug tolerance area. Risk assessment must be at the heart of risk management by 'identifying students' vulnerability and risk and adopt cost-effective methods to prevent drugs seeing the space around the university'.

After the presentations of the BWT and RMT, the strategic measures with all stakeholders must be developed and implemented, therefore, the PT was employed in this study, in terms of the PT, multidisciplinary approach in preventing illicit drugs and supply is crucial because all stakeholders involved such as the local SAPS, CPFs, and SRC, as the collaboratively fight drugs use and supply at UL, in essence they gets the opportunity to influence the buy-in against the use and supply of the illicit drugs, the system long been regarded as one of the elements of democratic decision-making processes (Maphazi et al. 2013:56). The multidisciplinary approach commended by Chapters 4 and 6 of the South African Constitution, respectively, the SAPS Act, 1995, and other approved policy documents which regards public participation as the cornerstone of democracy and providing service delivery, in this case, the prevention of use, and supply of illicit drugs.

3 Methodology

The qualitative research approach was adopted to set-out of strategies used by the researcher, this study involved smaller study sample, with an exception that only participants with in-depth experiences knowledge on this subject were selected. This research approach refers to the method generally associated with interpretative epistemology, including different forms of data collections and interpretations, to offer clear understanding of specific phenomenon identifying and applying emerging study themes and attached meanings, Minnaar, Mistry, Patel, and Rustin (2016). This research approach was supported by the phenomenological research design, which aims to describe the identified stakeholders' individuals' lived experiences (phenomena) to enrich the lived experiences of the targeted participants by drawing-out their respective meanings of the subject under research (Holloway, 2005). The research design in question shared the true meaning of the explored phenomena through the experiences of the chosen sample, the other goal of using this research design was to describe the lived experiences of the targeted sample, this was done by analysing their unique experiences on this subject (Burns & Grove 2001).

The Non-probability: Purposive sampling was employed to explore experiences of the targeted relevant stakeholders, including the SAPS officials, CPFs, Student Representative Council (SRC), all attached to UL, distributed as follows: three SAPS officials, three CPFs executive committee members, and three SRC members. They were subjected to the semi-structured KIIs. The collected data was analysed rough the application of the six-6 phases of inductive TCA (Maluleke 2020), the researcher adopted the 6-phase coding framework for the inductive TCA to identify the study themes in the collected empirical and the reviewed literature studies (Clarke & Braun 2013), namely: 1) Familiarisation of data, 2) Generation of codes, 3) Combining codes into study themes, 4) Reviewing study themes, 5) Determine significance of study themes, and 6) Reporting of study findings. The internal ethical approval was awarded to Dr. Tinyiku Ngoveni from University of South Africa Research Ethics Review Committee (UNISA RERC), in reference to the 2026 'policy on research ethics', approved from 11/07/2025 to 11/07/2028, with the reference number (8520) and the external approval of providing permission to conduct research in the SAPS, based on SAPS National Instruction 4/2022, with the following reference number (3/34/2).

4 Study Findings, Discussions, and Thematic Identifications

In understanding the existence and impact of the multidisciplinary approaches for the prevention illicit drug use and supply as per the case study, seven study themes emerged from the conducted experimental data through the semi-structured key informant interviews (KIIs). Their verbatim response was recoded as follows:

4.1 Theme 1: The Necessity of Establishing Stakeholders' Approach in the Prevention of Use of Illicit Drugs and Supply at University of Limpopo

- *It is necessitated by the fact that the supply chain of drugs is the most complex operation which requires different stakeholders* (Interviewee 1).
- *To prevent further illegal drugs activities continuation* (Interviewee 2).
- *Collective approach will bring different skills from different institutions* (Interviewee 3).
- *Too much consumers, high crime rate, and death of young people is high* (Interviewee 4).

- *To form partnership with various stakeholders and to share resources and knowledge. To identify high-risk groups and areas within the community* (Interviewee 5).
- *To prevent distribution and further harm on our society* (Interviewee 6).

4.2 Theme 2: The Policies Used to Apply the Stakeholders' Approach in the Prevention of Use of Illicit Drugs and Supply at University of Limpopo

- *Criminal Procedure Act [CPA] (No. 51 of 1977) and the policies in place* (Interviewee 1).
- *SAPS Act, 1995, CPA, 1977, and NCPS, 1996* (Interviewee 2).
- *SAPS Act, 1995, Constitution of Republic of South Africa, 1996 and NCPS, 1996* (Interviewee 3).
- *SAPS Act, 1995, National Health Act (No. 61 of 2003) and the Constitution of Republic of South Africa, 1996* (Interviewee 4).
- *Community engagement policy, drug abuse prevention, and education policy, zero tolerance policy* (Interview 5).
- *NCPS, 1996* (Interviewee 6).

4.3 Theme 3: Stakeholders' Approaches Used for the Prevention of Use of Illicit Drugs and Supply at University of Limpopo

- *The Constitution of Republic of South Africa, 1996 and SAPS Act, 1995 enables the stakeholders like CPF and others to report and alert criminals* (Interviewee 1).

- *It can be addressed through implementation of drugs team/task team* (Interviewee 2).
- *Civic society, faith-based organisation, youth formations* (Interviewee 3).
- *Awareness, workshops, CPF, civic churches, and youth structures* (Interviewer 4).
- *To form multidisciplinary operation to combat the use of illicit drugs use and supply. Community used approach-engaging with local communities and CPFs to address drug-related issues* (Interviewee 5).
- *CPF, South African National Council on Alcoholism and Drug Dependence (SANCA), Limpopo Anti-Drugs Gangsterism, and Crime (LADGAC) Rehabilitation Centre* (Interviewee 6).

4.4 Theme 4: Types of Drug-Related Crimes Experienced at University of Limpopo

- *Robberies, Assaults, Murder, and Gender-Based Violence (GBV)* (Interviewee 1).
- *Murders, assault with intent to inflict Grievous Bodily Harm (GBH)* (Interviewee 2).
- *Murder, robberies, theft, and rape* (Interviewee 3).
- *Distribution of illicit drug/selling drugs, possession of illicit drugs, and smoking in public* (Interviewee 5).
- *GBV, robberies, murder, and assault* (Interviewee 6).

4.5 Theme 5: Considerations of Types of Drug-Related Crimes Experienced and Measures in Place to Prevent the Occurrences at University of Limpopo

- *To activate informers' network, utilisation of technology, and to conduct disruptive operation to cut the chain of supply* (Interviewee 1).
- *Through awareness within the institution, involvement of SAPS* (Interviewee 2).
- *Awareness campaigns, door-to-door campaigns, visit to schools* (Interviewee 3).
- *Destructive operations, intelligent driven operation, and awareness campaigns* (Interviewee 4).
- *To conduct stop and search operations. To conduct awareness campaigns. To address the specific need and challenges of the community, as well as community policing strategies* (Interviewee 6).

4.6 Theme 6: Stakeholders' Involvement in the Prevention of Use of Illicit Drugs and Supply at University of Limpopo

- *Is very effective when measuring the arrest rate and conviction as compared to non-intelligence driven operations* (Interviewee 1).
- *Users will be aware and report the distribution and minimise the availability of illicit drugs* (Interviewee 6).

4.7 Theme 7: The Impact of Applying Multidisciplinary Approach in the Prevention of Use of Illicit Drugs and Supply at University of Limpopo

- *To conduct awareness campaign and avail rehabilitation centres* (Interviewee 1).
- *Different skills will be together to come with better solution to address the problem* (Interviewee 2).
- *Decline of crime, moral regeneration minimum usage of illicit drugs* (Interviewee 3).
- *Crime rate will go down, more arrest on the distributions, many people will be aware of the danger* (Interviewee 4).
- *To reduce the use of drug and Supply. To promote a safer community* (Interviewee 5).
- *Schools address, community meetings, roadshows* (Interviewee 6).

5 Conclusion and Recommendations

Taking into consideration the challenges of supply and use of the illicit drugs within the UL and the country at large, the number of crimes experienced by the Mankweng police station, the place in which the UL is allocated, the police station that is number 3 police station in the province behind Thohoyandou and Seshego, respectively, in terms of crime reported, further, the criminal activities such as murder, use and supply of drugs experienced and reported within the UL jurisdiction, it is very important that the multidisciplinary approach policing operation be prioritised. The following serves as the contributions, recommendations, and future research studies tailored to improve the fight against the use of illicit drugs and supply through the multidisciplinary approach:

- To supplement the police officers role and fighting crime and determine the effectiveness in fight against the use of illicit drugs and supply, within the UL and the surrounding, a multidisciplinary approach that involve various structures such as CPF, NPOs/Non-Governmental Organisations (NGOs) dealing with issues of drugs, UL protection services, and other UL formal structures such as the SRC, Religious structures, newly launched (2024) UL Campus Community Safety Forum, working in conjunction with the UL Department of Safety and Security and other recognised external relevant structures must be emphasised and monitored accordingly.

- The multidisciplinary approaches must initiate proactive and reactive drug use and supply techniques to offer effective prevention measures. The policy dealing with drugs must be established and made known to all within the premises. The issues of drugs and declaration statement must be included on the application forms.
- The live scree billboard to advertise issues of drugs and other tertiary matters must be activated as an awareness tool.

References

Africa Organised Crime Index.: Ranking by cocaine trade. Retrieved from: https://africa.ocindex.net/rankings/cocainetrade (2023)

Arria, A.M., Caldeira, K.M., Allen, H.K., Bugbee, B.A., Vicent, K.B., O'Grady, K.E.: Prevalence and incidence of drug use among college students: An 8-year longitudinal analysis. Retrieved from: https://pmc.ncbi.nlm.nih.gov/articles/PMC5638668/ (2017)

Balzacq, T.: Three faces of securitization: political agency, audience and context. European journal of international relations. **11**(2), 171–201 (2005)

Bullock, S.: Alcohol, drugs and student lifestyle: A study of the attitudes, beliefs and use of alcohol and drugs among Swedish university students. SoRAD Research Report No. 21. Centre for Social Research on Alcohol and Drugs, Stockholm (2004)

Burns, N., Grove, S.K.: The Practice of Nursing Research: Conduct, Critique and Utilisation. W.B. Saunders, Toronto (2001)

Clarke, V., Braun, V.: Teaching thematic analysis: overcoming challenges and developing strategies for effective learning. The Psychologist. **26**(2), 120–123 (2013)

Global Organised Crime Index: Ranking by criminality. Retrieved from: https://ocindex.net/rankings?f=rankings&view=List (2023)

Ipsos: Global attitudes to crime and law enforcement. Retrieved from: https://www.ipsos.com/en-us/global-attitudes-crime-and-law-enforcement# (2024)

Kabbash, I., Zidan, O., Said, S.: Substance abuse among university students in Egypt: prevalence and correlates. Eastern Mediterranean Health Journal**28**(1), 31–40 (2022)

Lombardo, R.M., Lough, T.: Community policing: broken windows, community building, and satisfaction with the police. Police Journal. **80**(2), 117–140 (2007)

Luvhengo, P.: Three suspects nabbed after University of Limpopo student murdered. Retrieved from: https://www.timeslive.co.za/news/southafrica/2024-05-14-three-suspects-nabbed-after-university-of-limpopo-student-murdered/#google_vignette (2024)

Maluleke, W.: The African scare of fall armyworm: are south African farmers immune? International Journal of Social Sciences and Humanity Studies. **12**(1), 207–221 (2020)

Maphazi, M., Raga, K., Taylor, J.D., Mayekiso, T.: Public participation: a south African local government perspective. African Journal of Public Affairs. **6**(2), 56–67 (2013)

Montesh, M., Sibanda, O.S., Basdeo, V., Lekubu, B.K.: Illicit drug use in selected high schools in Mamelodi Township. Retrieved from: https://journals.co.za/doi/abs/10.10520/EJC183378 (2015)

Mosel, S.: Substance abuse in college students: Statistics and rehab treatment. Retrieved from: https://americanaddictioncenters.org/rehab-guide/college (2024)

Motsila, P.: Gauteng unites against crime: Premier Lesufi leads historic safety agreement signing. Retrieved from: https://www.gauteng.gov.za/News/NewsDetails/%7B1a39950b-7ea8-44c7-bde1-5f20c37a0b32%7D (2025)

Mphahlele, R.S.: An exploratory study into the experience and behaviour(s) of students who use Nyaope/Whonga at the University of Limpopo. Retrieved from: http://ulspace.ul.ac.za/bitstream/handle/10386/4166/mphahlele_ls_2020.pdf?sequence=1&isAllowed=y (2020)

Mthembi, P.M., Mwenesongole, E.M., Cole, M.D.: Chemical profiling of the street cocktail drug 'nyaope' in South Africa using GC–MS I: Stability studies of components of 'Nyaope' in organic solvents. Retrieved from: https://www.sciencedirect.com/science/article/pii/S037907381830481X?via%3Dihub (2018)

Mutai, K.K.K., Stone, J., Scheibe, F., Fraser, H., Johnson, L.F., Vickerman, P.: Trends and factors associated with illicit drug use in South Africa: findings from multiple national population-based household surveys, 2002–2017.International Journal of Drug Policy. **125**, 1–12 (2024)

Mwanza, N., Mwale, G.: Students' perspectives on drugs and alcohol abuse at a public university in Zambia. HTS Teologiese Studies/Theological Studies. **79**(3), 1–8 (2023)

National Drug Intelligence Centre: The impact of drugs on society. Retrieved from: https://www.justice.gov/archive/ndic/pubs11/18862/impact.htm (2006)

Nel, E.: An overview of the management of the drug situation in South Africa. Retrieved from: http://www.unafei.or.jp/english/pdf/PDF_rms/no64/ch10.pdf (2023)

Ngoveni, T.D.: Investigating the interaction between South African Police Service and Community Policing Forums in ensuring community safety: A case study in Atteridgeville and Brooklyn police station. Unpublished Magister Technologiae: Policing, Tshwane University of Technology, Soshanguve South (2020)

Ngoveni, T.D.: A systematic study on the evolutions of community policing forums: a supplementary tool to crime prevention in South Africa. International Journal of Social Science Research and Review. **5**(11), 145–162 (2022)

Peltzer, L., Ramlagan, S., Johnson, B.D., Phaswana-Mafuya, N.: Illicit drugs use and treatment in South Africa. Retrieved from: https://pmc.ncbi.nlm.nih.gov/articles/PMC3010753/ (2010)

Phaladi, M.: Limpopo residents slam police for failing to curb crime. South African Broadcasting Corporation News. Retrieved from: https://www.sabcnews.com/sabcnews/limpopo-residents-slam-police-for-failing-to-curb-crime/#. (2023)

Phaladi, M.: University of Limpopo Raises Flags about Selling of Drugs Nearby. South African Broadcasting Corporation News. Retrieved from: https://www.google.com/search?q=THE+CHALLENGE+OF+DRUGS+IN+UNIVERSITY+OF+LIMPOPO (2025)

Ramdhani, S.: Demographic factors and depression as predictors of substance use among students at a South African university. Available from: https://humanities.uct.ac.za/sites/default/files/contentmigration/humanitiesuctacza/1233/files/Serita%2520Ramdhani.pdf (2025)

RehabHelper: How does drug abuse and addiction affect society in South Africa? Retrieved from: https://rehabhelper.co.za/blog/how-does-drug-abuse-and-addiction-affect-society-in-south-africa/ (2025)

Republic of South Africa, Department of Justice and Constitutional Development: Law to be amended to criminalise Nyaope. Retrieved from: https://www.gov.za/news/media-statements/law-be-amended-criminalise-%E2%80%98nyaope%E2%80%99-27-jan-2013 (2013)

Sekhwela, K.: Crime in Mankweng and Seshego: what the latest stats reveal. The Citizen.. Retrieved from: https://www.citizen.co.za/review-online/news-headlines/local-news/2025/03/16/crime-in-mankweng-and-seshego-what-the-latest-stats-reveal/. (2025)

South African Community Epidemiology Network on Drug Use: Monitoring alcohol, tobacco and other drug use trends in South Africa, July 1996–June 2023. Retrieved from: https://www.samrc.ac.za/sites/default/files/attachments/2024-07/SACENDU%20Brief%20Jan-June%202023.pdf (2024)
United Nations Office on Drugs and Crime.: The social impact of drug abuse. Retrieved from: https://www.unodc.org/pdf/technical_series_1995-03-01_1.pdf (1995)

Stakeholders' Crime Prevention Initiatives: A Rural-Based Phenomenological Study of Limpopo Province (South Africa)

Tinyiku David Ngoveni

Abstract Amongst the 104 police stations in the Limpopo Province, Thohoyandou police station, situated in Vhembe District Municipality of Limpopo Province (South Africa), occupied number one (1) provincially and 17 nationally, in reference to the 17-community reported serious crimes, as recorded in the third quarter South African Police Service (SAPS) crime statistics of 2024/2025, this remains a historical fate and it is currently worrisome, overall, Thohoyandou police station recorded 6.2 change/increase of the reported crimes. The researcher adopted the qualitative research approach, supported by the phenomenological research design to explore stakeholders' crime prevention initiatives. The study used the non-probability: Purposive sampling to solicit information from the selected participants, this was complimented by the unstructured Key Informant Interviews (KIIs) directed to the 23 participants comprising three SAPS members, coordinators of Community Policing Forums (CPFs), three CPF executive members, 12 CPF sub-forums committee members and five rural safety members, and one Vhembe United Group Against Crime. The inductive Thematic Content Analysis (TCA) was used to analyses the collected data. The findings of this study indicate that poor police visibility and ineffective partnership between the local SAPS and other identified stakeholders, based on identification and responses of well-known crime hotspot areas are inadequate. The witnessed established community crime prevention collaborations are currently unmonitored. This study concludes and recommends that the stakeholders' crime prevention in the study location should use the collaborative approach guided by Section 205(3) of the Constitution, 1996 of Republic of South Africa, 1996 and other developed integrated crime prevention strategies and policies.

Keywords Community policing forums · Rurality · South African police service · Thohoyandou policing area · South Africa

T. D. Ngoveni (✉)
University of South Africa (UNISA), Pretoria, South Africa
e-mail: ngovetd@unisa.ac.za

E. H. Redda (ed.), *Proceedings of the Social Sciences International Research Conference, 2025*, https://doi.org/10.1007/978-981-95-9489-4_25

1 Introduction

Crime in South Africa is regarded as a developmental issue, troubled by social and economic drivers and many consequences; furthermore, the cost of crime is reportedly direct and indirect. The direct costs of crime are exponential, including the expenditure on policing, prosecution, the judicial and the correctional system, as well as the 'health and social service expenditure, and private security'. The indirect costs refer to the economic and social, Inclusive Violence and Crime Prevention Programme (2020), amongst others. Moreover, South Africa continues prioritising crime prevention strategies as required by the Constitution, 1996 stipulating the responsibilities of the SAPS, including that the police should prevent, combat, and investigate crime; maintain public order; protect and secure the inhabitants of the Republic and their property; while upholding and enforcing the law (Republic of South Africa 1996). Although the SAPS work together with other legitimate structures, such as CPF to address the roots causes of crime, higher number of various crimes continue to be witnessed. Negatively, the South African crime situation results in unprecedented global attention, resulting in a meeting held by President Cyril Ramaphosa (South Africa) and President Donald Trump (United States), with the main focus placed on crime directed to White people, however; this was denied and justified by all South African representatives, indicating that crime in South Africa occurs without any discrimination (Lancaster and Gould 2025). The crime issue and related victimisations are concurred by Ngoveni (2021), indicating that in South African, crime does not discriminate, but everyone, irrespective of social status, can fall a victim of crime. This negatively impacts the lives of all citizens, enabling them to live in fear of various crimes.

The study by Human Science Research Council [HSRC] ([Sa]) (in Governance and Society, [Sa]) emphasises the crime, fear, and safety by South African citizens, coupled with complexities and detrimental effects on individuals, community members, and societal spheres, further affecting to quality of life resulting in racial stereotypes when discussing crime occurrences; limitations on their mobility and ability to socialise in public spaces, including the proliferation of gated communities; high walls; and the operations of private security measures (Governance and Society, [Sa]). The high volume of crime results in other members of community taking the law into their own hands (vigilantism), which is against the South African rule of law, often leading to harsh sentence, such life imprisonment to inform deterrence (South African High Court 2024).

Although the Limpopo Province witnessed decreased in some crimes, such as murder, which was reportedly decreased by 26% from 242 in 2024 to 179 in May 2025, this indicates that 3.0 ratio people per 100,000 gets killed daily in a province with 6,051.999 as per the 2021 mid-year population estimates, behind Eastern Cape with ratio of 15.4 murder rate per 100,000, Western Cape with a ratio of 14.2, and KwaZulu Natal, with a ratio of 10 murders per 100,000 people; the overall crime situation in the province remains a matter of grave concern as in the rest of the country (Mabeba 2025; Zoutnet 2025). The Thohoyandou area in the former Venda homeland, under the Thohoyandou police station, Limpopo Province, demographically highly rural is regarded as the most dangerous place in the province and struggling to decrease the crime. In the year 2019, the Safety and Justice Parliament Commitee raised concerns about the increase of Gender-Based Violence (GBV) reported cases, during this period, the police station's statistics revealed an increase of 19% to recorded cases related to sexual offences for the 2018/19 financial year, an increase of 12% to rape cases and an increase of 43.8% to cases related to sexual assault (Republic of South Africa 2019a, b). The situation that is persisting six years later, as the station is amongst the top 30 stations in the country to record various violent crime. It is perched at fifth spot, according to national statistics (Mabeba 2025). Thohoyandou has been plagued by a wave of contact crimes that include sexual assaults, murder, attempted murder, burglaries and armed robberies, carjacking and GBV, a huge concern to provincial commissioner (Mabeba 2025). According to Major General Eddie Van Der Walt, Vhembe district commissioner (2025), the high volume of liquor outlets is the contributory factors to crime in Vhembe district, Limpopo Province. South African perspective, the January–March 2025 crime statistics show the increase in violent crimes but decrease in key crime categories such as murder that is a thorn to South Africans, by 12.2% to 5,727 murders, or an average of 64 per day. Despite the fourth quarter crime statistics for the 2024/25 financial year revealing a decrease in contact crimes in Limpopo, 105 victims still reported cases including murder, assault GBH, rape, robbery, and theft to SAPS daily (South African Government 2026). If South Africa is indeed decreasing, what could be the reasons behind the positive result (Lancaster and Gould 2025). Thus, the multidisciplinary crime prevention should be prioritised.

According to Ngoveni (2021), to ensure effectiveness in crime prevention and decrease, the police and members of the community need to collaborate by using available platforms, such as the CPFs (Ngoveni 2021). Major General Jan Scheepers, Limpopo Province Deputy Commissioner (2025), concurs with this assertion, by stating that *the more community gets involved in fighting crime the more people see that when gets involved something happening, that install confidence to community. People keeps on trusting police as they see the action by the police based on the information provided by the community.* Mr Frans Kgasago, Limpopo Province CPF chairperson emphasised the importance of multidisciplinary approach indicating that CPF cannot be seen as the so-called *impimpi* (an informer) the apartheid terms used to identify the police informers against black democratic fighters to promote SAPS community approach in dealing with crimes within their area of jurisdiction and make sure that the victims receive justice (Bobodi 2025). Navarrete (2023) highlights that collaborative stakeholder's approaches encourage police officers to build positive relationships with the community, engage in open communication, and work together to identify and address crime and safety concerns. Overall, the reviewed studies highlighted stakeholders' crime prevention initiatives, however; the rural-based phenomenological study focuses on Limpopo Province (South Africa).

2 Literature Review

2.1 Thohoyandou Police Station Geographical Perspective

In responding to different forms of crime, which negatively impact South Africa, socio-economic and cultural dynamics should be prioritised. Rogers et al. (2017) contend that the predisposing factors (desire) such as environmental dimensions might be available to individual to commit crime that needs consideration and attention thereafter. Burger (2007) shares that in a country such as South Africa with high unemployment rate, socio-economic activities are amongst the contributing factors of crime and it is therefore the main role of both the SAPS, CPFs, and other relevant structures such as the business forums, traditional leaders, local councillors in collaboration, to consider this in order to fight and prevent crime as enshrined in the Constitution as the supreme law and the SAPS Act, 1995, effectively, Burger (2007). Therefore, the geographical location, size, and population of the municipality, and the basic socio-economic indicators of the area of Thohoyandou as per the case study were discussed in brief. The gathered information serves to provide a comprehensive and meaningful introduction of the research focus areas, as well as other information related to basic socio-economic activities (Ngoveni 2021).

The visit by the parliament selects committee members of safety and justice, by 2019, discovered that the station in question was providing a crime prevention service to the total number of estimated three hundred thousand (300,000) people with three hundred and ninety-one (391). Police officers work in shifts as normal (Republic of South Africa 2019a, b). Thohoyandou police station is situated in province of Limpopo of 6.5 million people. The province contributes 7.7% to the National Gross Domestic Product [GDP], with unemployment rate of 31.9% (Limpopo Province—State of the Province Address, SOPA 2025). By 2019, Thohoyandou urban population was estimated as 69,453 (Tsoriyo et al. 2021). By 2025, the Thohoyandou policing zone under Thulamela Municipality provides policing services to four hundred and sixty-one (461) villages, and hundred and ninety-eight (198) schools in 2025, with 391 police officers using 49 police vehicles (Republic of South Africa 2019a, b, 2025a). The Thohoyandou police station is the second and third quarters of the 2024/25 year, Thohoyandou, where crime has continued to rise. Thohoyandou is the only provincial station in the top 30 stations for all contact crimes, ranked 15th, and also appear in the top 30 stations for various crimes 13 times (Helm 2025). Despite Limpopo's recording an overall decline in contact crimes, Thohoyandou leads the country in both rape and sexual offence cases (Bele 2025). Just like the country in general in terms of crime, as it is country wise, the increase in seventeen (17) categories reported crime such as business robberies, murder, and sexual offences in Thohoyandou remains a challenge and it does not show any sign of improvement, the action that needs an urgent multidisciplinary stakeholder's approach, as supported by the depicted crime statistics of Thohoyandou in Fig. 1.

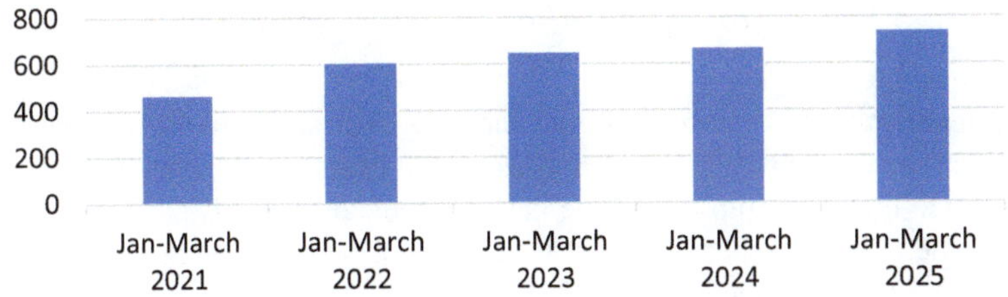

Fig. 1 Crime comparison, 4th quarter, January to March, 2021–2025: 4th Quarterly Crime Statistics (2021–2025), Thohoyandou police station. *Source* Republic of South Africa (2025a, b)

2.2 *Stakeholders' Crime Prevention Approaches and Policy Guidelines*

According to Ngoveni (2024), the legitimate documents and policies for promoting and emphasising safety and security on citizens are crucial to the operation of any company/organisation. As a result, policies and procedures along with the security measures promoting safety and security in relation to crime prevention must be established and adhered to by all crime prevention stakeholders (Van Jaarsveld 2011). The study finds that in Africa and Latin America poverty in the urban was identified as the perpetuator of crime and violence (Tsoriyo et al. 2021). Regarding safety and security in the context of crime prevention country perspective, the policies and documents act as a 'know how' platform for all Criminal Justice System (CJS) stakeholders involved in crime prevention. Accepted policies and documents provide strategic direction and enforce adherence to the subject of crime prevention, ensuring the constant maintenance of safety and security by all stakeholders and their existence and performance can be based as per the policies in place.

Various documents/policies, such as the Constitution of Republic of South Africa, 1996, NCPS of 1996, available to serve as the guidance to the stakeholders approach in the prevention of crime, the Local Government: Municipal Structures Act (No. 117 of 1998), SAPS Act, 68 of 1995, the White Paper on the Transformation of Public Service (*Batho Pele* Principles) of 1997, the White Paper on Safety and Security of 1998, White Paper on Safety and Security and Crime Prevention Measures, 2016, the Local Government: Municipal Systems Act (No. 32 of 2000) the SAPS Regulations for CPFs and Boards of 2001, the Intergovernmental Relations Framework Act (Act No. 13 of 2005), the SDG 2030, goal number 16 focusing on promoting peaceful and inclusive societies for sustainable development and Chapter 12 (building safer communities) of NDP, 2030, exist to compel and guide all involved in the crime prevention (Ngoveni 2024).

2.3 *Crime Prevention and Collaborative Approaches*

While members of the community find themselves in a position to prevent crime within their area and because of the non-existence of stakeholders multidisciplinary approach facilitator end-up taking the law into their own hands during the process violating the alleged suspect(s) rights leading them being perpetrator (Sithuga and Rakubu 2023), other members of Westbury in Gauteng avoid the importance of stakeholders collaborative approach find itself fighting with SAPS while dealing with crime within their area eNews Channel Africa [eNCA] (2025). In addressing the crime challenging issues within their cities respectively, United States of America (USA) and Australia are adopting principles of new-urbanism and engaging in integrated Public-Private Partnerships [PPPs] (Tsoriyo et al. 2021). In 2010, there was one police officer for every 250 people, and by 2022 the situation persists being dire even today with one (1) officer for every 450 people (Ngoveni 2024). Given the SAPS to community member ratio, it is evident that police cannot effectively address crime prevention on their own (Ngoveni 2024). Therefore, to be successful in crime prevention and promoting safety and security within the country at large legally so, the government must try to move beyond the mode of crisis management and reaction and be organised and apply a proactive approach (Ngoveni 2021). As part of transformation process to democratise the SAPS, from old apartheid system that was seen as anti-blacks in favour of the government of that day, soon after attaining democracy, all police stations were required as per the Constitution of the Republic of South Africa, 1996, which is the supreme law of the country, complimented by the NCPS, 1996, SAPS Act, 1995, and other established policies and documents to enact the multidisciplinary stakeholders approach that will directly involve members of the community in promoting safety and security within their areas of jurisdiction. The relationship between members of police and the community means a position or circumstance where the police and other role players destined to build a working relationship with the objective of jointly identifying and solving crime within their area of jurisdiction legally perspective, Sithuga and Rakubu (2023). This system brings together law enforcement, local government, community organisations, and residents to work collectively to reduce crime and improve neighbourhood safety (Navarrete 2023).

Dlamini (2023) reveals that with the collaborative stakeholders approach in crime prevention relying on CPF as the legitimate structure as per the established policies and documents where members of the community work with members of the police and other members from various structures to address crime problems in their respective communities. To realise the impact of stakeholders' collaborative approach in crime prevention, the togetherness must be promoted through other activities such public awareness campaigns and educational programmes (Navarrete 2023). This is emphasised by Kole (2015) by indicating that the legal relationship with positive buy-ins by various crime prevention stakeholders such as SAPS, CPFs, private security, business forum, traditional leaders, and the media can play a role in reducing crime in the community. The case study of patrols by members of CPFs and Community

Safety Forums (CSFs) in Dan Village around Tzaneen during identified peak crime periods led to the dismantling of the Boko Haram gangs (Ngoveni 2022).

In 2025, the President of South Africa, Honourable Cyril Ramaphosa indicated the network expansion by the South African Police Service (SAPS) by partnering with metropolitan municipalities and is working with communities to fight crime in the priority provinces where crime is particularly high. Various operations such as the 'operation *Shanela*' by Gauteng police are showing a positive outcome (Republic of South Africa 2025a, b). The stakeholders' collaborative approach in crime prevention deserves the commitment from various structures, such as the SAPS; Section 199 (1) of the Constitution, 1996, provides for the establishment of the country's security services that include a 'single police service' (Republic of South Africa 1996). The CPF as the 'umbrella' legally recognised unit that represents the policing interests of the local community and also intended to exert civilian oversight over the police at various levels, at the local police station level jurisdictional perspective (Dlamini 2023). The Private Security Personnel, regulated under the Private Security Industry Regulatory Authority (PSiRA) Act (No. 56 of 2001), can provide services of crime prevention within the community as per the agreement (Govender 2017). The Local Government Act, the White Paper on Safety and Security in 1998 supports the idea of empowering local government within its jurisdiction with the responsibility for implementing and coordinating social crime prevention programmes (Shawn 1998: 2). The Local Municipality, South Africa Local Government Association (SALGA) requires municipalities to consider the safety of the community as one of their key programmes that is given priority and effectively implemented (SALGA 2016). The CSF, evolved from the NCPS of 1996 and the White Paper on Safety and Security of 1998, meant to facilitate the delivery of a multi-sectoral governmental approach on safety in local communities (Singh and Zondi 2020). The School Governing Body (SGB), since schools fall victims of crime, is established under South African Schools Act [SASA] (No. 84 of 1996), as a framework to ensure that human rights were considered and adequately protected by all those involved in the formulation of school rules (Van Jaarsveld 2011). According to De Wet (2014), the media (print or online) platforms provide awareness of criminal activities in a communal setting, whether in the private or public sector is promoted mostly through the mass media such as radios, television, newspapers and social media, report daily about the scourge of violence that is taking place within South Africa at large (Khumalo 2019). The reset case of Westbury community members fighting members of SAPS while arresting a man for selling drugs must serve as a scenario on the importance of using media in understanding the positive impact of stakeholders' collaborative approach (eNCA 2025). The 59-year-old David *Nembahe* of *Makwarele* village under Thohoyandou policing states that 'as a community, we need to work together with the police in combating the crime in Thohoyandou by establishing the CPFs' (Monyela 2025).

Community policing is a policing strategy that focuses on police building ties and working closely with members of the community in combating crimes and criminals in society; their principle is to maintain close partnership between the police and the community in suppressing crimes and criminals in society. It is a philosophy founded by Sir Robert Peel in the nineteenth century in England, who argued by sharing the following:... *the police are the public and the public the police: The police being only members of the public who are paid to give full time attention to duties which are incumbent on every citizen in the interests of community welfare and existence and its forerunner was the London Metro Police Department around 1830* (Osborn 2018) Community policing is regarded as the effective strategies for preventing housebreaking in various studies (Timbury 2018). Engaging with communities proactively and using intelligence to target crime hotspots are key strategies.

Community policing focuses on fostering collaboration between law enforcement agencies and communities to solve local problems and improve public safety and it encourages police officers to engage directly with community members to establish trust, encourage participation, and strengthen resilience against crime. Consistent meetings, neighbourhood watch initiatives, and local events help cultivate a sense of responsibility amongst residents, leading to increased awareness and quicker responses to potential dangers (Goldstein 2018). By applying these approaches, the following three (3) strategies become essential: (1) improving community engagement is crucial. By engaging in regular meetings, crime prevention campaigns, and neighbourhood watch programmes, SAPS can enhance public trust and mobilise residents to report suspicious behaviour (Diphoorn and Berg 2014); (2) promoting information sharing is essential. Establishing hotlines, online platforms, and mobile applications enables residents to share intelligence efficiently with law enforcement; and (3) this strategy should be implemented more systematically, by analysing data on housebreaking incidents, and police can identify hotspots, modus operandi, and repeat offenders, enabling the strategic allocation of resources and targeted interventions (Ratcliffe 2016).

2.4 Crime Prevention Measures and Challenges

The importance of using stakeholders multidisciplinary approach for crime prevention is highly encouraged, while considering the challenges faced by the police stations in terms of crimes committed in Thohoyandou policing area, resulting in occupying position 1 and amongst the top 30 provincially and nationally respectively, on Friday, 16 May 2025, soon after the first quarterly crime report, the Civilian Secretariat for Police Service (CSPS) in conjunction with Department of Transport and Community Safety together with the SAPS in Limpopo under the leadership of Provincial Commissioner Lieutenant General, Thembi Hadebe, firstly with a courtesy visit to *Vho Thovhele Vho Gole Mphaphuli* Traditional Council led by the district commissioner, Major General Eddie Van Der Walt in company of the Thohoyandou SAPS Station Commander, Brigadier Mukwevho ([Sa]) (in Republic of South Africa 2025a, b) states that the stakeholder engagement at Vhembe district addressed the stubborn causes of crimes and the station conducted various operations resulting in arrest of 385 shebeen owners and 627 tavern owners for violations such as selling liquor to minors and failing to close on time.

Better working conditions and enough resources are crucial in being productive, and the absence of such might contribute to the opposite. While coming with various measures to address crime challenges, the station reported to the parliamentary monitoring groups various challenges hindering their daily services such as: university accommodation; service delivery protests; influx of illegal foreigners; lack of street lights, unnamed streets; un-rehabilitated roads; high unemployment rate; growing population due to business developments; low morale due to the unfortunate promotion of members; the SAPS Family Violence, Child Protection and Sexual Offences (FCS) facilities are not user friendly and they do not meet the minimum required standards for victim empowerment programmes. There are insufficient offices to accommodate members; none of the availability of psychologists which results in a high number of cases being withdrawn. There are also a high number of withdrawals of cases due to outstanding Deoxyribonucleic Acid (DNA) results from the SAPS Forensic Science Laboratory [FSL] (Republic of South Africa 2019a, b).

2.5 Theoretical Framework

Although there are various theories linked to the research topic, this study relied on the three (3) specific related theories, namely: (1) Broken Window Theory, (2) Normative Sponsorship Theory, and (3) Participative Theory.

The Broken Window Theory. The Broken Window Theory [BWT] (1981) is one of the theories identified in line when the study focus was introduced. The theory is based on the assumption that disorder and crime in the area are linked in a sequence of development. The theory herein, in line with the study focus, it is author's assumption, was that, in a situation where in use window gets broken and the one in charge decides to ignore the situation; there is a likelihood that all remaining windows that are intact are at a high risk of being broken thereafter. The BWT has been taken into consideration as it assists in highlighting and cautioning all stakeholders under the CPF umbrella and the SAPS in relation to the continuous various crime taking place within Thohoyandou area and the fact that, in case there is no partnership in combating such crime, criminals will see the opportunity and continue with their criminal act resulting in the station caring the brunt. Theory realises the importance of proactive action in fighting crime and the fact that everyone from various categories within the Thohoyandou area must be concerned of all the criminal activities and contribute legally as expected (Lombardo and Lough 2007: 122).

Normative Sponsorship Theory. The Normative Sponsorship Theory [NST] (1957) highlights the fact that, since crime affects everyone and the economy at large, most people have good will and are mostly willing to cooperate with other people with the aim of dealing with the challenge in place to satisfy their needs. The NST postulates that a community effort will only be supported if it is normative to all persons and interested groups proactively involved. To initiate community development and ensure that it is effectively accomplished, the initiator, in this case the SAPS must be knowledgeable on how two or more interest groups in relation to safety and security as per the case study can have enough unification of interest, or conscience concerning common goals to bring about effective implementation, to achieve these, various developed documents and policies serve as critical in imparting knowledge to all stakeholders (Mamosebo 2014).

Participatory Theory. The Participatory Theory [PT] (1970s), in terms of the PT, participation by all stakeholders in crime prevention is crucial because such collaboration provides members of the public with an opportunity to influence public decision and has long been regarded as one of the elements of democratic decision-making processes (Maphazi et al. 2013: 56). Chapters 4 and 6 of the Constitution of Republic of South Africa, 1996, respectively regard public participation as the cornerstone of democracy and providing service delivery, in this case the prevention of crime. To consult and to be allowed to participate are not privileges, but constitutional rights (Republic of South Africa 1996: 39; Nyati 2008: 1). The PT aims to assimilate and realise the ideals of direct democracy, which involves responsiveness and active citizenry, participation, and equality in the modern complex world of Nation States and it is important to provide opportunities to directly participate in decision-making at the local level for crime control and prevention interventions on daily basis. The platform to participate serves to educate people about key issues in resource creation and control and enables them to assess the performance of their crime prevention units (SAPS), to judge national questions, and to participate in national decisions [when the need arises] (Maphazi et al. 2013).

3 Methodology

For the purpose of the study, the qualitative research approach, that is classified by Minnaar et al. (2016) as the method generally associated with interpretative epistemology, which tends to be used to refer to different forms of collecting data and interpretation, which depend on understanding, where a prominence on meaning was chosen and applied. The chosen research approach further explained as the set of strategies the researcher use that most often include the involvement of smaller identified sample sizes, with an examination that is in depth of the stakeholders' participants' experiences. This research approach was supported by the phenomenological research design, which aims to describe the identified stakeholders' individuals' lived experiences (phenomena) to enrich lived experience by drawing out its meaning (Holloway 2005). In this context, this research design was employed to purposively explore experiences of the selected stakeholders (members of SAPS, CPFs, rural safety members, and Vhembe United Group Against Crime), respectively. This research design shares the true meaning of phenomena be explored through their sole experiences. It is inductive in nature; thus, in this study, the researcher adopted this research design with a goal to describe lived experiences of the targeted sample herein, and this was done by analysing their unique experiences on this subject (Burns and Grove 2001).

About 23 participants comprising of 3 SAPS coordinators of CPFs, 3 CPF executive members, 12 CPF sub-forums committee members and five rural safety members, and one Vhembe United Group Against Crime were identified and deemed necessary using the non-probability: purposive sampling. The unstructured KIIs and the literature studies on this subject were adopted for data collections. The collected data was analysed through the application of the inductive TCA (Maluleke 2020). The internal ethical approval was awarded to Dr Tinyiku Ngoveni from University of South Africa Research Ethics Review Committee (UNISA RERC), in reference to the 2026 'policy on research ethics', approved from 21/05/2025 to 21/05/2028, dated 21/05/2025, with the reference number (8517) and the external approval of providing permission to conduct research in the SAPS, based on SAPS National Instruction 4/ 2022, with the following reference number (3/34/2), dated 2025/12/31.

4 Study Findings and Discussion

In order to achieve the objective of this study, the 23 participants formed part of this study. The unstructured KIIs were conducted, and their verbatim responses and the following study themes emerged:

4.1 Theme 1: The Necessity of Establishing Stakeholders' Approach in the Prevention of Crime

Interviewee 1 (SAPS officials): Stakeholder approach is very important because it allows police and other stakeholders to work together in fighting of crime. Police will not fight crime alone.

Interviewee 2 (SAPS officials): It is necessary as it assists the police. The relationship with community members improves.

Interviewee 4 (CPF member): The establishment of stakeholders is important and must be done in all areas in our villages.

Interviewee 5 (Vhembe United Group Against Crime): Involvement of business forum in fighting crime is important. Business are at risk of theft, and some get killed.

Interviewee 6 (Rural safety member): It is very necessary. The relationship between the local police, community members, traditional leaders and business people is important to address matters affecting the immediate communities, as relevant crime information is clearly discussed and communicated between these stakeholders.

Interviewee 7 (CPF Sub-forum member): The multidisciplinary approach is necessary.

4.2 Theme 2: The Policies Used to Apply the Stakeholders' Approach in the Prevention of Crime

Interviewee 1 (SAPS member): Various polices are applied to promote stakeholder's approach in crime prevention. Policies such as South African Constitution, SAPS Act, 1995, NCPS, 1996, and other police guiding documents.

Interviewee 3 (CPF member): SAPS guide us in terms of policies. They tell us of what to do and how to perform of our work.

Interviewee 4 (Vhembe United Group Against Crime): We work with police. They talk to us about crime in our area. Police officer is given to us. We report to him when see crime.

Interviewee 5 (Rural safety members): SAPS policies guide us on the issues of dealing with crime. Police provide guidance always.

Interviewee 6 (CPF sub-forum member): We rely on police to teach and guide us. We also attend the workshops. Our coordinators play a role.

4.3 Theme 3: Types of Stakeholders' Approach Is Used in the Prevention of Crime

Interviewee 2 (SAPS member): We conduct campaigns and patrols with other stakeholders, such as CPF, provincial campaigns against various crime together with the Member of Executive Committee (MEC) and Deputy Minister.

Interviewee 4 (CPF member): Various methods are applied. Police work together with CPF to fight crime. We conduct patrol always and do roadblocks.

Interviewee 5 (Vhembe United Group Against Crime): We conduct meeting with police to talk about crime in our business. We also meet with station commander to talk about crime. We also participate in the campaigns in our areas.

Interviewee 6 (Rural safety member): Various stakeholders comprise the rural safety forum. The forum involves pastors, traditional leaders, various government departments, members of CPF, SAPS and business owners. Meetings are conducted to discuss issues of crime affecting members of the community.

Interview 7 (CPF sub-forum member): We work with police, CPF executive, traditional leaders, councillors in the village, and those who have shops.

4.4 Theme 4: Types of Crime(s) Experienced in Thohoyandou Policing Area

Interviewee 1 (SAPS member): We experience various crimes but robbery, drugs, house and business robbery, GBV and rape are a challenge.

Interviewee 4 (CPF member): We experience various crimes, and members of the community are at risk and living in fear. We experience use of drugs, robbery/ business robbery of local shops, murder, GBV and rape of young girls.

Interviewee 5 (Vhembe United Group Against Crime): We have challenges on crime. Our business is at risk. Shops are broken always. Murder, rapes, GBV, high jacking.

Interviewee 6 (Rural safety members): Our villages experience lots of crime, for example, house robbery, murder, rape, drugs, bullying at our schools, stock theft, GBV, illegal shebeens, violence in our school and alcohol abuse.

Interviewee 7 (CPF sub-forum member): Crime is a huge problem to the community. Girls and boys abuse alcohol, drugs, murder and people are not safe at home; businesses are robbed.

4.5 Theme 5: Considerations of Different Types of Crimes Experienced in Thohoyandou Policing Area and Measures in Place to Prevent Their Occurrences

Interviewee 2 (SAPS member): As police, we work together with all structures, and it shows progress. Although crime is still a challenge, but we also make number of arrests. The working together with CPF and campaigns plays a role.

Interviewee 4 (CPF member): Police are doing better in fighting crime, they talk about crime in local radios, and we engage schoolgirls and boys in secondary schools to talk about the danger of crime.

Interviewee 6 (Rural safety member): We always discuss in our monthly meeting of our challenges and try to come with solutions. We rely on our police to come with measures.

Interviewee 7 (CPF sub-forum member): Police are trying. The issue of unemployment is a challenge. School campaigns must continue.

4.6 Theme 6: Challenges of Using Stakeholders' Approach in the Prevention of Crime in Thohoyandou Policing Area

Interviewee 1 (SAPS member): The challenge is resources such as cars to make patrol and to transport members such as CPF after meeting and patrols. Shortage of staff is a big challenge.

Interview 4 (CPF member): The most challenge is shortage of cars within the station making patrol by police challenging.

Interviewee 5 (Vhembe United Group Against Crime): Shortage of police officers in our station. The distance to our station. The police complain about shortage of resources to our police station such as cars.

Interviewee 6 (Rural safety member): We have various challenges. Police does not have transport to visit the traditional leaders when needed in community meeting.

Interviewee 7 (CPF sub-forum member): we do not have transport to attend meetings and to patrol in our area even the bicycles. So, to conduct patrol is difficult.

5 Conclusion

Factual, South Africa is still lagging behind in terms of number of police officers versus the citizen population, with the number of crimes in various categories still persisting and reportedly high, to effectively respond to these crimes, all citizens of the selected study locations should work closely with the SAPS officials, using legitimate strictures, such as the CPFs. The nation as a whole need to be made aware of the importance of stakeholders' crime prevention initiatives and its impact. Stakeholders' development and effectiveness that is receiving criminal justice structures/ government attention must be prioritised and monitored.

The following serves as the main contributions, recommendations and future research studies to emphasise the use of the stakeholders' crime prevention initiatives, a rural-based phenomenological study of Limpopo Province (South Africa):

- A formalised stakeholders' crime prevention structure involving the identified participants of this study should be established, formalised and actively monitored.
- To maintain its core mandate, monthly or quarterly meeting should be introduced for public information, awareness and visibility.
- To emphasise the importance of multidisciplinary stakeholders' approach and instil confidence in crime prevention initiatives, the outcome should form part of the provincial quarterly crime reporting.
- The local SAPS should be provided with resources such as vehicles to interact and support other structures, such as the CPFs.
- Media coverage envisaging the importance of collaborative approach in crime prevention must be erected in those high number public space.

References

Bele, M.: Thohoyandou becomes South Africa's rape capital. Limpopo Mirror. https://www.limpopomirror.co.za/articles/news/62482/2025-05-30/thohoyandou-becomes-south-africas-rape-capital (2025). Last accessed 19 June 2025

Bobodi, L.: Crime stats for 2nd and 3rd quarter 2024/2025 [Video]. YouTube. https://www.youtube.com/watch?v=5JbyzeCGJqA (2025). Last accessed 19 June 2025

Burger, J.: Strategic perspectives on crime and policing in South Africa. Van Schaik, Hatfield (2007)

Burns, N., Grove, S.K.: The practice of nursing research: conduct, critique and utilisation. W.B. Saunders, Toronto (2001)

De Wet, C.: The *Herald*'s portrayal of school violence in the Eastern Cape Province, South Africa. https://www.researchgate.net/publication/274129641 (2014). Last accessed 19 June 2025
Diphoorn, T., Berg, J.: Typologies of partnership policing: case studies from urban South Africa. Polic. Soc. **24**(4), 425–442 (2014)
Diphoorn, T., Shearing, C.: Layers of security: the development of policing models in Southern Africa. Crime Law Soc. Chang. **77**(1), 53–69 (2019)
Dlamini, S.: Community police forums' future and legitimacy: redefining good community policing. Cogent Soc. Sci. **10**, 1–11 (2023)
eNews Channel Africa: Anti-gang unit police officers flee Westbury after being attacked. https://www.enca.com/top-stories/anti-gang-unit-police-officers-flee-westbury-after-being-attacked (2025). Last accessed 19 June 2025
Goldstein, H.: Improving policing: a problem-oriented approach. Crime Delinq. **25**(2), 236–258 (2018)
Govender, P.: Investigating serial murder: Case linkage methods employed by the South African Police Service. Master's dissertation, University of South Africa, Pretoria (2017)
Governance and Society: Governance and society: We investigate, evaluate, and assess governance and society. https://hsrc.ac.za/uploads/pageContent/625/Governance%20and%20society.pdf (n.d.). Last accessed 19 June 2025
Helm, M.: Latest crime stats reveal Thohoyandou as the rape capital of South Africa. Facebook. https://www.facebook.com/dalimpopo/posts/1012376847741953 (2025). Last accessed 19 June 2025
Holloway, I.: Qualitative research for nurses. Blackwell Science, Cambridge (2005)
Inclusive Violence and Crime Prevention Programme: Integrated crime and violence prevention strategy for consultation. Pretoria (2020)
Khumalo, S.S.: Implications of school violence in South Africa on socially just education. E-Bangi J. Soc. Sci. Humanit. **16**, 1–11 (2019)
Kole, O.J.: Partnership policing between the South African Police Service and the private security industry in reducing crime in South Africa. Doctoral thesis, University of South Africa, Pretoria (2015)
Lancaster, L., Gould, C.: Crime rates show hopeful decline: Is South Africa turning the corner on violence? Daily Maverick. https://www.dailymaverick.co.za/article/2025-05-28-crime-rates-show-hopeful-decline-is-south-africa-turning-the-corner-on-violence/ (2025). Last accessed 19 June 2025
Limpopo Province: State of the Province Address. https://www.gov.za/sites/default/files/2025-03/State-of-the-Province-Address-2025-26-February-2025_LMP.pdf (2025). Last accessed 19 June 2025
Lombardo, R.M., Lough, T.: Community policing: broken windows, community building, and satisfaction with the police. Police J. **80**(2), 117–140 (2007)
Mabeba, M.: Thohoyandou remains Limpopo's crime capital. Sunday World. https://sundayworld.co.za/news/thohoyandou-remains-limpopos-crime-capital/ (2025). Last accessed 19 June 2025
Maluleke, W.: The African scare of fall armyworm: are South African farmers immune? Int. J. Soc. Sci. Humanit. Stud. **12**(1), 207–221 (2020)
Maphazi, M., Raga, K., Taylor, J.D., Mayekiso, T.: Public participation: a South African local government perspective. Afr. J. Public Aff. **6**(2), 56–67 (2013)
Mamosebo, S.P.: A descriptive analysis of the implementation of community policing forums in the Lebowakgomo area. Master's dissertation, University of Limpopo, Sovenga (2014)
Minnaar, A., Mistry, D., Patel, C., Rustin, C.: Criminal justice research methodology. University of South Africa, Pretoria (2016)
Monyela, B.: Limpopo police lose the fight against crime in Thohoyandou. https://www.atnews.co.za/limpopo-police-lose-the-fight-against-crime-in-thohoyandou/ (2025). Last accessed 19 June 2025
Navarrete, R.: Collaborative approaches to crime prevention. https://ronnavarreta.co/collaborative-approaches-to-crime-prevention/ (2023). Last accessed 16 May 2026

Ngoveni, T.D.: Investigating the interaction between the South African Police Service and community policing forums in ensuring community safety. Unpublished MTech thesis, Tshwane University of Technology, Pretoria (2021)

Ngoveni, T.D.: The nexus of the South African Police Service and community policing forums in dismantling the "Boko Haram" gang in Tzaneen Township. Int. J. Soc. Sci. Res. Rev. **5**(11), 193–211 (2022)

Ngoveni, T.D.: Evaluating safety and security in the prevention of various crimes. Doctoral thesis, Tshwane University of Technology, Pretoria (2024)

Nyati, L.: Public participation: what has the Constitutional Court given the public? https://www.saflii.org/za/journals/LDD/2008/15.pdf (2008). Last accessed 19 June 2025

Osborn, B.: Crime prevention: methods to improve intelligence-led policing. Unpublished doctoral thesis, Colorado Technical University, Colorado (2018)

Ratcliffe, J.H.: Intelligence-led policing. Routledge, London (2016)

Republic of South Africa: The Constitution of the Republic of South Africa, 1996. Government Printer, Pretoria (1996)

Republic of South Africa: Committee notes challenges faced by Thohoyandou police station. https://www.parliament.gov.za/press-releases/committee-notes-challenges-faced-thohoyandou-police-station (2019a). Last accessed 19 June 2025

Republic of South Africa: Report of the select committee on security and justice on the oversight visit to Limpopo Province. https://pmg.org.za/tabled-committee-report/4031/ (2019b). Last accessed 19 June 2025

Republic of South Africa: Civilian Secretariat for Police Service stakeholder engagement in Vhembe District. https://www.saps.gov.za/newsroom/msspeechdetail.php?nid=60988 (2025a). Last accessed 19 June 2025

Republic of South Africa: President Cyril Ramaphosa: State of the Nation Address. https://www.gov.za/speeches/2025StateOfTheNation (2025b). Last accessed 19 June 2025

Rogers, C., van Schalkwyk, L., Ngoveni, T.D.: Security risk control measures. University of South Africa, Pretoria (2017)

Shawn, M.: The role of local government in crime prevention in South Africa. Institute for Security Studies. https://issafrica.s3.amazonaws.com/site/uploads/Paper33.pdf (1998). Last accessed 19 June 2025

Singh, S.B., Zondi, L.M.: Human beings and safety: the role of community safety structures in the fight against crime. Orient. Anthropologist **20**(1), 10–32 (2020)

South African Government. Minister Senzo Mchunu: Release of 2024/2025 fourth quarter crime statistics. https://www.gov.za/news/speeches/minister-senzo-mchunu-release-20242025-fourth-quarter-crime-statistics-23-may-2025 (2026). Last accessed 16 May 2026

South African High Court: In the matter between Piki Thabo Seopela. https://www.saflii.org/za/cases/ZAGPPHC/2024/447.pdf (2024). Last accessed 19 June 2025

Sithuga, N.P., Rakubu, K.: Community perspectives on public vigilantism in Limpopo Province. Int. J. Res. Bus. Soc. Sci. **12**(2), 1–14 (2023)

Timbury, A.: Intelligence-led and community-based policing: the effects on community relations. Doctoral thesis, Queen's University, Canada (2018)

Tsoriyo, W.W., Ingwani, J., Chakwizira, J., Bikam, P.: Street safety and security as spatial (in)justice in rural towns of Vhembe District. J. Inclusive Cities Built Environ. **1**(2), 1–14 (2021)

Van Jaarsveld, L.: An investigation of safety and security measures at secondary schools in Tshwane. Master's dissertation, University of South Africa, Pretoria (2011)

Zoutnet: Crime statistics nothing to celebrate. https://www.zoutnet.co.za/articles/news/62483/2025-05-30/crime-statistics-nothing-to-celebrate (2025). Last accessed 19 June 2025

Water, Energy, and Food Nexus in South Africa: Pathways to Social and Economic Resilience

Ogujiuba Kanayo and Maponya Lethabo

Abstract In South Africa, social stability and human well-being face growing risks due to ongoing issues in providing equitable access to essential resources, including energy, food, and water. Although the interconnection of these resources has been well documented, their impact on income inequality distribution has not been thoroughly investigated in the South African context. This research aims to investigate the impact of energy consumption, agricultural productivity, and water use on income inequality, as measured by the Gini coefficient, an indicator of social and economic resilience. Using the autoregressive distributed lag (ARDL) model for the years 1990–2022, the study examines dynamics in both the short and long run. The empirical evidence suggests that fossil fuel use significantly increases inequality in both the short and long terms. In contrast, water extraction intensity diminishes inequality over the long term but exacerbates it in the short term. Additionally, cereal production has a significant, long-term, inequality-increasing impact, indicating that agricultural benefits are not evenly distributed among households. The findings additionally validate a rapid adjustment rate towards long-term equilibrium, highlighting the enduring structural nature of inequality in resource distribution. These results underscore the urgent need for policies that promote a fair energy transition, equitable water management, and inclusive agricultural development. The research contributes to the broader conversation surrounding the food–water–energy nexus. It offers evidence-driven insights for developing resource governance strategies that can enhance resilience and reduce inequality in South Africa.

Keywords Economic · Energy · Water · Food Resilience · Social · Transition

O. Kanayo (✉) · M. Lethabo
Centre For Entrepreneurship and Rapid Incubator/School of Development Studies, University of Mpumalanga, Nelspruit, South Africa
e-mail: Kanayo.Ogujiuba@ump.ac.za

E. H. Redda (ed.), *Proceedings of the Social Sciences International Research Conference, 2025*, https://doi.org/10.1007/978-981-95-9489-4_26

1 Introduction

South Africa faces a critical juncture where limited resources, environmental harm, and deep-rooted inequality converge, endangering both human welfare and sustained stability. The availability of crucial resources like electricity, food, and clean water has become increasingly unpredictable for millions of families, highlighting the vulnerability of the nation's social and economic foundations. This issue is exacerbated by the complex interconnections of the food–water–energy (FWE) nexus, in which pressure in one area, whether agricultural production, water availability, or energy creation, ripples through others, destabilising communities and limiting economic progress (Essel-Gaisey et al. 2023; Kori et al. 2023). Although the connections within this nexus are widely acknowledged, the effects on inequality distribution in South Africa have not been thoroughly investigated.

The core of the issue presents a dual challenge. South Africa experiences significant resource instability. Annual rainfall averages only 464 mm, significantly below the global average of 860 mm, and almost one-third of the available water is wasted due to ageing infrastructure (Grewar 2019; Perera et al. 2021). At the same time, the economy's reliance on coal-fired plants exacerbates environmental damage and makes the transition to renewable energy more challenging, as it also requires considerable water resources (Reddy et al. 2023; Sparks et al. 2014). Droughts and climate fluctuations heighten the risk of food insecurity, placing pressure on agricultural livelihoods (Zhou et al. 2022). Conversely, there is an equally pressing issue of social equity in resource distribution that current policies fail to address adequately. Rural areas face unique challenges in accessing water (Magombeyi and Matambo 2023; Amoah and Thabisa 2022). Power outages and high rates hinder economic participation (Vasylyna 2023), while food scarcity continues to exacerbate poverty and malnutrition (Fagbemi et al. 2023). Consequently, disparities in access to essential resources enhance income gaps and diminish societal resilience.

These issues pose a vital research question: in what ways do fluctuations in food, water, and energy systems affect income disparities in South Africa, and what are the consequences for achieving a socially just and environmentally sustainable Just Transition? By emphasising this question, the research situates itself at the intersection of the FWE nexus and the concept of distributive justice, contributing to a body of literature that aims to secure resources and ensure their equitable distribution. The theoretical foundation underpinning this study combines the FWE nexus approach, emphasising systemic interconnections and compromises, with the concept of a Just Transition, which posits that sustainability must be aligned with equity (Giupponi 2023).

To address these issues, the research employs the autoregressive distributed lag (ARDL) model, which is particularly well suited for small-sample situations and allows for the simultaneous estimation of both short-term adjustments and long-term equilibrium dynamics. Using yearly data from 1990 to 2022, the model assesses the effects of fossil fuel use, cereal production, agricultural and livestock output, and water consumption on income disparity, as measured by the Gini coefficient. This method enables the evaluation of both short-term shocks and structural connections between resource dynamics and inequality, thereby providing insights into the rate and duration of adjustment.

The rest of the document continues in this manner. Section 2 examines the literature on the FWE nexus, inequality, and Just Transition; Sect. 3 details the data and methodology, focusing on variable creation and the ARDL specification. Section 4 reveals and examines empirical results. Section 5 concludes with policy implications, highlighting how combined resource management can promote environmental sustainability and social equity in South Africa.

2 Literature Review

2.1 Theoretical Framework

This research is grounded in two integrated theoretical viewpoints: the Water–Energy-Food (FWE) Nexus framework and the concept of a Just Transition. The FWE nexus model views food, water, and energy as an intertwined system in which stress on one resource affects the others, thereby impacting economic stability and social resilience (Scott et al. 2021). Instead of addressing resource challenges in isolation, the nexus approach offers a holistic perspective, highlighting synergies, trade-offs, and feedback loops that shape development outcomes. In this context, South Africa's socio-economic system is perceived as susceptible due to the structural interconnections among limited water resources, energy instability, and sensitive agricultural output.

Adding to this system's perspective, the idea of a Just Transition incorporates social equity. Rooted in discussions of labour and environmental justice, Just Transition theory emphasises that structural changes towards sustainability, such as the decarbonisation of energy systems, must also address concerns of equity in access, distribution, and involvement (Giupponi 2023). In the South African context, this perspective underscores how unequal access to water, energy, and food significantly impacts rural and low-income households, thereby exacerbating income inequality. It highlights that environmental sustainability policies must clearly include distributive justice to prevent worsening current social inequalities.

This study develops a framework to examine the dual challenges of resource scarcity and inequality by integrating the FWE nexus and Just Transition perspectives. The nexus framework situates the issue within the context of natural resources, while Just Transition theory ensures that equitable outcomes are not overlooked in the pursuit of sustainability. Together, these viewpoints provide the analytical framework for the study to explore how fossil fuel use, water consumption trends, and agricultural efficiency influence income inequality, as measured by the Gini coefficient. This combined theoretical framework enables the research not only to measure the relationships between resource use and inequality but also to situate the results within broader discussions of resilience, sustainability, and social justice.

2.2 Empirical Literature

The integration of food, water, and energy systems has become a key issue in discussions regarding sustainability and resilience. Worldwide, the water–energy–food (WEF) nexus is acknowledged as a framework for comprehending how strains in one sector impact others, thus influencing social stability and economic results. D'Odorico et al. (2010) demonstrate that globalisation of water, although alleviating immediate food deficits, could jeopardise long-term resilience by diminishing the ability to adapt during periods of drought. In a similar vein, Allouche (2011) emphasises how political power, trade, and institutions influence the emergence of water and food insecurities, whereas Linke (2014) points out the significance of low-impact water delivery systems in maintaining household food security within vulnerable ecosystems. Scott et al. (2021) build on this reasoning by urging the explicit incorporation of FWE nexus strategies into planning for earth-system resilience, given the increasing scarcity and climate-related disruptions. Together, these studies underscore the theoretical significance of perceiving resources as interrelated systems rather than separate sectors.

In emerging and developing economies, the risks associated with resource integration are particularly significant. Kamp and Østergård (2016) illustrate that resilience in localised energy and food generation is vital when external resource access is limited. In contrast, Spiegelberg et al. (2015) emphasise the importance of socio-ecological production landscapes in promoting sustainable rural livelihoods. Subedi et al. (2020) emphasise that cohesive strategies are vital for robust food systems in Nepal. In contrast, Zaman et al. (2017) highlight the trade-offs between resource consumption and ecological results in sub-Saharan Africa. Similarly, Ozturk (2017) finds that sustainable agricultural practices can alleviate food–energy–water poverty, while Ozturk (2015) demonstrates that increasing energy demand in BRICS nations jeopardises food security due to water and environmental limitations. Together, these contributions emphasise that the nexus challenge in developing contexts is not just technical but primarily socio-economic, with unequal access, poor governance, and climate vulnerability heightening risks.

In the South African setting, studies highlight the vulnerability of the relationship and its effects on distribution. Nhemachena et al. (2020) highlight the decreasing rainfall and diminished agricultural output resulting from climate change, whereas Mpandeli et al. (2018) contend that intersectoral approaches are crucial for enhancing resilience. Edokpayi et al. (2020) emphasise that adaptive water management is essential for lessening the socio-economic impacts of climate variability, while Juana et al. (2012) forecast welfare declines due to water scarcity caused by climate change. Mabhaudhi et al. (2019) propose a regional nexus framework, yet find that sustainability levels are still marginal, putting rural livelihoods at risk. Significantly,

Magombeyi and Matambo (2023) highlight the ongoing disparities in resource allocation between urban and rural families, whereas Amoah and Thabisa (2022) and Tantoh and McKay (2023) connect water shortages to health risks and economic insecurity. These results highlight a crucial issue: South Africa's resource insecurity is both environmental and profoundly inequitable, with marginalised communities facing unfair burdens.

Nonetheless, these insights reveal two significant gaps that persist. Firstly, although international and regional studies highlight the importance of integrated nexus strategies, empirical analysis is scarce regarding how the FWE nexus affects distributional outcomes, such as income inequality. Secondly, while South African literature emphasises environmental scarcity and governance shortcomings, it rarely measures the causal relationships between resource use and evolving inequality over time. Therefore, the issue of social equity in resource distribution remains inadequately addressed by existing policies, particularly in the context of South Africa's intended shift towards renewable energy and sustainable development.

This research helps to fill these gaps by using the ARDL model to evaluate the short- and long-term impacts of fossil fuel use, renewable energy implementation, water-use efficiency, and agricultural productivity on income inequality, represented by the Gini coefficient. Integrating the FWE nexus framework with the Just Transition concept enables the analysis to extend beyond environmental sustainability, directly addressing distributive justice and offering fresh empirical insights into how resource dynamics influence inequality in South Africa.

3 Data and Methods

This research uses annual time-series data for South Africa, spanning from 1990 to 2020, resulting in a total sample of 30 observations. Data were gathered from secondary sources to ensure reliability and consistency. Income inequality, represented by the Gini coefficient, is derived from the Standardised World Income Inequality Database (SWIID). In contrast, data on energy, water, and food resource dynamics come from the World Bank's World Development Indicators. The selection of this timeframe considers both the availability of data and the necessity to capture long-term structural dynamics, such as post-apartheid economic reforms, drought events, and the ongoing energy transition.

3.1 Justification of Variables

The dependent variable is income inequality, assessed through the Gini coefficient, which acts as an indicator of social and economic resilience. The explanatory variables are classified into three categories to represent the food–water–energy (FWE) nexus:

- Energy aspect: Usage of renewable energy (as a percentage of overall final energy) and consumption of fossil fuels (coal, oil, gas).
- Water aspect: Water efficiency (GDP generated per unit of freshwater extraction) and water extraction intensity (freshwater extraction as a percentage of total renewable resources).
- Food aspect: Crop output index, livestock output index, and cereal production (kg per hectare).

These variables are chosen to reflect both the access and use of resources, along with their distributional effects.

Table 1 summarises the variables, their proxies, and the anticipated signs grounded in theory and earlier research.

3.2 Model Specification

The empirical approach is based on the ARDL model developed by Pesaran, Shin, and Smith (2001). This method is especially appropriate because (i) it can be used when regressors are integrated of mixed order I(0) and I(1), given that none are I(2); (ii) it yields consistent estimates in small sample sizes; and (iii) it estimates both short-run and long-run dynamics at the same time.

The Mathematical Form of the Model is as follows:

$$GINI = f(SE, WA, FS) \tag{1}$$

where

$GINI$ = Social and economic resilience
SE = Sustainable Energy (Renewable Energy Consumption, Fossil Fuel Consumption)
WA = Water Availability (Water productivity, Water withdrawal intensity)
FS = Food Supplies (Crop production, livestock production, Cereal yield)

The Econometric Form of the Model is as follows:

$$\begin{aligned} GINI_t = \beta_0 + \beta_1 REC_t + \beta_2 FFS_t + \beta_3 WP_t + \beta_4 WWI_t \\ + \beta_5 CRP_t + \beta_2 LSP_t + \beta_6 CY_t + \mu_t \end{aligned} \tag{2}$$

where

$GINI_t$ =Social and economic resilience
REC_t= Renewable Energy Consumption
FFS_t= Fossil Fuel Consumption
WP_t= Water Productivity
WWI_t = Water Withdrawal Intensity

Table 1 Variables, proxies anticipated signs

Variable	Proxies	Details	Source
Social and economic resilience (Income Inequality)	Gini	Estimate of the Gini index of inequality in equivalised (square root scale) household disposable (post-tax, post-transfer) income	SWIID
Renewable Energy Consumption	REC	Renewable energy consumption is the share of renewable energy in total final energy consumption	World Bank
Fossil fuel Consumption	FFC	Fossil fuel comprises coal, oil, petroleum, and natural gas products	World Bank
Water productivity	WP_t	Water productivity is calculated as GDP in constant prices divided by annual total water withdrawal	World Bank
Water withdrawal intensity	WWI_t	The level of water stress is measured by the freshwater withdrawal ratio, which is the proportion of available freshwater resources used by all major sectors. This ratio is calculated by dividing the total freshwater withdrawn by all major sectors by the total renewable freshwater resources, taking into account environmental water requirements. Main sectors, as defined by ISIC standards, include agriculture; forestry and fishing; manufacturing; electricity industry; and services	World Bank
Crop production	CRP_t	The crop production index indicates agricultural production for each year in relation to the base period of 2014–2016. It includes all crops except those used for fodder. Regional and income group aggregates for the FAO's production indexes are calculated from the underlying values in international dollars, normalised to the base period 2014–2016	World Bank
Livestock production	LS_t	Livestock production index includes meat and milk from all sources, dairy products such as cheese, eggs, honey, raw silk, wool, and hides and skins	World Bank
Cereal yield	CY_t	Cereal yield, measured as kilogrammes per hectare of harvested land, includes wheat, rice, maize, barley, oats, rye, millet, sorghum, buckwheat, and mixed grains. Production data on cereals refer to crops harvested solely for dry grain. Cereal crops harvested for hay or harvested green for food, feed, or silage, and those used for grazing are excluded The FAO allocates production data to the calendar year in which the bulk of the harvest occurs, and these data are disseminated through the World Bank's World Development Indicators database	World Bank

Source Authors' compilation

CRP_t=Crop Production
LSP_t=Livestock production
CY_t=Cereal Yield
β_0 =Constant variable
β_{1-6} =Coefficients
μ_t=error term

The error term is often said to indicate the effect of variables that were omitted from the regression equation (Freedman and Drineas 2005).

3.3 Analytical Technique

The ARDL approach is suitable for investigating the effects of sustainable energy, water availability, and food supplies on social and economic resilience. The ARDL model is a valuable tool in time-series analysis, offering flexibility and the ability to test for cointegration (Natsiopoulos and Tzeremes 2022). It is beneficial in estimating long-run and short-run effects and can be used to fit models with optimal lags (Kripfganz and Schneider 2016).

The ARDL method, as discussed by Nkoro and Uko (2016), is a valuable tool for estimating long-run and short-run coefficients, particularly when dealing with variables of mixed order of integration. However, it is essential to note that the ARDL model is not suitable for non-stationary variables integrated at I (2) order of integration. To address this limitation, Harvey et al. (2010) proposes tests for the presence of nonlinear deterministic components in time series, which are robust to the order of integration. Tschernig et al. (2014) further emphasises the importance of considering the fractional integration order in VAR models, particularly in the context of long-run identification. Kripfganz and Schneider (2018) provides a practical tool for implementing the ARDL method, offering a command for estimating autoregressive distributed lag models and equilibrium correction models. The ARDL is specified as follows:

$$\begin{aligned}\Delta GINI_t = \alpha_0 &+ \sum_{i=0}^{p} \alpha_1 \Delta GINI_{t-i} + \sum_{i=0}^{p} \alpha_2 \Delta REC_{t-1} + \sum_{i=0}^{p} \alpha_3 \Delta FFS_{t-1} \\ &+ \sum_{i=0}^{p} \alpha_4 \Delta WP_{t-1} + \sum_{i=0}^{p} \alpha_5 \Delta WWI_t + \sum_{i=0}^{p} \alpha_6 \Delta CRP_t \\ &+ \sum_{i=0}^{p} \alpha_7 \Delta LSP_t + \sum_{i=0}^{p} \alpha_8 \Delta CY_t + \mu_t \end{aligned} \quad (3)$$

The empirical analysis is structured to sequentially assess the relationships among the study variables, ensuring robustness and alignment with the ARDL framework. The procedure is as follows:

- Preliminary Data Inspection: The dataset is first subjected to descriptive statistics and graphical analysis to evaluate distribution patterns, identify outliers, and examine trends over time. This step informs subsequent modelling decisions by highlighting potential structural breaks or anomalies.
- Stationarity Assessment: Each time series is tested for unit roots using the Augmented Dickey–Fuller (ADF) test, with inclusion of deterministic components (trend and/or intercept) as appropriate. Variables are retained in the ARDL framework only if they are I(0) or I(1), ensuring methodological validity, since the ARDL approach cannot accommodate I(2) series.
- Lag Length Determination: Optimal lags for each variable are identified using information criteria, namely the Akaike Information Criterion (AIC), Hannan–Quinn (HQ), and Final Prediction Error (FPE). In line with best practices for small-sample analyses, greater weight is given to AIC and FPE. This selection informs both short-term and long-term modelling.
- Bounds Testing for Cointegration: The existence of a long-run relationship among the variables is examined using the ARDL bounds testing procedure. Critical F-statistics are compared against the lower and upper bounds to establish whether a stable equilibrium relationship is present.
- Estimation of Long-Run and Short-Run Dynamics: Upon confirmation of cointegration, the unrestricted error correction model (UECM) is employed to estimate both long-run coefficients and short-run adjustments. The error correction term quantifies the speed of convergence towards the long-run equilibrium following a shock.

3.4 Diagnostic Tests

To guarantee the reliability of the estimated ARDL model, several post-estimation diagnostic evaluations were performed. Table 2 provides a summary of the tests, null hypotheses, and decision criteria.

These estimations together validate the dependability of the proposed model. The Jarque–Bera test evaluates the normality of residuals, whereas the Breusch–Godfrey and Breusch–Pagan tests examine autocorrelation and heteroskedasticity, respectively. Multicollinearity is assessed through the variance inflation factor (VIF), which verifies that explanatory variables do not exhibit concerning correlations. Ultimately, the CUSUM and CUSUMSQ tests provide a visual evaluation of parameter stability throughout time; model coefficients are considered stable if the plots stay within the critical limits.

Table 2 Diagnostic tests

Diagnostic test	Null hypothesis (H_0)	Method/test used	Decision rule
Normality	Residuals are normally distributed	Jarque–Bera test	Fail to reject H_0 if $p > 0.05$
Serial Correlation	No autocorrelation in residuals	Breusch–Godfrey LM test	Fail to reject H_0 if $p > 0.05$
Heteroskedasticity	Constant variance of residuals	Breusch–Pagan test	Fail to reject H_0 if $p > 0.05$
Multicollinearity	No high correlation among regressors	Variance inflation factor (VIF)	VIF < 10 indicates no concern
Stability	Model coefficients are stable	CUSUM & CUSUMSQ plots	Parameters are stable if plots remain within 5% significance bands

Source Author's own compilation

4 Findings and Discussion

4.1 Descriptive Statistics

Before estimating any econometric model, It is important to review the summary statistics for the variables before attempting to estimate any econometric model. Based on the findings of Blanca et al. (2013), it is concluded that normality is not a common property of real data, since only a small portion of real data distributions are close to a normal distribution. Bao (2013) derived approximate results for the skewness and kurtosis coefficients, which can be used to study the finite-sample behaviour of econometric estimators, as presented in (Table 3).

Before conducting the econometric estimation, it is crucial to consider the summary statistics of the variables to gain insight into their distribution and potential implications for model specification. In line with the findings of Blanca et al. (2013), who stated that real-world numbers are seldom normally distributed, and Bao (2013), who provided approximate values of skewness and kurtosis, this study reveals different distributional characteristics of the variables. The variables of renewable energy consumption (REC), water withdrawal intensity (WWI), crop production (CRP) and livestock production (LSP) represent a rightskewed distribution, where more cases have higher values than the average value of the distribution, as examples of higher renewable energy consumption may be associated with increased opportunities for adopting technologies, low energy costs, and possible exports. However, cases of lower renewable energy consumption values may indicate some measure of dependence on fossil fuel energy sources (Korpås et al., 2022; Lin, Schmid, and others). On the other hand, Gini, fossil fuel consumption (FFC) water productivity (WP) and cereal yield (CY) have negative skewness (NS), more cases have lower values than the average of the distribution,as examples of low fossil fuel consumption may be related to energy conservation measures, low water productivity values

Table 3 Descriptive statistics

Statistic	Gini	REC	FFC	WP	WWI	CRP	LSP	CY
Mean	61.9333	−1.5757	86.5167	0.6920	47.6550	2.4528	2.3699	8.0118
Median	62.2500	−2.1577	86.8931	1.6168	44.5202	0.4262	2.5457	7.9931
Maximum	63.5000	11.7382	88.1487	9.6676	65.0336	30.4054	25.2670	8.5814
Minimum	60.1000	−11.5257	84.2434	−12.0502	40.7717	−26.7101	−16.7283	6.8509
Std. Dev	1.1161	6.1791	1.1562	4.4099	8.0865	13.0824	6.8814	0.3918
Skewness	−0.3076	0.4142	-0.5359	−1.2581	1.1462	0.3982	0.3552	−0.8480
Kurtosis	1.6685	2.4762	2.0821	5.1296	2.7803	3.3331	6.9244	3.8871
Jarque–Bera	2.6895	1.2009	2.4891	13.5828	6.6297	0.9317	19.8825	4.5793
Probability	0.2606	0.5486	0.2881	0.0011	0.0363	0.6276	0.0000	0.1013
Sum	1858.000	−47.2717	2595.500	20.7609	1429.651	73.5828	71.0967	240.3539
Sum Sq. Dev	36.1267	1107.283	38.7670	563.9797	1896.375	4963.342	1373.261	4.4509
Observations	30	30	30	30	30	30	30	30

Source Author(s) computation using EViews 12

may be related to productivity constraints and economic stress in water-dependent sectors, and low cereal yield values may be related to less stable crop production and income uncertainty and vulnerability to food price fluctuations (Helsel & Ryker, 2002; Li et al., 2016). The majority of the variables are flatter-than-normal which is represented by kurtosis values below 3, and Jarque–Bera test shows that the null hypothesis of normality is not significant at the five per cent level, and there is no need in operating non-standard econometric techniques but admit a slight violation of normality. Such descriptive details provide a more detailed examination of the statistical features of the Water–Energy–Food nexus indicators in South Africa and their socio-economic resilience consequences.

4.2 *Preliminary Tests*

Below is the presentation of the preliminary results, which includes the unit root test/ stationarity test using Augmented Dickey–Fuller (ADF) and cointegration test using the ARDL cointegration bounds test, lag selection, which helps choose the max lags of dependent variable and regressors when running the ARDL, the ARDL for main results and the post-test results that include diagnostic tests.

Stationarity Test

a. *Visual Inspection.* For visual inspection, the study's variables: Gini, REC, FFC, WP, WWI, CRP, LSP, CY (Fig. 1).

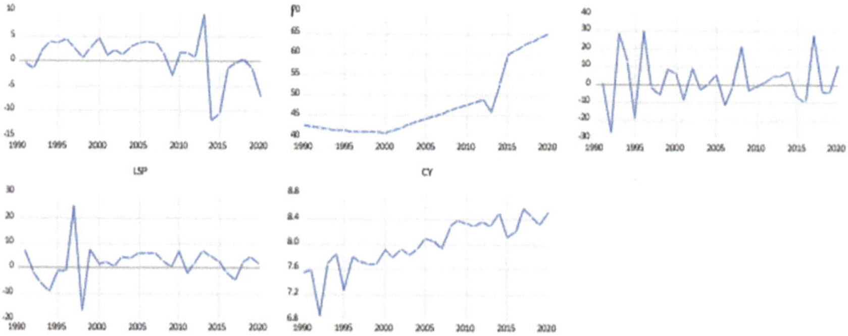

Fig. 1 Graphical representation of all variables at the level form. *Source* Author(s) computation using EViews 12

Table 4 Augmented Dickey–Fuller unit root tests

Variable	*p*-value (Level)	*p*-value (1st Diff.)	*t*-Statistic (Level)	*t*-Statistic (1st Diff.)	1% critical value	5% critical value	10% critical value	Order of integration
GINI	0.9285	0.0463	−1.0000	−5.1355	−4.3098	−3.5742	−3.2217	I(1)
REC	0.1611	0.0000	−2.9557	−9.2510	−4.3240	−3.5806	−3.2252	I(1)
FFC	0.0009	0.0000	−4.3774	−7.3010	−4.3098	−3.5742	−3.2217	I(0)
WP	0.0088	0.0200	−4.3655	−4.0830	−4.4163	−3.6220	−3.2486	I(0)
WWI	0.8980	0.0055	−1.1734	−4.5621	−4.3098	−3.5742	−3.2217	I(1)
CRP	0.0000	0.0001	−10.9335	−6.6875	−4.3943	−3.6122	−3.2431	I(0)
LSP	0.0000	0.0000	−7.2570	−11.2088	−4.3240	−3.5806	−3.2253	I(0)
CY	0.0001	0.0007	−6.4211	−5.5081	−4.3393	−3.5875	−3.2292	I(0)

Source Author(s) computation using EViews 12

b. *Augmented Dickey–Fuller Unit Root Tests.* The ARDL model is used only if none of the variables under study is I (2). All variables under the study must be either I (0) or I (1); otherwise, the ARDL approach is not applicable. Therefore, an ADF unit root test is used to verify that the data meet the requirements of the ARDL model. As exhibited in Table 4, none of the variables was I (2); all variables were at I (0) and at I (1). Therefore, ARDL could be used to test for cointegration between the study's variables.

When using the Augmented Dickey–Fuller test on the variables at the intercept level under the models, some variables tend to be stationary at the 1%, 5%, and 10% significance levels. At first, some variables become stationary with a trend and an intercept.

4.3 Lag Selection Criterion

There are six information criteria to choose from in Table 5. We look for the criterion with the lowest value, and in this case, we see that FPE has the lowest figure (18.09113). The optimum lag to use when running the model is 2.

4.4 ARDL Bounds Test

The ARDL procedure requires that the bound testing procedure be followed to investigate the presence of a long-run relationship among the variables. Furthermore, the Wald test is used to indicate the existence or absence of cointegration.

Table 5 Lag length selection

Lags	LogL	LR	FPE	AIC	SC	HQ
0	−500.5265	NA	822,863.7	36.32332	36.7010	36.4397
1	−315.1517	251.5800	170.6562	27.6537	31.0794	28.7010
2	−200.5962	90.000789*	18.09113*	24.04259*	30.51.33*	26.0208*

*Indicates lag order selected by the criterion
LR: Sequential modified LR test statistic (each test at 5% level) FPE: Final prediction error
AIC: Akaike information criterion SC: Schwarz information criterion
HQ: Hannan–Quinn information criterion
Source Author(s) computation using EViews 12

a. *Bounds test for cointegration.* The Wald test is used to compute the t-statistics, which are then compared to the specified critical bounds. The critical bounds and f-statistics for the model, taking into account lag 2, are presented in Table 6.

The F-statistic value of 6.2477, which exceeds the upper and lower bounds at significant levels of 10%, 5%, 2.5%, and 1%, suggests a long-run relationship between the variables.

b. *Long-run analysis.* After establishing that a co-integrating relationship exists between the variables, this indicates the presence of a long-run relationship. Table 7 presents the long-run relationship among the variables, and their explanation follows in detail.

Long-run interpretation. FFC: A statistically significant long-run relationship exists between FFC and the Gini coefficient. This implies that a 1% increase in FFC will, on average, result in a 26.05% increase in the Gini coefficient ceteris paribus. The relationship between FFC and the Gini coefficient is complex and multifaceted. Martins et al. (2021) found a positive causal relationship between FFC and carbon dioxide emissions, with a 1% increase in consumption resulting in a 0.4924% increase in emissions in the long run. This suggests that higher FFC can lead to greater environmental inequality. Papathanasopoulou and Jackson (2009) further support this, showing that the Gini coefficient for total fossil resource consumption increased by 24% in the UK between 1968 and 2000, driven by rising demand from high-income quintiles. However, Takada et al. (2020) introduce a different perspective,

Table 6 ARDL cointegration bounds test

Test Statistic	Value	Significance	I (0)	I (1)
F-statistic	6.2477	10%	2.03	3.13
K	7	5%	2.32	3.5
		2.5%	2.6	3.84
		1%	2.96	4.26

Source Author(s) computation using EViews 12

Table 7 Estimated long-run coefficients using ARDL and ECM short-run dynamic ARDL estimation

Dependent variable: Gini				
Included observations:30 after adjustments				
Long-run ARDL estimates				
Variable	Coefficient	Std. Error	*t*-Statistic	Prob
FFC	0.2605	0.0417	6.2482	0.0002
WWI	−0.0917	0.0094	-9.7714	0.0000
CY	3.8687	0.2271	17.003	0.0000
Short-run ARDL estimates				
D(FFC)	0.1821	0.0209	8.7312	0.0000
D(WP)	0.0319	0.0049	6.5515	0.0002
D(WWI)	0.0259	0.0102	2.5471	0.0343
D(CRP)	0.0077	0.0011	6.8077	0.0001
D(LSP(−1))	−0.0085	0.0014	−5.9845	0.0003
D(CY)	0.5208	0.0955	5.4518	0.0006
CointEq (−1)*	−0.5246	0.0542	0.0955	0.0000

Source Author(s) computation using EViews 12

highlighting a positive relationship between changes in the global spread of technological efficiency and the use of non-fossil fuels. This suggests that a shift towards non-FFC could reduce inequality.

Water withdrawal intensity. A statistically significant and negative long-run relationship exists between WWI and the Gini coefficient. This implies that a 1% increase in WWI will, on average, result in a 9.17% decrease in the Gini coefficient, ceteris paribus. A series of studies has explored the relationship between WWI and the Gini coefficient. Feng and Sun (2024) found that the water footprint intensity (WFI) in the water-receiving areas of the South-to-North Water Diversion Project in China fluctuated significantly, with a Gini coefficient range of 0.219 to 0.267. Sun et al. (2010) similarly observed a decline in WFI in China, indicating improved water-use efficiency. Haowen et al. (2019) noted a decrease in the Gini coefficient of water supply quantity in Yunnan Province, suggesting improved equilibration in water resource use.

Cereal yield. There is a statistically significant and positive long-run relationship between CY and the Gini coefficient. This implies that a 1% increase in CY will, on average, result in a 386,87% increase in the Gini coefficient, ceteris paribus. Several studies have investigated the relationship between CY and various factors. Simmonds (1995, 1996) found a strong negative correlation between grain yield and protein content in cereals, but a positive correlation between grain yield and protein yield per unit area. This suggests that an increase in CY may lead to a higher protein yield, which could have implications for the Gini coefficient. Engledow (1925) emphasised the complexity of yield in cereals, suggesting that it reflects the plant's vital processes.

Protić et al. (2009) further explored the components of grain yield in winter wheat varieties, finding positive correlations between grain yield and the number of plants and spikes per square metre. These findings collectively suggest that an increase in CY may potentially lead to a higher Gini coefficient; however, further research is needed to confirm this relationship.

After investigating the long-run relation of the specified model, we proceeded to examine the short-run dynamics using the ECM-ARDL model. The error correction term (CointEq (−1)) represents the speed of adjustment towards the long-run equilibrium, having one period of shock in the model. A stable model error correction term should satisfy two essential properties, proposed by Pahlavani et al. (2005), which are negative in sign and statistically significant, as shown in Table 8. The coefficient of CointEq (−1) is negative and less than the probability value of 0.0000, which is less than 0.05, showing that it is statistically significant. The coefficient of CointEq (−1) −0.5246 indicates that there is a high speed of adjustment from the short run to the long run. The variables FFC, WP, WWI, CRP, and LSP, as well as CY, have an impact on the Gini coefficient in the short run. There is a statistically significant short-run relationship between FFC and the Gini coefficient. This implies that a 1% increase in FFC will, on average, result in an 18.21% increase in the Gini coefficient ceteris paribus. There is a statistically significant short-run relationship between WP and the Gini coefficient. This implies that a 1% increase in WP will, on average, result in a 3.19% increase in the Gini coefficient ceteris paribus. There is a statistically significant short-run relationship between WWI and the Gini coefficient. This implies that a 1% increase in WWI will, on average, result in a 2.59% increase in the Gini coefficient ceteris paribus. There is a statistically significant short-run relationship between crop production and the Gini coefficient. This implies that a 1% increase in livestock production will, on average, result in a 0.77% increase in the Gini coefficient ceteris paribus. There is a statistically significant negative relationship between livestock production and the Gini coefficient in the short run. This implies that a 1% increase in livestock production will, on average, result in a 0.85% decrease in the Gini coefficient ceteris paribus. There is a statistically significant short-run relationship between CY and the Gini coefficient. This implies that a 1% increase in cereal will, on average, result in a 52.08% increase in the Gini coefficient ceteris paribus.

4.5 Post-Test Results

This study examines the assumptions of the classical linear regression model. The primary objective of a diagnostic test is to construct test statistics that correct for specification errors in an estimated model. The diagnostic tests performed here are those elucidated in number three, and they are subject to their own hypotheses.

The diagnostic analysis reveals that the estimated model meets the main econometric criteria of reliability, except for those of heteroskedasticity and serial correlation, which were corrected accordingly in the further estimations. The VIF values

Table 8 Diagnostic test

Diagnostic test	Purpose	Statistic/ Threshold	Decision	Inference
Variance inflation factor (VIF)	Detect multicollinearity among regressors	VIF < 5 indicates no multicollinearity	All VIF values < 5	Multicollinearity not detected
Normality test (Jarque–Bera / histogram)	Assess residual normality	$p > 0.05$ indicates normality	Accept H_0	Residuals are normally distributed
White heteroskedasticity test	Test for homoscedasticity	$p < 0.05$ indicates heteroskedasticity	Reject H_0	Heteroskedasticity present; corrected using robust standard errors
Breusch–godfrey serial correlation lm test	Detect serial correlation in residuals	$p < 0.05$ indicates serial correlation	Reject H_0	Serial correlation present; addressed through dynamic specification
CUSUM and CUSUMSQ Tests	Evaluate the structural stability of model parameters	The series must lie within 5% significance bounds	The model is stable over time	Parameter stability confirmed

Source Author(s) compilations

are significantly smaller than the traditional value of five, indicating the absence of multicollinearity and demonstrating that the regressors explain the changes independently. The p-value result of the residual normality test ($p = 0.69$) is not significant, indicating that the error term follows a normal distribution and therefore meets the assumptions of inferential validity.

The White test indicates statistically significant heteroskedasticity, whereas the Breusch–Godfrey test reveals the presence of serial correlation in the residual structure. These diagnostics necessitated heteroskedasticity- and autocorrelation-consistent (HAC) adjustments to draw strong inferences. Lastly, the statistics CUSUM and CUSUMSQ do not exceed the corresponding 5% confidence limits, which means that the model parameters are not structurally unstable during the sample period. Taken together, the results of the diagnostic procedures contribute to the validity of the estimated relationships, making the model suitable for analysing the underlying dynamics in the South African scenario.

5 Summary of Key Findings and Policy Recommendations

The empirical analysis reveals various meaningful long-term and short-term relationships among economic, environmental, and agricultural elements, as well as the Gini coefficient, in South Africa, offering insights into avenues for a Just Transition and cohesive resource management.

Determinants of income inequality over time. The consumption of fossil fuels, the intensity of water withdrawal, and CY stand out as significant long-term factors influencing income inequality. The use of fossil fuels exhibits a statistically significant positive correlation with the Gini coefficient, where a 1% rise in consumption results in a 26.05% rise in inequality, holding other factors constant. This suggests that dependence on fossil fuel-intensive industries unfairly benefits wealthier populations, exacerbating social inequalities. Within the framework of a Just Transition, this conclusion emphasises the need for targeted policies that redirect energy generation to renewable sources, while also establishing redistributive strategies, such as community-driven energy initiatives or job programmes, to ensure fair participation in the green economy.

On the other hand, the intensity of water withdrawal has a statistically significant negative long-term relationship with the Gini coefficient; a 1% rise leads to a 9.17% decrease in inequality. This suggests that enhancing access to water resources, especially for rural and agricultural communities, can reduce income inequality. These findings suggest that water management policies focusing on fair distribution and sustainable use could act as a vital tool for promoting socially inclusive development.

Cereal production exhibits a statistically significant long-term positive impact on inequality, where a 1% rise in yield results in a 386.87% increase in the Gini coefficient. This unequal advantage for large-scale producers highlights the potential injustices associated with agricultural intensification. Policies that support smallholder farmers through cooperative structures, access to technology, and inclusive market integration can help synchronise yield enhancements with fair outcomes.

The ECM-ARDL model also indicates a rapid adjustment to long-run equilibrium (CointEq $(-1) = -0.5246$), suggesting that short-run discrepancies from these relationships are corrected effectively, which enhances confidence in the stability of these results.

Short-term dynamics and transitional effects. In the short term, fossil fuel use shows a positive correlation with inequality (18.21% rise for every 1% change), highlighting the importance of gradual decarbonisation approaches that consider social equity. WP exhibits a favourable short-term impact on inequality (3.19%), indicating that early efficiency improvements could preferentially advantage more well-off participants. Likewise, the intensity of water withdrawal exhibits a short-term positive correlation with the Gini coefficient (2.59%), suggesting that the immediate benefits from resource use may not be distributed equally among individuals.

Crop production indicates a slight positive short-term effect on inequality (0.77%). In contrast, livestock production reveals a short-term adverse impact (−0.85%), implying that varied agricultural approaches, encompassing livestock initiatives, can help in reducing inequality while ensuring environmental sustainability.

The short-run yield of cereal still shows a positive correlation with inequality (52.08%), reflecting the long-term trend where advantages mainly favour larger producers. These findings collectively suggest that South Africa's Just Transition and integrated resource management approaches need to account for both time-related aspects and the effects of resource use distribution to avoid the accumulation of advantages among wealthier groups.

5.1 Policy Recommendations

Policy implications for Just Transition and integrated resource management, and results indicate that South Africa can promote a fair transition by:

- Emphasising the implementation of renewable energy and inclusive energy policies to decrease reliance on fossil fuels while generating fair job opportunities.
- Establishing fair water and land management systems that guarantee rural and disadvantaged communities achieve lasting access to productive resources.
- Assisting smallholder and medium-scale farming efforts, especially in cereal and crop cultivation, to reduce the unequal benefits enjoyed by large-scale producers.
- Orchestrating multi-sectoral planning to harmonise energy, water, and agricultural strategies, thus ensuring the management of resource interdependencies to attain environmental sustainability and social equity simultaneously.

These pathways demonstrate that South Africa can effectively manage the complex relationship among environmental sustainability, economic performance, and social equity, provided that integrated, evidence-driven, and participatory policies are implemented.

References

Allouche, J.: The sustainability and resilience of global water and food systems: political analysis of the interplay between security, resource scarcity, political systems and international trade. Food Policy **36**, S3–S8 (2011)

Amoah, L.N.A., Thabisa, J.: Small-scale farmers' experiences with the impact of water scarcity and coping strategies in the Eastern Cape, South Africa. Afr. J. Dev. Stud. **12**(4) (2022)

Bao, Y.: On sample skewness and kurtosis. Econometric Rev. **32**(4), 415–448 (2013)

Blanca, M.J., Arnau, J., López-Montiel, D., Bono, R., Bendayan, R.: Skewness and kurtosis in real data samples. Methodology (2013)

Boelee, E., Yohannes, M., Poda, J.N., McCartney, M., Cecchi, P., Kibret, S., Laamrani, H.: Options for water storage and rainwater harvesting to improve health and resilience against climate change in Africa. Reg. Environ. Change **13**, 509–519 (2013)

Brown, C., Lall, U.: Water and economic development: the role of variability and a framework for resilience. Nat. Resour. Forum **30**(4), 306–317 (2006)

Conradie, B.: A review of stochastic production functions with inefficiency models for extensive livestock systems with applications in the Karoo (2017)
D'Odorico, P., Laio, F., Ridolfi, L.: Does the globalisation of water reduce societal resilience to drought? Geophys. Res. Lett. **37**(13) (2010)
Durand-Lasserve, O., Cox, J., Engel-Cox, J., Igogo, T., Anand, P.K., Kumar, K., Saleh, L.: Integrated nexus policies for sustainable and resilient energy, water, and food systems (2020)
Edokpayi, J.N., Makungo, R., Mathivha, F., Rivers, N., Volenzo, T., Odiyo, J.O.: The influence of global climate change on water resources in South Africa: towards an adaptive management approach. Elsevier, pp. 83–115 (2020)
Engledow, F.L.: Investigations on yield in the cereals: II. A spacing experiment with wheat. J. Agric. Sci. **15**(2), 125–146 (1925)
Essel-Gaisey, F., Okyere, M.A., Forson, R., Chiang, T.F.: The road to recovery: financial resilience and mental health in post-apartheid South Africa. SSM Popul. Health **23**, 101455 (2023)
Fagbemi, F., Oke, D.F., Fajingbesi, A.: Climate-resilient development: an approach to sustainable food production in sub-Saharan Africa. Future Foods **7**, 100216 (2023)
Feng, Z., Sun, L.: Water conservation implications based on temporal-spatial characteristics of water footprint in the water-receiving areas of the South-to-North Water Diversion Project, China. Sustainability **16**(3), 1270 (2024)
Freedman, D., & Drineas, P.: Energy minimisation via graph cuts: Settling what is possible. In 2005 IEEE Computer Society Conference on Computer Vision and Pattern Recognition (CVPR'05) (Vol. 2, pp. 939–946). IEEE (2005, June)
Gallagher, P.W.: U.S. soybean yields: estimation and forecasting with nonsymmetric disturbances. Am. J. Agric. Econ. **69**, 796–803 (1987)
Gammans, M., Mérel, P.R., Ortiz-Bobea, A.: Negative impacts of climate change on cereal yields: statistical evidence from France. Environ. Res. Lett. **12** (2017)
Giupponi, B.O.: Just transition in the Global South: alternative approaches from Latin America. Int. J. Comp. Labour Law Ind. Relat. **39**(3–4) (2023)
Gracceva, F.: Fossil fuel sustainability from two different points of view. Miner. Energy Raw Mater. Rep. **17**, 25–37 (2002)
Grewar, T.: South Africa's options for mine-impacted water reuse: a review. J. S. Afr. Inst. Min. Metall. **119**(3), 321–331 (2019)
Haowen, G., Fei, W., Xiao, Y., Shixiang, G., Pingsen, M., Youyang, L.: Research on equilibration of water resources utilisation in the Yunnan Plateau by the Gini coefficient. IOP Conf. Ser. Earth Environ. Sci. **376**, 012065 (2019)
Harsono, O.A.: Sustainable and developmental energy resilience in a multidisciplinary perspective. RSF Conf. Ser.: Bus. Manag. Soc. Sci. **3**(3), 86–92 (2023)
Harvey, D.I., Leybourne, S.J., Xiao, L.: Testing for nonlinear deterministic components when the order of integration is unknown. J. Time Ser. Anal. **31** (2010)
Helsel, D.R., Ryker, S.J.: Defining surfaces for skewed, highly variable data. Environmetrics **13** (2002)
Hennessy, D.A.: Crop yield skewness under law of the minimum technology. IO: Productivity (2007)
Ho, M.: Descriptive statistics. In: Ownership Structure, Related Party Transactions, and Firm Valuation (2020)
Hu, J.L., Chang, T.Y.: Energy resilience: a cross-economy comparison. Energies **16**(5), 2214 (2023)
Juana, J.S., Mangadi, K.T., Strzepek, K.M.: The socioeconomic impacts of climate change on water resources in South Africa. Water Int. **37**(3), 265–278 (2012)
Kamp, A., Østergård, H.: Future scenario modelling and resilience indicators: a case study of small-scale food and energy production in a village in Ghana (2016)
Kaur, P., Stoltzfus, J.C., Yellapu, V.: Descriptive statistics. Int. J Acad. Med. **4**, 60–63 (2018)
Keskinen, M., Sojamo, S., Varis, O.: Enhancing security, sustainability and resilience in energy, food and water. Sustainability **11**(24), 7244 (2019)

Kori, D.S., Zhou, L., Sibanda, M.: A retrospective analysis of climate adaptation strategies and implications for food and nutrition security among small-scale farmers in South Africa. Front. Sustain. Food Syst. **7**, 1139078 (2023)

Korpås, M., Holttinen, H., Helistö, N., Girard, R., Koivisto, M., Frew, B., Söder, L.: Addressing market issues in electrical power systems with large shares of variable renewable energy. In: Proceedings of the 18th European Energy Market Conference (EEM), pp. 1–8. IEEE (2022)

Kraemer, W., Dette, H.: Beyond inequality: a novel measure of skewness and its properties. CESifo Work. Pap. (2016)

Kripfganz, S., Schneider, D.C.: ardl: Stata module to estimate autoregressive distributed lag models. Res. Pap. Econ. (2016)

Kripfganz, S., Schneider, D.C.: Ardl: estimating autoregressive distributed lag and equilibrium correction models. Stata J. **23**, 983–1019 (2018)

Kuye, O.O., Oludayo, O.O.: Effects of micro-credit on livestock production among smallholder livestock farmers in Yakurr LGA, Cross River State, Nigeria (2016)

Li, C.Z., Swain, R.B.: Growth, water resilience, and sustainability: a DSGE model applied to South Africa. Water Econ. Policy **2**(4), 1650022 (2016)

Li, M., Xu, W., Rosegrant, M.W.: Irrigation, risk aversion, and water rights under water supply uncertainty (2016)

Lin, C., Schmid, T., Weisbach, M.S.: Climate change, operating flexibility and corporate investment decisions. SSRN Electron. J. (2019)

Linke, P.: On the development of strategies for water and energy management in the context of the water-energy-food nexus. Comput.-Aided Chem. Eng. **34**, 196–201 (2014)

Lundqvist, J.: A high withdrawal ratio implies a changed water policy. Phys. Chem. Earth B **25**, 193–198 (2000)

Luptácik, M., Nežinský, E.: Measuring income inequalities beyond the Gini coefficient. Cent. Eur. J. Oper. Res. **28**, 561–578 (2020)

Mabhaudhi, T., Nhamo, L., Mpandeli, S., Nhemachena, C., Senzanje, A., Sobratee, N., ... & Modi, A. T. The water–energy–food nexus as a tool to transform rural livelihoods and well-being in Southern Africa. Int. J. Environ. Res. Public Health, **16**(16), 2970 (2019)

Magombeyi, M.S., Matambo, C.: Water Research Commission (2023)

Martins, T., Barreto, A.C., Souza, F.M., Souza, A.M.: Fossil fuel consumption and carbon dioxide emissions in G7 countries. Environ. Pollut. **291**, 118093 (2021)

Mpandeli, S., Naidoo, D., Mabhaudhi, T., Nhemachena, C., Nhamo, L., Liphadzi, S., Modi, A.T.: Climate change adaptation through the water-energy-food nexus in Southern Africa. Int. J. Environ. Res. Public Health **15**(10), 2306 (2018)

Natsiopoulos, K., Tzeremes, N.G.: ARDL bounds test for cointegration: replicating the Pesaran et al. (2001) results for the UK earnings equation using R. J. Appl. Econom. **37**(5), 1079–1090 (2022)

Nhemachena, C., Nhamo, L., Matchaya, G., Nhemachena, C.R., Muchara, B., Karuaihe, S.T., Mpandeli, S.: Climate change impacts on the water and agriculture sectors in Southern Africa. Water **12**(10), 2673 (2020)

Nkoro, E., Uko, A.K.: Autoregressive distributed lag cointegration technique: application and interpretation. J. Stat. Econom. Methods **5**, 1–3 (2016)

Oguge, N.O.: Building resilience to drought among small-scale farmers in Eastern African drylands. In: Curr. Dir. Water Scarcity Res. 2, 265–276. Elsevier (2019)

Okorie, V.O., Mphambukeli, T.N., Amusan, S.O.: Exploring the political economy of the water and food security nexus in BRICS. Africa Insight **48**(4), 21–38 (2019)

Orimoloye, I.R.: Livelihood vulnerability assessment and drought events in South Africa. In: Climate Change Impacts on Natural Resources, Ecosystems and Agricultural Systems, pp. 175–185. Springer, Cham (2023)

Ozturk, I.: Sustainability in the food-energy-water nexus: evidence from BRICS countries. Energy **93**, 999–1010 (2015)

Ozturk, I.: The dynamic relationship between agricultural sustainability and food-energy-water poverty in selected sub-Saharan African countries. Energy Policy **107**, 289–299 (2017)

Pahlavani, M., Wilson, E., & C Worthington, A.: Trade-GDP nexus in Iran: An application of the autoregressive distributed lag (ARDL) model. Am. J. Appl. Sci. **2**(7), 1158–1165 (2005)

Papathanasopoulou, E., Jackson, T.: Measuring fossil resource inequality: a case study for the UK between 1968 and 2000. Ecol. Econ. **68**(4), 1213–1225 (2009)

Perera, D., Smakhtin, V., Williams, S., North, T., Curry, A.: Ageing water storage infrastructure: an emerging global risk. UNU-INWEH Rep. Ser. **11**, 25 (2021)

Protić, R., Todorović, G., Protić, N.: Correlations of yield and grain yield components of winter wheat varieties. J. Agric. Sci. **54**(3), 213–221 (2009)

Reddy, K.V.N., Kumar, J.E., Kalita, M.: Droughts manifested by climate change in Southern Africa. Res. Highl. Agric. Sci. **6**, 117 (2023)

Scott, C.A., Albrecht, T.R., De Grenade, R.R., Zuniga-Teran, A., Varady, R.G., Thapa, B.: Water security and the pursuit of food, energy and earth systems resilience. In: Putting Water Security to Work, pp. 97–116. Routledge (2021)

Simmonds, N.W.: The relation between yield and protein in cereal grain. J. Sci. Food Agric. **67**(3), 309–315 (1995)

Simmonds, N. W.: Family selection in plant breeding. Euphytica, **90**(2), 201–208 (1996)

Sparks, D., Madhlopa, A., Keen, S., Moorlach, M., Dane, A., Krog, P., Dlamini, T.: Renewable energy choices and their water requirements in South Africa. J. Energy s. Afr. **25**(4), 80–92 (2014)

Spiegelberg, M., Hoshino, S., Hashimoto, S.: Serving the underserved: water-energy-food nexus in socio-ecological production landscapes. WIT Trans. Ecol. Environ. **193**, 181–191 (2015)

Subedi, R., Karki, M., Panday, D.: Food system and water-energy-biodiversity nexus in Nepal: a review. Agronomy **10**(8), 1129 (2020)

Sun, C.Z., Liu, Y.Y., Chen, L.X., Zhang, L.: Spatial-temporal disparities in water footprint intensity in China. Acta Ecol. Sin. **30**(5), 1312–1321 (2010)

Takada, H.H., Ribeiro, C.O., Costa, O.L., Stern, J.M.: Gini and entropy-based spread indices for primary energy consumption efficiency and CO_2 emissions. Energies **13**(18), 4938 (2020)

Tantoh, H. B., & McKay, T. J.: Utilising the water-land-food security nexus to review the underperformance of smallholder farmers in the Eastern Cape, South Africa. Frontiers in Sustainable Food Syst. **7**, 1143630 (2023)

Thiam, N.: Towards a sustainable energy system for Africa. In: Facing Global Environmental Change, pp. 445–456. Springer, Berlin Heidelberg (2009)

Tschernig, R., Weber, E., Weigand, R.: Long- versus medium-run identification in fractionally integrated VAR models. Econ. Lett. **122**, 299–302 (2014)

Vasylyna, O.: Influence of physical economy on energy security. Econ. Educ. **8**(1), 68–71 (2023)

Zaman, K., Shamsuddin, S., Ahmad, M.: Energy-water-food nexus under financial constraint environment in sub-Saharan Africa. Environ. Sci. Pollut. Res. **24**, 13358–13372 (2017)

Zhou, L., Kori, D.S., Sibanda, M., Nhundu, K.: Differences in vulnerability to climate change between rural and urban areas in South Africa. Climate **10**(8), 118 (2022)

Strategic Partnerships and Enterprise Supplier Development as Drivers of Performance in South Africa's Clothing Manufacturing Sector

Ntombizodwa J. Matsoma and Intaher M. Ambe

Abstract This study investigated the impact of strategic partnerships (SPs) and enterprise supplier development (ESD) on the performance of clothing manufacturers in South Africa. The research focused on how supplier-oriented procurement practices enhance operational efficiency, cost management, and competitiveness. Supplier-oriented procurement practices in the South African clothing industry face challenges due to a lack of support from local retailers and the government. Design/methodology/approach: The study employed structural equation modelling (SEM) to analyse data collected via a structured questionnaire from 621 manufacturers across Gauteng, KwaZulu-Natal, and the Western Cape. The results reveal that both SP and ESD significantly improve performance outcomes such as lead time reduction, cost efficiency, and product quality. The findings show that SPs between manufacturers and major retail buyers facilitate knowledge transfer, co-investment in upgrades, and contractual stability. Additionally, initiatives supporting ESD enable manufacturers to enhance production systems and compliance capabilities, ultimately improving their operational effectiveness and market readiness. However, challenges such as limited financial access, policy misalignment, and global competitive pressures persist. The findings offer actionable insights for policymakers, retail buyers, and supply chain managers, advocating for the institutionalisation of ESD and SP in public and private procurement strategies to promote industrial revitalisation. These require coordinated action among government, retailers, and suppliers to align procurement with localisation and inclusive development objectives. This research contributes to emerging literature on supplier development in developing economies by empirically demonstrating the performance benefits of relational procurement practices. It provides a practical framework for enhancing localisation, sustainability, and supply chain resilience.

N. J. Matsoma (✉)
University of South Africa (UNISA), Pretoria, South Africa
e-mail: matsonj@unisa.ac.za

I. M. Ambe
Department of Business Administration, Jackson State University, Jackson, MS, USA
e-mail: marcus.ambe@jsums.edu

E. H. Redda (ed.), *Proceedings of the Social Sciences International Research Conference, 2025*, https://doi.org/10.1007/978-981-95-9489-4_27

Keywords Strategic partnerships · Enterprise supplier development · Performance improvement

1 Introduction

Clothing and textile manufacturing remains a cornerstone of South Africa's industrial economy, contributing significantly to employment creation, domestic value chain integration, and inclusive economic development. However, despite its strategic importance, the sector continues to face entrenched structural challenges, including deindustrialisation, overreliance on imports, weak policy execution, and limited support for domestic manufacturers (CottonSA 2019; Jack 2020; Worku 2019). While the Retail-Clothing, Textile, Footwear and Leather (R-CTFL) Master Plan, Broad-Based Black Economic Empowerment (B-BBEE) codes, and Industrial Policy Action Plan (IPAP) aim to localise procurement and promote enterprise development, their implementation has yielded modest results. Within this policy landscape, enterprise supplier development (ESD) and strategic partnerships (SP) have emerged as critical instruments for revitalising domestic manufacturing and integrating historically disadvantaged suppliers into formal value chains. ESD initiatives are designed to build capacity, provide financial and technical support, and ensure long-term sustainability for local suppliers. In parallel, SP emphasises collaborative, trust-based buyer–supplier relationships that enhance agility, reduce costs, and foster innovation (Perry and Wood 2019; Singh and Chan 2022).

Despite the strategic intent of these policy-aligned mechanisms, limited empirical evidence exists on their joint impact on firm-level performance in the South African clothing sector. Most local manufacturers operate under the Cut-Make-Trim (CMT) model, lacking the skills, technology, and capital to compete globally. To address this gap, the current study investigated the extent to which ESD and SP influence procurement-driven outcomes, such as total cost of ownership (TCO), operational responsiveness, product quality, infrastructure development, and job creation. The study adopted a multidimensional theoretical framework grounded in the Resource-Based View (RBV), Transaction Cost Theory (TCT), and Stakeholder Theory (ST) to explore the relational and resource-based dynamics of procurement modernisation in developing contexts. Additionally, it used structural equation modelling (SEM) to empirically assess the relationship between ESD, SP, and performance among 621 clothing manufacturers across three industrial hubs, namely Gauteng, KwaZulu-Natal, and the Western Cape.

This research makes three key contributions. Firstly, it advances theory by integrating developmental procurement constructs, ESD, and SP into a performance model grounded in TCT, Resource Dependence Theory (RDT), ST, and RBV. Secondly, it delivers empirical insights based on robust SEM analysis in a developing country context, addressing a notable gap in procurement literature. Thirdly, it provides actionable policy recommendations for industry stakeholders, including government agencies, retail supply chain actors, and procurement professionals, on how localised sourcing strategies can drive competitiveness and inclusive economic development. The remainder of this paper is structured as follows. Section 2 presents a review of the relevant literature on supplier development, SP, and procurement performance. It also outlines the theoretical framework and hypotheses of the study. Section 3 describes the research methodology, including sampling, data collection, and SEM procedures. Section 4 reports on the empirical results and discusses key findings in relation to theory and practice. Section 5 concludes with a discussion of the practical implications, limitations, and directions for future research.

2 Literature Review and Conceptual Framework

This section synthesises the current literature on ESD, SP, and their combined effect on the performance of clothing manufacturers in South Africa. Drawing on both global insights and local policy contexts, the review identifies the theoretical and empirical underpinnings critical to understanding procurement-driven performance outcomes.

2.1 Enterprise Supplier Development, Strategic Partnership, and Performance

The concepts of enterprise supplier development, strategic partnership, and performance are discussed below.

Enterprise Supplier Development plays a pivotal role in South Africa's policy agenda in efforts to restructure its industrial base and support historically marginalised businesses. ESD initiatives aim to empower small, medium, and micro enterprises (SMMEs) by enhancing their technical capability, financial access, and integration into formal value chains, particularly in labour-intensive sectors such as clothing and textiles (DTIC 2021). With deindustrialisation and the dominance of offshore sourcing accelerating job losses and skill erosion, ESD has been positioned as a strategic intervention to revitalise domestic manufacturing and counter import dependency (Dos Santos 2020; Jenkin and Hattingh 2022). Despite supportive policies like tax incentives and machinery grants, the clothing sector continues to face barriers such as high input costs, outdated equipment, and limited financing (CottonSA 2019; Netshishivhe 2021). Scholars have argued that ESD must go beyond financial support to include technical training, digital integration, and strategic alignment with buyers (Joorst 2021; Perry and Wood 2019). To that end, this study frames ESD as a lever not only for compliance but also for industrial competitiveness, underpinned by constructs such as supplier upgrading, technology adoption, infrastructure readiness, and socio-economic impact (Khan and Yu 2019; Mercer 2020).

Strategic Partnerships are defined as long-term, trust-based collaborations between buyers and suppliers that facilitate innovation, agility, and shared value. In South Africa's clothing sector, these partnerships are underdeveloped, often replaced by short-term, cost-driven sourcing models that hinder transformation and sustainability (Perry and Wood 2019). Effective SPs require transparent communication, ethical practices, and alignment with standards like ISO 14001 to promote responsiveness, co-investment, and inclusive growth (Singh and Chan 2022; Utama et al. 2021). There are challenges, such as fragmented buyer–supplier relationships and transactional contracting, that undermine supply chain coherence and development goals (Mejías et al. 2019; Twyg 2020). Well-structured SPs can reduce lead times, stabilise demand, and build supplier capability, especially among vulnerable CMT manufacturers. Rooted in Relational Governance Theory, this study conceptualises SPs as enablers of mutual trust, regulatory compliance, and economic upliftment through shared risk and joint planning (Chen et al. 2023; Reis 2019).

Performance Improvement. Procurement-driven performance improvement encompasses operational, financial, and socio-economic outcomes. The intention of ESD and SPs is to enhance manufacturer performance across multiple dimensions, including lead time, cost efficiency, quality, compliance, and local economic impact (Dos Santos 2020; Shen et al. 2019). Collaborative procurement fosters transparency, better planning, and responsiveness, while digital integration and logistics agility are essential for competitive supply chains (Jin and Cedrola 2019; Swaminathan and Venkitasubramony 2024). Despite the availability of abundant resources, such as, mohair, local manufacturers face high production costs and global competition, limiting their ability to scale (InvestSA 2020; Netshishivhe 2021). According to the RBV, sustainable performance emerges from a blend of internal capabilities and external partnerships. Therefore, this research frames performance as a composite of operational efficiency, cost control, compliance, resource capacity, and local sourcing contribution (Braglia et al. 2020; Khan and Yu 2019).

2.2 *Hypotheses Development*

The theoretical foundation for this study integrates TCT, RDT, ST, and the RBV to explain how SP and ESD influence performance outcomes in South Africa's clothing manufacturing sector. The TCT theory, as developed by Williamson (1979, 1985), posits that firms adopt governance structures that minimise the costs associated with economic exchange, particularly under conditions of uncertainty and asset specificity. In procurement, SPs represent an institutional response to market failures such as opportunism, contractual instability, and coordination inefficiencies (Garfamy 2012; McIvor 2009). By fostering long-term collaboration, trust, and joint planning, SPs reduce transaction costs and enable adaptive responses to volatile demand, frequent design modifications, and logistics disruptions, which are common challenges in the clothing industry (Boateng 2016; Zijm et al. 2019). Thus, SPs enhance supply chain resilience and operational performance.

Table 1 Summary of hypotheses Source: Author's own complilation

Null hypothesis (H_0)	Alternative hypothesis (H_1)
H_{01}: Strategic partnerships do not significantly influence performance	H_1: Strategic partnerships have a significant and positive impact on firm performance
H_{02}: Enterprise supplier development does not significantly influence performance	H_2: Enterprise supplier development has a significant and positive impact on firm performance

Complementing this, RDT (Pfeffer and Salancik 1978) asserts that firms depend on external entities to access critical resources, making inter-organisational relationships a strategic necessity. In South Africa, small and medium-sized clothing manufacturers often lack advanced technology, access to markets, and bargaining power. SPs with dominant retail buyers can mitigate such dependencies by facilitating knowledge transfer, co-investment in upgrades, and contractual stability (Hillman et al. 2009; Petersen et al. 2008). Therefore, SPs are expected to contribute significantly to firm-level performance. Accordingly, the first hypothesis of the study was proposed as:

> **H1**: Strategic partnerships have a significant and positive impact on the performance of clothing manufacturers.

ESD is underpinned by ST, which views firms as accountable to a network of internal and external actors, including suppliers, regulators, communities, and customers, whose interests need to be considered for sustained value creation (Freeman 1984; Harrison et al. 2015). Within the South African policy context, ESD serves as a public-private mechanism to address inequalities and promote inclusive growth. It supports small manufacturers through technical training, access to finance, and infrastructure improvement, aiming to elevate their competitiveness and integration into mainstream supply chains (Dos Santos 2020; Johnson and Klassen 2022). The RBV further frames ESD as a catalyst for developing firm-specific resources and capabilities that are valuable, rare, and difficult to imitate, which are the key determinants of a competitive advantage (Barney 1991). Through ESD, manufacturers can enhance their production systems, human capital, and compliance capabilities, improving their operational effectiveness and market readiness (Table 1). Accordingly, the second hypothesis of the study was formulated as:

> **H2**: ESD has a significant and positive impact on the performance of clothing manufacturers.

2.3 Conceptual Framework

The conceptual framework guiding this study is built on the dual-theoretical axis of relational governance (via SP) and capability development (via ESD). It posits that SP

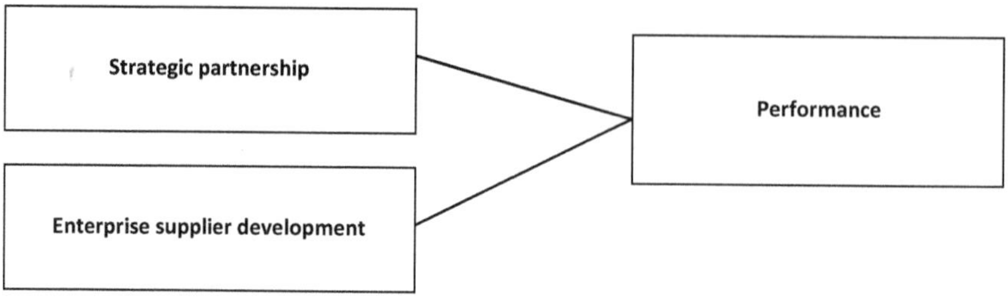

Fig. 1 Conceptual framework for the study

and ESD function as complementary mechanisms that influence the performance of clothing manufacturers through distinct but interconnected pathways. SPs, grounded in TCT and RDT, are modelled as mechanisms that reduce transaction inefficiencies, stabilise resource flows, and enhance coordination across the supply chain. These partnerships strengthen external linkages, enabling firms to benefit from collaborative forecasting, co-investment, and knowledge exchange, thereby driving operational efficiency and market agility. ESD, rooted in ST and RBV, is conceptualised as a capacity-building tool that enhances internal firm resources. ESD initiatives, such as technical training, financing support, and infrastructure upgrades, help build the foundational capabilities required for effective participation in competitive value chains. These internal improvements are theorised to have a direct and measurable impact on firm performance outcomes. Firm performance, the dependent variable, is conceptualised as a multidimensional construct encompassing operational efficiency, cost effectiveness, product quality, compliance, and socio-economic contribution. Based on the hypotheses and theoretical foundations, the study conceptualised a framework where SP and ESD function as independent variables that influence performance, the dependent variable. This integrative framework captures both the transactional and developmental aspects of procurement in emerging economies (Fig. 1).

3 Methodology

This section discusses the research methodology employed in the study.

3.1 Research Design and Philosophical Orientation

This study adopted a quantitative, cross-sectional research design underpinned by a positivist epistemological stance. The positivist paradigm supports objective measurement, hypothesis testing, and the generalisability of findings through empirical analysis (Saunders et al. 2019). The approach is suitable for establishing causal relationships among SP, ESD, and organisational performance within the South

African clothing manufacturing sector. A structured survey strategy was used to collect data for statistical analysis and model validation.

3.2 *Target Population, Sampling Strategy, and Unit of Analysis*

The target population comprised of all 800 registered clothing manufacturing firms in Gauteng, KwaZulu-Natal, and the Western Cape, South Africa's principal industrial provinces (Veitch 2021). Due to pandemic-related disruptions and limited accessibility, a non-probability snowball sampling technique was employed. Initially, respondents were purposively selected from regional industry directories and associations, and were asked to refer other eligible participants. This method yielded a final sample of 621 valid responses. The unit of analysis was the organisation, with procurement managers, supply chain officers, and general managers serving as key informants.

3.3 *Instrument Design and Operationalisation*

Data were collected through a structured, self-administered questionnaire. The instrument was grounded in prior empirical studies and theoretical constructs from the TCT, RDT, ST, and the RBV. It was divided into three sections:

- Section A: Organisational demographics (firm size, age, region, experience);
- Section B: ESD and SP constructs; and
- Section C: Organisational performance indicators.

Each construct was measured using validated multi-item scales on a four-point Likert scale. Items were adapted to the context of the South African clothing industry, and pre-tested for clarity and contextual relevance.

3.4 *Pilot Testing and Content Validity*

A pilot test was conducted with 15 industry professionals, including procurement specialists and supply chain managers. Their feedback informed revisions to item phrasing, response scaling, and questionnaire layout. Exploratory factor analysis (EFA) with Promax rotation was conducted to assess the dimensional structure of the measurement instrument. Items with factor loadings <0.40 and communalities <0.30 were eliminated to ensure construct validity (Hair et al. 2020).

Table 2 Reliability and validity metrics

Construct	No. of Items	Cronbach's α	Composite reliability	AVE
Strategic partnership	8	0.864	0.870	0.410
Enterprise supplier Development (ESD)	3	0.805	0.766	0.533
Performance	12	0.915	0.916	0.476

3.5 Reliability and Construct Validity

Cronbach's alpha and composite reliability (CR) metrics were used to assess internal consistency. All constructs surpassed the 0.70 threshold, indicating high reliability. Average variance extracted (AVE) values further confirmed convergent validity. Table 2 summarises the reliability statistics.

3.6 Data Collection Procedure

Data collection was conducted over six weeks using both electronic distribution (e-mail) and physical delivery of the questionnaires to manufacturing clusters. Ethical clearance was obtained from the relevant ethics committee of the university. Participants were informed of the study's purpose, assured of data confidentiality, and provided their informed consent. All responses were anonymised.

3.7 Data Analysis Strategy

SEM was applied using both AMOS 24.0 and JASP 0.16.4. The analysis proceeded in three phases:

- Confirmatory factor analysis (CFA) to validate the measurement model
- Structural path modelling to test the hypothesised relationships among SP, ESD, and performance
- Model fit assessment using Hu and Bentler's (1999) indices:
- Chi-square/df: 2.405 (acceptable range: 1–5)
- CFI: 0.916
- TLI: 0.910
- RMSEA: 0.042
- SRMR: 0.057

Ethical clearance for the study was granted by the College of Economic and Management Sciences, University of South Africa (UNISA).

Table 3 Model fit statistics

Fit index	Observed value	Acceptable threshold	Interpretation
Chi-square/df	2.405	Between 1–5	Acceptable model fit
CFI	0.916	≥0.90	Good incremental fit
TLI	0.910	≥0.90	Strong model fit
RMSEA	0.042	<0.08	Low approximation error
SRMR	0.057	<0.08	Acceptable residuals

4 Results

This section presents the empirical results derived from the SEM analysis that were used to assess the direct effects of SP and ESD on organisational performance.

4.1 Measurement Model Validation

Before evaluating structural relationships, CFA was conducted to assess the reliability and validity of the measurement model. All factor loadings exceeded 0.60, indicating strong item reliability. AVE values for all constructs ranged between 0.410 and 0.533, meeting the minimum threshold of 0.40–0.50 for convergent validity. Discriminant validity was confirmed by comparing the square roots of AVE to inter-construct correlations, with all constructs satisfying the Fornell–Larcker criterion.

4.2 Model Fit Indices

Table 3 presents the model fit statistics. The SEM results indicate an overall good fit between the hypothesised model and the observed data.

These values confirm the model's adequacy and robustness for hypothesis testing.

4.3 Structural Path Analysis

The standardised path coefficients (β), standard errors (SE), Z-values, and p-values are summarised in Table 4.

Both paths were statistically significant at $p < 0.001$, confirming that both SP and ESD significantly and positively influence performance. SPs ($\beta = 0.724$) exert a stronger influence than ESD ($\beta = 0.527$). The relationships between these variables are also presented in the SEM model illustrated in Fig. 2.

Table 4 SEM path coefficients

Path	β coefficient	SE	Z-value	p-value
Performance SP	0.724	0.0895	8.51	<0.001
Performance ESD	0.527	0.0791	6.06	<0.001

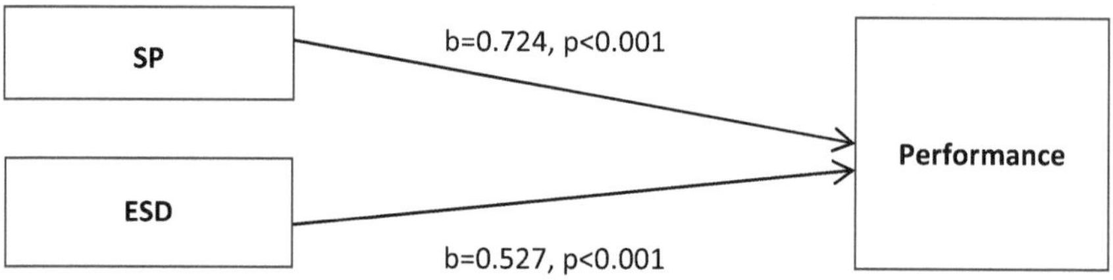

Fig. 2 SEM model Source: Author's own complilation

5 Interpretation and Discussion

This section interprets the SEM results in the light of the theoretical framework and broader empirical literature.

5.1 Strategic Partnerships and Performance (H1)

The finding that SPs significantly improve performance supports the propositions of the TCT and RDT. Long-term buyer–supplier collaborations reduce opportunism, renegotiation costs, and uncertainty in demand, thereby improving operational performance (McIvor 2009; Williamson 1985). These partnerships allow resource sharing, innovation, and co-investment, which are particularly vital for SMMEs in developing economies like South Africa. The results also align with prior studies showing that SPs enhance supply chain flexibility, reduce lead times, and optimise TCO (Chen et al. 2023; Singh and Chan 2022).

5.2 Enterprise Supplier Development and Performance (H2)

The significant influence of ESD on performance validates the theoretical perspectives of the RBV and ST. ESD initiatives strengthen the technical, financial, and operational capabilities of small suppliers, increasing their ability to meet buyer expectations and regulatory standards (Barney 1991; Freeman 1984). The results corroborate earlier findings that inclusive supplier development leads to improved local content integration, employment generation, and economic resilience (Dos Santos 2020; Perry and Wood 2019).

5.3 Sectoral Implications

Despite the positive impact of SP and ESD on firm performance, several systemic challenges persist in South Africa's clothing sector. For example, weak enforcement of supplier development policies and fragmented oversight mechanisms continue to hinder effective implementation. In addition, many small manufacturers face restricted access to credit, limiting their ability to invest in capacity or technology upgrades. Moreover, the sector's ongoing dependence on imports undermines national localisation goals. Digital integration across procurement systems also remains limited, reducing efficiency and transparency. To enhance outcomes, procurement frameworks must be more closely aligned with industrial policies, such as the R-CTFL Masterplan and B-BBEE codes, fostering a unified and developmental procurement environment.

5.4 Contribution to Theory and Practice

This study makes four notable contributions. Theoretically, it extends TCT, RDT, and ST into inclusive procurement models in emerging markets, showing how supplier development and partnerships drive performance. Empirically, it applied SEM to a large sample of South African manufacturers, offering robust evidence of the impact of SP and ESD. Practically, it provides guidance for procurement managers and policymakers on embedding local supplier development as a core performance strategy. From a policy standpoint, it underscores the importance of procurement in driving industrial transformation and supporting goals like employment growth, skills development, and supplier sustainability.

6 Conclusion and Recommendations

This study investigated the impact of SP and ESD on the performance of clothing manufacturers in South Africa, drawing data from firms in Gauteng, KwaZulu-Natal, and the Western Cape. Using SEM, the findings affirm that both SP and ESD significantly improve performance dimensions such as cost control, order accuracy, lead time, and compliance. SP yielded a slightly stronger effect, emphasising the value of sustained buyer–supplier collaboration in enhancing competitiveness. Nonetheless, systemic challenges, such as weak policy enforcement, limited financing, outdated technology, and reliance on imports, continue to constrain industrial transformation and undermine the goals of national policies like the IPAP, Broad-Based Black Economic Empowerment (B-BBEE), and the National Environmental Management Act (NEMA). Overcoming these constraints requires coordinated action among government, retailers, and suppliers to align procurement with localisation and inclusive development objectives. Theoretically, this study contributes to procurement and supply chain literature by reinforcing the applicability of the RBV and RDT to emerging economies. The study found that SP and ESD function not only as operational levers but also as sources of sustained advantage. The role of TCT was validated by demonstrating how relational governance lowers coordination and negotiation costs. Similarly, ST was extended by showing that procurement strategies aligned with national development goals generate both economic returns and social value. By grounding these theories in the South African context, the study illustrates how procurement can drive industrial upgrading and structural transformation.

Practically, the results offer clear guidance for stakeholders. Retailers are urged to invest in long-term partnerships with CMT manufacturers to enhance supply chain efficiency and support compliance with local content targets. Manufacturers can leverage such partnerships to access resources, build capacity, and stabilise production cycles. Policymakers should treat procurement as a strategic development tool by embedding performance metrics, incentivising local sourcing, and integrating supplier development into broader regulatory frameworks. Institutionalising these practices can enhance both firm-level performance and sector-wide resilience. Strategic actions should include formalised partnerships, co-investment in production infrastructure, and tailored enterprise development to transition CMTs towards full-package manufacturing. These interventions should be supported by performance monitoring systems and incentive mechanisms, such as grants or tax relief, for firms meeting localisation and sustainability targets. Alignment with national strategies like the IPAP and R-CTFL masterplan will be key to driving the systemic impact.

While the study provides valuable insights, it was limited by its cross-sectional design and focus on only three provinces. The reliance on procurement professionals may exclude relevant perspectives from retailers, policymakers, and financiers. Future research should adopt a longitudinal approach, incorporate multi-stakeholder views, and explore the impact of public incentives on supplier upgrading. Cross-sectoral and comparative studies across emerging markets are also recommended to deepen the understanding of procurement-driven industrial transformation.

References

Barney, J.B.: Firm resources and sustained competitive advantage. J. Manag. **17**(1), 99–120 (1991)

Boateng, D.: Executive insight series: compendium of supply chain management terms. 2nd edn. Panavest International & Partners, South Africa (2016)

Braglia, M., Marrazzini, L., Padellini, L., Rinaldi, R.: Managerial and Industry 4.0 solutions for fashion supply chains. J. Fash. Mark. Manag. 25(1), 184–201 (2020)

Chen, T., Shan, F., Yang, F., Xu, F.: Online retailer bundling strategy in a dual-channel supply chain. Int. J. Prod. Econ. **259**, 1–14 (2023)
CottonSA: Outlook on the South African cotton textile industry. Cluster Report No. 1/2019. http://www.cottonsa.org.za (2019). Last accessed 19 June 2025
Department of Trade, Industry and Competition: A guide to the DTIC incentive schemes 2020/21. http://www.thedtic.gov.za/wp-content/uploads/the-dti-incentive-schemes-guide-2021.pdf (2021). Last accessed 19 June 2025
Dos Santos, P.: The impact of manufacturing incentives in the clothing and textile industry in KwaZulu-Natal: a case study of Durban. Master's dissertation, Durban University of Technology, Durban (2020)
Freeman, R.E.: Strategic management: a stakeholder approach. Pitman, Boston (1984)
Garfamy, R.M.: Supply management: a transaction cost economics framework. S. East Eur. J. Econ. Bus. **7**(2), 139–147 (2012)
Hair, J.H., Page, M., Brunsveld, N.: Essential of business research methods. 4th edn. Routledge, New York (2020)
Harrison, J.S., Freeman, R.E., Abreu, M.C.S.D.: Stakeholder theory as an ethical approach to effective management: applying the theory to multiple contexts. Rev. Bras. Gest. Neg. **17**(55), 858–869 (2015)
Hillman, A.J., Withers, M.C., Collins, B.J.: Resource dependence theory: a review. J. Manag. **35**(6), 1404–1427 (2009)
InvestSA: Investing in South Africa's clothing, textile, footwear and leather sector. http://www.investsa.gov.za/wp-content/uploads/2020/02/FACT-SHEET_TEXTILES_2020.pdf (2020). Last accessed 19 June 2025
Jack, S.T.: The slow movement in the garment manufacturing industry of South Africa. Master's dissertation, University of Johannesburg, Johannesburg (2020)
Jenkin, N., Hattingh, E.: Designing climate-compatible industrial strategies for South Africa: The textiles value chain. Trade and Industrial Policy Strategies (TIPS). https://www.tips.org.za/images/TIPS_Designing_climate-compatible_industrial_strategies_for_South_Africa_-_The_textiles_value_chain_August_2022.pdf (2022). Last accessed 19 June 2025
Jin, B., Cedrola, E.: Product innovation in the global fashion industry. Springer, New York (2019)
Johnson, P.F., Klassen, R.D.: New directions for research in green public procurement: the challenge of inter-stakeholder tensions. Cleaner Logist. Supply Chain **3**, 1–7 (2022)
Joorst, G.: Among petrochemical companies in South Africa. Master's dissertation, University of Johannesburg, Johannesburg (2021)
Khan, S.A.R., Yu, Z.: Strategic supply chain management. Springer, Cham (2019)
McIvor, R.: How the transaction cost and resource-based theories of the firm inform outsourcing evaluation. J. Oper. Manag. **27**(1), 45–63 (2009)
Mejías, A.M., Bellas, R., Pardo, J.E., Paz, E.: Traceability management systems and capacity building: new approaches for improving sustainability in the multi-tier fashion supply chain. Int. J. Prod. Econ. **217**, 143–158 (2019)
Mercer, S.: An investigation of employment outcomes in South African manufacturing. Master's dissertation, University of Cape Town, Cape Town (2020)
Netshishivhe, T.C.: The challenges impacting the growth of small, micro- and medium-sized textile enterprises (SMMEs): Ekurhuleni Metropolitan Municipality. Master's dissertation, University of the Free State, Mangaung (2021)
Perry, P., Wood, S.: Exploring the international fashion supply chain and corporate social responsibility: cost, responsiveness and ethical implications. In: Fernie, J., Sparks, L. (eds.) Logistics and retail management, pp. 77–100 (2019)
Petersen, K.J., Handfield, R.B., Lawson, B., Cousins, P.D.: Buyer dependency and relational capital formation: the mediating effects of socialization processes and supplier integration. J. Supply Chain Manag. **44**(4), 53–65 (2008)
Pfeffer, J., Salancik, G.R.: The external control of organizations: a resource dependence perspective. Harper & Row, New York (1978)

Reis, T.L.: Validation of a scale to measure consumers' engagement in voluntary simplistic clothing consumption behaviour in South Africa. Unpublished master's dissertation. University of Pretoria, Pretoria, South Africa (2019)

Saunders, M.N.K., Lewis, P., Thornhill, A.: Research methods for business students. 8th edn. Pearson Education, Harlow (2019)

Shen, B., Gu, Q., Yang, Y.: Fashion supply chain management in Asia: concepts, models and cases. Springer, Singapore (2019)

Singh, P.K., Chan, S.W.: The impact of electronic procurement adoption on green procurement towards sustainable supply chain performance: evidence from Malaysian ISO organizations. J. Open Innov. Technol. Mark. Complex. **8**(2), 1–27 (2022)

Swaminathan, K., Venkitasubramony, R.: Demand forecasting for fashion products: a systematic review. Int. J. Forecast. **40**(1), 1–21 (2024)

Twyg: Assessing the economic value of the designer fashion sector in South Africa: Executive summary. https://twyg.co.za/wp-content/uploads/2020/08/SACC004-A-SACC-THE-ECONOMIC-VALUE-OF-THE-DESIGNER-FASHION-SECTOR-RESEARCH-REPORT11.pdf (2020). Last accessed 19 June 2025

Utama, D.M., Maharani, B., Amallynda, I.: Integration DEMATEL and ANP for supplier selection in the textile industry: a case study. J. Ilm. Tek. Ind. **20**(1), 119–130 (2021)

Veitch, C.: The clothing industry in South Africa. Who Owns Whom. https://www.whoownswhom.co.za/store/clothing-industry-south-africa (2021). Last accessed 19 June 2025

Williamson, O.E.: Transaction cost economics: the governance of contractual relations. J. Law. Econ. **22**(1919), 233–261 (1979)

Williamson, O.E.: The economic institutions of capitalism: Firms, markets and relational contracting. Free Press, New York (1985)

Worku, Z.: Risk factors of failure in start-up textile enterprises. Afr. J. Sci. Technol. Innov. Dev. **11**(1), 97–105 (2019)

Zijm, W.H.M., Klumpp, M., Heragu, S., Regattieri, A.: Operations, logistics and supply chain management: definitions and objectives. In: Zijm, W.H.M., Klumpp, M., Heragu, S., Regattieri, A. (eds.) Operations, logistics and supply chain management. Springer, Cham (2019)

Key Determinants of Youth Unemployment in Regions D and G of Johannesburg, Gauteng Province

Ntombifuthi Winnie Gamede

Abstract This study investigates the principal determinants of youth unemployment in Gauteng Province, with a focused examination of regions D and G. Recognising youth unemployment as a persistent barrier to economic growth and sustainable development in South Africa, the research addresses its role as a driver of poverty, inequality, and social insecurity. The analysis identified several core factors underlying elevated youth unemployment rates. Key determinants include the persistent mismatch between educational attainment and labour market requirements, insufficient entrepreneurial and job-creation skills, lack of formal work experience, and the enduring structural legacies of apartheid within post-1994 governance. Additionally, disparities in the provision of basic services and unequal access to socio-economic opportunities were found to exacerbate youth unemployment. Methodologically, the study employed a quantitative approach to capture the multifaceted nature of the challenge. Stepwise logistic regression analysis was applied, enabling identification and validation of statistically significant predictors among a range of socio-demographic variables. The results highlight that the economic system does not adequately account for the diversity of the workforce, which includes individuals with low, intermediate, high, and professional skill levels. Although educational attainment is essential, during periods of economic stagnation, unemployment affects all. In response to these findings, the study recommends targeted policy interventions to reform service delivery and address the lingering systemic effects of apartheid. Addressing these critical determinants is essential to alleviate chronic poverty, strengthen socio-economic inclusion for youth, and facilitate sustainable economic growth in South Africa.

Keywords Youth unemployment · Economic instability · Stepwise regression

N. W. Gamede (✉)
University of South Africa (UNISA), Pretoria, South Africa
e-mail: gamednw@unisa.ac.za

E. H. Redda (ed.), *Proceedings of the Social Sciences International Research Conference, 2025*, https://doi.org/10.1007/978-981-95-9489-4_28

1 Introduction

Youth unemployment has emerged as a persistent global challenge, deepening poverty and signalling broader threats to social and economic stability worldwide (Zinhnwe 2012; Yarima 2014). According to the International Labour Organisation (ILO 2010; 2011), high youth unemployment rates affect both developed and developing countries, often leaving many young people facing severe economic hardship—circumstances that have far-reaching repercussions for global order. As Pandey (2005) emphasises, youth unemployment is uniquely problematic, as it drives unacceptably high levels of poverty and inequality. Difficulties in securing employment lead to frustration and social unrest among young people, resulting in heightened risks of crime, violence, drug abuse, and other adverse social outcomes that exacerbate human insecurity (Malik 2014). Moreover, youth unemployment diminishes economic output and undermines livelihoods, deepening both individual and societal vulnerability.

In South Africa, this problem remains especially pronounced, serving as a root cause of social insecurity, rising violence, and even suicide, particularly within Gauteng Province. Bernstein (2008) attributes these outcomes to enduring policy challenges. Despite the scale of the issue, underlying and region-specific determinants of youth unemployment have been underexplored, particularly within Johannesburg's regions D and G, where this study is focused. Previous studies have largely centred on broader themes of human capital and labour supply and demand but have rarely examined the intricate local factors that perpetuate unemployment despite available workforce resources. It is argued here that mismatches between education and job market demand, limited entrepreneurial skills, lack of experience, historical legacies of apartheid, and persistent inequalities in the provision of basic services and opportunities are among the most significant contributors to youth unemployment rates in South Africa.

Prior studies have typically addressed national trends or macro-level factors, often neglecting the complex intersection of local structural challenges. Issues such as persistent skills mismatches, limited job-creation capacity, historical legacies of apartheid, and enduring inequalities in access to essential services and opportunities have not received sufficient attention. Since these localised dynamics shape outcomes within communities, a significant gap persists in understanding the proper drivers of youth unemployment at this level.

To address these limitations, the present study focuses explicitly on identifying the key determinants of youth unemployment in regions D and G of Johannesburg, Gauteng Province. The guiding research question is: What are the key determinants of youth unemployment in regions D and G, Johannesburg, Gauteng Province? Stepwise logistic regression has been adopted to rigorously isolate and analyse the most influential factors within these communities. This provides a much-needed empirical foundation for understanding both regional specificities and the statistical relationships that national-level research may obscure.

Ultimately, this study aims to provide policymakers and scholars with targeted insights to develop effective interventions and policies that foster economic growth and support social stability. The paper is organised as follows: Sect. 2 reviews the relevant literature; Sect. 3 outlines the research methodology; subsequent sections present the results, discuss their implications, and conclude with key recommendations.

2 Literature Review

The youth population is vital to the economic structure of a country. A young population forms a crucial part of the labour force and makes a significant contribution to economic growth (Writer 2025). One could argue that when young people enter the workforce, they expand the labour supply, boost productivity, and drive innovation. They also tend to be more receptive to new ideas and technologies. Moreover, youth are central to entrepreneurship, as they have the potential to establish new industries and create job opportunities (Writer 2025).

Despite their importance in strengthening the labour force, young people frequently face periods of unemployment. This is reflected in labour markets worldwide, where youth unemployment rates are typically higher than those of adults (O'Higgins 1997). Ekanem (2025) notes that youth unemployment remains one of the most pressing challenges in Africa. According to the African Development Bank (2016), the majority of young people on the continent lack stable economic opportunities. Unemployment is therefore a major continental challenge with wide-reaching consequences.

Baah-Boateng (2015) highlights that in Africa, countries such as Lesotho, South Africa, Botswana, Namibia, Zambia, Sudan, Tunisia, and Algeria have unemployment rates ranging between 10 and 25%, while other countries, such as Benin and Madagascar, record rates below 3%. Approximately 420 million youth aged 15 to 35 years, one-third are unemployed or discouraged, another third are engaged in precarious employment, and only one in six has a salaried job. The same report indicates that around 10 to 12 million young people enter the labour force each year, but only 3.1 million jobs are created, leaving a substantial number without secured employment (African Development Bank 2016).

Baah-Boateng (2015) asserts that, in terms of institutional and systemic terms, nearly nine out of every ten jobs in Africa are organised informally, with only 13% of workers employed in the formal sector. Furthermore, Baah-Boateng (2015) links the incidence of unemployment directly to poverty rates, noting that youth are particularly affected. The unemployment rate is consistently higher among youth than adults, which correlates with elevated poverty levels in younger age groups. Baah-Boateng and Turkson (2005) maintained that unemployment predominantly affects youth more than adults and declines with advancing age. Supporting this, Baah-Boateng's (2015) findings reveal that the average youth unemployment rate in Africa is 12.6%, compared to 4.5% for adults. Moreover, youth unemployment in North Africa is three times higher than adult unemployment, while in Sub-Saharan Africa, the youth rate is almost double that of adults. Comparable patterns are evident in South Africa, Botswana, and Ghana (Baah-Boateng 2015).

The situation becomes more acute during economic downturns, where the 'last-in, first-out' approach to hiring and layoffs disproportionately impacts young workers (UNECA 2005). In South Africa, the crisis has been exacerbated by the inability of multiple municipalities to address key determinants of unemployment. Many have failed to develop effective centres for job creation, skills development, and employment placement, leaving young people with limited opportunities for gaining work experience.

In South Africa, those aged 15 to 34 represent approximately 50.2% of the working-age population, or roughly 20.9 million people (Stats SA 2024). This demographic is central to driving innovation and entrepreneurship, helping establish new industries and create employment opportunities. Young citizens typically demonstrate higher consumption rates, especially regarding entertainment, education, and technology—activities that further stimulate economic growth by boosting consumer demand (Writer 2025). Nonetheless, youth unemployment remains persistently high, rising from 36.8% in 2014 to 45.5% in 2024. Recent data from the first quarter of 2025 indicate that the expanded unemployment rate for individuals aged 18–34 stands at approximately 56.3% (Stats SA 2025).

Foundational economic thinkers such as Smith (1976), Von Thünen, and Marshall (1920) recognised that educated individuals possess economic value because they embody capital comprising knowledge and skills. In the 1960s, Shultz (1961) and Becker (1964) advanced the concept of human capital, positing that investment in education, training, and skill development promotes national productivity, enhances individual earning potential, and widens employment opportunities (Becker 1964). Ehrenberg and Smith (1988) supported this framework, describing human capital as the collective skills within individuals that constitute a valuable stock of productive assets, which can be marketed to employers for earnings. Consequently, the prospect of higher lifetime earnings and a higher standard of living is a significant incentive for individuals to pursue education and training. This theory is particularly relevant in the South African context for understanding, as it offers insight into rising youth unemployment and underscores the importance of sustained investment in human capital to improve long-term economic outcomes.

2.1 Unemployment

A fundamental objective of every economic system is to optimise the use of limited factors of production to maximise the output of goods and services (Kaufman 1989). Unemployment, however, signals that a portion of available labour is being underutilised, resulting in a loss of potential production (Morh and Fourie 2008). The magnitude of this economic loss can be estimated using Okun's law, which states that each percentage-point increase in the unemployment rate corresponds to a 2.5

percentage-point decrease in gross national product. Therefore, sustained economic expansion is essential to prevent skill erosion and maintain productivity (Okun 1962).

Despite ongoing efforts to stimulate growth, youth unemployment rates consistently exceed those of adults in many countries. Youth unemployment typically refers to individuals aged 15–24 in the labour force, although age classifications vary by country. In South Africa, for example, youth refers to those aged 15–34, representing roughly one-third of the national population and being the group most acutely impacted by unemployment (Stats SA 2024).

Unemployment refers to a situation in which individuals who are capable and willing to work cannot secure employment at the current wage rate (Van Zyl et al. 2002). The root causes of unemployment are complex and dynamic. For some, unemployment is brief and occurs amid normal transitions between individual jobs or when entering or exiting the workforce. Because information in the job market is imperfect, finding each other typically takes time for job seekers and employers. Due to imperfect information in the labour market, job seekers and employers require time to connect. If labour market information were perfect and mobility costs were negligible, unemployment would be eliminated (Kaufman 1989). As such, some degree of frictional unemployment is unavoidable, and full employment (i.e. where all willing workers are employed) is unattainable in practice (Morh and Fourie 2008).

However, for many, unemployment persists for more extended periods and is not simply a consequence of imperfect information. Instead, it reflects a fundamental mismatch between the types of available jobs and the characteristics of job seekers, such as skills, education, location, and age (McConnell and Brue 2002). Structural unemployment arises when job openings require specific skills or qualifications that job seekers lack. For instance, positions in skilled occupations such as teaching may be available. However, the candidates are either younger people with inapplicable qualifications, or those lacking the necessary experience or education, or recently laid off from unskilled or semi-skilled jobs (Kaufman 1989).

Structural unemployment may also develop when job vacancies are concentrated in rural areas, whereas job seekers reside mainly in urban centres. In such situations, unemployment results not from imperfect labour market information but rather from mobility barriers and a reluctance to relocate (Sapsford and Tzannatos 1993). This form of unemployment leads to the simultaneous existence of job vacancies and job seekers unable to access those positions. A subtle distinction exists between frictional and structural unemployment, since both involve individuals seeking jobs. The key difference lies in the relative speed of the job search process (Kaufman 1989).

Both frictional unemployment and structural unemployment stem from the challenges of effectively matching job seekers with available positions. In contrast, some instances of unemployment are attributable to insufficient economic demand (Kaufman 1989). When the economy cannot generate enough jobs for job seekers due to demand shortfalls, cyclical unemployment arises. This type of unemployment is closely related to fluctuations in the business cycle, namely periods of economic growth and contraction. If an economy endures prolonged stagnation, this phenomenon is termed secular stagnation and is characterised by sustained low growth (Summers 2014, 2020). Cyclical unemployment exhibits significant year-to-year variation, reflecting macroeconomic expansions and contractions. While it shares with frictional and structural unemployment an impact across numerous sectors, the duration of cyclical unemployment generally falls between the short-lived joblessness typical of frictional unemployment and the longer spells associated with structural unemployment (Kaufman 1989; Sapsford and Tzannatos 1993).

2.2 Factors Contributing to Youth Unemployment in South Africa

Youth unemployment in South Africa is driven by a diverse and persistent set of factors, many of which remain inadequately addressed. Unemployment is defined as the condition in which an individual who is both willing and able to work is unable to secure paid employment (Van Zyl et al. 2002). The unemployment rate, which reflects the proportion of unemployed individuals within the labour force, requires careful classification, as it includes both those currently employed and those actively seeking work. In the South African context, youth unemployment arguably represents the most formidable barrier to poverty reduction and socio-economic advancement (Ligthelm 2006). The burden of unemployment is borne disproportionately by young South Africans, who typically lack skills and formal work experience (National Treasury 2011). The National Planning Commission has highlighted the extreme difficulty young people face in escaping unemployment, warning that a significant share of today's unemployed youth may never find formal employment (National Planning Commission 2012).

The South African labour market exhibits unemployment levels far exceeding those of other emerging economies (National Planning Commission 2011). Although South Africa's labour force constitutes a mere 0.5% of the global total, it remarkably accounts for 2% of worldwide unemployment (National Treasury 2011). Furthermore, the country's labour force participation rate stands at 56.5%, substantially below that of comparable nations, while formal sector labour absorption rates have declined steadily since 1985 (Ligthelm 2006).

Distinguishing between the employed and unemployed often necessitates subjective judgements, such as the minimum threshold of paid work required to classify an individual as employed, alongside statistical counts of jobholders and job seekers. Addressing youth unemployment demands an in-depth understanding of its underlying drivers. In South Africa, many contributing factors stem from persistent market distortions. Policies of import substitution industrialisation have fostered an uncompetitive, capital-intensive economy, largely inaccessible to most South Africans due to historical racial and spatial exclusion (National Planning Commission 2011). The inherited legacies of apartheid, including uncompetitive primary and secondary industries, continue to impede broad-based job creation (National Planning Commission 2011).

A further major contributor to unemployment is the prevalence of a largely unskilled and spatially misaligned workforce, a consequence of long-standing barriers to quality education (National Planning Commission 2012). Economic growth in recent decades has been concentrated predominantly in skills-intensive sectors, such as finance and business services (Hausmann 2008). This dynamic has precipitated a significant structural mismatch between the skills demanded by the modern South African economy and those possessed by the workforce (National Planning Commission 2011; Hausmann 2008). Moreover, robust labour force growth since the end of apartheid has intensified competition for a limited pool of suitable jobs (Hausmann 2008). South African youth, in particular, are affected by this mismatch, as they are generally low-skilled and often excluded from relevant employment opportunities.

Employers frequently cite education's failure to serve as a reliable signal of skill level, deeming unskilled and inexperienced youth a risky investment and thereby raising their effective employment costs (National Treasury 2011). As a result, youth unemployment arises from both a deficient demand for labour, driven by the increasingly skills-oriented nature of the economy, and a weak supply of adequately prepared entrants, many of whom are seen as high risk due to inadequate education and work experience. Additional factors exacerbating youth unemployment include corruption and nepotism, as well as a scarcity of real-time labour market information, all of which further marginalise young job seekers (ManPowerGroup 2012).

3 Research Methodology and Data Collection

Rajasekar et al. (2006) assert that research constitutes a logical and systematic pursuit of new and relevant information concerning a particular subject. Within the present study, research was conceptualised as a rigorous academic inquiry seeking solutions to youth unemployment in the D and G regions of Gauteng Province. Research methodology, therefore, denotes a structured approach employed to address a research problem, informed by the need to describe, explain, and predict patterns of youth unemployment in South Africa. A quantitative methodology was adopted, selected for its capacity to generate findings that are generalisable to larger populations. This approach facilitated the extrapolation of concise conclusions regarding youth unemployment, using representative samples and statistical analysis (Kothari 2004). Consequently, the study's findings provide inferences that extend beyond the immediate determinants of youth unemployment in South Africa.

3.1 Sample

The research used data collected in 2021 from the black townships in regions D and G, located within Johannesburg, Gauteng, South Africa. The sample comprised 600 black respondents, inclusive of both male- and female-headed households. Participants represented a spectrum of residential circumstances, including informal settlements, hostels, older four-roomed dwellings, RDP (Reconstruction and Development) houses, and mortgage-financed homes. For ethical reasons, the precise names of locations within these townships have been deliberately withheld. Respondents' identities were further protected by omitting names from questionnaires. Verbal consent was obtained from all participants in compliance with Level 3 COVID-19 regulations, which enforced social distancing and discouraged the handling of physical documents to limit transmission risk. As a result, written consent was replaced by verbal consent to protect both participants and researchers while maintaining research integrity and compliance. Survey data were analysed using the Statistical Package for the Social Sciences (SPSS) software.

3.2 Descriptive Statistics of the Age Distribution of the Sample

Table 1 presents a frequency distribution of the sample population's age groups. The sample comprised 600 respondents, with the most significant proportion identified as young adults aged 25–34 (29.5%), followed by middle-aged individuals aged 35–44 (26.8%), and participants aged 45–54 (18.7%). Nair (2011) emphasises that the age structure of a population is fundamental to socio-economic and demographic analysis. The distribution of age within a population significantly influences key societal domains, including the allocation of housing, educational and healthcare services, infrastructure planning, labour force participation, and overall economic growth (Bloom et al. 2010; Harper 2014). Furthermore, the age profile of a population is crucial for assessing variations in unemployment rates, educational attainment, and poverty levels (Wang 2019).

Table 2 illustrates the relationship between age and respondents' employment status, revealing that approximately 44% of young adults aged 18 to 24 are unemployed. This means that out of every 10 youth in the sampled population, four are unemployed. This number is slightly lower than Stats SA's (2025) figure of 62.2% for national youth unemployment among those aged 18–24. This indicates that, at the national level, six out of every ten youths are unemployed.

Furthermore, the unemployment rate among young adults aged 25 to 34 remains high at 44.1%. This result corresponds with the findings of Stats SA (2025), which report a similarly elevated unemployment rate of 40.4 per cent within this age group. These figures indicate that employment opportunities for new entrants to the labour market are diminishing, leaving young people increasingly vulnerable in their efforts

Table 1 Sample age distribution

		Frequency	Percentage	Valid Percentage	Cumulative Percentage
Valid	18–24	75	12.5	12.5	12.5
	25–34	177	29.5	29.5	42.0
	35–44	161	26.8	26.8	68.8
	45–54	112	18.7	18.7	87.5
	55–64	41	6.8	6.8	94.3
	65 and above	34	5.7	5.7	100.0
	Total	600	100.0	100.0	

Source Data Survey 2021

to secure work. Given that youth constitute at least one-third of the total population, persistent unemployment among this demographic is likely to remain a growing concern, as their numbers already surpass those of older age groups (Graham and De Lannoy 2016). Table 2 further shows that there is a high percentage of middle-aged adult unemployment between the ages of 35–44 and 45–54 at 42.2% and 42%, respectively. The unemployment rate for these age groups indicates a significant shift in the dependency ratio across age categories. The dependency burden ratio is the percentage of the working-age population that supports the non-working population. It comprises children under 15 and adults aged 65 or older (Todaro and Smith 2015; Hayes 2022).

According to Gonzo and Platter (2003), employment fulfils a range of social and interpersonal functions by enabling individuals to meet various psychological needs. Work plays a pivotal role in facilitating young people's transition into the labour market, enabling them to apply the skills and knowledge they have acquired through investments in human capital. This, in turn, enables youth to make meaningful contributions to national economic performance and societal well-being. Given that humans are inherently social, the workplace also serves as a setting for forging friendships, fostering connections, delivering mutual support, and providing avenues for personal growth and recognition. Employment offers individuals a sense of purpose and self-worth, which significantly influences self-esteem and personal identity (Moller 1993; Gonzo 2003).

Conversely, unemployment imposes considerable psychological stress on those who are willing and able to work but remain unable to secure employment. Such individuals frequently report reduced overall life satisfaction and may be prone to depressive symptoms, particularly when unemployment is prolonged despite active job searching efforts (Carl and Sloan 2003). The psychological consequences often extend to feelings of pessimism about the future and a sense of inadequacy (Carl and Sloan 2003). Youth unemployment, in particular, deprives young people of crucial labour market entry points, discourages further investment in human capital, erodes existing skills through disuse, and impedes the ability to capitalise on economic

Table 2 Results of cross-tabulation between age and employment status

Age		Employment status					Total
		Employed	Unemployed	Informal employed	Student	Pensioner	
18–24	Count	3	33	23	16	0	75
	% Within Age Classification	4.0%	44.0%	30.7%	21.3%	0.0%	100.0%
	% Within Employment Status	3.2%	14.0%	11.3%	66.7%	0.0%	12.5%
	% of Total	0.5%	5.5%	3.8%	2.7%	0.0%	12.5%
25–34	Count	41	78	52	6	0	177
	% Within Age Classification	23.2%	44.1%	29.4%	3.4%	0.0%	100.0%
	% Within Employment Status	44.1%	33.1%	25.5%	25.0%	0.0%	29.5%
	% of Total	6.8%	13.0%	8.7%	1.0%	0.0%	29.5%
35–44	Count	26	68	66	1	0	161
	% Within Age Classification	16.1%	42.2%	41.0%	0.6%	0.0%	100.0%
	% Within Employment Status	28.0%	28.8%	32.4%	4.2%	0.0%	26.8%
	% of Total	4.3%	11.3%	11.0%	0.2%	0.0%	26.8%
45–54	Count	18	47	46	1	0	112
	% Within Age Classification	16.1%	42.0%	41.1%	0.9%	0.0%	100.0%
	% Within Employment Status	19.4%	19.9%	22.5%	4.2%	0.0%	18.7%
	% of Total	3.0%	7.8%	7.7%	0.2%	0.0%	18.7%

Source Survey data (2021)

opportunities (Writer 2025). This not only hampers their immediate contribution to the economy but also undermines long-term wealth accumulation and retirement prospects.

3.3 Model Specification

Stepwise logistic regression was employed to estimate the probability of youth unemployment by analysing its relationship with a range of independent variables, including dwelling type, highest educational attainment, duration of residence, spoken language, occupational position, marital status, and gender. This stepwise selection approach identified the most significant predictors in the dataset. The regression equation applied in this analysis is presented in Eq. 1.

$$P(Y = 1|X) = \frac{1}{1 + e\,(\beta 0 + \beta_1 X_1 + \beta_2 X_2 + \quad + \beta_n X_n)} \tag{1}$$

where

- P(Y = 1|X) is the probability of the dependent variable (Y) being 1, given the values of the independent variables (X).
- where e is the base of the natural logarithm.
- X1, X2,…., Xn the number of independent variables.

The odds of youth unemployment are expressed in Eq. 2 as follows:

$$\frac{\mathrm{p}}{1 - \mathrm{p}} = (\beta_0 + \beta_1 X_1 + \beta_2 X_2 + \quad + \beta_n X_n) \tag{2}$$

4 Regression Results

A further analysis was conducted using stepwise logistic regression to ascertain the factors influencing the probability of youth unemployment, and the results are presented in Table 3.

Upon initial implementation of the stepwise regression model, all variables—namely, type of dwelling, the highest educational attainment, duration of residence, speaking ability, occupational position, marital status, and gender—were included in the analysis. Subsequent refinement of the model resulted in only the significant variables from step 4 being retained. Household dependents, highest level of education, occupational position, and gender emerged as significant predictors of youth unemployment.

The results from the stepwise logistic regression indicate that the likelihood of being dependent on others for welfare is 0.45, highlighting how unemployment exacerbates the dependency burden within households and across the broader economy. Unemployed individuals frequently lack economic support beyond their immediate family (Cloete 2015). Additionally, the odds of the head of the household being unemployed are less than 0.01 in comparison with other family members, while

Table 3 Stepwise logistic regression of youth unemployment factor

Variables in equation

		B	S.E	Wald	of	Sig	Exp(B)
Step 1[a]	Gender (1)	1.268	0.298	18.131	1	<0.001	3.554
	Constant	–0.940	0.202	21.580	1	<0.001	0.391
Step 2[b]	Position of the Respondent			11.964	2	0.003	
	Position of the Respondent (1)	0.382	0.655	0.341	1	0.559	1.466
	Position of Respondent the (2)	1.117	0.325	11.801	1	<0.001	3.056
	Gender (1)	1.289	0.319	16.334	1	< 0.001	3.631
	Constant	– 1.574	0.293	28.908	1	< 0.001	0.207
Step 3[c]	Highest level of education			9.530	3	0.023	
	Highest level of education (1)	–0.421	0.895	0.222	1	0.638	0.656
	Highest level of education (2)	–0.327	0.899	0.132	1	0.716	0.721
	Highest level of education (3)	1.263	0.993	1.619	1	0.203	3.538
	Position of the Respondent			13.188	2	0.001	
	Position of the Respondent (1)	0.465	0.675	0.476	1	0.490	1.593
	Position of the Respondent (2)	1.237	0.342	13.055	1	<0.001	3.446
	Gender (1)	1.364	0.331	16.939	1	<0.001	3.911
	Constant	– 1.496	0.883	2.873	1	0.090	0.224
Step 4[d]	Number of dependents in the household	–0.168	0.084	4.001	1	0.045	0.845
	Highest level of education			8.201	3	0.042	
	Highest level of education (1)	–0.314	0.894	0.123	1	0.726	0.731
	Highest level of education (2)	–0.143	0.902	0.025	1	0.874	0.867
	Highest level of education (3)	1.281	0.991	1.671	1	0.196	3.599
	Position of the Respondent			14.293	2	<0.001	

(continued)

Table 3 (continued)

Variables in equation

		B	S.E	Wald	of	Sig	Exp(B)
	Position of the Respondent (1)	0.421	0.680	0.384	1	0.536	1.524
	Position of the Respondent (2)	1.317	0.351	14.091	1	<0.001	3.734
	Gender (1)	1.350	0.335	16.236	1	<0.001	3.858
	Constant	−1.216	0.888	1.875	1	0.171	0.296

a. Variable(s) entered in Step 1: Gender
b. Variable(s) entered in Step 2: Respondent's position
c. Variable(s) entered in step 3: highest education level
d. Variable(s) entered in Step 4: Number of household dependents

the likelihood of male youth being unemployed is also less than 0.01 compared to their female counterparts. This suggests that young males are more likely to secure employment than young females.

Although educational attainment remains a significant variable, the results do not differ substantially, with a coefficient of 0.42. This finding indicates that during periods of secular stagnation—a prolonged state of economic inactivity—unemployment persists regardless of educational level. Consequently, individuals with advanced educational qualifications may remain unemployed and encounter considerable difficulty finding work, while those with lower or semi-skilled education also experience substantial barriers to employment.

5 Conclusion

The study's findings suggest that the South African economy is experiencing sustained economic stagnation, marked by diminishing employment opportunities for job seekers. The economic system does not adequately account for the diversity of the workforce, which includes individuals with low, intermediate, high, and professional skill levels. This stagnation has led to unemployment across all skill levels, affecting youth with varying educational backgrounds. The impact of youth unemployment is felt by those with limited education, individuals who have completed basic education, and graduates holding post-matric qualifications.

Addressing these challenges requires developing and implementing public policies specifically designed to reduce non-cyclical unemployment and stimulate higher economic growth. In addition to policy interventions, governmental agencies can improve labour market transparency by enhancing the dissemination of vacancy information to job seekers. Also, distributing job listings at local community centres, shopping malls, health clinics, and places of worship, ensuring broad access to employment information, may be among the methods to disseminate information about the job market.

References

Baah-Boateng, W.: Unemployment in Africa: How appropriate is the global definition and measurement for policy purpose. Int. J. Manpow. **36**(5), 650–667 (2015)

Baah-Boateng, W., Turkson, E.F.: Employment in globalisation, employment and poverty reduction: A case study of Ghana. In: Aryeetey, E. (ed.) Employment and Poverty Reduction, pp. 104–139. Institute of Statistical, Social and Economic Research (ISSER), University of Ghana, Legon, Accra (2005)

Becker, G.S.: Human Capital: A Theoretical and Empirical Analysis with Special Reference to Education. National Bureau of Economic Research, New York (1964)

Bernstein, A.: Youth Unemployment Interventions Neither Attacking Root Causes Nor Going to Scale. Centre for Development and Enterprise, Johannesburg (2008)

Bloom, D.E., Canning, D., Fink, G.: Implications of population ageing for economic growth. Oxf. Rev. Econ. Policy **26**(4), 583–612 (2010)

Carr, S.C., Sloan, T.S. (eds.).: Poverty and Psychology: From Global Perspective to Local Practice, pp. 285–299. Kluwer Academic/Plenum Publishers (2003). https://doi.org/10.1007/978-1-4615-0029-2_15

Cloete, C.A.: Youth unemployment in South Africa: A theological reflection through the lens of human dignity. Missionalia **43**(3), 513–525 (2015)

Kaufman, B.E.: The Economics of Labour Market. Dryden Press, Orlando, Florida (1989)

Graham, L., De Lannoy, A.: Youth unemployment: What can we do in the short run. Date of access: 08 February 2025 (2016)

Graham, L., Mlatsheni, C.: Youth unemployment in South Africa: Understanding the challenge and working on solutions. South African Child Gauge, pp. 51–59 (2015)

Gonzo, W.: Unemployment in an African Country: A Psychological Perspective. University of Namibia (2003)

Gonzo, W., Platter, I.E.: Unemployment in an African Country: A Psychological Perspective. University of Namibia Press, Namibia (2003)

Hayes, A.:. What is the dependency ratio, and how do you calculate it? (2022). https://www.investopedia.com/terms/d/dependencyratio.asp. Accessed 10 October 2022.

Hausmann, R.: Final recommendations of the International Panel on Growth. Harvard University Center for International Development, Working Paper 161 (2008)

Harper, S.: Economic and social implications of ageing societies. Science **346**(6209), 587–591 (2014)

ILO.: Global Employment Trends for Youth: A Generation at Risk. International Labour Office, Geneva, Switzerland (2010)

Ligthelm, A.: An evaluation of the role and potential of the informal economy for employment creation in South Africa. South African Journal of Labour Relations **30**(1), 30 (2006)

Malik, K.: Human Development Report: Sustaining Progress: Reducing Vulnerabilities and Building Resilience. United Nations Development Program (2014)

ManpowerGroup.: How Policymakers Can Boost Youth Employment. ManpowerGroup (2012)

Mbulaheni, A.D., Agyapong, G.: Determinants of youth unemployment in South Africa: Evidence from the Vhembe district of Limpopo province. Environ. Econ. 7(4) (2016). https://doi.org/10.21511/ee.07(4).2016.06

Marshall, A.: Principles of Economics, 8th edn. Macmillan, London (1920)

Møller, V.: The Quality of Life in Unemployment. HSRC Publishers, Pretoria (1993)

Nair, P.S.: Age structural transition in South Africa. Afr. Popul. Stud. **25**(2), 194–214 (2011)

O'Higgins, N.: The Challenges of Youth Unemployment. Labour Market Organisation, United Kingdom (1997)

Okun, A.M.: Potential GNP: Its Measurement and Significance. Cowles Foundation (1962)

Sapsford, D., Tzannatos, Z.: The Economics of the Labour Market. Macmillan Press, Hampshire (1993)

Smith, A.: An Inquiry into the Nature and Causes of the Wealth of Nations. University of Chicago Press, Chicago (1976)
Statistics South Africa.: Quarterly Labour Force Survey. Pretoria: Government Printers (2024)
Statistics South Africa.: South African youth continue to bear the burden of unemployment. Pretoria: Government Printers (2025)
Summers, L.H.: US economic prospects: Secular stagnation, hysteresis, and the zero lower bound. Bus. Econ. **49**, 65–73 (2014). https://doi.org/10.1057/be.2014.13
Summers, L.H.: Accepting the reality of secular stagnation (2020). https://www.imf.org/en/Publications/fandd/issues/2020/03/larry-summers-on-secular-stagnation. Accessed 7 June 2025.
Todaro, M.P., Smith, S.C.: Economic Development, 12th edn. Pearson Education, New York (2015)
Van Zyl, J.S.H., Swanepoel, D.J.: New Economics 2000. Nasou, Cape Town (2002)
Wang, L.: Analysis of population structure. Department of Population and Employment Statistics, National Bureau of Statistics (2019)
Writer, S.: Why focus on the youth? (2025). https://nedbank-private-wealth.nedbank.co.za/learn/blog/economic-opportunities-for-youth.html. Accessed 2 June 2025.
Yarima, Y.A.: Socio-economic effects of unemployment among the youth in Nigeria. Journal of Contemporary Issues in Business Research **3**(5), 240–249 (2014)
Zinhunwe, C.: The youth and unemployment in Zimbabwe. Paper presented at the National Conference "Building a Future with Decent Work for Young People", Rainbow Towers, Harare, 29 March 2012 (2012)

The Relationship Between Entry Qualification and Academic Performance of Accountancy Students at a Selected University of Technology

Gauda Johannes Maseko and Mothemba Cecilia Mokoena

Abstract Globally, the admission requirements and academic performance in accounting programmes have received attention at tertiary institutions. The purpose of this study was to establish whether entry qualifications determine the academic performance of accountancy students at a university of technology (UoT) in South Africa. The study only focused on students' performance in accounting as their major. Against this background, the intention of this study was to assist policymakers at universities and UoTs to make informed decisions regarding admission criteria, specifically admission requirements and the admission points score (APS). The study compared two groups of students: those who met the requirements and those who do not. Document analysis was employed to examine the records of first-year accountancy students in 2019. Correlation and regression analyses established the relationship between the variables, while a t-test was used to determine differences between groups. There was no correlation found between APS and students' academic performance. Regression analysis found that mathematics and english positively influence performance while accounting was found to negatively affect students' performance. According to the knowledge of the authors, this is the first study conducted in South Africa to establish the impact of APS and admission requirements on accountancy students' academic performance at a UoT and therefore contributes to the body of knowledge. This study may assist university policymakers in practically determining admission criteria, particularly for accounting qualifications. This study recommends that there should be separate subject delivery for students who meet admission requirements and those who do not as they need more attention.

Keywords Academic performance · Accountancy · Accountancy students · Admission requirements · Universities · Admission point score

G. J. Maseko (✉) · M. C. Mokoena
Department of Accountancy, Vaal University of Technology, Vanderbijlpark, South Africa
e-mail: johannes@vut.ac.za

M. C. Mokoena
e-mail: ceciliap@vut.ac.za

E. H. Redda (ed.), *Proceedings of the Social Sciences International Research Conference, 2025*, https://doi.org/10.1007/978-981-95-9489-4_29

1 Introduction

The former President, Mr Jacob Zuma (2017) announced that the government will introduce free higher education and training for all South African students from the poor and working-class backgrounds, starting in 2018 for first-year students (Zuma, 2017). After 1994, the number of students from previously disadvantaged backgrounds admitted to South African universities increased (Cloete and Moja 2005; Mentz 2012), North West University (NWU) (North-West University 2025) had 70 104 students in 2017 and 70,493 in 2018 (North-West University: 2025), Rhodes University (RU) (Rhodes University 2017) had 7 005 students in 2017 and 8 100 in 2018 (Rhodes University 2025) and Witwatersrand University (WU) (University of the Witwatersrand 2017) had 38 353 students in 2017 and 39 953 in 2018 (University of the Witwatersrand: Quick stats about Wits 2025). However, when looking at high schools the student enrolment rates in 2018 and in 2019 have seen a decline of 11 596 and 12 972, respectively, in the number of students enrolling to write accounting for National Senior Certificate (often referred to as matric or grade 12) examinations. This indicates that in the long run, universities will face challenges in maintaining enrolment levels in accounting-related fields.

Apart from the declining enrolment rates, the Department of Basic Education (DoE) (Department of Basic Education 2025) has lowered the pass mark to 30%, which was enforced in 2014 (Department of Basic Education, 2014). This implies that a learner may pass with knowledge of only three out of ten answers. However, the Senior Certificate is not awarded if a candidate passes all subjects with 30% (The Citizen 2022). The number of students who enrolled in 2018 is reflected in Fig. 1. The percentages of students passing accounting between 2014 and 2018 at the 30% pass mark and at 40% and above are shown in Table 1. The Minister of Basic Education, Mrs Angie Motshekga, reported that the overall 2019 matric pass rate is 83.1% (Department of Basic Education 2019). Conversely, Banton (Banton 2019) argues that the 'real' matric pass rate for South Africa in 2019 is 38.9%.

After completing matric, these students apply for enrolment at the institutions of higher learning and training. Traditional universities and universities of technology (UoTs) have admission requirements for each course offered as stated by the Council of Higher Education (CHE) (Council on Higher Education 2025) (Higher Education Monitor 9, 2010). The admission requirements are based on the academic achievement in matric (grade 12) and Technical and Vocational Education and Training (TVET) colleges (Diploma and Bachelor's Degree Programmes Requiring a National Senior Certificate 2025). Admission requirements are referred to as permission to register for a specific course at the institution of higher learning and training. They are used by tertiary institutions to regulate or limit the number of students enrolled in specific course due to large number of applicants each year (Ellett 2025). Tables 2 and 3 indicate the admission requirements for students who are eligible to register in the Department of Accountancy Diplomas at a selected UoT in South Africa (NQF level 6) in Cost and Management Accounting, Internal Auditing and Financial Information System.

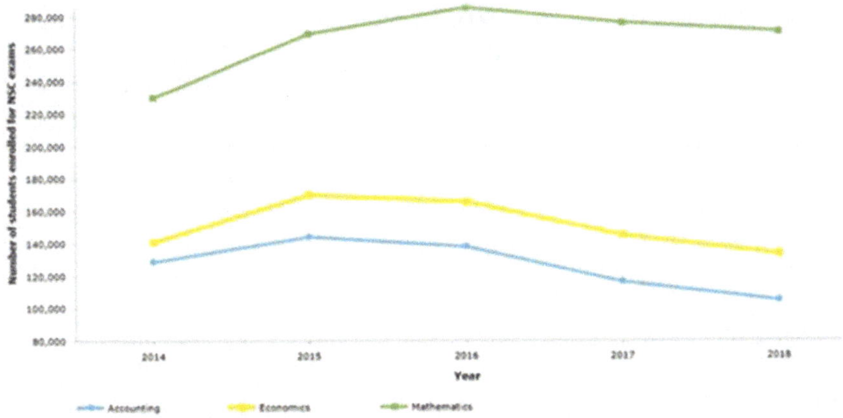

Fig. 1 Declining number of students enrolled for accounting. *Source* Businesstech, 2020

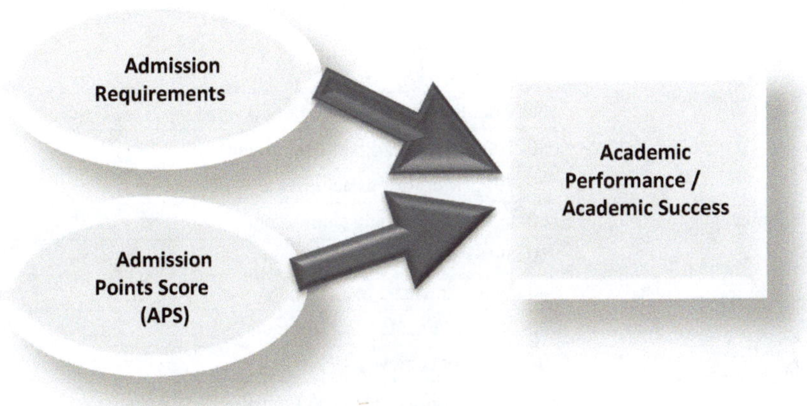

Fig. 2 Research model. *Source* Authors' illustration

Table 1 Number of matriculants who wrote and percentage that passed accounting

Year	No. that wrote	No. achieved at 30%	% achieved at 30%	No. achieved at 40% and above	% achieved at 40 and above
2014	125 897	85 681	68.0	55 837	44.3
2015	140 474	83 746	59.6	50 906	36.2
2016	128 853	89 507	69.5	57 914	44.9
2017	103 427	68 318	66.1	44 041	42.6
2018	90 278	65 481	72.5		
2019	80 110	62 796	78.4		

Source Department of Basic Education, 2020

Table 2 Admission Requirements Accountancy Department

Subject	Cost and Management Accounting	Internal Auditing	Financial Information System
NSC Endorsement		Eligibility for National Diploma	
Compulsory Subjects		Level	
English	4	4	4
Accounting	4	4	4
Mathematics (Maths) or	3	3	3
Mathematics literacy (Maths Lit.)	6	6	6
Remaining three subjects (Min three per subject)	11	11	11

Source Author's illustration based on the website of the UoT under study

Table 3 Admission points score

National Senior Certificate (NSC) scale	Level
80–100% (Outstanding achievement)	7
70–79% (Meritorious achievement)	6
60–69% (Substantial achievement)	5
50–59% (Moderate achievement)	4
40–49% (Adequate achievement)	3
30–39% (Elementary achievement)	2
0–29% (Not achieved – Fail)	1

Source Department of Basic Education, 2020

Table 2 presents the minimum requirements for admission to a UoT, while Table 3 reflects admission points that are used in admitting the students in the Department of Accountancy as well as the percentage requirements at each level. Based on the admission requirements shown in Table 2, the UoT under study requires 50–59% in accounting, 40–49% in mathematics, or 60–69% in mathematics literacy, 50–59% in English, and the remaining three subjects must total a minimum of 11 points.

Universities assume that students with higher academic grades will perform better academically as opposed to those who obtained average grade 12 results. This was confirmed and substantiated by Fraser and Killen (Fraser and Killen 2003) that grade 12 learners with high marks will succeed academically at the Higher Education Institutions (HEIs). Furthermore, Alias and Zain (2006), as well as Zezekwa and Mudavanhu (2011), indicate that there is a positive association between entry requirements and the academic performance of students.

Ganyaoupfu (2013) is of the opinion that teaching methods that lecturers use to conduct their classes can contribute towards students' academic performance. Mlambo (2011) endorsed this by reporting that lecturers' teaching methods influence the students' academic performance in a positive way. Adeyele and Yusuf (2012) also acknowledged that knowledge that academic staff have in the subjects they teach plays an important role in the student's academic performance. According to Hanson (2000) academic performance of students can be affected by factors such as learning abilities, gender and background. Academic performance is defined as the result or proof that the lecturer, student and the university have accomplished their educational objectives (Annie et al. 1996). However, academic staff knowledge, teaching methods, learning abilities and students' backgrounds are not measured in this study.

A paramount concern for South African academic institutions is the decreasing number of high school students matriculating in accounting as well as the number of students who meet the admission requirements of these institutions. This is a concern for these institutions as they offer a number of programmes majoring in accounting and this has resulted in an inadequate number of students enrolling in the institutions of higher learning leaving these institutions with a question on how to fill up the lecture halls and meet their enrolment targets. Different universities have different admission criteria and the degree of admission adversity will differ among universities.

The Department of Basic Education in South Africa considers 30% to be a pass mark for matriculants. However, universities and universities of technology (UoTs) require a 50% pass mark in accounting for those who seek admission in the courses majoring in accounting. The overall pass rate from 2015 to 2018 in accounting ranges between 40 and 44%. This is a clear indication that the pass mark of 40% for accounting is far below the 50% threshold considered for admission by universities and UoTs making it difficult for students to gain admission to these academic institutions and preventing them from pursuing the careers of their choice. Students are forced to rewrite the subject or enrol for preparatory programmes (extended programmes) at universities and UoTs, which are quite expensive because they are not financed by the National Student Financial Aid Scheme (NSFAS).

2 Literature Review

2.1 Theoretical Framework

To acquire adequate background and structure for this study, Tinto's (Tinto 1975) Student Integration Model, Elger's (Elger 2007) Theory of Academic Performance and Astin's (Astin 1984) Inputs-Environment-Outputs (I-E-O) model were used in this study.

Student Integration Model (SIM). According to Tinto (Tinto 1975), the main features of the SIM are goal commitment and institutional commitment. Goal commitment is the degree of persistence a student has towards the attainment of their degree while institutional commitment refers to the extent to which a student is devoted to the university or college that the student attends. However, with enough goal commitment, a student may remain at the institution they may have little commitment to. The above features are affected by pre-college characteristics like family background which refers to the social status of an individual; personal attributes which include gender, race and academic ability; and pre-enrolment experiences like school grade point average. The aforementioned are the factors of performance discussed in detail in Sect. 3.5.

Tinto's theory asserts that students from higher socio-economic backgrounds have a greater likelihood of graduating than those from lower backgrounds. Furthermore, Tinto (1975) reported that for students to be retained by the university on their chosen field of study it depends on the attitude a student has on his/her academic aims and career objectives. Hence, the theory highlights the association between academic persistence, high school experience and student's point of view. Consistent with Tinto's theory, Tinro (2004) proposed that for students not to drop out, HEIs must make sure that students have no problems in accessing academic staff and student support services. Universities must provide the academic and social support to their students for them to perform well in their studies. Berzensk (2019) reported that for students to succeed academically academic staff must provide support, which can help students to graduate.

Theory of Academic Performance (TOP). The study also employed the TOP to enhance understanding of the phenomenon. The choice was informed by Elger (2007) who specified that performance has three attributes, namely performer's mindset, environment and reflective practice. Performer's mindset is an individual's attitude towards the subject or an activity. Environment refers to one's surroundings which may either be physical, social or intellectual and may have a positive or a negative influence on an individual's performance. Reflective practice refers to paying attention to and learning from past experiences (Balogun et al. 2020). For example, a student who performed poorly in matric may consciously decide to improve their results when at the university or college, or a student who achieved outstanding results may then adapt their learning styles to cope with tertiary work.

Inputs-Environment-Output Theory (I-E-O). In developing a conceptual model on the correlation of admission requirements and academic performance of the students, the model by Astin (Tinto 1975) I-E-O was used in the study. The model focuses on how student input (student's background) interrelates with the university environment (student's participation in their academics and social events) in order to achieve the desired output (perseverance and academic performance). First-year students find it difficult to belong in the university surroundings and that can also affect their ability to succeed academically and socially (Yeager and Walton 2011).

Most HEIs admit students based on different entry qualifications in such a way that students can do the same course but differ in prior information (Mlambo 2011).

The admission process is crucial for the universities as it can determine student's academic success in the chosen field of study. Adeyemi (2001), Agboola (2011), Allen and Sherry (2008) are of the opinion that universities set their admission requirements in a way that students can succeed academically and graduate. It has been confirmed to have value in relation to academic performance, student retention and graduation rates. As a result, the way universities admit students plays a vital role. Tinto (1975) stated that good relationship with the faculty and classmates influences students to reach their personal goals and succeed academically. This can result in quality education and quality graduates (Fajar et al. 2019).

Engler (Engler 2010) explained that there is a strong correlation between the admission requirements and academic performance but not in all courses. Ali, Haider, Munir, Khan and Ahmed (Ali et al. 2013) reported that academic performance of students is influenced by the kind of education they receive from the university. Based on the aforementioned, the following hypotheses are formulated:

H1: There is a relationship between admission points score and academic performance.

H2: There is a relationship between admission requirements and academic performance.

2.2 *Definition of Concepts*

Concepts used in the research model illustrated above are briefly defined as follows:

Admission requirements. Admission requirements are calculated based on learner's grade 12 final marks and universities use different requirements to admit students as stated by the CHE (Council on Higher Education 2025). The admission requirements are designed in a way that students can cope with the modules of the university to succeed academically. A study conducted by Ayyaf and Magzoub (2014) in the Kingdom of South Arabia found that there is a positive relationship between the admission grades and academic performance among nursing students. On the other hand, some universities use entry tests for students to be admitted and the tests include reading comprehension and problem solving (Sulphey and Alkahtani 2018).

Admission point score (APS). The matric subjects' marks are graded into a hierarchy of points/levels to the maximum of seven. That is, they are converted into a point system (see Table 3 above). These marks are then added together to give an APS. It determines the eligibility for a specific tertiary education. Universities use a total APS point to determine if the student qualifies to study for the specific course that the student applied. Each university allocates the number of APS points needed to study a course differently. This means that a course at one university will not need the same APS as the same course at another university (Fundi Connect: What is APS (Admission Point Score) 2025).

University. University is an institution of learning of the highest level that has the right to award degrees. A place where academic research is done. It can also develop its students through a high level of education in arts and science (Alemu 2018). A university is also referred to as students' environment, which is the surroundings where students will be studying. Environment is what is around us or what surrounds an individual's life which consists of living and non-living components that include physical elements. It can also be something that one is used to Nguyen (2025).

Accountancy students. Accountancy students are the ones that pursue a career in the accounting field. Students who have secondary school accounting and mathematics or mathematics literacy are likely to succeed academically (Gloeck 2025). Accountancy entails broader learning concerning subjects such as accounting, auditing, financial management and taxation (Gloeck 2025).

Accountancy. Buys (Buys 2010) described accountancy as application of skills to be used in the organisational atmosphere and such application depends on the accountant's knowledge of theory. Accountancy is therefore a profession of people who are working or doing accounting work in the organisation referred to as accountants to develop strategies for how they can advise the organisation to best achieve its business objectives. According to Collin, York and Joliffe (Collin et al. 2001) accountancy is the work that accountants do on a daily basis. The profession of accountancy helps with business approach and also assists in cutting costs that are not necessary (International Federation of Accountants 2025).

Academic performance. Student output is what they achieve after teaching and learning has taken place whether they succeed academically or not, that is, the academic performance. Academic performance has been examined from the level of information that is determined by graduation numbers and is measured by teaching and learning process (Granados et al. 2019).

Factors that affect the academic performance of students are discussed below.

Student's learning abilities. Academic performance of students can be affected by factors such as learning abilities, gender and their background (Hanson 2000). Student's learning abilities differ from one student to another. Some develop the skill to study at the early age, as early as grade 8, while others start in grade 12 because they want to further their studies or prove to the society in which they live that they can pass grade 12. On the other hand, some students would develop the skill of learning at the institution of higher learning because they want to obtain their degrees and begin earning a living. Students' actions are key in determining what has been studied in class as opposed to the educator's actions (Shuell 1986). Learning style is defined as methods that are used by students to process what has been learned while learning strategies are concerned with the tactics students use in approaching a certain learning condition (Conti and Kolody 1995).

Students' gender. It is believed that most female students perform better academically in theory-based subjects such as communications studies as compared to their male counterpart, simply because female students like to read magazines and are used to reading (Volchok 2019). While on the other hand, their male counterparts perform well in calculation-based subjects such as accounting, mathematics and science simply because males like the application. Iroegbu (2000) discovered that boys perform better than girls in high school science and he further explained that there are factors that affect the way each gender processes the information learned. Penner and Willer (2019) are of a view that boys achieve good mathematics marks as compared to girls while Wrigley-Asante et al. (2023) reported that male students rate more marks than female students in science at the intermediate phase. Nonetheless, Holmlund and Krister (2008) revealed that girls perform better than boys at the school level and rate higher marks on reading assessments.

Student's background. Students' backgrounds can affect his or her academic performance in most cases students from poor backgrounds struggle to cope with the modules or subjects in the HEIs. Jeffery and Johnson (2019) denote that dispositions and backgrounds are essential to equip learners for tertiary education. This is based on the fact that learners are from different family backgrounds with different learning skills and diverse cultures (Elemam and Saide 2023). Shi and Ko (2023) reported that some students attend school not receiving enough support and inspiration from their families. On the other hand, some receive the necessary support and motivation and have information that is beyond their grade. Studies revealed that there can be so many issues that disturbing learners' academic performance and some are outside the school premises (Bowen 2009).

Socio-economic issues, such as poverty, teacher's abilities, insecurity, negative attitude towards their studies and parents working long hours, can also affect the student's academic performance in South African schools (Geske and Ozola 2008).

3 Methodology

The study is situated within a mixed-methods research paradigm. It used qualitative data generated through document analysis to contextualise the quantitative data. Document analysis was employed in the study to review or evaluate documents, both printed and electronic materials. It entails studying existing documents to understand their substantive content (Bowen 2009). Documentary analysis included the examination of matric results of extended programme students to establish why they were admitted into the extended programme (admission requirements and APS), as well as their gender and academic performance. This information was obtained from the UoTs Integrated Tertiary System (ITS).

Students (all first-year 2019 accountancy students) were divided into two groups: students who do not meet the admission requirements and the required APS. This group consisted of 49 extended programme students. The second group was made up of 94 students who met the admission requirements and obtained the required APS. These two groups of students were exposed to the same learning conditions: the same environment, the same content, they were taught by the same lecturer, they wrote the same assessments and the duration of assessments was the same. A tutor was appointed, working together with the lecturer to assist extended programme students and, as a result, these students had more contact hours.

Data obtained through document analysis was used to produce numerical data (frequency tables) indicating the number and percentage of female and male students, students who passed or failed the subject as well as class attendance which was then analysed to establish correlation between admission requirements/APS and academic performance. Students' commitment and gender were compared to academic performance to determine the relationship as well as the strength of the relationship between variables (Creswell 2010). T-tests were used to determine differences between gender and the two groups of students indicated above.

4 Results and Discussion

The results discussed in this section were derived from responses to the questionnaire, after statistical analysis.

4.1 Frequency Tables

Table 4 shows the total number of students who met the admission requirements, labelled 'Normal' and the number of students who did not meet the admission requirements and were admitted into the extended or preparatory programme, labelled 'Extended'. It reflects 49 students in the extended programme, of which 24 were female and 25 were male while the normal group consisted of 61 females and 33 males, with the total number of 94 students in the normal group. The last column presents a total of 85 female students and 58 males for both groups combined.

Students who seek admission to the university should have the subjects indicated in Tables 2 and 5 above. The remaining three subjects must total 11 points, indicated as other subjects in Table 5. Examples of other subjects include History, Geography and Life Orientation. Students who are one point short in other subjects and in one

Table 4 Gender

Cross tabulation		Extended	Normal	Total
Female	Count	24	61	85
	% within Gender	28.2%	71.8%	100%
Male	Count	25	33	58
	% within Gender	43.1%	56.9%	100.0%
Total	Count	49	94	143
	% within Gender	34.3%	65.7%	100%

Source Authors' illustration

of the subjects shown above were admitted into the extended programme. Table 5 reveals that 31% of students were admitted to the extended programme because they were one point short in accounting, 6% in English, 43% in mathematics (mathematics and mathematics literacy combined) and 20% in other subjects.

Table 6 shows the average APS for both groups. The normal group recorded the highest mean. This could be attributed to the fact that these are the students who obtained APS levels between 4 (50%) and 8 (100%) in accounting, mathematics and English and a minimum of 11 points in other subjects, which explains the relatively high standard deviation ($\overline{x} = 32.43$; $\sigma = 5.013$). The extended group recorded a lower mean. These are the students who obtained APS levels between 1 (0%) and 3 (49%) in at least one of the subjects mentioned above ($\overline{x} = 30.08$; $\sigma = 3.796$).

Table 7 indicates students' performances at the end of the semester in accounting as a major subject. The pass mark for the subject is 50% and above. 41 (84%) of 49 extended programme students passed the subject and only eight students (12%) failed, while 12 (13%) of 94 students failed in the normal group with the pass rate of 87%.

Table 8 shows a higher mean for normal group than the extended programme group, indicating that students in this group are stronger with regard to academic performance ($\overline{x} = 66.32$; $\sigma = 17.740$) and ($\overline{x} = 59.20$; $\sigma = 13.075$), respectively.

Table 5 Reason for admittance into the extended programme

Extended	Accounting (Level 4)	English (Level 4)	Maths (Level 3)/ Maths Lit (Level 6)	Other subjects	Total
Number	15	3	21	10	49
Percentage	30.6%	6.1%	42.9%	20.4%	100%

Source Authors' illustration

Table 6 Group statistics for admission points score

	Groups	N	Mean	Std. Deviation	Std. Error Mean
APS	Extended	49	30.08	3.796	0.542
	Normal	94	32.43	5.013	0.517

Source Authors' illustration

Table 7 Students' academic performance

		Extended		Normal	
Academic Performance	Pass	41	84%	82	87%
	Fail	8	16%	12	13%
	Total	49	100%	94	100%

Source Authors' illustration

Table 8 Group statistics for academic performance

		N	Mean	Std. Deviation	Std. Error Mean
Academic Performance	Extended	49	59.20	13.075	1.868
	Normal	94	66.32	17.740	1.830

Source Authors' illustration

Table 9 Equality of variance and means for admission points score

	Levene's test for equality of variances		t-test for equality of means				
APS	F	P-value/ Sig	t	df	Sig. (2-tailed)	Mean Difference	Std. Error Difference
Equal variances assumed	5.657	0.019	-2.870	141	0.005	-2.344	0.817
Equal variances not assumed			-3.128	122.630	0.002	-2.344	0.749

Source Authors' illustration

4.2 *T-Test*

The t-test was used to establish differences between the group means. The results are highlighted in Table 9, which exhibits statistically significant differences between the APS of the normal group and the extended programme group. This is indicated by the p-value, which was < 0.05 ($p = 0.002$) and is consistent with the results shown in Table 6.

Statistically significant differences were also established between the groups regarding academic performance. The p-value was < 0.05 ($p = 0.014$) as shown in Table 10. These findings are consistent with the results presented in Table 8.

4.3 *Correlation Analysis*

Pearson correlation was used to examine the relationships between constructs and the strength of those relationship. Correlation coefficients range from -1 to + 1, where -1 indicates perfect negative correlation and + 1 indicates perfect positive correlation.

A markedly low and negligible positive linear relationship is observed between APS and students' academic performance as shown in Table 11. However, the p-value reflected in the table is greater than the level of significance chosen for the study as highlighted above. Based on the fact that these results are drawn from the

Table 10 Equality of variance and means for academic performance

	Levene's Test for Equality of Variances		t-test for Equality of Means				
Academic performance	F	P-value/ Sig	t	df	Sig. (2-tailed)	Mean Difference	Std. Error Difference
Equal variances assumed	2.824	0.095	-2.477	141	0.014	-7.115	2.872
Equal variances not assumed			-2.721	124.939	0.007	-7.115	2.615

Source Authors' illustration

Table 11 Correlations between APS and academic performance

APS – Academic performance	APS	Academic performance
Pearson correlation	1	
Sig. (2-tailed)		
N	143	
Pearson correlation	0.089	1
Sig. (2-tailed)	0.289	
N	143	143

** Significant at 0.05 (2-tailed). *Source* Authors' illustration

sample, there is not enough evidence to suggest that this correlation exists in the population; therefore, we conclude that it likely occurred by chance (r 141), 0.089, p > 0.001. The hypothesis (H1) is rejected and there is a relationship between APS and academic performance. This is also confirmed by the results highlighted in Table 7 and consistent with Adewale and Adhuze's (2014) study. The purpose of the study was to establish the relationship between entry qualification and academic performance of architecture students in Nigerian polytechnics. The results revealed a low correlation was found between students' academic performance and their basic knowledge in physics and mathematics. However, this contrasts with Windle et al. (2018), who found that high school qualifications were statistically significant predictors of academic success in a pharmacy programme.

4.4 *Regression Analysis*

A regression model was built with academic performance as a dependent variable and admission requirements at the UoT, namely English, accounting and mathematics as independent variables to establish whether they are good predictors of academic performance. The model summary was used to establish the relative contribution of each variable to the variance explained by the model. This is highlighted in Tables 12, 15 and 18.

The model summary in Table 12 indicates that 3% of the variance in students' academic performance can be explained by English. ANOVA was used to determine whether the model is statistically significant and whether the predictor variable meaningfully predicts of the outcome variable as reflected in Table 13, $F (1, 141) = 4.945$, $p < 0.001$, $R^2 = 0.027$.

The regression analysis reveals a significant positive correlation between English and academic performance ($p = 0.028$; $\beta = + 0.184$) shown in Table 14. This indicates that English predicts academic performance. As the APS in English increases by one point, academic performance is predicted to increase by 0.082%.

The model summary in Table 15 reveals that mathematics contributes 5% of variance in students' academic performance and the model is statistically significant as highlighted in Table 16 below, $F (1, 141) = 4.945$, $p < 0.001$, $R^2 = 0.052$.

A significant positive influence is unveiled between mathematics and performance ($p = 0.003$; $\beta = + 0.243$). As mathematics points score increases by one unit, performance is expected to increase by 0.188 units (Table 17).

Accounting accounts for 3% of variance in academic performance. The significance of this model was determined in the table below and was found to be significant at $F (1,141) = 5.071$, $p < 0.001$, $R^2 = 0.028$ as reflected in Table 18.

The regression table above (Table 20) discloses a negative association between accounting and academic performance. As the APS in accounting decreases by one point, students' performance is expected to decrease by 0.066 units ($p = 0.026$; $\beta = -0.186$). The level of significance is below 0.05 which indicates that accounting significantly predicts academic performance.

Table 12 Model summary showing the contribution of English on academic performance

Model	R	R Square	Adjusted R Square	Std. Error of the Estimate	Change Statistics		
					R Square Change	F Change	df1
1	0.184	0.034	0.027	0.343	0.034	4.945	1

Source Authors' illustration

Table 13 ANOVA highlighting the significance of the contribution of English

Model		Sum of Squares	df	Mean Square	F	Sig
1	Regression	0.583	1	0.583	4.945	0.028
	Residual	16.620	141	0.118		
	Total	17.203	142			

Source Authors' illustration

Table 14 Regression analysis of English with academic performance

Model	Unstandardised Coefficients		Standardised Coefficients	t	Sig
	B	Std. Error	Beta		
1 (Constant)	0.769	0.169		4.547	0.000
English	0.082	0.037	0.184	2.224	0.028

Source Authors' illustration

Table 15 Model summary showing the contribution of mathematics on academic performance

Model	R	R Square	Adjusted R Square	Std. Error of the Estimate	Change Statistics		
					R Square Change	F Change	df1
1	0.243	0.059	0.052	0.339	0.059	8.834	1

Source Authors' illustration

Table 16 ANOVA highlighting the significance of the contribution of Mathematics

Model		Sum of Squares	df	Mean Square	F	Sig
1	Regression	1.014	1	1.014	8.834	0.003
	Residual	16.189	141	0.115		
	Total	17.203	142			

Source Authors' illustration

Table 17 Regression analysis of mathematics with academic performance

Model		Unstandardised Coefficients		Standardised Coefficients	t	Sig.
		B	Std. Error	Beta		
1	(Constant)	0.900	0.086		10.510	0.000
	Mathematics	0.188	0.063	0.243	2.972	0.003

Source Authors' illustration

Table 18 Model summary showing the contribution of accounting on academic performance

Model	R	R Square	Adjusted R Square	Std. Error of the Estimate	Change Statistics		
					R Square Change	F Change	df1
1	0.186	0.035	0.028	0.343	0.035	5.071	1

Source Authors' illustration

Table 19 ANOVA highlighting the significance of the contribution of Accounting

Model		Sum of Squares	df	Mean Square	F	Sig
1	Regression	0.597	1	0.597	5.071	0.026
	Residual	16.606	141	0.118		
	Total	17.203	142			

Source Authors' illustration

Table 20 Regression analysis of accounting with academic performance

Model		Unstandardised Coefficients		Standardised Coefficients	t	Sig.
		B	Std. Error	Beta		
1	(Constant)	1.436	0.135		10.658	0.000
	Accounting	-0.066	0.029	-0.186	-2.252	0.026

Source Authors' illustration

The frequency table showing students' academic performance (final mark) did not reveal many differences between the 'normal' group (87%) and the 'extended' group (84%) as the pass rates of the two groups were above the university threshold of 60%. This is an indication that with proper support students who do not meet the required APS can still perform well academically. This was supported by Pearson's correlation results which did not establish any relationship between students' academic performance and the overall APS.

The above results are consistent with the three theories guiding this study. SIT requires persistence towards the attainment of goals and institutional support, TOP requires positive attitude towards the subject or an activity and learning from past experiences while I-E-O requires participation, perseverance and enabling environment for performers to achieve desired outcomes. Although the students in the extended programme did not meet admission requirements, they used the opportunities and the support provided by the UoT through tutorials and extra lessons. Students' participation in the extra activities especially in the extended group, perseverance, positive attitude and students' decision to improve their academic performance led to the attainment of desired results with the throughput rate above the UoTs 60% threshold. Hence, this study contributed to the three theories, which are related and can be used interchangeably.

Regression analysis shed more light on these three compulsory subjects required for admission, namely mathematics, English and accounting. It was evident in the results that mathematics and English positively affect academic performance, the higher the score the better the academic performance. This could be attributed to better understanding of the content and instructions as well as adequate numeracy skills. Tewary's (2014) study conducted at the University of KwaZulu-Natal (2014) also found mathematics to be a significant variable which determines academic performance in the first-year university modules. However, accounting was found to negatively affect performance highlighting that students with lower score in accounting struggle in the subject. Hence the study accepted the hypothesis (H2) that there is a relationship between admission requirements and academic performance. Henceforth, this is a practical contribution to the body of knowledge.

Based on the aforesaid, it is recommended that there should be separate subject delivery for students who meet admission requirements and those who do not meet admission requirements as they need more attention. Mathematics classes should be offered and extra classes in accounting should be provided for these students to assist them to improve their numeracy and accounting skills. English is not included as Communication Skills is part of the curriculum at the UoT. However, students should be encouraged to make use of the Writing Laboratories available at the university to improve their English proficiency.

Policymakers should consider changing the admission criteria in the Accountancy Department. Students who meet admission requirements in English, mathematics and accounting but are one point short in their other subjects should be admitted in the normal group. The above-mentioned subjects were found to be key determinants of academic performance in the Accountancy Department. Potential students are denied the opportunity to further their studies because they are one point short in Geography or History, subjects that are unrelated to accounting. 20% of the students in this study were admitted into the extended programme due to other subjects although they met the requirements for three compulsory subjects mentioned above.

Extended programmes are often costly and only a few can afford them as they are not financed by the NSFAS. Students in the extended group have to complete their studies in four years instead of three years resulting in unnecessary costs and accumulation of debt for students. Furthermore, the university should do away with APS as it does not influence students' academic performance and this will enable Accountancy Department at the university to fill up the lecture halls and meet their enrolment requirements.

More tutors should be appointed to assist students who struggle in accounting and their performance should be monitored regularly through class tests and quizzes to enable lecturers to come up with intervention strategies to assist underperforming students.

5 Conclusion

It can therefore be concluded that APS does not have any influence on Accountancy students' academic performance at the university. However, students with higher scores in Mathematics and English perform better, while those with lower scores in accounting struggle in the subject.

The study was conducted at a single UoT in South Africa and therefore the results may not be generalised. For future research, the study may be replicated and conducted at other universities to enable the generalisation of findings.

References

Adewale, P.O., Adhuze, O.B.: Entry qualifications and academic performance in architecture. Front. Architect. Res. **3**(1), 69–75 (2014)

Adeyele, J.S., Yusuf, Y.S.: Effect of teaching method, discipline choice and student–lecturer relationship. Journal of Economics and Sustainable Development **3**(7), 1–7 (2012)

Adeyemi, J.K.: Equality of access and catchment area factor in university admissions. High. Educ. **42**(3), 307–332 (2001)

Agboola, B.M.: Admission criteria and student retention in federal universities. Unpublished doctoral thesis, University of Benin (2011)

Alemu, S.K.: The meaning, idea and history of university education in Africa. Forum Int. Res. Educat. **4**(3), 210–227 (2018)

Ali, S., Haider, Z., Munir, F., Khan, H., Ahmed, A.: Factors contributing to students' academic performance. American Journal of Educational Research **1**(8), 283–289 (2013)

Alias, M., Zain, A.F.M.: Relationship between entry qualification and performance in graduate education. Int. Electron. J. **7**(3), 371–378 (2006)

Allen, D.T., Sherry, W.: Admission criteria as predictors of business student graduation. College Student Journal (2008)

Annie, W., Howard, W.S., Mildred, M.: Achievement and ability tests: definition of the domain. In: Educational Measurement, vol. 2, pp. 2–5. University Press of America (1996)

Astin, A.W.: Student involvement: a developmental theory. J. Coll. Stud. Pers. **25**(4), 297–308 (1984)

Ayyaf, A.H.M., Magzoub, M.E.: Admission criteria as predictors of academic performance. In: Proceedings of the First International Conference on Assessment and Evaluation. Riyadh (2014)

Balogun, O.S., Olaleye, S.A., Ibidoja, O.J.: University students' academic performance: a tau statistic approach. In: Proceedings of the 36th IBIMA Conference. Granada (2020)

Banton, V.: DA, EFF dispute 2019 matric pass rate, say 'real' number is much lower. News24 (8 Jan 2020). https://www.news24.com/news24/southafrica/news/da-eff-dispute-2019-matric-pass-rate-say-real-number-is-much-lower-20200108, accessed 29 Nov 2025

Berzenski, S.R.: The when and who of graduation and dropout predictors. Journal of College Student Retention 0(0), 1–25 (2019)

Bowen, G.A.: Document analysis as a qualitative research method. Qual. Res. J. **9**(2), 27–40 (2009)

BusinessTech: South Africa's matric pass rate: 2008 to 2019. https://businesstech.co.za/news/government/364476/south-africas-matric-pass-rate-2008-to-2019/, accessed 29 Nov 2025

Buys, P.W.: The legitimacy predicament of accounting theory. Unpublished PhD thesis, North-West University (2010)

Cloete, N., Moja, T.: Transformation tensions in higher education: equity, efficiency and development. Soc. Res. **72**(3), 693–772 (2005)

Collin, P.H., York, D., Jolliffe, A.: Dictionary of Accounting, 2nd edn. Peter Collin, London (2001)

Conti, G.J., Kolody, R.C.: Learning strategies: an international perspective. In: Proceedings of the Adult Education Research Conference, pp. 77–89. University of Alberta, Edmonton (1995)

Council on Higher Education: Higher Education Monitor 9: Access and throughput in South African higher education. https://www.che.ac.za/sites/default/files/publications/Higher_Education_Monitor_9.pdf, accessed 29 Nov 2025 (2010)

Creswell, J.W.: Planning, Conducting, and Evaluating Quantitative and Qualitative Research. Sage, London (2010)

Department of Basic Education: National policy pertaining to the programme and promotion requirements of the National Curriculum Statement grades R–12. https://www.education.gov.za/Portals/0/Documents/Policies/PolicyProgPromReqNCS.pdf, accessed 29 Nov 2025

Department of Basic Education: Report on the 2019 National Senior Certificate examination. https://www.education.gov.za/Portals/0/Documents/Reports/2019%20NSC%20Examination%20Report.pdf, accessed 29 Nov 2025

Department of Higher Education and Training: Minimum admission requirements for Higher Certificate, Diploma and Bachelor's degree programmes requiring a National Senior Certificate. https://www.dhet.gov.za/HED%20Policies/Minimum%20Admission%20Requirements%20for%20Higher%20Certificate,%20Diploma%20and%20Bachelor%27s%20Degree.pdf, accessed 29 Nov 2025

Elemam, S.M., Saide, A.: Education across cultural differences. Res. Educat. Rehabilit. **6**(2), 166–174 (2023)

Elger, D.: Theory of performance. In: Beyerlein, S.W., Holmes, C., Apple, D.K. (eds.) Faculty Guidebook, 4th edn., pp. 19–22. Pacific Crest, Lisle (2007).

Ellett, P.: The real story behind university entry requirements. https://university.which.co.uk/advice/ucas-application/the-real-story-behind-entry-requirements, accessed 29 Nov 2025

Engler, R.: School subjects and university performance. Tertiary Sector Performance Analysis and Reporting, p. 2. Ministry of Education, New Zealand (2010)

Fajar, S., Hussain, M., Sarwar, H., Afzal, M., Gilani, S.A.: Factors affecting academic performance of undergraduate nursing students. International Journal of Science and Management **6**(1), 1–16 (2019)

Fraser, J.W., Killen, R.: Factors influencing academic success or failure of first-year and senior university students. South African Journal of Education 23(4) (2003)

Fundi Connect: What is APS (Admission Point Score)?. https://fundiconnect.co.za/know-aps-score/, accessed 29 Nov 2025.

Ganyaupfu, E.M.: Teaching methods and students' academic performance. International Journal of Humanities and Social Science Invention **2**(9), 29 (2013)

Geske, A., Ozola, A.: Factors influencing reading literacy. Probl. Educat. 21st Century 6(6), 71–77 (2008)

Gloeck, D.: Is accountancy education paying its social debt?. Auditing SA. https://www.saiga.co.za/index_htm_files/1%20Gloeck%20Editorial%20Accountancy%20education.pdf, accessed 29 Nov 2025

Granados, D.E., Olivares, G., De la Hoz, R., Figueroa, S.: Academic performance, intelligence and personality. Int. J. Arts Commerce **8**(1), 30–39 (2019)

Güngörmüş, H.A., Uyar, A.: Factors associated with student performance in financial accounting. Eur. J. Econ. Polit. Stud. **4**(2), 139–154 (2011)

Hanson, J.B.: Student performance and student growth as measures of success. In: Proceedings of the American Educational Research Association Annual Meeting, New Orleans (2000).

Holmlund, H., Sund, K.: Gender gap and teacher sex. Labour Econ. **15**(1), 37–53 (2008)

International Federation of Accountants: Roles and importance of professional accountants in business. https://www.ifac.org/news-events/2013-10/roles-and-importance-professional-accountants-business, accessed 29 Nov 2025.

Iroegbu, T.O.: Exploring hands-on investigation in wave and optics. African J. Educat. Res. **6**, 1–2 (2000)

Jeffery, D., Johnson, D.: Whose fault is failure? Res. Comp. Int. Educ. **14**(3), 376–393 (2019)

Mentz, M.: Measuring and using pre-university levels of student engagement at a South African university. Unpublished PhD thesis, University of the Free State (2012)

Mlambo, V.: Factors affecting student academic performance in an introductory biochemistry course. Caribbean Teaching Scholar **1**(2), 79–92 (2011)

Nguyen, B.: What is environment class?. https://www.quora.com/What-is-the-environment1, accessed 20 Jan 2025

North-West University: Student profile. https://www.nwu.ac.za/sites/www.nwu.ac.za/files/files/institutionalinformation/Corporateprofile/2017-2018/NWU-Profile-2017-18.pdf, accessed 29 Nov 2025.

North-West University: Student profile: student headcount enrolments and graduates. https://143.160.39.37/PowerHEDA/Dashboard.aspx

Penner, A.M., Willer, R.: Men's overpersistence and the gender gap. Socius **5**, 2378023118821836 (2019)

Rhodes University: Admission gateway: facts and figures. https://www.ru.ac.za/admissiongateway/whyrhodes/factsandfigures/, accessed 29 Nov 2025

Rhodes University: Rhodes University. In: Wikipedia (10 Dec 2017). https://en.wikipedia.org/wiki/Rhodes_University, accessed 29 Nov 2025

Shi, Y., Ko, Y.C.: Family and school psychological environment. Participatory Educat. Res. **10**(1), 106–121 (2023)

Shuell, T.J.: Cognitive conceptions of learning. Rev. Educ. Res. **56**, 411–436 (1986)

Sulphey, M.M., Alkahtani, N.S.: Academic excellence of business graduates through nudging. Int. J. Innovat. Learn. **24**(1), 98–114 (2018)

Tewary, D.D.: Is matric mathematics a good predictor of first-year performance? Int. J. Educat. Sci. **6**(2), 233–237 (2014)

The Citizen: Maimane's claims 'misleading' – no one can pass matric with 30% says DBE. The Citizen (10 Jan 2022). https://www.citizen.co.za/news/south-africa/education/2954769/education-department-pass-mark-misleading-10-january-2022, accessed 29 Nov 2025

Tinto, V.: Dropouts from higher education: a theoretical synthesis. Rev. Educ. Res. **45**, 89–125 (1975)

Tinto, V.: Student retention and graduation. Pell Institute, Washington, DC (2004)

University of the Witwatersrand: Quick stats about Wits. https://www.wits.ac.za/about-wits/quick-stats/, accessed 29 Nov 2025

University of the Witwatersrand: University of the Witwatersrand. In: Wikipedia (8 Dec 2017). https://en.wikipedia.org/wiki/University_of_the_Witwatersrand, accessed 29 Nov 2025

Volchok, E.: Gender differences in partially online courses. Commun. College J. Res. Pract. (2019)

Windle, J.M., Spronken-Smith, R.A., Smith, J.K., Tucker, I.G.: Preadmission predictors of academic performance. Curr. Pharm. Teach. Learn. **10**, 842–853 (2018)

Wrigley-Asante, C., Ackah, C.G., Frimpong, L.K.: Gender differences in STEM academic performance. SN Soc. Sci. **3**(1), 12 (2023)

Yeager, D.S., Walton, G.M.: Social-psychological interventions in education. Rev. Educ. Res. **81**(2), 267–301 (2011)

Zezekwa, N., Mudavanhu, Y.: Effects of entry qualification on students' performance in university science courses. African Journal of Education and Technology **1**(3), 32–39 (2011)

Zuma, J.: Zuma announces free higher education for poor and working class students. News24 (16 Dec 2017). https://www.news24.com/SouthAfrica/News/zuma-announces-free-higher-education-for-poor-and-working-class-students-20171216, accessed 29 Nov 2025

Determining Perceptions and Attitudes of Non-Accounting Majors Towards Accounting: A Case Study of a Selected University of Technology in South Africa

Mothemba Cecilia Mokoena and Gauda Johannes Maseko

Abstract This paper contributes to the continuing debate regarding the perceptions of non-accounting majors in higher education institutions. Many non-accounting majors at a university of technology in South Africa struggle with accounting due to lack of exposure to the subject, which resulted in a high failure rate. An investigation was undertaken to examine students' perceptions and attitudes towards accounting in order to enable lecturers to improve the throughput rate, keep students focused and interested in accounting and establish relationships between perceptions, attitudes, lecturer capabilities, students' struggles and academic performance. 453 structured questionnaires were collected and the results indicated that the majority of students displayed positive perceptions and attitudes towards Accounting. Students concluded that lecturers are knowledgeable and easy to understand. Students' struggles and poor academic performance were attributed to the fast pace of lecturers, lack of Accounting exposure, inadequate mathematical background and problems with motivation and focus. Positive correlations were revealed between perceptions and attitudes; lecturer capabilities and performance. However, negative correlations were found between students' struggles and perceptions; attitudes and lecturer capabilities. This study could lead to curriculum changes and encourage policymakers to apply constructive interventions to improve performance. The study contributes to the body of knowledge by highlighting the difficulties faced by non-accounting majors as limited research has been conducted on this topic in South Africa.

Keywords Accounting · Non-accounting majors · Perceptions · Attitudes · Academic performance · Lecturer capabilities · Students' struggles

M. C. Mokoena (✉) · G. J. Maseko
Department of Accountancy, Vaal University of Technology, Vanderbijlpark, South Africa
e-mail: ceciliap@vut.ac.za

G. J. Maseko
e-mail: johannes@vut.ac.za

E. H. Redda (ed.), *Proceedings of the Social Sciences International Research Conference, 2025*, https://doi.org/10.1007/978-981-95-9489-4_30

1 Introduction

Over the past few years, student perceptions and attitudes have been researched from various perspectives. Several studies involving Accounting as a subject have been widely studied and researched. For example, Yu (Yu 2011) investigated how study habits, skills and attitudes affect student performance in Accounting courses. Accounting is a skills course that requires constant practice. It exposes students to a system of analysing transactions, recording, evaluation of financial information and, moreover, understanding the process of input and output in business decisions.

Knight, Braun, Simpson and Le Guyader (Knight et al. 2021) found that generally students see the importance of accounting in their profession; however, they have negative attitudes towards accounting and thereby affect their academic performance. This is due to lack of an accounting background, insufficient mathematical skills and perceived complexity of Accounting by non-accounting majors; non-Accounting majors have negative perceptions and attitudes towards Accounting. As a result, the capability of Accounting lecturers is considered to be poor. The perceptions and attitudes of students towards lecturer capability and the challenges faced by non-accounting majors are expected to negatively influence students' academic performance in this study.

2 Literature Review

Over the years, non-accounting students have perceived Accounting to be a difficult subject to pass (see (Doran et al. 1991; Elias 2005; Lane and Porch 2002)). This phenomenon has been found globally. For example, (Roy 2019), in a study in Fiji, also revealed that students perceive Accounting as a difficult subject. Najid et al. (2016) in a study completed in the United States of America (USA) suggested that games could be used to motivate non-accounting students to develop an interest in financial accounting courses. (Tun 2019), who conducted a study in Malaysia, established that lack of prior Accounting knowledge can affect the perception of non-accounting students regarding the disclosure of financial and non-financial information by Zakat institutions. Lloyd and Abbey (2009) found in their study at a university in the USA that students with an Accounting background and strong mathematical skills perform well in Accounting. Students who are not equipped in Accounting and who may not have a strong background in mathematics, as is often the case in non-accounting majors, may need remedial work in the areas of mathematics and accounting. Consistent with the above studies, Goh and Scerri (Goh and Scerri 2016) concur that poor numeracy skills negatively influence performance.

Huang and Si's study (Huang and Si 2019) in China found that it is imperative to reform accounting courses for non-accounting students to simplify the subject matter for students. This study concluded that lecturers fail to take into account the characteristics of non-accounting majors; instead, lecturers teach these students in the same way they teach accounting majors, despite the former not being trained to become accountants. This may affect their confidence in their academic strength and abilities, which have been found to be good predictors for non-accounting students' interest in studying accounting, as suggested by Owusu et al. (2019), who conducted a study on non-accounting students at the University of Ghana Business School. In 2009, the study by Steenkamp and Baard (Steenkamp and Baard 2009) at a South African university determined the factors influencing success in first-year accounting. The study highlighted that students considered the subject to be boring and that this led to poor academic performance. Lack of motivation and negative attitudes towards the subject were established. This may have been due to the fact that lectures were presented in large classes to both accounting and non-accounting majors. Lloyd and Abbey (2009) argue that lecturers can unintentionally make learning difficult for non-accounting majors, as they tend to explain concepts in ways that are optimal for accounting majors, given that most lecturers are accountants themselves.

Studies, such as those by Bahar (2016), Naing and Thein (2018) and (Longobardi et al. 2020), have associated student perceptions with academic achievement. However, these studies investigated student perceptions and academic performance in general. This study will focus specifically on non-accounting majors. Academic performance can be measured based on students' grades (Vergara et al. 2015). An important aspect that can affect student performance is the lecturer. Lecturers are regarded as the most significant factor influencing students' achievement levels (Muzenda 2013). The lecturer can change students' attitudes and help them to achieve better performance, as attitude determines an individual's behaviour. Attitude is a combination of trust, a person's feelings towards an object and the tendency to act in relation to the object (Weli 2015). In order to accomplish this, lecturers must have suitable capabilities or competencies.

Lecturer capability refers to the ability to deliver information clearly and to equip students with the skills and knowledge required in the subject. Subject knowledge enables lecturers to develop strong teaching competencies and to deliver quality teaching (Weli 2015). This enables lecturers to create an atmosphere that is conducive to learning and where students and the lecturer alike enjoy the process of teaching and learning. Ganyaupfu (Long et al. 2014) and (Sultan and Shafi 2014) found that lecturer competence has a significant influence on students' academic performance. This is consistent with (Akiri and Ugborogbo 2009), who indicated that lecturer competence in the learning process will drive and help the student to achieve better performance. From a western cultural point of view, (Yoon et al. 2007) found that teachers who receive substantial professional development can boost their students' achievement.

(Shamsudin et al. 2020) established that non-accounting students perceive Accounting as complicated, due to its technical difficulties; however, the intervention by application of simplified accounting templates in teaching and learning can assist students. (Ismail et al. 2017) observed that the appropriateness of teaching techniques employed in class for the Accounting subject is important for student success. These authors have emphasised that the success or failure of a lecturer in the university or higher education institution depends highly on students' perceptions of the lecturer's performance in the lecture hall or class.

The current study was conducted at a selected university of technology located in Gauteng province, South Africa. Universities of technology are universities that offer mainly diploma and certificate courses but also offer degree courses that take about four years to complete. A university of technology places emphasis on innovative problem-solving and vocationally-oriented courses, while universities offer academic and professional programmes. The courses at universities of technology are designed to be career oriented and practical, with experiential learning forming part of the courses (Department of Higher Education and Training: Registering a private higher education institution 2025).

In 2019, the university of technology described above had approximately 2000 students registered as non-accounting majors under ten different courses. These students were taught by 10 lecturers who hold accounting qualifications. A high proportion of non-accounting majors who enrol every year at the university of technology in this study have no accounting background and are mostly doing business courses. The number of students has increased over the years as more university courses require students to have exposure in Accounting. The failure rate in the subject also has increased. Only 54% of registered students passed in the first semester of 2019, which is below the university threshold of 60%. Based on the aforementioned, Accounting was listed as an "At-Risk Subject" at the university. The academic success or throughput rate has also deteriorated due to the rising failure rate.

The failure rate of non-accounting majors is especially worrying and has become an escalating concern for the department, faculty and the university. Repeating modules has negative financial implications for both students and the institution as more resources are needed to cater for repeating students in relation to lecturers and facilities. It also affects the number of students graduating, resulting in reduced government subsidy (approximately R30 000 per student graduating). Repeating modules has a psychological impact on students as it affects their attitudes towards the subject and therefore leads to high dropout rates. (Moodley and Singh 2015) found that dropout rates at some institutions of higher learning in South Africa are as high as 80%, with a dropout cost of R4.5 billion to the National Treasury in grants and subsidies to higher education institutions without a return on investment (Council on Higher Education: Access and throughput in South African higher education 2010).

From the literature above, it can be concluded that lecturer capability influences performance and that a change in students' attitudes and lead to improved academic performance. The studies above have also highlighted the association between students' perceptions and performance. The current study was based on the framework outlined in Fig. 1.

Based on the above, the broader aim of this study was to establish the perceptions and attitudes of non-accounting majors towards Accounting and the impact of these on students' academic performance at a university. The study gives students a "voice" to indicate why they struggle with the subject. The intention of this was to make visible the areas of improvement or gaps relating to subject delivery and lecturer capabilities that may enhance students' academic performance. This study

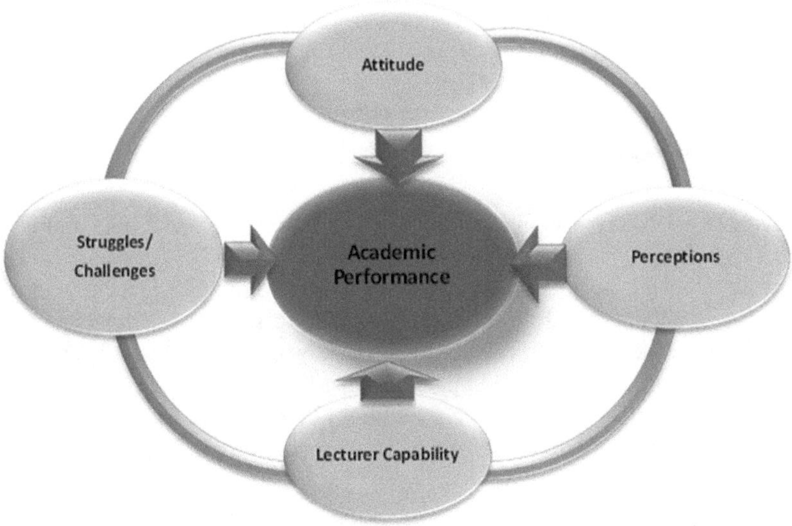

Fig. 1 Conceptual framework. *Source* authors' illustration

makes a theoretical contribution to the body of knowledge by highlighting the difficulties faced by non-accounting majors, which could lead to lecturers modifying their teaching methodologies. This, in turn, could change the perceptions and attitudes of students. It could also enable curriculum developers, designers of university programmes and policymakers to apply constructive interventions to improve the academic performance of non-accounting majors. With this in mind, the research questions guiding this paper are:

(i) What are the perceptions and attitudes of non-accounting majors towards Accounting at a university of technology in Gauteng province, South Africa?
(ii) What are the perceptions of non-accounting majors regarding lecturer capabilities at the university?
(iii) What are the challenges or struggles faced by non-accounting majors at the university of technology in relation to Accounting?
(iv) How is academic performance influenced by perceptions, attitudes, lecturer capabilities and challenges or struggles faced by non-accounting majors at the university of technology mentioned above?

To answer the above research questions, the following objectives were formulated:

(i) To determine the perceptions and attitudes of non-accounting students towards Accounting.
(ii) To establish lecturer capabilities as perceived by non-accounting majors.
(iii) To ascertain the challenges or struggles faced by non-accounting majors at the university.
(iv) To examine the performance of non-Accounting majors during the semester and expected performance at the end of the semester.

(v) To assess whether students' perceptions, attitudes, lecturer capabilities and students' struggles in Accounting influence academic performance.

3 Methodology

A quantitative research paradigm was used to produce numerical data and enable statistical analysis, as suggested by Zikmund and Babin (2007). As a result, the researcher was able to determine the perceptions and attitudes of students towards Accounting, lecturer capabilities, and the challenges faced by non-accounting students and to establish the relationship between these variables and academic performance. A structured questionnaire was distributed to non-accounting majors.

The questionnaire consisted of 24 five-point Likert scale questions. It contained modified questions/items, ranging from 1, "strongly disagree", to 5, "strongly agree" and was adapted from the questionnaire used by Zikmund and Babin (2013). Geiger and Ogilby (2000) have noted that 2- to 3-point Likert scales lack discrimination ability and therefore make respondents uncomfortable due to limited freedom of expression. Increasing categories in a Likert scale increases discriminatory power and improves reliability. Clow and James (2014) concurred that a Likert scale questionnaire makes it easy for the respondents to read and understand the questions and it is generally used to assess attitudes. It was thus found suitable for the study at hand as it determined the attitude of non-accounting majors towards Accounting as a subject.

The questionnaire had five sections (A-E), with section A requesting the participants' general information. Section B determined students' perceptions towards Accounting; section C established students' attitude towards the subject; section D determined lecturer capability, as perceived by students; and section E consisted of self-evaluation questions where students indicated the challenges that they are faced with in Accounting. They also indicated their performance during the semester and their expected performance at the end of the semester. A pilot test was undertaken with 20 students who were later excluded from the main survey, to establish the initial reliability of the questionnaire. Cronbach's alpha was generated using SPSS version 20 to further determine the reliability of the scale and yielded values of 0.72, 0.76, 0.80 and 0.71 for sections B, C, D and E, respectively. All these values were above the acceptable range, as the value of 0.60 or less designates inadequate reliability (Goddard and Melville 2009).

The questionnaire was assessed for validity to establish whether it measures what it is supposed to measure. Content validity was ensured by confirming that the instrument used takes into account all the meanings of the concept being measured and enabled the researcher to identify and eliminate problematic questions with regard to wording and arrangement of questions. After careful modification and the exclusion of ambiguous questions, the final questionnaire was distributed to 600 students in the second semester of 2019.

The questionnaire was administered by lecturers offering the subject and peer collaborators across the ten courses with non-accounting majors at the university to enable generalisation of the findings. To ensure the best possible response rate, questionnaires were distributed at the beginning of accounting lectures to students who voluntarily chose to participate. A total of 453 usable questionnaires were received, yielding a 76% response rate. Due to constraints of time, as the questionnaires were completed during classes, a structured format was used that made data collection easier and quicker. Questionnaires also made it possible to avoid bias and error and ensured anonymity of participants. A letter was handed out with the questionnaire indicating the purpose of the study and instructions on how to complete the questionnaire. The study was conducted at one university of technology in Gauteng province, South Africa (defined in Sect. 1), rather than at multiple universities and was therefore a case study. Case studies enable exploration of an event and provide thorough and detailed investigations. However, case studies are unique, cannot be generalised and are prone to researcher bias and ethical issues. The target population was restricted to one university consisting of 2000 students. Purposive sampling was adopted to ensure that only registered non-accounting majors participated in the study and to ensure that the sample consisted of students from all courses with non-accounting majors with at least 70% attendance rate. Purposive sampling is not without limitations, as it is prone to researcher bias arising from subjective judgements. The following data analysis techniques were applied:

(i) Frequency tables classified various responses into categories in Sections A and E. The counts were then expressed as percentages to produce percentage frequency tables, as these are easy to understand and interpret (Churchill et al. 2010). Frequency tables were used to determine gender, year of study, whether students had an accounting background before enrolling at the university, students' academic performance as well as their expected performance at the end of the semester in order to achieve the fourth (iv) objective of the study.
(ii) Descriptive statistics permitted the researcher to describe and compare variables numerically (Saunders et al. 2009). (O'Leary 2004) has noted that the main function of descriptive statistics is to provide measures of central tendency, dispersion and distribution shape. They were employed to determine the perceptions and attitudes of students towards Accounting, to establish lecturer capability as perceived by students and to highlight participants' struggles with the subject in order to achieve the first three objectives of the study.

In order to achieve the fifth objective of the study, correlation and regression analyses were used. Correlation analysis established the relationship between variables and analysed the direction and magnitude of the relationship between variables in the current study (Diamantopoulos and Siguaw 2000). Pearson's and Spearman's correlation procedures were used to explore these relationships. Regression analysis narrated independent and dependent variables in a model that takes mathematical inter-correlation into account (Malhotra 2010). Regression analysis was used to predict the values of a dependent variable given the values of one or more independent variables (Saunders et al. 2009). For this study, regression analysis was used to establish whether there was any relationship that existed between students' academic performance and expected performance.

4 Results and Discussion

The results discussed in this section were derived from the questionnaire subsections after statistical analysis.

4.1 *Frequency Tables*

Table 1 presents the composition of the sample. The sample consisted of 261 (57.6%) females and 192 males (42.4%). A total of 356 students (83%) did not have an accounting background prior to their enrolment at the university. Accounting, as a non-core subject, is taken by all students in their first year of study. The table indicates that 140 students out of 453 are repeating the subject as they are in their second year of study and upwards (31% indicates cumulative percentage from second year to other. Other refers to students who are in their fifth year of study or higher at the university).

Table 1 General information

Gender	Number of participants	Percentage
Female	261	57.6%
Male	192	42.4%
Total	453	100%
Did you matriculate in accounting?		
Yes	77	17.0%
No	356	83.0%
Total	453	100%
Year of study		
1st year	313	69.1%
2nd year	125	27.6%
3rd year	11	2.4%
4th year	1	0.2%
Other	3	0.7%
Total	453	100%

Source Authors' illustration

4.2 Descriptive Statistics

The first column in Tables 2 to 5 indicates the section and questions under each section, while the second column depicts the number of participants, the minimum values are in the third column and the maximum values are in the fourth column on a five-point Likert scale ranging from 1, "strongly disagree", to 5, "strongly agree". These tables present the summary of results from the questionnaire based on the measures of central tendency in the fifth column (mean) and dispersion in the last column (standard deviation). The tables also show the mean and standard deviation of each item found in the questionnaire. Since mean scores are not whole numbers but rather fractions, the following scales are used to interpret the mean scores of items in Sections B to E. Mean scores of 4 or above indicate "strongly agree"; scores between 3 and 4 indicate "agree"; scores between 2 and 3 indicate "disagree" while less than 2 indicates "strongly disagree".

In Table 2, the average scores range between $\overline{x} = 3.61$ ($\sigma = 1.058$) and $\overline{x} = 4.15$ ($\sigma = 0.884$), which is an indication that, on average, participants agree with the positive statements given under Section B about perceptions towards Accounting. Participants are in agreement that the subject will help them in their careers, that Accounting is useful in day-to-day life and that they learn a lot in accounting class. The standard deviation is within the acceptable range of $\pm$ 2 from the mean, showing that the results are close to the mean (Manoharan 2010).

Table 3 contains statements under Section C of the questionnaire, which assessed students' attitudes towards the subject. The participants agreed with the positive statements under this section. Students indicated that they attend lessons in order to do well in the subject ($\overline{x} = 4.13$; $\sigma = 0.988$), do all activities ($\overline{x} = 3.90$; $\sigma = 0.987$), that they are active participants in class ($\overline{x} = 3.28$; $\sigma = 1.074$) and that they enjoy the subject ($\overline{x} = 3.83$; $\sigma = 0.990$). As a result, students study the subject continuously,

Table 2 Perceptions of students towards Accounting

Section B	N	Min	Max	Mean	Standard deviation
The subject will help me to do well in my career (B1)	453	1	5	3.91	0.920
Doing well in this subject would be personally rewarding (B2)	453	1	5	4.15	0.884
The subject is useful in day-to-day life (B3)	453	1	5	3.99	0.934
I am highly motivated and do well in this subject (B7)	453	1	5	3.61	1.058
I learn a lot in accounting class (B8)	453	1	5	4.04	0.990
I have passed the test(s) in this subject (B10)	453	1	5	3.78	1.179
Valid N (listwise)	453				

Source Authors' illustration

Table 3 Students' attitude towards Accounting

Section C	N	Min	Max	Mean	Standard deviation
I attend all my accounting lessons to do well in the subject (C2)	453	1	5	4.13	0.988
I do all activities required (C3)	453	1	5	3.90	0.987
I am an active participant in accounting class (C4)	453	1	5	3.28	1.074
I learn accounting continuously and not only before the test (C5)	453	1	5	3.41	1.038
I enjoy accounting class (C8)	453	1	5	3.83	0.990
Valid N (listwise)	453				

Source Authors' illustration

Table 4 Lecturer capabilities

Section D	N	Min	Max	Mean	Standard deviation
Lecturer always comes to class prepared (D1)	453	1	5	4.82	0.493
Lecturer explains concepts clearly/ lecturer easy to understand (D2)	453	1	5	4.49	0.792
Lecturer shows knowledge of the subject (D3)	453	1	5	4.74	0.571
Lecturer uses examples to clarify theory (D4)	453	1	5	4.57	0.743
Valid N (listwise)	453				

Source Authors' illustration

Table 5 Struggles/challenges

Section E	N	Min	Max	Mean	Standard deviation
I struggle in the subject due to lack of related previous study (E2)	453	1	5	3.84	1.216
I struggle in the subject due to lack of numeracy skills (E3)	453	1	5	3.27	1.011
I struggle in the subject due to lack of motivation and focus (E4)	453	1	5	3.39	1.082
I struggle in the subject due to unpreparedness of lecturers (E5)	453	1	5	1.60	0.759
I struggle in the subject due to the fast pace of lecturers (E6)	453	1	5	3.03	1.071
Valid N (listwise)	453				

Source Authors' illustration

rather than only before assessments ($\overline{x} = 3.41$; $\sigma = 1.038$). In Table 3, a positive attitude towards the subject is evident.

Section D established lecturer capabilities. Students agreed that lecturers come to class prepared ($\overline{x} = 4.82$; $\sigma = 0.493$) and that concepts are clearly explained by lecturers and are easy to understand ($\overline{x} = 4.49$; $\sigma = 0.792$). Students stated that lecturers use examples to clarify theory ($\overline{x} = 4.57$; $\sigma = 0.743$) and display knowledge of the subject ($\overline{x} = 4.74$; $\sigma = 0.571$). A low standard deviation confirms that the majority of participants were in agreement with these statements.

Table 5 highlights that challenges faced by students in the subject are due to lack of previous related study ($\overline{x} = 3.84$; $\sigma = 1.26$), inadequate mathematical skills ($\overline{x} = 3.27$; $\sigma = 1.011$), problems with motivation and focus ($\overline{x} = 3.39$; $\sigma = 1.082$) as well as the fast pace of lecturers ($\overline{x} = 3.03$; $\sigma = 1.071$). However, students strongly disagreed that the challenges faced by students in the subject are due to unpreparedness of lecturers ($\overline{x} = 1.60$; $\sigma = 0.759$). This is consistent with the results shown in Table 4 regarding lecturer capabilities, which indicate that lecturers come to class prepared and show knowledge of the subject. The results demonstrate that non-accounting majors at the university have positive perceptions and attitudes towards the subject. Positive feedback was given by students regarding lecturer capabilities. The students in the sample struggle with the subject due to lack of related previous study, as well as inadequate numeracy skills and problems with motivation and focus. The fast pace at which lecturers teach was also identified as a challenge.

The high failure rate is presented in Table 6. However, students still have a positive attitude, perceive the subject as important for their future careers and are determined to perform better at the end of the semester (Table 7).

Table 6 Academic performance

Performance	Number of participants	Percentage %
Pass	185	40.8
Fail	268	57.6
Total	453	100

Source Authors' illustration

Table 7 Expected performance

Expected performance	Number of participants	Percentage %
Pass	342	65.6
Fail	111	34.4
Total	453	100

Source Authors' illustration

4.3 Correlation Analysis

Spearman correlation examined monotonic relationships between constructs. Correlation coefficients range between −1 and + 1, where −1 indicates perfect negative correlation and + 1 indicates perfect positive correlation. The test was two-tailed (non-directional) because no specific direction of the relationship between constructs was hypothesised. The strength of the relationship was determined using the guidelines suggested by Choudhury (Choudhury, 2025): 0–0.1 (weak or no relationship); 0.1–0.3 (weak); 0.3–0.5 (moderate); and 0.5–1.0 (strong relationship).

Based on the correlations shown in Table 8, there is a positive (strong) correlation between perceptions of students and their attitude towards the subject, r_s (451) = 0.639, $p < 0.001$. This is consistent with the results of a study conducted by Grunert, Bredahl and Scholderer (Grunert et al., 2003), which found that individuals' attitudes guide their subsequent perceptions. Lecturer capabilities show a weak positive relationship with perceptions, rs (451) = 0.253, p < 0.001, while a moderate positive correlation is observed between the attitude of students towards the subject and lecturer capabilities, rs (451) = 0.319, p < 0.001. Challenges faced by students reflected negative (moderate to weak) association with perceptions of students towards Accounting, rs (451) = −0.440, p < 0.001, attitude of students towards the subject, rs (451) = −0.389, p < 0.001 and lecturer capabilities, rs (451) = −0.288, p < 0.001.

Table 8 Correlation between perception, attitude, lecturer capabilities and students' struggles

			Perception	Attitude	Lecturer
Spearman's rho	Perception	Correlation coefficient	1.000		
		Sig. (2-tailed)			
	Attitude	Correlation coefficient	0.639**	1.000	
		Sig. (2-tailed)	0.000		
	Lecturer capabilities	Correlation coefficient	0.253**	0.319**	1.000
		Sig. (2-tailed)	0.000	0.000	
	Struggles/ challenges	Correlation coefficient	−0.440**	−0.389**	−0.288**
		Sig. (2-tailed)	0.000	0.000	0.000
		N	453	453	453

Table 8 Correlation between perception, attitude, lecturer capabilities and students' struggles

			Perception	Attitude	Lecturer
Spearman's rho	Perception	Correlation coefficient	1.000		
		Sig. (2-tailed)			
	Attitude	Correlation coefficient	0.639**	1.000	
		Sig. (2-tailed)	0.000		
	Lecturer capabilities	Correlation coefficient	0.253**	0.319**	1.000
		Sig. (2-tailed)	0.000	0.000	
	Struggles/ challenges	Correlation coefficient	−0.440**	−0.389**	−0.288**
		Sig. (2-tailed)	0.000	0.000	0.000
		N	453	453	453

4.4 Regression Analysis

Regression analysis is a technique that formulates a model in order to establish relationships amongst constructs in order to determine the degree of relationships between two or more variables and the relative contribution of each of the variables that make up the model (Pallant, 2011). Students' perceptions and attitudes, lecturer capabilities, students' struggles and students' performance were compared to determine relationships using Pearson correlation.

A positive correlation was observed in this study between students' performance (in the assessments written during the semester) and students' perception, $r = (451)$, 0.521, $p < 0.001$. (Ahmed et al. 2018) made the same finding that students with more positive perceptions have higher academic achievements. A positive correlation was found between academic performance and attitude, $r = (451)$, 0.402, $p < 001$, which confirms (Bakar et al. 2010)' finding that motivation and attitude are the best predictors of academic performance. Correlation between academic performance and lecturer capabilities was found to be positive in this study, (Table 9), $r = (451)$, 0.101, $p < 0.016$, while a negative correlation between performance and challenges faced by students, $r = (451)$, −0.315, $p < 0.001$ was established in the study. This supports Sultan and Shafi's (Sultan and Shafi 2014) argument that there is a negative relationship between performance and challenges faced by students, a finding also confirmed by Steenkamp and Baard (Steenkamp and Baard 2009). Steenkamp and Baard's (Steenkamp and Baard, 2009) study maintains that a significant correlation was not found between having Accounting as a subject at high school and achieving success in Accounting.

The regression model summary was used to determine the overall fit of the model and the relative contribution of each variable to the total variance explained. This is highlighted in Table 10. The table indicates that independent variables (as a group)

Table 9 Correlations between performance, attitude, lecturer capabilities and struggles

		E7	Perception	Attitude	Lecturer
Pearson correlation	E7—Academic performance	1.000			
	Perception	0.521	1.000		
	Attitudes	0.402	0.620	1.000	
	Lecturer capabilities	0.101	0.269	0.368	1.000
	Struggles/ challenges	−0.315	−0.389	−0.321	−0.245
Sig. (1-tailed)	E7		0.000**	0.000**	0.016**
	Perception	0.000**		0.000	0.000**
	Attitude	0.000**	0.000**		0.000**
	Lecturer	0.016**	0.000**	0.000**	
	Struggles/ challenges	0.000**	0.000**	0.000**	0.000**
N	E7	453	453	453	453

a Correlation is significant at the 0.05 level (1-tailed). *Source* Authors' illustration

contribute approximately 30% of variance in the dependent variable. ANOVA was used to determine the significance of the contribution of these variables, as reflected in Table 11.

Based on the results in Table 11, it was concluded that the model is appropriate in predicting the outcome as the overall regression model is statistically significant, $F(4,448) = 48.25$, $p < 0.001$, $R^2 = 0.30$.

Table 10 Regression model summary showing the contribution of independent variables

R	R Square	Adjusted R Square	Std. Error of the Estimate	Change Statistics		
				R Square Change	F Change	df1
0.549	0.301	0.295	14.553	0.301	48.253	4

Source Authors' illustration

Table 11 ANOVA indicating the significance of the contribution

Model		Sum of Squares	df	Mean Square	F	Sig.
1	Regression	40875.660	4	10218.915	48.253	0.000
	Residual	94875.656	448	211.776		
	Total	135751.316	452			

Source Authors' illustration

The regression analysis, as shown in Table 12, highlighted that students' perceptions and attitude towards Accounting have a significant positive relationship with students' performance ($p = 0.000$; $\beta = +0.406$) and ($p = 0.007$; $\beta = +0.141$), respectively, while lecturer capabilities ($p = 0.030$; $\beta = -0.093$) and challenges faced by students ($p = 0.002$; $\beta = -0.135$) showed a negative relationship with academic performance. Adediwura and Tayo (Adediwura and Tayo 2007) concurred that no relationship exists between academic performance and perceptions of teachers' knowledge.

The Pearson correlation was further used to establish whether students' performance during the semester was a predictor for students' expected performance at the end of the semester. Results are shown in Table 13.

A strong positive correlation between academic performance and expected performance is evident, $r = (451), 0.587, p < 0.001$.

The model summary in Table 14 indicates that students' performance explains approximately 35% of students' expected performance with ANOVA (Table 15) showing a statistically significant influence ($F (4,451) = 237.10, p < 0.001, R^2 = 0.35$).

The regression analysis in Table 16 reveals that performance has a significant positive relationship with students' expected performance ($p = 0.000$; $\beta = +0.587$). This indicates that performance significantly predicts expected performance. As the

Table 12 Regression analysis of performance with perception, attitude, lecturer capabilities and struggles

Dependent variable: Performance					
Independent variables	Unstandardised coefficients		Standardised coefficients	T	Sig.
	B	Std. Error	Beta		
Perception	11.130	1.428	0.406	7.793	0.000
Attitude	3.303	1.227	0.141	2.693	0.007
Lecturer capabilities	−3.100	1.427	−0.093	−2.172	0.030
Struggles/challenges	−3.288	1.063	−0.135	−3.093	0.002

Source Authors' illustration

Table 13 Correlations between academic performance and expected performance

E8—Expected performance		E8	E7
Pearson correlation	E8	1.000	
	E7	0.587	1.000
Sig. (1-tailed)	E8	.	0.000
	E7	0.000	.
N	E8	453	453

Source Authors' illustration

Table 14 Model summary showing the contribution of academic performance on expected performance

Model	R	R Square	Adjusted R Square	Std. Error of the Estimate	Change Statistics		
					R Square Change	F Change	df1
1	0.587	0.345	0.343	11.585	0.345	237.101	1

Source Authors' illustration

Table 15 ANOVA highlighting the significance of the contribution

Model		Sum of Squares	Df	Mean Square		F	Sig.
1	Regression	31819.787	1	31819.787		237.101	0.000
	Residual	60525.868	451	134.204			
	Total	92345.656	452				

Source Authors' illustration

performance increases by 1%, the expected performance is predicted to increase by 0.484%.

Regression analysis provided enough evidence that perceptions and attitude towards the subject positively influence academic performance. Lack of an accounting background, numeracy skills, focus and motivation as well as the fast pace of lecturers were found to negatively affect academic performance. The study also revealed that students' performance during the semester was positively correlated with students' expected performance at the end of the semester.

The study further established whether students' perceptions towards the subject influence their attitude and their conclusion regarding lecturer capabilities as well as the challenges faced by students. It was found that the way in which students perceive the subject determines their attitude and perceptions regarding lecturer capabilities.

It is recommended that there should be more contact hours with students as it was indicated in Sect. 1 that Accounting is a skills subject and requires considerable practice. Separate subject delivery for students with and without accounting exposure is recommended, as students without accounting exposure need more attention in order to address skills gaps, build confidence and develop an interest in the

Table 16 Regression analysis of academic performance with expected performance

Model		Unstandardised Coefficients		Standardised Coefficients	T	Sig.
		B	Std. Error	Beta		
1	E8	45.777	1.988		23.032	0.000
	E7	0.484	0.031	0.587	15.398	0.000

Source Authors' illustration

subject; lecturers also need more time to motivate those who are struggling and keep students focused in the subject. Lectures dedicated to equipping students with basic mathematics skills are recommended.

It is also recommended that students' progress be monitored through the use of quizzes to enable lecturers to identify students who are struggling in order to decide on intervention strategies and implementation before formal assessments are written. The appointment of tutors, more peer assistants or mentors is recommended to assist lecturers with marking and to help students who fail assessments or quizzes during the semester, as this may demoralise students. The training of lecturers is also recommended to enable lecturers to cope with non-accounting majors and be patient with the students, encourage lecturers to take into account the characteristics of non-accounting majors and to not teach non-accounting majors in the same way that accounting majors are taught.

5 Conclusion

It can be concluded that students at the university have positive perceptions and attitudes towards Accounting and that lecturers show knowledge of the subject. The high failure rate of non-accounting majors can be attributed to the absence of an accounting background, inadequate mathematical skills, the fast pace of lecturers, as well as insufficient motivation and focus.

The study was not without limitations. The study was conducted at a single university of technology in South Africa; therefore, the results may not be generalised to broader South African higher education institutions, which may have different dynamics. A recommendation for future research is to conduct similar studies at other universities to determine differences and whether the results would facilitate a generalisation of the findings.

References

Adediwura, A.A., Tayo, B.: Teachers' competence as predictor of academic performance. Educational Research and Review. **2**(7), 165–171 (2007)

Ahmed, Y., Taha, M.H., Alneel, S., Gaffar, A.: Learning environment and academic performance. Int. J. Med. Educ. **9**, 145–150 (2018)

Akiri, A.A., Ugborogbo, N.M.: Teachers' effectiveness and students' academic performance. Stud. Home Community Sci. **3**(2), 107–113 (2009)

Bahar, M.: Student perception of academic achievement factors. European Journal of Educational Research. **5**(2), 85–100 (2016)

Bakar, K.A., Tarmizi, R.A., Malhyuddin, R., Elias, H., Luan, W.S., Ayub, A.F.M.: Achievement motivation and academic performance. Procedia–Social and Behavioral Sciences 2, 4906–4910 (2010)

Choudhury, A.: Spearman rank correlation coefficient. http://explorable.com/spearman-rank-correlation-coefficient, Accessed 29 Nov 2025

Churchill, G.A., Brown, T.J., Suter, T.A.: Basic Marketing Research. South-Western Cengage Learning, Mason (2010)

Clow, K.E., James, K.E.: Essentials of Marketing Research. Sage, Thousand Oaks (2014)

Council on Higher Education: Access and throughput in South African higher education. Higher Education Monitor No. 9 (2010). https://www.che.ac.za, Accessed 29 Nov 2025

Department of Higher Education and Training: Registering a private higher education institution. https://www.gov.za/services/tertiary-education/register-tertiary-institution, Accessed 29 Nov 2025

Diamantopoulos, A., Siguaw, J.A.: Introducing LISREL. Sage, London (2000)

Doran, M., Bouillon, M., Smith, C.: Determinants of student performance in accounting principles I and II. Issues Account. Educ. **6**(1), 74–84 (1991)

Elias, R.Z.: Students' approaches to study in introductory accounting courses. J. Educ. Bus. **80**(4), 194–199 (2005)

Ganyaupfu, E.M.: Factors influencing academic achievement in quantitative courses. J. Educ. Pract. **4**(15), 57–65 (2013)

Geiger, M.A., Ogilby, S.M.: First accounting course and choice of major. J. Account. Educ. **18**, 63–78 (2000)

Goddard, W., Melville, S.: Research Methodology: an Introduction. Juta, Lansdowne (2009)

Goh, E., Scerri, M.: I study accounting because I have to. J. Hosp. Tour. Educ. **28**(2), 85–94 (2016)

Grunert, K.G., Bredahl, L., Scholderer, J.: European consumers' attitudes to genetic modification. Innov. Food Sci. Emerg. Technol. **4**(4), 435–445 (2003)

Huang, X., Si, Y.: Reform of accounting courses for non-accounting majors. In: Proceedings of the 5th International Conference on Social Science and Higher Education (ICSSHE 2019). Xiamen (2019)

Ismail, S., Jamaludin, N., Zakaria, W.Z.W., Navi, N.M.: Accounting students' perceptions of teaching quality. International Journal of Accounting, Finance and Business. **2**(5), 85–98 (2017)

Knight, E., Braun, R.L., Simpson, W.R., Le Guyader, L.P.: Attitudes of non-accounting majors toward an introductory accounting course. e-Journal of Business Education and Scholarship of Teaching. **15**(3), 16–24 (2021)

Lane, A., Porch, M.: Computer-aided learning and its impact on non-specialist accounting undergraduates. Acc. Educ. **11**(3), 217–233 (2002)

Lloyd, C.B., Abbey, A.: Teaching elementary accounting to non-accounting majors. Am. J. Bus. Educ. **2**(2), 23–30 (2009)

Long, C.S., Ibrahim, Z., Kowang, T.O.: Relationship between lecturers' competencies and students' satisfaction. Int. Educ. Stud. **7**(1), 37–46 (2014)

Longobardi, C., Settanni, M., Lin, S., Fabris, M.A.: Student–teacher relationship quality and prosocial behaviour. Br. J. Educ. Psychol. (2020)

Malhotra, N.K.: Marketing Research: an Applied Orientation. Prentice Hall, New Jersey (2010)

Manoharan, P.K.: Research Methodology. A.P.H. Publishing, New Delhi (2010)

Moodley, P., Singh, R.J.: Addressing student dropout rates at south African universities. Alternation. **17**, 91–115 (2015)

Muzenda, A.: Lecturers' competences and students' academic performance. International Journal of Humanities and Social Science Invention. **3**(1), 6–13 (2013)

Naing, M.M., Thein, H.H.M.: Impact of classroom perception on student academic achievement. Universities Research Journal. **11**(5), 239–256 (2018)

Najid, N.A., Tahir, W.M.M.W., Pauzi, S.F.M., Rahim, N.M., Dangi, M.R.M.: Use of Amazacc Brain Teaser Card for financial accounting classrooms. In: Proceedings of the Regional Conference on Science, Technology and Social Sciences (RCSTSS 2014), pp. 19–26. Springer, Singapore (2016). doi:10.1007/978-981-10-1458-1_2

O'Leary, Z.: The Essential Guide to Doing Research. Sage, London (2004)

Owusu, G.M.Y., Bekoe, R.A., Okyere, S.A., Welbeck, E.E.: Course major decision of accounting and non-accounting students. J. Int. Educ. Bus. **12**(1), 26–42 (2019)

Pallant, J.: SPSS Survival Manual, 4th edn. Allen & Unwin, Crows Nest (2011)

Roy, S.: Development of students' conception about accountancy. International Journal of Business, Accounting and Management. **4**(1), 56–64 (2019)

Saunders, M., Lewis, P., Thornhill, A., Wang, C.L.: Analysing quantitative data. In: Saunders, M., Lewis, P., Thornhill, A. (eds.) Research Methods for Business Students, 5th edn, p. 444. Prentice Hall, Harlow (2009)

Shamsudin, A., Pauzi, N.F.M., Karim, M.S., Roslan, N., Ahmad, K.: Measuring students' understanding of financial statements. Jurnal Dinamika Akuntansi. **12**(1), 24–33 (2020)

Steenkamp, L.P., Baard, R.S.: Factors influencing success in first-year accounting. S. Afr. J. Account. Res. **23**(1), 113–140 (2009)

Sultan, S., Shafi, M.: Impact of perceived teachers' competence on students' performance. J. Educ. Psychol. **8**(1), 10–18 (2014)

Tun, K.O.: Zakat disclosure of zakat institutions: perceptions of accounting and non-accounting students. Master's thesis, International Islamic University Malaysia (2019). http://studentrepo.iium.edu.my/handle/123456789/9395, accessed 29 Nov 2025

Vergara, J.P., Cabalonga, M.L.M., Genove, E.J.T., Lucero, E.J.C., Valmores, M.V.S., Villena, M.N.: Accountancy students' attitudes, methods and academic performance. In: Proceedings of the National Business and Management Conference. Philippines (2015)

Weli, I.: Accounting students' attitudes towards computers and accounting information systems. Procedia–Social and Behavioral Sciences 172, 18–25 (2015)

Yoon, K.S., Duncan, T., Lee, S.W.-Y., Scarloss, B., Shapley, K.: Reviewing the Evidence on Teacher Professional Development. U.S. Department of Education, Washington, DC (2007)

Yu, D.D.: How much do study habits, skills, and attitudes affect student performance in introductory college accounting courses? New Horizons in Education. **59**(3), 1–15 (2011)

Zikmund, W.G., Babin, B.J.: Exploring Marketing Research. Thomson South-Western, Cincinnati (2007)

Zikmund, W.G., Babin, B.J.: Essentials of Marketing Research. Cengage Learning, Mason (2013)

Effect of Personal Traits on Conflict Management Strategies in Family Business Succession Planning in South Africa

Victoria Adekomaya and Chris Schachtebeck

Abstract Family businesses constitute a vital component of South Africa's economic and social landscape, playing a critical role in job creation and community development. However, one of the most pressing challenges facing these enterprises is succession planning, a process often marked by significant conflict due to intertwined family relationships and differing individual interests. This study explores the influence of personal traits such as openness, conscientiousness, extraversion, agreeableness, and emotional stability on the conflict management strategies adopted during succession planning in family businesses. Using a qualitative research design, data were collected through semi-structured interviews with founders, successors, and key family members across multiple family-owned enterprises in South Africa. Thematic analysis was employed to interpret the data and identify patterns in how individual personality traits influence the choice of conflict management strategies, including collaboration, accommodation, avoidance, competition, and compromise. Using a qualitative research design, the study employs semi-structured interviews with founders, successors, and key family stakeholders across selected South African family businesses. Thematic analysis was used to interpret the data and uncover patterns in how individual personality traits shape negotiation styles, communication approaches, and decision-making processes. Preliminary findings suggest that personal traits significantly influence whether conflicts are approached collaboratively, avoidantly, or confrontationally, with implications for succession outcomes and family cohesion. This study contributes to the growing body of literature on family business succession by emphasising the role of individual personality in conflict resolution. It also offers practical implications for family business consultants, leadership coaches, and policymakers seeking to design more effective succession planning interventions that are attuned to the psychological dynamics within family businesses.

V. Adekomaya (✉) · C. Schachtebeck
DHET-NRF SARChI in Entrepreneurship Education College of Business & Economics, University of Johannesburg, Johannesburg, South Africa
e-mail: ceciliap@vut.ac.za

E. H. Redda (ed.), *Proceedings of the Social Sciences International Research Conference, 2025*, https://doi.org/10.1007/978-981-95-9489-4_31

Keywords Family business · Succession planning · Conflict management · Personal traits · South Africa · Personality · Family dynamics

1 Introduction

Family businesses are a dominant force in both developed and emerging economies, contributing substantially to employment, GDP, and economic resilience (Discua Cruz et al. 2019). In South Africa, family-owned enterprises play a particularly vital role in job creation and local economic development (Birdthistle and Hales 2023). However, despite their significance, many of these businesses face a common and persistent challenge: succession planning. The transition of leadership from one generation to the next is often fraught with emotional tension, misaligned expectations, and interpersonal conflict, which can threaten both business continuity and family harmony (Mzid et al. 2019). While various structural, financial, and governance mechanisms have been proposed to facilitate succession, less attention has been given to the psychological and interpersonal factors that underpin conflict during this critical process.

In the South African context, succession planning is further complicated by unique socio-cultural dynamics, including collectivist family norms, historical inequities, and a socio-political legacy that affects intergenerational business practices (Amann and Jaussaud 2014). These contextual factors add layers of complexity to the way personal traits influence conflict management. For example, cultural expectations of respect for elders and communal decision-making may clash with younger successors' desires for autonomy and innovation, resulting in strategic disagreements that are emotionally charged (Zapata-Cantu et al. 2023).

Moreover, empirical literature suggests that conflict in family business succession is not inherently negative; when managed effectively, it can lead to growth, innovation, and stronger family cohesion (Nölke et al. 2015). However, the ability to manage such conflict constructively often hinges on the individual personalities of those involved. For instance, successors with high emotional stability may be better equipped to handle the stress of succession discussions, while founders with low openness may resist necessary changes, perceiving them as threats rather than opportunities (Pont and Simon 2025). Despite the importance of personality in conflict dynamics, research specifically examining how personal traits shape conflict management strategies in South African family business succession remains scarce. Most existing studies focus either on Western contexts or emphasise structural variables such as ownership, legal frameworks, and governance models (Nölke et al. 2015). This leaves a gap in understanding the micro-level psychological mechanisms that influence succession outcomes in diverse cultural settings. To address this gap, this study adopts a qualitative approach to explore how personal traits influence conflict management strategies during succession planning in South African family businesses. Through in-depth interviews with founders, successors, and key family members, the research seeks to uncover the nuanced ways in which individual differences shape interpersonal interactions, negotiation styles, and ultimately, succession outcomes. The goal is to contribute a more holistic understanding of succession planning by integrating psychological insights into existing models of family business transition.

This study research is timely and relevant, particularly in the context of South Africa's evolving entrepreneurial landscape. As more family businesses seek to institutionalise their operations and ensure long-term sustainability, understanding the human factors that support or hinder succession becomes critical. Furthermore, the findings have practical implications for family business consultants, leadership coaches, and policymakers, who must design succession frameworks that are not only structurally sound but also emotionally intelligent. The remainder of this paper is organised as follows: the next section provides a detailed literature review on family business succession, personal traits, and conflict management. This is followed by a description of the research methodology, including sampling, data collection, and analytical procedures. The findings section presents key themes that emerged from the interviews, and the discussion interprets these results in light of existing theory. Finally, the conclusion offers recommendations for future research and practical applications for family business succession planning in South Africa.

2 Literature Review

Family businesses are a dominant force in both developed and emerging economies, contributing substantially to employment, GDP, and economic resilience (Nölke et al. 2015). In South Africa, family-owned enterprises play a particularly vital role in job creation and local economic development (Zapata-Cantu et al. 2023). However, despite their significance, many of these businesses face a common and persistent challenge: succession planning. The transition of leadership from one generation to the next is often fraught with emotional tension, misaligned expectations, and interpersonal conflict, which can threaten both business continuity and family harmony (Pont and Simon 2025). While various structural, financial, and governance mechanisms have been proposed to facilitate succession, less attention has been given to the psychological and interpersonal factors that underpin conflict during this critical process.

2.1 Theoretical/Conceptual Framework

The theoretical framework guiding this study is grounded in the Trait Theory of Personality, integrated with principles of Conflict Management Theory, and contextualised within the dynamics of family business succession. This framework provides a basis for understanding how individual personality traits influence the choice and application of conflict management strategies during succession planning in family-owned businesses in South Africa. Trait Theory posits that individuals possess consistent, stable characteristics referred to as traits that influence their behaviour across different situations (Zaccaro 2007: Novikova 2013; Derue et al. 2011). The Five-Factor Model (FFM), also known as the Big Five Personality Traits, is widely accepted in psychological and organisational research. These five traits include openness to experience, imaginative, open to new ideas organised, assertive, cooperative, and emotionally unstable (Buss 1989). These traits influence how individuals perceive conflict and respond to it. For example, a highly agreeable individual may prioritise harmony, leading to a preference for accommodating strategies, while a conscientious person may take a more structured, problem-solving approach to manage disputes. In the context of family business succession, where interpersonal relationships, generational expectations, and emotional investments converge, these personality traits play a pivotal role in shaping the nature of interactions and the approach to conflict resolution.

The Five-Factor Model (Big Five Personality Traits). Personal traits particularly those defined by the FFM, including openness, conscientiousness, extraversion, agreeableness, and emotional stability (neuroticism) are crucial in shaping how individuals perceive and manage interpersonal conflict (Soto and Jackson 2013). These traits influence communication patterns, decision-making styles, and emotional regulation, all of which are central to the dynamics of family business succession. Founders and successors often bring distinct personalities into the succession discourse, and these traits may either facilitate collaborative dialogue or escalate tensions (Antonioni 1998). The FFM, also known as the Big Five Personality Traits, is one of the most widely accepted frameworks for understanding human personality. Developed by McCrae and Costa (2008), the model proposes that personality can be understood across five broad dimensions: openness to experience, conscientiousness, extraversion, agreeableness, and neuroticism. These dimensions have been empirically validated across cultures and contexts and are particularly useful for examining behavioural tendencies in interpersonal and organisational settings (Soto and Jackson 2013). In the context of family business succession planning, particularly in South Africa, the Big Five traits offer a useful lens for examining how personal attributes influence the style, nature, and outcomes of conflict management strategies during transitions of leadership and ownership.

In South Africa, conscientious successors may clash with more informal or relational leadership styles common among older founders, leading to procedural conflicts unless mutual expectations are clarified early. Extraversion encompasses traits such as sociability, assertiveness, and emotional expressiveness. In family business settings, extraverts may assume leadership roles more readily, initiate conversations about succession, and manage conflict more openly. Extraverted individuals tend to confront conflict directly, often using competitive or collaborative conflict management styles (McCrae 2010). In hierarchical South African family firms, extraverted successors may challenge authority more readily, leading to power-based conflict if not managed through clear governance mechanisms. Founders who are also extraverted may engage in power struggles with similarly assertive successors unless a collaborative dialogue is maintained.

In family business contexts, high agreeableness is often linked to emotional intelligence, effective communication, and family cohesion (Chamorro-Premuzic and Furnham 2009). Agreeable individuals are likely to adopt accommodating, obliging, or compromising conflict styles to maintain relationships. While this trait may reduce open conflict in the short term, excessive agreeableness may lead to suppressed concerns, avoidance of difficult conversations, or reluctance to challenge suboptimal decisions. In South African businesses shaped by the Ubuntu philosophy, agreeableness aligns with cultural values of harmony, respect, and consensus, reinforcing the use of relational conflict resolution mechanisms (Boyle 2008). Neuroticism refers to emotional instability, anxiety, and a tendency towards negative emotions. Low neuroticism is considered essential for calm and rational decision-making, especially under stress (Thompson et al. 2002). Individuals high in neuroticism may adopt avoidant or passive-aggressive conflict strategies due to fear of confrontation or emotional reactivity (Rahim 2002). In high-stakes succession situations, especially those involving wealth and legacy, neuroticism can contribute to escalated family tensions, misinterpretation of motives, and prolonged disputes. In the South African socio-cultural landscape, where emotional restraint is sometimes encouraged in formal contexts, neurotic tendencies may be suppressed only to resurface later in destructive ways if succession plans are not openly discussed.

2.2 *Conflict Management Theories*

In South Africa, where power distance and collectivist values are often prominent (Kassim and Ibrahim 2014), obliging or avoiding styles may be more prevalent among younger successors. The TKI also identifies five conflict management modes and overlaps conceptually with Rahim's model (Thomas 1992; Rahim 2003). The relevance of TKI lies in its applicability in organisational development and leadership training. The use of TKI in South African family businesses may help understand how founders and successors deploy different conflict strategies, particularly during succession phases (Rahim et al. 2001). Family businesses in South Africa face unique challenges, including socio-cultural expectations, legacy concerns, and the need for economic resilience (Awalluddin and Maznorbalia 2023). Succession planning in such contexts often involves multi-generational stakeholders with diverse personality profiles. Rahim et al. (2000) refers personal traits has how to determine the stakeholders engage with conflict situations in a highly conscientious successor might take proactive steps to engage family members constructively, a neurotic predecessor might resist succession due to fear of losing control, an open family member may welcome innovation in leadership transition. These personality-driven tendencies shape the conflict resolution landscape and influence succession outcomes. Understanding these dynamics is particularly crucial in South Africa's collectivist and family-oriented cultural setting, where interpersonal relationships and respect for elders may mediate or escalate conflicts during succession (Afzalur Rahim et al. 2001). This theoretical framework integrates Trait Theory and Conflict Management Theory to explore how personality traits influence conflict resolution during family business succession. It provides a psychological lens for examining succession challenges in South African family businesses, highlighting how individual differences can shape conflict dynamics and succession outcomes. This framework is critical for both academic understanding and practical intervention, offering insights that can guide leadership development, succession planning, and family governance practices.

The FFM provides a powerful framework for understanding the behavioural patterns and interpersonal dynamics that characterise conflict during succession in family businesses (Berings et al. 2004). In the South African context, where succession planning is influenced not only by organisational strategy but also by culture, history, and interpersonal relationships, the model offers valuable insights for predicting conflict behaviour, designing interventions, and facilitating smoother transitions. Personality assessments based on the Big Five can be integrated into succession planning protocols, helping to identify potential sources of conflict and appropriate mediation strategies tailored to individuals' dispositions (McCrae and Sutin 2007). Future research should examine how these personality traits intersect with cultural values, governance mechanisms, and generational identity, especially in emerging economies like South Africa, where family businesses form a crucial part of socio-economic development.

2.3 *Cultural and Contextual Frameworks in South Africa*

In South Africa, the Ubuntu philosophy emphasises collectivism, community, and shared humanity (Nussbaum 2003). The philosophy supports conflict resolution through dialogue, respect, and community harmony. Individuals who exhibit high agreeableness and conscientiousness may align more closely with Ubuntu-oriented conflict resolution. Theoretical models that ignore Ubuntu values may misinterpret avoidance or obliging strategies as weakness rather than culturally contextual behaviour (Swanson 2007). Understanding personality traits within the Ubuntu framework can thus provide a more nuanced interpretation of conflict strategies in South African family businesses.

According to Molose et al. (2018), South Africa reflects medium to high power distance, where authority and hierarchy are respected. Senior family members with dominant personalities may suppress conflict, while successors with higher openness or extraversion may challenge traditional norms, creating intergenerational conflict. Applying personality theory within this socio-cultural frame allows researchers to account for the generational and hierarchical nuances that impact conflict management styles in succession processes (Ge & Campopiano 2022). From the reviewed theories, an integrative conceptual framework emerges. Personal traits, as described by the Big Five and Thematic Apperception Test (TAT), interact with conflict management styles such as those described by Rahim and the Thomas–Kilmann

Conflict Mode Instrument (TKI). These interactions are moderated by family business dynamics (Systems Theory, SEW, and Stewardship Theory) and shaped by South Africa's cultural milieu (Ubuntu, power distance, collectivism). This conceptual synthesis offers a multidimensional view of how succession-related conflict can be managed or exacerbated by the personalities of the involved parties, especially in the South African context.

The reviewed theoretical and conceptual frameworks offer critical insight into how personality traits influence conflict management strategies during succession planning in family businesses. In South Africa, where culture, family legacy, and economic uncertainty intersect, these frameworks offer a lens to understand and potentially mitigate conflict during succession transitions. Future research should empirically test these frameworks in South African family businesses, exploring how specific personality traits correlate with conflict management strategies and succession outcomes. Mixed-methods studies, cultural adaptations of assessment tools, and longitudinal designs could all enhance understanding and inform practical interventions.

3 Methodology

This section outlines the methodological approach adopted to investigate the influence of personal traits on conflict management strategies during family business succession planning in South Africa. The research design, population and sample, data collection methods, data analysis techniques, and ethical considerations are discussed.

3.1 Research Design

The study adopts a qualitative research design, grounded in the interpretivist paradigm, to explore participants' subjective experiences, meanings, and interpretations of succession planning, conflict, and personality. This design is appropriate because it enables an in-depth understanding of social interactions, relational dynamics, and the nuanced role of personality traits within family business contexts (Tisdell et al. 2025). A multiple case study approach (Kumar and Prameswari 2018) is employed to examine family businesses of different sizes, generations, and sectors. This allows for rich, contextual comparisons and the identification of both shared and unique patterns across cases. A purposive sampling strategy is used to select information-rich participants who are directly involved in succession processes within South African family businesses. The sample includes current successors (next-generation leaders), senior family members, and non-family managers involved in succession-related decisions. To ensure maximum variation, cases are selected from various provinces (Gauteng, Western Cape, KwaZulu-Natal), business sectors and generational stages. Approximately 12–15 participants across 5–7 family businesses. This is adequate for thematic saturation (Guest et al. 2006), where no new insights emerge from additional interviews.

3.2 Data Collection

Data will be collected through in-depth, face-to-face or virtual interviews using Zoom, MS Teams, or phone calls, depending on participants' availability and location. Interviews will be conducted in English, but accommodations will be made for participants preferring to speak in other South African languages, with the help of a translator if needed. Each interview is expected to last 45–60 min and will be audio-recorded with consent. The recordings will be transcribed verbatim for analysis. Field notes will also be kept to capture non-verbal cues and contextual information. To enhance credibility and reflexivity, the researcher will maintain a research diary to document personal reflections, decisions made during data collection, and emerging insights.

3.3 Data Analysis

A thematic analysis approach (Braun and Clarke 2006) will be employed. The steps includes familiarisation with the data, coding of text segments, generating initial themes, reviewing and refining themes, and defining and naming final themes. NVivo software will assist in coding and organising data. The analysis will seek to uncover patterns of meaning, contradictions, and context-specific interpretations of how personality traits influence conflict management.

4 Findings

This section presents the core findings of the study, which explored how personal traits influence conflict management strategies during succession planning in South African family businesses. Thematic analysis of the interview data revealed five overarching themes: (Table 1).

Table 1 Themes for the effect of personal traits on conflict management strategies in family business succession planning

Theme	Subtheme
Personal traits on conflict management strategies in family business succession planning	1. Dominance of personality in leadership conflict dynamics. 2. Trait-driven conflict management styles. 3. Generational differences in conflict responses. 4. Personality traits and succession readiness. 5. Cultural and familial norms as moderators of trait expression.

4.1 Dominance of Personality in Leadership Conflict Dynamics

Participants highlighted how dominant personality traits often shaped leadership styles and influenced the dynamics of conflict within the family business. Those with strong, assertive personalities were perceived as taking control during disagreements, sometimes silencing alternative views or creating power struggles. While such traits were seen as effective for decisive leadership, they also contributed to tension and resistance, particularly when collaboration was needed. Participants emphasised that the dominance of certain personalities could either strengthen or destabilise leadership structures, depending on how conflict was managed and how inclusive decision-making processes were.

Participant 9: *Even my siblings have withdrawn from discussions because they feel caught in the middle. My younger brother, who is more accommodating and avoids conflict, just agrees with my dad to keep the peace, even when he doesn't really support the decisions being made. It's frustrating, because you can clearly see that the personality dynamics are preventing the family from planning effectively for the future. There's a fear of confrontation because of how emotional and unpredictable some people can become when their authority is challenged.*

Participant 14: *I strongly believe that if we had more self-awareness as a family—if we understood how our personalities influence the way we communicate and handle disagreements—we would avoid a lot of the conflict. Personality isn't just about who's right or wrong; it shapes how we relate to one another, especially in leadership. And in a family business, those relationships are everything. Without addressing that, succession planning becomes almost impossible.*

4.2 Trait-Driven Conflict Management Styles

Participants described how individual personality traits played a key role in shaping their approaches to managing conflict within the family business. Traits such as assertiveness, empathy, patience, and openness were frequently linked to collaborative and constructive conflict resolution. In contrast, traits like dominance or emotional reactivity were often associated with more confrontational or avoidance-based styles. Many participants emphasised that understanding these trait-driven tendencies helped them navigate interpersonal challenges and improve communication, particularly during high-stakes moments like succession planning.

Participant 4: As *for me, I've always seen myself as someone in the middle. I tend to compromise I try to bring people together and find common ground. Sometimes that works, especially when both sides are willing to listen, but other times, it leaves me feeling like I'm carrying the burden of keeping everyone happy. What I've learned is that none of these styles are right or wrong. They're just shaped by who we are. But when it comes to family business, especially succession planning, it's important to be aware of these tendencies. Otherwise, we keep repeating the same patterns without realising that our personalities are driving our reactions.*

Participant 8: *If we had been more intentional about understanding our conflict styles maybe even doing some kind of personality or leadership assessment it would have helped us navigate succession more smoothly. Because sometimes, it's not the issue itself that causes the breakdown, but how we each deal with disagreement based on our traits. Recognising that could change the way we handle difficult conversations moving forward.*

4.3 Generational Differences in Conflict Responses

Participants shared insights on how different generations within family businesses respond to conflict, highlighting notable contrasts in communication styles, decision-making approaches, and emotional expression. Older generations were often described as favouring authority, tradition, and indirect communication, while younger members tended to adopt more open, collaborative, and assertive conflict resolution strategies. These generational differences sometimes led to misunderstandings or tension but also created opportunities for dialogue and learning. Overall, the responses emphasised that generational perspectives significantly influence how conflict is perceived and managed in succession planning.

Participant 14: *There's also the emotional aspect. The older generation often holds in their feelings; they rarely express frustration or vulnerability. But in our generation, we're more open about emotions. We want to understand not just the business, but the people behind it. This difference means we don't always interpret silence the same way. They think silence equals peace. We often see it as avoidance or unresolved tension. So there's this invisible wall between generations not because we don't care about the business, but because we don't speak the same conflict language.*

Participant 20: *If we're going to move forward successfully, especially with succession planning, we need to bridge that gap. I truly believe both generations have wisdom, but we must acknowledge that our responses to conflict come from different places (different upbringings, different worldviews). Until we start respecting those differences and adapting how we communicate, succession will continue to be a point of tension rather than a smooth transition.*

4.4 Personality Traits and Succession Readiness

Participants reflected on how specific personality traits influenced an individual's readiness for succession in family businesses. Traits such as responsibility, adaptability, confidence, and emotional intelligence were frequently mentioned as indicators of potential successors' preparedness. Many noted that successors who demonstrated initiative, resilience, and a strong work ethic were viewed more favourably, while those perceived as passive or overly dependent struggled to gain trust. The responses revealed that personality traits were seen not only as personal attributes but also as critical factors shaping leadership potential and succession outcomes.

Participant 10: *What complicates things is that some traits are not always visible unless there's tension or pressure. It's easy to look ready when everything's going smoothly. But when the business hits a rough patch or when there's conflict between family members, that's when your true leadership personality shows. That's why I think personality assessments or leadership simulations should be part of the succession planning process. It's not just about preparing someone for the job; it's about evaluating whether they're personally ready to handle all that comes with it.*

Participant 21: *If we had paid more attention to personality traits earlier in our succession planning, I believe we would have avoided some of the missteps we made appointing people who couldn't handle the interpersonal side of leadership, or overlooking others who had the right temperament but weren't as vocal. Succession readiness is as much about who the person is as it is about what they know.*

4.5 Cultural and Familial Norms as Moderators of Trait Expression

Participants highlighted how cultural values and family expectations often shaped the way personality traits were expressed within the family business setting. Many noted that even when individuals possessed strong leadership traits or assertiveness, these were sometimes subdued out of respect for elders or adherence to traditional roles. Others emphasised how family hierarchy, gender roles, and cultural beliefs influenced decision-making, communication, and conflict management. The responses underscored the complex interplay between individual traits and the socio-cultural environment in which they are enacted.

Participant 7: *The result is that people end up being judged not for their true capabilities or leadership potential, but for how well they conform to the family script. In our succession discussions, the family tends to gravitate toward the person who fits the cultural image of a leader usually older, male, quiet, composed even if that person isn't necessarily the best fit for the business. Meanwhile, someone who may be more innovative, outspoken, or emotionally intelligent might be sidelined because they don't 'act the part' as defined by the family culture.*

Participant 13: *That's why I believe succession planning must go deeper than just observing behaviour. We need to ask: Are we seeing the real person, or just the version they feel safe to show in this family setting? Without that awareness, we risk making decisions based on appearances, not true leadership readiness.*

5 Discussion of the Findings

This study explored the relationship between personal traits and conflict management strategies during family business succession planning in South Africa. Grounded in the FFM of personality (Soto and Jackson 2013) and Rahim's Conflict Management Styles Theory (Rahim 2002), the findings contribute to the understanding of how individual dispositions shape conflict navigation within intergenerational leadership transitions. Five key themes emerged from the data, offering new insights into the psychological, interpersonal, and socio-cultural dimensions of conflict in family business succession (McCrae 2010). The discussion below integrates these findings with existing literature and highlights their theoretical and practical implications. The study confirms that personal traits substantially influence how individuals perceive, engage in, and manage conflict during succession. Founders exhibiting high conscientiousness and low openness were described as controlling and resistant to change, creating power struggles and communication barriers. This aligns with findings by (Moberg 2001) and, who emphasise that founder-centric leader ship often complicates succession due to resistance to relinquishing control.

Conversely, successors with high openness to experience, emotional stability, and agreeableness employed collaborative approaches, preferring integrative conflict management strategies. These findings affirm research by Chiaburu et al. (2011), who argue that personality is a strong predictor of cooperative conflict resolution in organisational settings. This evidence supports the argument that the psychological composition of key actors significantly shapes succession outcomes. The FFM proved useful in explaining how underlying traits translate into behaviours during conflict situations. An important contribution of this study is the identification of trait–strategy alignment in managing succession-related conflict. For instance, agreeable individuals preferred compromising or integrating styles, consistent with Rahim's (2002) model, which links cooperative behaviour with prosocial traits. Highly neurotic individuals adopted avoiding strategies, echoing findings by Soomro et al. (2023), who found that emotional instability correlates with conflict avoidance. Extraverted successors or founders often resorted to dominating strategies, particularly in high-stakes negotiations. This link highlights the predictive power of personality profiles in succession planning and points to the need for incorporating psychological assessments into leadership transition planning processes. It also contributes to conflict management literature by extending its application to family business environments, where emotional intensity and relational complexity are elevated.

A central theme in the findings was the intergenerational personality gap and its contribution to conflict. Founders were often perceived as risk-averse, emotionally reserved, and authoritative traits associated with high conscientiousness and low openness while successors expressed values aligned with openness, flexibility, and innovation. These generational contrasts frequently led to misinterpretations and value-based conflict, particularly around decision-making and strategic vision. This finding echoes McCrae and Löckenhoff (2010) work, which identified intergenerational value misalignment as a core challenge in family business succession. In South Africa, this dynamic is further intensified by social transformation, where younger generations are shaped by democratic values, global exposure, and digital fluency, while older generations maintain traditional leadership norms. The study affirms that generational differences are not merely about age, but also about divergent personality profiles influenced by lived experience.

While trait theory emphasises internal disposition, the study also revealed that cultural and familial expectations moderate trait expression. South African values rooted in Ubuntu emphasising respect, harmony, and community encouraged individuals to suppress conflict and avoid direct confrontation, even when their personality would naturally support assertive behaviour (Ngubane and Makua 2021). Female participants, in particular, felt constrained by patriarchal norms, often suppressing traits such as extraversion or assertiveness to conform to familial expectations. This supports the argument by Udah et al. (2025) that trait expression is culturally embedded, and must be interpreted through a contextual lens. These findings indicate that while personality traits offer predictive value, trait-based models are incomplete without accounting for cultural and social influences. For family business scholars, this calls for an integration of personality and socio-cultural theories in future models.

The findings insight from the study is that personality traits not only affect conflict styles but also influence readiness and openness to succession. Founders with high emotional intelligence, openness, and agreeableness were more likely to initiate succession plans and support the empowerment of the next generation. This aligns with Berings et al. (2004), who found that self-aware and emotionally stable incumbents are more successful in transitioning leadership. Similarly, successors with traits like resilience, adaptability, and low neuroticism were perceived as better prepared for leadership roles. This insight extends existing literature by showing that succession planning is not just a structural issue, but a psychological readiness process, influenced by both individual disposition and relational trust. In sum, the study demonstrates that personal traits significantly affect conflict management approaches during family business succession. The findings validate existing personality and conflict theories while contributing new, context-specific knowledge relevant to South Africa. Importantly, they suggest that successful succession planning must incorporate psychological, relational, and cultural dimensions, not just procedural planning. By bridging trait theory, conflict management, and family business literature, the study presents a multidimensional understanding of succession conflict. Future research should consider longitudinal studies to explore how personality traits evolve across the succession cycle, and how targeted interventions (coaching, mediation, emotional intelligence training) can improve intergenerational transitions.

6 Theoretical Implications

The study offers several theoretical implications that enhance the understanding of how personal traits influence conflict management strategies during family business succession planning, particularly within the South African context. By employing the FFM of personality as its conceptual anchor, this study contributes to expanding the boundaries of personality theory, Conflict Management Theory, and family business succession frameworks. This study reinforces the utility of the FFM (Costa and McCrae 1992) in predicting individual behaviour in complex interpersonal settings, such as family business succession. While the FFM is well-established in general organisational behaviour research, its application in intergenerational, emotionally charged, and kinship-bound environments like family businesses remains underexplored. The findings suggest that agreeableness and emotional stability contribute to constructive conflict handling and smoother succession. High conscientiousness and low openness, especially in founders, can hinder collaborative succession planning and extraversion, when coupled with emotional intelligence, may facilitate assertive but effective leadership transitions. Thus, the study contributes to contextualising trait theory within relational and affective domains, showing that trait expression is shaped by not only the individual's disposition but also by familial, generational, and cultural dynamics.

In another way, the findings bridge personality psychology and Conflict Management Theory by empirically linking traits to preferred conflict styles, consistent with Rahim's (2002) model. For instance, avoidance strategies were associated with high neuroticism, integrative and compromising approaches were employed by those high in agreeableness and openness, and dominating styles were prevalent in individuals with high extraversion or low agreeableness. This cross-theoretical synthesis shows how trait-based tendencies manifest in actual conflict resolution behaviours, particularly in family enterprises where personal history and emotional ties amplify conflict intensity. The study thereby extends Conflict Management Theory to family business scholarship by highlighting how dispositional variables rather than only structural or situational ones determine the effectiveness and choice of conflict strategies in succession planning.

6.1 Theoretical Contribution to Family Business Succession Literature

Family business succession models traditionally emphasise structural factors, such as governance, succession planning stages, or business continuity frameworks (Bizri 2016; Aladejebi 2021; Campopiano 2022). However, this study brings a psychological and interpersonal lens to bear, emphasising that succession failure is often rooted in personality misalignments, poor emotional intelligence, and conflicting personal styles rather than procedural gaps alone. The study thus contributes to a more human-centric model of succession, in which personality dynamics are considered foundational to succession success or failure. This has theoretical implications for developing personality-based readiness assessments for successors and incumbents, designing intervention models that integrate personality coaching and conflict mediation within succession planning processes, expanding succession theory to include psychosocial variables as critical to family firm longevity. While the FFM is often regarded as universal, this study highlights how cultural, gendered, and familial norms in South Africa mediate the expression of personality traits. For instance; Ubuntu values may moderate the expression of extraversion and dominance, patriarchal norms may suppress assertiveness in female successors, regardless of their natural trait inclinations, historical and economic factors may amplify control and anxiety among founders from previously disadvantaged groups. These findings imply that trait theory must account for cultural conditioning and role expectations in collectivist or post-colonial societies. This theoretical implication supports a more contextualised and culturally sensitive understanding of personality theory, as previously advocated by cross-cultural scholars like (Nussbaum 2003; Swanson 2007).

6.2 Towards an Integrated Theoretical Framework for Family Business Conflict

Based on the study's insights, a new integrated framework is suggested that combines the FFM (individual traits), Rahim's conflict management styles (behavioural responses), generational role theory (intergenerational dynamics), and socio-cultural theory (contextual influence). This integrated model offers a multidimensional understanding of conflict in succession, where individual personality traits are shaped by and enacted within broader systemic and cultural contexts. The implication is that family business conflict theory should move beyond dyadic conflict models to include intrapersonal and cultural drivers of succession challenges. In sum, the study advances theoretical knowledge on the FFM to emotionally charged, culturally complex family business environments. It integrates trait theory with conflict management frameworks to better understand behaviour during succession. It highlights the limitations of universal trait models and argues for culturally embedded, relationally informed approaches to family business theory. It calls for more explored research on interdisciplinary models that blend psychology, sociology, and organisational studies in examining succession conflict.

7 Conclusion

This study set out to explore the influence of personal traits on conflict management strategies during family business succession planning in South Africa. Guided by the FFM of personality and Conflict Management Theory, and grounded in qualitative data from South African family business stakeholders, the research offers rich insights into the psychological and interpersonal dimensions of succession planning. The findings clearly demonstrate that personality traits significantly shape both the experience of conflict and the strategies employed to manage it. Traits such as openness, conscientiousness, agreeableness, emotional stability, and extraversion influence not only individual responses to conflict but also broader relational dynamics between incumbents and successors. These traits contribute to conflict prevention, escalation, or resolution, depending on how they interact with contextual and generational variables. Moreover, the study reveals that generational differences in personality expression often manifest as conflict during succession processes. Founders tend to exhibit traits associated with authority, control, and structure, while successors are more inclined towards openness, innovation, and collaboration. This trait mismatch frequently leads to resistance, miscommunication, and delayed succession planning.

Importantly, the research highlights the cultural and contextual moderation of personality traits in the South African setting. Ubuntu values, patriarchal norms, and the historical burden of economic exclusion play pivotal roles in shaping how personality traits are expressed and perceived. These socio-cultural factors must be acknowledged when applying universal trait theories in African family business contexts. From a theoretical perspective, the study contributes to expanding the FFM by illustrating its application in emotionally complex, intergenerational, and culturally nuanced environments. It also bridges the gap between trait theory and conflict management frameworks, offering an integrated approach to understanding succession-related conflict in family enterprises.

In practical terms, the study points to the importance of personality assessments, emotional intelligence training, and culturally sensitive conflict resolution mechanisms in improving succession outcomes. South African family businesses are encouraged to integrate psychosocial dynamics into their planning processes to ensure a smoother leadership transition and preserve both family harmony and business continuity. In conclusion, this study affirms that succession planning in family businesses is as much a human challenge as it is a structural one. Addressing the psychological undercurrents especially the influence of personal traits is vital for mitigating conflict and ensuring generational continuity in family-owned enterprises. Future research should further investigate these dynamics across diverse cultural settings and explore how personality development interventions might enhance succession preparedness and conflict resilience in family firms.

References

Aladejebi, O.: Succession planning: a key to sustainable family business. Journal of Business and Social Science Review. **2**(7), 26–38 (2021)

Amann, B., Jaussaud, J.: Family and non-family business resilience in an economic downturn. In: Japan's Lost Decade, pp. 69–89. Routledge, London (2014)

Antonioni, D.: Relationship between the big five personality factors and conflict management styles. Int. J. Confl. Manag. **9**(4), 336–355 (1998)

Awalluddin, M.A., Maznorbalia, A.S.: A systematic literature review of preferred organizational conflict management styles in Malaysia. Environment and Social Psychology. **8**(2), 1–13 (2023)

Berings, D., De Fruyt, F., Bouwen, R.: Work values and personality traits as predictors of enterprising and social vocational interests. Personal. Individ. Differ. **36**(2), 349–364 (2004)

Birdthistle, N., Hales, R.: The family business: Meaning and contribution to global economies. In: Attaining the 2030 Sustainable Development Goal of Good Health and Well-Being, pp. 13–25 (2023)

Bizri, R.: Succession in the family business: drivers and pathways. Int. J. Entrep. Behav. Res. **22**(1), 133–154 (2016)

Boyle, G.J.: Critique of the five-factor model of personality. In: Boyle, G.J., Matthews, G., Saklofske, D.H. (eds.) The SAGE Handbook of Personality Theory and Assessment, vol. 1, pp. 295–312. Sage, London (2008)

Braun, V., Clarke, V.: Using thematic analysis in psychology. Qual. Res. Psychol. **3**(2), 77–101 (2006)

Buss, A.H.: Personality as traits. Am. Psychol. **44**(11), 1378–1388 (1989)

Chamorro-Premuzic, T., Furnham, A.: Mainly openness: the relationship between the big five personality traits and learning approaches. Learn. Individ. Differ. **19**(4), 524–529 (2009)

Chiaburu, D.S., Oh, I.S., Berry, C.M., Li, N., Gardner, R.G.: The five-factor model of personality traits and organizational citizenship behaviors: a meta-analysis. J. Appl. Psychol. **96**(6), 1140–1166 (2011)

Costa Jr., P.T., McCrae, R.R.: The five-factor model of personality and its relevance to personality disorders. J. Personal. Disord. **6**(4), 343–359 (1992)

Derue, D.S., Nahrgang, J.D., Wellman, N.E., Humphrey, S.E.: Trait and behavioral theories of leadership: an integration and meta-analytic test of their relative validity. Pers. Psychol. **64**(1), 7–52 (2011)

Discua Cruz, A., Basco, R., Parada, M.J., Malfense Fierro, A., Alvarado-Alvarez, C.: Resilience and family business groups in unstable economies. In: The Family Business Group Phenomenon: Emergence and Complexities, pp. 315–352 (2019)

Ge, B., Campopiano, G.: Knowledge management in family business succession: current trends and future directions. J. Knowl. Manag. **26**(2), 326–349 (2022)

Guest, G., Bunce, A., Johnson, L.: How many interviews are enough? An experiment with data saturation and variability. Field Methods. **18**(1), 59–82 (2006)

Kassim, M.A.M., Ibrahim, H.I.: Conflict management styles and organizational commitment: a study among bank employees in Penang. International Journal of Business, Economics and Law. **4**(1), 45–53 (2014)

Kumar, S., Prameswari, Q.: Determinants of successful family business succession in Indonesia: a multiple case study. International Journal of Family Business Practices. **1**(2), 87–103 (2018)

McCrae, R.R.: The place of the five-factor model in personality psychology. Psychol. Inq. **21**(1), 57–64 (2010)

McCrae, R.R., Costa, P.T.: Empirical and theoretical status of the five-factor model of personality traits. In: Boyle, G.J., Matthews, G., Saklofske, D.H. (eds.) The SAGE Handbook of Personality Theory and Assessment, vol. 1, pp. 273–294. Sage, London (2008)

McCrae, R.R., Löckenhoff, C.E.: Self-regulation and the five-factor model of personality traits. In: Hoyle, R.H. (ed.) Handbook of Personality and Self-Regulation, pp. 145–168. Wiley, New York (2010)

McCrae, R.R., Sutin, A.R.: New frontiers for the five-factor model: a preview of the literature. Soc. Personal. Psychol. Compass. **1**(1), 423–440 (2007)

Moberg, P.J.: Linking conflict strategy to the five-factor model: theoretical and empirical foundations. Int. J. Confl. Manag. **12**(1), 47–68 (2001)

Molose, T., Goldman, G., Thomas, P.: Towards a collective-values framework of Ubuntu: implications for workplace commitment. Entrepreneurial Business and Economics Review. **6**(3), 1–18 (2018)

Mzid, I., Khachlouf, N., Soparnot, R.: How does family capital influence the resilience of family firms? J. Int. Entrep. **17**(2), 249–277 (2019)

Ngubane, N., Makua, M.: Ubuntu pedagogy: transforming educational practices in South Africa through an African philosophy. Inkanyiso: Journal of Humanities and Social Sciences. **13**(1), 1–12 (2021)

Nölke, A., Ten Brink, T., Claar, S., May, C.: Domestic structures, foreign economic policies and global economic order. Eur. J. Int. Rel. **21**(3), 538–567 (2015)

Novikova, I.: Trait, trait theory. In: Encyclopedia of Cross-Cultural Psychology, pp. 1–2 (2013)

Nussbaum, B.: Ubuntu: reflections of a south African on our common humanity. Reflections. **4**(4), 21–26 (2003)

Pont, A., Simon, A.: Nurturing family business resilience through strategic supply chain management. Journal of Family Business Management. **15**(1), 68–101 (2025)

Rahim, M.A.: Toward a theory of managing organizational conflict. Int. J. Confl. Manag. **13**(3), 206–235 (2002)

Rahim, M.A.: Managing conflict in organizations. In: Construction Conflict Management and Resolution, pp. 370–379. Routledge, London (2003)

Rahim, M.A., Antonioni, D., Psenicka, C.: A structural equations model of leader power, subordinates' styles of handling conflict, and job performance. Int. J. Confl. Manag. **12**(3), 191–211 (2001)

Soomro, B.A., Saraih, U.N., Ahmad, T.S.T.: Personality traits and conflict management styles via job performance in higher education. Journal of Applied Research in Higher Education. **15**(4), 1069–1094 (2023)

Soto, C.J., Jackson, J.J.: Five-factor model of personality. J. Res. Pers. **42**, 1285–1302 (2013)

Swanson, D.M.: Ubuntu: an African contribution to the search for a humble togetherness. Journal of Contemporary Issues in Education. **2**(2), 53–67 (2007)

Thomas, K.W.: Conflict and conflict management: reflections and update. J. Organ. Behav. **13**, 265–274 (1992)

Thompson, R.L., Brossart, D.F., Carlozzi, A.F., Miville, M.L.: Five-factor model personality traits and universal-diverse orientation in counselor trainees. J. Psychol. **136**(5), 561–572 (2002)

Tisdell, E.J., Merriam, S.B., Stuckey-Peyrot, H.L.: Qualitative Research: a Guide to Design and Implementation. Wiley, Hoboken (2025)

Udah, H., Tusasiirwe, S., Mugumbate, R., Gatwiri, K.: Ubuntu philosophy, values, and principles: an opportunity to do social work differently. J. Soc. Work. (2025)

Zaccaro, S.J.: Trait-based perspectives of leadership. Am. Psychol. **62**(1), 6–16 (2007)

Zapata-Cantu, L., Sanguino, R., Barroso, A., Nicola-Gavrilă, L.: Family business adapting to a new digital-based economy: opportunities and challenges for future research. J. Knowl. Econ. **14**(1), 408–425 (2023)

Balancing Autonomy and Oversight: Strategies for Improving Delegation in South African

Victoria Adekomaya and Chris Schachtebeck

Abstract Delegation is a critical managerial function that influences organisational performance, employee development, and operational efficiency. However, in the dynamic context of South African small and medium-sized enterprises (SMEs), leaders often struggle to strike an effective balance between granting employees sufficient autonomy and maintaining necessary oversight. This study investigates the challenges and strategic approaches to improving delegation practices within South African SMEs, focusing on how leaders navigate the tension between control and trust. Drawing on theoretical frameworks such as agency theory, empowerment theory, and stewardship theory, the research adopts a qualitative approach. Qualitative data were gathered through semi-structured interviews to provide deeper insights into contextual influences, leadership styles, and cultural factors affecting delegation. Using a qualitative methods approach, the research examines how SME owner-managers implement delegation, the barriers they encounter, and the impact of delegation practices on business performance and workforce productivity. NVivo analysis was adopted for the findings. Through case studies, the study identifies key factors influencing delegation effectiveness, including leadership styles, communication mechanisms, and accountability structures. Preliminary findings suggest that poor delegation is linked to unclear expectations, lack of training, and over-centralisation of decision-making. The study aims to identify practical, scalable strategies for improving delegation, including competency-based task assignment, transparent communication, and adaptive feedback loops. By generating evidence-based recommendations, the research seeks to inform leadership development programmes and policy frameworks that support sustainable growth and innovation in the SME sector. Ultimately, the study contributes to enhancing managerial effectiveness and employee engagement in South Africa's emerging economy, while addressing broader challenges of organisational resilience and human capital development.

Keywords Delegation · Autonomy · Leadership · SMEs · South Africa

V. Adekomaya (✉) · C. Schachtebeck
DHET-NRF SARChI in Entrepreneurship Education College of Business & Economics, University of Johannesburg, Johannesburg, South Africa
e-mail: vicfunsolutions@gmail.com

E. H. Redda (ed.), *Proceedings of the Social Sciences International Research Conference, 2025*, https://doi.org/10.1007/978-981-95-9489-4_32

1 Introduction

Effective delegation is widely recognised as a fundamental component of leadership and organisational performance. It involves the transfer of responsibility and authority from managers to subordinates, allowing employees to make decisions and execute tasks within a defined framework (Kusworo 2020). In practice, however, achieving an optimal balance between granting autonomy and maintaining oversight is complex particularly in contexts where trust, capacity, and accountability are unevenly distributed. In South Africa, the challenge is further compounded by diverse workplace cultures, historical management legacies, and varying leadership competencies across sectors (Sev 2017; Stea et al. 2015; Ugoani 2020). Many South African organisations, especially in the public sector and among small and medium enterprises (SMEs), face challenges related to either micromanagement or weak accountability structures (Alasomuka 2020). Excessive oversight may stifle innovation and reduce employee engagement, while unstructured autonomy can lead to poor performance, errors, or lack of strategic alignment (Stea et al. 2015). Moreover, the country's transformation agenda and evolving organisational dynamics call for more inclusive and participatory management styles, yet practical strategies for achieving this remain underexplored (Nor 2023).

The existing literature points to factors such as trust, communication, leadership style, and employee competence as central to effective delegation (Sebakamotse and Khahliso 2020). However, in South Africa, little empirical work has been done to examine how these factors interact within local organisational settings or how leaders can strike a practical balance between employee autonomy and necessary oversight. As such, there is a critical need to identify context-specific delegation strategies that enhance productivity, empower employees, and ensure managerial accountability. This study seeks to address this gap by exploring how South African managers and organisational leaders navigate the autonomy–oversight continuum. The research will examine strategies currently in use, evaluate their effectiveness, and propose frameworks tailored to the South African organisational environment. In doing so, the aim of this study is to explore and identify effective strategies for improving delegation practices in South African organisations by examining how managers balance employee autonomy with oversight. The research seeks to understand the challenges and enablers of delegation in diverse organisational contexts and to offer context-specific recommendations that enhance trust, accountability, and professional development within the South African workforce, and contribute to leadership development practices and improve delegation outcomes within both public and private sector institutions.

2 Literature Review

Delegation is a crucial leadership function that contributes to organisational effectiveness, employee empowerment, and leadership development. At its core, delegation involves the assignment of responsibility and authority to subordinates while the delegator retains accountability (Sebakamotse and Khahliso 2020). The literature has emphasised the importance of balancing two often competing dynamics in delegation: autonomy, which fosters trust, innovation, and employee motivation; and oversight, which ensures accountability, direction, and quality control (Nor 2023). Effective delegation hinges on the ability of leaders to navigate this balance, particularly in contexts of diverse workforce capabilities and rapidly changing organisational environments.

2.1 Theoretical Perspectives on Delegation Autonomy, and Oversight

Delegation theory highlights the relational and contextual nature of delegation. According to Downey (2021), successful delegation requires leaders to assess the readiness, competence, and motivation of employees to handle assigned tasks. The Situational Leadership Theory of Mahapatra and Jain (2023) suggests that leadership and by extension, delegation should be adapted to the maturity level of followers. Employees with high competence and confidence can handle greater autonomy, while those with lower readiness require closer supervision. Autonomy in the workplace has been associated with enhanced job satisfaction, intrinsic motivation, and increased innovation. However, autonomy without appropriate oversight can lead to inconsistent performance and strategic misalignment. Oversight mechanisms such as progress monitoring, feedback loops, and reporting structures are therefore essential for risk mitigation and quality assurance (Cortes 2023). The key challenge is ensuring that oversight does not evolve into micromanagement, which can demoralise employees and stifle initiative. Understanding the balance between delegation, autonomy, and oversight requires grounding in relevant theoretical frameworks that explain how power, responsibility, and accountability are distributed within organisations. This section explores key theories that inform the study's conceptual foundation.

Agency Theory. Agency Theory provides a foundational framework for understanding the complex relationship between delegation, autonomy, and oversight in organisational management. At its core, Agency Theory examines the relationship between a principal and an agent to whom authority is delegated to act on the principal's behalf (Bjurstrøm 2020). The theory assumes that agents may not always act in the best interests of the principal due to differing goals, preferences, or access to information creating what is known as an agency problem (Schillemans and Bjurstrøm 2020). In this context, delegation involves the transfer of responsibility and decision-making power from the principal to the agent. However, this transfer is not without risk. The agent may shirk responsibilities, make self-serving decisions, or underperform, especially when information asymmetry exists for example the agent knows more about the task than the principal. Consequently, oversight mechanisms such as performance monitoring, accountability systems, and incentive structures become necessary to reduce the likelihood of opportunistic behaviour and align the agent's interests with those of the principal.

While autonomy is increasingly recognised as essential for motivation, innovation, and job satisfaction (Al-Faryan 2024). Agency Theory cautions that unchecked autonomy without adequate oversight can increase agency costs and reduce organisational control. Therefore, effective delegation requires a delicate balance: giving employees enough freedom to make decisions while implementing structured oversight mechanisms to mitigate potential risks (Bjurstrøm 2021). In the South African organisational context, where workplace dynamics are often shaped by historical power hierarchies, skills gaps, and diverse organisational cultures, Agency Theory helps to explain managers' often cautious approach to delegation. It suggests that trust, clarity of expectations, and performance-based incentives are critical to minimising agency problems and maximising the benefits of delegation. Furthermore, Agency Theory emphasises the role of contractual clarity and outcome measurability in delegation. When roles, tasks, and outcomes are clearly defined and measurable, oversight becomes more effective, and agents are more likely to align their actions with organisational objectives (Drolc and Keiser 2021).

Empowerment Theory. Empowerment Theory provides a compelling lens through which to examine the relationship between delegation, autonomy, and oversight, especially within organisational settings seeking to enhance employee engagement, motivation, and productivity. At its core, Empowerment Theory emphasises the importance of giving individuals the authority, resources, and confidence to take initiative, make decisions, and act independently within their roles (Shuhaimi et al. 2023). In the context of management, this theory supports the idea that effective delegation is not just a transfer of tasks but a strategic tool for enabling autonomy and fostering a culture of trust, accountability, and personal growth. Delegation, when viewed through Empowerment Theory, is a means of transferring power from supervisors to subordinates in a way that enhances their sense of ownership, competence, and impact (Peng et al. 2023). Empowered employees are more likely to be proactive, innovative, and committed to organisational goals (Železnik and Fink-Hafner 2023). This perspective aligns closely with contemporary views on leadership that prioritise transformational and participative approaches, where autonomy is seen not as a risk to control but as a catalyst for productivity and organisational learning.

Oversight, within the empowerment framework, does not imply strict monitoring or micromanagement, but rather the provision of support, feedback, and development opportunities. Managers are seen as facilitators or coaches who help employees build the skills and confidence needed to handle greater responsibilities. This shift from control to support is crucial for fostering psychological empowerment, which includes four dimensions: meaning, competence, self-determination, and impact (Venkataramani et al. 2022). In South African organisational contexts especially in diverse and historically hierarchical environments empowerment-based delegation is particularly significant. Many participants in empowerment-oriented workplaces report higher job satisfaction, greater innovation, and a deeper sense of inclusion and value. Moreover, empowerment is viewed as an essential strategy in addressing inequities in skills development and leadership representation, particularly among historically marginalised groups. Thus, Empowerment Theory suggests that effective delegation requires more than the assignment of tasks; it involves creating conditions that support autonomy while maintaining oversight through mentorship, encouragement, and constructive feedback, rather than punitive control.

Stewardship Theory. Stewardship Theory provides an alternative viewpoint to traditional agency-based models by emphasising the intrinsic motivation of managers and employees to act as stewards of the organisation's interests rather than as self-interested agents (Obermann et al. 2020). Unlike Agency Theory, which assumes that individuals require strict oversight due to potential conflicts of interest, Stewardship Theory posits that employees are naturally aligned with organisational goals and seek to contribute positively through responsible behaviour and commitment (Torfing and Bentzen 2020). From the perspective of delegation, Stewardship Theory suggests that granting autonomy fosters a sense of ownership and responsibility, encouraging employees to act in the best interest of the organisation without the need for excessive monitoring or control (Rouault and Albertini 2022). This theory highlights that when employees perceive their roles as meaningful and trust is established between managers and subordinates, oversight can shift from formal control mechanisms to supportive, relational approaches.

In the South African context, where organisational cultures are increasingly embracing participative and transformational leadership styles, Stewardship Theory resonates well. It promotes delegation practices that empower employees, minimise hierarchical constraints, and build trust-based relationships crucial for navigating the complexities of diverse workforces and socio-economic challenges (Bjurstrøm 2020). Therefore, Stewardship Theory advances the argument that effective delegation balances autonomy with oversight through trust and shared values rather than rigid controls. This perspective encourages managers to view employees as partners in achieving organisational objectives, emphasising relational governance over transactional oversight. Together, these theories underscore that delegation is not merely a function of assigning tasks but a strategic leadership process that requires careful balancing of trust, control, context, and motivation. Integrating these perspectives allows South African managers to move beyond simplistic models and develop nuanced delegation practices that reflect organisational goals, employee capacity, and socio-cultural dynamics.

2.2 *Delegation Challenges in the South African Context*

In the South African organisational landscape, delegation presents unique challenges shaped by socio-cultural, historical, and economic dynamics. Legacy issues from apartheid-era hierarchical management systems have contributed to control-oriented leadership cultures in many sectors, particularly in government and state-owned enterprises (Mathebula and Barnard 2020). While transformation policies have sought to democratise management and promote inclusivity, implementation has often been inconsistent, and delegation practices remain uneven across sectors (Du Plessis 2020). Research suggests that many South African managers are reluctant to delegate effectively due to a lack of trust in subordinates, fear of losing control, and a shortage of leadership development training (Nzimakwe and Biyela 2021). Conversely, in some settings, especially among SMEs, delegation is informal or poorly structured, resulting in insufficient oversight and accountability (Thatcher et al. 2023). These patterns point to a need for tailored delegation strategies that consider organisational culture, leadership style, employee capacity, and sectoral differences.

Furthermore, historical legacies of hierarchical and authoritarian management styles persist in many South African organisations, particularly in public sector institutions and traditional corporate environments (Dlamini et al. 2020). These entrenched leadership paradigms often promote centralised decision-making and micromanagement, creating an environment where delegation is either superficial or avoided altogether due to fears of losing control. Cultural diversity and communication barriers also present significant challenges. South Africa's multicultural workforce comprises diverse languages, values, and work ethics, which can complicate clarity in task assignments and expectations (Mathebula and Barnard 2020). Miscommunication or misunderstandings arising from these differences may lead to delegation failures and decreased trust between managers and subordinates.

In addition, organisational trust issues frequently undermine delegation efforts. Due to a combination of past experiences with corruption, nepotism, and lack of accountability in some organisations, managers may exhibit reluctance in granting full autonomy, preferring stringent oversight to mitigate risks (Munzhedzi 2021). This cautious approach, while understandable, can stifle innovation and demotivate employees. Lastly, the impact of rapid technological change and digital transformation has introduced new complexities in delegation. While technology can facilitate remote supervision and information sharing, it also requires employees to possess digital literacy and adaptability, which are not uniformly present across the workforce (Dlamini et al. 2020).

2.3 *Strategies for Improving Delegation*

Effective delegation is essential for organisational success, enhancing productivity, employee development, and managerial efficiency. Improving delegation requires deliberate strategies that address both the human and structural aspects of the process. A fundamental strategy to improve delegation is investing in employee training and development to build the necessary skills and confidence for autonomous task execution. When employees are competent and prepared, managers can delegate more effectively, reducing the need for excessive oversight (Mathebula and Barnard 2020). Skills development programmes, mentoring, and job rotation are proven methods to enhance readiness. Sarigiovannis et al. (2021) posits that trust between managers and employees forms the foundation of effective delegation. Managers should foster open communication channels to clarify expectations, provide feedback, and encourage dialogue, which reduces misunderstandings and enhances accountability (Chen 2021). Establishing a culture of transparency helps to create an environment where delegation flourishes. Adopting adaptive leadership approaches, such as Situational Leadership Theory, allows managers to tailor their delegation style according to the employee's competence and commitment levels (Fenn III and Plake 2020). By adjusting the balance of autonomy and oversight, leaders can provide the right amount of support while encouraging independence.

Clear delegation requires unambiguous communication about what tasks are delegated, the scope of authority, and accountability measures. Documentation of delegated tasks and performance expectations supports this clarity. Organisations should promote cultures that value empowerment, learning, and shared accountability. When employees feel supported and recognised, they are more willing to accept responsibility and perform delegated tasks diligently (Ugoani 2020). Snowdon et al. (2020) states that leadership development programmes that emphasise empowerment and trust contribute to this cultural shift for a combination of leadership development, trust-building, and structured oversight mechanisms as critical to improving delegation outcomes. Leadership training focused on emotional intelligence, communication, and coaching can help managers develop the confidence to delegate more effectively (Ali et al. 2021). Furthermore, building trust through transparency, competence recognition, and feedback supports a culture where autonomy can thrive (Fenn III and Plake 2020). In the South African context, Campbell et al. (2020) emphasise the importance of context-sensitive delegation frameworks that incorporate mentorship, skills development, and clear role definitions. Effective delegation strategies also require the integration of performance management systems that align delegated tasks with organisational goals, supported by monitoring tools that do not undermine employee autonomy (Ugoani 2020).

3 Methodology

3.1 Research Design

This study adopts a qualitative research design to explore how South African managers and organisational leaders balance autonomy and oversight in delegation. The qualitative approach is appropriate as it enables an in-depth understanding of participants' lived experiences, perceptions, and strategies within their specific organisational contexts (Creswell and Poth 2018). An exploratory case study design were used to generate rich contextual data across selected sectors, including public institutions, corporate organisations, and SMEs. The study is grounded in an interpretivist paradigm, which assumes that reality is socially constructed and best understood through the subjective experiences of individuals. This paradigm allows the researcher to interpret meanings and insights that managers attach to their delegation practices, challenges, and strategies.

3.2 *Population and Sampling*

The target population includes middle to senior-level managers in South African organisations who are directly involved in delegation and supervision processes. A purposive sampling technique was used to select participants from SMEs especially those operating in urban townships. Approximately 15–20 participants were selected to ensure diversity in organisational type, leadership experience, and management level. The sampling was guided by the principle of data saturation, where no new insights emerged from additional interviews.

3.3 *Data Collection Methods*

Primary data will be collected through semi-structured interviews to allow for flexibility in probing and follow-up questions while maintaining a focus on key thematic areas. Each interview will last approximately 45–60 min and will be audio-recorded with participant consent. The interview guide will include questions related to experiences with delegation, perceptions of autonomy and oversight, strategies used to balance control and independence, organisational culture, and leadership development practices. Data will be analysed using thematic analysis as outlined by Braun and Clarke (2006). The process involves six steps: familiarisation with data, generating initial codes, searching for themes, reviewing themes, defining and naming themes, and producing the report. NVivo software will be used to facilitate coding and categorisation. The goal is to identify common themes and divergent experiences across different organisational contexts.

3.4 *Ethical Considerations*

Ethical clearance was obtained from the University of Johannesburg ethics committee. Participants were provided with informed consent forms outlining the study's purpose, confidentiality provisions, and voluntary nature. Pseudonyms will be used to protect participants' identities, and data will be stored securely.

4 Findings

Using Braun and Clarke's (2006) six-step thematic analysis process, five core themes emerged from the data. These themes reveal how South African managers and leaders navigate the tension between giving employees autonomy and maintaining necessary oversight in delegation. The findings highlight that effective delegation in South African organisations is not just a technical function, but a relational and contextual process. Trust, capacity building, leadership intention, and organisational culture all play pivotal roles in shaping delegation practices. A balanced approach where oversight supports rather than controls was found to be most effective (Table 1).

Theme 1: Trust as the Cornerstone of Delegation. Trust emerged as a central theme in understanding how delegation is enacted and experienced in South African organisational contexts. Participants consistently emphasised that effective delegation is grounded in a trusting relationship between the manager and the subordinate. Without trust, managers reported feeling reluctant to delegate meaningful tasks, often resulting in micromanagement or delayed decision-making. Conversely, when trust is present, employees are empowered, take ownership of their responsibilities, and feel motivated to perform at higher levels. Trust was thus viewed not only as a facilitator of delegation but as its foundational pillar.

Participant 2: *If I don't trust the person, I tend to hold on to the task myself. Delegation becomes very difficult when you constantly fear that things will not be done right. Trust gives me the confidence to let go.*

Participant 6: *In a small business, trust is everything. I don't have time to double-check everything. I train my staff and show them that I trust them. That's how we grow together. Sometimes, we delegate because we have to. But when trust is there, it feels like a partnership, not just a task being handed over. It makes a big difference.*

Theme 2: Limited Skills and Readiness as a Barrier to Autonomy. While delegation is often seen as a tool to promote autonomy and leadership development, many participants in this study highlighted that limited employee skills, experience, and readiness can significantly hinder the effectiveness of delegation. Managers reported

Table 1 Summary of key themes

Theme	Insight
Trust as the cornerstone of delegation	Delegation depends on mutual trust between leaders and employees.
Skills and readiness limit autonomy	Skill gaps hinder confidence in giving full autonomy to employees.
Accountability vs. micromanagement	Oversight is necessary but must not stifle employee motivation.
Delegation as growth strategy	Used as a tool to develop leadership skills and promote internal talent.
Culture shapes delegation practices	Organisational values determine the balance between autonomy and oversight.

that in some cases, employees lacked the technical expertise, critical thinking ability, or confidence to work independently, resulting in reluctance to delegate key tasks. This skill gap often led to increased oversight, repeated mistakes, and a slowdown in performance. The findings suggest that effective delegation must be accompanied by capacity building and mentorship to ensure employees are adequately prepared for autonomous responsibilities.

4.1 Participant Comments

Participant 1: *I've had cases where I delegated a task and had to redo it because the person just wasn't ready. Not everyone is equipped to handle responsibility without close supervision. Sometimes we want to delegate, but the staff don't have the skills yet. You can't leave them to sink. You have to guide them step by step until they build confidence.*

Participant 5: *I often have to spend more time fixing errors than it would have taken me to do the task myself. That's when I realise that the person I gave the task to wasn't ready for that level of autonomy. Delegation is not just about handing over work it's about making sure the person can actually do it well. If they lack the skills or understanding, it puts the whole project at risk.*

Theme 3: Balancing Accountability with Supportive Oversight. This theme highlights the tension managers face in ensuring delegated tasks are completed responsibly while avoiding micromanagement. Participants emphasised that delegation is not about relinquishing responsibility entirely but rather about finding the right balance between holding employees accountable and providing guidance where needed. Effective oversight, when delivered in a supportive and non-intrusive manner, was seen as critical in building confidence, improving performance, and fostering professional growth. Many participants also expressed the view that overly rigid monitoring could stifle innovation and autonomy, while too little oversight could lead to costly errors. Striking this balance was regarded as a key skill in leadership and delegation practice.

Participant 7: *Some employees thrive when you trust them, but even then, they still need your guidance now and then. The oversight should help, not hinder. If something goes wrong, as a manager, I'm still accountable. That's why I ensure regular progress reviews but I also try to empower staff to make decisions along the way.*

Participant 14: *Too much oversight and people feel suffocated. Too little, and mistakes happen. I try to stay visible but not overbearing. It's a constant balancing act. Delegation doesn't mean disappearing. I stay involved just enough to guide and course-correct if needed. It keeps everyone focused and accountable.*

Theme 4: Delegation as a Tool for Professional Growth. This theme reflects participants' recognition of delegation not just as a managerial necessity but as a strategic tool for developing employee potential and leadership capacity. Delegation was widely viewed as a mechanism for building confidence, exposing staff to new responsibilities, and preparing them for future roles. Managers highlighted how entrusting employees with meaningful tasks fostered learning-by-doing, encouraged critical thinking, and signalled confidence in their abilities. The theme underscores that when done purposefully, delegation becomes a form of on-the-job training, helping employees grow professionally and aligning their development with organisational succession goals.

Participant 4: *When we delegate strategically, it becomes a development opportunity. I always explain the bigger picture, so they understand the value of the task and their role in it. In small companies, delegation is essential for growth. I can't do everything myself, and my staff need to be ready to lead someday. I use delegation to expose them to leadership challenges.*

Participant 9: *Delegation is how you build confidence. When staff see they can handle something important, it builds their self-belief and encourages initiative. For me, delegation is about sustainability. I want my business to thrive even without me. That means helping my staff grow into their full potential through real responsibility.*

Theme 5: Organisational Culture Influences Delegation Style. This theme explores how organisational culture shapes managers' approaches to delegation. Participants reported that values, norms, and historical practices within their organisations often determined the extent to which they could delegate, how they delegated, and to whom. In hierarchical or rigid environments, delegation was often limited or controlled tightly, with decision-making power concentrated at the top. In contrast, organisations with more collaborative or developmental cultures encouraged flexible delegation and empowered staff at multiple levels. The findings reveal that organisational culture can either enable or constrain effective delegation by influencing trust, communication flow, and attitudes towards employee autonomy.

Participant 7: *In our company, there's a top-down approach. You don't question authority, and delegation only happens if the manager is overloaded. It's not about development it's about survival. We have a culture of mentorship, so delegation is part of how we build capacity. Senior staff are expected to involve juniors and coach them along the way.*

Participant 13: *In my business, I created a culture where everyone is involved. We discuss tasks together, and I delegate based on interest and strength. That's how we stay agile. The culture here is quite academic and individualistic. People tend to hold onto responsibilities and knowledge. Delegation happens, but it's not encouraged as a way to build teams.*

5 Discussion of the Findings

This section interprets and contextualises the study's findings in relation to the existing literature and theoretical frameworks. The discussion highlights the implications of each theme for leadership and delegation practices within South African organisations. The study confirms that trust is a crucial enabler of effective delegation, aligning with Stea et al. (2015). who assert that trust between leaders and subordinates reduces the need for control and enhances performance. In South Africa, trust is particularly significant due to the socio-political history of inequality and hierarchical management structures, especially in the public sector (Alasomuka 2020). The findings indicate that in environments with high interpersonal trust, leaders are more willing to delegate tasks with less intrusive oversight. This underscores the need for leaders to intentionally foster trust through open communication, reliability, and respect, particularly in diverse and transitional workplace cultures. The second major theme that skills gaps hinder delegation resonates with the Situational Leadership Model Mahapatra and Jain (2023). which posits that delegation should be matched with the competence and maturity level of the employee. The findings suggest that in many South African SMEs and under-resourced sectors, leaders refrain from full delegation due to perceived or real limitations in employee readiness. This reflects a structural issue within the South African labour market: insufficient upskilling, lack of mentorship, and inconsistent employee development. Addressing these gaps through training, coaching, and structured feedback could empower leaders to delegate with greater confidence, while preparing employees to take on more responsibility.

The findings highlight a key tension: balancing accountability with autonomy. Leaders must monitor progress to meet organisational goals, but excessive oversight can demotivate employees and undermine innovation. This aligns with Bjurstrøm (2021) who argued that effective leaders clarify roles and expectations while avoiding over-control. The study suggests that the use of supportive oversight mechanisms such as milestone-based reporting, check-ins, and team reflections can help maintain accountability while preserving autonomy (Downey 2021). The implication is that oversight should be perceived as supportive rather than punitive, fostering a delegation was also found to be a tool for grooming future leaders, consistent with Sev's (2017) view that strategic delegation contributes to employee engagement and leadership pipeline development. In South African companies that invest in internal talent, delegation is not just a means to get work done it is a leadership strategy. This insight suggests that leaders should be trained not only in task delegation but also in developmental delegation, where responsibilities are assigned based on growth potential, not just operational need (Sebakamotse and Khahliso 2020). This approach has significant implications for leadership succession planning in both private and public institutions.

Finally, the study reinforces that organisational culture profoundly shapes delegation practices. In hierarchical or compliance-driven cultures, oversight is prioritised, whereas in agile or entrepreneurial cultures, autonomy is celebrated. Nor (2023) structural configurations support this: mechanistic organisations tend towards tight control, while organic structures enable autonomy. This suggests that cultural transformation is necessary for organisations seeking to improve delegation. Leaders must intentionally align their delegation practices with the desired culture promoting initiative, psychological safety, and accountability (Ugoani 2020). Overall, the findings indicate that effective delegation in South Africa is context-dependent, requiring a thoughtful balance between structure and freedom, competence and trust, control and empowerment. Organisations that succeed in this balance are those that build trusting relationships between leaders and teams, invest in continuous employee development, design transparent oversight processes that do not demoralise staff, using delegation to develop leadership capacity, and align delegation practices with an enabling organisational culture. These insights not only contribute to the theory of delegation but also have practical relevance for improving leadership effectiveness, organisational performance, and employee development in diverse sectors.

5.1 Managerial/Practical Implication

The findings of this study suggest that effective delegation is a critical leadership competency that directly impacts organisational performance, employee motivation, and leadership development. Practically, managers must adopt a strategic approach to delegation that balances autonomy with accountability. This means building a culture of trust—Managers should foster open communication, transparency, and fairness to create psychologically safe environments where employees feel empowered to take ownership of delegated tasks. Delegation should be based on continuous assessment of employee skills and readiness. Providing mentorship and training to close skill gaps will enhance delegation outcomes. Instead of micromanagement, managers should use structured check-ins, performance metrics, and feedback loops that ensure progress without stifling initiative. Assigning tasks that challenge and stretch employees helps in grooming future leaders and improving retention. Managers must integrate delegation into talent development and succession planning strategies. Aligning delegation practices with organisational values. Delegation should reflect and reinforce a company's broader culture and strategy, promoting agility, innovation, and shared accountability. By applying these practical strategies, South African organisations can enhance employee engagement, optimise leadership effectiveness, and contribute to more resilient and adaptive workplace cultures in both public and private sectors.

6 Conclusion

This study explored the complex but critical challenge of balancing autonomy and oversight in delegation practices within South African organisational contexts. The findings, derived through thematic analysis, reveal that effective delegation is not merely a functional task, but a strategic leadership practice that requires high levels of trust, employee competence, structured oversight, and alignment with organisational culture. The study established that while managers often recognise the value of delegation, they struggle with releasing control due to competency gaps, fear of failure, and organisational constraints. However, when trust is cultivated, and employees are empowered and supported, delegation becomes a powerful tool for enhancing productivity, developing future leaders, and fostering a culture of shared responsibility. Ultimately, improving delegation practices in South Africa requires a multifaceted approach one that includes targeted leadership development, continuous employee upskilling, and the institutionalisation of supportive oversight mechanisms. This balance of autonomy and control is essential for organisations aiming to be agile, innovative, and sustainably high-performing in an increasingly complex business environment. By integrating the study's insights into practice, South African managers can shift from reactive task assignment to strategic delegation that empowers teams, builds leadership capacity, and drives long-term organisational success.

6.1 Contribution

This study makes several significant contributions to both academic knowledge and practical leadership within the South African context. First, it addresses a critical gap in the literature by providing an in-depth exploration of how South African managers negotiate the balance between granting autonomy and maintaining oversight during delegation an area often overlooked in research focused predominantly on Western organisational settings. By centring the unique socio-economic, cultural, and organisational dynamics of South Africa, the study offers contextually relevant insights that advance understanding of delegation practices in emerging economies. Second, the research highlights the interplay between trust, employee competence, and organisational culture as pivotal factors influencing delegation effectiveness. Unlike prior studies that treat delegation largely as a transactional process, this study frames it as a relational and developmental leadership strategy. This shift expands theoretical perspectives on delegation by emphasising the importance of emotional intelligence and cultural sensitivity in leadership practices.

Third, the study provides practical frameworks and actionable strategies tailored for South African organisations, including SMEs and public sector entities, which face distinct challenges such as skills shortages and hierarchical cultures. These strategies support managers in designing delegation systems that enhance accountability without stifling innovation and employee empowerment. Lastly, by integrating thematic analysis of real-world managerial experiences, the study contributes methodological value to leadership research, demonstrating how qualitative approaches can uncover nuanced understandings of complex managerial phenomena. In summary, this study enriches the discourse on delegation by bridging theory and practice with a strong emphasis on contextual relevance, thereby offering valuable guidance for both scholars and practitioners aiming to improve leadership effectiveness in South Africa and comparable settings.

7 Limitations and Future Research

While this study provides valuable insights into balancing autonomy and oversight in delegation within South African organisations, several limitations should be acknowledged. First, the study's qualitative design and thematic analysis, while rich in depth, limit the generalizability of findings across all sectors and organisational sizes in South Africa. The sample may have been influenced by specific industries or regional contexts, potentially restricting the applicability of results to broader or different organisational settings. Second, the study primarily captures the perspectives of managers and supervisors, which may omit important views from employees who experience delegation firsthand. Including employee perspectives could provide a more holistic understanding of delegation dynamics. Third, the complex socio-cultural and economic context of South Africa means that delegation practices may be influenced by factors such as labour regulations, union presence, and historical inequalities, which were not extensively explored in this study. Finally, the cross-sectional nature of the research limits insight into how delegation practices evolve over time or in response to organisational change.

References

Alasomuka, V.A.: Delegation: a tool for leadership effectiveness in higher education institutions. Int J Inst Leadersh Policy Manag. **2**(1), 1–16 (2020)

Al-Faryan, M.A.S.: Agency theory, corporate governance and corruption: an integrative literature review approach. Cogent Soc Sci. **10**(1), 2337893 (2024)

Ali, A.H., Jafar, S.M., Abbas, N.H.: Delegation of authority and its impact on the development of administrative skills of staff in the student activities section. Rev Iberoam Psicol Ejerc Deporte. **16**(4), 6 (2021)

Bjurstrøm, K.H.: Principal–agent or principal–steward: how ministry–agency relations condition the impact of performance management in the steering of government agencies. Public Perform. Manag. Rev. **43**(5), 1053–1077 (2020)

Bjurstrøm, K.H.: How interagency coordination is affected by agency policy autonomy. Public Manag. Rev. **23**(3), 397–421 (2021)

Braun, V., Clarke, V.: Using thematic analysis in psychology. Qual. Res. Psychol. **3**(2), 77–101 (2006)

Campbell, A.R., Layne, D., Scott, E., Wei, H.: Interventions to promote teamwork, delegation and communication among registered nurses and nursing assistants: an integrative review. J. Nurs. Manag. **28**(7), 1465–1472 (2020)

Chen, Y.: The impact of responsibility delegation on policy and practice implementation: a contingency approach. Public Perform. Manag. Rev. **44**(4), 842–867 (2021)

Cortes M (2023) The Employees' Perspective: Situational Leadership Style Flexibility and Effectiveness and Employee Performance in a Technological Organization. PhD thesis, Walden University

Creswell, J.W., Poth, C.N.: Qualitative inquiry and research design: Choosing among five approaches. Sage publications (2016)

Dlamini, N., Mazenda, A., Masiya, T., Nhede, N.T.: Challenges to strategic planning in public institutions: a study of the Department of Telecommunications and Postal Services South Africa. Int J Public Lead.Sh. **16**(1), 109–124 (2020)

Downey, L.: Delegation in democracy: a temporal analysis. J Polit Philos. **29**(3), 305–329 (2021)

Drolc, C.A., Keiser, L.R.: The importance of oversight and agency capacity in enhancing performance in public service delivery. J. Public Adm. Res. Theory. **31**(4), 773–789 (2021)

du Plessis, A.: The emergence of decentralised centralism in the south African education governance system. J. South. Afr. Stud. **46**(1), 165–183 (2020)

Fenn III, N.E., Plake, K.S.: Developing delegation skills. Am. J. Health Syst. Pharm. **77**(4), 254–258 (2020)

Kusworo, K.: The importance of delegation of authority, budget allocation and leadership in improving performance. J Lead.Sh Organ Stud. **2**(1), 221–228 (2020)

Mahapatra, R.R., Jain, A.K.: Situational leadership theory: an extension and a test of global organizational prescriptions. In: Globalization, Human Rights and Populism: Reimagining People, Power and Places, pp. 623–640. Springer, Cham (2023)

Mathebula, B., Barnard, B.: The factors of delegation success: accountability, compliance and work quality. Expert. J Bus Manag. **8**(1) (2020)

Munzhedzi, P.H.: An evaluation of the application of the new public management principles in south African municipalities. J. Public Aff. **21**(1), e2132 (2021)

Nzimakwe, T.I., Biyela, A.C.: Exploring the procurement challenges in the south African public sector. In: Public Procurement, Corruption and the Crisis of Governance in Africa, pp. 73–92. Springer, Cham (2021)

Nor MI (2023) Optimizing public financial management in Somalia: delegation, autonomy, and strategic focus for sustainable economic growth. Working paper, July 8

Obermann, J., Velte, P., Gerwanski, J., Kordsachia, O.: Mutualistic symbiosis? Combining theories of agency and stewardship through behavioral characteristics. Manag. Res. Rev. **43**(8), 989–1011 (2020)

Peng, J., Nie, Q., Chen, X.: Managing hospitality employee cyberloafing: the role of empowering leadership. Int. J. Hosp. Manag. **108**, 103349 (2023)

Rouault, J., Albertini, E.: Reconciling the social sector with external accountability requirements: lessons from stewardship theory. J. Bus. Res. **142**, 485–498 (2022)

Sarigiovannis, P., Jowett, S., Saunders, B., Corp, N., Bishop, A.: Delegation by allied health professionals to allied health assistants: a mixed methods systematic review. Physiotherapy. **112**, 16–30 (2021)

Schillemans, T., Bjurstrøm, K.H.: Trust and verification: balancing agency and stewardship theory in the governance of agencies. Int. Public Manag. J. **23**(5), 650–676 (2020)

Sebakamotse J, Khahliso T (2020) The Development of an Oversight and Accountability Framework for Municipalities in the Free State Province. PhD thesis, Central University of Technology, Free State

Sev, J.T.: Effective delegation of authority as a strategy for task accomplishment and performance enhancement in business organizations: an empirical survey of flour mills of Nigeria plc, Lagos. Bus Manag. Rev. **8**(4), 138 (2017)

Shuhaimi, N.A., Md Noor, A., Wan Jaafar, W.M.: Concepts and theories of empowerment and competency. Int J Acad Res Bus Soc Sci. **13**, 18 (2023)

Stea, D., Foss, K., Foss, N.J.: A neglected role for organizational design: supporting the credibility of delegation in organizations. J. Organ. Des. **4**(3), 3–17 (2015)

Thatcher, M., Sweet, A.S., Rangoni, B.: Reversing delegation? Politicization, de-delegation, and non-majoritarian institutions. Governance. **36**(1), 5–22 (2023)

Torfing, J., Bentzen, T.Ø.: Does stewardship theory provide a viable alternative to control-fixated performance management? Adm Sci. **10**(4), 86 (2020)

Ugoani, J.: Effective delegation and its impact on employee performance. Int. J. Econ. Bus. Adm. **6**(3), 78–87 (2020)

Venkataramani, V., Bartol, K.M., Zheng, X., Lu, S., Liu, X.: Not very competent but connected: leaders' use of employee social networks as prisms to make delegation decisions. J. Appl. Psychol. **107**(3), 458–473 (2022)

Železnik, A., Fink-Hafner, D.: Theoretical perspectives on bureaucrats: a quest for democratic agents. Adm. Soc. **55**(7), 1432–1456 (2023)

Exploring Student Capital as a Competitive Advantage in Academic Excellence for Private Higher Education Institutions

Henk de Jager, Mark Bussin, and Calvin Mabaso

Abstract In the context of private higher education institutions (PHEIs), student capital refers to the skills, knowledge and attributes (competencies) students acquire throughout their academic journey. It requires exceptional levels of teaching excellence to develop students for the future world of work who can also become responsible citizens, meaningfully contributing to society. Student capital directly impacts the student's academic performance, satisfaction, career opportunities, overall personal development, potential for success and employability. This study is grounded in an integrated theoretical framework that draws on human capital theory. The study followed a constructivist and positivist stance within a mixed-methods approach. As part of the qualitative approach, thirty experts from six private higher education institutions in South Africa were selected through purposive sampling to identify key performance indicators (KPIs) that align with the vision and mission of PHEIs. The quantitative approach purposively selected one hundred experts from six PHEIs to rate the KPIs, while the analytical hierarchy process was applied to rank the KPIs in terms of importance. The study identified four distinctive key perspectives: human, institutional, student and brand capital. In terms of student capital, PHEIs need to focus on producing high-quality academic graduates who are employable and can make a positive contribution to society and industry. They need to rethink their curriculum, teaching, and assessment policies and practices to ensure their relevance to the world of work. This must be supported by academics who need to develop students' competencies with a passion and desire to make a difference. To further enhance academic excellence, institutions must provide student-centred education, create a positive academic environment, a favourable staff-to-student ratio and academics with a passion for academia and developing students' potential. The broader implications for PHEIs are to adopt business principles for sustainability and react promptly to market and student demands.

H. de Jager (✉)
STADIO Higher Education, Johannesburg, South Africa
e-mail: henkd@stadio.ac.za

M. Bussin · C. Mabaso
University of Johannesburg, Johannesburg, South Africa

E. H. Redda (ed.), *Proceedings of the Social Sciences International Research Conference, 2025*, https://doi.org/10.1007/978-981-95-9489-4_33

Keywords Student capital · Teaching excellence · World of work · Key performance indicators · Higher education institutions · Human capital · Competitiveness

1 Introduction

Over the past three decades, higher education (HE) globally has faced enormous challenges regarding its accountability to various stakeholders. This includes stakeholders such as governments, regulating authorities, industry, society, and students as concerned clients. These challenges have ranged from the increased demand for access to HE, as indicated by Stander and Herman (2017), to the exponential growth in emerging technologies (Eager & Brunton, 2023; Du Toit & Verhoef, 2018). The demand for leadership and good governance within these institutions ensures the institution's sustainability and adequate response to external environmental demands (Kok & McDonald, 2017; Welch, 2020). However, more recent and pressing issues have arisen due to the changing societal expectations of the educational outcomes of higher education institutions (HEIs) (Cheah et al., 2023). This societal expectation has placed pressure on HEIs to produce employable students in today's limited and demanding job market (Hernandez-Diaz et al. 2020).

The primary aim of this study was to develop a novel theoretical framework for measuring the holistic organisational performance of private higher education institutions (PHEIs) in South Africa. The framework is designed to serve as a strategic alignment tool, integrating institutional strategies, structures, people, and processes with the institution's overarching vision and mission. In doing so, it aims to support the development of sustainable, agile, and high-performing PHEIs.

2 Literature Review

2.1 Introduction

Due to the limited capacity in the public HE space, PHEIs can no longer be ignored as key role players in addressing the skills shortage in South Africa. The role can be broader or niche market skills development (Dirkse van Schalkwyk 2011). The limited capacity of public HEIs is highlighted by Badat (2016) and Maharaj (2016), who indicate that a lack of funding, insufficient resources and an infrastructure and skills shortage are among the challenges that negatively impact public HEIs' ability to accommodate the growing demand for quality HE.

2.2 Developing Students' Competencies

It is widely accepted that, within a global context, HE is essential for providing quality education that equips students with the skills and capabilities required for the world of work (Fernandes & Singh 2021). It underscores the need for society to change for the better and the expectation that HE will provide graduates with the required competencies, which include skills, knowledge, and the right attitudes (Houston 2007). Therefore, PHEIs must acknowledge this need as part of their strategy and develop students' competencies for the current and future world of work. Leiber (2019) argues that developing competencies in HEI ensures that students are empowered to develop academic knowledge, methodologies, and skills that can be used directly for their profession and employment. PHEIs must emphasise developing students' competencies, including effective communication, teamwork, leadership, problem-solving and creative thinking. These renewed efforts to develop people (students) is supported by human capital theory Becker (1964), and Schultz (1961), which posits that individuals possess a range of skills, competencies, and knowledge that can be enhanced through education and training, ultimately contributing to increased organisational productivity (Becker 1964; Schultz 1961).

The Economist Intelligence Unit (2019) suggests that evidence indicates that the competency level of new graduates from HEIs often falls short of meeting the industry's expectations. It may result from inadequate curriculum design, a high staff-to-student ratio, unnecessary programmes, and inefficient use of resources, which can significantly impact the quality of education (Fernandes & Singh 2021). If not addressed adequately, this may substantially affect how HE can contribute to national socio-economic growth, access, equity and inclusion (Koch & Fisher 1998). PHEIs need to be proactive in their strategy to ensure they send out quality individuals into the workplace by consistently assessing and updating their programmes to meet the future needs of industries.

Creating adequate institutional support structures can enhance students' academic preparedness, supporting their aspirations for obtaining a qualification. It can prepare students for the world of work, supporting students' goals and career aspirations. It has the potential to enable students to build social networks and relationships that impact their success and develop the opportunity to become responsible citizens. A culture of inclusivity created by institutional capabilities can enhance the students' ability to embrace diversity and build a strong sense of community and belonging. This supportive environment fosters collaboration and mutual respect among students from various backgrounds, ultimately enriching the educational experience and promoting personal and academic growth.

The institution's human and institutional capabilities must provide the opportunity for frequent student input with adequate feedback. This can create a responsive educational environment that embraces students' individuality and development. It has the potential to develop a collaborative learning environment that meets the students' needs and places their voices at the institution's centre. This can create a sense of belonging, improve students' perceptions of the institution, and contribute to brand capital. It must be acknowledged that students' word of mouth can attract new students to the institution, thereby impacting the institution's bottom line.

2.3 *Industry Requirements*

Duderstadt (2000) argued that a knowledge-driven economy's increasingly sophisticated labour market needs advanced education and training. This is further amplified by the rapid changes brought about by the fourth Industrial Revolution, establishing technology as a key component to be considered in how HEIs develop and present education to meet the demands of the future world of work. Considering the broad diversity and ever-evolving nature of society, with increased market competition, PHEIs will need to demonstrate their ability to adapt and modify their traditional educational services to be more flexible and agile.

Developing a responsive curriculum that resonates with students' aspirations and career goals is important in becoming an institution of choice. This can further enhance students' opportunities for employment and readiness for the world of work. Plewa et al. (2015) argue that HEIs should engage with businesses in their curriculum design, which can directly impact the alignment of the curriculum with business needs. Further, providing students with experiential learning opportunities and mentorship programmes can enhance their capabilities for employment. Allowing students to be part of the community can build social skills that prepare them for the world of work, where relationships are key to building successful organisations. Building student capital will enrich students' educational journey and prepare them for their future endeavours, benefitting both the student and the institution.

The ability of students to complete their qualifications with a set of competencies sought after by industry is shaped by both the institutions and human capital. Fernandes and Singh (2021) argue that HE is necessary to provide quality education that equips students with the skills and capabilities required for the world of work. Understanding students' needs and individuality can help the institution develop strategies that improve student retention and success. At the same time, academics play a crucial role in creating a conducive learning environment that supports students' success throughout their academic journey. As the most valuable asset, the academic team has the potential to produce outstanding students who drive innovation, leading to the transformation of businesses and the creation of innovative new companies (Patil et al., 2023). It can ensure that students are developed with competencies that adequately prepare them for the world of work.

Lastly, acquiring this practical knowledge and skills should have a meaningful and positive effect on the economy and society in which these students function (Brosky 2011). PHEIs need to acknowledge that their role is not only about developing students' competencies but also about how these new competencies will make a difference in the economy and society in which they function. Human and institutional capabilities impact students' academic performance and overall personal development, providing them with an improved chance to gain employment. At the same time, showcasing students' success through various platforms can enhance their motivation and commitment to their studies and the institution. Institutions that enhance student capital are likely to produce well-rounded graduates who can succeed in the world of work.

3 Methodology

3.1 Research Philosophy

The philosophical foundation of the study was informed by both epistemological and ontological perspectives. The research paradigm adopted a positivist and constructivist stance within a case study design. Positivism, as an epistemological stance, employs natural science methods in social research, emphasising objective, observable, and measurable phenomena (Bryman & Bell 2017). In contrast, ontology concerns the nature of reality, with **constructivism asserting that social phenomena and their meanings are continuously shaped through individual and collective interactions** (Bryman & Bell 2017).

3.2 Research Design

Qualitative Research Design. A multiple case study design was employed for the qualitative portion of the study. This design facilitates an in-depth understanding of the meaning participants attribute to their life experiences (Fouche et al., 2021).

Quantitative Research Design. The design for the quantitative research was non-experimental. A non-experimental design is primarily used in descriptive studies, where selected units are measured across relevant variables without the aim of conducting rigorous hypothesis testing (Fouche et al., 2021). The non-experimental design is useful for descriptive purposes and involves collecting and analysing numerical data to identify patterns or trends (Bryman & Bell, 2017). Non-experimental studies are often used to build theories, models, or frameworks.

3.3 Research Approach

A qualitative and quantitative approach was applied, also known as the mixed-methods approach.

Qualitative Approach. According to Bryman and Bell (2017), De Vos et al. (2017), and Fouche et al. (2021), the qualitative approach aims to understand reality through a systematic, interactive, and methodical approach. The qualitative phase employed one-on-one interviews with open-ended questions; experts were free to add any KPIs most suitable for PHEIs in South Africa.

Quantitative Approach. The quantitative research approach was non-experimental. Within a non-experimental design, experts rated the KPIs identified in the qualitative phase of the study, through a net promotor score (NPS) questionnaire using a 10-point Likert scale. The results from the NPSs were then analysed through the analytical hierarchy process (AHP), providing a final ranking of the KPIs based on their priority.

3.4 Population

This study was conducted in the private HE setting in South Africa. The total population consisted of 87 DHET (2024) fully registered institutions that employ approximately 15,857 staff (Sithole, 2024), representing 18% of total student enrolment in South Africa (SAPHE, 2023).

3.5 Sampling

Qualitative Sampling. A non-probability, purposive convenience sampling technique was applied to select 30 experts who assisted in identifying the KPIs through one-on-one interviews with open-ended questions. Including participants (such as expert panels or designated locations) in a study is based on their capacity to provide expert opinions that answer the research question. Experts from six PHEIs in South Africa who have spent significant time in private HE and, in some cases, public HEIs were selected.

The sample of 30 experts who spent a significant part of their career in six PHEIs was selected for the survey. Of the 30 experts selected for the qualitative portion of the study, 70% held a PhD or doctorate, and 30% had a master's degree. This included 3% executive managers, 40% senior managers and 50% professional academics with an average of 12.7 years of experience in PHE and 21.9 years of experience in HE. Many of these participants are recognised for their research contribution and for developing frameworks and policies for quality teaching and assessment. A substantial number of the participants were involved in setting up PHEIs in South Africa or significantly contributing towards the strategic direction of these institutions.

Quantitative Sampling. For the quantitative sampling, 100 experts were selected through a non-probability, purposive convenience sampling technique. These 100 experts were chosen to rate the KPIs identified in the previous step. The sample of 100 experts also spent a significant part of their career in six leading PHEIs and were selected guided by the following inclusion criteria and availability:

Of the 100 experts for the quantitative section of the study, 43% had a PhD or doctorate, 43% a master's degree and 14% an honours degree. The number of years of experience in HE was distributed as follows: 15% more than 15 years, 16% between 11 and 15 years, 28% between 7 and 10 years, and 10% below 6 years.

3.6 Data Collection Method

Qualitative Data Collection. One-on-one interviews with open-ended questions conducted with 30 experts were done to identify a list of KPIs for PHEIs. Due to the open-ended questions, the experts could freely explain and express their opinions, experience and understanding of the KPIs.

Quantitative Data Collection. Data were collected through a cross-sectional questionnaire within a descriptive context with 100 experts.

3.7 Data Analysis

Qualitative Data Analysis. All the information from the one-on-one interviews was analysed through inductive thematic analysis (ITA) using ATLAS.ti v23. These analyses provided a set of KPIs and rich data that described the KPIs. The ITA framework, which draws on inductive analytical methods that involve identifying and coding emerging themes within data, is also referred to as content analysis (Bryman & Bell 2017).

Quantitative Data Analysis. An analysis of the quantitative data was done through descriptive statistics. Fouche et al. (2021) indicate that descriptive statistics describe numerical data by organising, summarising, and interpreting sample data univariately, looking at the frequency and central tendency.

The data from the NPS questionnaire that provided the rating of the KPIs were analysed through SPSS to establish the reliability of the questionnaire and the mean scores of each KPI. The diagnostic evaluation of the questionnaire indicated the reliability of Cronbach's Alpha, and the descriptive analysis establishes the mean scores for each KPI, as reported in Table 1. Through descriptive statistics, a pairwise comparison was done using the AHP to determine the relative weight of each KPI, as indicated in Table 1. This provided a final ranking of all the KPIs regarding their priority.

The principle of AHP, introduced by operational researcher Saaty, is that it can calculate relative weights (importance) for the different key organisational performance indicators (Saaty 2008). Zhilong (2014) further indicates that AHP is a practical and flexible multi-criteria decision-making method that uses quantitative analysis to analyse a qualitative problem.

Table 1 KIPs ranked by importance

c	KPIs	Mean scores	AHP ranking priority
Student capital	High importance		
	Developing student's skills	9.12	0.954
	Experience a positive learning environment	8.98	0.941
	Student employability	8.87	0.930
	Positive perception of the institution	8.85	0.924
	Average importance		
	Student's ability to do postgraduate studies	8.59	0.896
	Student satisfaction	8.54	0.891
	Quality of students	8.30	0.870
	Student situation, circumstances and needs change	8.14	0.850
	Student individuality and potential	7.86	0.802

3.8 *Validity and Reliability*

Methodological triangulation was achieved using qualitative and quantitative methodology in the study, which included multiple case studies. Data triangulation was achieved through the use of multiple data layers, including document analysis, one-on-one interviews and a survey that included a much larger population sample, capturing experts' perspectives and experiences. Theoretical triangulation consists of a range of theoretical and contextual perspectives to frame the study's topic within its context.

3.9 *Ethics*

All ethical protocols were adhered to during the conduct of this study. Ensuring ethical compliance was a priority throughout the research process to uphold integrity and protect participants.

4 Results and Discussion

Four key performance indicators (KPIs) of significant importance were identified, supporting the conceptualisation of student capital as presented in Table 1. KPI evaluation j.

4.1 Student Capital

Student capital in the context of PHEIs refers to the skills, knowledge, and attributes (competencies) students acquire during their academic journey as supported by the human capital theory (Becker 1964; Schultz 1961). It includes the student's overall satisfaction and perception of the institution. Student capital has a direct impact on a student's academic performance, career opportunities, overall personal development, and potential for success and employability. Kinash et al. (2014) refer to the educational outcome influenced by the institution's capacity to develop students' knowledge, skills and attributes for employment and career success after graduation. It underscores the role of organisational and human capital as enablers for student capital. Ayuk and Jacobs (2018) support this and argue that the institution's capacity can equip students with the competencies to access and sustain employment. Kaplan and Norton (1996) refer to the customer perspective, which includes customer retention, satisfaction and acquisition.

Student Employability and Skills Development. Previous studies by Crouchley and Taylor (2004) and Leiber (2019) highlighted the importance of student employability as a KPI, while Ewell, as cited in Ruben (1999), and Stewart and Carpenter-Hubin (2001), refer to students' placement. The KPI of skills development, as highlighted in this study, is notably not explicitly mentioned in previous studies, which mainly refer to staff's skill development as an important KPI (Beard 2009; Zaki and Zaki Rashidi 2013).

Although earlier studies mentioned students' employability as an important KPI, it is noticeable that later studies do not mention this pertinently as a KPI. This contradicts the latest thinking on the accountability of HEIs to developing students fit for employment, as highlighted in this study and others (Hernandez-Diaz et al. 2020; Kinash et al. 2014; Sahney & Thakkar. 2015). The aim of PHEIs must be to produce quality graduates who are employable and can positively contribute to society and industry.

However, social scientists and education researchers postulate that HEIs may not provide the relevant knowledge and basic skills required to adequately and efficiently position graduates in the world of work and society (Johnson & Stage 2018; Kilgo et al. 2015). MoEYS (2016) confirms the ongoing challenge of student performance, revealing a gap between market needs and students' skill sets.

Participant 8: *Employability: Is he/she actually contributing to the labour market or the profession that he is studying in?*

Participant 10: *I have seen it in so many of the schools where they have this absolute fixation with what quality is, but those are also the schools where we see the lowest employability; I think we missed the point: employability in any education programme must be right up there. Actually, industry and the workplace valuing the type of graduate that you produce, that is the ultimate indicator of quality.*

Academics need to develop students' competencies with a passion and desire to make a difference. In the late 1990s, the Boyer Commission (1998) argued that employability could be improved by integrating theoretical, practical, and self-regulatory knowledge. It is acknowledged by Ahmad and Soon (2015), who argue that HEIs must be able to provide relevant knowledge and skills through their teaching, learning, and research processes and practices. The interconnectedness between the student, HEIs and the world of work is a critical aspect of increasing the relevance of HE (Wall 2017).

Participant 1: *I want to go back to the curriculum because, you know, in a way, there is always something about the curriculum. So, maybe the relevance of the curriculum, the regular review of the curriculum, how you review the curriculum.*

It can be achieved by providing a student-centred education, a positive internal work environment, a favourable staff-to-student ratio and academics with a passion for academia. The knowledge assets of HEIs can improve the quality of education and produce graduates who can actively compete in today's world of work (Yoeung et al., 2017).

Participant 10: *You will then, at the same table, have an academic really believing in teaching and learning and what takes place in the classroom, that means quality because their viewpoint would be, you would deliver somebody to the workplace that can actually do the job. Alignment with the world of work, what I mean by that, is we deliver students that are in demand in the workplace, being student-centred because you need to also attract a student to your, to your, institution and make the student happy to be in the institution.*

HEIs have to better understand the world of work's needs by developing close relationships with industries and businesses that can influence the design of programmes relevant to the market (Yoeung et al., 2017). A recent review confirms the collaboration between HE and industry in the Australian teaching and learning domain, highlighting the need for HE and industry to form a partnership to address the development of skills relevant to the world of work (Bean & Dawkins 2021). A good reputation among employers and within the industry, as well as building long-lasting relationships with students, industry professionals, and the community through community engagement, can contribute to the sustainability of PHEIs.

Participant 6: *Our brand reputation, the one that I did not say, was student employability, and we have now started what we call the graduate desirability surveys, and that is to look at how employers perceive graduates from the institutions, so after you have employed somebody from our institution, what do those employers say about it?*

Gready (2014) argues that students must adequately deal with the complexities and realities of today's life, including high levels of uncertainty and a particular distinctiveness and conflict in values. This requires a conducive relationship between PHEIs and industry to develop students for the future world of work. It can provide a platform to maintain a positive community and industry perception while simultaneously equipping students with the ability to navigate complex changes in the future world of work.

Participant 6: *We start looking at student employability indicators, and I think that will be important. I think I have said to you responsiveness to the twenty-first century and agility.*

Participant 12: *I wanted to say something and that is also linked to academic citizenship: academic citizenship should not only be focused on the university community, it should be focused on the industry also, and that links well to tuition also, so one of the things you should make sure of is that your lecturers are in contact with someone, institutions or parties in the specific industry.*

In an ever-evolving and complex world, it seems more important for students to have a broad set of competencies and graduate attributes to easily transition into the world of work and perform effectively in their workplace. In this regard, An and Loes (2023) indicate that employers seek workers with analytical and practical skills that are transferable across various domains, which are essential for adaptability and innovation.

Participant 6: *I am looking at what are the students, the institution's graduate attributes, and their focus on preparing students for the twenty-first century and the world of work.*

The importance of the graduation rate of students, as a percentage of those who continue their studies and ultimately graduate from the institution, needs to be highlighted in the context of employability. It directly reflects the PHEI's ability to provide quality education and its reputation of excellence, ultimately leading to employability. In this process, it is essential to continuously analyse the pass rate of students and identify reasons why students are not passing to improve support for at-risk students. Additionally, measuring the dropout rate of students is crucial for understanding student satisfaction and making the necessary improvements to retain students.

Participant 26: *A successful institution must have good pass rates for their students; the through-put must be good, of good quality, and there must be employed, there must be high employability of the graduates when they go out, and then also a good private institution also must be able to fulfil the three core responsibilities of a university or higher institution, which is research, teaching and learning and community engagement.*

The graduation rate, employability, and the quality of graduates should be the ultimate measure of success for PHEIs in South Africa. Graduates are extremely concerned about what the job market can offer them and the time it takes to complete a qualification, underscoring the relevance of the qualification for employability and the need for student support during the learning journey.

Participant 7: *I think people do study because they want to get employed, and it is really important that we properly prepare them for the world of work so that they get jobs, but that also industries want to employ our students.*

Student Satisfaction and Experience: A Positive Learning Environment and Perception of the Institution. Student satisfaction is a critical KPI, as highlighted in the current study, and is widely supported by previous studies (Ashraf & Javed, 2012; Gordon & Fischer 2008; Martin & Sauvageot 2011). Student satisfaction supports the value-for-money experience and must be driven by a student-centred approach that acknowledges students' individuality. Satisfied students can positively influence student engagement and retention (Cheah et al. 2023), leading to more productive, confident, and capable students (Alzamel 2014).

The rising expectations of students for quality services and value for money within a student-centric approach have become a key focus point for many HEIs in their quest to provide a unique learning experience and support (Biswas et al. 2022). Students have options in today's competitive academic environment (Chawla & Sharma 2014), and PHEIs must develop a clear strategy to retain and attract students. A consumerist society seeks modern facilities with quality education and a wide range of academic components (Pandey & Solanki 2021).

Participant 7: *I do think feedback from students is critical in understanding their experience and their satisfaction in terms of their studies.*

Participant 13: *I think client satisfaction, obviously, students, we need to know well how happy are students with the current stance of processes, systems and things, overall, I think student satisfaction.*

Biswas et al. (2022) indicate that there is a positive relationship between student satisfaction and HEI performance, which can further contribute to the institution's reputation. A PHEI with a student-centred approach in its strategy, policies, and academic and administrative practices can make a significant difference in how various stakeholders perceive the institution (Camilleri 2021). Besides these efforts, it will require that the staff of PHEIs change their mindset and behaviour to acknowledge the students as paying customers, as this can positively influence the institution's sustainability. In this regard, Donia et al. (2020), Frick et al. (2020) and Gibbons et al. (2015), argue that this can ultimately retain students while at the same time attracting new students.

Participant 10: *Being student-centred because you need to also attract a student to your institution and make the student happy to be in the institution.*

Participant 18: *It is the student, that student-centredness and a student that is really competent and that gets all the assistance, all the necessary support that they can get to enable them to have the knowledge and skills as they enter the work environment; that they can really make a change there, that they can do the work and that they really know what to do, not just a theoretical background that you have given them, but that you have really equipped them to go and do their work.*

Participant 23: *You will never succeed if your students are treated like a number; it needs to be very student-centred.*

HEIs need to establish whether learners are satisfied with their learning environments and engaged in the learning process (Jankowski 2017), supported by the academic teaching attitude, teaching materials, and technology (Guo 2016). This underscores the importance of PHEIs, providing quality education, support services, and an engaging learning environment that is student-centred in all its processes and practices. The emotional connection between students and the institution, motivation from academics, mental support, realistic expectations, and success are all contributing factors to the satisfaction of students (Wilcox & Nordstokke 2020).

Participant 27: *The students that they are producing, whether now when they step into the door of the workplace, they are seen to be having that badge, that label, that says: I am from Environment A and the Environment A is flourishing, is producing the cream of the crop.*

Measuring students' satisfaction and wellness, as well as tracking their alumni's success, is pivotal for the success of PHEIs. The feedback from both students and lecturers, as well as the subsequent positive response from the leadership and management of PHEIs, is critical for the institution's growth. It can substantially improve the retention of students who ultimately graduate from the institution.

It is a well-known phenomenon in HEIs to continually use student satisfaction surveys that will enable the institution to assess the quality of their teaching and the use of resources to create a unique learning experience for students that can ultimately improve student satisfaction (Gannaway et al. 2018; Oliver et al. 2018). Student surveys can also provide valuable information that informs the design and development of innovative curricula, with the potential to produce graduates who are ready for jobs sought after by employers (Gannaway et al. 2018; Wilkins et al., 2012).

Participant 13: *What things do you believe are important that we need to measure in private higher education institutions? So you have mentioned client satisfaction, staff wellness, your student wellness is something that needs to be measured.*

Participant 18: *Doing surveys at the workplaces where students are currently working to determine the satisfaction of employers; it would be the student satisfaction, but also the lecturer's satisfaction with the institution. We get the opinion of everybody, but we have never asked the lecturers what are their opinions and I think it is very important to include them as well.*

Culture significantly influences student and staff satisfaction, including the institute's beliefs, values, and attitudes, and how this influences the behaviour of its employees and students. Culture can significantly affect how students and staff experience the institution. Building a positive organisational climate and culture for PHEIs is crucial, leading to motivated staff and students. A positive culture can further attract students, contributing to the sustainability of PHEIs.

A student-centric culture can facilitate active academic engagement, provide a positive learning experience and equip students with employable skills sought after by employers (Biswas et al. 2022). Creating a student-centric culture will require the voices of students and academics to continually improve the quality offered by PHEIs. The student's voice must be viewed within the context of the HEI as a business, with the student as the primary customer, considering the student as a valued partner in the institution (Canning 2017).

Participant 2: *But if you do not have a healthy organisational culture, then you can do all of that, and it is not going to work.*

Participant 26: *A private higher institution must have a good culture where academics know it is a business, but I feel that they need a workforce that is motivated and enthusiastic and happy to deliver, because even if the lecturers are not doing their jobs or not teaching properly, it is going to show in the results and in terms of marketing the marks will also help; so, if the students are good and happy and the turnout is great there will be more… there will be more students that will be attracted to that institution, and we can say they are successful.*

5 Recommendations

This study presents several strategic recommendations aimed at strengthening student capital, thereby enhancing academic success, engagement, and the overall institutional reputation, with a focus on various stakeholders.

- The student voice must be prominent in the institution.
- Provide a variety of educational practices that support students' individuality.
- Measure students' satisfaction with strategic interventions to address any shortcomings.
- Allow students to co-create the curriculum.
- Developing students' attributes relevant to the world of work.
- Embracing technology as a vehicle to change how education is done and prepare students for this new reality in the world of work.
- Provide students with the opportunity to attain micro-credentials that support their development for the world of work.
- Develop quality graduates who can make a difference in society and industry.
- Re-align their teaching and learning practices to prepare students for a digital future.
- Ensure that students are equipped with industry-desired competencies, enhancing their employment ability.
- Ensure that students experience value for money when engaging with the institution.

6 Conclusion

This study highlights the significance of student capital, which fosters academic excellence in developing students who can make a substantial contribution to organisations and the community as responsible citizens, thereby contributing to the country's socio-economic development.

References

Ahmad, A.R., Soon, N.K.: Balanced scorecard in higher education institutions: What should be considered? In: Proceedings of the International Symposium on Technology Management and Emerging Technologies, Langkawi, 25–27 August 2015, pp. 64–68 (2015). https://doi.org/10.1109/ISTMET.2015.7359002

Alzamel, S.: Factors that influence student satisfaction with international programs in institutions of higher learning: a proposed case study of University of Dayton. International Journal of Global Business. **7**(1), 15 (2014)

An, B.P., Loes, C.N.: Participation in high-impact practices: considering the role of institutional context and a person-centered approach. Research in Higher Education. **64**(4), 520–546 (2023). https://doi.org/10.1007/s11162-022-09715-6

Ayuk, P.T., Jacobs, G.J.: Developing a measure for student perspective on institutional effectiveness in higher education. SA Journal of Industrial Psychology. **44**, a1485 (2018). https://doi.org/10.4102/sajip.v44i0.1485

Badat, S.: Deciphering the meanings and explaining the south African higher education student protests of 2015/16. Pax Academica Afr. J. Acad. Freedom. **1**, 71–106 (2016)

Bean, M., Dawkins, P.: Review of University–Industry Collaboration in Teaching and Learning. Australian Government, Department of Education, Skills and Employment (2021)

Beard, D.: Successful applications of the balanced scorecard in higher education. Journal of Education for Business. **84**(5), 275–282 (2009). https://doi.org/10.3200/JOEB.84.5.275-282

Becker, G.S.: Human Capital: a Theoretical and Empirical Analysis, with Special Reference to Education, 3rd edn. University of Chicago Press, Chicago (1964)

Biswas, K.S., Bose, M., Shams, C.S.: Determinants and consequences of student satisfaction in Australian universities: evidence from QILT surveys. Account. Finance. **63**(1), 1821–1850 (2022). https://doi.org/10.1111/acfi.12930

Boyer Commission: Reinventing Undergraduate Education: a Blueprint for America's Research Universities. Stony Brook, NY (1998)

Brosky, D.: Micropolitics in the school: teacher leaders' use of political skill and influence tactics. Int. J. Educ. Leadersh. Prep. **6**(1), 1–11 (2011)

Bryman, A., Bell, E.: Business Research Methods. Oxford University Press, Oxford (2017)

Camilleri, M.A.: Evaluating service quality and performance of higher education: a systematic review and a post-COVID-19 outlook. International Journal of Quality and Service Sciences. **13**(2), 268–281 (2021). https://doi.org/10.1108/IJQSS-03-2020-0034

Canning, J.: Conceptualising student voice in UK higher education: four theoretical lenses. Teaching in Higher Education. **22**(5), 519–531 (2017). https://doi.org/10.1080/13562517.2016.1273207

Chawla, M., Sharma, T.: Service quality in higher education: a comparative study of management and education institutions. NMIMS Rev. **24**, 59–73 (2014)

Cheah, L.F., Cheng, M.Y., Hen, K.W.: The effect of quality management practices on academics' innovative performance in Malaysian higher education institutions. Studies in Higher Education. **48**(4), 643–656 (2023). https://doi.org/10.1080/03075079.2022.2160702

Crouchley, R., Taylor, J.: Higher education performance indicators: invited comments. J. R. Stat. Soc. A. **167**(3), 497–499 (2004). https://doi.org/10.1111/j.1467-985X.2004.a2054.x

De Vos, A.S., Strydom, H., Fouché, C.B., Delport, C.S.L.: Research at Grass Roots: for the Social Sciences and Human Service Professions, 4th edn. Van Schaik, Pretoria (2017)

Department of Higher Education and Training: Register of Private Higher Education Institutions. DHET, Pretoria (2014)

Dirkse van Schalkwyk, R.: The Impact of Leadership Practices on Service Quality in Private Higher Education in South Africa. Master's thesis,. University of South Africa, Pretoria (2011)

Donia, M., Mach, M., Ronen, S., Brutus, S., O'Neill, T.A.: Student satisfaction with use of a peer feedback system. Acad. Manag. Proc. **2020**, 20759 (2020). https://doi.org/10.5465/AMBPP.2020.20759

Du Toit, J., Verhoef, A.H.: Embodied digital technology and transformation in higher education. Transform. High. Educ. **3**, a52 (2018). https://doi.org/10.4102/the.v3i0.52

Duderstadt, J.J.: A University for the 21st Century. University of Michigan Press, Ann Arbor (2000)

Eager, B., Brunton, R.: Prompting higher education towards AI-augmented teaching and learning practice. J. Univ. Teach. Learn. Pract. **20**(5) (2023). https://doi.org/10.53761/1.20.5.02

Fernandes, J.O., Singh, B.: Accreditation and ranking of higher education institutions: review and recommendations for India. TQM J. **34**(5), 1013–1038 (2021). https://doi.org/10.1108/TQM-04-2021-011

Fouché, C.B., Strydom, H., Roestenburg, W.J.H.: Research at Grass Roots for the Social Sciences and Human Service Professions. Van Schaik, Pretoria (2021)

Frick, H., Birt, J., Waters, J.: Enhancing student engagement in large management accounting lectures. Account. Finance. **60**(1), 271–298 (2020). https://doi.org/10.1111/acfi.12318

Gannaway, D., Green, T., Mertova, P.: Investigating the impact of class size on student evaluations. Assess. Eval. High. Educ. **43**(2), 175–184 (2018). https://doi.org/10.1080/02602938.2017.1317327

Gibbons, S., Neumayer, E., Perkins, R.: Student satisfaction, league tables and university applications. Econ. Educ. Rev. **48**, 148–164 (2015). https://doi.org/10.1016/j.econedurev.2015.07.002

Gordon, T.P., Fischer, M.: Communicating performance in U.S. colleges and universities. J. Public Budget. Account. Financ. Manag. **20**(2), 217–255 (2008)

Gready, P.: First encounters: early career researchers and fieldwork. J. Hum. Rights Pract. **6**(2), 195–200 (2014). https://doi.org/10.1093/jhuman/huu013

Guo, K.: Empirical study on factors of student satisfaction in higher education. Rev. Ibér. Sist. Tecnol. Inf. **11**, 344–355 (2016)

Hernandez-Diaz, P.M., Polanco, J., Escobar-Sierra, M.: Building a measurement system of higher education performance. Int. J. Qual. Reliab. Manag. **38**(6), 1278–1300 (2020). https://doi.org/10.1108/IJQRM-04-2020-0133

Houston, D.: TQM and higher education: a critical systems perspective. Qual. High. Educ. **13**(1), 3–17 (2007). https://doi.org/10.1080/13538320701272672

Jankowski, N.A.: Unpacking Relationships: Instruction and Student Outcomes. American Council on Education, Washington, DC (2017)

Johnson, S.R., Stage, F.K.: Academic engagement and student success. J. High. Educ. **89**(5), 753–781 (2018). https://doi.org/10.1080/00221546.2018.1441107

Kaplan, R.S., Norton, D.P.: The Balanced Scorecard: Translating Strategy into Action. Harvard Business School Press, Boston (1996)

Kilgo, C.A., Sheets, J.K.E., Pascarella, E.T.: High-impact practices and student learning. High. Educ. **69**(4), 509–525 (2015). https://doi.org/10.1007/s10734-014-9788-z

Kinash, S., Crane, L., Schulz, M., Dowling, D., Knight, C.: Improving Graduate Employability. Ireland International Conference on Education, In (April 2014) (2014)

Koch, J.V., Fisher, J.L.: Higher education and total quality management. Total Qual. Manag. **9**(8), 659–668 (1998). https://doi.org/10.1080/0954412988136

Kok, S.K., McDonald, C.: Leadership and management behaviours in high-performing academic departments. Stud. High. Educ. **42**(2), 210–231 (2017). https://doi.org/10.1080/03075079.2015.1036849

Leiber, T.: Performance indicators for higher education institutions. Qual. High. Educ. **25**(1), 76–97 (2019). https://doi.org/10.1080/13538322.2019.1594030

Maharaj, A.: Academic mobility and immigration trends in South Africa. J. Econ. Behav. Stud. **8**(4), 52–66 (2016)

Martin, M., Sauvageot, C.: Constructing an Indicator System or Scorecard for Higher Education. UNESCO IIEP, Paris (2011)

Oliver, B., de St, J., Jorre, T.: Graduate attributes for 2020 and beyond. High. Educ. Res. Dev. **37**(4), 821–836 (2018). https://doi.org/10.1080/07294360.2018.1446415

Pandey, Y.V., Solanki, N.: Student satisfaction towards virtual learning. Ilkogretim Online. **20**(6), 87–93 (2021) https://doi.org/10.17051/ilkonline.2021.06.011

Plewa, C., Galan-Muros, V., Davey, T.: Engaging business in curriculum design. High. Educ. **70**(1), 35–53 (2015). https://doi.org/10.1007/s10734-014-9822-1

Ruben, B.D.: Towards a balanced scorecard for higher education. High. Educ. Forum. **99**(2), 99–115 (1999)

Saaty, T.L.: Decision making with the analytic hierarchy process. Int. J. Serv. Sci. **1**(1), 83–98 (2008). https://doi.org/10.1504/IJSSCI.2008.017590

Sahney, S., Thakkar, J.: Performance assessment of higher education institutes in India. Qual. Assur. Educ. **24**(2), 278–302 (2015). https://doi.org/10.1108/QAE-02-2015-0006

Schultz, T.W.: Investment in human capital. Am. Econ. Rev. **51**(1), 1–17 (1961)

Stander, E., Herman, C.: Quality assurance challenges in south African private higher education. S. Afr. J. High. Educ. **31**(5), 206–224 (2017) https://doi.org/10.28535/31-5-1481

Stewart, A.C., Carpenter-Hubin, J.: The balanced scorecard beyond reports and rankings. Plan. High. Educ. **29**(2), 37–42 (2001)
Wall, T.: A manifesto for higher education, skills and work-based learning. High. Educ. Skills Work-Based Learn. **7**(3), 304–314 (2017). https://doi.org/10.1108/HESWBL-06-2017-0036
Welch, A.: Governance and corruption in east and southeast Asian higher education. Stud. High. Educ. **45**(10), 2073–2081 (2020). https://doi.org/10.1080/03075079.2020.1823643
Wilcox, G., Nordstokke, D.: Predictors of first-year university student satisfaction. Can. J. High. Educ. **49**(1), 104–124 (2020) https://doi.org/10.47678/cjhe.v49i1.188230
Wilkins, S., Balakrishnan, M.S., Huisman, J.: Student satisfaction at international branch campuses. J. High. Educ. Policy Manag. **34**(5), 543–556 (2012). https://doi.org/10.1080/1360080X.2012.716003
Zaki, S., Zaki Rashidi, M.: Parameters of quality in higher education. Int. J. Soc. Sci. Educ. **3**(4), 1098–1105 (2013)
Zhilong, C.: Using Delphi and AHP for kindergarten teacher evaluation. J. Chem. Pharm. Res. **6**(4), 6–13 (2014)

Lecturers' Perspectives on the Use of ChatGPT in Accountancy Education: Opportunities and Threats

Lilian Ifunanya Nwosu, Veruschka Pelser-Carstens, and Anina Marx

Abstract The use of ChatGPT in educational settings often faces resistance due to concerns about academic integrity and the potential misuse of artificial intelligence (AI) technology. Nevertheless, rather than opposing these emerging AI tools, it is critical for educational institutions to consider their adoption, given their ability to enhance the educational experience. The purpose of this study is to explore accounting lecturers' views on the use of ChatGPT in accounting education, with a focus on the potential challenges and opportunities that ChatGPT presents for pedagogical practices. A qualitative research methodology was employed to answer the research questions. Qualitative data were collected through in-depth interviews with purposively selected accounting lecturers from various universities in South Africa. The authors used data triangulation techniques and performed a thematic analysis, which involved stages of data reduction, presentation, and drawing conclusions. The findings reveal that ChatGPT is a widely used AI tool in accounting education, offering weighty advantages in improving the efficiency and accessibility of information for both accounting lecturers and students. Accounting lecturers, in particular, have recognised the ability of ChatGPT to enhance plagiarism detection, increase the precision of academic work, assist students in completing tasks, and provide timely feedback. However, the study also identified several challenges, including ethical dilemmas related to academic integrity, concerns about data accuracy, and the risk of excessive dependence on ChatGPT. The findings of this study offer several implications for accounting lecturers and higher education institutions (HEIs) regarding the effective implementation of AI tools, such as ChatGPT.

Keywords Accounting education · ChatGPT · Lecturers · Opportunities · Teaching and Learning · Threats

L. I. Nwosu (✉) · V. Pelser-Carstens · A. Marx
TELIT-SA, North-West University, Potchefstroom, South Africa
e-mail: lilian.nwosu@nwu.ac.za

A. Marx
e-mail: Anina.Marx@nwu.ac.za

E. H. Redda (ed.), *Proceedings of the Social Sciences International Research Conference, 2025*, https://doi.org/10.1007/978-981-95-9489-4_34

1 Introduction

The advent of artificial intelligence (AI) in educational settings has brought transformative changes, thereby revolutionising how knowledge is delivered (Mohd Amin et al. 2025). Among the most prominent developments in AI is the emergence of advanced state-of-the-art language models, such as ChatGPT (a sibling model to InstructGPT), which has garnered significant attention for its ability to generate human-like text (Roumeliotis and Tselikas 2023). ChatGPT can understand and respond to natural language queries, thereby enabling it to assist with various tasks, including tutoring, content creation, and research (Menon and Shilpa 2023). In the academic setting, research has shown that ChatGPT has the potential to enhance teaching and learning experiences by providing immediate, personalised support and assisting both students and instructors with a wide range of educational activities (Deng et al. 2025; Lo et al. 2024; Menon and Shilpa 2023; Mohd Amin et al. 2025; Naznin et al. 2025; Roumeliotis and Tselikas 2023; Wang and Fan 2025).

However, the integration of ChatGPT in education is not without concerns, given that instructors in fields that require critical thinking and ethical reasoning, such as accounting, have raised concerns about the potential for misuse of these skills (Ballantine et al. 2024; Filasari and Suranto 2025; Henry et al. 2025). While some researchers perceived ChatGPT as a resource to aid students in better understanding complex concepts (Bikanga Ada 2024; Hasanein and Sobaih 2023), others have raised concerns about issues such as academic dishonesty, over-reliance on technology, and the diminishing of essential cognitive skills (Adel et al. 2024; Dempere et al. 2023; Jaleel et al. 2025; Perse 2025). In the South African context, where higher education institutions (HEIs) are constantly adapting to advance technology, understanding the perspectives of lecturers on the use of ChatGPT in accounting education is of utmost importance (Maistry and Singh 2025; Phokoye et al. 2025; Sokhulu et al. 2025). Therefore, it is crucial to examine the perspectives of South African accounting lecturers on the role of ChatGPT in enhancing or hindering the quality of teaching and learning in accounting education, considering both its opportunities and threats.

1.1 Background

Accounting education has long prioritised developing analytical, problem-solving, and ethical decision-making skills among students (Anis 2025; Oben and Van Rooyen 2023). The accountancy field requires students to grasp complex financial concepts, engage in ethical reasoning, and apply their knowledge to real-world situations (Rossouw and Steenkamp 2025). Traditionally, these skills have been achieved through a combination of lectures, case studies, and practical assignments that test the student's ability to think critically and apply what they have learnt.

The integration of ChatGPT in HEIs presents a new dimension in accounting education. On the one hand, ChatGPT can serve as an invaluable supplementary resource, assisting students to quickly access information, clarify their doubts, and deepen their understanding of complex accounting principles (Dosumu et al. 2025). ChatGPT can also support lecturers by automating repetitive tasks, such as grading, providing feedback, answering frequently asked questions, or providing personalised tutoring to students (Naznin et al. 2025; Nwosu 2025). This could lead to a more efficient and effective use of classroom time, as well as increased student engagement.

Conversely, the use of ChatGPT in accounting education may present several challenges. One significant concern is that ChatGPT may undermine the development of critical thinking and problem-solving skills among accounting students (Dosumu et al. 2025). If students rely heavily on ChatGPT-generated responses, they may bypass the rigorous cognitive processes necessary to understand and apply accounting concepts in the real-world context (Filasari and Suranto 2025). Furthermore, ChatGPT could intensify academic integrity, as students might misuse it to complete assignments without engaging with other pedagogical materials such as textbooks and notes (Deng et al. 2025; Lo et al. 2024). Another possible challenge is the ethical concern that may arise regarding the accuracy of ChatGPT-generated content, particularly if students rely on ChatGPT for information without critically evaluating the quality and credibility of the sources (Hasanein and Sobaih 2023; Lo et al. 2024).

Recent research has highlighted the dual nature of ChatGPT's impact on education (Dempere et al. 2023; Jaleel et al. 2025). For example, Jarrah et al. (2023) suggest that while AI tools like ChatGPT can support scholarly writing and improve efficiency, they may also increase plagiarism rates and reduce the authenticity of student work. Equally, Mlambo (2024) emphasises the need for lecturers to guide students in using AI tools ethically, ensuring that they complement instead of replacing cognitive engagement. These observations highlight the importance of examining accounting lecturers' perspectives to understand how ChatGPT can be effectively incorporated into accounting education, considering both its potential benefits and risks.

1.2 Problem Statement

The rapid adoption of AI tools, such as ChatGPT, in educational settings has outpaced the development of comprehensive guidelines and pedagogical strategies, particularly in specialised fields, including accounting (Ballantine et al. 2024; Filasari and Suranto 2025; Henry et al. 2025). Although ChatGPT offers the potential to enhance teaching efficiency, provide personalised learning experiences, and improve access to information, its unregulated use could pose significant challenges to accounting education (Maistry and Singh 2025; Phokoye et al. 2025; Sokhulu et al. 2025). Some of the potential risks include, but are not limited to, academic dishonesty, the erosion of critical thinking skills, and the spread of inaccurate or misleading information (Dosumu et al. 2025; Hasanein and Sobaih 2023; Lo et al. 2024). Despite the growing interest in the use of ChatGPT in education, limited research has been conducted on accounting lecturers' perspectives regarding the integration of ChatGPT into the teaching and learning of accounting, especially in an emerging economy such as South Africa.

The lack of a clear understanding of how ChatGPT is perceived by accounting lecturers hinders the formulation of effective implementation strategies for its use (Haugland Sundkvist and Kulset 2024). Without addressing these concerns, educational institutions can struggle to balance the benefits of AI integration with the maintenance of academic integrity and the development of essential cognitive skills among students. This research problem underscores the need for a targeted examination of the opportunities and challenges associated with the use of ChatGPT in accounting education.

1.3 Research Questions and Objectives

To address the research problem for this study, the following research questions and objectives were formulated, as shown in Table 1.

2 Literature Review

Over the years, accounting education has incorporated specific information and communication technology (ICT) tools into its curricula and degree programmes, often reflecting their practical applications in contemporary accounting practice (Ballantine et al. 2024; Du Toit et al. 2024). For example, South African accounting education has integrated elements of accounting information systems into its curriculum, which requires accredited accounting degree programmes to cover topics

Table 1 Research questions and objectives

Research Question	Research Objective
What are the perceptions of accounting lecturers in South Africa regarding the opportunities presented by integrating ChatGPT into accounting education?	To explore the perspectives of accounting lecturers in South Africa on the potential benefits of incorporating ChatGPT into accounting education
What are the perceptions of accounting lecturers in South Africa regarding the challenges presented by integrating ChatGPT into accounting education?	To identify the perceived challenges associated with the use of ChatGPT in accounting education in South Africa
What strategies do accounting lecturers propose to mitigate potential risks while leveraging the benefits of ChatGPT in education?	To explore strategies proposed by accounting lecturers to mitigate the potential risk associated with the use of ChatGPT in accounting education

related to the use of ICT spreadsheets, expert systems, and audit software (Du Toit et al. 2024; Osmani et al. 2020; Stumke 2023).

ChatGPT has become an increasingly transformative force in higher education, facilitating personalised learning, automating administrative tasks, and enhancing student engagement (Zawacki-Richter et al. 2019). Research has shown that AI-driven adaptive learning systems can tailor content delivery to individual needs, resulting in improved learning outcomes and higher retention rates (Chassignol et al. 2018; Filasari and Suranto 2025; Menon and Shilpa 2023; Mohd Amin et al. 2025; Perse 2025; Roumeliotis and Tselikas 2023). This foundational shift towards AI-enhanced education sets the stage for more specialised applications, including the use of large language models such as GPT-3 and ChatGPT, which excel in generating human-like text, answering domain-specific queries, and providing real-time feedback to students (Brown et al. 2020). In the academic context, ChatGPT has functioned as an on-demand tutor, assisting students in drafting essays and providing answers to multiple-choice questions (Li et al. 2023). Nevertheless, the use of ChatGPT in education is not without challenges, as concerns about plagiarism and the authenticity of AI-generated content continue to provoke academic debate (Hutson 2024; Adel et al. 2024; Mohd Amin et al. 2025).

The implications of ChatGPT in education extend to specialised fields such as accounting, where analytical reasoning, ethical judgement, and the application of complex financial principles are dominant (Du Toit et al. 2024; Neu et al. 2013). Traditional accounting pedagogies, including lectures, case studies, and problem-based learning, aim to foster critical thinking but often struggle with scalability due to large class sizes and delays in personalised feedback (Brivot and Gendron 2013; Dosumu et al. 2025). ChatGPT presents a dual-edged opportunity: while it can automate routine tasks such as grading and generating questions and answers—freeing instructors to focus on higher-order teaching activities (Chasokela and Hlongwane 2025), ChatGPT can provide instant clarifications on accounting principles and simulate real-world accounting scenarios for student practice (Henry et al. 2025). Nevertheless, over-reliance on AI risks undermining students' problem-solving abilities and intensifying concerns about academic integrity (Jones and Prieto 2023). Thus, as AI continues to redefine higher education, its integration must be carefully balanced to preserve pedagogical rigour while harnessing efficiency and adaptability.

3 Theoretical Framework

The theoretical framework employed in this study is the Technology Acceptance Model (TAM) (Figure 1), which posits that perceived usefulness (PU) and perceived ease of use (PEOU) are the primary determinants of technology adoption (Davis 1989).

In this study, the acceptance of ChatGPT will be interpreted using the TAM constructs to understand their Behavioural Intentions (BI) towards its integration (Figure 2).

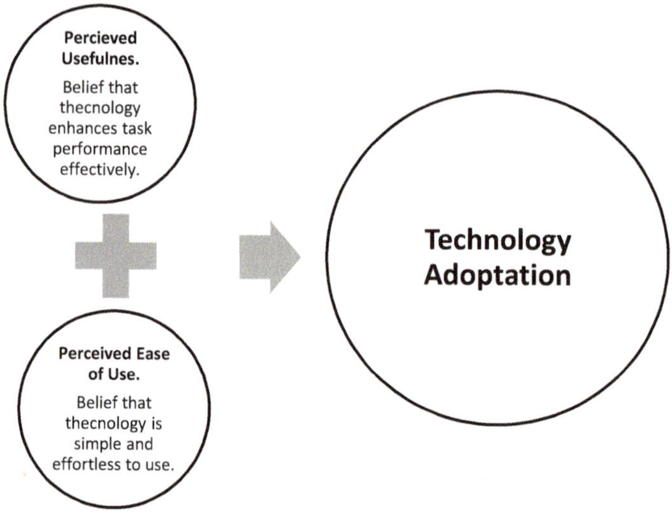

Fig. 1 The TAM for technology adaptation (Davis 1989).

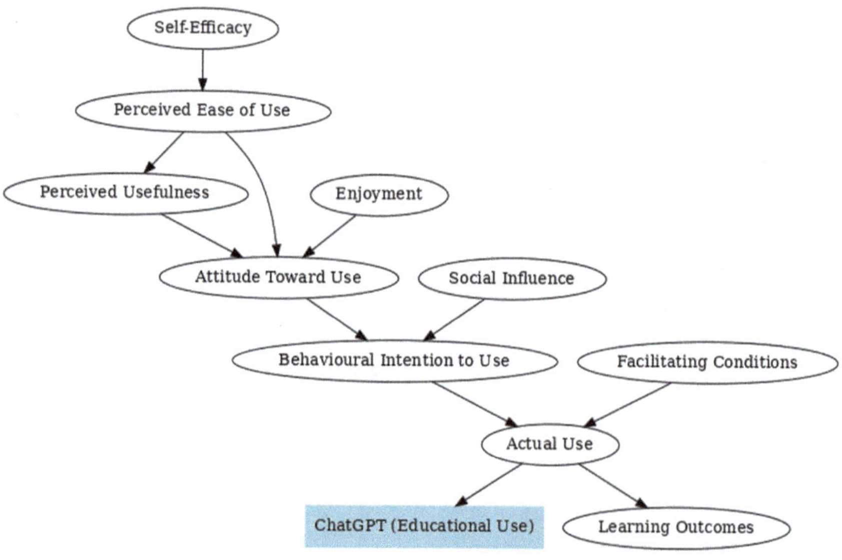

Fig. 2 An expanded TAM conceptual model integrating ChatGPT for educational use (Davis 1989)

In this model, PEOU and PU determine the user's Attitude Toward Use (ATU), which, in turn, influences their BI to use ChatGPT. The intention then drives the actual use of ChatGPT in learning activities. ChatGPT is depicted at the centre of the tool being adopted, mediating between user intentions and educational outcomes. In addition to the core constructs, several other factors relevant to educational technology adoption are incorporated into the model:

1. Self-Efficacy: User confidence in their ability to use ChatGPT; greater self-efficacy leads to higher PEOU and a greater likelihood of adoption.
2. Enjoyment: The fun or pleasure derived from using ChatGPT; higher perceived enjoyment can improve the user's attitude towards ChatGPT (though studies suggest it is less critical than practical benefits).
3. Social Influence: The influence of peers, instructors, or social circles; strong social influence (for example, believing that others expect one to use ChatGPT) can directly increase one's intention to use it.
4. Facilitating Conditions: The availability of resources, support, and infrastructure for ChatGPT; sufficient facilitating conditions (e.g. training, technical support) directly enable the actual use of ChatGPT.
5. Learning Outcomes: The educational results or performance gains from using ChatGPT; effective use of ChatGPT can lead to improved student learning outcomes and academic performance.

The researchers also adopted the ethical framework proposed by McCabe et al. (2012) to assess how AI tools intersect with academic honesty, focusing on the principles of responsibility, fairness, and respect for intellectual property (Figure 3).

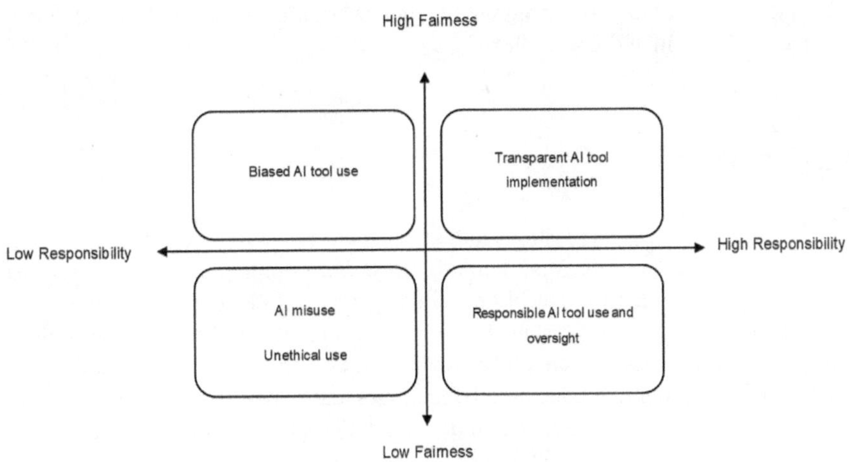

Fig. 3 Ethical framework for assessing AI tools (Peters et al. 2020).

4 Methodology

In this study, a specific research methodology was employed to address both the research questions and objectives.

4.1 *Research Design*

A qualitative design was used to gain in-depth insight into the perspectives (Creswell and Poth, 2018) of accounting lecturers teaching final-year students, who were subsequently interviewed. These lecturers were purposefully selected from the Department of Accounting Sciences at 25 accredited universities. Selection ensures variation in experience, institution type, and familiarity with AI tools (Palinkas et al. 2015). Thus, a total of 25 lecturers were sampled until saturation was reached. Semi-structured interviews (lasting 45-60 minutes each) were conducted via videoconference (Microsoft Teams). The interview guide covered perceptions of benefits, concerns, and mitigation strategies related to the use of ChatGPT. The interviews were transcribed verbatim and analysed using Braun and Clarke's (2006) six-phase thematic analysis, which includes familiarisation, coding, theme development, review, defining, and documentation. Credibility was addressed through member checking, transferability through in-depth description, dependability through an audit trail, and confirmability by triangulating interview data with institutional policy documents (Lincoln and Guba 1985).

5 Findings

As ChatGPT becomes increasingly integrated into educational environments, this paper explores the perspectives of South African accounting lecturers on the use of ChatGPT in accounting education, focusing on the potential opportunities and threats that this integration presents for pedagogical practices. The thematic analysis conducted in this study revealed three main themes: the recognition of the benefits of ChatGPT in accounting education by lecturers, their concerns about the potential threats associated with its use, and their recommended strategies to address the threats related to ChatGPT, as shown in Table 2.

5.1 *Perceived Opportunities*

The findings from the interviews with accounting lecturers reveal that a majority of the lecturers recognise several significant advantages associated with using ChatGPT

Table 2 Interview themes

Themes	Sub-themes
Perceived Opportunities	- Efficiency - Accessibility - Personalised support
Perceived Threats	- Academic integrity - Over-reliance - Data accuracy and referencing
Proposed Mitigation Strategies	- Clear guidelines - Pedagogical integration - Training and awareness

in educational contexts, particularly for accounting students. The findings suggest that ChatGPT has become a vital AI tool, providing substantial benefits in improving efficiency, accessibility, and personalised support for both learners and lecturers. Accounting lecturers have highlighted ChatGPT's ability to improve plagiarism detection, enhance the accuracy of academic work, assist students in completing tasks, and provide prompt responses to questions.

In terms of efficiency, most of the participating accounting lecturers noted that ChatGPT allows for quick and effective access to information. One lecturer stated, *ChatGPT enhances my productivity as a lecturer, particularly when drafting and editing academic documents, assessments, and professional emails.* Furthermore, ChatGPT saves time by offering quick phrasing suggestions and streamlining written communication, which is especially helpful when working under tight deadlines. This highlights the importance of ChatGPT as a significant tool in the learning process, particularly for students struggling with complex accounting concepts. Thus, ChatGPT serves as a reliable resource for clarifying students' understanding of both fundamental principles and more complex accounting topics.

Based on the responses from the participants, the study revealed that accounting lecturers generally view ChatGPT as a valuable and user-friendly tool that improves academic accuracy and efficiency. The findings suggest that ChatGPT offers several advantages for accounting lecturers, particularly in enhancing the engagement of accounting students through appropriate guidance on its use. Specifically, ChatGPT can automate tasks for students, generate personalised feedback, and assist with grading and assessments. This allows accounting lecturers to focus more on student interaction and providing comprehensive instruction. Additionally, students can use ChatGPT as a resource to explore complex concepts, practice their language skills, and prepare for tests.

From an accessibility perspective, most of the participating accounting lecturers observed that ChatGPT plays a crucial role in helping students write, structure arguments, and generate innovative ideas. For example, one of the participants noted, *As English is my second language, I consider ChatGPT to be a readily accessible support tool for me to use. ChatGPT helps improve my academic writing and bridges any language gaps that might otherwise hinder clear and professional communication.*

Participants' responses indicate that ChatGPT fosters innovative thinking that may not have been previously considered. In essence, ChatGPT serves not only as an information retrieval tool but also as a source of inspiration for developing new concepts in academic work, especially assignments (Filasari and Suranto 2025). These findings align with a similar study by Halaweh (2023), which infers that ChatGPT is currently more effective in improving the comprehension of English writing and editing than in generating or synthesising new ideas. Therefore, the accounting curriculum should prioritise student engagement in their own synthesis by placing them in real-world contexts that draw on their unique experiences and perspectives. AI technologies may significantly limit this process.

For personalised support, most participants are of the view that ChatGPT encourages the development of independent thinking skills. For example, one participant remarked, *ChatGPT provides personalised, real-time support. I can receive instant feedback on grammar, tone, and structure, tailored to the context of higher education and accounting discourse, which contributes to my ongoing professional development.* In the digital landscape, where students encounter a multitude of information sources, the ability to navigate and select appropriate information is crucial (Filasari and Suranto 2025). ChatGPT empowers users to search for solutions independently and retrieve information, thereby enhancing their confidence in the learning process. However, it is essential to recognise that while ChatGPT provides information quickly, users should critically evaluate and consult additional resources to ensure accuracy and comprehension.

5.2 *Perceived Threats*

A significant concern in higher education is academic integrity, which encompasses not only the creation of content but also issues related to content privacy and the sharing of information with third-party entities that prioritise business objectives over educational ones (Memarian and Doleck 2023). Although participants recognised the potential benefits of ChatGPT, they highlighted the importance of critical evaluation, responsible use, and appropriate citation practices to maintain academic integrity and minimise the risk of misinformation. However, this perspective also aligns with the view that AI tools like ChatGPT can help improve academic integrity by helping detect plagiarism (Arabyat et al. 2022; Chan et al. 2023; Zhou and Li 2023) and verifying the originality of student submissions (Gavilán et al. 2022; Fajt and Schiller 2025).

Many participants expressed concerns about over-reliance on AI. They mentioned that lecturers who depend on AI to evaluate students' work undermine their professional insights and experiential knowledge. One participant remarked, *There is a risk that both students and some academics may become overly dependent on the tool.* If used naively, this could limit problem-solving, independent writing skills, and critical thinking—essential skills in accountancy. Another participant added,

Excessive reliance on AI can hinder the development of critical thinking and independent problem-solving skills, leading to a passive and 'indoctrinated' learning style. These responses suggest that excessively relying on technology could lead to a decline in critical and analytical thinking skills, which are deemed crucial in the accounting profession. Thus, users need to use ChatGPT carefully. Maintaining academic integrity is crucial in this context; students should revise content or sources obtained from ChatGPT in their own words and with their own understanding and incorporate relevant references as necessary. This practice helps avoid plagiarism and enhances comprehension and communication skills.

For data accuracy and reference, there is significant distrust about the reliability of the references that ChatGPT generates. One participant noted, *In my experience, it sometimes fabricates sources or provides incorrect bibliographic information. This is particularly problematic in academic contexts, where accurate reference is essential for maintaining credibility and integrity.* Another participant expressed concerns about students' potential misuse of ChatGPT, specifically noting plagiarism and unauthorised aid with assessments. They stated, *without proper guidance, it can be used to generate content that students submit as their own, undermining their learning outcomes.* ChatGPT facilitate various forms of plagiarism. Based on the participant's feedback, it can thus be suggested that this tool poses a high academic misconduct risk, as it provides prompts that foster critical thinking. This raises ethical concerns about the integrity of scholarly work. A recent study by Klyshbekova and Abbott (2024) highlighted that ChatGPT raises ethical concerns in educational settings, with a particular focus on academic integrity and plagiarism.

The findings reveal that lecturers hold varying viewpoints on whether using ChatGPT constitutes 'cheating'. Some lecturers believe that using ChatGPT is not dishonest if it is employed as an educational tool. However, if used as a replacement for independent thinking, it can violate academic integrity. Additionally, the lecturers acknowledge the ethical concerns associated with the use of this technology. Thus, it is recommended that ChatGPT be used to enhance understanding instead of providing answers. In summation, while ChatGPT facilitates learning, it is important to recognise that misuse may compromise academic integrity. Therefore, its use in the educational context must be considered carefully (Dwivedi et al. 2023).

5.3 Proposed Mitigation Strategies

The findings reveal that the participants proposed several strategies to mitigate concerns surrounding generative AI in higher education. These strategies include implementing training and awareness opportunities, establishing clear guidelines, and integrating pedagogical approaches. One participant emphasised the need for HEIs to set forth policies that clearly define the acceptable versus unacceptable uses of generative AI tools. Such policies would aid in educating students about these technologies as supportive tools for original thought or academic integrity, rather than replacements.

To mitigate concerns regarding ChatGPT, such as over-reliance, academic integrity challenges, and concerns about data accuracy and referencing, it is recommended that HEIs define ethical standards and improve the transparency of AI applications (Ahmad et al. 2023). However, a potential oversight can involve the construction of ethics and bias, as well as the collaborative determination by humans and AI regarding the criteria used to differentiate between biased and unbiased practices (Alabool 2023; Anders 2023; Bahrini et al. 2023).

For pedagogical integration, it is suggested that ChatGPT be systematically incorporated into instructional methods. For instance, students could be equipped to use AI tools for idea generation, grammar improvement, or study support, while also learning the importance of critically evaluating and verifying their outputs.

For training and awareness, both educators and students alike require guidance on the ethical and practical use of AI tools. This training should cover the limitations of these tools, including issues related to bias and the potential for generating inaccurate content, as well as promoting appropriate academic citation practices and cross-verification of information. Participant feedback suggests a need for educational programmes that prioritise student instruction in the proper and ethical use of ChatGPT. This would enable them to use its capabilities to enhance their learning while maintaining academic integrity (Ali et al. 2024).

Figure 4 illustrates an expanded TAM that embeds the three main themes of a qualitative study on ChatGPT use in accountancy education—perceived opportunities, perceived threats, and proposed mitigation strategies—into the TAM constructs. The classical elements of TAM (PEOU, PU, ATU, BI and AU) are augmented with additional educational factors, including self-efficacy, social influence, facilitating conditions, learning outcomes, and enjoyment (intrinsic motivation). This framework shows how ChatGPT's perceived benefits (e.g. improved efficiency, greater accessibility, personalised support) and perceived risks (e.g. academic integrity concerns, over-reliance, data accuracy issues) influence the acceptance of ChatGPT and how strategic interventions (clear guidelines, pedagogical integration, training & awareness) can mediate negative impacts. In the study, the lecturers found ChatGPT to be highly 'beneficial and user-friendly' for accounting instruction (reflecting high PU and PEOU) while simultaneously expressing concerns about academic integrity, excessive reliance and data accuracy—precisely the balance this model captures.

Extended TAM-based model integrating ChatGPT adoption themes in accountancy education. Perceived opportunities (dotted ellipses), like efficiency, accessibility, and personalised support, positively influence users' perceptions of usefulness and ease of use. These, in turn, improve attitude (ATU) and intention to use ChatGPT. Perceived threats (dotted ellipses)—academic integrity issues, over-reliance on AI and data accuracy concerns—act as negative factors, hindering positive attitudes or outcomes (denoted by dashed negative (–) arrows). For example, a fear of plagiarism or 'cheating' can undermine favourable attitudes towards ChatGPT, and over-reliance on the tool might 'limit the development of critical thinking … leading to a passive learning style', ultimately undermining learning outcomes. Mitigation strategies (dashed ellipses) are linked by dotted arrows to these threat factors, indicating how interventions can moderate or reduce the negative effects. Establishing clear

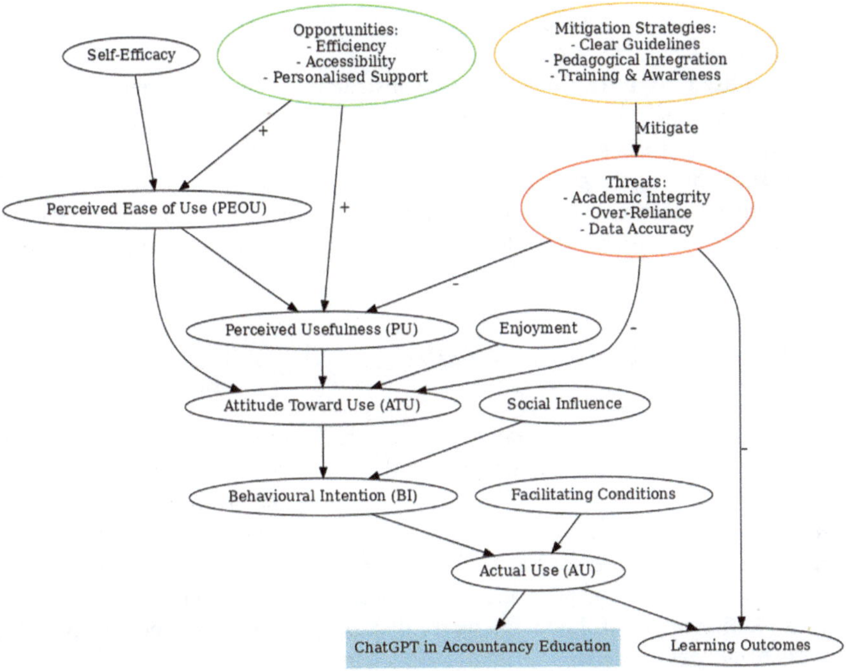

Fig. 4 Extended TAM framework for ChatGPT adoption in accountancy education

guidelines (policies on acceptable use) helps preserve academic integrity, pedagogical integration of ChatGPT into teaching/assessment design prevents over-reliance by ensuring AI is used to complement (not replace) cognitive effort, and targeted training and awareness (for both faculty and students) addresses data accuracy and ethical use by highlighting ChatGPT's limitations, proper citation, and critical evaluation of its outputs (In the figure, '+' and '–'on the arrows denote positive versus negative influence, and 'mitigate' denotes a moderating effect of a strategy on a given threat.)

5.4 Perceived Opportunities Driving Adoption

Efficiency. ChatGPT enables *quick and effective access to information,* helping lecturers and students save time drafting content and finding answers. This productivity boost increases PU; this is where users believe the tool significantly improves their performance (a core TAM driver of adoption).

Accessibility. Participants noted that ChatGPT is a *readily accessible support tool* available on demand, which *bridges gaps* (for example, assisting non-native English speakers with writing). This ubiquitous availability and ease of getting help contribute to PEOU and an inclusive user experience, thus encouraging a positive attitude towards using the technology.

Personalised Support. ChatGPT offers instant, personalised guidance, 'personalised, *real-time support'* with feedback on queries, writing, and problem-solving. This one-on-one assistance can enhance learners' confidence and independence, thereby strengthening the perceived value of ChatGPT as a learning aid. In the TAM terms, these opportunities collectively increase users' perceptions of utility and ease, leading to more favourable attitudes (ATU) and stronger intentions to use the tool.

5.5 Perceived Threats and Challenges

Academic Integrity. A prominent concern among lecturers is that unchecked use of ChatGPT could facilitate plagiarism or *provide immediate answers* that students might pass off as their own work. This ethical dilemma negatively skews the attitude towards use, as using AI may be considered *cheating* or undermining the integrity of learning. The model reflects this with a negative influence on ATU: if users (or institutions) doubt the ethical fit, they will be less inclined to adopt it.

Over-Reliance on AI. Lecturers worry that excessive dependence on ChatGPT could *limit the development of critical thinking, problem-solving, and independent writing skills* in students. Within this framework, this threat is linked to learning outcomes as a negative influence—over-reliance might increase usage in the short term but diminish actual learning gains. In fact, participants cautioned that without careful use, technology could *undermine … learning outcomes* by replacing genuine effort. This highlights that the actual use (AU) of ChatGPT in class must be weighed against its impact on skill development.

Data Accuracy. Many users are *sceptical of ChatGPT's reliability*, noting that *it sometimes fabricates sources or provides incorrect information,* including faulty references. Such accuracy issues erode trust and PU—if the information quality is questionable, users will not find the tool useful for academic work. This is shown as a negative path to the PU in the model. Users must double-check AI output, which adds friction and can also weaken the ease-of-use perception. Overall, these threats can introduce perceived risk that counterbalances the positive drivers of adoption, explaining any hesitancy in BI or constrained AU.

5.6 Mitigation Strategies to Address Threats

Clear Guidelines. Lecturers emphasised the need for institutional policies and honour codes that explicitly define *acceptable* versus *unacceptable* uses of generative AI. By clarifying that ChatGPT is a *supportive tool* (not a replacement for original student work), clear guidelines can alleviate academic integrity fears. In the model, this strategy mitigates the negative impact of integrity concerns on attitude: When users are aware of the 'rules of engagement' for AI, they are more comfortable and positive about using it. For example, making it clear that using ChatGPT for idea generation or grammar help is allowed (with proper citation), but not submitting AI-written answers helps maintain trust in the tool's educational use.

Pedagogical Integration. Rather than banning ChatGPT, participants suggest integrating it into teaching and assessment in structured ways. This could involve designing assignments where students use ChatGPT for specific tasks (like brainstorming or getting feedback) *while also learning to critically evaluate its outputs.* This integration ensures the AI *complements, rather than replaces* students' own thinking. Within this framework, effective pedagogical integration moderates the risk of over-reliance—students learn with the AI but also practise their skills, thereby safeguarding learning outcomes. For example, an assignment might require students to improve a ChatGPT-generated solution or to compare their work with AI's, leveraging the tool to deepen learning rather than shortcut it.

Training and Awareness. Both faculty and students require training to use ChatGPT effectively and an understanding of its limitations. The study's participants call for workshops or guidance that cover AI's potential biases, the fact that it *may generate inaccurate content,* and provide instructions on proper citation and verification of AI-provided information. By improving user self-efficacy with the technology (i.e. their confidence and know-how in using it appropriately), such training can enhance PEOU (people find it easier to use the tool correctly). Training also directly mitigates data accuracy concerns—informed users will double-check references and facts, reducing the chance of blindly trusting incorrect outputs. In the model, this is shown as the training and awareness strategy attenuating the negative link between accuracy issues and PU. Building digital literacy around AI generally empowers users to explore its benefits without succumbing to its pitfalls.

5.7 Extended TAM Elements in the Model

In addition to qualitative themes, the diagram incorporates key TAM extensions relevant to educational settings. Self-efficacy (user confidence in their ability to use ChatGPT) is positioned as an antecedent of PEOU: Higher self-efficacy 'significantly enhances perceived ease of use', making users more likely to engage with the tool. Social influence reflects peer or institutional pressure/support. In studies, it plays a vital role in shaping students' intentions to use AI-based technologies. Thus, if colleagues or authorities endorse ChatGPT, a user's behavioural intention can increase. Facilitating conditions represent the resources and support—adequate resources, training, and technical support—that enable usage. The model illustrates facilitating conditions that feed into AU (and indirectly BI), recognising that even a willing user needs access and infrastructure to actually use ChatGPT (e.g. availability of the tool, internet access, institutional IT support). Enjoyment (or hedonic motivation) is also included: The extent to which using ChatGPT is fun or satisfying can positively influence attitudes and intentions. If students find ChatGPT engaging (e.g.

as an interactive 'study buddy'), this intrinsic motivation can enhance their attitude towards the tool and increase their willingness to use it. Finally, learning outcomes are highlighted as the ultimate educational goal—the model explicitly links AU to learning results. Ideally, if ChatGPT is used appropriately (capitalising on opportunities and mitigating risks), it can enhance learning outcomes (e.g. through efficient practice and personalised feedback). On the contrary, misuse (such as unmitigated threats like plagiarism or over-reliance) can harm learning outcomes. By including learning outcomes, the framework aligns technology acceptance with pedagogical success, emphasising that the true measure of ChatGPT adoption is its impact on student learning.

This integrated model offers a comprehensive view of the acceptance of ChatGPT in accountancy education, illustrating how user perceptions of usefulness and ease of use are shaped by distinct opportunities (efficiency, accessibility, support) and tempered by concerns (integrity, reliance, precision). The traditional TAM sequence, from positive perceptions (PU, PEOU) to favourable attitude (ATU), intention (BI), and actual usage, is thus expanded to account for the dual nature of AI in education: Its capacity to significantly assist learning and the challenges it poses. Inclusion of self-efficacy, social influence, facilitation conditions, and enjoyment acknowledges personal and environmental factors that also mediate the adoption of technology in this context. By visualising where ChatGPT's influence is beneficial and where it is potentially detrimental, the figure underscores the importance of proactive strategies (clear policies, integration into the curriculum, and user training) to maximise opportunities while mitigating the threats. This extended TAM framework can guide educators and institutions in crafting informed approaches to adopt ChatGPT—leveraging its strengths (e.g. improved efficiency and personalised learning support) to enhance educational outcomes, while implementing safeguards to uphold academic integrity and critical skill development. Such a balanced and well-managed adoption of ChatGPT aligns with the dual mandate of innovation and integrity of higher education, ensuring that the technology serves as a valuable learning aid rather than a hindrance in accounting education.

6 Discussion

Research provides valuable insights into the perceived utility, ease of use, attitudes towards use, and behaviour intentions associated with the use of ChatGPT in accounting education. The discussion highlights both the possible advantages and disadvantages of its application.

6.1 Lecturers' Perspectives on Integrating ChatGPT in Accounting Education

In today's technology-advanced era, it showcases the transformative nature of the Fourth Industrial Revolution. The critical convergence of AI, data, and automation has ushered in an era where machines command and engage in interactive learning, resembling human behaviour (Zahrani and Haqque 2024). ChatGPT's ability to provide immediate responses and information, as well as offer intelligent responses, has sparked interest among accounting lecturers. Its ability to produce human-quality text positions it as a valuable tool for content creation (Dwivedi et al. 2023; Fajt and Schiller 2025; Kothari 2023).

Accounting lecturers found that ChatGPT serves as an engaging conversational partner, particularly when exploring unknown information. Although it cannot replace human interaction, it offers diverse perspectives, provides attentive listening, and generates ideas. Its strengths include personalised support, accessibility, efficiency, and timely responses, which help users feel recognised and esteemed. In some contexts, ChatGPT can serve as a neutral mediator, particularly when individuals encounter conceptual challenges or uncertainty about their viewpoints.

Given the subjective nature of balancing constructivist principles with AI assistance, what encapsulates both potential negative and positive outcomes simultaneously represents an opportunity in another when engaging with ChatGPT. Implementing this technology could enhance motivation among accounting students. ChatGPT can be classified as a form of e-learning, as it facilitates learning at any time, reduces costs, offers flexible schedules, and expands the range of available educational resources (Mustofa and Riyanti 2019). It can serve various functions, including assisting with assignment completion and retrieving educational materials. It is widely recognised as a valuable tool in accounting education, supporting students with multiple tasks such as providing sample problems and solutions related to journals, a statement of financial position, and a statement of comprehensive income.

6.2 Opportunities and Threats in Using ChatGPT in Accounting Education

ChatGPT presents numerous opportunities to enhance accounting education by improving the overall learning experience. Research indicates that ChatGPT's capabilities include facilitating personalised and complex learning, designing specific teaching and learning activities, administering assessments, enabling asynchronous communication, and providing feedback mechanisms. It can also enhance the research accuracy, facilitate persona development, support task delegation, and promote cognitive off-loading.

However, the findings highlight several challenges that ChatGPT currently faces or may encounter in an educational context. These challenges include issues related to plagiarism, deception, misuse, inadequate learning, accountability, and privacy concerns. A comparable examination by Memarian and Doleck (2023) confirms these findings, highlighting the potential to support complex and personalised learning experiences, as well as its role in various educational activities and assessments. Their research identifies similar challenges, including misuse, plagiarism, and issues related to accountability.

While ChatGPT offers advancements in identifying instances of plagiarism, elevating the quality of academic work, facilitating students' task completion, and providing timely feedback, the study also highlights significant challenges. These include academic misconduct, concerns about data accuracy, potential for misuse, and over-reliance on technology. Additional studies (Dai et al. 2023; Duha 2023; Elder et al. 2023) suggest that although tools like ChatGPT can critically evaluate student-generated content, they may also foster over-reliance on technology, ultimately undermining students' critical thinking skills and self-assessment abilities.

6.3 *Ethical Implications and Policy Considerations*

Accounting educators have raised a significant concern about the use of ChatGPT, which involves potential ethical dilemmas associated with its application. According to the TAM proposed by Davis (1989), this study's findings suggest that lecturers view ChatGPT as user-friendly and beneficial for accounting education. However, they also expressed concerns about academic integrity, over-reliance on the tool, the accuracy of the data, possible misuse, and overall reliability. The intention to use ChatGPT is closely linked to the willingness or planned decision to adopt this technology in accounting education (Fajt and Schiller 2025).

According to the ethical framework established by McCabe et al. (2012), accounting educators recognise the advantages of ChatGPT in completing academic tasks. However, integrating ChatGPT into educational settings requires a comprehensive examination of the ethical implications, particularly concerning plagiarism. The study's findings suggest that students who actively use ChatGPT are more likely to commit plagiarism and academic dishonesty. Furthermore, students who perceive greater risks associated with AI tend to identify more factors that contribute to plagiarism and regard its consequences as more severe. This highlights the complex ethical considerations related to the use of AI-generated content (Dillon et al. 2024; Macdonald et al. 2023).

These results align with a systematic literature review conducted by Jarrah et al. (2023), which highlights concerns raised in numerous studies indicating that students who use ChatGPT are more likely to submit plagiarised work. This raises ethical concerns related to authorship (Anderson et al. 2023; Bom 2023; Gao et al. 2023; Grassini 2023; Kumar 2023; Mijwil et al. 2023; Perkins 2023; Urban et al. 2024).

This research provides valuable insights into how accounting lecturers perceive the use of ChatGPT in education, which can be better understood when interpreted through the lens of the TAM (Davis 1989). According to the TAM, the adoption of technology is influenced by PU and PEOU, which shape ATU and, ultimately, BI and AU. The study's findings align with this theoretical structure: Lecturers who identified ChatGPT's personalised support, accessibility, and efficiency demonstrated strong associations with PU and PEOU, supporting earlier work that depicts these constructs as critical in determining the adoption of AI tools in higher education (Zawacki-Richter et al. 2019; Chassignol et al. 2018). These positive perceptions were identified in favourable attitudes towards ChatGPT as a supportive educational tool, supporting research by Filasari and Suranto (2025), who emphasise the transformative role of ChatGPT in enhancing students' productivity.

Nonetheless, the findings illustrate how perceived threats, such as concerns about academic integrity, data accuracy, and over-reliance on AI, intersect with TAM by negatively shaping attitudes towards use and perceptions of usefulness. For example, plagiarism and non-existent references impact the value of PU, in line with Hutson's (2024) argument that AI-generated content impacts academic integrity. Similarly, fears of over-reliance suggest that while ChatGPT may initially facilitate AU, it risks diminishing learning outcomes, a construct often considered as an extension of TAM in educational settings (Jones and Prieto 2023). Thus, the threats identified in this research off-set the positive drivers of adoption, confirming the double-sided nature of AI highlighted in previous literature (Mlambo 2024; Memarian and Doleck 2023).

Notably, the mitigation strategies proposed by lecturers, such as clear guidelines, awareness/training, and pedagogical integration, act as enabling conditions that map onto TAM extensions. For example, clear guidelines explain and simplify, while experiential learning aligns with the idea of facilitating conditions that support AU (Venkatesh and Bala 2008). Furthermore, awareness and training initiatives enhance the academic conviction of both lecturers and students, thereby reinforcing PEOU and mitigating the risks of misuse. These findings align with Alabool's (2023) SWOT analysis, which emphasises that structured institutional support is crucial to maximising AI opportunities while mitigating its threats.

This study indicates that lecturers' perspectives are concerned with the practical benefits of ChatGPT and the conditions under which it can be responsibly integrated into accounting education. By locating these insights within the TAM and related extensions, this study infers that ChatGPT adoption depends on balancing perceived opportunities with perceived threats, mediated by institutional strategies that maintain academic integrity while fostering innovation. This discussion builds upon previous work by highlighting the integration of ChatGPT in a South African context, exemplifying both the opportunities and threats of AI, and providing a foundation for future policy and pedagogical decisions.

7 Practical Implications

The study's findings suggest that integrating ChatGPT into accounting pedagogy presents both advantages and disadvantages. ChatGPT offers the potential for a transformative shift in learning, promoting a deeper comprehension of accounting principles and preparing students to meet the evolving nature of the digital era.

These findings have implications for accounting educators and HEIs in effectively using AI tools, such as ChatGPT, for educational contexts. This study contributes to the scant body of knowledge on the application of ChatGPT in accounting education by integrating knowledge and experiences from accounting educators in South Africa. It examines both the potential opportunities and threats of incorporating ChatGPT in teaching and learning, providing a novel approach in the field.

The authors are of the view that the current challenges related to implementing ChatGPT in accounting education present a unique opportunity to move beyond the traditional focus on technical, managerial, and financial aspects that have long defined the discipline. Furthermore, emerging AI technologies have generated a crisis that accounting educators must address to ensure their sustainability in the long-term.

8 Limitations and Future Research

Like most research, this study has limitations that require recognition. Firstly, the sample comprised 25 South African university accounting lecturers who teach final-year students. This may limit the generalisability of the findings to other contexts and populations. Furthermore, relying on self-reported data presents a risk of social desirability bias, where participants may exaggerate positive behaviours and attitudes or minimise negative ones, such as their willingness to engage in plagiarism (Fajt and Schiller 2025).

Another limitation is the study's cross-sectional approach, which captures perceptions and attitudes at a single point in time (Bowen and Wiersema 1999). To address this limitation, future research should use mixed methods approaches that combine quantitative surveys with qualitative interviews or focus groups, incorporating input from accounting lecturers and students. This method would provide a deeper comprehension of integrating ChatGPT into accounting academic settings and improve the generalisability of the findings. Furthermore, conducting longitudinal studies could aid in monitoring potential changes in participants' perspectives and attitudes over time.

9 Conclusion

The discussion on prospects and threats examines perspectives on integrating ChatGPT into accounting teaching. Thus, potential opportunities and challenges, as well as broader implications for pedagogical practices within accounting education were outlined. The findings indicate a significantly positive attitude towards using ChatGPT in educational settings, while also highlighting concerns about its potential for abuse.

The benefits of using ChatGPT in accounting include various applications, such as facilitating personalised and complex learning experiences, designing targeted teaching and learning activities, administering assessments, enabling asynchronous communication, and providing feedback mechanisms. Additionally, it enhances research accuracy, facilitates persona development, supports task delegation, and promotes externalisation. Accounting lecturers recognise the possibility of improving plagiarism detection, enhancing the rigour of academic output, supporting students in completing tasks, and responding in a timely manner.

Conversely, several issues were recognised in implementing ChatGPT in accounting education in South Africa. These issues include plagiarism, misuse, deception, unsatisfactory learning outcomes, and concerns regarding data protection. Profound ethical conflicts arise regarding academic integrity, including concerns about data accuracy, possible misuse, and over-reliance on technology.

Although studies on the impact of AI in accounting education have increased over the past two decades, they have primarily focused on ICTs, including intelligent knowledge systems, expert systems, and decision support systems (Damerji and Salimi 2021; Enholm Papagiannidis Mikalef and Krogstie 2022). This exploratory study focuses on recent advancements in integrating ChatGPT within accounting education, expanding the developing area of scholarship that examines the possible implications of AI on accounting education and practices.

However, integrating ChatGPT into accounting education also raises concerns of academic integrity, such as plagiarism (Klyshbekova and Abbott 2024). The study found that students who regularly use ChatGPT are more likely to engage in academic misconduct. Furthermore, students who perceive risks associated with AI are more likely to identify the factors contributing to academic dishonesty and perceive its consequences as serious. This observation underscores the complex nature of ethical considerations in this context.

References

Adel, A., Ahsan, A., Davison, C.: ChatGPT promises and challenges in education: Computational and ethical perspectives. Educ. Sci. **14**(8), 814 (2024). https://doi.org/10.3390/educsci14080814

Ahmad, N., Murugesan, S., Kshetri, N.: Generative artificial intelligence and the education sector. Computer. **56**(6), 72–76 (2023)

Alabool, H.M.: ChatGPT in education: SWOT analysis approach. In: 2023 International Conference on Information Technology (ICIT), pp. 184–189. IEEE (2023)

Anders, B.A.: Is using ChatGPT cheating, plagiarism, both, neither, or forward thinking? Patterns. **4**(3), 100694 (2023). https://doi.org/10.1016/j.patter.2023.100694

Anderson, N., Belavy, D.L., Perle, S.M., Hendricks, S., Hespanhol, L., Verhagen, E., Memon, A.R.: AI did not write this manuscript, or did it? Can we trick the AI text detector into generated texts? The potential future of ChatGPT and AI in sports & exercise medicine manuscript generation. BMJ Open Sport. & Exerc. Med. **9**(1), e001568 (2023). https://doi.org/10.1136/bmjsem-2023-001568

Anis, A.: Auditing ethics education, moral intensity and accounting students' ethical decision-making: An experimental study. Account. Educ. **1–34** (2025). https://doi.org/10.1080/09639284.2025.2547623

Arabyat, R.M., Qawasmeh, B.R., Al-Azzam, S.I., Nusair, M.B., Alzoubi, K.H.: Faculty members' perceptions and attitudes towards anti-plagiarism detection tools: Applying the theory of planned behavior. J. Empir. Res. Hum. Res. Ethics **17**(3), 275–283 (2022). https://doi.org/10.1177/15562646221078655

Bahrini, A., Khamoshifar, M., Abbasimehr, H., Riggs, R.J., Esmaeili, M., Majdabadkohne, R.M., Pasehvar, M.: ChatGPT: Applications, opportunities, and threats. In: 2023 Systems and Information Engineering Design Symposium (SIEDS), pp. 274–279. IEEE (2023)

Ballantine, J., Boyce, G., Stoner, G.: A critical review of AI in accounting education: Threat and opportunity. Crit.Al Perspect. Account. **99**, 102711 (2024). https://doi.org/10.1016/j.cpa.2024.102711

Bikanga Ada, M.: It helps with crap lecturers and their low effort: Investigating computer science students' perceptions of using ChatGPT for learning. Educ. Sci. **14**(10), 1106 (2024). https://doi.org/10.3390/educsci14101106

Bom, H.S.H.: Exploring the opportunities and challenges of ChatGPT in academic writing: A roundtable discussion. Nucl. Med. Mol. Imaging. **57**, 165–167 (2023). https://doi.org/10.1007/s13139-023-00809-2

Bowen, H.P., Wiersema, M.F.: Matching method to paradigm in strategy research: Limitations of cross-sectional analysis and some methodological alternatives. Strat. Manag. Journal. **20**(7), 625–636 (1999). https://doi.org/10.1002/(SICI)1097-0266(199907)20:7<625::AID-SMJ45>3.0.CO;2-V

Brivot, M., Gendron, Y.: Beyond panopticism: On the ramifications of surveillance in a contemporary professional setting. Account., Organ. Society. **38**(8), 583–594 (2013). https://doi.org/10.1016/j.aos.2011.03.003

Chan, C.K.Y., Lee, K.K.W.: The AI generation gap: Are Gen Z students more interested in adopting generative AI such as ChatGPT in teaching and learning than their Gen X and Millennial generation teachers? Smart Learning Environments **10**. Article. **60** (2023). https://doi.org/10.1186/s40561-023-00269-3

Chasokela, D., Hlongwane, J.: Assessing higher-order and critical skills in the era of artificial intelligence. In: Educational assessments in the age of generative AI, pp. 285–314. IGI Global Scientific Publishing (2025). https://doi.org/10.4018/978-8-3693-6351-5.ch010

Chassignol, M., Khoroshavin, A., Klimova, A., Bilyatdinova, A.: Artificial intelligence trends in education: A narrative overview. Procedia Comput. Sci. **136**, 16–24 (2018). https://doi.org/10.1016/j.procs.2018.08.233

Davis, F.D.: Technology acceptance model: TAM. In: Al-Suqri, M.N., Al-Aufi, A.S. (eds.) Information seeking behavior and technology adoption: Theories and trends, pp. 205–219, IGI Global (1989). https://doi.org/10.4018/978-1-4666-8156-9.ch013

Dempere, J., Modugu, K., Hesham, A., Ramasamy, L.K.: The impact of ChatGPT on higher education. Front. Educ. **8**, 1206936 (2023). https://doi.org/10.3389/feduc.2023.1206936

Deng, R., Jiang, M., Yu, X., Lu, Y., Liu, S.: Does ChatGPT enhance student learning? A systematic review and meta-analysis of experimental studies. Computers & Education. **227**, 105224 (2025). https://doi.org/10.1016/j.compedu.2024.105224

Dosumu, O., Porumb, V.A., Stafford, A., Zimmer, A.: In the wake of ChatGPT: Early reflections on marking open-book online accounting assessments. Account. Education. **1–32** (2025). https://doi.org/10.1080/09639284.2025.2487487

Du Toit, E., Marx, B., Smith, R.: A model to develop integrated thinking skills of prospective professional accountants. Advances in Accounting Education: Teaching and Curriculum Innovations (2024)

Dwivedi, Y.K., Kshetri, N., Hughes, L., Slade, E.L., Jeyaraj, A., Kar, A.K., Baabdullah, A.M., et al.: "So what if ChatGPT wrote it?" Multidisciplinary perspectives on opportunities, challenges and implications of generative conversational AI for research, practice and policy. Int. J. Inf. Manag. **71**, 102642 (2023). https://doi.org/10.1016/j.ijinfomgt.2023.102642

Fajt, B., Schiller, E.: ChatGPT in academia: University students' attitudes towards the use of ChatGPT and plagiarism. J. Acad. Ethics. (2025). https://doi.org/10.1007/s10805-025-09603-5

Filasari, I., Suranto, S.: Transformative effect of ChatGPT on accounting education: A new era AI chatbot. J. Paedagogy. **12**(1), 185–195 (2025). https://doi.org/10.33394/jp.v12i1.13709

Gao, C.A., Howard, F.M., Markov, N.S., Dyer, E.C., Ramesh, S., Luo, Y., Pearson, A.T.: Comparing scientific abstracts generated by ChatGPT to real abstracts with detectors and blinded human reviewers. NPJ Digit. Medicine. **6**, 75 (2023). https://doi.org/10.1038/s41746-023-00819-6

Gavilán, J.C.O., Díaz, D.Z., Huallpa, J.J., Cabala, J.L.B., Aguila, O.E.P., Puma, E.G.M., Vasquez-Pauca, M.J., Mansilla, E.B.R., Laura, P.A.S., Hoces, W.B.: Technological social responsibility in university professors. Eurasian J. Educ. Research. **100**, 104–118 (2022)

Grassini, S.: Shaping the future of education: Exploring the potential and consequences of AI and ChatGPT in educational settings. Educ. Sciences. **13**, 692 (2023). https://doi.org/10.3390/educsci13070692

Hasanein, A.M., Sobaih, A.E.E.: Drivers and consequences of ChatGPT use in higher education: Key stakeholder perspectives. Eur. J. Investig. Health, Psychol. Educ. **13**(11), 2599–2614 (2023). https://doi.org/10.3390/ejihpe13110181

Haugland Sundkvist, C., Kulset, E.M.: Teaching accounting in the era of ChatGPT: The student perspective. J. Account. Education. **69**, 100932 (2024). https://doi.org/10.1016/j.jaccedu.2024.100932

Henry, T.F., Rosenthal, D.A., Weitz, R.R.: ChatGPT-4: Can it handle real-world accounting cases? J. Emerg. Technol. Account. **1–12** (2025). https://doi.org/10.2308/JETA-2024-028

Hutson, J.: Rethinking plagiarism in the era of generative AI. Journal of Intelligent. Communication. **4**(1) (2024). https://doi.org/10.54963/jic.v4i1.220

Jaleel, A., Aziz, R., Farid, G., Bashir, M.Z.: The impact of ChatGPT on academic integrity in medical education: A developing nation perspective. Front. Educ. **10** (2025). https://doi.org/10.3389/feduc.2025.1554444

Jarrah, A.M., Wardat, Y., Fidalgo, P.: Using ChatGPT in academic writing is (not) a form of plagiarism: What does the literature say. Online J. Commun. Media Technologies. **13**(4), e202346 (2023). https://doi.org/10.30935/ojcmt/13572

Jones, A., Prieto, L.P.: The double-edged sword of AI in education: Efficiency vs. critical thinking erosion. Computers & Education. **194**, 104700 (2023). https://doi.org/10.1016/j.compedu.2022.104700

Klyshbekova, M., Abbott, P.: ChatGPT and assessment in higher education: A magic wand or a disruptor? Electron. J. E-Learning. **22**(2), 30–45 (2024). https://doi.org/10.34190/ejel.21.5.3114

Kumar, A.: Analysis of ChatGPT tool to assess the potential of its utility for academic writing in biomedical domain. Biol. Eng. Med. Sci. Reports. **9**(1), 24–30 (2023). https://doi.org/10.5530/bems.9.1.5

Lo, C.K., Hew, K.F., Jong, M.S.Y.: The influence of ChatGPT on student engagement: A systematic review and future research agenda. Computers & Education. **219**, 105100 (2024). https://doi.org/10.1016/j.compedu.2024.105100

Maistry, S.M., Singh, U.G. : Faculty perspectives on the role of ChatGPT-4.0 in higher education assessments. J. Educ. (Univ. Kwa Zulu-Natal) **98**, 86–102 (2025). https://doi.org/10.17159/2520-9868/i98a05

Memarian, B., Doleck, T.: ChatGPT in education: Methods, potentials, and limitations. Comput. Hum. Behav.: Artif. Humans. **1**(2), 100022 (2023). https://doi.org/10.1016/j.chbah.2023.100022

Menon, D., Shilpa, K.: "Chatting with ChatGPT": Analysing the factors influencing users' intention to use OpenAI's ChatGPT using the UTAUT model. Heliyon. **9**(11), e20962 (2023). https://doi.org/10.1016/j.heliyon.2023.e20962

Mijwil, M.M., Hiran, K.K., Doshi, R., Dadhich, M., Al-Mistarehi, A.H., Bala, I.: ChatGPT and the future of academic integrity in the artificial intelligence era: A new frontier. Al-Salam J. Eng. Technology. **2**(2), 116–127 (2023). https://doi.org/10.55145/ajest.2023.02.02.015

Mlambo, P.B.: Exploring the adoption of ChatGPT in higher education: A case of lecturers in a university of technology in KwaZulu-Natal, South Africa. In: EDULEARN24 Proceedings, pp. 1826–1832. IATED (2024)

Mohd Amin, M.R., Ismail, I., Sivakumaran, V.M.: Revolutionising education with artificial intelligence (AI)? Challenges and implications for open and distance learning. Soc. Sci. and Hum.Ities Open. **11**, 101308 (2025). https://doi.org/10.1016/j.ssaho.2025.101308

Naznin, K., Al Mahmud, A., Nguyen, M.T., Chua, C.: ChatGPT integration in higher education for personalised learning, academic writing, and coding tasks: A systematic review. Computers. **14**(2), 53 (2025). https://doi.org/10.3390/computers14020053

Neu, D., Warsame, H., Pedwell, K.: Exploring the role of accounting in the construction of ethical subjects. Crit.Al Perspect. Account. **24**(2), 141–157 (2013). https://doi.org/10.1016/j.cpa.2012.09.004

Nwosu L.I.: AI Proficiency and accounting exam papers. In Artificial intelligence and accounting 2025 Aug 8 (pp. 137–155) (2025). Routledge.

Oben, J.A., Van Rooyen, A.: Social cognitive career theory and rural high school learners' intentions to pursue an accounting career. Account. Education. **32**(4), 445–478 (2023). https://doi.org/10.1080/09639284.2022.2076564

Osmani, M., Hindi, N., Weerakkody, V.: Incorporating information communication technology skills in accounting education. Int. J. Inf. Commun. Technol. Education. **16**(4), 100–110 (2020). https://doi.org/10.4018/IJICTE.2020100107

Perkins, M.: Academic integrity considerations of AI large language models in the post-pandemic era: ChatGPT and beyond. J. Univ. Teach. and Learn. Practice. **20**(2), Article 7 (2023). doi:https://doi.org/10.53761/1.20.02.07

Perse, R.J.P.: The pros and cons of ChatGPT: Incorporating AI literacy into the curriculum. Preprints. (2025). https://doi.org/10.20944/preprints202504.1014.v1

Peters, D., Vold, K., Robinson, D., Calvo, R.A.: Responsible AI: Two frameworks for ethical design practice. IEEE Trans. Technol. Soc. **1**(1), 34–47 (2020). https://doi.org/10.1162/99608f92.8cd550d1

Phokoye, S.P., Dlamini, S., Mthalane, P.P., Luthuli, M., Moyane, S.P.: A comprehensive review of ChatGPT in teaching and learning within higher education. Informatics. **12**(3), 74 (2025). https://doi.org/10.3390/informatics12030074

Rossouw, M., Steenkamp, G.: Developing the critical thinking skills of first-year accounting students with an active learning intervention. Int. J. Manag. Educ. **23**(1), 101086 (2025). https://doi.org/10.1016/j.ijme.2024.101086

Roumeliotis, K.I., Tselikas, N.D.: ChatGPT and OpenAI models: A preliminary review. Futur Internet. **15**(6), 192 (2023). https://doi.org/10.3390/fi15060192

Sokhulu, L.H., Zulu, M.W., Lott-Naidoo, D.: Artificial intelligence in higher education in South Africa: The role of ChatGPT in students' learning J. Educ. (Univ. KwaZulu-Natal) **99**, 142–166 (2025). doi:https://doi.org/10.17159/2520-9868/i99a08

Stumke, O.: Information communication technology integration via Bloom's taxonomy in the accountancy curriculum. Transform. High. Educ. **8**, a283 (2023). https://doi.org/10.4102/the.v8i0.283

Zou, L., Huang, L., Yang, X.: ChatGPT and academic writing: Opportunities and threats. J. Acad. Writ. **13**(1), 101–118 (2023). https://doi.org/10.18552/joaw.v13i1.995

Zuo, L., Liu, P., Wang, J.: The influence of ChatGPT on academic integrity: Evidence from Chinese universities. Asian Educ. Dev. Stud. **13**(4), 512–529 (2023). https://doi.org/10.1108/AEDS-05-2023-0143

Income Tax Treatment of Assessed Losses: A Three-Way Comparative Analysis Based on the Fundamental Principles of Good Tax Policy

Cayla Vijoen and Herman Albertus Viviers

Abstract More restrictive tax treatment was recently imposed that prohibits South African corporate taxpayers to fully use their balances of assessed losses against their taxable income. This paper evaluates this restrictive tax treatment and its predetermined policy objectives against the fundamental principles of good tax policy. The restrictive South African income tax treatment of corporate taxpayers' assessed losses is found to lack in its adherence to the fundamental principles of taxation and in meeting its policy objectives. These outcomes are conceptually analysed and compared against that of other jurisdictions. A combination of a literature review and doctrinal analysis is employed to identify and conceptualise the fundamental principles of taxation and the income tax treatment of assessed losses in the hands of South African corporate taxpayers. The findings are then qualitatively analysed and evaluated in terms of how this tax treatment reflects the fundamental principles of taxation and achieves its policy objectives. A comparative analysis is applied to compare misalignments identified with their tax treatments in the United Kingdom and Canada. The evaluation shows that the limitation of assessed losses in the hands of South African corporate taxpayers does not fully reflect the fundamental principles of taxation and its policy objectives. The paper recommends that the tax treatments and practices in the United Kingdom and Canada serve as valuable guidance for proposed amendments and improvements to the South African corporate income tax context. This paper contributes towards better aligning the income tax treatment of corporate taxpayers' assessed losses in South Africa with the fundamental principles of taxation.

Keywords Assessed loss · Balance of assessed loss · Company · Fundamental principles of taxation · Income tax · Section 20 · South Africa

C. Vijoen (✉) · H. A. Viviers
North-West University, WorkWell Research Unit, Potchefstroom, South Africa

H. A. Viviers
e-mail: herman.viviers@nwu.ac.za

E. H. Redda (ed.), *Proceedings of the Social Sciences International Research Conference, 2025*, https://doi.org/10.1007/978-981-95-9489-4_35

1 Introduction

Section 20(1)(a) of the South African Income Tax Act No. 58 of 1962 ('the Act') was amended for tax years ending on or after 31 March 2023. A limitation was placed on corporate taxpayers' use of their balance of assessed losses carried forward from a previous tax year (National Treasury 2022). This limitation entails that the offset of a qualifying balance of assessed losses carried forward to a current tax year is limited to the higher of either R1 million or 80% of the amount of the current year's taxable income determined before any balance of assessed loss from a prior tax year is considered (South Africa 1962).

The term "assessed loss" is defined in an income tax context under section 20(2) of the Act as: "…any amount by which the deductions admissible under section 11 exceeded the income in respect of which they are so admissible". The unused portion of an assessed loss of the preceding tax year that meets the requirements to be carried forward to a subsequent tax year(s), is referred to as the "balance of assessed loss". Although the latter term is not formally defined or described in the Act, it was used and described as per the latter description in the CIR v Louis Zinn Organization (Pty) Ltd. (1958) case. South African taxpayer companies with years of assessments that ended before 31 March 2023 were allowed to carry forward their full balance of assessed losses from previous tax years for income tax purposes. This effectively meant that such corporate taxpayers were able to offset their full balance of assessed loss against their taxable income, provided that all the requirements for the use of such an assessed loss had been met (KPMG 2021).

Tax statistics published by the National Treasury and the South African Revenue Service (SARS) on 10 December 2024 reveal that 25,4% of all South African taxpayer companies (1,166,692 in total) as assessed by SARS by 31 August 2024 reported an assessed loss position for income tax purposes (National Treasury & SARS 2024). These statistics elucidates the dire negative consequences of the new limitation on the use of assessed losses for the income tax positions and tax liabilities of just over a quarter of all South African corporate taxpayers. The limitation on a company's use of an assessed loss could result in lesser tax savings, which may in time lead to higher cash outflow for companies trying to survive in a struggling South African economy with poor growth (African Development Bank Group 2024).

Navigating weak economic conditions strains any company's ability to generate income and to realise profits. Taxpayers' operations are negatively affected by adverse economic conditions that contribute to larger assessed losses. The fact that assessed losses can reduce a taxpayer's taxable income is perceived as tax beneficial, as it results in an overall reduction in such a taxpayer's income tax liability. Lower income tax liabilities due to the full use of previously suffered assessed losses could help corporate taxpayers grow as it alleviates cash flow pressures, which in turn could contribute to more job opportunities and overall economic growth. Therefore, the new limitation on companies' use of assessed losses could contribute to an even heavier burden on the cash flow positions of South African corporates, many of which are already under severe pressure due to a frail South African economy. Therefore, it is questionable if the recently introduced limitation on the use of assessed losses by South African corporate taxpayers is fair, reasonable and justifiable against the backdrop of the continuous weak economic conditions prevailing in South Africa.

2 Purpose and Contribution

The main research objective of this paper is to evaluate if the income tax treatment of assessed losses in the hands of South African corporate taxpayers reflect the principles of good tax policy when measured against its tax policy objectives. The paper also endeavoured to benchmark the identified misalignments to international income tax treatment under the tax systems of the United Kingdom (UK) and Canada (see the comparative analysis—section 6 for a motivation for the selection of these countries for comparison to South Africa).

A conceptual comparison to the other two countries' tax legislative regulations may inform and guide the South African income tax regulations of assessed losses for corporate taxpayers by showing how it measures up to its international counterparts against the backdrop of international tax policy objectives.

3 Methodology

McKerchar (2008) highlights the social importance of taxation by describing it not only as a discipline, but rather a social phenomenon that could be interpreted through various lenses. This research was positioned within the interpretivist paradigm using a qualitative research strategy. The research followed inductive reasoning to subjectively evaluate whether the restrictive use of corporate taxpayers' assessed losses reflects the fundamental principles of taxation. This evaluation relied on a combination of research methods, namely a literature review and doctrinal analysis, to analyse the income tax treatment of assessed losses in the hands of South African taxpayer companies. This treatment is measured against the fundamental principles of good tax policy in the context of the tax policy objectives with these income tax provisions. The identified misalignments were further benchmarked by way of a conceptual comparison to its income tax treatment in the UK and Canada.

The remainder of this paper is structured as follows: First, a literature review is conducted to identify the characteristics and fundamental principles of a good, sound and well-formed tax policy that is commonly recognised and accepted. This is followed by a conceptualisation and doctrinal analysis of the income tax treatment of assessed losses in the hands of South African corporate taxpayers. Next, this restrictive income tax treatment is evaluated against the fundamental principles of a good tax policy. The identified instances of misalignment between the South African income tax treatment of assessed losses of corporate taxpayers and the fundamental principles of good tax policy are then compared to the income tax treatment of assessed losses in the hands of corporate taxpayers in the UK and Canada. Finally, the paper concludes with recommendations and an acknowledgement of its research limitations.

4 Literature Review

4.1 History behind the more Restrictive Tax Treatment of Assessed Losses

The South African Minister of Finance initially outlined the proposal to introduce a limitation on the use of assessed losses during the 2020 South African National Budget Review. This was followed by the publication of the 2021 Draft Explanatory Memorandum on the Taxation Laws Amendment Bill (TLAB) that first indicated a proposed reduction in the South African corporate tax rate (National Treasury 2021a).

Broadening the tax base and reducing the corporate income tax rate was advocated as ways to improve the country's competitiveness and promote economic growth (National Treasury 2021a). The South African government initially proposed to restrict South African corporate taxpayers from offsetting their balance of assessed losses to be carried forward to only 80% of its taxable income. This proposal was motivated by the view that such a limitation will provide the South African government with the fiscal space to lower the corporate tax rate from 28% to the proposed 27% of taxable income (National Treasury 2021a).

This legislative proposal became a reality when the new limitation on the use of assessed losses was enacted as part of the 2021 legislative cycle (National Treasury 2021c). During the 2022 South African National Budget Speech, it was announced that the South African corporate tax rate will be lowered from 28% to 27% for all South African corporate taxpayers with effect from 31 March 2023, and that it will apply to all tax years for companies that end on or follow this date (National Treasury 2022). Aligned with this reduction in the corporate tax rate, the amendments limiting the use of assessed losses suffered by companies during previous tax years also took effect on the same date.

However, the originally proposed rule to limit assessed losses was amended again. This amendment determined that a company's balance of assessed loss could be carried forward to be set off against its current taxable income, however, now not to exceed the higher of R1 million or 80% of such company's taxable income amount during the tax year to which the balance of assessed loss is to be carried forward (National Treasury 2021c).

Limitations on the use of assessed losses are considered a global trend. The new legislative restriction makes South Africa one of approximately 33 countries worldwide where there are no group tax regimes and where assessed losses are being restricted (National Treasury 2021b). The UK amended its tax legislative requirements for companies in 2017 to broadly restrict the offsetting of operating (trading) losses to 50% of "taxable income" exceeding five million British Pound Sterling (GBP) (PricewaterhouseCoopers (PwC), 2021a). The Canadian tax legislation also restricts the offsetting of tax losses, but only in terms of the number of years that such a tax loss is allowed to be carried forward (PwC 2021b).

4.2 Tax Policy Objectives Behind the More Restrictive Tax Treatment

The South African government aimed to improve the country's competitiveness and promote economic growth by restructuring corporate taxpayers' use of assessed losses to reduce the corporate income tax rate (National Treasury 2021a). Despite National Treasury's intent to restructure the corporate income tax system by broadening the tax base and reducing the corporate income tax rate in a revenue-neutral manner, it remains uncertain whether these income tax policy objectives fall within the purview of the fundamental principles of sound tax policy and similar tax systems' treatment of assessed losses. Since the income tax treatment of assessed losses for South African taxpayer companies was not limited for years of assessments ending prior to 31 March 2023, it is questionable if this limitation reflects the principles of good tax policy given the tax policy aims and objectives behind this limitation provision. If not, it is further questionable how this tax treatment compares to international tax legislation and tax policy objectives.

4.3 Fundamental Principles of Good Tax Policy

Smith (1784) famously introduced his four fundamental principles of a good, sound and well-formed tax system in 1776: (i) equity; (ii) certainty; (iii) convenience; and (iv) economic efficiency. These principles are commonly applied as a theoretical framework and measuring instrument in tax and legal research (Du Preez and Stiglingh 2018). Although Smith's maxims of taxation formed an important cornerstone in the development of tax systems globally (Bandelj and Sowers 2010), several different combinations of principles have been considered and applied over the years to stay relevant in light of modern business practices (Alley and Bentley 2005). For the context of this paper, the fundamental principles of taxation reflect the tax policies embedded within the tax system, synonymous to characteristics of a good, sound and well-formed tax system. Table 1 (see below) contains an overview of burgeoning literature that conceptualises and describes the fundamental principles of good tax policy.

Table 1 Conceptualisation of the fundamental principles of good tax policy

Fundamental tax princple	Summary of the meaning, characteristics, and the tax policy objective of each fundamental principle
Equity and fairness *Also referred to as "neutrality"*	• Contributions should be in proportion to the ability to pay according to one's income (Smith 2000). • Tax should produce the right amount of tax at the right time (OECD 1998). • Taxpayers in similar situations should be taxed similarly, linking the concepts of horizontal equity and vertical equity (American Institute of Certified Public Accountants (AICPA) 2017). • The ability-to-pay principle should be encouraged as a clear, reliable, and simple way to calculate equitable income for tax purposes (Davis Tax Committee 2016).
Efficiency *Also referred to as "convenience of payment"*	• Taxes should be levied in a convenient manner (Smith 2000). • Compliance costs for taxpayers and administrative costs for tax authorities should be kept to a minimum (OECD 1998). • Cost of collecting taxes should be minimised for all parties to achieve economic efficiency (AICPA 2017). • Should create sufficient revenue while limiting the impact on the economy (Davis Tax Committee 2016).
Certainty and simplicity	• Taxes and tax laws ought to be certain, clear and simple to comply with (Smith 2000; OECD 1998). • Tax rules should specify when, where and how taxes should be paid (AICPA 2017). • The collection and calculation of taxes should be certain (Davis Tax Committee 2016).
Transparency and accountability *Also referred to as "transparency and visibility"*	• Transparent process of tax collection to ensure that all parties involved could be held accountable (AICPA 2017). • Tax rules and procedures should be transparent and applied consistently (Davis Tax Committee 2016). • Accountability is promoted when policy setting and approval processes are accessible and available to taxpayers (Lombard 2018).

5 Doctrinal Analysis

Limiting South African corporate taxpayers who operate in dire economic conditions from using their full balances of assessed losses, gives rise to the possibility that the income tax treatment of assessed losses under section 20 of the Act may not conform with the fundamental principles of a good tax policy and the objectives it aims to achieve. The next part of this paper examines the income tax treatment of assessed losses in the hands of South African corporate taxpayers by analysing the relevant sections of the Act, case law, Interpretation Note 33 as issued by SARS (2025), and other commentary on the income tax treatment of assessed losses.

5.1 *Income Tax Regulation of Assessed Losses in South Africa*

The income tax treatment of assessed losses by South African taxpayers is regulated in terms of sections 11A, 20, 20A and 103(2) of the Act (South Africa, 1962). However, the scope of this research is limited to companies only, excluding companies that carry on mining operations as referred to in section 15 of the Act (South Africa 1962) as special limitation rules of assessed loss treatment apply.

Corporates taxpayers (akin to other South African taxpayers) may use a balance of assessed loss to reduce their taxable income for a specific tax year. An assessed loss may only be set off against taxable income if the requirements as stipulated in section 20 of the Act are met. The term "taxable income" is formally defined in section 1(1) of the Act (South Africa 1962) to mean:

> the aggregate of –-
>
> the amount remaining after deducting from the income of any person all the amounts allowed under Part I of Chapter II to be deducted from or set off against such income; and
>
> all amounts to be included or deemed to be included in the taxable income of any person in terms of this Act.

This definition clarifies that a South African taxpayer will only be liable for income tax if their income is more than (or exceeds) the amount of deductions allowed during a specific tax year, meaning that excess income is taxed as taxable income. In contrast, if a taxpayer's allowable deductions exceed its income during a specific tax year, the excess will result in an "assessed loss" position (see Introduction).

5.2 *Section 20 of the Act*

Section 20 of the Act is concerned with the income tax treatment of assessed losses. This provision was first introduced to provide income tax relief to taxpayers who are not making taxable income (i.e. profits). In terms of proviso (b) to section 20(1) of the Act (South Africa 1962), taxpayers may not reduce any amount they derived from a source in the Republic by any assessed loss they suffered from a trade carried on outside the Republic during the same/previous year(s) of assessment. The definition of an assessed loss in section 20(1) of the Act refers to the requirement of adhering to a "trade" or "income from trade" to apply the assessed loss. The term "trade" is defined in section 1(1) of the Act (South Africa 1962), and it includes:

> … every profession, trade, business, employment, calling, occupation or venture, including the letting of any property and the use of or the grant of permission to use any patent as defined in the Patents Act or any design as defined in the Designs Act or any trade mark as defined in the Trade Marks Act or any copyright as defined in the Copyright Act or any other property which is of a similar nature.

Based on the definition included in the Act, it is evident that the term "trade" could have a wide scope and interpretation. Any type of enterprise carried on from a business perspective with the intention to generate a profit, would constitute a "trade" for income tax purposes (Burgess v CIR 1993; Sub-Nigel Ltd. v CIR 1948). A South African taxpayer company can only carry forward its assessed loss from the preceding year of assessment (i.e. the "balance of assessed loss") if it carried on a trade during the current year of assessment (SARS 2025). If the company fails to carry on a trade for an entire tax year, it will forfeit the right to carry forward its balance of assessed loss in terms of section 20(1)(a) of the Act (South Africa 1962). This principle was also established in the SA Bazaars (Pty) Ltd. v CIR (1952) case.

Therefore, if a company does not carry on an active trade during a specific tax year, it loses its right to carry forward its balance of assessed loss due the failure to meet the "trade" requirement. Effectively, section 20(1) of the Act (South Africa 1962) has two requirements for companies before an assessed loss may be carried forward, namely that a taxpayer must: (i) carry on a trade during the current year of assessment (described as the "trade" requirement); and (ii) to have derived income from the carrying on of that trade (the "income from trade" requirement).

5.3 Assessed Loss Income Tax Treatment Features Identified from Doctrinal Analysis

Seven key features emerge from the doctrinal analysis of the legislative provisions and the relevant case law, interpretation notes and literature on the income tax treatment of assessed losses in the hands of South African taxpayers. These should be further evaluated against the fundamental tax principles and the predetermined policy objectives behind the introduction of the new assessed loss limitation rule for corporate taxpayers. The seven income tax treatment features are as follows:

Pre-trade expenditure: Companies are not allowed to create an assessed loss position for income tax purposes during any pre-trade tax year since section 11A of the Act (South Africa 1962) defers the use of deductions to be set off against taxable income during the tax year that trade commences. However, any pre-trade ring-fenced portion of allowable deduction can be carried forward under section 11A to a future tax year during which a trade is carried on and income is realised.

Aggregation of trades—Single net figure (taxable income or assessed loss) from all trades: Apart from foreign trades, income and expenditure from various trades carried on by a single company taxpayer are aggregated to establish a single taxable income or assessed loss position for the taxpayer for each tax year. These different incomes and expenses are also grouped together when declared to SARS on the company's income tax return (ITR14) and are not classified according to each type of trade. If the final net outcome of all aggregated amounts results in an assessed loss position, this assessed loss position is measured on its own (once-off) against the income tax regulations to determine the income tax treatment of such single assessed loss figure from all the company's trades.

Balance of assessed loss retention in circumstances where trade temporarily ceased: A company taxpayer will lose/forfeit its balance of assessed loss if it did not carry on a trade for a full year of assessment (SA Bazaars (Pty) Ltd. v CIR (1952)).

Setting off foreign trade assessed losses against local trade taxable income (profits): Any amount derived by a company from a source within SA may not be reduced by any assessed loss/balance of assessed loss the company suffered from a trade carried on outside the Republic during the same year of assessment/any previous years of assessment.

Anti-avoidance on takeover of other loss-making trades: Section 103(2) of the Act (South Africa 1962) provides clear anti-avoidance provisions aimed at all companies to prohibit any company from acquiring an assessed loss (or balance of assessed loss) from other trade(s) via an entity/business takeover with the main or sole aim to reduce its overall income tax liability.

Taxable income levels linked to restrictions on use of full/entire balance of assessed loss: The level of a company's taxable income is not a determining factor for the limitation on the use of the balance of assessed losses to become applicable. Irrespective of the level of the company's taxable income, the assessed loss carried forward will always be limited under section 20(1)(*a*)(i) of the Act. In contrast, the taxable income level of natural person taxpayers first has to be considered before the ring-fencing provisions of assessed losses are triggered in terms of section 20A of the Act (South Africa 1962) since only natural persons taxed at the maximum marginal rate of tax could face ring-fencing of assessed losses.

The revenue authority's guidance to the taxpayer on income tax treatment of assessed losses: SARS (2025) issued additional guidance to company taxpayers by means of Interpretation Note 33 titled: *Assessed losses: companies: the "trade" and "income from trade" requirements.* This interpretation note is currently in its sixth issue and was recently revised on 4 July 2025).

6 Comparative Analysis and Findings

An evaluation of the seven key income tax treatment features of assessed losses against the principles of a good tax policy and the envisioned tax policy objectives, reveals misalignments. Table 2 contains an overview of these misalignments.

The discussion subsequently explains each misalignment and compares it to its tax treatment in the UK and Canada.

The UK and Canada were selected for comparison to South Africa since they both have certain restrictions on the use of assessed losses to protect their tax bases. They are also both members of the Organisation for Economic Co-Operation and Development (OECD), where governments work together to find solutions to mutual challenges, develop global standards, share their experiences, and identify the most appropriate practices to promote improved policies for better lives (Bainbridge 2000). In 2007, South Africa became one of the five key partners of the OECD that contribute to the OECD's work on issues of fiscal policy (OECD 2018). South Africa, the UK and Canada are considered comparable as they all evaluate their fiscal policies with the aim to promote and strengthen inclusive growth by weighing their policies against the standards established and discussed in the OECD and with their international members and partners.

Table 2 Misalignments identified

Assessed loss income tax treatment features seen from the doctrinal analysis	Fundamental tax principles			
	Equity and fairness/ neutrality?	Efficiency/ convenience of payment?	Certainty and simplicity?	Transparency and accountability/ visibility?
Pre-trade expenditure	No	No	No	Yes
Aggregation of trades	Yes	Yes	Yes	No
Retention of losses in instances where trade temporarily ceases	Yes	Yes	No	No
Setting off foreign trade assessed losses	No	No	No	No
Anti-avoidance provisions on the takeover of other loss-making trades	Yes	Yes	Yes	No
Taxable income levels linked to the use of assessed losses	Yes	Yes	Yes	No
Guidance from the revenue authority on the income tax treatment of assessed losses	Yes	Yes	Yes	Yes

The UK's tax system is considered an important benchmark and comparative measuring instrument when analysing South African tax policy as the UK formed the cornerstone for the tax models of many former British colonies, including South Africa (Mostert 2021). South Africa was a British colony until 1961, with its tax legislation borrowing principles from English law. A survey performed in the UK in mid-February 2024 showed that 26% of actively trading businesses in the UK has experienced a decrease in turnover (Office for National Statistics 2024). It is therefore evident that businesses in the UK are also experiencing a decline in their business profits and financial strains as an aftershock of the COVID-19 pandemic, with more businesses showing lower confidence in their ability to survive (Office for National Statistics 2021). A study published in 2021 analysed the impact of the COVID-19 pandemic on Canadian businesses' revenue. In comparison to 2019, 31% of enterprises in Canada experienced a loss in revenue of 30% or more during 2020, while 29,5% reported a 1% to less than 30% decline in revenue (Statista Research Department 2022).

Pre-trade expenditure: An evaluation of the fundamental tax principle of *transparency* against the income tax treatment of pre-trade expenditure shows that the tax treatment seems to be misaligned. The value/worth of the pre-trade expenditure amounts could be lower by the time it is eventually allowed to be set off against income during a future tax year in which trade commences or during a tax year when sufficient trade income is generated. Transparency is diluted due to the time value of money decreasing over time. This negatively affects the corporate taxpayer.

The UK only compensates for the risk of a loss in value due to the time value of money in respect of natural person taxpayers. They may make use of the "early trade loss relief" under section 72 of the UK Tax Act (United Kingdom 2007) in respect of losses incurred in the first four tax years of trading. This is not available to UK companies. The tax treatment in Canada is regarded as more reflective of the principle of *transparency* since Canada compensates for the risk of a loss in value by allowing all taxpayers to carry back their losses for a period of up to three years (therefore a retrospective use of losses) in terms of subsection 111(1) of the Canadian Tax Act (1985, c. 1). This enables taxpayers to use losses incurred during a current tax year at its current value to be set off against income generated during a previous tax year. Canadian taxpayers receive an immediate cash refund of previously paid taxes.

Aggregation of trades—Single net figure (taxable income or assessed loss) from all trades: The reflection of the principles of *equity and fairness*, *efficiency* as well as *certainty and simplicity* were found to be lacking and misaligned when evaluating the differences in the income tax treatment of South African taxpayer companies and natural persons. Companies are allowed to aggregate all its expenses and income from different types of trade to derive a single taxable income or assessed loss. For natural persons, the aggregation of amounts from all types of trades is not allowed and each type of trade must be kept separate to determine either a taxable income or assessed loss for each type of trade.

These tax treatments in South Africa are similar for companies and natural persons in both the UK and Canada. It is therefore concluded that the fundamental principles of *equity and fairness*, *efficiency* as well as *certainty and simplicity* regarding the aggregation of income and expenses in terms of different types of trades are non-reflective when evaluated in an international context.

Retention of losses in instances where trade temporarily ceases: The tax principles of *equity and fairness* as well as *efficiency* were found to be poorly reflected and misaligned with respect to the differences in the income tax treatment of the retention of the carry forward of balances of assessed losses. Companies will forfeit their right to retain their balance of assessed loss from a prior tax year in instances where they did not carry on a trade for a full tax year. In comparison, a natural person's right to carry forward the balance of assessed loss is not affected by whether or not trade temporarily ceased for a full tax year.

In contradiction, the UK does allow companies who cease trading to have access to the "Terminal Loss Relief". This relief allows a company to carry back the full trading loss amount incurred over the last preceding tax year in which trade was carried on, to be set off against the profits realised during the three preceding tax years. The order of the relief provided is first applicable to the most recent of the three tax years in which a profit was realised. This relief is also subject to the condition that the company still carried on a trade during those three preceding tax years. This tax treatment is more reflective of the fundamental principles of *equity and fairness* as well as *efficiency.*

Canada also allows company taxpayers to only use any previously unused loss subject to their business being conducted with the aim to realise profits, or with a reasonable expectation of realising profits over a particular tax year. Also, this loss can be carried forward for a maximum period of 20 years and/or can be carried back for a limited period of three years. Consequently, this tax treatment is considered to better reflect the fundamental tax principles of *equity and fairness* as well as *efficiency* when compared to that of South Africa

Anti-avoidance provisions on the takeover of other loss-making trades: The principles of equity and fairness, efficiency as well as certainty and simplicity were found to be poorly reflected and misaligned when evaluating the anti-avoidance measures in place to prevent taxpayers from using an assessed loss acquired from another person via a restructuring or business takeover transaction. The Act (Republic of South Africa 1962) contains a specific anti-avoidance provision aimed only at companies (namely, section 103(2)), while no specific provision aimed at natural persons exists. This allows more subjectivity when sections 80A to 80 L of the Act (Republic of South Africa 1962) are applied to natural persons. The application of section 103(2) of the Act could negatively affect companies during periods of weak economic conditions where the need for company restructuring could be a strategy for survival.

These tax treatments in South Africa are in alignment with international practices in both the UK and Canada. The measures also apply to companies and not to natural person taxpayers in terms of Part 14 of the Corporation Tax Act (United Kingdom, 2010, c.4) and subsections 111(4) to (5.5) of the Canadian Tax Act (1985, c. 1), respectively. The fundamental tax principles of equity and fairness, efficiency as well as certainty and simplicity regarding the specific and strict anti-avoidance provisions applicable to company taxpayers compared to the general and more subjective anti-avoidance provisions applicable to natural persons, are not reflected in the international tax treatment of the UK and Canada.

Taxable income levels linked to the use of assessed losses: All four the fundamental tax principles under consideration in this paper were found to be poorly reflected and misaligned when considering the differences in the income tax treatment of assessed losses for companies and natural persons based on different levels of taxable income. It either triggers the restrictions on the use of the balance of assessed loss carried forward by companies, or the ring-fencing provisions to be applied to natural persons.

For UK company taxpayers, the maximum operating (trading) loss that is allowed to be carried forward is broadly limited to GBP5 million plus 50% of the current tax year's profits exceeding the GBP5 million in terms of sections 45, 45A and 45B of the Corporations Tax Act (United Kingdom 2010). These rules apply to all company taxpayers of the UK. The UK Act does not provide for any ring-fencing provisions that target high net worth individuals.

Since Canada does not place any restrictions on or provide for ring-fencing of assessed losses linked to specific levels of taxable income in terms of the number of tax years in which losses may be carried back or forward, no equal comparison can be made for these tax treatments between the three counties.

Guidance provided by the revenue authority on the income tax treatment of assessed losses: Finally, all four the fundamental principles under consideration in this paper were found to be poorly reflected and misaligned when evaluating the scope and relevance of the guidance and practical examples provided to companies as opposed to natural persons on the income tax treatment of assessed losses.

SARS (2025) issued Interpretation Note 33 (Issue 6) to provide guidance to SA company taxpayers on the income tax treatment of assessed losses and this document was recently updated. However, the updated document only contains a single example serving as the only guidance on the new limitation placed on companies in using their full balance of assessed losses carried forward from a previous tax year with no other detailed illustrative examples.

His Majesty's Revenue and Customs (HMRC) does not seem to have issued any guidance or supporting explanatory documents on the tax treatment of assessed losses for either type of taxpayer in the UK. In this instance, the income tax treatment in SA seems to be better than the UK, as SARS has issued a guideline document to assist SA taxpayers in this regard, although outdated.

The Canada Customs and Revenue Agency (CCRA 1997) issued a highly detailed Interpretation Bulletin (IT) 232R3 that provides technical interpretations and positions regarding the tax treatment of losses for Canadian taxpayers. This bulletin document contains practical illustrative examples and guidance on the different types of losses and their tax deductibility in the loss-making year as well as in other tax years. This bulletin document on the tax treatments strengthens the reflection of the fundamental principle of *certainty and simplicity*. Since this guidance addresses losses for both company and natural person taxpayers, it is also more reflective of the fundamental tax principles of *equity and fairness*, *efficiency*, and *transparency and accountability*.

7 Conclusion and Recommendations

There may be room for improvement in the design and application of the South African legislative provisions that regulate the income tax treatment of assessed losses, especially with reference to how it reflects the fundamental tax principles when measured against its tax policy objectives. Based on the comparison with the UK, it is recommended that South Africa consider adopting a mechanism for the carry-back of assessed losses, similar to the UK's approach. This would address the impact of the "time value of money" on assessed losses and their balances, enhancing the effectiveness of the tax treatment in South Africa.

In relation to anti-avoidance measures concerning the acquisition of loss-making businesses, it is recommended that section 103(2) of the Act (South Africa 1962) be repealed. This would only leave sections 80A to 80 L of the Act to apply uniformly to both companies and natural persons and would ensure that the same level of subjectivity is applied when the existence of an "impermissible tax avoidance arrangement" need to be determined. These changes would promote greater *equity, fairness, certainty, transparency*, and *accountability* in the income tax treatment of acquired assessed losses. Finally, it is suggested that the newly introduced restrictions under section 20(1)(*a*) of the Act (South Africa 1962) apply only to companies taxed at the full corporate tax rate of 27%, prior to the carry-forward of any assessed loss. This would align the income tax treatment of companies more closely with that of natural persons, whereby only those individuals subject to the highest marginal tax rate could be affected by the ring-fencing provisions for their assessed losses. Such a regulation would uphold the principles of *equity, fairness*, and *efficiency*, ensuring that taxpayers in similar circumstances are treated similarly. Additionally, it would limit the economic impact by exempting small- and medium-sized businesses classified as "small business corporations" in terms of the Act from assessed loss restrictions, thereby fostering tax relief and stimulating economic growth.

It is acknowledged that this paper is subject to the limitation that the findings were solely and inductively deduced from the views and interpretations of the researchers based on the review and doctrinal analysis. In summary, it is advocated that some of the tax treatments and tax practices employed by the UK and Canada may serve as valuable guidance for proposed amendments and improvements to the South African income tax treatment regulations of assessed losses. A suggestion for future research is to investigate the impact of the restricted use of assessed losses by South African corporate taxpayers by attracting or discouraging foreign investment to South Africa.

References

African Development Bank Group: African Economic Outlook 2024. https://www.afdb.org/sites/default/files/2024/06/06/aeo_2024_-_country_notes.pdf (last accessed 19 June 2025)

Alley, C., Bentley, D.: A remodelling of Adam Smith's tax design principles. Aust. Tax Forum. **20**(4), 579–624 (2005)

American Institute of Certified Public Accountants: Concept statement no. 1 (updated version of March 2001): Guiding principles of good tax policy: a framework for evaluating tax proposals. Healthc. Financ. Manag. **49**(8), 72–78 (2017)

Bainbridge, T.: A brief history of the OECD. OECD Obs. **2000**(2), 1–124 (2000)

Bandelj, N., Sowers, E.: Economy and State: a Sociological Perspective. Polity Press, Cambridge (2010)

Burgess v Commissioner for Inland Revenue 1993 (4) SA 161 (AD)

Canada: Income Tax Act (1985). https://laws-lois.justice.gc.ca/eng/acts/i-3.3/page-1.html (last accessed 19 June 2025)

Canada Customs and Revenue Agency: Income Tax Act: Losses—their deductibility in the loss year or in other years. Interpretation Bulletin IT-232R3. https://www.canada.ca/en/revenue-agency/services/forms-publications/publications/it232r3.html (last accessed 19 June 2025) (2025)

Commissioner for Inland Revenue v Louis Zinn Organisation (Pty) Ltd 1958 (4) SA 478 (A), 22 SATC 85

Davis Tax Committee: Final report on macro analysis on the tax system and inclusive growth in South Africa: Towards an analytical framework. https://www.taxcom.org.za/docs/20160421%20DTC%20Macro%20Analysis%20Final%20Report%20-%20Full%20Report.pdf (last accessed 19 June 2025)

Du Preez, H., Stiglingh, M.: Constructing the fundamental principles of taxation through triangulation. ATA J. Leg. Tax Res. **16**(2), 25–45 (2018)

KPMG: South Africa: Proposal to limit use of losses by companies. https://home.kpmg/us/en/home/insights/2021/11/tnf-south-africa-proposal-to-limit-use-of-losses-by-companies.html (last accessed 19 June 2025)

Lombard, M.: South Africa's sugar tax system: a taxation perspective. Master's dissertation, University of the Free State, Bloemfontein (2018)

McKerchar, M.A.: Philosophical paradigms, inquiry strategies and knowledge claims: applying the principles of research design and conduct to taxation. eJ. Tax Res. **6**(1), 5–22 (2008)

Mostert, T.: An Interpretation of the deeming provisions in legislation in the context of a good tax system: a South African perspective. Doctoral thesis,. Rhodes University, Grahamstown (2021)

National Treasury: Draft explanatory memorandum on the Taxation Laws Amendment Bill, 2021a. http://www.treasury.gov.za/public%20comments/tax%20bills%202021%20draft/2021%20draft%20explanatory%20memorandum%20on%202021%20tlab%20-%2028%20July%202021.pdf (last accessed 19 June 2025) (2021a)

National Treasury: Draft response document on the Taxation Laws Amendment Bill, 2021. http://www.treasury.gov.za/documents/mtbps/2021/Draft%20Response%20Document%20on%20the%202021%20Draft%20Tax%20Bills-10%20November%202021.pdf (last accessed 19 June 2025) (2021b)

National Treasury: Taxation Laws Amendment Act No. 20 of 2021. Government Printer, Pretoria (2021c)

National Treasury: Taxation Laws Amendment Bill, 2022. https://www.sars.gov.za/wp-content/uploads/Legal/Bills/Legal-LPrep-Bills-2022-02-Taxation-Laws-Amendment-Bill-B26%E2%80%942022.pdf (last accessed 19 June 2025)

National Treasury (South Africa), South African Revenue Service: 17th annual edition of tax statistics. https://www.sars.gov.za/media-release/joint-media-statement-national-treasury-and-south-african-revenue-service-on-the-release-of-the-17th-annual-edition-of-tax-statistics/ (last accessed 19 June 2025) (2024)

Office for National Statistics: Business insights and impact on the UK economy, 18 November 2021. https://www.gov.uk/government/statistics/business-insights-and-impact-on-the-uk-economy-18-november-2021 (last accessed 19 June 2025)
Office for National Statistics: Business insights and impact on the UK economy, 22 February 2024. https://www.gov.uk/government/statistics/business-insights-and-impact-on-the-uk-economy-22-february-2024 (last accessed 19 June 2025)
Organisation for Economic Co-operation and Development: South Africa and the OECD. https://www.oecd.org/southafrica/south-africa-and-oecd.htm (last accessed 19 June 2025) (2018)
Organisation for Economic Co-operation and Development: Electronic commerce: Taxation framework conditions. OECD Ministerial Conference, Paris (1998)
PricewaterhouseCoopers: Canada: Corporate—taxes on corporate income. https://taxsummaries.pwc.com/canada/corporate/taxes-on-corporate-income (last accessed 19 June 2025) (2021a)
PricewaterhouseCoopers: United Kingdom: Corporate—income determination. https://taxsummaries.pwc.com/united-kingdom/corporate/income-determination (last accessed 19 June 2025) (2021b)
Smith, A.: An inquiry into the nature and causes of the wealth of nations. Strahan and Cadell, London (1784)
Smith, A.: An inquiry into the nature and causes of the wealth of nations. Modern Library, New York (2000) (original work published 1776)
SA Bazaars (Pty) Ltd v Commissioner for Inland Revenue 1952, 18 SATC 240
South Africa: Income Tax Act No. 58 of 1962. Government Gazette Extraordinary, 29 May (1962)
South African Revenue Service: Interpretation Note 33 (Issue 6): Assessed losses—companies: The "trade" and "income from trade" requirements. https://www.sars.gov.za/wp-content/uploads/Legal/Notes/Legal-IntR-IN-33-Assessed-Losses-Companies-Trade-and-income-from-trade-requirements.pdf (last accessed 19 June 2025) (2017)
Statista Research Department: Business revenue from 2020 compared with 2019 in Canada, by employment size. https://www.statista.com/statistics/1227187/business-revenue-from-2020-compared-with-2019-in-canada/ (last accessed 19 June 2025) (2022)
Sub-Nigel Ltd v Commissioner for Inland Revenue 1948 (4) SA 580 (A), 15 SATC 381
United Kingdom: Income Tax Act 2007 (c. 3). https://www.legislation.gov.uk/id/ukpga/2007/3 (last accessed 19 June 2025)
United Kingdom: Corporation Tax Act 2010 (c. 4). https://www.legislation.gov.uk/id/ukpga/2010/4 (last accessed 19 June 2025)

Tracing the Career Paths of Thuthuka Bursary-Funded Students: A South African Universities' Perspective

Lemlem Amhatsion and Ndivhuwo Sirembe

Abstract The persistent shortage of Black and Coloured Chartered Accountants (CAs) in South Africa remains a pressing concern, even after three decades of democracy. As of 2024, only 20.2 percent of registered CAs are Black, and 5.1 percent are Coloured, highlighting the continuing challenge of achieving demographic transformation within the profession. To address this, the South African Institute of Chartered Accountants (SAICA) launched the Thuthuka programme, a key initiative designed to improve representation by providing funding and comprehensive social as well as emotional support to students. This study explores the career paths of students who benefited from the Thuthuka bursary, with a focus on how such support influences their journey towards becoming CAs (SA). The motivation is to assess the bursary's effectiveness in addressing demographic disparities and in promoting upward mobility for previously disadvantaged individuals. Employing a quantitative research design based on secondary data analysis, the study investigates the career trajectories of bursary recipients between 2013 and 2017, drawing on enrolment and throughput statistics as well as information from the SAICA website. The analysis focuses on students who were expected to have completed their studies and attained professional qualifications by 2025. The findings indicate positive trends in increasing diversity among South African CAs; however, slow growth rates and retention challenges among bursary students suggest that the journey towards transformation remains complex. Continued, targeted interventions are necessary to address academic barriers, improve retention, and support career progression for historically disadvantaged students.

Keywords Thuthuka Bursary · CA(SA) Profession · South Africa · Career Paths · SAICA

L. Amhatsion (✉) · N. Sirembe
School of Accountancy, University of the Free State, Bloemfontein, South Africa
e-mail: AmhatsionLG@ufs.ac.za

E. H. Redda (ed.), *Proceedings of the Social Sciences International Research Conference, 2025*, https://doi.org/10.1007/978-981-95-9489-4_36

1 Introduction

The persistent shortage of Black and Coloured Chartered Accountants (CAs) in South Africa remains a pressing concern, even after three decades of democracy. Despite the country's political transformation in 1994, systemic disparities continue to impede progress. As of 2024, only 20.2 percent (11,924 of 58,962) of registered CAs are Black, and 5.1 percent (3002 of 58,962) are Coloured, highlighting the extent of continuing challenges (SAICA 2024). This is particularly notable given that Black people make up 81.8 percent of the South African population, followed by 8.4 percent Coloured (Stats SA 2024). Considering that these groups constitute the majority of the population, one would expect greater representation of Black and Coloured professionals in the chartered accountancy field, especially given the high regard in which the profession is held both within South Africa and internationally.

During apartheid, Black and Coloured South Africans, now often referred to as 'previously disadvantaged,' faced systemic oppression under white minority rule (Pellicer and Ranchhod 2023). Following the end of apartheid in 1994, the South African government implemented measures to address these historical inequalities, particularly within the education sector, which remains a significant barrier to racial representation in the accounting profession (Mdepa and Tshiwula 2012). Although higher education has been undergoing transformation since 1994, substantial disparities persist, prompting the South African Institute of Chartered Accountants (SAICA) to introduce various initiatives, most notably the Thuthuka project, one of the most important skills development programmes aimed at increasing participation of previously disadvantaged groups (Musundwa and Hammond 2024). A recent study indicates that while progress has been made, more targeted interventions are needed to achieve meaningful change (Musundwa and Moses 2024).

The Thuthuka Bursary Fund (TBF), established by SAICA in 2005 and operating under the Thuthuka Education Upliftment Fund (TEUF), assists disadvantaged Black and Coloured students to become CAs (SA) in order to transform the profession. The Thuthuka project invests in human and institutional capacity development, specifically to address South Africa's critical skills shortages (SAICA 2020). According to SAICA (2020), 'Thuthuka' is a IsiZulu verb meaning 'to develop', reflecting the programme's action-oriented approach to transformation. The TBF offers financial support for three years to full-time undergraduate students pursuing a Bachelor of Accounting (BAcc) or an equivalent qualification, and for one year to postgraduate students enrolled in the Postgraduate Diploma in Chartered Accountancy (PGCA) (UFS SOA 2024), all studying towards becoming CA(SA) at SAICA-accredited universities.

Beyond financial support, students also need social and emotional assistance. Thuthuka bursary programme goes further than providing funding, offering comprehensive support throughout students' academic and professional development. In addition to covering expenses, recipients benefit from valuable non-financial 'wrap-around' support, including academic support through interventions such as additional tutorials, study sessions, study skills workshops, and individual consultations; social and emotional support; work-readiness and life skills training; mentoring and counselling; assistance with administrative matters such as registration and accommodation; and exposure to experienced CAs and business leaders (UFS SOA 2024). All graduates are also given the opportunity to complete internships at reputable companies. The primary goal of the Thuthuka programme is to foster workplace readiness while providing the non-academic support necessary for students to succeed in becoming CA(SA) (Barac 2015).

This study contributes valuable insights by exploring the career pathways of Thuthuka bursary-funded students within South African universities. It highlights how bursary support influences students' progression into the accounting profession, particularly among previously disadvantaged groups. The findings enhance understanding of the factors that facilitate or hinder the successful transition from bursary recipient to professional accountant, providing a foundation for developing more targeted and effective strategies to increase diversity and representation within the CA(SA) community. Ultimately, this study advances knowledge on the impact of bursary programmes on career development and social mobility in the context of South African higher education and professional fields. To explore the career paths of students who benefited from the Thuthuka bursary in their journey towards becoming CAs, this study examines the following objectives:

To determine how the Thuthuka bursary can contribute to transformation and promoting growth in the accounting profession.

To provide an overview of the University of the Free State Thuthuka Bursary Fund Programme.

The presentation of this study consists of five sections. The first provides the introduction. The second reviews relevant literature, examining existing research and theories that underpin the investigation. The third outlines the methodology employed. The fourth presents an interpretation of the empirical findings. The fifth and final section summarises the conclusions drawn from the study, suggests potential avenues for future research, and provides a comprehensive reference list.

2 Literature Review

Extensive research has been conducted to explore the various factors that hinder Black Africans from qualifying as CAs. Carmel et al. (2025) emphasises that systemic barriers, such as financial limitations, unequal access to quality education, and a shortage of professional role models, persist as significant obstacles (Carmel et al. 2025). Barac (2015) highlights that socio-economically disadvantaged students, including those benefiting from the Thuthuka bursary programme, may require tailored support and interventions to succeed within higher education institutions.

2.1 Access to Accredited Institutions

For students aspiring to become CAs, access to accredited universities is a critical first step. In South Africa, SAICA-accredited institutions are largely concentrated in urban centres such as Johannesburg, Cape Town, and Durban, which poses significant challenges for students originating from rural and township areas (Chiramba and Ndofirepi 2023). Limited transportation, financial constraints, and lack of local tertiary institutions often hinder equitable access, perpetuating disparities within the profession. The geographical concentration of SAICA-accredited universities results in an uneven distribution of opportunities for aspiring Black and Coloured students, who are disproportionately represented in rural settings (Davids and Waghid 2016; De Jager and Bitzer 2018).

2.2 A History of Inequality

South Africa's history of racial segregation has left a lasting imprint on the accounting profession. During apartheid, Black and Coloured South Africans faced systemic exclusion from many professional fields, including accountancy. It was only in 1976 that the first Black individual qualified as a CA (SAICA 2024), followed by the first Black woman in 1987 (SAICA 2013). This historical marginalisation has contributed to a significant underrepresentation of Black professionals in the sector. Addressing this inequality remains a challenge, with ongoing efforts to redress past injustices and promote diversity through programmes such as Thuthuka (De Jager 2014; Barac 2015).

2.3 High Failure Risks and Scholarship Pressures

Despite the availability of merit-based scholarships such as Thuthuka, many recipients face significant risks of failure due to the strict conditions attached to the funding. Scholarships typically require students to maintain high academic standards and meet specific performance benchmarks; failure to do so often results in the withdrawal of financial support. This creates intense pressure on students to perform consistently at a high level, often without sufficient support to address personal, academic, or socio-economic challenges.

Recent cases highlight that students can lose their scholarships after failing a year, sometimes due to circumstances beyond their control. For example, Ontong and Smit (2023) reported on a student who lost her Thuthuka bursary after failing an academic year following the death of her father and health complications. Such cases illustrate how personal hardships can jeopardise scholarship eligibility, disrupting educational and career trajectories.

This rigid approach to scholarship management not only increases the risk of academic failure but also imposes significant psychological and financial stress on students, many of whom are already vulnerable due to socio-economic disadvantages (Chiramba and Ndofirepi 2023). The fear of losing financial support can lead to heightened stress, anxiety, and reluctance to seek help, further impairing academic performance and increasing dropout rates Ontong and Smit (2023). As recent research indicates, the high failure risks associated with scholarships like Thuthuka highlight the urgent need for comprehensive support systems, including academic mentoring, counselling, and financial guidance, to help students overcome personal and academic challenges. Without such interventions, the pressure to succeed may become overwhelming, undermining the scholarship's purpose of promoting equitable access and success in the accounting profession Ontong and Smit (2023)

2.4 *Educational Gaps*

The persistent educational inequalities in South Africa, rooted in its apartheid-era legacy, significantly hinder Black students' ability to qualify as CAs. Many students enter tertiary education underprepared due to gaps in foundational skills, limited exposure to relevant industries, and inadequate support in their schooling. Recent studies, such as those by Ontong and Smit (2023) highlight that students struggle with industry-specific case studies because they lack practical experience and contextual knowledge. Additionally, disparities in school resources, including shortages of qualified teachers and basic facilities like internet access, further diminish students' readiness for university and professional careers (Dlamini 2024).

Schools from disadvantaged backgrounds often focus on high matric results rather than developing critical skills such as language proficiency, which are essential for higher education success (Chiramba and Ndofirepi 2023). Research continues to show that systemic inequalities and resource deficiencies hinder Black students' academic progression and contribute to their underrepresentation in the accounting profession (Dlamini 2024). The widening gap between students from historically disadvantaged schools and those from better-resourced institutions highlights the need for targeted interventions. Addressing these barriers is essential for creating equitable pathways for bursary-funded students, such as Thuthuka recipients, to pursue careers in accounting and contribute to a more diverse professional landscape.

2.5 *Language Barrier*

Language remains significant challenge for Black accountancy students, impacting both their written and spoken communication skills essential for academic success and professional competence. Language difficulties hinder students' ability to fully engage with complex accounting concepts, participate actively in classroom discussions, and perform well in assessments that demand high levels of literacy and numeracy.

Studies consistently demonstrate that limited language proficiency correlates with poor academic performance, particularly in disciplines requiring precise communication and analytical thinking (Koch and Burkett 2006). More recent research emphasise that language proficiency is a critical factor in student retention and success, directly influencing their ability to comprehend course material, articulate ideas clearly, and participate meaningfully in academic and professional settings (Koch and Burkett 2006).

Addressing language barriers is therefore vital for fostering an inclusive and equitable learning environment that recognises the diverse linguistic backgrounds of Black students. Creating supportive interventions such as language development programmes, academic literacy workshops, and culturally responsive teaching practices can significantly enhance students' communication skills and confidence. Such measures not only improve academic performance but also boost self-efficacy, motivation, and engagement. In doing so, they enable more Black students to overcome barriers and succeed in qualifying as CAs, ensuring that language does not remain an insurmountable obstacle to professional and socio-economic advancement in the accounting field.

3 Methodology

3.1 Research Design

This study adopted a quantitative research design based on secondary data analysis to investigate the career paths of students funded by the Thuthuka bursary. The focus was on interpreting existing numerical data related to enrolment, throughput, and professional qualification outcomes (CA(SA)). This approach allowed an objective examination of the impact of bursary support on students' progression into the accounting profession over time.

3.2 Sampling Method

The study focuses on students who received the Thuthuka bursary between 2013 and 2017. This cohort was purposively selected because, based on their intake year, they are expected to have completed their academic programmes and obtained their professional qualifications by 2025. The sampling sought to include all eligible bursary recipients within this period to ensure comprehensive coverage of career outcomes related to bursary support.

3.3 Research Instrument

The primary research instruments consisted of academic data summaries obtained from the University of the Free State's School of Accountancy and publicly available information from the SAICA website. These sources provided quantitative data on students' academic performance, progression, and qualification status, which formed the basis for analysing the impact of bursary support on career advancement.

3.4 Data Collection

Data collection involved extracting relevant academic records and qualification status information from the University of the Free State and SAICA's online resources. The data encompassed demographic details, academic progress, completion status, and professional qualification attainment. Ethical considerations, such as confidentiality and data privacy, were upheld throughout the process.

3.5 Data Analysis

The collected data was organised and visually displayed using Excel spreadsheets. Descriptive statistics summarised enrolment and throughput rates of Black and Coloured students funded through Thuthuka. Visual representations, including bar graphs and pie charts, were used to illustrate key themes and findings. To evaluate the impact of bursary support on professional qualification attainment, the study analysed the correlation between enrolment data, throughput rates, and successful registration as CAs(SA). This analysis provided insights into the effectiveness of the bursary programme in promoting diversity and career progression within the accounting profession.

4 Results and Discussion

This section presents the results derived from the academic data summaries provided by the University of the Free State's School of Accountancy, as well as publicly available information obtained from the SAICA website.

Figure 1 shows the distribution of 178 Thuthuka bursary recipients from 2013 to 2017 based on their race classification. The analysis indicates that the majority of students during this period were Black, comprising 92%, while Coloured students accounted for 8%.

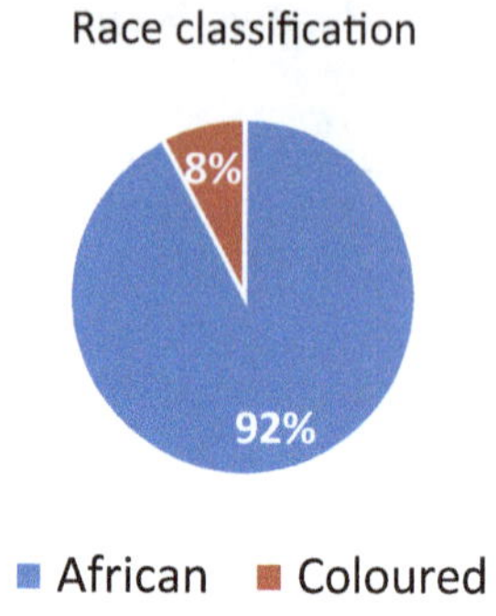

Fig. 1 Race classification. *Source* School of Accountancy, 2024

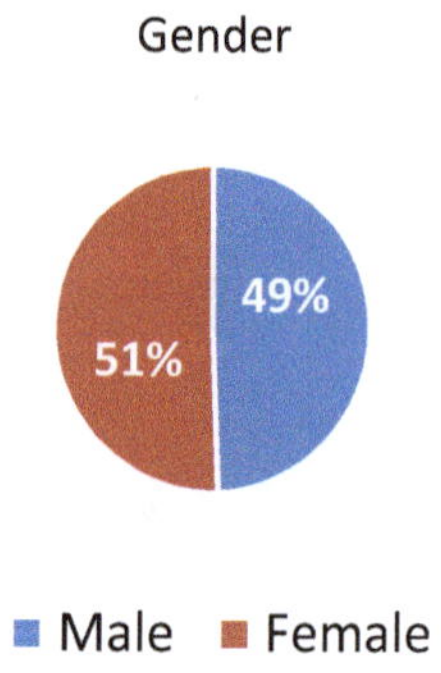

Fig. 2. Gender. *Source* School of Accountancy, 2024

Figure 2 shows the distribution of 178 Thuthuka bursary recipients from 2013 to 2017 by gender. The data indicates that during this period, 51% of the students were female, and 49% were male. While this reflects a relatively balanced gender distribution, it is important to acknowledge the need for continued support for women, especially those from previously disadvantaged backgrounds, to ensure sustained progress towards gender equity in the accounting profession.

Figure 3 displays the Black and Coloured CAs (SA) from 2022 to 2024. Although the percentage of Black CAs (SA) has increased from 17.7% in 2022 to 20.2% in 2024, representing a total rise of 2.5 percentage points over two years, the pace of growth remains slow. The average annual increase for Black CAs (SA) over this period is approximately 1.25%. Similarly, Coloured CAs (SA) recorded a slight increase from 4.8% in 2022 to 5.1% in 2024, a total increase of 0.3 percentage points. The average annual increase for Coloured CAs (SA) is about 0.15%.

Figure 4 displays the mid-year population estimates for South Africa in 2024, broken down by population group. The data shows that 81,7% of the population is Black, followed by 8,5% Coloured, 7,2% White, and 2,6% Indian/Asian. Despite Black South Africans comprising the majority of the population, with Coloureds as the next largest group, the representation of CAs remains relatively low compared to their population share. This highlights the ongoing shortage of Black professionals in

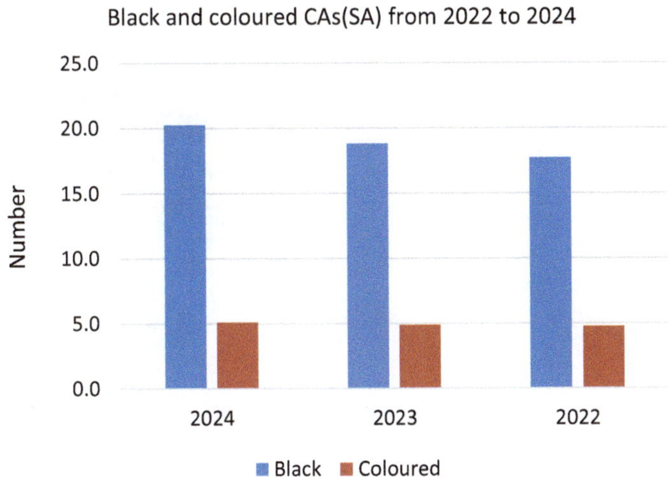

Fig. 3 Black and coloured CAs (SA) from 2022 to 2024. *Source* SAICA integrated Report, 2024

the accounting field and emphasises the need for targeted efforts to promote greater inclusion and representation within the profession.

Figure 5 illustrates the distribution of 178 Thuthuka bursary recipients from 2013 to 2017 based on their year of intake. The proportions of recipients were 21% in 2013, 23% in 2014, 16% in 2015, 17% in 2016, and 23% in 2017. Notably, the years 2014 and 2017 each accounted for the highest proportion of recipients at 23%.

Figure 6 displays the distribution of 178 Thuthuka bursary recipients from 2013 to 2017 based on the students' year of study at the time of receiving the bursary at

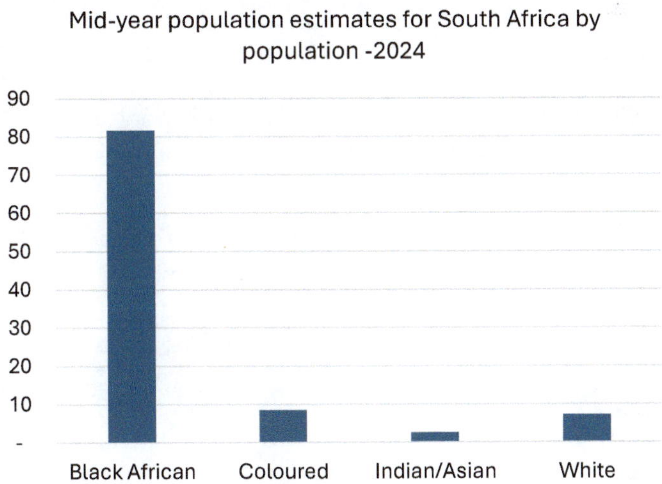

Fig. 4 2024 SA population. *Source* Stats SA, 2024

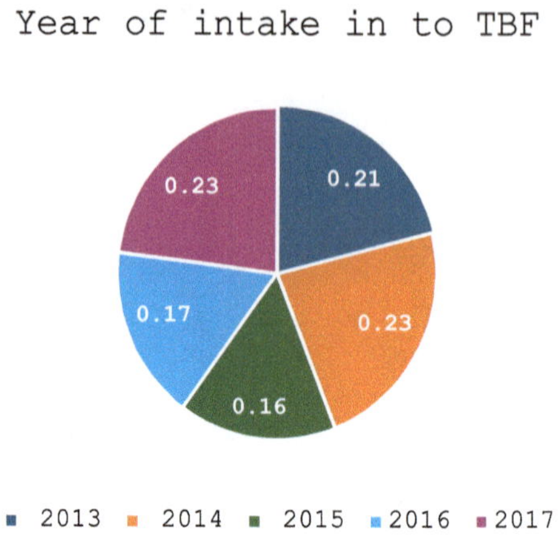

Fig. 5 Year of intake into TBF. *Source* School of Accountancy, 2024

the University of the Free State (UFS). Intake is primarily conducted at 1st year of study; however, some intakes also include students who are in the 2nd year or 3rd year or honours level of study. The data shows that the majority of bursaries were awarded to students in their first year of study, accounting for 73% of all intakes. This was followed by 22% of students in their second year, 3% in their honours year, and 2% in their third year. Overall, most bursary allocations occurred at the first-year level.

Figure 7 illustrates the distribution of 178 Thuthuka bursary recipients from 2013 to 2017 based on bursary retention during their studies. The data indicates that 63% of students lost their bursary due to academic performance issues, such as failing one or more core accounting modules, whereas 37% did not lose their bursary. The majority of students faced the risk of losing their bursary, highlighting the challenges associated with maintaining the strict conditions tied to the funding. These conditions typically require students to uphold high academic standards and meet specific performance benchmarks; failure to do so often results in the withdrawal of financial support.

Figure 8 illustrates the distribution of 113 Thuthuka bursary recipients from 2013 to 2017 who either lost the bursary or subsequently regained it during their BAcc

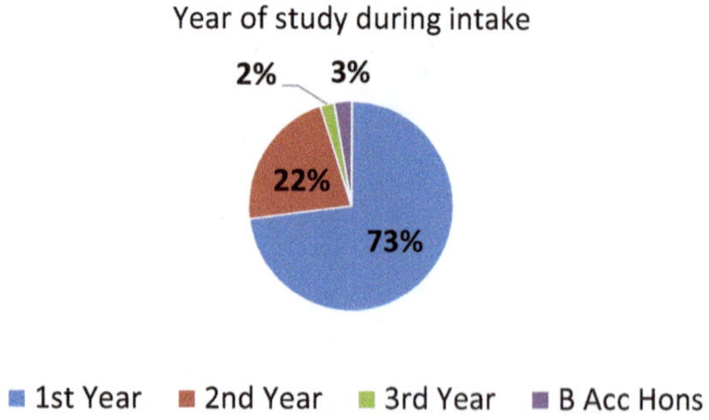

Fig. 6 Year of study during intake. *Source* School of Accountancy, 2024

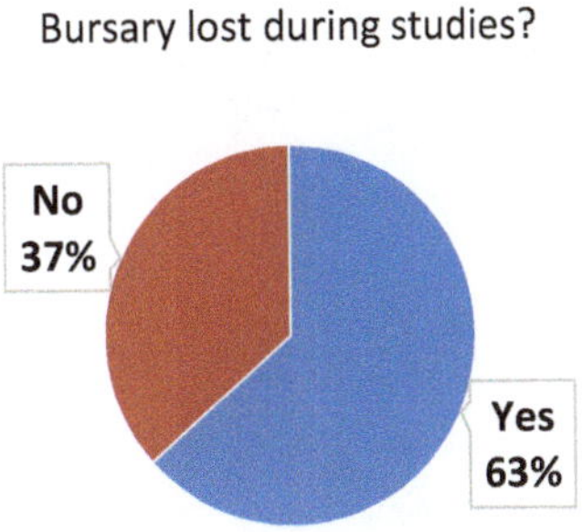

Fig. 7 Bursary lost during studies. *Source* School of Accountancy, 2024

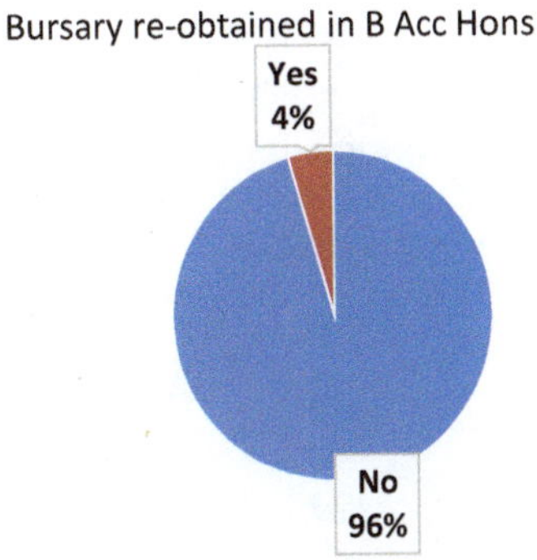

Fig. 8 Bursary re-obtained in BAcc Hons. *Source* School of Accountancy, 2024

Honours studies. Students who lost the bursary during their undergraduate studies are permitted to re-obtain it if they qualify for BAcc Honours through their own funding or alternative sources. Among these students, only 4% managed to re-obtain the bursary, while the remaining 96% did not.

Figure 9 displays the distribution of 178 Thuthuka bursary recipients from 2013 to 2017 based on their highest qualification obtained at the UFS. These qualifications were earned during their studies at the UFS and do not include any qualifications obtained at other institutions. The data shows that 11% of the students do not have UFS qualifications, likely due to deregistration. Additionally, 3% obtained other qualifications within the Economics and Management Faculty. Regarding specific degrees, 20% earned a Bachelor of Accounting (BAcc), and 10% earned a BCom in Accounting. During the 2013 to 2017 period, students who failed under the BAcc programme were automatically transferred to the BCom Accounting programme. Furthermore, 12% obtained a BCom Accounting Honours or PGDipGA, while 44% achieved a BAcc Honours or PGCA.

Figure 10 illustrates the distribution of 178 Thuthuka bursary recipients from 2013 to 2017, based on the SAICA qualifications recorded as of 19 May 2025, according to the SAICA member verification information available online. The data shows that 39% of the students are qualified as CA(SA), and 2% as an Associate General Accountant (South Africa) (AGA(SA)). Additionally, 59% of the recipients are not members of SAICA. Both CA(SA) and AGA(SA) are professional designations

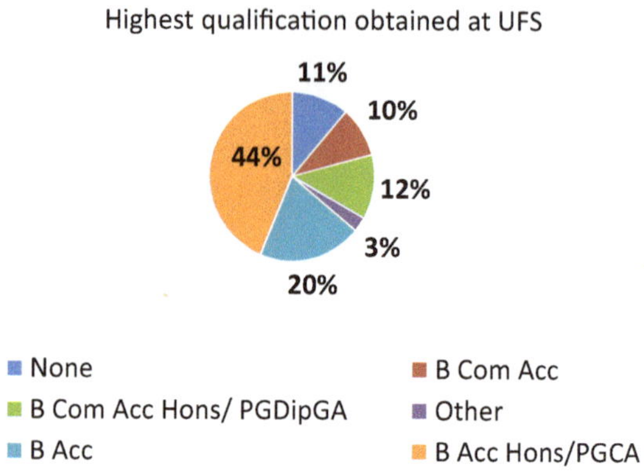

Fig. 9 Highest qualification obtained at the UFS. *Source* School of Accountancy, 2024

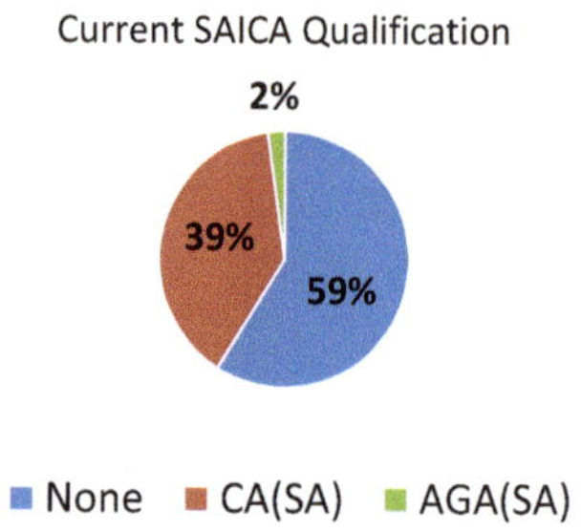

Fig. 10 SAICA qualification. *Source* School of Accountancy, 2024

under the SAICA umbrella. The primary distinction is that CA(SA) is a full Chartered Accountant designation, signifying a higher level of qualification, expertise, and membership status within SAICA. Conversely, AGA(SA) is a recognised qualification for accountants who have met specific educational and experience criteria but are not necessarily full members with the CA(SA) designation.

Discussion. The analysis of the data reveals both progress and persistent challenges in the journey of Thuthuka bursary-funded students towards becoming **CAs** and diversifying the profession. According to Figure 3, the percentage of Black CAs (SA) increased from 17.7% in 2022 to 20.2% in 2024, reflecting a total growth of 2.5 percentage points over two years. The average annual increase of approximately 1.25% indicates a slow but steady movement towards greater representation. Similarly, Coloured CAs (SA) experienced a modest rise from 4.8% in 2022 to 5.1% in 2024, with an average annual increase of about 0.15%.

While these upward trends are positive, they highlight the slow pace of change within the profession, despite SAICA's concerted efforts to promote diversity. The findings demonstrate a strong positive correlation between bursary support and successful qualification as a CA(SA). However, the gradual growth indicates that, despite some progress, significant barriers persist, requiring more targeted and sustained initiatives to accelerate the inclusion of Black and Coloured professionals into the CA(SA) community. Without intensified efforts, the profession risks stagnating in its diversity goals, which is particularly concerning given its stated commitment to transformation. The results resonate with the work of Frans (2022) at Stellenbosch University, who emphasised the transformative role of the TBF in recipients' academic and professional trajectories. Their study demonstrated that financial, academic, and psychosocial support not only facilitates access to higher education but also enables successful degree completion and entry into the profession. They recommended further interventions such as workshops to support students' emotional and mental well-being, additional academic support, greater integration of real-world applications into the curriculum, and enhanced assistance in navigating professional examinations. Such interventions are vital to dismantle the barriers still impeding progress and to accelerate meaningful transformation within the CA(SA) community.

Further insights are gained from Fig. 7, which illustrates the retention challenges faced by Thuthuka bursary recipients. Out of 178 students from 2013 to 2017, 63% lost their bursary primarily due to academic performance issues, such as failing core accounting modules. This high attrition rate highlights the difficulties students encounter in meeting the rigorous academic requirements tied to bursary conditions. The pressure to maintain high standards, while essential for professional qualification, can pose significant obstacles, especially for students from historically disadvantaged backgrounds.

Finally, Fig. 10 provides a broader view of the qualification outcomes of these bursary recipients. Of the 178 students, 39% qualified as CA(SA), 2% as AGA(SA), while 59% are not members of SAICA. The fact that a significant proportion of bursary recipients have not yet achieved full membership accentuates the long and demanding pathway to professional qualification and highlights potential areas for intervention to support student progression.

5 Conclusion

In conclusion, while the data demonstrates positive trends in increasing diversity among CAs (SA), the slow rates of growth and the retention challenges faced by bursary students reveal that the journey towards transformation remains complex. Continued, targeted enhancement are necessary to address academic barriers, improve retention, and support the career advancement of historically disadvantaged students. Achieving meaningful change will require a collaborative approach that combines academic support, mentorship, and systemic reforms within the profession and higher education sectors.

The study recommends an increase in the intake of Thuthuka bursary recipients across all participating universities in South Africa, as the current allocation at the UFS is restricted to approximately 30 students per year. Expanding the bursary nationally would enhance access to accounting education for a broader and more diverse pool of talented students, thereby advancing inclusion and equity within the profession. Furthermore, the bursary scheme should introduce an appeals mechanism for students who encounter setbacks arising from external factors beyond their control. For instance, as highlighted by Ontong and Smit (2023), personal hardships such as bereavement or health-related crises may threaten students' bursary support despite their academic capability. Providing opportunities for appeal in such cases would help ensure that deserving students are not unfairly disadvantaged and are able to continue pursuing their educational and professional aspirations in accounting.

Furthermore, it is recommended that the bursary scheme incorporate a one-year bridging programme, serving as a transitional pathway from the BCom Accounting degree to the SAICA-accredited Postgraduate Diploma (PGDip) in Accounting. Funding this bridging year would support a more seamless progression for bursary recipients, thereby enhancing their likelihood of success in obtaining the Chartered Accountant qualification.

This study has certain limitations that present avenues for future research. It was conducted exclusively at the UFS, a single South African university. The choice of this institution was deliberate to facilitate an in-depth exploration of how bursary support influences students' progression into the accounting profession, providing detailed insights into the specific challenges and practices within this context. However, this focused approach limits the generalisability of the findings to other institutions or broader student populations. Additionally, the study primarily focused on students who pursued a CA (SA) qualification, leaving the career paths of students who chose alternative professional designations less understood.

Notwithstanding the above limitations, findings emerging from this study could have an impact on the future transformation initiatives of the South African accounting profession, as they will assist SAICA and TBF trustees to review the current programme, especially since it does not appear to equip Thuthuka students with the skills and confidence they require to overcome social closure based on race (and class) within the profession. The findings are of interest to other parties involved in the introduction or maintenance of support programmes for socio-economically disadvantaged students, to either benchmark the features of their existing programmes or to consider the features suggested by this study for possible inclusion in new programmes.

References

Barac, K.: Helping disadvantaged students: findings from the Thuthuka programme. Accounting Education. **24**(2), 75–101 (2015). https://doi.org/10.1080/09639284.2015.1010171

Carmel, J.M., Chapman, S., Wright, P.: Disability justice: the challenges of inclusion in everyday life. Int. J. Disabil. Dev. Educ. **72**, 1–14 (2025). https://doi.org/10.1080/1034912X.2025.2456288

Chiramba, O., Ndofirepi, E.: Access and success in higher education: disadvantaged students' lived experiences beyond funding hurdles at a metropolitan South African university. S. Afr. J. High. Educ. **37**(6), 56–75 (2023) https://doi.org/10.20853/37-6-6021

Davids, N., Waghid, Y.: #FeesMustFall: history of South African student protests reflects inequality's grip. Mail & Guardian. (2016) https://mg.co.za/article/2016-10-10-feesmustfall-history-of-south-african-student-protests-reflects-inequalitys-grip, last accessed 23 Sept 2023

De Jager, E.: Thuthuka students' perceptions of factors influencing success. J. Econ. Financ. Sci. **7**(1), 53–72 (2014). https://doi.org/10.4102/jef.v7i1.130

De Jager, E., Bitzer, E.: The views of commerce students regarding 'free' higher education in South Africa. S. Afr. J. High. Educ. **32**(4), 12–36 (2018) https://doi.org/10.20853/32-4-2436

Dlamini, P.H.: Offenders' non-participation in freely provided correctional educational programmes: an investigative approach. Master's thesis. University of South Africa, Pretoria (2024) https://hdl.handle.net/10500/32258

Frans, I.: Transforming the accounting profession through Thuthuka: an outcome-based evaluation of the Thuthuka Bursary Fund programme at Stellenbosch University. Master's thesis,. Stellenbosch University, Stellenbosch (2022) https://scholar.sun.ac.za/items/8b894008-af3b-4447-934b-58fd4a7f12b6

Koch, E., Burkett, B.: Making the role of African languages in higher education a reality. S. Afr. J. High. Educ. **19**(6), 1089–1107 (2006). https://doi.org/10.4314/sajhe.v19i6.25547

Mdepa, M., Tshiwula, M.: Student diversity in South African higher education. Widening Particip. Lifelong Learn. **13**(1), 19–33 (2012). https://doi.org/10.5456/WPLL.13.S.19

Musundwa, S., Hammond, T.A.: Black chartered accountants in post-apartheid South Africa: global implications for the profession. Accounting Horizons. **38**(1), 111–122 (2024). https://doi.org/10.2308/HORIZONS-2022-071

Musundwa, S., Moses, O.: The lived experiences of South African Black accountants within the framework of the chartered accountancy profession sector code. Account. Audit. Account. J. **37**(7–8), 1726–1760 (2024). https://doi.org/10.1108/AAAJ-10-2023-6690

Mzangwa, S.T.: The effects of higher education policy on transformation in post-apartheid South Africa. Cogent Educ. **6**(1), 1–15 (2019). https://www.researchgate.net/publication/331946890_The_effects_of_higher_education_policy_on_transformation_in_post-apartheid_South_Africa

Ontong, J.M., Smit, S.: Reflecting on why transformation students in a comprehensive bursary programme may be academically unsuccessful. Perspect. Educ. **41**(4), 342–355 (2023) https://journals.ufs.ac.za/index.php/pie/article/view/7221

Pellicer, M., Ranchhod, V.: Understanding the effects of racial classification in apartheid South Africa. J. Dev. Econ. **160**, 1–13 (2023). https://inq-dp.eu/wp-content/uploads/2023/07/understanding-the-effects-of-racial-classification-in-apartheid-south-africa.pdf

South African Institute of Chartered Accountants (SAICA): Winning women: Nonkululeko Gobodo (CA) SA. https://www.accountancysa.org.za/cover-story-winning-women-nonkululeko-gobodo-casa/, last accessed 19 June 2025 (2013)

South African Institute of Chartered Accountants (SAICA): Thuthuka Bursary Fund overview. https://www.saica.co.za/, last accessed 19 June 2025 (2020)

South African Institute of Chartered Accountants (SAICA): Annual integrated report. https://www.saica.org.za/, last accessed 19 June 2025 (2024)

Statistics South Africa (Stats SA): Population demographics and professional statistics. https://www.statssa.gov.za/, last accessed 19 June 2025 (2024)

University of the Free State, School of Accountancy: Thuthuka programme. https://www.ufs.ac.za/econ/faculty-of-economic-and-management-sciences-home/departments-and-divisions/school-of-accountancy-home/thuthuka/thuthuka-programme, last accessed 19 June 2025 (2024)

A Systematic Literature Review on Strategic and Performance Information Disclosure by Universities

Lemlem Amhatsion and Ndivhuwo Sirembe

Abstract The study undertook a systematic literature review of universities' strategic and performance information disclosures. It reviewed existing literature to assess the theoretical frameworks used and examined annual patterns, geographical variations, and research contexts, to suggest areas for future research. The Preferred Reporting Items for Systematic Reviews and Meta-Analyses (PRISMA) framework was used to examine articles from peer-reviewed journals published between 2015 and 2025. Using PRISMA, 38 articles were identified from the SCOPUS and Web of Science (WoS) databases and subsequently assessed using a framework developed by the researchers, which comprised 12 categories. The results highlighted the predominance of research in the Global North, and the widespread use of quantitative content analysis as a research method, and the preference for cross-sectional study designs. The study provided insights into the examination of strategic and performance information disclosure within universities and offered guidance for future research by outlining potential directions for investigation. It also found that the most applied theories were stakeholder theory, legitimacy theory, and institutional theory. Future research should consider employing alternative methodologies, such as surveys and questionnaires, and place greater emphasis on studies conducted in the Global South to promote equity and the inclusion of diverse knowledge systems. Additionally, there is a need for more longitudinal research, which would facilitate the analysis of emerging trends and the evolution of strategic and performance information reporting. Furthermore, future studies should examine the extent to which official frameworks for strategic and performance information reporting exist within universities and explore the relationship between strategic plans, institutional performance indicators, targets, and actual achievements disclosed.

Keywords Strategic and Performance Information · SCOPUS · Web of Science · Disclosure · Universities · PRISMA · Systematic Literature Review

L. Amhatsion (✉) · N. Sirembe
School of Accountancy, University of the Free State, Bloemfontein, South Africa
e-mail: AmhatsionLG@ufs.ac.za

E. H. Redda (ed.), *Proceedings of the Social Sciences International Research Conference, 2025*, https://doi.org/10.1007/978-981-95-9489-4_37

1 Introduction

Universities frequently develop strategic and performance information disclosure practices to comply with the regulatory frameworks applicable to their operating environment. In such cases, disclosures are often structured in a manner similar to the annual reports of leading universities or aligned with requirements set by higher education authorities. In the absence of regulatory frameworks, universities may disclose strategic and performance information according to sectoral norms, that is, by following the practices of peer institutions.

Strategic Planning in universities involves setting specific goals, identifying opportunities, and allocating resources to accomplish institutional objectives, which serves as a guide for the future growth of institutions (Creatix Campus 2025). Strategic information relates to the data and insights used to inform this planning process.

Within universities, common areas of strategic information include performance information, risk management, sustainability related information, regulatory and political information, social and cultural information, corporate governance, and intellectual capital information (HEA 2025; Ziemba 2025). Such information is typically found in annual reports, strategic plans, sustainability reports, university websites, reports to regulatory bodies, and integrated reports (Larrán Jorge et al. 2019; Manes-Rossi et al., 2018 as cited on Nicolò G et al., 2020).

Performance reports serve as the essential instrument for assessing and confirming an institution's success in carrying out its strategic plan (Hanover Research, 2014). Performance data allows stakeholders to evaluate how effectively and efficiently these public entities use their resources to deliver services and meet their goals, which makes the measurement and reporting of this performance more vital than ever due to severe financial limitations (Tiron-Tudor et al. 2022).

This study presents a systematic literature review (SLR) with the objective of answering the following questions:

- What are the trends and themes in the existing literature on the strategic and performance information reporting in universities?
- What gaps exist in the current literature regarding strategic and performance information reporting in universities?

2 Literature Search and Sample Development

This study adopted a SLR approach to extract data on the literature concerning strategic and performance information disclosure in universities. An SLR provides an authoritative account of existing evidence, employing reliable, objective, thorough, and reproducible research practices (University of Wollongong Library 2025). In an SLR, reviewers systematically screen and analyse published evidence from multiple sources to identify relevant studies and extract data (Mtech Access, 2024).

The database search was conducted using the SCOPUS and Web of Science (WoS) databases. These databases are recognised as leading search engines for citations, covering more than 330 disciplines, including business, management and accounting, economics, and social sciences (Samsuddin et al. 2020, as cited in Ah Choi 2020).

The Preferred Reporting Items for Systematic Reviews and Meta-Analyses (PRISMA) framework was employed to examine articles from peer-reviewed journals published between 2015 and 2025. Although the PRISMA 2020 framework was originally designed for systematic reviews of studies on the effects of health interventions, its checklist items can be adapted for systematic reviews of non-health-related studies (Page et al. 2021). The PRISMA guidelines allow for a thorough search of terms related to the factors that influence strategic and performance information disclosure by universities. The flow of the study can be seen in Fig. 1.

As the first step of the systematic review process, the following search strings were developed using keywords and relevant terms identified from previous literature searches. These were applied in the SCOPUS and WoS databases to locate articles:

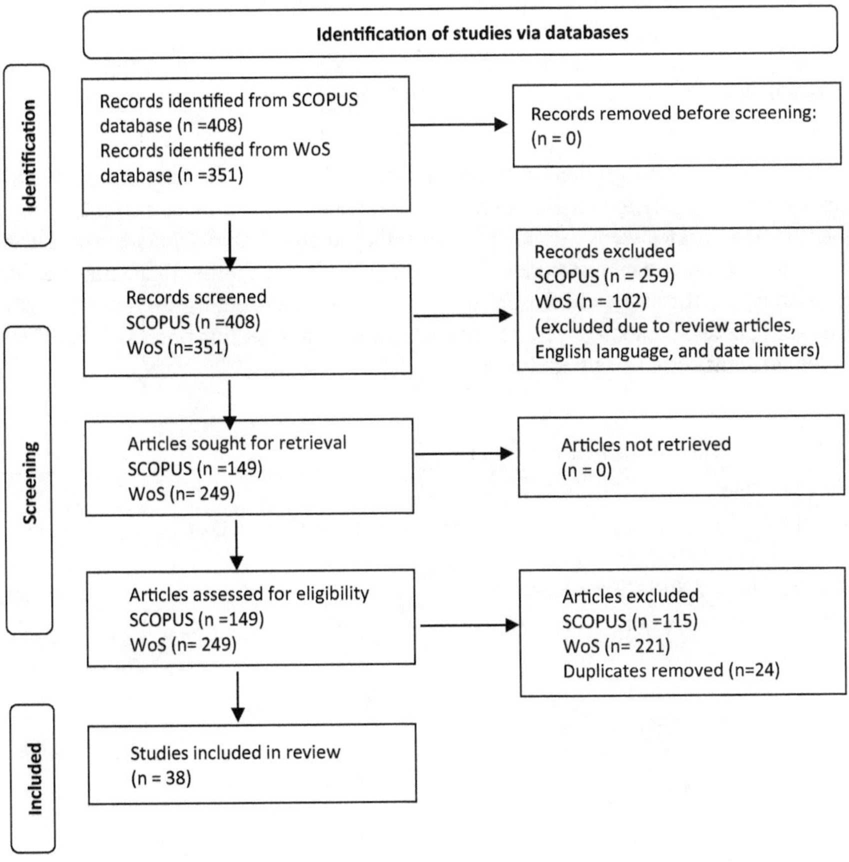

Fig. 1 Flow diagram of the study adapted from PRISMA 2020

TITLE-ABS-KEY ((*"Strategic disclosure"* OR *"Strategic information"* OR *"Integrated plan disclosure"* OR *"Performance disclosure"* OR *"Sustainability disclosure"* OR *"Integrated report* disclosure"* OR *"Corporate social responsibility disclosure"* OR *"Accountability disclosure"* OR *"Performance report*"* OR *"Corporate governance disclosure"* OR *"Governance disclosure"* OR *"Risk management disclosure"* OR *"Social responsibility disclosure"* OR *"Information systems disclosure"* OR *"Financial performance disclosure"* OR *"intellectual capital disclosure"* OR *"financial sustainability disclosure"* OR *"financial viability disclosure"* OR *"Information technology disclosure"*) AND (*"Higher education"* OR "*universities* OR *colleges*"))

The search yielded 759 documents. All identified documents were then screened based on the inclusion and exclusion criteria presented in Table 1.

Based on the screening process, 361 articles were excluded for the following reasons:

- Firstly, documents published before 2015 were excluded, as the last 10 years were considered to represent the most recent studies in this area.
- Secondly, in terms of literature type, documents such as conference papers, book chapters, books, and review articles were excluded, as they were not considered primary sources.
- Lastly, only articles written in English language were included. Articles in other languages were excluded due to resource limitations in translating them into English.

To ensure the appropriateness of the articles for this study, the remaining 398 articles were thoroughly examined based on titles, abstracts, and main content. Consequently, 336 articles were excluded because they focused on disclosures by private companies or on enhancing student academic performance rather than strategic and performance information disclosure by universities. Additionally, 24 duplicate articles from the two databases were removed as part of the selection process. Therefore, the final sample comprised 38 articles.

Table 1 The inclusion and exclusion criteria

No	Criterion	Inclusion	Exclusion
1	Literature type	Research articles	Review articles, books, book chapters, conference proceedings, editorial note
2	Timeline	2015–2025	Before 2015
3	Language	English	Non-English

3 The Analytical Framework and Data Analysis

The remaining 38 articles were assessed and examined to answer formulated research questions and, information was extracted by examining the abstracts as well as reading the full articles to identify appropriate categories using a framework developed by the researchers. The analytical framework comprised 12 categories: (1) year of publication, (2) country of research, (3) subject area, (4) affiliation, (5) number of citations, (6) research theme, (7) theoretical frameworks adopted, (8) research methodology, (9) data collection methods used, (10) time horizon, and (11) nature of the study. To ensure reliability, the researchers independently tested the framework on five articles and compared their results. This was followed by the researcher performing manual coding of the assigned articles and recording the process in an Excel document.

4 Results and Discussion

4.1 General Sample Characteristics

Analysis by Year of Publication. Figure 2 shows the distribution of the 38 articles by year of publication. The analysis indicates that the majority of the articles (61%) were published in the last five years, indicating a general increase on the number of publications relating to strategic and performance information disclosure.

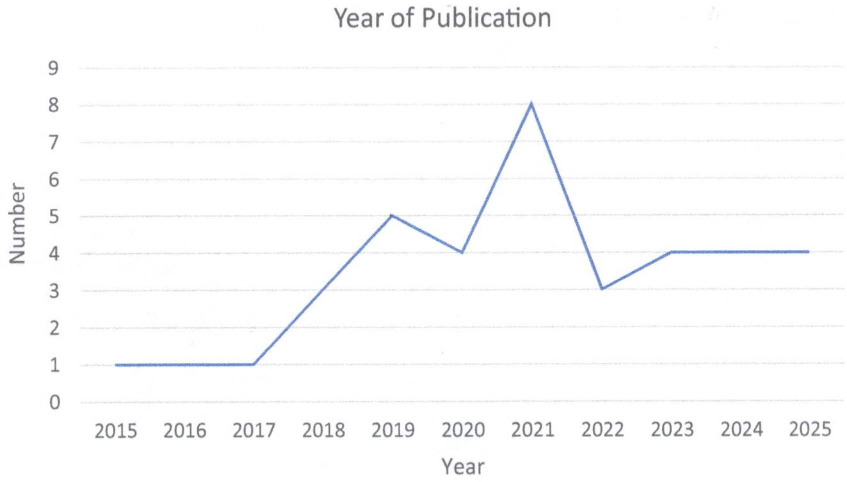

Fig. 2 Classification by year of publication

Analysis by Country of Publication. Figure 3 shows the distribution of the 38 articles by country. The analysis indicates that the articles are widely distributed, being based on 17 countries, with Italy having the highest percentage of 29%. It also indicates a dominance of research (71%) by countries in the Global North. Future research should focus more on conducting research in Global South to address issues of equity and diverse knowledge systems.

Analysis by Subject Area Figure 4 shows the distribution of the 38 articles by subject area. The analysis indicates that the majority of the articles (61%) were related to business, management and accounting and economics, with social sciences following at 34%. This demonstrates a general dominance of the two subject areas in the number of publications that relate to strategic and performance information disclosure.

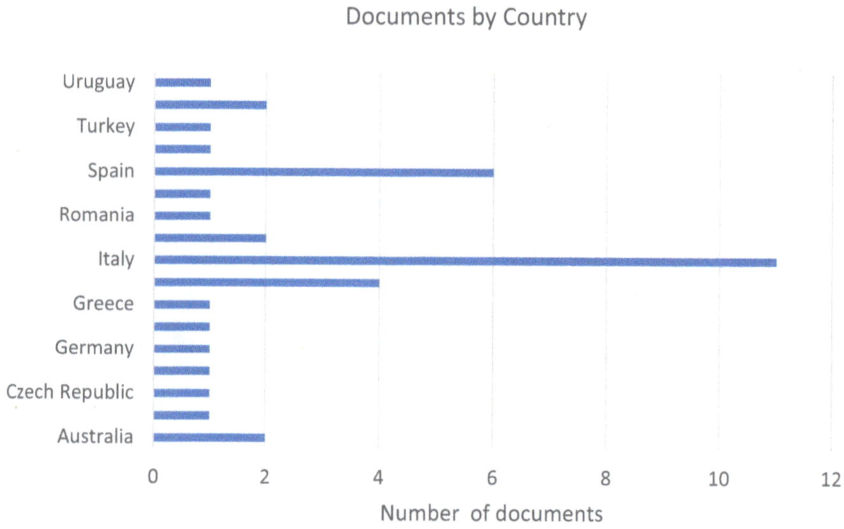

Fig. 3 Classification by country

Fig. 4 Classification by subject area

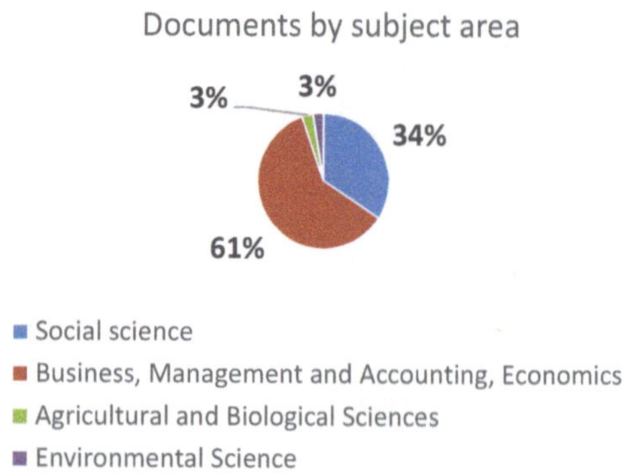

Analysis by affiliation Figure 5 shows the distribution of the 38 articles by affiliation. The analysis indicates that the articles are widely distributed among 27 higher education institutions, with the University of Salerno having the highest percentage of 24% (9 publications). No dominance by any institute was indicated.

Analysis by Number of Citations Figure 6 shows the distribution of the 38 articles by number of citations. The analysis indicates that 37% of the articles were cited less than 10 times, with four articles having no citation at the time this study was conducted. It was also noted that three of the four articles with no citation were published in 2024/2025. On the other hand, eight publications (21%), that had the highest number of citations of 40 and above, were all published at least five years ago.

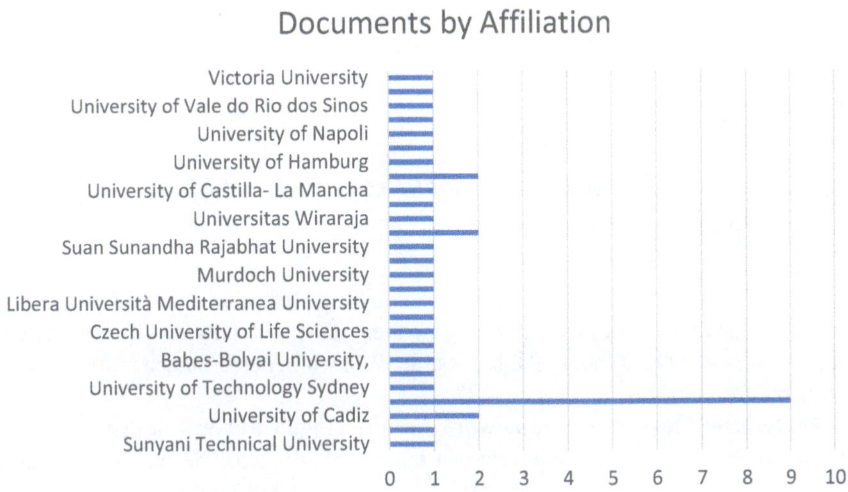

Fig. 5 Classification by affiliation

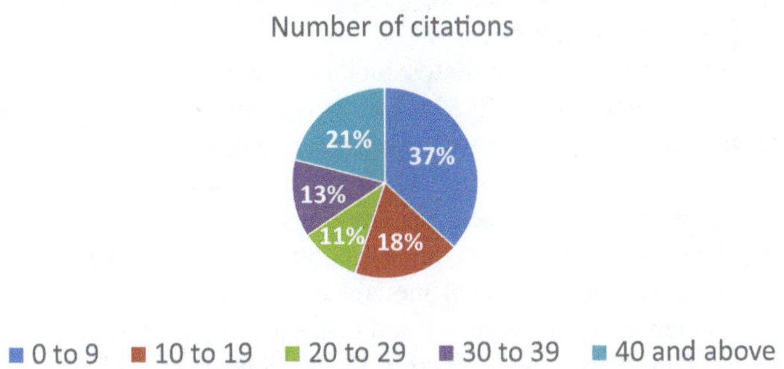

Fig. 6 Classification by citation

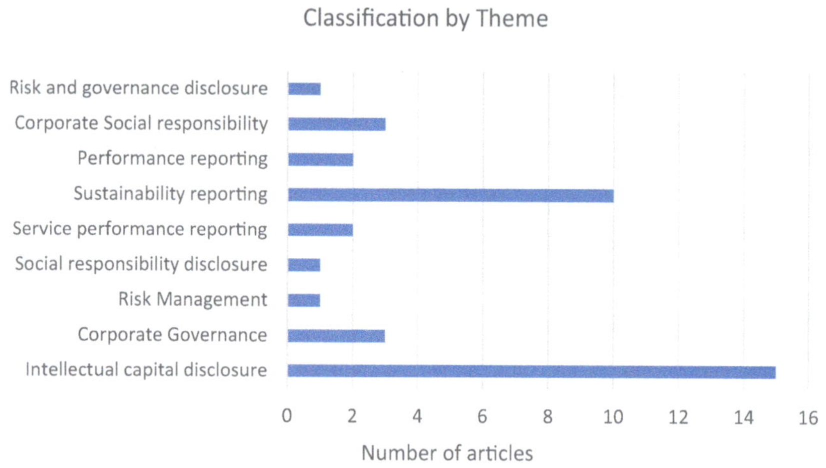

Fig. 7 Classification by theme

4.2 Research Theme, Theoretical Framework, Research Methodology Classifications

Analysis by Research Theme Figure 7 shows the distribution of the 38 articles by theme. The analysis indicates that nine main themes were identified, with intellectual capital disclosure (ICD) being the highest at 39%, while performance reporting was the theme for only four articles i.e. 10%.

Analysis by Theoretical Framework Adopted Figure 8 shows the distribution of the 38 articles by the theoretical framework that was adopted. The analysis indicates that seven theories were cited, and stakeholder theory and legitimacy theory were the most adopted theories. Stakeholder theory was cited in 19 articles and legitimacy theory was cited in 16 articles.

Analysis by Research Methodology Figure 9 shows the distribution of the 38 articles by research methodology used. The analysis indicates that in the majority of the articles (28 articles; 74%) quantitative methods were used, whereas in six articles qualitative methods were used. Mixed methods were used in four articles only. The study shows the dominance of the use of quantitative methods.

Analysis by Data Collection Method Figure 10 shows the distribution of the 38 articles by data collection methods used. The analysis indicates that the majority of the articles (32 articles; 84%) used content analysis for data collection, confirming that content analysis is the dominant method in this field. Future research should focus on non-content analysis research methods, such as surveys and questionnaire.

Analysis by Time Horizon Covered in the Articles Figure 11 shows the distribution of the 38 articles by time horizon covered. The analysis indicates that in the majority of the articles (29 articles; 76%) a cross-sectional study was conducted, whereas longitudinal study was conducted in only nine articles, indicating the need

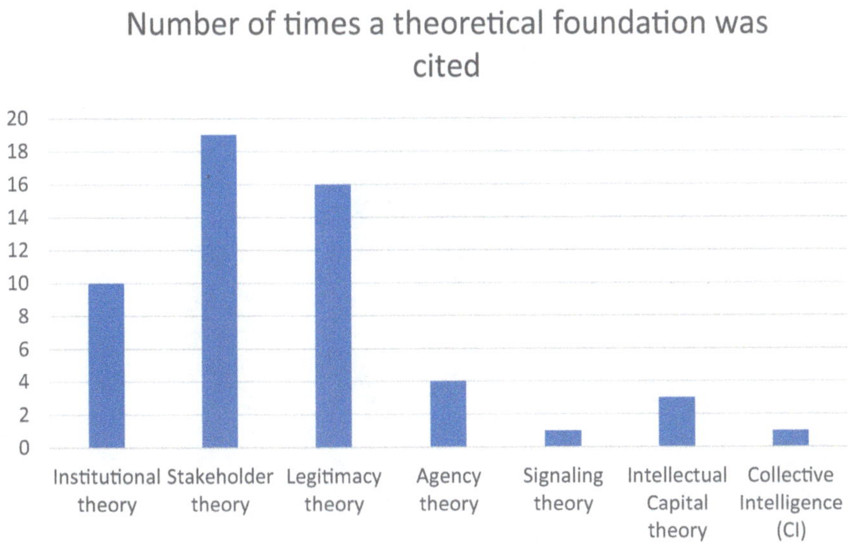

Fig. 8 Classification by theoretical framework adopted

Fig. 9 Classification by research methodology used

to increase longitudinal research which would allow the analyses new trends on strategic and performance information reporting.

Analysis by Nature of the Study of the Articles. Figure 12 shows the distribution of the 38 articles by nature of study. The analysis indicates that 15 articles (39%) focused on the disclosure made by universities or higher education institutions, in comparison with the characteristics of the institutions and the leadership thereof. Eight articles (21%) related to studies conducted by comparing disclosure with frameworks developed by the researchers or in previous studies. Seven articles (18%) related to studies conducted by comparing disclosure with official guidelines or frameworks. Where are another seven articles (18%) related to studies that compared disclosure with the interest of various stakeholders at the institutions. The remaining one article related to the comparison of disclosure with strategic plans, performance

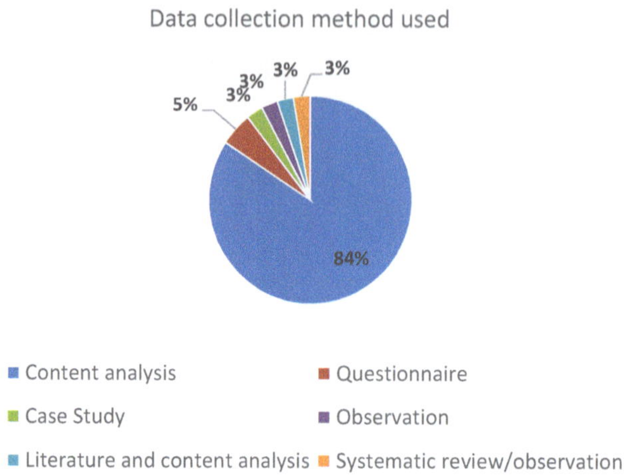

Fig. 10 Classification by data collection methods used

Fig. 11 Classification by time horizon of articles

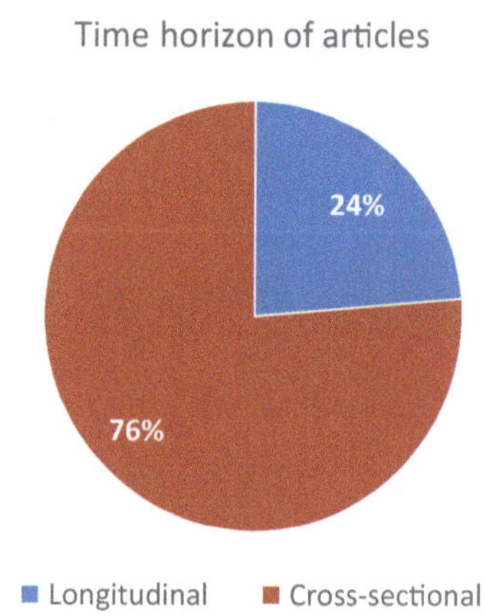

indicators of the institutions, including the targets that were set, actual achievements disclosed by the institutions. Future research should focus on the extent of the existence of official strategic and performance information reporting frameworks that are used by universities.

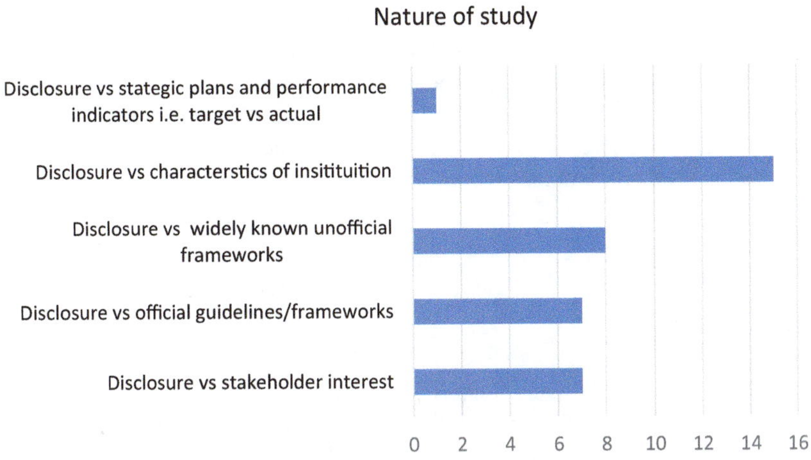

Fig. 12 Classification by nature of study

5 Conclusion

An organisation's mandatory reporting and voluntary disclosure are influenced by the coercion of legislation, reporting and funding regulations, and other stakeholder requirements for information (Tiron-Tudor et al. 2022). This study provided a SLR on strategic and performance information disclosure by universities. A total of 38 articles, identified using PRISMA from the SCOPUS and WoS databases, were assessed. The results show that in the last five years (2021–2025), there is an increase in the number of articles published. The articles were mainly from Global North with business management and accounting and social sciences being the main subject areas of the articles.

Nine main themes were identified: intellectual capital disclosure; corporate governance disclosure; risk management disclosure; social responsibility disclosure; service performance reporting disclosure; sustainability reporting disclosure; performance reporting disclosure; corporate social responsibility disclosure; and risk and governance disclosure. Quantitative research methods, particularly content analysis, were dominant. Future research should focus on non-content-analysis methods, such as surveys and questionnaires.

It was also found that the most adopted theories were stakeholders, legitimacy, and institutional theories, alongside a dominance of research conducted in the Global North. Future research should focus more on conducting research in Global South to address issues of equity and diverse knowledge systems.

The study also identified that in the majority of the articles a cross-sectional study design was employed, indicating the need to increase longitudinal research, which would allow the analysis of new trends, and a better understanding of the evolution of strategic and performance information reporting.

The study also identified that the studies mainly focused on the comparison of disclosure with frameworks developed by the researchers or previous studies, stakeholders' interest, or characteristics of the institutions. Future research should focus on the extent to which official strategic and performance information reporting frameworks are used by universities, as well as examine the linkages between strategic plans, institutional performance indicators, targets set, and the actual achievements disclosed.

References

*Amoako, K.O., Marfo, E.O., Akwaa-Sekyi, E.K., Amaning, N., Yankey, N.: Stakeholder perceptions of sustainability reporting on the websites of technical universities in Ghana. BIJ. **30**(10), 4248–4284 (2023). https://doi.org/10.1108/BIJ-02-2022-0104

*Andrades, J., Martinez-Martinez, D., Larrán Jorge, M.: Corporate governance disclosures by Spanish universities: how different variables can affect the level of such disclosures? Meditari Accountancy Research. **29**(1), 86–109 (2021). https://doi.org/10.1108/MEDAR-02-2020-0766

*Aversano, N., Nicolò, G., Sannino, G., Tartaglia Polcini, P.: The integrated plan in Italian public universities: new patterns in intellectual capital disclosure. Meditari Accountancy Research. **28**(4), 655–679 (2020). https://doi.org/10.1108/MEDAR-07-2019-0519

*Brusca, I., Cohen, S., Manes-Rossi, F., Nicolò, G.: Intellectual capital disclosure and academic rankings in European universities: do they go hand in hand? Meditari Accountancy Research. **28**(1), 51–71 (2020). https://doi.org/10.1108/MEDAR-01-2019-0432

Creatix Campus: Strategic planning in higher education: path to success. Accessed 20 June 2025. https://www.creatixcampus.com

*Elmagrhi, M.H., Ntim, C.G., Wang, Y., Elamer, A.A., Crossley, R.: The effect of vice-chancellor characteristics and internal governance mechanisms on voluntary disclosures in UK higher education institutions. J. Int. Account. Audit. Tax. **45**, 100428 (2021). https://doi.org/10.1016/j.intaccaudtax.2021.100428

*Galip, Y., Gönen, S.: Examination of sustainability disclosures of public sector institutions in Türkiye within the scope of public sector sustainability reporting standards. EDPACS. **1–19** (2025). https://doi.org/10.1080/07366981.2025.2500213

*Gamage, P., Sciulli, N.: Sustainability reporting by Australian universities. Aust. J. Public Adm. **76**(2), 187–203 (2017). https://doi.org/10.1111/1467-8500.12215

Hanover Research: Strategic planning in higher education: best practices and benchmarking. Accessed 4 September 2025. https://www.hanoverresearch.com

*Herli, M., Tjahjadi, B.: The role of social media as a voluntary intellectual capital disclosure in university: evidence from Indonesia. Intangible Capital. **18**(2), 247–262 (2022). https://doi.org/10.3926/ic.1505

Higher Education Authority (HEA): Governance oversight framework—higher education institutions. Accessed 20 June 2025. https://www.hea.ie

*Hsiao, P.C.K., Low, M., Scott, T.: Service performance reporting and principles-based authoritative guidance: an analysis of New Zealand higher education institutions. Meditari Account. Res. **32**(2), 367–395 (2024). https://doi.org/10.1108/MEDAR-10-2022-1825

Ah Choi, K.: A systematic review: determinants of forward-looking information disclosure by Malaysian companies. Int. J. Serv. Manag. Sustain. **5**(2), 13–36 (2020) https://doi.org/10.24191/ijsms.v5i2.11709

*Kuralova, K., Margarisova, K.: Intellectual capital disclosure at Czech public universities in relation to the stakeholder information need. Acta Universitatis Agriculturae et Silviculturae Mendelianae Brunensis. **64**(6), 1989–1998 (2016) https://doi.org/10.11118/actaun201664061989

*Larrán Jorge, M., Andrades Peña, F.J., Herrera Madueño, J.: An analysis of university sustainability reports from the GRI database: an examination of influential variables. J. Environ. Plan. Manag. **62**(6), 1019–1044 (2019). https://doi.org/10.1080/09640568.2018.1457952

*Ma'wa, A., Setiawan, D.: Online disclosure of intellectual capital and the determinants: evidence from higher education in Indonesia. Journal of Applied Research in Higher Education. (2024) https://www.emerald.com/insight/2050-7003.htm

*Mafa, M.M., Visser, C.T., Avenant, W.V.E.: Risk management practices by south African universities: an annual report disclosure analysis. J. Manag. Gov. **21**(2), 435–460 (2017). https://doi.org/10.3390/JRFM14050195

*Manes Rossi, F., Nicolò, G., Tartaglia Polcini, P.: New trends in intellectual capital reporting: exploring online intellectual capital disclosure in Italian universities. J. Intellect. Cap. (2018). https://doi.org/10.1108/JIC-09-2017-0119

*Elmagrhi, M.H., Ntim, C.G.: Non-financial reporting in non-profit organisations: the case of risk and governance disclosures in UK higher education institutions. Account. Forum. **47**(2), 223–248 (2023). https://doi.org/10.1080/01559982.2022.2148854

Mtech Access: 7 steps of a systematic literature review. Accessed 6 June 2024. https://mtechaccess.co.uk/7-steps-systematic-literature-review/

*Ndou, V., Secundo, G., Dumay, J., Gjevori, E.: Understanding intellectual capital disclosure in online media: an exploratory case study in a university. Meditari Accountancy Research. **26**(3), 499–530 (2018). https://doi.org/10.1108/MEDAR-03-2018-0302

*Nicolò, G., Aversano, N., Sannino, G., Tartaglia Polcini, P.: Online sustainability disclosure practices in the university context: the role of the board of directors. Corp. Gov. **23**(4), 800–826 (2023). https://doi.org/10.1108/CG-05-2022-0224

*Nicolò, G., Raimo, N., Vitolla, F., Aversano, N.: Unveiling universities' sustainability disclosure antecedents: an empirical analysis during the COVID-19 pandemic. Journal of Financial Reporting and Accounting. (2024) https://www.emerald.com/insight/1985-2517.htm

*Nicolò, G., Manes-Rossi, F., Aversano, N., Christiaens, J.: Unveiling the link between performance and intellectual capital disclosure in the context of Italian public universities. J. Intellect. Cap. **22**(6), 1215–1240 (2021) https://doi.org/10.1016/j.evalprogplan.2021.101969

*Nicolò, G., Manes-Rossi, F., Christiaens, J., Aversano, N.: Accountability through intellectual capital disclosure in Italian universities. J. Manag. Gov. **24**, 1055–1087 (2020). https://doi.org/10.1007/s10997-019-09497-7

Page, M.J., Moher, D., Bossuyt, P.M., Boutron, I., Hoffmann, T.C., Mulrow, C.D., Shamseer, L., Tetzlaff, J.M., Tricco, E.E., Turner, L.C., Ware, L.A., Wardle, M., Wright, A.J., Berlin, J.A., Wells, G.A.: PRISMA 2020 explanation and elaboration: updated guidance and exemplars for reporting systematic reviews. BMJ. **372**, n160 (2021) https://www.bmj.com/content/372/bmj.n160

*Raimo, N., Vitolla, F., Marrone, A., Esposito, P.: Corporate governance disclosure by Italian universities: an empirical analysis of the determinants. Corporate Governance (Bingley). **25**(3), 567–585 (2025). https://doi.org/10.1108/CG-05-2023-0217

*Ramírez, Y., Manzaneque, M., Merino, E.: Factors influencing online sustainability disclosure in Spanish universities. Int. J. Sustain. High. Educ. **26**(2), 315–333 (2025). https://doi.org/10.1080/13504622.2014.977231

*Ramírez-Córcoles, Y., Manzaneque-Lizano, M.: The relevance of intellectual capital disclosure: empirical evidence from Spanish universities. Knowl. Manag. Res. Pract. **13**(1), 31–44 (2015). https://doi.org/10.1057/kmrp.2013.27

*Rana, T., Ahmed, Z.U., Narayan, A., Zheng, M.: An institutional theory perspective on public sector reform and service performance reporting by New Zealand universities. Journal of Accounting &

Organizational Change. **18**(3), 461–484 (2022) https://www.emerald.com/insight/1832-5912.htm
*Sanchez, R.G., Bolívar, M.P.R., Hernandez, A.M.L.: Which are the main factors influencing corporate social responsibility information disclosures on universities' websites. Int. J. Environ. Res. Public Health. **18**, 524 (2021). https://doi.org/10.3390/ijerph18020524
*Sassen, R., Dienes, D., Wedemeier, J.: Characteristics of UK higher education institutions that disclose sustainability reports. Int. J. Sustain. High. Educ. **19**(7), 1279–1298 (2018). https://doi.org/10.1108/IJSHE-03-2018-0042
*Seibert, R.M., Macagnan, C.B.: Social responsibility disclosure determinants by philanthropic higher education institutions. Meditari Account. Res. **27**(2), 258–286 (2019). https://doi.org/10.1108/MEDAR-04-2018-0328
*Son-Turan, S., Lambrechts, W.: Sustainability disclosure in higher education: a comparative analysis of reports and websites of public and private universities in Turkey. Int. J. Sustain. High. Educ. **20**(7), 1143–1170 (2019). https://doi.org/10.1108/IJSHE-02-2019-0070
*Sun, Y., Ip, P.S.: The transformation of intellectual capital disclosures: exploring the online intellectual capital disclosure by Hong Kong universities. J. Intellect. Cap. **21**(6), 1109–1127 (2025). https://doi.org/10.1108/QRAM-09-2022-0143
*Suwanajote, N., Daiporn, A.: A guideline of performance report by indicators on the requirement of Suan Sunandha Rajabhat university. Int. J. Learn. Teach. **6**(2), 2 (2020)
*Tiron-Tudor, A., Nistor, C.S., Fekete, S., Alexandru, A.: Factors influencing public higher education institutions' performance reporting in the Romanian context. Administrative Sciences. **12**, 163 (2022). https://doi.org/10.3390/admsci12040163
*Trireksani, T., Zeng, Y.-T., Djajadikerta, H.G.: Extent of sustainability disclosure by Australian public universities: inclusive analysis of key reporting media. Aust. J. Public Adm. **80**(4), 830–848 (2021). https://doi.org/10.1111/1467-8500.12469
*Ulum, I., Harviana, R.R., Zubaidah, S., Jati, A.W.: intellectual capital disclosure and prospective student interest: an Indonesian perspective. Cogent Bus. Manag. 6:1, 1707041 (2019). https://doi.org/10.1080/23311975.2019.1707041
University of Wollongong Library: Systematic review: what is a systematic review? LibGuides. Accessed 12 June 2025. https://uow.libguides.com/systematic-review/what-is-a-systematic-review (2025)
*Vitolla, F., Raimo, N., Nicolò, G., Ricciardelli, A.: Exploring intellectual capital disclosure and its determinants in knowledge-based institutions: empirical evidence from international universities. Meas. Bus. Excell. **27**(2), 227–245 (2023). https://doi.org/10.1108/MBE-01-2022-0020
*Wahyudin, A., Utaminingsih, N.S., Yulianto, A., Sari, M.P., Putri, W.A.: The influence of intellectual capital (IC) disclosure and geographic location on the sustainability of higher education: a case in southeast Asian universities. Journal of Infrastructure Policy and Development. **8**(8), 3186 (2024) https://doi.org/10.24294/jipd.v8i8.3186
Ziemba, D.: Strategic planning in higher education: a comprehensive guide. Accessed 13 June 2025. https://www.ziemba.com (2025)

Teaching and Learning of Accounting in Higher Education Institutions in South Africa Through the Lens of the Fourth Industrial Revolution: A Systematic Literature Review

Tlotlo Segotso, Lilian Nwosu, Musili Oladele, and Innocent Mashava

Abstract The Fourth Industrial Revolution (4IR) has tremendously reshaped higher education institutions' (HEIs) teaching and learning approaches. As a result, the learning experiences and pedagogical approaches have significantly transformed due to the impact and implications of the 4IR tools incorporated in these processes, including in accounting and accounting education. However, students and lecturers are uncertain about how the adoption of these tools aligns with the institutional goals of the 4IR. This review intends to explore the impact and implications of integrating the 4IR technology tools in the teaching and learning of accounting in HEIs, with a specific focus on South Africa. The study employs a qualitative approach involving a systematic literature review (SLR), with the use of the Preferred Reporting Items for Systematic and Meta-Analysis (PRISMA) to analyse available data.in order to achieve the set objectives. Peer-reviewed articles published between 2020 and 2024, which were considered relevant to the study were analysed in a bid to further address the study objectives. Data was extracted from databases such as Google, Google Scholar, EBSCO host, JSTOR, Research Gate, and Science Direct during the research process. A search procedure, research keywords, exclusion and inclusion criteria, and quality assessments of chosen literature were used in conjunction with the SLR approach. Thirty-three documents were finally analysed after selecting them according to their themes and outline objectives. The study found that, while some South African HEIs have successfully adopted 4IR virtual and contact instruction technologies, more still has to be done to bridge the extant digital divide or gap. The study also found that, despite the substantial benefits that come with the adoption of 4IR in accounting pedagogy and learning, these technologies pose significant barriers for developing countries, due to inadequate infrastructure and lack of skilled and trained lecturers to adapt and use these emerging technologies that enhance

T. Segotso
University of Kwa Zula Natal, Durban, South Africa

L. Nwosu (✉) · M. Oladele · I. Mashava
North West University, Potchefstroom, South Africa
e-mail: lilian.nwosu@nwu.ac.za

E. H. Redda (ed.), *Proceedings of the Social Sciences International Research Conference, 2025*, https://doi.org/10.1007/978-981-95-9489-4_38

teaching and learning in the context of 4IR. Further, there are still misconceptions about the concepts of emerging technologies and digitalisation in accounting pedagogy. The study also found that, students and lecturers are still uncertain about the adoption of emerging technologies in the teaching and learning of accounting. The study concludes by observing that, while the 4IR is recognised to have obvious transformative effects on educational pedagogy and learning including in the accounting sector, an ongoing digitalisation training that outlines the need for these technologies in the 4IR should be implemented. The study further suggests that these technologies be properly adopted by HEIs in South Africa in the teaching and learning of accounting, and that proper investment in infrastructure and comprehensive training be regularly provided for academic staff.

Keywords Fourth Industrial Revolution · Benefits and challenges · Impact · Implications · Higher education · Higher education institutions · Accounting · Accounting education · Teaching and learning · Technology integration · Systematic review · Educational technology

1 Introduction

1.1 Introduction and Background

The 4IR was coined by Klaus Schwab as the big idea that builds on the previous digital revolutions, but is very distinct in the sheer speed, magnitude and disruptive impact it potentially could have on current emerging business models based on a sharing or on-demand economy (Schwab 2016; Yang 2018). The 4IR requires the blending of emergent technologies such as artificial intelligence, automation, and robotics, with the ever-increasing access and use of data and information learners have due to developments in mobile device technologies (Mbizi et al. 2022).

The 4IR has tremendously transformed and reshaped HEIs' teaching and learning approaches. As a result, the pedagogical approaches and learning experiences have witnessed significant changes due to the impact and implications of incorporating the 4IR tools in these processes, including in accounting and accounting education. However, students and lecturers are uncertain about how the adoption of these tools aligns with the current curricula demands or requirements and institutional goals of the 4IR (Tsiligris and Bowyer 2021). Further, the adoption and full implementation of 4IR technology is a cause for global economic concern. Hence, countries such as South Africa has been found to be lacking behind in terms of the country's preparedness for the 4IR against the background of the World Economic Forum's (WEF) benchmarking framework, as it is still in its nascent stage (Mthimkhulu et al. 2024).

Through the use of a SLR this paper examines the teaching and learning of accounting in HEIs in South Africa. This is particularly relevant as it speaks to accounting graduates' employability, which presents challenges and opportunities for HEIs to restructure their curricula and for accounting educators in South Africa to enhance their teaching methods. The relevance and value of this study is further underscored by the rapid transformation and adoption of 4IR technology and or tools such as virtual or hybrid classrooms, artificial intelligence (AI), and the Internet of Things (IoT) in the global pedagogical practise as well as in accounting education in HEIs in South Africa. The study concedes that, there is a dearth of information and study on the impact of the 4IR on the teaching and learning of accounting and its implications thereof, with most literature focusing on the impact of the 4IR on the broader economic and or educational sector. This study aims to bridge this gap by exploring the teaching and learning of accounting in the context of the 4IR, the benefits and challenges thereof, and the need for a paradigm shift by HEIs from the conventional to modern pedagogical practices ushered in by the 4IR.

The Industrial Revolutions including the 4IR and their effects on education, specifically HEIs must therefore be explained as it enables a background for better appreciation, contextualisation and understanding of the study. The first industrial revolution began in 1790 when the steam engine was developed and employed for industry in feudal society and for agricultural use by farmers. With the development of the internal combustion engine in 1900, the Second Industrial Revolution began, ushering in a period of electricity and oil powered machinery. When Information Technology (IT) and electronics were developed in 1960 to automate production, the Third Industrial Revolution got underway. The early 2003 saw the beginning of the.41R (Nwosu et al. 2023). The 4IR includes computer-generated data and products including algorithms, AI, 3D printing, cyber-physical systems, digital mobile services and the internet, machine learning, management of big data, distributed control systems such as smart electricity grids and manufacturing systems (Vurayai 2025).

The significant technical developments brought about by the 41R are changing a number of industries, including higher education. Yende (2021) asserts that the 4IR is now the primary engine propelling global improvements in higher education in the twenty-first century, as a result effective teaching and learning environment is seen to have been generated by the rapid technological developments. Gleason (2018) and Penprase (2018) emphasise how critical it is for higher education to keep up with the latest advancements in technology. In order to satisfy the demand for highly qualified graduates who can thrive in the 41R, they advised that universities enhance their current curricula programmes and launch new ones. To succeed in the 41R, HEIs are required to produce graduates with advanced abilities including digital literacy, problem-solving, and critical thinking (Gleason 2018). Generally, the education industry including HEIs in South Africa have been slow to implement 41R technology and only a small percentage of public universities have done so (Lubinga et al. 2023).

Gleason (2018) further asserts that without the required infrastructure educational institutions cannot be transformed for growth in the 41R. This entails sophisticated technology (internet, powerful computers), data (stream, unstructured, structured, big, and audio, picture, text, and sensor), institutional setup (curriculum), and knowledgeable professionals (formal, non-formal, and field or practical).

Furthermore, Landberg (2021) is of the view that, HEIs adapt their Accountancy curricula in order to equip students with the required skills for the changing technological environment. This is because; the Accountancy curriculum aims to bridge the expectations of the labour market and the competencies and capabilities of graduates to ensure employability. This means that HEIs must provide the required hardware, software, and technical assistance since students will be able to properly use 41R related technologies efficiently if HEIs upgrade their technological resources.

1.2 Problem Statement

According to Lubinga et al. (2023) several industries (including the accounting industry) do not fully grasp the 4IR, despite its advancement in various sectors of the economy. This has led to various stakeholders such as academics, business professionals, and legislators now giving more attention to the 4IR. HEIs therefore play a key role in prompting the required changes to adapt to the demands of the 4IR for enhanced growth and development. This is due to the fact that, higher education was previously created to cater for the demands of earlier Industrial Revolutions which prioritised mechanised production (Gleason 2018). Consequently, it is imperative that HEIs bridge the gap between the initial Industrial Revolutions and the 4IR by integrating the use of 4IR tools to enhance teaching and learning (Nwosu et al. 2023), even in accounting.

The benefits of integrating 4IR tools in the teaching and learning of accounting includes the automation of work tasks, reduced human error and enhanced productivity of accountants in the workplace (Van dan berg and Rothman 2024). Furthermore, the benefits of time and place convenience, cost effectiveness, student engagement, and increased access to education have been associated with the integration of 4IR tools in accounting education (Mthimkhulu et al. 2024). However, in order to maximise the full potential of the 4IR, student's capabilities and competencies need to meet the demands of the twenty-first century. Specific challenges include ensuring data security, complying with global regulatory requirements, managing ethical dilemmas, integrating big data analytics, fostering continuous learning and maintaining relevance as automation and AI increase (Van dan berg and Rothman 2024).

Further, one of the biggest obstacles to higher education students' ability to take advantage of 41R prospects particularly in Southern African nations is access to infrastructure and technology (Almutairi et al. 2025). In addition to students, institutions are also impacted by limited access. Furthermore, it is observed that, the majority of the region's universities use conventional, teacher-centred teaching methods. This means that, due to excessively large class sizes in most cases, students have little to no opportunity to engage in meaningful interactions with their teachers or with each other (Vurayai 2025). Lecturers and students, therefore, remain uncertain about how the adoption of these 4IR technologies or tools enhances teaching and learning, as well as how it aligns with the institutional goals of HEIs of the 4IR. This review intends to explore the benefits, challenges, and implications of integrating 4IR technology in the teaching and learning of accounting in HEIs in South Africa.

1.3 Objectives

1. To examine the impact of 41R in the teaching and learning of accounting in HEIs in South Africa.
2. To identify the benefits and challenges 41R integration in accounting pedagogy HEIs and how it can be resolved.
3. Examine accounting students' readiness and preparedness for the 41R.

1.4 Questions

1. What are the impacts and implications of the integration of 4IR tools in the teaching and learning of accounting in HEIs in South Africa?
2. What are the challenges facing the integration of 4IR in HEIs?
3. How prepared are accounting students for the demands of the 4IR?

2 Literature Review

The 4IR was coined by Klaus Schwab as the big idea that builds on the previous digital revolutions, but is very distinct in the sheer speed, magnitude and disruptive impact it potentially could have on current emerging business models based on a sharing or on-demand economy (Schwab 2016; Yang 2018). The 4IR requires the blending of the emergent technologies such AI, automation and robotics with the ever-increasing access and use of data and information learners have, due to developments in mobile device technologies (Mbizi et al. 2022).

The 4IR has therefore, transformed and reshaped HEIs' teaching and learning approaches. As a result, the didactic methods and learning experiences of both educators and students have witnessed significant changes due to the impact and implications of incorporating the 4IR tools in these processes, including in accounting and accounting education. However, students and lecturers are uncertain about how the adoption of these tools aligns with the current curricula demands or requirements and institutional goals of the 4IR (Tsiligris and Bowyer 2021). Further, the adoption and full implementation of 4IR technology is a cause for global economic concern. Hence, countries such as South Africa have been found to be lacking behind in terms of the country's preparedness for the 4IR against the background of the WEF's benchmarking framework, as it is still in the early developmental stage (Mthimkhulu et al. 2024).

The 41R technologies which includes AI, block chain, machine learning, robotics., cloud computing, data science, virtual reality, and 3D printing, has emerged as a result of technological advancements (Vurayai 2025). The WEF predicts that these technologies will cause significant shifts in people's social and economic lives, leading to the introduction of new employment and the loss of some. In order to work and preserve job security in the 41R era, people must enhance their skills. This is where HEIs come in, helping to build the abilities of the next generation and bridge the extant digital or technological gap. This will allow the HEIs to create and start training courses and educational programmes, as well as evaluate students' readiness, in order to promote their understanding of 41R and its applications (Mthimkhulu et al. 2024).

Cele et al. (2023) are of the view that the rise of cloud-hosted Information Communication Technology (ICT) infrastructure present, myriad opportunities in teaching and learning. Moreover, much emphasis is placed on low-cost, mobile, adaptable and widespread technology solutions, developed and frequently provided by academics and students. This implies that individual students and academics currently possess a significant amount of control over their learning and teaching. As automation becomes more sophisticated, modern skills are becoming increasingly important for young people entering the labour market. However, Almutairi et al. (2025) with reference to the WEF 2017 Report, observe that technologies may lead to increased inequality with regard to social stability, despite the fact that they have also been shown to increase access to education and equality. The problem in this case, is that, while new technologies bring about a growth in employment and entrepreneurship, they also pose new threats, such as the dominance of multinational technological corporate giants, which exacerbates social fragmentation and inequality (World Economic Forum 2017).

2.1 Challenges and Opportunities

According to Van dan berg and Rothman (2024) the benefits of integrating 4IR tools in accounting pedagogy includes the automation of work tasks, reduced human error, and enhanced productivity of accountants in the workplace. Also, the benefits of time and place convenience, cost effectiveness, student engagement and increased access to education have been associated with the integration of 4IR tools in accounting education (Mthimkhulu et al. 2024). However, in order to maximise the full potential of the 4IR, student's capabilities and competencies need to match the demands of the twenty-first century. Particular challenges include ensuring data security, complying with global regulatory requirements, managing ethical dilemmas, integrating big data analytics, fostering continuous learning, and maintaining relevance as automation and AI increase (Van dan berg and Rothman 2024).

Dangi et al. (2023) highlighted the need for educators and HEIs to integrate relevant 4IR technologies into their teaching and learning practices to meet various stakeholder expectations and keep abreast with the accounting professions advancement in the 4IR. In order for HEIs to further enhance their productive capacity which as has been observed, needs to shift from the conventional to the latest 4IR technologies to best equip graduates, in this case, accounting graduates with the requisite skills to thrive in the dynamic work environment ushered in by the 4IR, a prompt re-evaluation and revamping of the current accounting curricula, is therefore imperative (Landberg 2021). Karsten et al. (2020) observe that the present emphasis on the 4IR and debates on twenty-first century skills, promotes the complexity that HEIs are facing in terms of their curricular planning, teaching and learning practices. This further adds another dimension to the support required, if their aim is to cultivate students who possess the knowledge, skills and attributes, seen to be vital for the future of the accounting profession as envisioned by the South African Institute of Chartered Accountants (SAICA), amid the changing circumstances.

Vurayai (2025) further observe that one of the biggest obstacles to higher education students' ability to take advantage of 41R prospects particularly in Southern African nations is access to infrastructure and technology. In addition, students, institutions are also impacted by limited access. This speaks to the inequalities challenge mentioned by Almutairi et al. (2025) brought about by the digital or technological gap in society. This is because a lot of HEIs in the South African context lack the necessary infrastructure and technology due to limited national or state funding. The cost of setting up relevant 4IR technology and infrastructure is therefore enormous and one which has to be given more attention by the various governments in Subs-Saharan Africa, in order to bridge the extant technological divide or gap and inequalities as a result of the 4IR.

2.2 *Skills and Preparedness of Accounting Students*

A few studies have looked at how prepared HEI accounting students are for the 41R in general, and specifically South Africa. For example, Dangi et al. (2023) employing a mixed methodology, highlighted the slow pace of technology adoption, and the need for educators and HEIs to rapidly integrate relevant 4IR technologies into their teaching and learning practices to meet various stakeholder expectations, as well as keep abreast with the accounting professions advancement in the 4IR. This is in order for HEIs to further enhance their productive capacity which as has been observed, needs to shift from the conventional to the latest 4IR technologies to better equip graduates, in this case, accounting graduates with the requisite skills to thrive in the dynamic work environment ushered in by the 4IR.

Mbizi et al. (2022) study identified the competency and skills required by accountants to effectively use 4IR technologies at the workplace in the African context. With the use of a qualitative approach, the study found that, accountants on the continent are found wanton and mostly require computer skills and critical thinking skills to navigate the 4IR. The study also found that, accountants are expected to have data management skills, to effectively execute their duties in the 4IR era. Further skills that accountants need to possess include IT skills, data analytics, and complex problem-solving skills, specialised software and business acumen.

Van dan berg and Rothman (2024) study aimed at determining which twenty-first century skills were essential for financial accounting students in South African HEIs from the perspective of students, their educators and professionals, as well as other stakeholders. Using a quantitative approach and cross-sectional survey, the study revealed that thinking and learning to learn were the most critical twenty-first century skills for the financial accounting profession. Further, the study indicated that there exists a divergence among stakeholders concerning the second-highest ranked skill, with professionals preferring entrepreneurial and working skills, while educators and lecturers opted for communication and IT skills. For students, it was self-care and managing of daily life activities (self-preservation), while cultural and building a sustainable future were rated very minimally by the three stakeholder groups. The study concludes by proposing a new conceptual framework on the twenty-first century competencies and capabilities as perceived by different stakeholders.

Landberg (2021) study, investigated whether the current accounting curricula provide graduates (in this case Accountancy graduates in South Africa), the needed competencies and skills required in the twenty-first century. The study employed a systematic literature review (SLR), which involved a comparing of the proposed competency framework 2025CF, with the current competency framework 2021/2CF in order to identify additional skills. The study found that the current 2021/2CF, does not adequately meet the demands of the 4IR. The study concludes by recommending a prompt re-evaluation and revamping of the current 2021/2CF, in order to align with the 2025CF ushered in by the 4IR.

Tsiligris and Bowyer (2021) identified certain skills necessary for accounting graduates and practitioners in the current 4IR. These include but are not limited to adaptability and a lifelong approach to continuous personal and professional development, business skills, digital skills, ethical skills, and soft skills. The study embarked on a SLR of recent study available by accounting professional organisations delineating the impact of digital technologies on the accounting profession. These skills are lacking, and therefore necessary for accounting graduates and practitioners in order to successfully navigate the dynamic teaching, learning, and working environment as a result of the 4IR.

Finally, in a similar study, which examined the impact of the 4IR on the development of critical cross-field outcomes (CCFOs) among accounting students in South African HEIs, as well as, how to enhance students' autonomy and motivation to maximise CCFOs development. It was discovered that, there exist a substantial increase in relationship between the evolutionary trajectory of CCFOs among accounting students and the 4IR. The study indicated that there is a need for accountants to acquire the 4IR aligned CCFOs. The study also found that, there is a rising demand in the job market for accountants who possess the 4IR aligned CCFOs. With the use of a SLR, the study among others, indicated that there exists a noticeable gap as students lack the CCFOs highly sort after by employers, and recommends a multifaceted approach by all stakeholders including HEIs to bring about a change in the education curricula to accommodate the development of CCFOs in the context of the broader 4IR (Vorster 2023).

3 Methodology

The study employs a qualitative approach involving a SLR, with the use of the preferred reporting items for systematic review and meta-analysis (PRISMA) to analyse available data.in order to achieve the set objectives. Peer-reviewed articles published between 2020 and 2024, which were considered relevant were analysed in order to address the study objectives. Data was extracted from databases such as Google, Google Scholar, EBSCOhost, and Research Gate during the research process. A search procedure, research keywords, exclusion and inclusion criteria, and quality assessment of selected literature publications were also employed in combination with the SLR approach. The SLR offers current data and evidence on a specific study area and explains future research directions (Nwosu et al. 2023). According to Nwosu et al. (2023) the SLR is also considered as a method that assists researchers in answering the research question by evidencing a clear systematic proof based on the contemporary literature. The SLR process for this study used the PRISMA against the background of 4IR practices to improve the growth and development of accounting pedagogy in South African HEIs. This methodology was considered the best for the current study as it allowed the researchers to provide a detailed and systematic analysis of the bulk of available data concerning the use of 4IR technology tools as a teaching and learning resource in South African HEIs.

The databases for this study were chosen as they are appropriate for investigating the benefits and challenges of integrating the 4IR into the higher education accounting curricula in South Africa. For example, according to Olaitan et al. (2024) the Google Scholar provides a broad access to diverse scholarly works, including those from the Global South, while EBSCOhost covers various subject-specific databases, enabling detailed analysis of educational research, including studies from South Africa (Chisango and Lesame 2017). Collectively, these sources enabled a detailed blend of high-quality literature on the subject matter, with specific focus on study pertinent to Sub-Saharan Africa.

3.1 Search Strategy, Exclusion and Inclusion Criteria

The literature data used in this study was extracted from four databases including Google, Google Scholar, EBSCOhost, and Research Gate. The literature search used specific keyword or terms designed to encapsulate a broad range of information and material on the subject matter. The literature includes studies on the 4IR technologies, teaching and learning, teaching and learning of accounting, and South African HEIs. The following search keywords or terms were used to extract literature data: '4IR and teaching and learning in higher education institutions', '4IR and teaching and learning, challenges and benefits', '4IR teaching and learning of accounting in higher education institutions in South Africa'. An exclusion and inclusion determination was also applied after the initial search process, to further identify relevant information

Table 1 Summary of search results

Source	Number of records
Google	11
Google Scholar	10
EBSCOhost	9
ResearchGate	3
Total identified	33
Excluded	20
Included	13

or literature data on the theme of the study. The search databases were selected for their prevalence in the publication of academic research material focusing on the development and improvement of the use of 4IR technologies and tools in HEIs.

Exclusion Criteria. Certain exclusion criteria were applied to the study. Firstly, only literature published in the English Language was considered. No time limit was applied, however, only literature publications between 2020 and 2025, were considered relevant for a final analysis based on the objectives of the study. An initial number of sixty documents were extracted, after the search process was completed, and duplicated documents including articles, books or book chapters as well as conference contributions were excluded, a total of six duplicated documents were eliminated, resulting in thirty-three (33) literature publications.

Inclusion Criteria. In order to ensure that only relevant publications dealing with the subject matter were included, the researchers perused the abstracts of all 33 publications and classified them as either excluded or included in this review. As a result, twenty among the 33 research papers were deemed irrelevant to the current study; and as such, they were excluded. This led to the inclusion of thirteen publications which were considered pertinent to the study. This stage was conducted by two authors and further verified by two authors to reinforce the assessment, reducing the number of documents and avoiding excluding relevant and non-related articles (Fig. 1). A summary of the search results is presented in Table 1.

3.2 *Methodological Quality Assessment*

The methodological quality of the 13 literature publications used was evaluated by means of a six-question checklist propounded and developed by Pitchforth et al. (2017) and Nwosu et al. (2023). The objective was to assess whether these papers meet the study's exclusion or inclusion criteria or requirements. The checklist evaluates six study criteria or requirements by means of a 'yes' or 'no' score. Three authors independently conducted the quality assessment, which was further confirmed by one author to reinforce the assessment. An agreement on the relevance of the 13 literature publications for this study was therefore reached among the four authors regarding

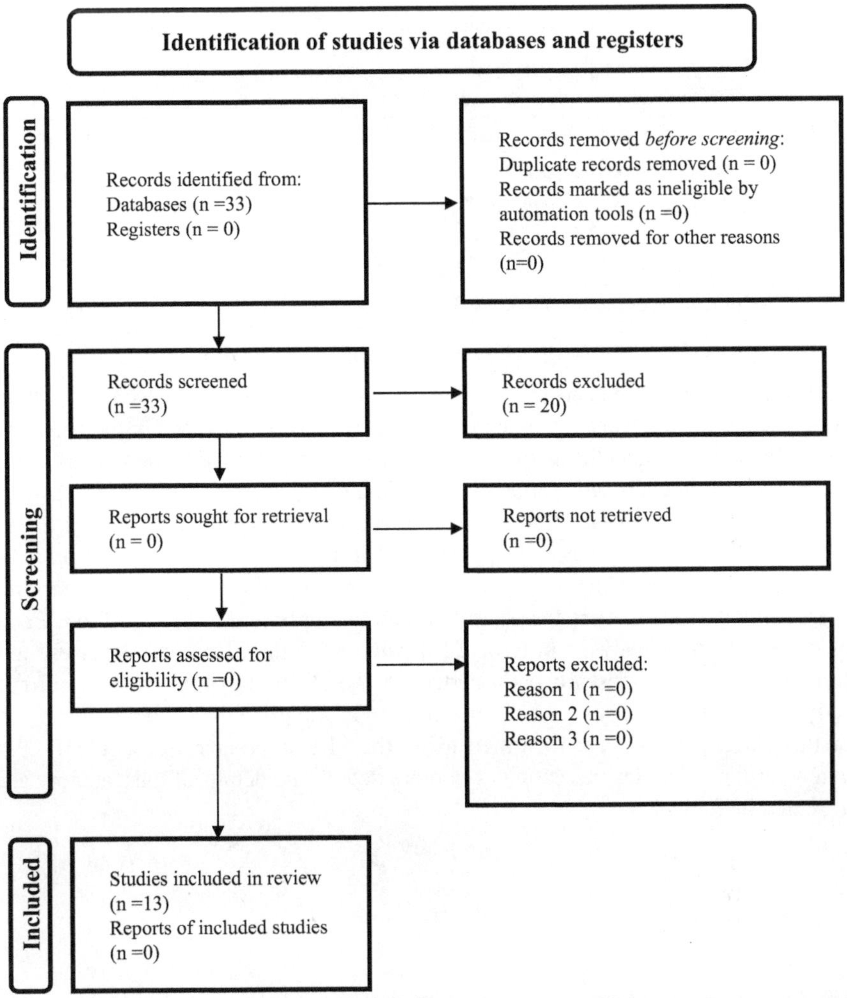

Fig. 1 Identification of studies via databases and registers. *Source* Adapted from Page et al. (2021)

the 13 documents employing the Pitchforth et al. (2017) checklist mentioned above (Table 2).

4 Results

In total, 13 literature publications focusing on the theme of the study were analysed based on content analysis and guidelines proffered by (Jarvis et al. 2003; Aarseth et al. 2017). Perusal of the article publications was implemented applying the exclusion and

Table 2 Methodological quality assessment criteria

Quality assessment question	Consensus
Q1 Are the objectives of the study clearly identified?	Yes
Q2 Is the context of the study clearly stated?	Yes
Q3 Do the research methods support the aims of the study?	Yes
Q4 Does the study provide a comprehensive description of the instruments or scales used?	Yes
Q5 Is there a clear statement of the findings?	Yes
Q6 Are the limitations of the study discussed extensively and explicitly?	Yes

inclusion criteria on the basis that only reviews focusing on the 4IR technology and tools in teaching and learning of accounting in HEIs were relevant. On the contrary, reviews that did not meet the inclusion criteria were considered irrelevant to this study and consequently excluded. Table 1 outlines the details of all the reviews analysed during the search process. They are categorised under separate headings as either 'Excluded in the current review' or 'Included in the current review'. The reviews focusing on the 4IR technology and tools to improve the teaching and learning of accounting in HEIs were therefore retained by applying the SLR method. Each review was further analysed in terms of the authors, title, theme, weaknesses, and strengths to allow the researchers ascertain the key factors that address the current study's research objectives. The themes used to accomplish the research objectives for this current study were identified using the SLR approach. Subsequently, the review further discusses these themes in more detail. The details of this category are presented in Table 3.

4.1 Findings

This review among others indicate that despite the substantial benefits that come with the adoption of 4IR, these technologies pose significant barriers for developing countries, due to insufficient infrastructure, and a lack of skilled and trained professionals to adapt and use these emerging technologies that enhance teaching and learning within the 4IR revolution road map (Almutairi et al. 2025; Vurayai 2025). The review also discovered that, there are still some misconceptions about the concepts of emerging technology and digitalisation in the teaching and learning of accounting, and that students and lecturers are still uncertain about the adoption of emerging technologies in teaching and learning of accounting (Skhephe et al. 2020; Tsiligris and Bowyer 2021).

Van dan berg and Rothman (2024), indicate that, the benefits of integrating 4IR tools in accounting pedagogy includes the automation of work tasks, reduced human error and enhanced productivity of accountants in the workplace. Further, Mthimkhulu et al. (2024) observe that, the benefits of time and place convenience,

Table 3 Categorisation of reviewed studies

Author	Title	Benefits of 4IR tools in the teaching and learning of accounting in HEIs in South Africa:	Challenges of 4IR tools in teaching and learning in HEIs in South Africa
Cele, W.S. (2022)	The applicability of Accounting Information Systems for the Financial Accounting curriculum in the Fourth Industrial Revolution era: A case study at the Durban University of Technology (DUT)	The benefits of using accounting software packages are numerous and significant. Essentially, accounting software packages aid businesses in a variety of processes. When compared to computerised accounting systems, manual accounting systems take longer to process information. Furthermore, manual systems require more personnel	A large number of students have never heard of accounting software packages or had awareness but little or no real experience with accounting software packages; graduates that are incompetent in the use of accounting software packages
Dangi, M.R.M. Saat, M.M., and Saad, S. (2023)	Teaching and learning using twenty-first century educational technology in accounting education: Evidence and conceptualisation of usage behaviour	Nowadays, technology plays an immense role in humans' lives, revolutionising many fundamental activities and penetrating most areas, including education, transforming research activities, scholarship and services to society in the academic field and enhancing teaching and learning. Educational technology refers to technology integration into the curriculum, altering the educational process. Students develop new skills and enhance their learning experience through various technology platforms technology supplements teaching materials, offering students a new customisable learning experience with a self-paced, problem based, interactive learning environment, improving their knowledge and understanding Past research; shows educational technology improves students' motivation, satisfaction, attitudes, interaction, engagement and academic performance	The World Economic Forum (2018) reported several declining occupations such as accounting, bookkeeping, accountants and auditors due to technology's advancement; the accounting curriculum neglecting technology aggravates the situation

(continued)

Table 3 (continued)

Author	Title	Benefits of 4IR tools in the teaching and learning of accounting in HEIs in South Africa:	Challenges of 4IR tools in teaching and learning in HEIs in South Africa
Landberg, E. (2021)	Investigating the preparedness of the Accountancy curricula for the 4th industrial revolution	The purpose of the Accountancy curriculum aims to bridge the expectations of the labour market and the competencies of graduates to ensure employability	The business environment is rapidly changing while the Accounting curricula are too slow to keep up. Literature on specific and detailed guidance on how to develop the accounting curricula for the 4IR is lacking. Literature indicates that the accounting curricula are not currently up to pace to deliver 4IR ready students for the labour market and guidance and development is required to achieve the goal
Mbizi, R., Sifile, O., Gasheja, F., Twesige, D., Gwangava, E., Makurumidze, S., Matowanyika, K., Chinofunga S., and Sunday, K. (2022)	Accountants in Africa and the evolving Fourth Industrial Revolution (4IR): Towards a competency framework	Digitalisation is affecting organisations on how they organise, process and evaluate the financial data which improves productivity and saves costs and time. The block chain is changing the accounting and auditing processes in the world. It is creating new opportunities and challenges in accounting and auditing. Block chain technology can be applied in property records, banking, supply chain, auditing, anti-money laundering, customer survey, stock trading, smart customers. IT offers opportunities to accounting professionals and practitioners to explore the environment and claim the benefits related to digitalisation. Digitalisation is making machines perform accounting tasks better and faster than humans	In spite of the importance of the accounting profession in ensuring development and achieving the SDGs, the accounting professional in Africa is faced with a number of challenges. Previous research show that the accounting professional in Africa is faced with a lack of capacity to implement accounting reforms, poor technological advancement as well as political and institutional interference at the workplace

(continued)

Table 3 (continued)

Author	Title	Benefits of 4IR tools in the teaching and learning of accounting in HEIs in South Africa:	Challenges of 4IR tools in teaching and learning in HEIs in South Africa
Mthimkhulu, T.T., Gamede B.T., and Ezenia, J.C. (2024)	Impact of 4IR On Accounting Education in Higher Education	The benefits of time and place convenience, cost effectiveness, student engagement and increased access to education have been associated with the integration of 4IR tools in accounting education (Mthimkhulu et al. 2024)	Further, the adoption and full implementation of 4IR technology is a cause for global economic concern. Hence, countries such as South Africa have been found to be lacking behind in terms of the country's preparedness for the 4IR against the background of the WEF's benchmarking framework, as it is still in the early developmental stage
Nabi, N. (2024)	Digital Literacy in Accounting Education: Navigating the Fourth Industrial Revolution (4IR) Landscape	Several studies have highlighted the need for accounting education to adapt to rapid technological changes and integrate digital literacy, technology literacy, and human literacy to prepare students for work readiness in the 4IR era (Rosi and Mahyuni 2021; Tran and Phan 2019)	With the future of 4IR, Internet of Things, big data, artificial intelligence, autonomous technologies etc. are being adopted by the individuals and organisation that are performing human jobs (Santoso and Lestari 2019). In consequence, the 4IR passes opportunities and threats for existing and future accountants and other jobs in the accounting field (Mohamed 2018). Jobs like accountant and auditor, bookkeeping, accounting and auditing clerks, revenue agents, and budgets analysts, tax examiners and collectors will be replaced by technology (Kroon et al. 2021)

(continued)

Table 3 (continued)

Author	Title	Benefits of 4IR tools in the teaching and learning of accounting in HEIs in South Africa:	Challenges of 4IR tools in teaching and learning in HEIs in South Africa
Papageorgiou, E., Van der Merwe, N., Terblanche, W., and Lubbe, I. (2024)	Perceptions of Using E-learning Applications in Accounting At Residential Universities In South Africa: A social Justice Lens	Typically, the internet. E-learning applications are interactive online services that provide learners with information, tools, and resources that support and enhance educational delivery and management (Kumar Basak et al. 2018; Terblanche et al. 2023). Students who experience e-learning as a replacement for or addition to face-to-face learning, report benefits around accessing course information, administration, and support	Prior studies of student experiences of online distance learning identified self-regulation skills such as time management, balancing academic and personal life, and taking personal responsibility for learning; developing an online community including accessibility of lecturers and connection with peers; and support from family and friends were needed to deal with the lack of structure and proximity of an online learning environment (Becker and Schad 2022; Blackmon and Major 2012)
Tavares, M.C. Azevedo, G., Marques, R.P., and Bastos, A. (2023)	Challenges of education in the accounting profession in the Era 5.0: A systematic review	This context of accelerated change is an opportunity for accounting and accounting professionals. Accounting is more influential in our lives than many of us think or can imagine. Accounting, beyond a professional technique, is a social and moral practice that can help move capabilities and influence the creation of a better world (Carnegie 2022), consistent with the principles of sustainability and the Era 5.0	However, accounting education has not kept pace with the technological innovation of multinational accounting firms, despite these firms' large investment in technological innovation. In the face of changes in the industry, there is a need for existing accounting curricula at universities to prepare and 'equip' students with the necessary technological skills for successful careers in accounting and auditing (Damerji and Salimi 2021). With the new needs for information literacy, whether in the field of digitisation or big data, accounting (and finance) can take on a more strategic role and help to shape the future. It is accounting education that must produce accountants who match the needs of industry and society, with a curriculum that covers these areas of study (Surianti 2020)

(continued)

Table 3 (continued)

Author	Title	Benefits of 4IR tools in the teaching and learning of accounting in HEIs in South Africa:	Challenges of 4IR tools in teaching and learning in HEIs in South Africa
Tsiligris, V. and Bowyer, D. (2021)	Exploring the impact of 4IR on skills and personal qualities for future accountants: a proposed conceptual framework for university accounting education	The benefits of using accounting software packages are numerous and significant. Essentially, accounting software packages aid businesses in a variety of processes. When compared to computerised accounting systems, manual accounting systems take longer to process information. Furthermore, manual systems require more personnel	There was consensus among accountants that their very existence was being threatened by technologies that surround 4IR e.g., Smart Contracts through block chain technology, artificial intelligence and machine learning. In order to accomplish accounting tasks, accountants had to possess both technical and social skills
Vorster (2023)	The Influence of the Fourth 4IR on the development of critical cross-field outcomes (soft skills) in Accounting Students in higher education institutions in South Africa: A systematic literature review	By incorporating CCFOs, HEIs can promote lifelong learning, personal development, ethical business judgement, critical thinking, and social responsibility (Tsiligris and Bowyer 2021). According to them accounting students need to apply their knowledge to demonstrate competencies that require a wide range of abilities and attributes. HEIs must expose students to CCFOs at the higher education level, providing learning experiences that facilitate their application and achievement of these skills (Tsiligris and Bowyer 2021)	Business faculties are under pressure, according to De Villiers (2010), to provide accounting graduates with the orientation, attitudes, qualities, and CCFOs they require. According to Berry and Routon (2020), the debate over the CCFOs needed by accountants is not wholly new; academics have long argued that in order to sustain a sustainable profession, accountants need to properly deploy critical abilities

(continued)

Table 3 (continued)

Author	Title	Benefits of 4IR tools in the teaching and learning of accounting in HEIs in South Africa:	Challenges of 4IR tools in teaching and learning in HEIs in South Africa
Van den Berg, E., and Rothman, S. (2024)	Twenty-first-century competencies and capabilities for financial accounting students	In the Fourth Industrial Revolution, pertinent individual competencies are not clearly defined for accounting (Jarosz et al. 2020). However, the shift to multidisciplinary teams, the ability to navigate the integration between humans and machines and the management of an instant yet colossal amount of available information is not merely predicted but gradually advancing into the workforce (Espahbodi 2020; Jarosz et al. 2020; Mah'd and Mardini 2020; Prifti et al. 2017)	Bowyer and Ghalayini (2022) identified a discrepancy between employers' expectations of soft skills and students' belief that technical skills and grades are the most important factors for employment. They concluded that soft skills (competencies B and G) are more critical for post-degree employability than technical skills. These findings have significant implications for accounting graduates' job market readiness. A study by Schwinger et al. (2022) confirmed the importance of students' competency in self-care and managing everyday life (competency C). Learning environments that foster academic motivation and decrease concerns about failure are needed to reduce students' self-handicapping tendencies. The most powerful predictors of self-handicapping were emotional instability, conscientiousness and fear of failure
Skhephe, M., Ntombekhaya, P.C., and Boadzo, R.K. (2020)	Accounting Teachers' Readiness for e-learning in the Fourth Industrial Revolution: A case of selected High Schools in the Eastern Cape, South Africa	With the inclusion of technology within the subject (accounting), the methods of getting cutting edge information around the subject has improved as a result and teachers and learners are able to go beyond the limit and are able to receive all the information they want with the help of e-learning resources	Accounting classrooms are not designed in a way that supports e-learning. Another finding was that accounting teachers do not understand e-learning or the benefit associated with an e-learning classroom

(continued)

Table 3 (continued)

Author	Title	Benefits of 4IR tools in the teaching and learning of accounting in HEIs in South Africa:	Challenges of 4IR tools in teaching and learning in HEIs in South Africa
Peens, S. (2021)	Higher education institutions' Need To Initiate Change To The Accounting Curriculum During the Fourth Industrial Revolution (4IR)	Secondary school (FET) Accounting is beneficial to the performance of students in their first year of accounting studies if there is a close association between the two curricula	In South Africa, the Further Education and Training phase in accounting faces a major decline in learner numbers. The current format of FET Accounting serves little purpose in preparing learners for accounting courses at tertiary level, if FET Accounting is not a precondition to Professional and Chartered Accountant courses

cost effectiveness, student engagement and increased access to education have been associated with the integration of 4IR tools in accounting education. Furthermore, several studies have highlighted the need for accounting education to adapt to rapid technological changes and integrate digital literacy, technology literacy, and human literacy to prepare students for work readiness in the 4IR era (Nabi 2024; Rosi and Mahyuni 2021; Tran and Phan 2019). The study therefore aligns with the view that the aim or purpose of the Accountancy curriculum is to bridge the expectations of the labour market and the competencies of graduates to ensure employability, even in the current 4IR era (Landberg 2021).

However, this study observed that, accounting education has not kept abreast with the developments ushered in by the 4IR despite large investment in technological innovation. It therefore suffices that, due to innovations in the industry, there is a need for existing accounting curricula at HEIs to prepare and train students with the necessary technological skills for successful careers in accounting and auditing (Damerji and Salimi 2021).

Therefore, while the 4IR is recognised to have potential transformation in all sectors of life and the economy including accounting, an ongoing digitalisation training that outlines the need for these technologies and how they can be adequately adopted in teaching and learning of accounting, and proper investment in infrastructure as well as comprehensive training should be provided for the academic staff and other accounting professionals on a regular basis.

4.2 Discussion

This review further, offers insight into the integration of 4IR technology and tools for promoting the growth and development of accounting pedagogy in HEIs in South Africa, while ascertaining relevant themes through a SLR of studies (as outlined in Table 3). The themes identified therefore address the challenges and opportunities in the integration or use of 4IR technology and tools in promoting the teaching and learning of accounting in HEIs, and ways to combat these challenges. These themes address the study's main objectives.

4.3 The Impact and Implications of the Integration of 4IR Technology and Tools in the Teaching and Learning of Accounting in Higher Education Institutions in South Africa

The important technological advancements ushered by the 41R are transforming a number of sectors, including higher education. Yende (2021) states that the 4IR is now the main engine propelling global improvements in HE in the twenty-first century, as such effective teaching and learning environment is seen to have been generated by the rapid technological developments. Gleason (2018) and Penprase (2018) emphasise how critical it is for higher education to keep up with the latest advancements in technology. In order to satisfy the demand for highly qualified graduates who can thrive in the 41R, they advised that universities enhance their current curricula programmes and launch new ones. To succeed in the 41R, HEIs are required to produce graduates with advanced abilities including digital literacy, problem-solving, and critical thinking (Gleason 2018). Generally, the education industry including HEIs in South Africa have been slow to implement 41R technology and only a small percentage of public universities have done so (Lubinga et al. 2023).

The 4IR has therefore, transformed and reshaped HEIs' teaching and learning approaches. As a result, the didactic methods and learning experiences of both educators and students have witnessed significant changes due to the impact and implications of incorporating the 4IR tools in these processes, (including in accounting and accounting education). However, students and lecturers are uncertain about how the adoption of these tools aligns with the current curricula demands or requirements and institutional goals of the 4IR (Tsiligris and Bowyer 2021). Further, the adoption and full implementation of 4IR technology is a cause for global economic concern. Hence, countries such as South Africa have been found to be lacking behind in terms of the country's preparedness for the 4IR against the background of the WEF's benchmarking framework, as it is still in the early developmental stage (Mthimkhulu et al. 2024).

As a result, while the 4IR positions to transform all sectors of life and the economy including accounting, an ongoing training that outlines the need for these 4IR technologies and tools and how they can be adequately adopted in teaching and learning of accounting, and proper investment in infrastructure as well as comprehensive training should be provided for the academic staff and other accounting professionals on a regular basis. Further, a speedy re-evaluation and overhauling of the current 2021/2CF, in order to align with the 2025CF ushered in by the 4IR is therefore necessary in HEIs in South Africa Landberg (2021).

4.4 Benefits of 4IR Technology and Tools Integration in the Teaching and Learning of Accounting in Higher Education Institutions in South Africa

The benefits of using accounting software packages are numerous and significant. Essentially, accounting software packages aid businesses in a variety of processes. When compared to computerised accounting systems, manual accounting systems take longer to process information. Furthermore, manual systems require more personnel (Cele 2022). It, therefore, implies that, for accounting students and graduates to be able to effectively use these software packages even in the workplace, HEIs are expected to provide the required training and practical experience to students, by integrating the use of 4IR technology and tools in teaching and learning. A curricula alignment is also imperative in this regard and should be implemented in order to meet the competency and skills demand of the twenty-first century. Further, the 41R technologies such as AI, block chain, machine learning, robotics., cloud computing, data science, virtual reality, and 3D printing, has emerged as a result of technological advancements (Vurayai 2025). The integration of these 4IR technology and tools in the teaching and learning of accounting HEIs is therefore imperative.

According to Papageorgiou et al. (2024), the Internet and other E-learning applications are interactive online services that provide learners with information, tools, and resources that support and enhance educational delivery and management (Kumar Basak et al. 2018; Terblanche et al. 2023). They further assert that, students who experience e-learning as a replacement for or addition to face-to-face learning, report benefits around accessing course information, administration, and support. The benefits of time and place convenience, cost effectiveness, student engagement and increased access to education have been associated with the integration of 4IR tools in accounting education (Mthimkhulu et al. 2024).

Furthermore, in addition to 4IR technology competencies and skills, HEIs can promote lifelong learning, personal development, ethical business judgement, critical thinking, and social responsibility by incorporating CCFOs (Vorster 2023; Tsiligris and Bowyer 2021). According to Tsiligris and Bowyer (2021) accounting students need to apply their knowledge to demonstrate competencies that require a wide range of abilities and attributes. HEIs must therefore expose students to CCFOs at the higher education level, providing learning experiences that facilitate their application and achievement of these skills (Tsiligris and Bowyer 2021).

Finally, for this purposes of the section and in the context of a high schools studies, it was found that, with the integration of technology within the subject (accounting), the methods of getting cutting edge information around the subject has improved as a result and teachers and learners are able to go beyond the limit and are able to receive all the information they want with the help of e-learning resources (Skhephe et al. 2020). The latter is an encouraging development as it indicates that, high school learners including in the subject of accounting are being giving the necessary education, exposition and preparation for the use of the 4IR technology and tools prior to their progression into HEIs.

4.5 Challenges of 4IR Technology and Tools Integration in the Teaching and Learning of Accounting in Higher Education Institutions in South Africa

Despite the importance of the accounting profession in ensuring development and achieving the Sustainable Development Goals (SDGs), the accounting professional in Africa is faced with a number of challenges. Initial study indicates that the accounting professional on the continent is faced with a lack of capacity to implement accounting reforms, poor technological advancement as well as political and institutional interference at the workplace. The adoption and full implementation of 4IR technology is therefore a cause for global economic concern. Hence, countries such as South Africa have been found to be lacking behind in terms of the country's preparedness for the 4IR against the background of the WEF's benchmarking framework, as it is still in the early developmental stage (Mthimkhulu et al. 2024). The WEF (2018) reported several declining occupations such as accounting, bookkeeping, accountants, and auditors due to technology's advancement; the accounting curriculum neglecting technology aggravates the situation.

In South Africa, accounting classrooms are not designed in a way that supports e-learning. Another finding was that accounting teachers do not understand e-learning or the benefit associated with an e-learning classroom (Skhephe et al. 2020). Also, the Further Education and Training phase (FET) in accounting faces a major decline in learner numbers, this is because the current format of FET Accounting serves little purpose in preparing learners for accounting courses at tertiary level, if FET Accounting is not a precondition to Professional and Chartered Accountant courses Peens (2021). Further, a large number of students have never heard of accounting software packages or had awareness but little or no real experience with accounting software packages; which results in graduates that are incompetent in the use of accounting software packages. As a result, business faculties are under pressure, according to De Villiers (2010), to provide accounting graduates with the orientation, attitudes, qualities, and CCFOs they require.

Furthermore, there was consensus among accountants that their very existence was being threatened by technologies that surround 4IR for example, Smart Contracts through block chain technology, AI, and Machine Learning. In order to accomplish accounting tasks, accountants had to possess both technical and social skills (Tsiligris and Bowyer 2021). However, accounting education has not kept pace with the technological innovation of multinational accounting firms, despite these firms' large investment in technological innovation. There is a need therefore for existing accounting curricula at universities to prepare and equip students with the necessary technological skills for successful careers in accounting and auditing in the face of changes in the industry (Damerji and Salimi 2021). With the new requirements for information literacy, whether in the field of digitisation or big data, accounting (and finance) can take on a more strategic role and help to shape the future. It is accounting education that must produce accountants who match the needs of industry and society, with a curriculum that covers these areas of study (Surianti 2020). In order to enhance the accounting curricula to cater to the demands of the 4IR, a prompt re-evaluation and revamping of the current 2021/2CF, so as to align with the 2025CF ushered in by the 4IR is imperative in HEIs in South Africa Landberg (2021).

4.6 Accounting Students' Readiness and Preparedness for the 4IR

As indicated above, the study identified that in South Africa; accounting classrooms are not designed in a way that supports e-learning. Another finding was that accounting teachers do not understand e-learning or the benefit associated with an e-learning classroom (Skhephe et al. 2020). Also, the FET phase in accounting faces a major decline in learner numbers; this is because the current format of FET Accounting serves little purpose in preparing learners for accounting courses at tertiary level, if FET Accounting is not a precondition to Professional and Chartered Accountant courses Peens (2021). This situation is untenable as it defeats the purpose of students aiming to pursue a career in accounting by doing so at any of the FET colleges or institutions across the country, despite the position of FETs being recognised as HEIs in South Africa. It further reduces the moral of accounting students in FETs, knowing that their qualifications are not worth its value and as such prepare them for the accounting profession or workplace. This is even more concerning in the context of the study, as the study observes that a change in the status quo is necessary, as the teaching and learning of accounting in FETs, should also align with the demands of the 4IR in order to enhance students' readiness and preparedness for the demands of the twenty-first century on the accounting profession. Further, the study finds that, a large number of students have never heard of accounting software packages or had awareness but little or no real experience with accounting software packages, which results in graduates that are incompetent in the use of accounting software packages. As a result, business faculties are under pressure according to De Villiers (2010), to provide accounting graduates with the orientation, attitudes, qualities, and CCFOs they require to succeed in the 4IR era.

The study furthermore, observed that, notwithstanding the benefits of integrating 4IR technology tools in the teaching and learning of accounting, such as the automation of work tasks, reduced human error, and enhanced productivity of accountants in the workplace, time and place convenience, cost effectiveness, student engagement and increased access to education; accounting students and lecturers still remain in the dark on how the adoption of these technological tools aligns with the current curricula demands or requirements and institutional goals of the 4IR (Mthimkhulu et al. 2024; Tsiligris and Bowyer 2021).

Finally, the study found that, one of the biggest obstacles to higher education students' inadequate preparedness and readiness for the twenty-first century era is access to infrastructure and technology. In addition, students, institutions are also impacted by limited access, which further highlights the inequality challenges brought about by the digital or technological gap in society due to the 4IR era. This is because a lot of HEIs in the South African context still lack the necessary infrastructure and technology due to limited national or state funding. The cost of setting up relevant 4IR technology and infrastructure is therefore enormous and one which needs to be given more attention by the various governments in Subs-Saharan Africa in order to bridge the extant technological divide and inequalities as a result of the 4IR (Vurayai 2025; Almutairi et al. 2025).

5 Contribution of the Study

Considering, the dearth of information and study on the impact of the 4IR on the teaching and learning of accounting and its implications thereof, with most literature focusing on the impact of the 4IR on the broader educational sector. This study attempted to fill this gap by exploring the teaching and learning of accounting in the context of the 4IR, the benefits and challenges, and the need for a paradigm shift by HEIs from the conventional to modern pedagogical practices ushered in by the 4IR. The relevance and contribution of this study is further underscored by the rapid transformation and adoption of 4IR technology and or tools such as virtual or hybrid classrooms, AI, and the IoT in the global pedagogical practise as well as in accounting education in HEIs in South Africa. The study and its findings therefore hope to shed more light on the impact, implications, challenges, and benefits of integrating the use of 4IR technology and tools in the teaching and learning of accounting in HEIs in South Africa. Through the use of a SLR this paper propounds a change in the current HEIs curricula, and official and national policy framework in accounting teaching and learning by implementing as well as using the 4IR technological tools. This study has the potential to influence both policy and practice within South African higher education. Through systematically reviewing teaching and learning approaches in accounting through the lens of the Fourth Industrial Revolution, the findings can provide evidence-based insights for higher education institution to align national curriculum standards with the digital competencies demanded by industry.

Furthermore, the results may guide HEIs in adopting innovative pedagogical strategies, enhancing lecturer training, and integrating technology to better prepare students for a rapidly evolving professional landscape. Explicitly connecting these findings to policy frameworks and institutional practices underscores the broader impact and relevance of this research.

6 Conclusion and Recommendations

This study set out to examine how integration of the 4IR technology and tools can be used to advance the growth and development of accounting pedagogy in South African HEIs. Through employing the SLR methodology, this study proffers in-depth information on the theme of the study being the impact and implications of the 4IR technology and tools integration and application in teaching and learning of accounting in HEIs in South Africa. The choice on methodology was based on the consideration that the SLR method was relevant to the study. This study found that South African higher educational institution faces significant challenges, including limited infrastructure, inadequate lecturer training, outdated curricula, and unequal access to technology, which hinder students' readiness for 4IR demands. To address these challenges, the study recommends that relevant stakeholders such as government, HEIs, academics and employers should play their role by among others, implementing policies that promote the integration of 4IR technologies and tools in teaching and learning. Secondly, that continuous digitalisation training that outlines the need for these technologies in the 4IR be provided on a regular basis to the academics, students and other professionals, on how these technologies can be adequately adopted in teaching and learning of accounting, as well as proper investment in infrastructure by all relevant stakeholders. Thirdly, the study recommends that, HEIs' curricula including that of the FETs be re-valuated and the necessary changes implemented to align with the demands of the 4IR, in order to enhance students' preparedness and readiness to function in the twenty-first century work environment as accounting professionals. Finally, further study on the theme of this review paper is recommended. This could be done using an empirical approach sampling from a few or more HEIs in South Africa. A further approach could be a mixed method involving SLR and results from empirical survey, as this would provide a more in-depth analysis on the challenges of the 4IR and teaching and learning of accounting in the country.

References

Aarseth, W., Ahola, T., Aaltonen, K., Økland, A., Andersen, B.: Project sustainability strategies: a systematic literature review. Int. J. Project Manage. **35**(6), 1071–1083 (2017). https://doi.org/10.1016/j.ijproman.2016.11.006

Almutairi, A., Al-Saad, A., Elmelegy, A., Bakr, A., Abdallah, A., Almotaire, A.: Systematic literature review of the Fourth Industrial Revolution in higher education: implications for sustainability (2025)

Berry, R., Routon, W.: Soft skill change perceptions of accounting majors. J. Account. Educ. **53**, 100691 (2020)

Becker, J., Schad, M.: Understanding the lived experience of online learners: towards a framework for phenomenological research on distance education. Online Learn. **26**(2), 296–322 (2022)

Blackmon, S., Major, C.: Student experiences in online courses a qualitative research synthesis. Q. Rev. Distance Educ. **13**(2), 77–85 (2012)

Bowyer, D., Ghalayini, G.: Future proofing tomorrow's accounting graduates: skills, knowledge and employability. Australas. Account. Bus. Finance J. **16**(3), 55–72 (2022). https://doi.org/10.14453/aabfj.v16i3.05

Carnegie, G.D.: Accounting 101: redefining accounting for tomorrow. Acc. Educ. **31**(6), 615–628 (2022). https://doi.org/10.1080/09639284.2021.2014915

Cele, W.: The applicability of accounting information systems for the Financial Accounting curriculum in the Fourth Industrial Revolution era: a case study at the Durban University of Technology (DUT) (Doctoral dissertation) (2022)

Cele, S., Bhana, A., Matli, W.: Exploring undergraduate students' perception of 4IR digital era at a higher education institution in South Africa. J. Account. Res. Organ. Econ. **6**(1), 49–62 (2023). https://doi.org/10.24815/jaroe.v6i1.32046

Chisango, G., Lesame, C.: Challenges of ICT policy implementation in rural South Africa. Communitas **22**, 48–61 (2017). https://doi.org/10.18820/24150525/comm.v22.4

Damerji, H., Salimi, A.: Technology readiness and adoption of artificial intelligence in accounting. Acc. Educ. **30**(2), 107–130 (2021). https://doi.org/10.1080/09639284.2021.1872035

Dangi, M.R.M., Saat, M.M., Saad, S.: Teaching and learning using 21st century educational technology in accounting education: evidence and conceptualisation of usage behaviour. Australas. J. Edu. Technol. **39**(1), (19–38) (2023)

De Villiers, R.: Incorporating soft skills into accounting curricula. Meditari Accountancy Res. **18**(2), 1–22 (2010)

Espahbodi, L.: Multistakeholder alliances and equity in higher education for the Fourth Industrial Revolution. Amstat News **520**, 8–9 (2020)

Gleason, N.W. (ed.): Higher Education in the Era of the Fourth Industrial Revolution. Springer Nature, Singapore (2018). https://doi.org/10.1007/978-981-13-0194-0

Jarvis, C.B., MacKenzie, S.B., Podsakoff, P.M.: Construct indicators and measurement model misspecification. J. Consum. Res. **30**(2), 199–218 (2003). https://doi.org/10.1086/376806

Jarosz, S., Soltysik, M., Zakrzewska, M.: Fourth Industrial Revolution and competence change. Eur. Res. Stud. J. **23**(1), 530–548 (2020). https://doi.org/10.35808/ersj/1776

Karsten, J., Van der Merwe, N., Steenkamp, L.: Empowering accounting students for self-determination skills in the Fourth Industrial Revolution (2020)

Kroon, N., do Céu Alves, M., Martins, I.: The impacts of emerging technologies on accountants' role and skills: connecting to open innovation—a systematic literature review. J. Open Innov.: Technol. Mark. Complex. **7**(3), 163 (2021)

Kumar Basak, S., Wotto, M., Bélanger, P.: E-learning, m-learning and d-learning. E-Learning Digital Media **15**(4), 191–216 (2018). https://doi.org/10.1177/2042753018785180

Landberg, E.: Investigating the preparedness of accountancy curricula for the Fourth Industrial Revolution (2021)

Lubinga, E., Maramura, T., Masiya, T.: Adoption of the Fourth Industrial Revolution in South African higher education institutions. J. Cult. Values Educ. **6**(2), 1–17 (2023). https://doi.org/10.46303/jcve.2023.5

Mah'd, O.A., Mardini, G.H.: Quality of accounting education and international education standards. Acc. Educ. **31**(2), 113–133 (2020). https://doi.org/10.1080/09639284.2020.1790020

Mbizi, R., Sifile, O., Gasheja, F., Twesige, D., Gwangava, E., Makurumidze, S., Matowanyika, J., Chinofunga, L., Sunday, K.: Accountants in Africa and the evolving Fourth Industrial Revolution. Towards a competency framework (2022)

Mohamed, M.: Challenges and benefits of industry 4.0: an overview. Int. J. Supply Oper. Manage. **5**(3), 256–265 (2018)

Mthimkhulu, T.T., Gamede, B.T., Ezenia, J.C.: Impact of the Fourth Industrial Revolution on accounting education in higher education (2024)

Nabi, N.: Digital literacy in accounting education and the Fourth Industrial Revolution (2024)

Nwosu, L., et al.: Fourth Industrial Revolution tools for teaching and learning in higher education. A systematic literature review (2023)

Olaitan, O., Vijayalekshmi, R., Kumar, P.: Integrating Fourth Industrial Revolution technologies into higher education in South Africa (2024)

Page, M.J., Moher, D., Bossuyt, P.M., Boutron, I., Hoffmann, T.C., Mulrow, C.D., Shamseer, L., Tetzlaff, J.M., Tricco, E.E., Turner, L., et al.: PRISMA 2020 explanation and elaboration: updated guidance and exemplars for reporting systematic reviews. BMJ **372**, n160 (2021)

Papageorgiou, E., Van der Merwe, N., Terblanche, W., Lubbe, I.: 2024. Perceptions of using e-learning applications in accounting at residential universities in South Africa: a social justice lens. S. Afr. J.High. Educ. **38**(6), 203–225 (2024)

Penprase, B.E.: Fourth Industrial Revolution and higher education. In: Gleason, N.W. (ed.) Higher Education in the Era of the Fourth Industrial Revolution, pp. 207–229. Palgrave Macmillan, Singapore (2018)

Pitchforth, E., et al.: Community hospitals and their services in the NHS. NIHR J. Libr. **5**, 1–220 (2017)

Prifti, L., Knigge, M., Kienegger, H., Krcmar, H.: A competency model for Industrie 4.0 employees. In: Proceedings of the 13th International Conference on Wirtschaftsinformatik. St. Gallen (2017)
Rosi, N.M.K., Mahyuni, L.P.: Future of the accounting profession in Industry 4.0. E-Jurnal Akuntansi **31**(4) (2021)
Santoso, A., Lestari, S.: The roles of technology literacy and technology integration to improve students' teaching competencies. KnE Soc. Sci. **3**(11), 243 (2019) [online]
Schwab, K.: The Fourth Industrial Revolution: what it means and how to respond. World Economic Forum (2016). https://www.weforum.org/agenda/2016/01/the-fourth-industrial-revolution-what-it-means-and-how-to-respond/
Schwinger, M., et al.: Academic self-handicapping strategies. J. Educ. Psychol. **114**(3), 576–596 (2022). https://doi.org/10.1037/edu0000706
Skhephe, M., Ntombekhaya, P.C., Boadzo, R.K.: Accounting teachers' readiness for e-learning in the Fourth Industrial Revolution (2020)
Surianti, M.: Accounting curriculum model based on industrial revolution approach. Res. J. Finance Acc. **11**(2), 116–123 (2020). https://doi.org/10.7176/RJFA/11-2-12
Tavares, M.C., Azevedo, G., Marques, R.P., Bastos, M.A.: Challenges of education in the accounting profession in the era 5.0. A systematic review (2023)
Terblanche, W., Lubbe, I., Papageorgiou, E., Van der Merwe, N.: Acceptance of e-learning by accounting students. South Afr. J. Acc. Res. **37**(1), 35–61 (2023). https://doi.org/10.1080/10291954.2022.2101328
Tran, T.K.L., Phan, T.T.H.: Impact of Industry 4.0 on the accounting industry in Vietnam. Int. J. Manag. Sci. Bus. Res. **8**(5), 85–92 (2019)
Tsiligris, V., Bowyer, D.: Impact of the Fourth Industrial Revolution on skills for future accountants. Acc. Educ. **30**(6), 621–649 (2021). https://doi.org/10.1080/09639284.2021.1938616
Van den Berg, E., Rothman, S.: Twenty-first-century competencies for financial accounting students (2024)
Vorster, J.: Influence of the Fourth Industrial Revolution on critical cross-field outcomes in accounting students. A systematic literature review (2023)
Vurayai, S.: Impact of the Fourth Industrial Revolution on Southern African higher education. Int. J. Knowl. Content Dev. Technol. **15**(1), 51–64 (2025). https://doi.org/10.5865/IJKCT.2025.15.1.051
World Economic Forum: The Global Competitiveness Report (2017)
World Economic Forum (WEF): The future of jobs report 2018. World Economic Forum (2018)
Yang, P.A.: Educational mobility and transnationalization. In: Gleason, N.W. (ed.) Higher Education in the Era of the Fourth Industrial Revolution, pp. 39–65. Palgrave Macmillan, Singapore (2018)
Yende, S.J.: Transition towards the Fourth Industrial Revolution in South African education. Afr. J. Dev. Stud. **11**(2), 55–75 (2021). https://doi.org/10.31920/26343649/2021/v11n2a3

Identifying the Project Management Success Factors in the Mauritian Construction Industry

M. Lallbeeharry, R. Sannegadu, and S. Pudaruth

Abstract This study aims to assess the project management factors influencing project success within the Mauritian construction industry. Employing dimensions of the European Foundation for Quality Management (EFQM) business excellence model, the research assesses six independent variables (stakeholder management, project lifecycle processes, leadership, employee performance, strategy, and partnership) and their relative importance regarding project success within the Mauritian context. From a project management standpoint, understanding the factors that contribute to the failure or success of construction projects is fundamental. Data collected from 110 professionals from the construction industry was analysed using the Statistical Package for the Social Sciences (SPSS). The results identify all factors as significantly important, with high mean scores ranging between 4.1 and 4.6 on a scale of 1–5 (1 being 'Not important at all' and 5 being 'Very important'). The findings contribute to the existing literature by providing valuable insights into the dynamics that shape successful projects in Mauritius. Consequently, to achieve the effective delivery of project deliverables, it is recommended that (i) clear objectives and deliverables are defined at the start of the project, (ii) all the stakeholders are engaged throughout the project lifecycle, (iii) effective leadership is applied to guide decision-making, and (iv) training and professional development programmes are dispensed to employees to strengthen their capacity.

Keywords Project management success factors · Construction industry · Mauritius · EFQM business excellence

M. Lallbeeharry · R. Sannegadu (✉) · S. Pudaruth
University of Mauritius, Moka, Mauritius
e-mail: r.sannegadu@uom.ac.mu

S. Pudaruth
e-mail: sh.pudaruth@uom.ac.mu

R. Sannegadu
North-West University, Potchefstroom, South Africa

E. H. Redda (ed.), *Proceedings of the Social Sciences International Research Conference, 2025*, https://doi.org/10.1007/978-981-95-9489-4_39

1 Introduction

In the world of business and construction, projects are the driving force behind progress and economic growth. Projects are meticulously planned endeavours with specific goals, allocated resources, and stringent deadlines. Traditionally, project success was measured using the *iron triangle* model, which focused on cost, quality, and schedule (De Wit 1988). However, contemporary research has expanded our understanding, highlighting that project success extends beyond this and includes criteria such as stakeholder satisfaction, among others. The factors contributing to the project's success are multifaceted and include effective management of the projects requiring skills and competencies, including effective planning, well-defined responsibilities, strong leadership, governance and effective communication (Amoah et al. 2021). Moreover, with the intense competition in the construction industry and the increase in the price of construction materials, it is becoming increasingly difficult to manage construction projects in markets to avoid cost overruns, delays in completing the construction and meeting the satisfaction of all stakeholders. These changing contexts have also redefined the parameters of project success within the context of developing countries, expanding the criteria beyond traditional factors such as cost, quality, and time. Success in contemporary project management in developing countries encompasses broader considerations beyond the existing literature, mostly based on developed countries, including strategic alignment, effective leadership within project teams, sustainability practices, innovation, environmental responsibility, and societal well-being (Chen et al. 2004).

It is evident from the literature that projects are becoming more complex day by day, and the success factors need to be explored (Azmat and Siddiqui 2023). However, figuring out what makes a construction project a success or a failure is a complicated question. There are many factors that can lead to success, and just as many that can lead to failure. The effective management of construction projects has been one fundamental factor that contributes to project success (Azmat and Siddiqui 2023). Researchers have focused their attention on this subject for over 50 years and despite all the studies, this topic still fascinates both academics and professionals. Numerous studies have been conducted in the past to design an efficient performance management system for the construction industry in developed countries, addressing shortcomings in existing global practices for developing countries (Parikh and Phugat 2019; Amoah et al. 2021).

Mauritius is no different and has a dynamic construction industry covering a wide range of activities, from building new homes to improving public infrastructure. Construction projects are pivotal to the nation's economic growth, contributing significantly to its gross domestic product (GDP). In 2017, the construction sector contributed approximately 7.5% to the GDP (Competition Commission of Mauritius 2018). Beyond economic benefits, the construction industry generates employment opportunities and income for the population. According to Statistic Mauritius (2023), the construction industry is poised for substantial growth, with a projected 28.6% expansion in 2023, driven by both public projects (e.g. Metro Express extension, road network expansion) and private sector endeavours (e.g. smart city development, hotel construction). Notably, the construction sector in Mauritius employs more than one-third of the country's workforce, engaging around 46,300 workers, making it a significant contributor to the labour force (Dudley et al. 2022). The landscape of the construction sector in Mauritius is characterised by unique environmental challenges, evolving sustainability requirements, and heightened expectations regarding social responsibility. Several instances of incomplete projects in Mauritius exist, such as Le Merritt up by Trianon—today two abandoned towers in the eye line of the population across the Motorway M1, the Heritage City project—improper site selection, the Jin Fei Industrial Town now called The Chinese Ghost Town, the Riviere des Anguilles dam initially planned to start in 2014 and not yet started in 2023, the Mon Tresor Smart City planned by Omnicane Ltd in 2016, abandoned and now sold to the Mauritius Investment Corporation, the Grand Baie and Pailles Guibies sewer project, interrupted during construction in 2023, highlight the need for effective project management and project success.

As the construction industry contributes significantly to the socio-economic development of Mauritius, which is classified as a Small Island Developing State (SIDS) with limited resources, it is crucial to ensure that projects are delivered on time, within budget, with high quality and to the satisfaction of all stakeholders. Nevertheless, even when project managers appear to be applying the *correct* methodologies, instances of project failure continue to persist in Mauritius, often without clear explanations. The enigmatic failures previously detailed in the research background underscore the challenges in comprehending the complexity of project management within the Mauritian context, necessitating a holistic perspective. Success criteria should encompass more than cost, quality, and time; it should also encompass broader objectives, and the success factors, such as the management of these projects, sustainability, human resources, social responsibility, and innovation, should be investigated (Al-Nabae and Sammani 2021).

Performance management in the construction sector is a continuous process that involves recognising, quantifying, and improving organisational performance by coordinating individual efforts in line with the organisation's overarching goals and objectives (Parikh and Phugat 2019). As part of its budget speech of 2023, the finance minister has recognised that construction is one of the industries 'in need of more employees' and that 'it needs to reinforce its capacity and improve the skills of its workforce'. While extensive research exists on the critical success factors affecting project success and project management at the global level in developed countries, there is a noticeable gap in research that specifically addresses the management of projects in developing countries, such as in the Mauritian construction industry. Although there is consensus among researchers on the critical success factors affecting project success, their relative influence on the success of the projects varies from context to context. The industry's unique dynamics, success factors, and challenges have not been comprehensively explored, leading to the need for in-depth research tailored to the Mauritian context. The presence of unexplained project failures, despite the diligent application of seemingly appropriate methods, highlights the necessity of adopting a comprehensive perspective. This gap in research is a clear sign of the need to study the management of construction projects within the Mauritian context more deeply to identify its link with project success.

Against this backdrop, this study aims to identify the critical success factors for project completion in the Mauritian construction industry. Consequently, the following research questions are set for this study:

- What factors contribute to the successful completion of construction projects?
- How important do construction professionals perceive the relative importance of these factors to the successful completion of construction projects?

2 Literature Review

2.1 Theoretical Underpinning

The theoretical foundation of the study is rooted in several theories, which stress the importance of identifying and managing factors contributing to the success of projects. These theories include the project management triangle (iron triangle) (Barnes 1969), the critical success factors (CSFs) theory (Rockart 1979), the stakeholder theory (Freeman 1984) and the success dimensions framework (Shenhar et al. 2001).

Also referred to as the triple constraints, the iron triangle depicts three main elements, including time, cost and scope that determine project success (Pollack et al. 2018). These interrelated elements are constantly negotiated throughout the project depending on the changing requirements (Fayomi et al. 2023). However, a change in one dimension influences the other dimensions. For example, increasing the scope of the project will lead to an increase in cost and time. On the other hand, a reduction in time may lead to an increase (i) in cost due to the need to hire more people to complete the project or (ii) a reduction in scope. However, van der Hoorn and Whitty (2015) consider that the iron triangle oversimplifies project practice and is based on the task and psychosocial success criteria as evident in managers' experience rather than the psychosocial criteria such as customer and stakeholders' satisfaction and team member relationships.

On the other hand, the CSFs theory posits that competitive performance and the success of an organisation are achieved when the key areas, essential to its success, are identified and effectively managed (Rockart 1979). By focusing on key critical areas, managers can effectively execute their plans by optimising limited resources. The principles of the CFS theory can also be applied within the project management context. For example, practitioners should focus on a limited number of factors necessary to guarantee the completion of projects within time and budget and to the satisfaction of the stakeholders. However, the CSF theory has been criticised by Azimi and Manesh (2010) for its simplicity in the way the CSFs are identified and applied within different contexts. They claim that CSFs cannot be universally applied since they are based on existing reality rather than being created.

Developed by Freeman (1984), the stakeholder theory holds that the management of stakeholders is critical to the success of any project. A stakeholder is defined as any individual or group who can affect or is affected by an organisation's activity. Since projects are multifaceted in nature, they engage several stakeholders, including the client, team members, suppliers, and the community at large, each having different influence over the outcome of the project, depending on their expectations and interests.

Finally, the success dimensions framework, by Shenhar et al. (2001), is considered a more comprehensive framework for evaluating project success since it applies a multidimensional approach instead of focusing on a limited number of factors as prescribed by traditional models such as the iron triangle model. Moreover, it acknowledges that the success of a project depends on context-specific factors, thereby broadening the scope to include other factors in assessing a project.

2.2 *Project Success Factors*

The above theories lead us to conclude that the management of project success factors is key to the effective completion of projects to the satisfaction of all the stakeholders. One of the widely used frameworks to improve the performance and success of projects is the European Foundation of Quality Management's Business Excellence Model (EFQM).

The EFQM excellence model. Having its origin in total quality management (TQM), the EFQM business excellence model was developed in 1992 and subsequently revised in 2013. Driven by quality, efficiency, and sustainability, it offers a holistic approach to pursue excellence (Olaru et al. 2010). The application of the EFQM to the management of projects has proved to be successful in various contextual settings, including Pakistan (Qureshi et al. 2009) and the United Arab Emirates (UAE) (Mir and Pinnington 2014), as well as in diverse fields including education, management, project design, social corporate sustainability, strategy, construction, and tourism. According to Paraschi et al. (2019), the EFQM business excellence model provides a comprehensive organisational management philosophy that empowers management to focus on stakeholders and achieve superior performance. Abenet and Maniha (2023) have also confirmed the positive causal relationships between EFQM factors and project quality and success. The dimensions/enablers of the EFQM excellence model contributing to the success of projects are explained in the following paragraphs.

Leadership. Leadership in project management (PM) is a multifaceted and pivotal aspect, integral to achieving project success. Defined as an influencing process within the organisational hierarchy, leadership involves not only the delegation of tasks and responsibilities but also requires decision-making processes, a sense of duty, and the ability to motivate followers towards organisational objectives (Islam et al. 2017). Dynamic leadership is emphasised as indispensable for effective PM, ensuring compliance with procedures, establishing relationships, and communicating organisational values, ultimately contributing to project management success (PMS) (Zada et al. 2024). Leadership methodologies in PM, characterised by effective communication, cooperation, teamwork, and trust, significantly affect PMS (Kerzner 2017). The project manager's attention to both management and leadership roles is underscored as crucial for project success (Pandya 2014; Siddiqui et al. 2024). As construction projects in Mauritius are becoming more complex due to rapid changes in the environment, it is essential that the right leadership style is adopted in the planning and execution phase of the projects to ensure successful project delivery.

Employee performance. Employee performance (EP) plays a pivotal role in determining project success, serving as the conduit through which organisational goals are realised, and interpersonal activities align with established standards (Osman et al. 2016; Watanabe et al. 2024). Within the project context, EP encompasses skills, behaviours, knowledge, and values, contributing significantly to organisational productivity. The costs incurred in training and development are viewed as a long-term investment with the potential for substantial benefits (Rafiei and Davari 2015). Recognising that EP is a key factor in assessing organisational performance and productivity, it becomes synonymous with the organisation's capacity to achieve its goals (Muda et al. 2014). Several studies, such as Salman et al. (2021) and Gul et al. (2025), have demonstrated that a direct relationship exists between project success and employee efficiency. As new techniques are employed in the Mauritian construction industry, including sustainable building technologies, the development of new skills through training and development programmes is essential for project success.

Policy and strategy. While policies refer to the set of rules, values and principles guiding the organisation's actions, strategy relates to the long-term vision of the organisation. The critical importance of strategy, governing policies and procedures, an effective communication management plan, and a collaborative project culture in enhancing project performance and success is widely acknowledged among researchers and construction professionals (Rehan et al. 2024). For example, Anantatmula (2015) emphasises the need for a well-established framework that encompasses these elements to ensure project success. Another study, conducted by Qureshi et al. (2009), revealed the promotion of PM benefits, the establishment of a company-wide PM system with centralised control, the redefinition of jobs through organisation-wide PM methods, and the accreditation of a quality management system. To ensure that there is consistency in the way construction projects are managed in Mauritius, it is imperative that well-established policies and standard procedures are set in place to provide clear guidance on the planning, execution and management of the key performance indicators (KPIs).

Partnerships and resources. Partnerships and resources emerge as critical determinants of successful PM, as highlighted by both empirical studies and the EFQM model. Coordination factors within partnerships are recognised for their pivotal role in ensuring the effective implementation of project stages (Almarri and Boussabaine

2025). Oliveira and Lumineau (2017) describe these factors as dynamic activities that wield a substantial influence on project performance. In line with this, studies by Qureshi et al. (2009) and Busco et al. (2024) accentuate the superior value delivered by project alliances, particularly those characterised by open two-way partnerships with customers, suppliers, and both internal and external stakeholders. This underlines the significance of fostering collaborative relationships to enhance value and project success, a theme echoed in the EFQM model. For example, knowledge sharing and specialised expertise from both the internal and external stakeholders contribute to the smoother execution of the project. As partnerships and alliances are becoming increasingly popular, it is important to investigate the extent to which they can contribute to project success in the Mauritius construction industry, particularly in terms of resource sharing, knowledge transfer, and risk mitigation.

Project lifecycle processes. Abenet and Maniha (2023) highlight the significance of processes within a company, organised by interrelated actions that transform input into output and consequences. Well-defined processes within each phase of the project lifecycle help in reducing risks and improving planning (Ahmed 2024). The effectiveness of the quality planning process is crucial for bridging the gap between a firm's quality improvement aspirations and the realisation of benefits (Shanmugapriya and Subramanian 2015). Quality planning, integral to TQM and International Standardisation of Organisations (ISO), should be integrated with strategic planning within companies (Inyang-Udoh 2002). The EFQM model underscores that customers and stakeholders benefit from firms' efforts to enhance internal processes and procedures, with process enablers being critical for successful framework deployment (Chinda and Mohamed 2008). Therefore, by establishing well-structured processes across all phases of the project lifecycle, firms in the Mauritian construction industry can enhance planning and mitigate risks.

Stakeholder management. Stakeholder management plays a crucial role in determining the success of a project, extending beyond traditional criteria such as time, cost, and quality (Dugbartey and Kehinde 2025). While these factors remain vital, recognising the importance of stakeholder satisfaction has gained prominence in the project management domain (PMBOK 2017). Stakeholders are integral to successful PM, and their engagement is a key aspect of stakeholder management, impacting both the research domain and the industry concerned (Oppong et al. 2017; El Khatib et al. 2020). According to Shaukat et al. (2022), project managers should promote stakeholders' engagement to create meaningful value propositions for each category of stakeholders. As projects often involve diverse stakeholders such as government, suppliers, contractors and consultants, among others, it is apparent that firms within the Mauritian construction industry ensure continuous engagement with the stakeholders to minimise conflicts and build trust, which are essential for the smooth implementation of projects (Fig. 1).

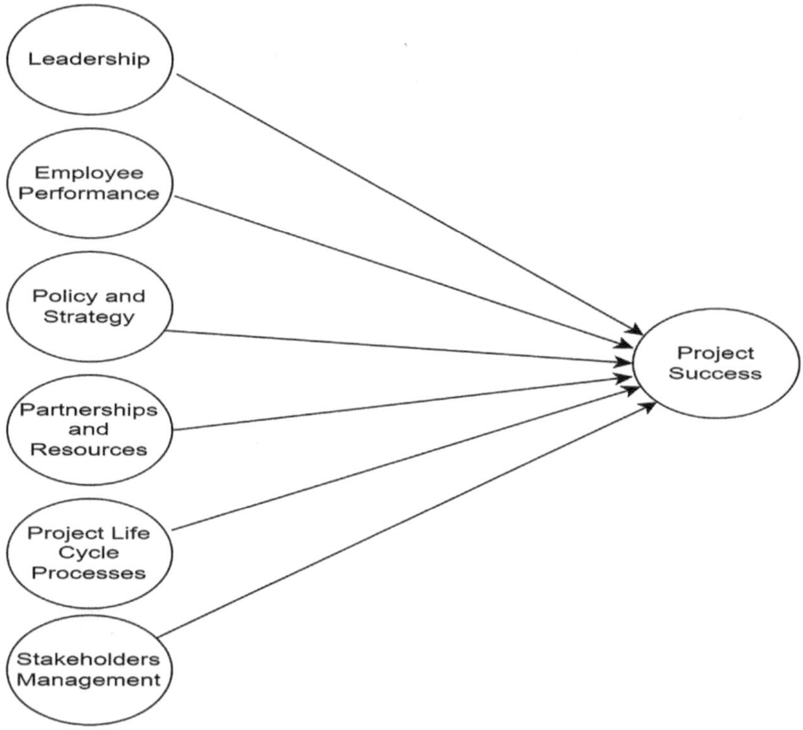

Fig. 1 The conceptual framework of the study

3 Methodology

Guided by the positivist paradigm, this study employed quantitative research to address the research questions. To develop a foundational knowledge base of the phenomenon being investigated, secondary data was reviewed from relevant academic literature. To ensure reliability of the literature reviewed, the researchers retrieved the secondary data from reputable sources such as Google Scholar, ProQuest, and other academic databases and included peer-reviewed journals and other official publications. Moreover, quantitative data relevant to the research questions was collected through a structured survey method from professionals who were actively involved in the management of construction projects in Mauritius. This combined methodology provided a comprehensive and in-depth analysis, leveraging both existing knowledge and direct perspectives from individuals engaged in construction PM.

Sample size and strategy. The researchers faced some difficulties in obtaining information on the number of project managers and other professionals operating in the construction industry. According to official records from the Council of Registered Professional Engineers (CRPE), there have been approximately 1850 registered engineers since its establishment in 1966. However, this figure excludes other professionals such as architects and quantity surveyors. Based on the information gathered from the experts of the construction industry, the population of professionals in the construction industry is around 1500. Therefore, the target sample size was set at approximately 306, based on a 95% confidence interval with a 5% margin of error (The Research Advisor 2006). The estimated sample size of the study is within the range of 30–500 as recommended by Cavana et al. (2001). In the absence of a sampling frame, the respondents were selected through a non-probability technique, namely the purposive sampling based on their relevance to the study. However, considerations of convenience also influenced the selection of the respondents. This methodological choice, that is, the combined approach, is consistent with a similar study by Sarvari et al. (2021).

To answer the research questions, a survey was designed and distributed to project managers and engineers working on projects in the Mauritian construction industry. A five-point Likert scale was used to measure the respondents' perceived level of importance for each statement of the questionnaire. The survey was divided into three parts:

- Part 1 pertains to details regarding the participant's profile.
- Part 2 focused on their perception of PM factors affecting project success. The respondents were asked to rate the importance of each factor on a five-point Likert-type scale (5 = Very important, 4 = Important, 3 = Neutral, 2 = Low importance, 1 = Not important at all). The averages of the numerical scores provided by the respondents are used for ranking the most important factors.
- Part 3 focused on confirmatory questions to capture the level of agreement (5 being strongly agree and 1 being strongly disagree) with regard to the presence of success factors in the Mauritian construction industry.

To ensure internal validity of the questionnaire, the selection of initial measurement items was grounded in a comprehensive review of theoretical and empirical literature. The questionnaire was pilot tested among eight potential participants at their workplace to evaluate clarity, relevance, and reliability of the survey items. Based on the feedback from the pilot survey, adjustments were made to the final questionnaire, including the transition from the initial seven-point Likert scale (based on importance) to a five-point Likert scale. To ascertain the reliability of the collected data, Cronbach's alpha test values were examined. Ethical considerations were strictly adhered to throughout the research process. Respondents were informed that participation in the survey was voluntary and that they have the right to withdraw at any stage without any negative consequences. They were also ensured that their anonymity and confidentiality will be preserved. They were further informed that the data would be used exclusively for academic purposes. These measures were implemented to protect the rights of the respondents.

3.1 Relative Importance Index to Rank the Project Success Factors

The relative importance index (RII) was applied to gauge the relative significance of various factors influencing project success. The computation involves weights assigned by respondents (ranging from 1 to 5), the highest weight (A), and the total number of respondents (N). The resulting RII, on a scale from 0 to 1, provided insights into the importance of each factor within the context of project success. Similar methods have been used by Digehsara et al. (2018) and Abal-Seqan et al. (2023) to identify and rank project success factors.

4 Findings

4.1 Sample and Response Rate

The main survey sample consisted of 111 participants from the Mauritian construction industry. Data collection took place through the administration of an online questionnaire. One questionnaire was excluded from the analysis, resulting in a final dataset of 110 completed questionnaires ($N = 110$) used for the analysis of the main survey. This screening process ensures the reliability and integrity of the data used in drawing conclusions and insights from the study.

4.2 Respondents' Profile

Age and gender. The sample size was dominated by respondents aged between 21 and 30 years (49.1%, $n = 54$), followed by respondents aged between 31 and 40 years (23.6%, $n = 26$), while the remaining participants were above 40 years old (27.3%, $n = 30$). This age distribution shows that a significant proportion of respondents in construction-related studies fall within the relatively younger age brackets, reflecting the demographic characteristics of professionals in the industry. Hoxha and McMahan (2019) demonstrate that project managers' age did not significantly predict project success. The ratio of male to female respondents was 63.6% to 36.4%. The higher percentage of male respondents can be attributed to the historical dominance of the construction sector by the male gender. However, the survey results indicate a noteworthy shift, suggesting an increase in gender equality practices and greater involvement of the female gender in the construction sector, reflecting a positive trend towards a diverse and inclusive workforce in the industry. Rodriguez et al. (2017) state that the male gender adopts a more authoritarian managerial style and that gender influences the project success (Fig. 2).

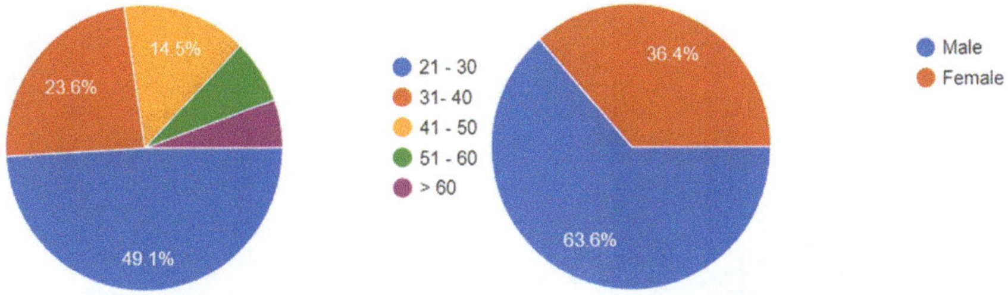

Fig. 2 Diagram showing gender and age distribution of survey respondents

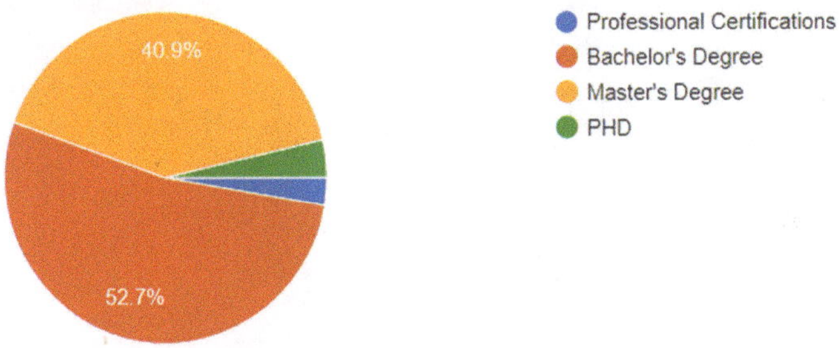

Fig. 3 Diagram showing level of education of survey respondents

Level of education. The data collected revealed that 52.7% ($n = 58$) of the respondents had a bachelor's degree, and 40.9% ($n = 45$) had a master's degree. This educational distribution shows a good educational background of the respondents. The level of education and its correlation with project success have been extensively explored in past research, providing a valuable context for interpreting the diverse educational backgrounds among the surveyed individuals. Semerádová and Mrazek (2015) state that education and experience to some extent determine the project's outcomes (Fig. 3).

Occupation and years of experience. The respondents in this study were exclusively drawn from the Mauritian construction industry, representing a diverse range of occupations. Most participants were identified as civil engineers, comprising 57.3% ($n = 63$) of the sample. The remaining respondents belonged to various other key roles such as mechanical engineers, electrical engineers, quantity surveyors, and architects, among others.

In terms of professional experience, the dataset was characterised by a notable distribution of respondents across different tenure brackets. The majority, 31.8% ($n = 35$), reported six to 10 years of experience, closely followed by 30% ($n = 33$) with one to five years of experience. A significant proportion, 38.2% ($n = 42$), possessed 11 or more years of experience in the Mauritian construction industry. This distribution underscores the presence of respondents with substantial expertise,

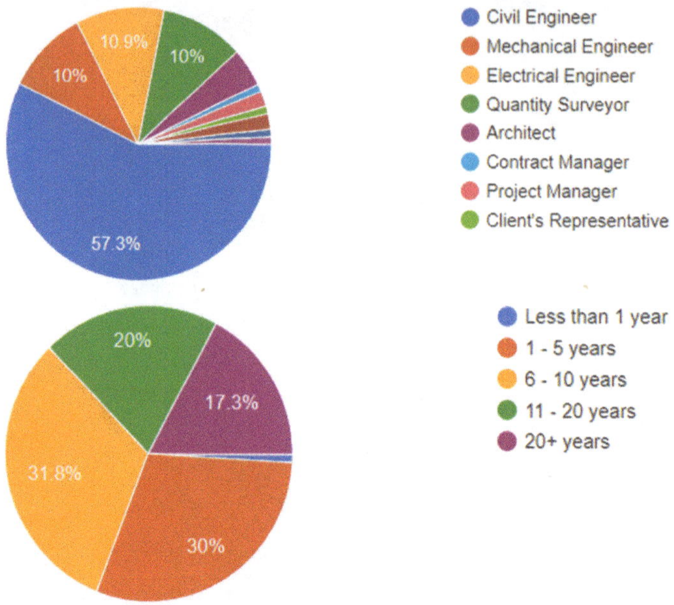

Fig. 4 Diagram showing occupation and years of experience of survey respondents

contributing to the richness of insights drawn from individuals with varied levels of experience in the field (Semerádová and Mrazek 2015) (Fig. 4).

4.3 *Company Details*

The Mauritian construction industry exhibits a clear distinction between two primary sectors: consultancy services and contracting services. This distinction is highlighted by the requirement for professionals to be registered with the Construction Industry Development Board (CIDB) under the CIDB Act of 2008. Item 8 of the Act explicitly states, *No person shall offer his services as a consultant or as a contractor unless he is registered under this Act.* An analysis of the respondent data reveals that 47.3% ($n = 52$) of participants were affiliated with the consultancy sector, while 23.6% ($n = 26$) were associated with the contracting sector. A subset of respondents was engaged in both sectors simultaneously. Further insights into the organisational landscape of the participants indicate that 36.4% ($n = 40$) worked in organisations comprising 51 to 250 employees, while 31.8% ($n = 35$) were employed by organisations with 11–50 employees. This organisational distribution provides an understanding of the workforce composition within different-sized entities in the Mauritian construction industry (Fig. 5).

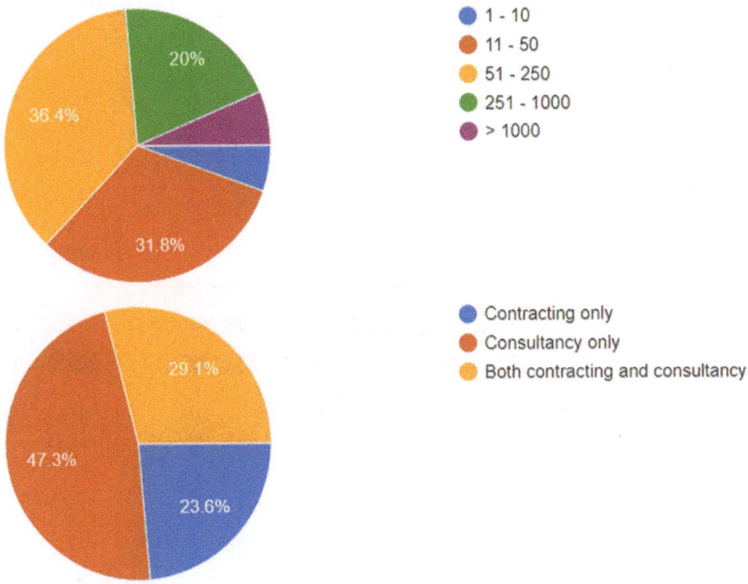

Fig. 5 Diagram showing details of organisational landscape

Project involvement. The survey respondents were drawn from individuals involved in both private and public sector projects, constituting the majority at 73.6% (n = 81). Smaller segments of the participants were exclusively engaged in private sector projects at 15.5% (n = 17) and public sector projects at 10.9% (n = 12). This distribution reflects a representation of professionals with varied project involvements across different sectors. Insights into the nature of projects undertaken by the respondents were also captured, with 88.8% (n = 97) reporting experience in infrastructure construction works (roads, drains, and utilities). Moreover, 61.8% (n = 68) and 65.5% (n = 72) indicated involvement in residential and commercial buildings, respectively. Additionally, 32.7% (n = 36) and 29.1% (n = 32) of respondents shared their experience in institutional constructions, projects such as hospitals, schools, and water and wastewater treatment plants. This diversity in project experience underscores the wide-ranging expertise of the survey participants within the Mauritian construction domain (Fig. 6).

4.4 Statistical Analysis

Descriptive analysis. In this section, the results of the statistical analysis are comprehensively detailed. The mean scores of the constructs for both the dependent and independent variables are reported, offering insights into the central tendencies of the measured variables. Furthermore, the reliability of the measurement instruments

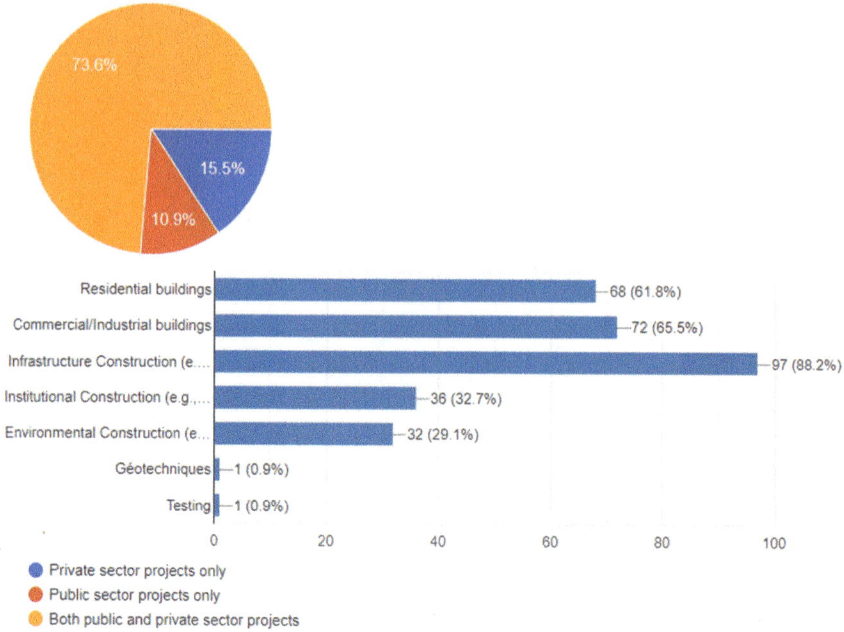

Fig. 6 Diagram showing distribution of projects

was assessed using the Cronbach's alpha test. Notably, only one variable reported a coefficient of less than 0.7. However, the decision was made to retain all items despite the low alpha value. This choice was informed by the understanding that the low alpha coefficient might be attributed to the limited number of questions rather than an intrinsic issue with the construct's reliability, as suggested by Tavakol and Dennick (2011). Therefore, all items were retained to maintain the integrity and completeness of the measurement instruments. Overall Cronbach's alpha of variables = 0.845.

Reliability. To ensure the reliability of the collected data, Cronbach's coefficient alpha (α) was employed to assess the internal consistency among items within each factor. Generally, an instrument's internal consistency should be at least 0.70 (Zikmund 2013). As presented in Table 1, six out of the seven variables demonstrated alpha (α) values equal to or exceeding the recommended threshold. Notably, the independent variable *partnerships and resources* reported a coefficient slightly below the 0.70 threshold ($\alpha = 0.669$). Despite this, a strategic decision was made to retain all items associated with this variable, even with the lower alpha value. This choice was guided by the recognition that the diminished alpha coefficient might be attributed to the limited number of questions rather than an inherent flaw in the construct's reliability, as proposed by Tavakol and Dennick (2011). Retaining all items was deemed crucial to maintaining the integrity and completeness of the measurement instruments.

Table 1 Cronbach's alpha for variables

Constructs	Indicator	Cronbach's alpha
Leadership	LD1: Leadership skills and behaviours	0.715
	LD2: Experience in project management	
	LD3: Openness to creativity, innovation and new ideas	
	LD4: Adaptability to unforeseen challenges	
Employee performance	EP1: Training and development	0.752
	EP2: Performance appraisal practices	
	EP3: Providing feedback	
	EP4: Adoption of new industry practices by employees	
Policy and strategy	ST1: Communication management plan	0.699
	ST2: Alignment of project goals with organisational goals	
	ST3: Effective communication with project stakeholders	
	ST4: Quality management system accreditation	
Partnerships and resources	PA1: External alliances for enhancing technical expertise	0.669
	PA2: Allocation of resources for implementation	
	PA3: Availability of hardware and software	
	PA4: Alliance with industry experts	
Project lifecycle processes	PR1: Definition of aim, objectives and deliverables	0.782
	PR2: Project planning	
	PR3: Workflow management	
	PR4: Closing and handing over of project deliverables	
Stakeholder management	SM1: Understanding stakeholder needs and expectations	0.841
	SM2: Addressing stakeholder needs and expectations	
	SM3: Communication with stakeholders	
	SM4: Stakeholder satisfaction and engagement	
Project success	PS1: Stakeholders' satisfaction	0.756
	PS2: Delivery within time	

(continued)

Table 1 (continued)

Constructs	Indicator	Cronbach's alpha
	PS3: Delivery with budget	
	PS4: Abiding to technical requirements	

Measure of constructs. A five-point Likert scale was adopted for measuring each indicator of the independent and dependent variables. Figure 7 illustrates the mean score for each individual indicator. The Likert scale used was based on the level of importance, with 1 indicating 'Not important at all' and 5 representing 'Very important'. This scaling approach provides a complex understanding of the respondents' perceptions and emphasises the degree of importance attributed to each indicator in the study.

Mean, median, and variance. The mean, standard error of the mean, standard deviation, variance, skewness, and kurtosis were calculated for each variable, and the results are presented in Table 2.

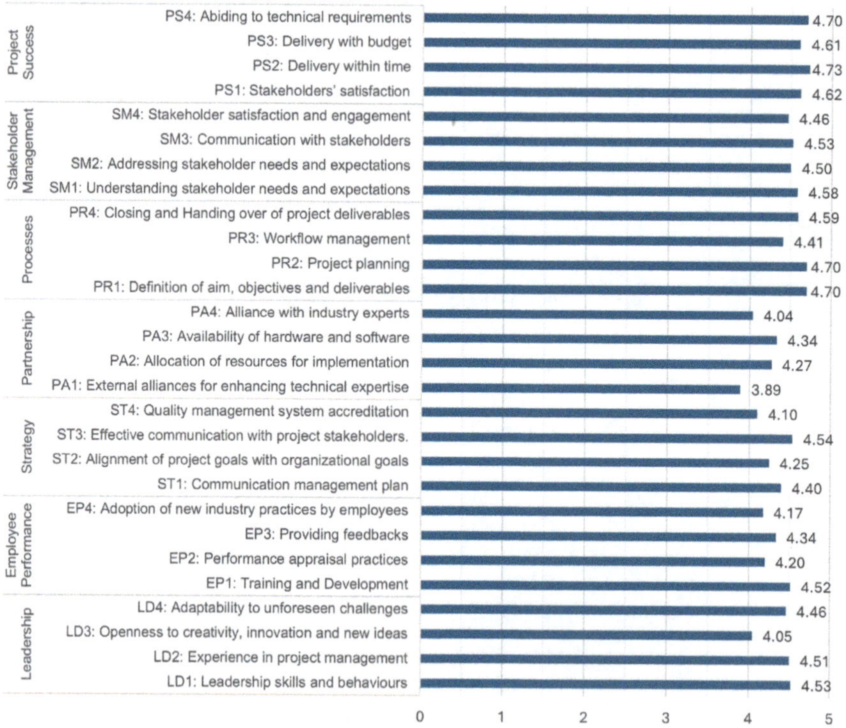

Fig. 7 Measurement scale for independent and dependent variables

Table 2 Mean, standard error of the mean, standard deviation, variance, skewness, and kurtosis

	Mean		Std. deviation	Variance	Skewness	Kurtosis
	Statistic	Std. error				
Leadership	4.386	0.039	0.409	0.167	−1.173	2.773
Employee performance	4.386	0.051	0.536	0.287	−0.845	0.305
Strategy	4.321	0.046	0.486	0.236	−0.422	−0.386
Partnership	4.134	0.044	0.459	0.211	−0.213	−0.167
Process	4.600	0.041	0.430	0.185	−2.395	10.972
Stakeholder management	4.518	0.049	0.512	0.262	−1.509	4.37
Project success	4.664	0.035	0.362	0.131	−0.701	−0.866

Ranking of factors using the RII. Table 3 outlines the RII of key factors influencing project success, providing a comprehensive ranking based on the perceived importance of each independent variable by the respondents. Topping the list is the *definition of aim, objectives, and deliverables within the project lifecycle processes*, securing the first rank with an RII of 0.940. *Understanding stakeholder needs and expectations* under stakeholder management is ranked second with an RII of 0.916. In the third position, *effective communication with project stakeholders* within policy and strategy achieved an RII of 0.907. Leadership skills and behaviours, represented by *leadership skills and behaviours*, ranked fourth with an RII of 0.905. The *training and development* aspect of EP claimed the fifth rank with an RII of 0.904. Lastly, *availability of hardware and software* within partnerships and resources achieved an RII of 0.867, placing it in the sixth rank. These rankings, derived from the RII values, indicate the perceived importance of each factor in contributing to project success in the Mauritian construction industry, thereby possibly guiding strategic decision-making and resource allocation as detailed in the literature review. These findings are in conformity with empirical studies highlighting the importance of having a clear definition of project aims, objectives and deliverables (Ahmed 2024); understanding stakeholders' needs and expectations (Shaukat et al. 2022), effectively communicating with project stakeholders (Rehan et al. 2024), applying effective leadership skills (Zada et al. 2024), developing new skills though relevant training (Rafiei and Davari 2015) and securing the required hardware and software through the collaborations with other partners (Almarri and Boussabaine 2025) to the successful planning and implementation of projects in the Mauritian construction industry.

Regarding measurement of the dependent variable, project success, *delivery within time* had the highest RII at 0.946, followed by *abiding with technical specifications* at 0.940. *Stakeholder satisfaction* and *delivery within budget* were ranked third and fourth with RII values of 0.924 and 0.922, respectively.

Table 3 Relative importance index of factors

Constructs	Indicator	RII	Rank
Project lifecycle processes	PR1: Definition of aim, objectives and deliverables	0.940	1
Stakeholder management	SM1: Understanding stakeholder needs and expectations	0.916	2
Policy and strategy	ST3: Effective communication with project stakeholders	0.907	3
Leadership	LD1: Leadership skills and behaviours	0.905	4
Employee performance	EP1: Training and development	0.904	5
Partnerships and resources	PA3: Availability of hardware and software	0.867	6

5 Conclusions and Recommendations

This study explored the impact of PM factors, using the EFQM business excellence model, on project success within the Mauritian construction industry. Six independent variables (stakeholder management, project life cycle processes, leadership, EP, strategy, and partnership) were tested to identify their impact on the dependent variable, i.e. project success. The RII of key factors influencing project success, providing a comprehensive ranking based on the perceived importance of each independent variable, was computed. Findings of the study revealed that these factors are multifaceted and context specific. Based on the ranking of the key factors, recommendations are provided to shape successful construction projects in Mauritius. Firstly, it is essential that project managers clearly define the objectives and deliverables of the project at the start of the project. A clear articulation of these elements into a project governance framework ensures a better alignment between strategy and project execution. Secondly, since understanding and addressing stakeholder needs and expectations are critical factors that influence project outcomes, the policy framework should be designed in such a way that multi-stakeholder engagement is provided at the very start of the project lifecycle. Regular meetings and feedback loops should also be held at all stages of the project lifecycle to reduce resistance and enhance the long-term sustainability of the project outcomes. Thirdly, as effective leadership has emerged as one of the key project success factors, project leaders should demonstrate both technical and soft skills in managing projects. Management development programmes embedding leadership skills should be offered to project leaders. Moreover, to ensure that project leaders are applying effective leadership skills, clearly defined leadership performance indicators can be used to ensure that key leadership competencies are demonstrated throughout the project lifecycle. Fourthly, given that up-to-date technical, managerial and interpersonal skills are key factors driving project success, effective training and professional development programmes should be given to all the concerned employees. For example, project-specific training to understand the project scope, complexity and objectives is essential in ensuring that the organisation has capable team members to deliver results with consistency and excellence. Fifthly, organisations should ensure that they have the right mix of resources to plan, execute, monitor, and deliver project deliverables effectively. These resources include finance, technology and technical support policies, among others.

Although the study highlights the relative importance of project success factors in the Mauritian construction industry, it is recommended that firms do not consider each factor in isolation, as the interplay among them generates synergistic effects that intensify the overall benefits. For example, effective leadership is key in supporting stakeholders' engagement, employees' performance and partnerships' engagement. It is, therefore, recommended that projects are managed using an integrated and holistic approach whereby firms leverage the dynamic relationships between the success factors to maximise performance and realise long-term benefits.

6 Limitations and Future Research

Although this study provides valuable insights into the factors contributing to the success of projects within the Mauritian construction industry, it is subject to certain limitations. Firstly, this study is limited by its adoption of a single-stakeholder perspective, focusing exclusively on project managers and excluding other stakeholders who can significantly affect the outcomes of projects. Accordingly, future research should adopt a multi-stakeholder approach to provide a more comprehensive understanding of the CSFs of PM. Secondly, given that the study focuses solely on the construction industry, the generalisability of the findings to other industries is limited. Therefore, future studies can extend the scope of research to include comparative analyses across various industries. This would allow for a better understanding of how factors influencing project success vary or remain consistent across different sectors, providing valuable insights for industry-specific best practices. Thirdly, this study adopts a cross-sectional view of the situation at a specific point in time, excluding other emerging factors that may potentially affect project success. Moreover, the impact of these factors on project success is likely to vary over time, highlighting the need for longitudinal studies. Longitudinal studies would offer a dynamic perspective on project success by tracking and analysing specific projects over an extended period and using cases. This approach would facilitate a more comprehensive understanding of the temporal dynamics and the evolution of success factors. Fourthly, given the limited scope of the EFQM, the authors acknowledge that certain critical factors, such as the role of information technology, are not captured in the study. Therefore, given the rapid advancements in technology, future research could explore how emerging technologies, such as artificial intelligence, big data analytics, and digital PM tools, impact project success in the construction industry. Fifthly, since the study was conducted within a specific geographical context, the generalisability of the findings to other locations and industries is limited. To enhance the generalisability of the findings, similar studies should be conducted in other regions to confirm whether the observed factors have similar weights on project success. Finally, future studies can employ more robust statistical software to validate and strengthen the findings.

References

Abal-Seqan, M.H., Pokharel, S., Naji, K.K.: Key success factors and their impact on the performance of construction projects: case in Qatar. Sustainability **15**(4), 3700 (2023)

Abenet, Y., Maniha, N.: Factors influencing the quality of construction projects in public universities. Int. J. Bus. Manag. Econ. **4**(1), 13–31 (2023). https://doi.org/10.47747/ijbme.v4i1.990

Ahmed, B.A.: Optimizing project management methodologies to enhance efficiency and success in new product development. Doctoral dissertation, Politecnico di Torino (2024)

Almarri, K., Boussabaine, H.: Critical success factors for public–private partnerships in smart city infrastructure projects. Constr. Innov. **25**(2), 224–247 (2025)

Al-Nabae, M., Sammani, D.: Factors influencing project management performance: a review. Int. J. Acad. Res. Bus. Soc. Sci. **11**(8), 628–643 (2021)

Amoah, A., Berbegal-Mirabent, J., Marimon, F.: Making the management of a project successful: case of construction projects in developing countries. J. Constr. Eng. Manag. **147**(12), 04021166 (2021)

Azimi, A., Manesh, F.S.: A new model to identify and evaluate critical success factors in IT projects: case study using RFID technology in the Iranian fuel distribution system. Int. J. Inf. Sci. Manag. (Special Issue), 99–112 (2010)

Azmat, Z., Siddiqui, M.A.: Analyzing project complexity, its dimensions and their impact on project success. Systems **11**(8), 417 (2023)

Busco, C., Walters, J., Provoste, E.: Stakeholder management within PPP-arranged civil engineering megaprojects: a systematic literature review. Int. J. Public Sect. Manag. **37**(5), 649–671 (2024)

Chen, Z., Li, H., Hong, J.: An integrative methodology for environmental management in construction. Autom. Constr. **13**(5), 621–628 (2004)

Chinda, T., Mohamed, S.: Structural equation model of construction safety culture. Eng. Constr. Archit. Manag. **15**, 114–131 (2008)

De Wit, A.: Measurement of project success. Int. J. Project Manage. **6**(3), 164–170 (1988). https://doi.org/10.1016/0263-7863(88)90043-9

Digehsara, A., Amin, R., Hassan, R., Soleimani, M.: Performance evaluation of project management systems based on EFQM and QFD. J. Proj. Manag. **3**, 171–182 (2018). https://doi.org/10.5267/j.jpm.2018.4.003

Dudley, T.K., et al.: An assessment of the factors leading to project success and failure in the Mauritian building construction industry. Int. J. Soc. Sci. Hum. Res. **5**(11) (2022). https://doi.org/10.47191/ijsshr/v5-i11-47

Dugbartey, A.N., Kehinde, O.: Optimizing project delivery through agile methodologies. World J. Adv. Res. Rev. **25**, 1237–1257 (2025)

El Khatib, M., Alabdooli, K., AlKaabi, A., Al Harmoodi, S.: Sustainable project management: trends and alignment. Theor. Econ. Lett. **10**(6), 1276 (2020)

Fayomi, F.O., Onifade, M.K., Adeoye, A.B.: Iron-triangle framework and the efficient management of public building projects in Nigeria. J. Innov. **71**, 12–14 (2023)

Freeman, R.E.: Strategic Management: A Stakeholder Approach. Pitman, Boston (1984)

Gul, J., Rasheed, R., Usman, M., Majid, A.: Impact of communication on project success: mediating role of employee efficiency. Rev. Appl. Manag. Soc. Sci. **8**(1), 35–52 (2025)

Islam, M.A., Jantan, A.H., Saimoon, A.: Factors affecting females' participation in leadership positions in the RMG industry in Bangladesh. Int. J. Sci. Res. Publ. **7**(10) (2017)

Kerzner, H.: Project Management: A Systems Approach to Planning, Scheduling, and Controlling, 12th edn. Wiley, New York (2017)

Mir, F.A., Pinnington, A.H.: Exploring the value of project management: linking project management performance and project success. Int. J. Proj. Manag. **32**(2), 202–217 (2014)
Muda, I., Rafiki, A., Harahap, M.R.: Factors influencing employees' performance in Islamic banks in Indonesia. Int. J. Bus. Soc. Sci. **5**(2), 73–80 (2014)
Olaru, M., Dinu, V., Stoleriu, G., Sandru, D., Dinca, V.: Responsible commercial activity of SMEs and sustainable development values. Econ. **12**(27), 10–26 (2010)
Oliveira, N., Lumineau, F.: How coordination trajectories influence the performance of interorganizational project networks. Organ. Sci. **28**, 965–1167 (2017)
Osman, S., Shariff, S.H., Lajin, M.N.A.: Does innovation contribute to employee performance? Procedia Soc. Behav. Sci. **219**, 571–579 (2016). https://doi.org/10.1016/j.sbspro.2016.05.036
Pandya, K.D.: The key competencies of project leaders beyond technical capabilities. IUP J. Knowl. Manag. **12**(4), 39–48 (2014)
Paraschi, E.P., Georgopoulos, A., Kaldis, P.: Airport business excellence model: a holistic performance management system. Tourism Manag. **72**, 352–372 (2019)
Parikh, D.V., Phugat, S.: Performance management in road construction projects. Int. Res. J. Eng. Technol. **6**(4) (2019)
Pollack, J., Helm, J., Adler, D.: What is the iron triangle, and how has it changed? Int. J. Manag. Proj. Bus. (2018). https://doi.org/10.1108/IJMPB-09-2017-0107
Qureshi, T.M., Warraich, A.S., Hijazi, S.T.: Significance of project management performance assessment (PMPA) model. Int. J. Proj. Manag. **27**(4), 378–388 (2009)
Rafiei, N., Davari, F.: Role of human resource management in enhancing faculty teaching skills. Materia Socio Medica **27**(1), 35–38 (2015)
Rehan, A., Thorpe, D., Heravi, A.: Project success factors for leadership practices and communication. Int. J. Manag. Proj. Bus. **17**(3), 562–590 (2024)
Rockart, J.F.: Chief executives define their own data needs. Harv. Bus. Rev. **57**, 81–93 (1979)
Salman, A., Jaafar, M., Malik, S., Mohammad, D., Muhammad, S.A.: Communication, employee motivation and project success in agile software development. Bus. Perspect. Res. **9**(1), 46–61 (2021)
Sarvari, H., Chan, D.W., Alaeos, A.K.F., Olawumi, T.O., Aldaud, A.A.A.: Critical success factors for managing construction SMEs in the Middle East. J. Build. Eng. **43**, 103152 (2021)
Shanmugapriya, S., Subramanian, K.: Structural equation modelling of quality performance in Indian construction projects. Sadhana **40**(6), 1975–1987 (2015)
Shaukat, M.B., Latif, K.F., Sajjad, A., Eweje, G.: Sustainable project management and project success. Sustain. Dev. **30**(1), 58–75 (2022)
Shenhar, A.J., Dvir, D., Levy, O., Maltz, A.Z.: Project success: a multidimensional strategic concept. Long Range Plan. **34**(6), 699–725 (2001)
Siddiqui, A., Qureshi, B., Shaukat, M.: Project manager competencies and project success. Int. J. Organ. Leadersh. **13**(Special Issue) (2024)
Tavakol, M., Dennick, R.: Making sense of Cronbach's alpha. Int. J. Med. Educ. **2**, 53–55 (2011)
van der Hoorn, B., Whitty, S.: Signs to dogma: a Heideggerian view of project artefacts. Int. J. Proj. Manag. **33**(6), 1206–1219 (2015)
Watanabe, W.C., Shafiq, M., Nawaz, M.J., Saleem, I., Nazeer, S.: Emotional intelligence and project success. Int. J. Eng. Bus. Manag. **16**, 18479790241232508 (2024)
Zada, M., Khan, J., Saeed, I., Zada, S., Yong Jun, Z.: Sustainable leadership and sustainable project performance. J. Knowl. Manag. **28**(6), 1588–1608 (2024)

Kolibri 360 GmbH
Ferdinand-Jühlke-Straße 7
99095 Erfurt, Deutschland
produktsicherheit@kolibri360.de